S0-BJU-143

REMOTE SENSING FOR RESOURCES DEVELOPMENT
AND ENVIRONMENTAL MANAGEMENT

VOLUME 2

PROCEEDINGS OF THE SEVENTH INTERNATIONAL SYMPOSIUM ON REMOTE SENSING FOR RESOURCES DEVELOPMENT AND ENVIRONMENTAL MANAGEMENT
ISPRS COMMISSION VII / ENSCHEDE / 25-29 AUGUST 1986

Remote Sensing for Resources Development and Environmental Management

Edited by
M.C.J.DAMEN / G.SICCO SMIT / H.TH.VERSTAPPEN
ITC, Enschede, Netherlands

VOLUME TWO

5 *Non-renewable resources*
6 *Hydrology*
7 *Human settlements*
8 *Geo-information systems*

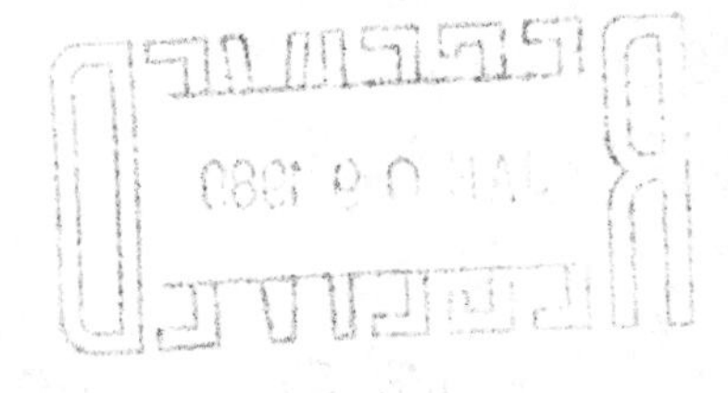

A.A.BALKEMA / ROTTERDAM / BOSTON / 1986

Scheme of the work

VOLUME ONE

VOLUME TWO

VOLUME THREE

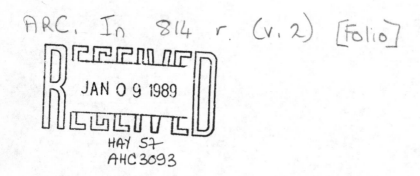

Photograph on the cover
Image of SPOT satellite taken on 16th May 1986: Coast of the Netherlands (IJmuiden up to just north of Hook of Holland). Approximate scale 1:200.000.

The texts of the various papers in this volume were set individually by typists under the supervision of each of the authors concerned.

For the complete set of three volumes, ISBN 90 6191 674 7
For volume 1, ISBN 90 6191 675 5
For volume 2, ISBN 90 6191 676 3
For volume 3, ISBN 90 6191 677 1

Distributed in USA & Canada by: A.A.Balkema Publishers, P.O.Box 230, Accord MA 02018

Printed in the Netherlands

Table of contents

5 *Non-renewable resources: Geology, geomorphology and engineering projects*

6 *Hydrology: Surface water, oceanography, coastal zone, ice and snow*

7 *Human settlements: Urban surveys, human settlement analysis and archaeology*

8 *Geo-information systems*

5 Non-renewable resources: Geology, geomorphology and engineering projects

Chairman: J.V.Taranik
Liaison: B.N.Koopmans

Symposium on Remote Sensing for Resources Development and Environmental Management / Enschede / August 1986

Application of stereo-terrestrial photogrammetric technique to varied geoscientific investigations

N.K.Agarwal
Geological Survey of India

ABSTRACT : Stereo-terrestrial photogrammetry provides distinct advantages over conventional techniques in large scale topographic and thematic mapping of selected areas. During last six years this technique has been advantageously applied to various geoscientific investigations in the Geological Survey of India, important amongst them being the Nilgiris landslide investigation, Tamil Nadu; escarpment slope mapping of Supa Dam, Karnataka; mapping of Coconut Island, off west coast, Karnataka; and mapping of a glacier in north-west Himalayas.

The methodology consisted of taking terrestrial stereo-pairs from 'UMK-100' Camera; surveying of control points with the help a theodolite or 'Distomat' and plotting of details on the Unviersal Anologue Stereo-plotter 'Topocart'. Topographic and photo-interpreted thematic maps of these areas were prepared on scales between 1:500 and 1:2000, except for the glacier which was covered on 1:5000 scale. The experience obtained in applying this technique in diverse terrains has helped in gradually selecting the most suitable methodology.

Results clearly demonstrate the potential of this advanced technique in large scale mapping, for a comprehensive evaluation of selected areas with speed, economy and precision. Further, the technique offers exciting possibilities of digital terrain modelling for multitheme mapping by making use of 'Analytical Stereo--plotter'.

1. INTRODUCTION

Geoscientific investigations of project areas require preparation of large scale topographic base map and relevant thematic maps of the area. Normally, this work is being carried out with the help of conventional plane table or theodolite survey technique.

Aerial photogrammetric mapping has already been established as a powerful tool for preparation of topogrphic base maps with speed, economy and precision, and the potential of photo-interpretation technique is increasingly being exploited for thematic mapping.

Terrestrial photogrammetry permits acqusition of controlled stereo-photographs of selected areas at the desired time. The stereo-pairs are used for the simultaneous preparation of large scale topographic base map and relevant thematic maps of the area. Further, computer assisted analytical stereo-plotters, which are making a breakthrough in photogrammetric mapping, provide scope for simultaneous digitization of topographic and thematic data of the area. This in turn would make morphometric and thematic mapping, either separately or in combination an easy proposition.

Terrestrial photogrammetry was introduced in Geological Survey of India in 1978. Since then this technique has been advantageously applied to a variety of geoscientific projects involving diverse terrains, varying from Coastal tracts to high mountaneous regions. Some of the important investigations in which this technique has been applied include Nilgiris landslide investigation, Tamil Nadu; escarpment slope mapping of Supa Dam, Karnataka; geo-environmental study of Coconut Island, off West Coast, Karnataka; and mapping of a glacier in Northwest Himalayas.

The experience obtained while applying the technique in diverse terrains has progressively helped in adopting the most suitable methodology.

2. METHODOLOGY AND INSTRUMENTS

The technique of terrestrial photogrammetry involves establishment and survey of four to five control points in the area; photography for obtaining stereo-pairs with the help of a terrestrial phorogrammetric Camera unit; and finally map preparation in the laboratory on a stereo-plotter.

In the present surveys, a wide angle terrestrial photogrammetric plate Camera 'UMK-100' (plate size 13 cm x 18 cm) of 100 mm focal length was used for photography. Survey of control points was intially done with help of a theodolite and subsequently by an electronic distance measuring unit Distomat. Map preparation was undertaken on the analogue stereo-plotter Topocart.

2.1 Control point survey and establishment

Establishment and survey of control points in the area is an important part of the photograpmmetric survey. For an expeditious and accurate survey of the control points an electronic distance measuring unit Distomat has been found to be highly convenient and handy. The coordinates of the surveyed points can be obtained instantaneously in the field.

In case of present surveys two different types of orientation systems were used, one for the Camera and the other for the Distomat. However, if the Camera and the Distomat can be mounted on the same orientation system then the field unit becomes quite compact. It is particularly significant if the unit has to be transported on head loads for survey in remote inaccessible areas.

During application of the technique in different areas various objects have been tried as control points. Ultimately, it has been found that white painted circles, of appropriate dimensions (Manual of photogrammetry : 837), on suitable faces of either rock outcrops, hut or tree serve best as the control points.

These, besides being very convenient to establish in the field, also provide accuracy and consistency during model orientation in the stereo-plotter. The ground survey of these permanent control points can be undertaken independent of photography. Further, in case a fresh photography of the area becomes necessary, repeat survey of such control points is not required.

If the area to be surveyed does not have suitably located objects for establishing permanent control points, then temporary control points need to be established. A white painted round ball fixed on the mounting rod of the reflector of the Distomat (Fig.1) forms a suitable temporary control point. An advantage with such an arrangement is that the photography and the survey of control point can be done simultaneously. Even a temporary control point should be located, to the extent possible, over some fixed object with a centering mark, so that if repeat photography becomes necessary, the temporary control point can be placed at the same location, thereby avoiding a fresh ground survey.

In case the area to be mapped falls in more than one stereo-pair then it is required to have at least two control points common in adjoining stereo-pairs. The above described temporary control point, being round in shape is equidimensional and if located at vantage open places would be visible from different places and hence would form an ideal tie point.

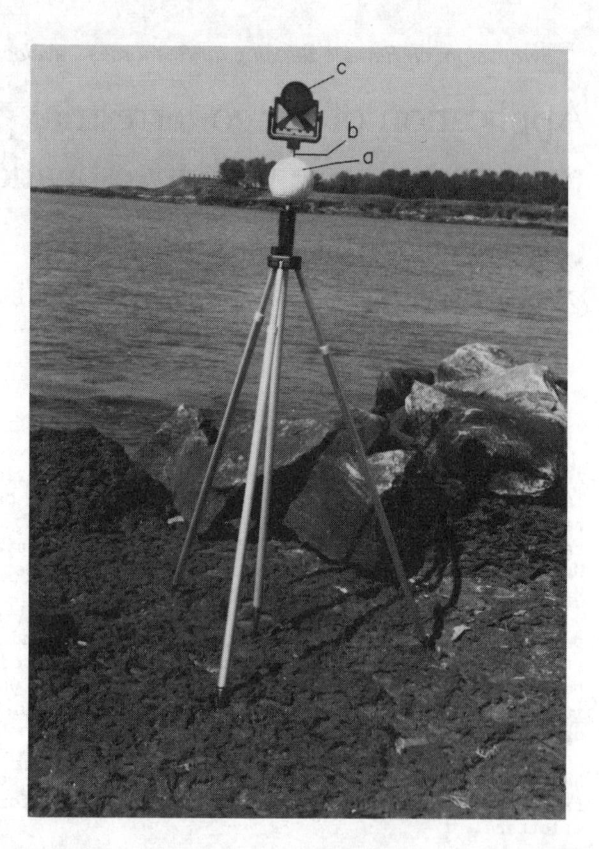

Figure 1. Temporary control point (a), fixed on mounting rod (b), of reflector prism (c)

2.2 Photographic Camera Choice

In course of photography it has been noted that in some cases a camera of 200 mm focal length would have been appropriate in place of the available Camera of 100 mm focal length. Since on account of topographic restrictions in field, the camera stations had to be located at a farther distance than required. As a result the intended area was covered only in a part of the photoframe. In such cases if a larger focal length Camera is used then the required area would be covered in full photoframe, providing a larger scale photography, which in turn would result in higher resolution and greater accuracy. Therefore, it is desirable that the photographic unit contains Cameras of different focal lengths so that depending upon the available locations for photography, the appropriate camera may be used. Since the orientation system remains the same for these Cameras, only one oreintation system is sufficient for the unit.

In most of the cases only one Camera was used for photography. The stereo-pairs were obtained from two Camera stations by interchanging the position of the target and the Camera. However, in case of glacier surveys two Cameras were used for obtaining the stereo-pairs at the same instant of time, since on glaciers the weather conditions fluctuate rather rapidly. Unmounting, packing and shifting of the Camera from one station to the other, through rugged glaciated terrain, and resetting it at the second station involves good amount of time resulting in variations in photographic conditions.

However, it was noted that orientation of model on the stereo-plotter was much easier and relatively more accurate when only one Camera was used for the photography. It was perhaps due to the fact that the elements of inner orientation remain same when only one Camera is used.

2.3 Photographic prints in field

It is desirable to get photographic prints of the stereo-pair in the field itself so that control points are properly identified and numbered to avoid any confusion during model orientation in the laboratory. Further stereo-pairs can be interpreted in the area itself supported by field checks and collection of ground truth data. Thus revisit to the area for the purpose can be avoided. For photo-interpretation in field a pocket mirror-stereoscope has been found to be very convenient and handy.

In remote, inaccessible areas it is not possible to have the facility of a photolab, therefore, an improvised, handy photo-printing technique was adopted. The standard ready-made re-agents were used for the photographic work. Developing of negatives is a simpler work, but the main problem comes in making prints which require a controlled diffused light source. For making prints the photographic paper and the negative were placed between two ground glass plates, it was exposed by a light source from a three cell torch whose transparent glass was replaced by a ground glass. After one or two trials it became possible to get sufficiently good quality prints. Duplicate exposures of each area was obtained, one for developing in the field and the other for processing in the laboratory.

3. SELECTED CASE STUDIES

The main consideration for applying terrestrial photogrammetric technique for the selected case studies has been the constraints inherent in conventional survey techniques for accurate mapping of such areas. Besides, the technique has distinct advantages of speed, economy and precision over conventional survey techniques.

3.1 Escarpment slope mapping of Supa dam

A concrete dam was under construction across a

Figure 2. Terrestrial stereo-pair of left abutment of Supa Dam, a - excavated weathered basic dyke, x-x' fold axis, b - shear zone.

200m deep gorge section of Kali river at Supa, Karnataka. The left abutment of the Dam formed an escarpment in the upper portion (Fig.2). Being inaccessible it could not be mapped by conventional methods. Utilizing terrestrial photogrammetric technique maps of the area could easily be prepared on 1:500 scale with a contour interval of 5 m. Various significant geological details that could be mapped out included weathered basic dykes, prominent shear zones and joint planes, traces of bedding, faults and joints, and attitude of planar surfaces. Folding pattern of the rocks was well deciphered by correlating precise levels of the fold axes on the left and the right abutment areas (Agarwal 1985). Thus terrestrial photogrammetry not only made it possible to accurately map an inaccessible zone but also helped in the expeditious completion of the geological exploration.

3.2 Geo-environmental study of Coconut island

Coconut Island off the west coast of India in the Arabian sea, has been declared as a 'National Geological Monument' for preservation of its natural exposures of columnar lava. The Island exhibits highly irregular topographic features marked by conspicuous scarps. This coupled with the remoteness and accessibility constraints rendered accurate mapping of the island through conventional methods a tedious preposition. In this case also, terrestrial photogrammetric technique was applied to map an important portion of the island on 1:500 scale with a contour interval of one metre. The area was mapped with the help of three stereo-pairs taken from three sides of the island. Apart from topographic map, a geo-environmental map of the area was also prepares simultaneously. Various geo-environmental features recognised include wave cut platform, dissected marine terraces, scarps, sea side cliffs, gently sloping surface, stacks, master joints, sandy and shingle beaches, bushes and tree covered areas (Agarwal & Srivastava 1982)

The island exhibits a conspicuous erosional regime. A comprehensive geo-environmental evaluation of the area could easily be brought out with the help of the stereo-pairs.

3.3 Glacier mapping

Under International Hydrological Programme, a long-term, detailed glaciological study was initiated on a glacier in NW Himalayas. The existing large scale map of the glacier was on 1:50,000 scale only.

An accurate, large scale topographic base map of the glacier was required for the project. It was rather impossible to prepare the same through the conventional plane table or theodolite survey, mainly on account of following :

i) inaccessibility of many portions of the glacier, especially the upper reaches,

ii) large survey time required, during which period the glacier surface would have modified due to ablation and glacier movements, and

iii) rugged and dissected nature of lower portion of the glacier requiring considerable interpolation and consequent generalisation.

Taking recourse to terrestrial photogrammetric technique an accurate large scale topographic map of the glacier could easily be prepared on 1:5000 scale with a contour interval of 10m. Five stereo--pairs taken from vantage points selected around the lower portion of the glacier were used for the purpose. A photo-interpreted glaciological map was also prepared simultaneously showing moraine ridges, bergschrund, transient snowline. supraglacial channels, ice faces, crerasses and ice falls. Results have clearly brought out that terrestrial photogrammetry is the most appropriate technique for glacier mapping.

3.4 Nilgiris landslide investigation

Nilgiris district of Tamil Nadu was ravaged by numerous landslides on an unprecedented scale, causing considerable damage to life and property, following heavy rains during the winter monsoon of 1978-79. This necessitated a quick appraisal of the affected areas. As the conventional survey technique is a time consuming process, terrestrial photogrammetry was resorted, initially to map four important landslides in the area. The maps were prepared on 1:500 or 1:2000 scale with contour intervals of 2 m and 4 m, respecitvely. Apart from topographic details various cultural and landslide details were also mapped. These included old and fresh landslide scars, fracture lines, fissures, broken ground, houses, vegetation. The stereo-pairs were found to be highly useful for a comprehensive evaluation of the land-slides and significantly helped in a quick assessment of the slide areas.

4. DIGITAL TERRAIN MODELLING

The significance of morphomertic and thematic maps in the comprehensive evaluation of the project areas, particularly those related to geo-environmental aspects, hardly need any emphasis. In case of Nilgiris landslide investigation the 'Surface Area Ratio' maps of the two areas were prepared. These established a strong correlation between the slope parameters and the incidence of landslides and thereby helped in identifying safe and hazardous areas. Further, erosion, geomorphological and existing landuse maps of the areas were also prepared. Integration of the morphometric and the thematic maps led to a comprehensive evaluation of the areas (Agarwal & Sharma 1982).

Morphometric mapping requires topographic data in a digital form at desired intervals, depending upon the accuracy requirements and the surface irregularities of the terrain. Computer assisted analytical stereo-plotters in fact record the topographic data in a digital form. Also, they hold great promise in simultaneously recording the thematic data in a coded form, as well. Thus if an analytical stereo-plotter is used in place of an analogue stereo-plotter then apart from topographic base maps, morphometric and thematic maps can also be produced simultaneously, either independently or in combination.

DISCUSSIONS

Though the science of photogrammetry owes its origin to the terrestrial photogrammetry, it was soon overtaken by aerial photogrammetry due to its obvious advantages in quickly covering larger areas. However, terrestrial photogrammetry maintained its relevance and advantages in certain disciplines like glacier mapping, and surveys of steep inaccessible or remote areas. In snow mapping the significant advantages of the technique has clearly been demonstrated by Blyth et.al. (1974).

Inspite of the increased applications of terrestrial photogrammetry, particularly in Architectural engineering and various other fields, its application potential in geo-scientific investigations has not yet been fully exploited. In this context the case histories cited, amply demonstrate the potential and advantages of this technique, in geo-scientific studies. On account of the numerous advantages of this technique over conventional survey, the latter appears obsolete in comparison. The technique holds great merit in monitoring dynamic features, as repeat photography can easily be carried out from fixed camera stations, at desired time invervals.

While preparing maps by this technique on stereo - plotters the ratio between the model scale to the map scale is so selected that the maps prepared have a plan accuracy of 0.5 mm on the map scale, and the height accuracy lies within a quarter of the contour interval. This is as per the accuracy limits followed by the Survey of India (Agarwal 1974:78). Thus, if features to be monitored show variations larger than 1 mm on plan scale, these can be resolved through comnparison of different maps. In case monitoring of smaller variations is required then either an analytical or semi-analytical approach (Marzan & Karara 1976) is to be applied.

An analytical stereo-plotter, unlike an analogue one, has no constraints of photographic parameters, and hence it is more versatile, besides being more accurate too. Another aspect where it has significant advantage over analogue stereo-plotter is in digital terrain modelling which holds great promise for simultaneous topographical, morphometric and thematic mapping. However, it is much more sophisticated, costly and would involve the interaction with experts for its operation. Accounting various aspects, Marzan & Karara (1976) have opined that if the maps produced by analogue stereo-plotter meet the accuracy requirements for a particular work, then the anologue method is the simplest, most straightforward and perhaps the cheapest method to employ.

It may be summarised that the technique of terrestrial photogrammetry is unique for preparation of large scale topographic and thematic maps, for a comprehensive evaluation of the project areas in the shortest possible time. Simultaneous photo-interpretation, field checks and ground truth data collection considerably expedites the work.

However, it would be relevant to mention that the suitability of the technique for a given area largely depends upon the availability of suitable locations from where stereo-photographic coverage of the area could be obtained.

REFERENCES

Agarwal, G.C. 1974. Photogrammetric surveys, their planning and costing, Tech. Public No.7401, Survey of India, Hyderabad.

Agarwal N.K. 1985. Mapping of inaccessible escarpment slope of left abutment of Supa Dam, Karnataka, utilizing stereo-terrestrial photogrammetry, Jr. of Engineering Geology Vol.XIV, Nos. 1 & 2, ISEG.

Agarwal, N.K. & Srivastava, G.S. 1982. Geomorphology and Environment of Coconut Island, South Kanara distt., Karnataka, Proc. Symposium on Resources surveys for Landuse planning and Environmental Conservation, ISPI & RS, Dehradun, India.

Agarwal, N.K & Sharma. R.P. 1982. Terrestrial photogrammetric evaluation of slide areas, Nilgiris dist., Tamil Nadu, Proc. 4th International Congress, IAEG, India, Vol. III.

Blyth, K. Cooper, M.A.R., Lindsey, N.E. Painter, R.B. 1974. Snow depth measurements with terrestrial photos. Photogrammetric Engineering, Vol. XL, No.8.

Marzan, G.T. & Karara, M.M. 1976. Rational design for close range photogrammetry, photogrammetry series No. 43, Deptt. of Civil Engineering, Univ. of Illinois, Urbana, Illinois, 61891.

Manual of Photogrammetry 1980, Fourth Ed. Am. Society of Photogrammetry, Falls Church, U.S.A.

Symposium on Remote Sensing for Resources Development and Environmental Management / Enschede / August 1986

Regional geologic mapping of digitally enhanced Landsat imagery in the southcentral Alborz mountains of northern Iran

Sima Bagheri
New Jersey Institute of Technology, Newark, USA

Ralph W.Kiefer
University of Wisconsin-Madison, USA

ABSTRACT: This study evaluates the utility of using Landsat MSS data in regional geologic mapping of lineaments. Both conventional image interpretation techniques and digital enhancement techniques were utilized. The presence and orientation of lineaments can have great structural significance and a correlation may exist between them and zones of weakness characterized by seismic activity and mineral concentrations. The lineaments detected on computer-enhanced imagery of the study area exhibited definite trends providing a regional view of the geological "grain" of the area. When lineament alignments located by Landsat image analysis were plotted, a correlation was found between lineaments detected on the enhanced scene and earthquake epicenters, as well as the mapped location of phosphate deposits of the study area.

1. INTRODUCTION

Landsat images are widely used in regional geologic studies. They are especially useful for displaying extended structural elements such as intrusive bodies, domes, folded mountain belts, and fault and fracture zones. This study is concerned with the utilization of Landsat data for the detection of geologic lineaments in the southcentral Alborz Mountains of northern Iran. The term "lineament", as used here, is defined by Siegal and Gillespie (1980), as follows:

> Lineament: a two-dimensional geomorphological term referring to a mappable, simple or composite linear feature of a surface, whose parts are aligned in a rectilinear or slightly curvilinear relationship and which differs distinctly from the patterns of adjacent features and presumably reflects a subsurface phenomenon.

The study area is located in the southcentral part of the Alborz Mountains of northern Iran (Figure 1). The region is one of high to moderate relief which consists of marine sedimentary rocks of Paleozoic and Mesozoic ages and volcanic rocks of Tertiary age.

The primary objective of this research was to delineate geological structures by means of digital image processing techniques, since the presence and orientation of these features may have great structural significance and correlation may exist between them and seismic activities, as well as mineral concentrations.

The data used here include both film products and digital data of four Landsat MSS bands imaged in November 1976. The scene is characterized by a relatively low sun elevation angle (31°) and a light continuous snow cover in the mountainous areas, characteristics that are advantageous for mapping and interpreting the structural features. A lineament map and a rock type classification map were produced from the computer enhanced MSS data. These maps were compared with an earthquake epicenter map and the mapped location of ore deposits (mainly phosphate minerals) in order to investigate relationships between them.

2. REGIONAL GEOLOGIC SETTING

Recent discoveries indicate that the traditional concept of Iran as a pair of orogenic belts (the Alborz and Zagros Mountains) with a median mass between them has been invalid. Nor can central Iran be regarded as an eugeosyncline. The only distinction between the Alborz Mountain and central Iran is one of relief and the two areas are structurally and stratigraphically very similar (Stocklin, 1968). The Bouguer gravity anomaly map conforms with the regional structural pattern revealing no separation of the Alborz as an independent zone from Central Iran.

The Alborz Mountains are continuation of the Alpine type mountains, which are a complex, asymmetric belt of folded and faulted rocks. The Alpine orogeny of the Iranian ranges is characterized by the absence of nappes present in the European Alps which extend over a width of 1200 km (Stocklin, 1968). The compressive stress of Alpine orogeny results in thrusts and high-angle reverse faults in the central Alborz and a wide range of fold systems throughout the mountain chain. These structural features are surface manifestations of underlying faults, joints, folds, lithologic contacts or other geologic discontinuities and are expressed as lineaments of different dimensions on Landsat imagery. An understanding of these features and their significance in the regional geologic framework is essential in analyzing both earthquake prone areas and mineral potential locations of the Alborz Mountains.

Physiographically the range is divided into three segments--Western, Central and Eastern Alborz--with the central part subdivided into seven lithostructural units (Gansser & Huber, 1962). The study area is 3500 sq-km in size and is located between Karaj River to the west and Damavand volcano to the east and includes the following units of the southern Central Zone: (1) the Tertiary Central Zone, (2) the Southern Paleozoic-Mesozoic Zone and (3) the Southern Tertiary Zone.

3. METHODOLOGY

Two methods of analysis were utilized, visual image interpretation and computer-assisted interpretation.

Figure 1. The Study Site (southcentral Alborz Mountains of northern Iran) Landsat MSS Band 5 image - scale about 1:1,600,000.

The visual image interpretation was accomplished by means of a color additive viewer using 70mm black and white film positives of the four Landsat MSS bands. The interpretation depended upon the evaluation of image tone, texture, fabric and relief. Another visual image interpretation method was the use of Ronchi rulings for analyzing images to identify linear features. The Ronchi ruling used in this study is a diffraction grating having a spacing of 79 line pairs per centimeter. When the ruling is rotated between the eyes and the Landsat image, lines on the image that are perpendicular to the direction of ruling are enhanced (by diffraction), and lines in other directions become diffuse (Pohn, 1978).

The second method of analysis was computer-assisted image processing of Landsat digital data in which the lineaments were enhanced for the purpose of interpretation. In geological analysis, enhancement techniques are often performed on band 7 which is the preferred near-infrared band. The enhancement routines used here are suitable for diverse topography and complex structural geology. The enhancement is a form of digital image processing and involves the adjustment of brightness value for each individual pixel. Potentially useful enhancements include contrast stretching (linear, non-linear), band ratioing, high pass filtering and diagonal derivative processing. The most effective enhanced product for this study was computer-enhanced high-pass-filtered, contrast-stretched image of MSS band 7, as shown in Figure 2. This enhancement facilitated the interpretation and was especially useful in distinguishing between structural features and artifacts.

The computer-assisted image interpretation techniques provided superior results to the visual image interpretation techniques.

Spectral band ratioing was attempted, using blue for the MSS band ratio 5/4, green for MSS 7/6, and red for MSS 6/5. This technique did not enhance faults and other lineaments. It did, however, provide a clearer picture of alluvial deposits of different ages.

The lineament map (Figure 3) was produced from the computer-enhanced Landsat data and is based on the application of the following criteria for lineament identification (Short & Lowman, 1973): (1) lines of variable length, straightness and continuity which are differentiated by tonal contrast in images; (2) tonal discontinuities; (3) bands of variable width which contrast in tone to the area immediately adjacent; (4) alignment of topographic forms; (5) alignment of drainage patterns; (6) association of vegetation along linear trends; and (7) co-alignment of cultural features (e.g., farms, roads pattern, etc.) with underlying structural and/or surrounding topographical control.

In addition to enhancement algorithms, an attempt was made to obtain a lithological classification using computer-based spectral pattern recognition. The classification routine was applied to the Landsat MSS data set in order to discriminate rock types by focusing on both sides of major lineaments and by attempting to identify whether the structural break occurs at near, or at some depth beneath, the surface. This discrimination provides an alternative

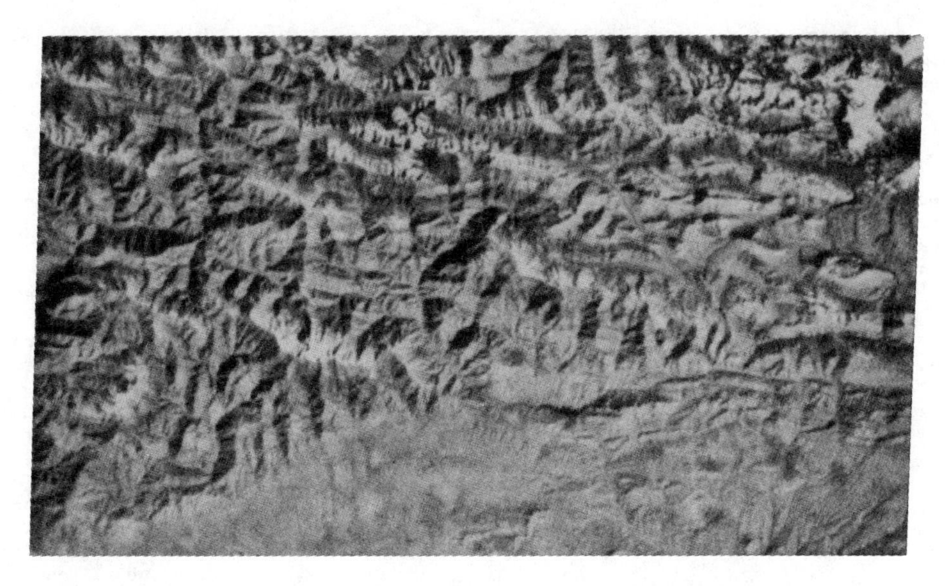

Figure 2. Landsat MSS Band 7 image of a subscene of the study site. High-pass filtered and stretched data - scale about 1:750,000.

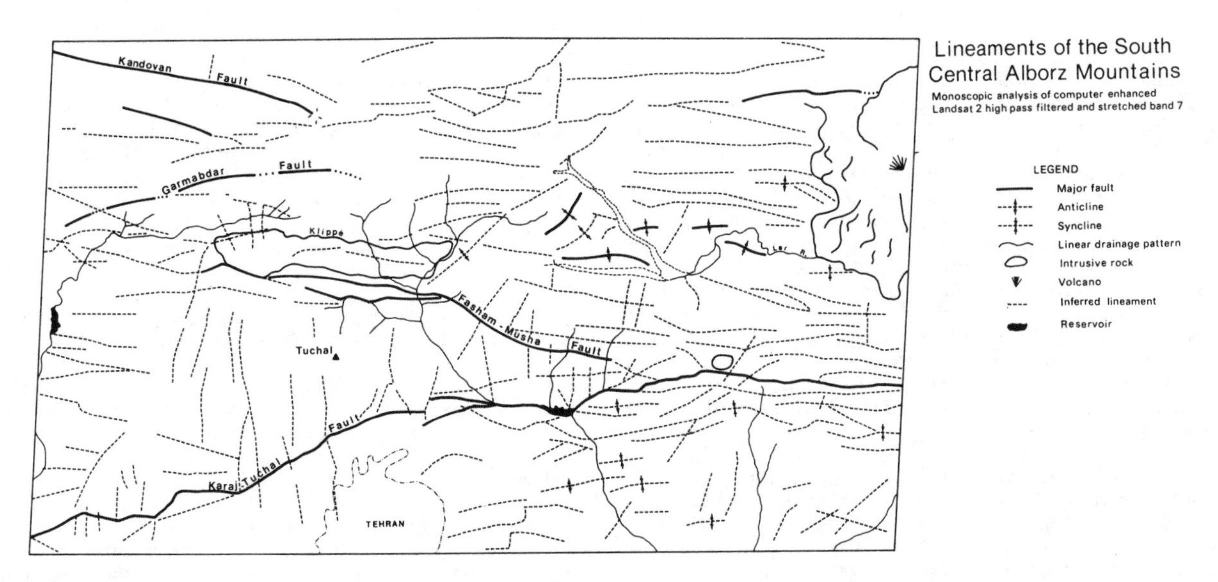

Figure 3. Lineaments of the southcentral Alborz mountains - scale about 1:750,000.

concept to guide mineral resource exploration and can be used as a complementary procedure to geophysical techniques for the purpose of exploration. Additionally, this technique of rock discrimination could be utilized in order to extend geological mapping into unmapped and inaccessible areas within the region.

It should be noted that image classification techniques have not been as widely used for geologic applications as enhancement techniques. This is due to the fact that classification provides information on cover conditions and is affected largely by non-homogenity of geologic units as well as similarity of spectral signatures of different rock types.

A supervised classification--maximum likelihood classifier--was used to identify the individual pixels in the scene. Training sets were delineated on the computer's color display screen with polygon programs. The selection of training sets was aided by field geology information and topographic maps. Data on the sample means, and variance-covariance matrices were derived from training set statistics. Several statistical programs, including training set check and training set divergence programs, were used to evaluate spectral separability by creating confusion matrices and computing transformed divergence for each pair of training sets. The transformed divergence analysis procedure is described by Haack (1984) as follows:

> Transformed divergence, which is calculated from the means and covariance matrices of each spectral class or training site, is a measure of the statistical distance between class or site pairs of interest and provides information on their "separability". This separability is an indirect estimate of the likelihood of correct classification between groups of different band combinations. Such an estimate provides information usually obtained by the time consuming and expensive process of actual classification and accuracy evaluations. Transformed divergence can

also be used to examine intraclass variability, to examine how separability may change with an increase or decrease in the number of channels utilized, and to determine the spectral channels most useful for classifying specific class pairs.

The transformed divergence analysis identified the optimum subset of two MSS bands as bands 5 (.5 to .6 μm) and 7 (.8 to 1.1 μm). Note that the selected bands are those that have been generally recommended for geologic analysis. The transformed divergence algorithms allow the analyst to determine if information classes are spectrally unique or should be modified by merging or deleting classes.

The resulting classification of rock types (not illustrated here), based on the procedures described above, is estimated to be about 80% accurate based on field data.

4. STRUCTURAL CHARACTERISTICS OF LINEAMENTS DETECTED FROM ENHANCED LANDSAT IMAGERY

The major structural elements of the southcentral Alborz Mountains are faults. These faults trend east-west, causing the Precambrian formations drop from the height of Alam Kuh (4800 m) to an unknown depth below Kavir plain (Asserto, 1966, Stocklin, 1974).

The small scale of Landsat images makes the alignment of individual features across large areas more obvious. Within the study area there are several lineaments that possess characteristics normally associated with faults, which are classified as such and marked by names (Figure 3). The main fault traces correspond to Kandovan, Garmabdar, Fasham-Musha and Tuchal-Karaj faults. These fault traces appear on the enhanced Landsat images as terrain scars, and are easily recognizable over a distance of many kilometers. The continuity of the Garmabdar Fault, particularly at its western end, is surprising when compared with information previously acquired. On a geology map (Asserto, 1966), the location of the Garmabdar Fault is in part assumed, and shows a bend toward north that can be confirmed by examination of the Landsat images.

Despite the linearity of the Fasham-Musha Thrust, there are minor bends in the fault that may, in fact, be small offsets at its intersection with several small transverse faults. This suggests that tectonic forces have continued to be active in recent times. Based on geological mapping, this lineament is the trace of a high angle fault, with evidence of fault breccia and stratigraphic displacement on the ground. It is clearly depicted on the Landsat MSS images that all the fold axes here are parallel to the trace of this thrust. In addition to the trace of the fault, the outline of Shahrestanak Klippe, a calcareous allochthonous, mass is traceable.

In the case of Tuchal-Karaj Upthrust, the line of structure seems to be fairly clear indication of a thrust marking the Alborz front as a continuous scar. The trace of this thrust is marked by springs and patches of vegetation that appear in dark tone on the Landsat images. This fault separates the gentle morphology of short ridges and elongated hills to its southern side from the intensity folded zone to its northern side.

In addition to fault lineaments, several fracture zones were identified on the enhanced Landsat imagery of the study area. These features include linear features attributable to terrain discontinuities that are arranged in sets having a common orientation. The fracture traces of Qazvin Plain located near the far west boundary of the study area are typical expressions of such structure and trend ESE to WNW parallel to the main trends of the area. Some east-striking structural elements interfere with this strike direction and are visible in the southern border of Alborz Mountains. It is postulated that the interplay of these trends contributed to the occurrence of Boyin Earthquake.

Topographic lineaments are additional features depicted on Landsat scene of the study area. They are characterized by the alignment of topographic features and tonal alignments indicated by specific soil, moisture, vegetation and relief.

On the basis of analysis of the enhanced Landsat MSS images of the study area, most of the topographic lineaments, faults, and other lineaments seem to be related to a system of compressional stress that extends from the south or central Iranian Plateau to the north. The tectonic activity along this stress zone has been going on since Cretaceous time (Stocklin, 1974). The most recent activity was manifested by numerous earthquakes, suggesting that the evolution of the range is still in progress.

5. CORRELATION OF LINEAMENTS WITH EARTHQUAKE AND MINERAL OCCURRENCES

The identification of previously unknown lineaments and fracture systems has been a major contribution of Landsat image analysis in mineral exploration and the identification of earthquake occurrences. Indications of the relationship between lineaments and zones of seismic activities, as well as mineralization areas, can be detected by comparative analysis of lineament maps, maps of the regional distribution of mineral deposits, and epicenter maps of historical earthquakes.

The detection of such information on Landsat imagery could provide a better understanding of seismic risk hazards or mineral concentration locations for regional development schemes. The information derived from an analysis of lineament concentrations can indicate where the crust is the weakest and, therefore, most likely to be mineralized or susceptible to seismic activity.

For the purpose of this study, overlays were prepared depicting (1) the occurrence of phosphate deposits and (2) the epicenters of historic earthquakes within the study area. These overlays were compared with the orientation of the lineaments interpreted from enhanced and classified Landsat images.

A correlation was found between the distribution of phosphate deposits in central Alborz (Geological Survey of Iran, 1974), and lineament trends expressed on the Landsat images. The comparison also demonstrated the dependence of seismically active zones on the regional structural features and showed a good agreement between the general distribution of epicenters and active faults.

The significance of these correlations in future mineral exploration and earthquake risk hazard assessment is of invaluable importance.

6. SUMMARY

Landsat images are being used operationally in the completion of geologic maps of different parts of the world. Geologic maps based on Landsat-derived information are especially useful in providing information in areas where field geological mapping is sparse or lacking.

In this study, the interpretation of enhanced Landsat MSS imagery revealed many previously unmapped lineaments and led to the discovery of several formerly undetected active faults responsible for much of the the seismic activity in the southcentral Alborz region. For example, the study showed that the rocks on both sides of major lineaments in the Tuchal-Karaj area are not the same in age and lithology and appear to have been significantly deformed along the fault. This finding suggests that this lineament represents a deep-seated structure rather than a structural break near the surface as is the case for other major lineaments.

The detection of lineaments also helped to identify areas that have the potential for discovery of phosphate, providing an accessible source of agricultural fertilizer. The development of this source could be a great step toward the mitigation of Iran's production problems.

In addition to mineral identification, Landsat lineament analysis facilitates the assessment of the risk of damage from earthquakes. Such assessments are essential to providing for the safety and well-being of the area's inhabitants.

WORKS CITED

Assereto, R. 1966. Geology of Jadjerad and Lar Valley. Univ. of Milan, Milan, Italy.

Gansser, A., and H. Huber 1962. Geological observations in the Central Alborz, Iran. Schwerz Min. Petr. Mitt. vol. 42, no. 2: 583-630.

Geological Survey of Iran 1974. Reports no. 10 and 11, Tehran, Iran.

Haack, B.N. 1984. L- and X-Band Like- and Cross-Polarized Synthetic Aperture Radar for Investigating Urban Environments. Photogrammetric Engineering and Remote Sensing, vol. 50, no.3: 331-340.

Pohn, H.A. 1978. The Use of Ronchi Rulings in Geologic Lineaments Analysis. Proc. 3rd International Conference on Basement Tectonics, Durango, Colorado: 250-254.

Siegal, B.S. and A.R. Gillipsie (eds.) 1980. Remote Sensing in Geology. Wiley, New York.

Stocklin, J. 1968. Structural History and Tectonics of Iran, A Review. AAPG Bull., vol. 52: 1229-1258.

Stocklin, J. 1974. Northern Iran; Alborz Mountains. Scottish Academic Press, London, pp. 213-234.

SELECTED BIBLIOGRAPHY

American Society for Photogrammetry and Remote Sensing 1984. Multilingual Dictionary of Remote Sensing and Photogrammetry. Falls Church, VA.

American Society for Photogrammetry and Remote Sensing 1985. Special Issue on Landsat Image Data Quality Analysis. Photogrammetric Engineering and Remote Sensing, vol. 51, no. 9. Falls Church, VA.

American Society of Photogrammetry 1983. Manual of Remote Sensing. Second Edition.

Bagheri, S. 1984. Utilization and Application of Landsat CCT's (Computer Compatible Tapes) in the Detection of Lineaments in the Southcentral Alborz Mountains of Northern Iran. Ph.D. Thesis, University of Wisconsin-Madison.

Barzegar, F. 1979. Rock Type Discrimination Using Enhanced Landsat Imagery. Photogrammetric Engineering and Remote Sensing, vol. 45, No. 5: 605-610.

Jensen, J.R. 1986. Introductory Digital Image Processing: A Remote Sensing Perspective. Prentice-Hall.

Lamar, J.V. 1974. Applications of ERTS images to study of active faults. USGS Tech. Rept. 74-3.

Lillesand, T.M. and R.W. Kiefer 1986. Remote Sensing and Image Interpretation, Second Edition. Wiley, New York.

Lyon, R.J.P. 1977. Mineral Exploration Applications of Digitally Processed Landsat Imaging. Proceedings: 1st William T. Pecora Memorial Symposium, USDA Paper 1015, Government Printing Office, Washington, D.C.

Moik, J.G. 1980. Digital Processing of Remotely Sensed Images. NASA SP-431, Government Printing Office, Washington.

Rieben, H. 1955. The Geology of Tehran Plain. Amer. Jour. Sci., New Haven, Conn., vol. 253: 617-639.

Sabins, F.F., Jr. 1978. Remote Sensing: Principles and Interpretation. Freeman, San Francisco.

Schowengerdt, R.A. 1983. Techniques for Image Processing and Classification in Remote Sensing. Academic Press, New York.

Short, N.M. and P.D. Lowman 1973. Earth Observations from Space: Outlook for Geological Sciences. NASA X-650-73-316, Greenbelt, MD.

Sheffield, C. 1981. Earthwatch--A Survey of the World from Space. Macmillan, New York.

Sheffield, C. 1985. Selecting Band Combinations from Multispectral Data. Photogrammetric Engineering and Remote Sensing, vol. 51, no. 6: 681-687.

Shih, E.H., and R.A. Schowengerdt 1983. Classification of Arid Geomorphic Surfaces Using Landsat Spectral and Textural Features. Photogrammetric Engineering and Remote Sensing, vol. 49, no. 3: 337-347.

Short, N.M. 1982. The Landsat Tutorial Workbook. NASA Ref. Pub. 1078, Government Printing Office, Washington.

Slater, P.N. 1980. Remote Sensing: Optics and Optical Systems. Addison-Wesley, Reading, MA.

Smith, W.L. (ed.) 1977. Remote Sensing Applications for Mineral Exploration. Dowden, Hutchinson & Ross, Stroudsburg, PA.

Swain, P.H., and S.M. Davis (eds.) 1978. Remote Sensing: The Quantitative Approach. McGraw-Hill, New York.

USGS 1979. Landsat Data Users Handbook Revised.

USGS and NOAA 1984. Landsat 4 Data Users Handbook.

Symposium on Remote Sensing for Resources Development and Environmental Management / Enschede / August 1986

Operational satellite data assessment for drought/disaster early warning in Africa: Comments on GIS requirements

Hubertus L.Bloemer & Scott E.Needham
Ohio University, Athens, USA

Louis T.Steyaert
NOAA, NESDIS/AISC, Columbia, Mo., USA

ABSTRACT: The National Oceanic and Atmospheric Administration (NOAA/National Environmental Satellite Data and Information Service (NESDIS) Assessment and Information Service Center (AISC) has developed operational climate impact assessment for improved drought/disaster early warning in semi-arid regions of Africa. The system is based on daily United States NOAA polar orbiting, Advanced Very High Resolution Radiometer (AVHRR) satellite data used in combination with ten-day rainfall reports from ground stations throughout the region.

The current assessments are prepared using a "light table" G.I.S. approach for map/image overlay and statistical time series analysis. As part of the United States Agency for International Development's (AID) funded project, NOAA/AISC and Ohio University developed and defined requirements for a cost effective, reliable computer based G.I.S. In addition to describing the assessment process, computer hardware and software considerations to meet the needs of the spatial analysts are discussed. Remote sensing data processing and G.I.S. capabilities are assessed according to various data handling proficiency and applicability of data. This includes considerations of a variety of computer systems currently available, including "turn-key" stations with G.I.S. packages as well as a comparison of the obtainable G.I.S. software packages for different types of data sets.

1 INTRODUCTION

The National Oceanic Atmospheric Administration (NOAA), National Environmental Satellite Data, and Information Service (NESDIS) Assessment and Information Service Center (AISC) has developed a reliable and cost-effective program to support disaster early warning and technical assistance objectives of the Agency for International Development (AID). The operational Early Warning Program was developed at the request of the AID Office of U.S. Foreign Disaster Assistance (OFDA) and in cooperation with the University of Missouri-Columbia, US/AID Missions, and host countries. Qualitative climatic impact assessments routinely provide early warnings of weather impacts on subsistence agricultural crops and the potential for drought caused food shortages with a lead-time of 3-6 months before socio-economic impacts occur. Recent advances in the operational use of daily NOAA polar orbiting meteorological satellite data for agricultural assessment have significantly upgraded the existing system which is based primarily on daily weather station reports. Global meteorological satellite information increases the spatial resolution for analysis and presents an opportunity to incorporate a Geographic Information Systems approach into the assessments. This paper comments on some of the GIS requirements for the upgraded assessment system.

2 BACKGROUND ON EARLY WARNING PROGRAM

Since July 1979, AISC has issued weekly and monthly assessments of climatic impacts on food security for developing countries in the Caribbean Basin, Africa, South and Southeast Asia, and more recently, the South Pacific Islands, Central America, and the Andes countries of South America. Decision makers, planners, and economists are provided with timely, reliable information based on continuous monitoring of environmental conditions. The AISC assessment cables are sent through NOAA international communication facilities to American Embassies, AID including overseas missions, the U.S. Department of Agriculture, and various United Nations agencies (e.g., Food and Agriculture Organization). Climatic impact on potential food supplies, subsistence crop conditions and field operations are assessed. Information on crop calendars, estimated soil moisture, and unusual or severe weather events such as flooding, high winds and other meteorological extremes are reported. If available, information from ancillary sources is included to supplement AISC analysis. The result is a cost-effective increase in the lead-time for planning of food assistance strategies and mitigation measures to reduce the adverse impact of climate.

In addition, AISC prepares special assessment reports for AID, e.g.: the 1981/82 drought impact in Bangladesh, Sri Lanka, Malaysia, Tanzania, Botswana; the 1981/83 drought problems in Nepal, Philippines, Ethiopia, Somalia, Uganda, Sudan, Sahel Region, southern Africa, Haiti, NE Brazil and the South Pacific; and more recently Kenya, Tanzania and Sahel countries. These were used by AID as one input to estimate the magnitude of drought impact, potential food shortage deficits and/or disaster assistance needs. The Early Warning Program is based on weekly rainfall/weather analysis and climatic impact assessment models for more than 350 agroclimatic regions (i.e., regions which are generally homogeneous with respect to agricultural crops and climatic type). Regional rainfall estimates are determined from ground station reports received through an international communications network. Satellite cloud data (e.g., METEOSAT) are used to improve the accuracy of precipitation estimates, particularly in those regions where weather data are sparse and unreliable. Weather data are then interpreted by regional agroclimatic indices which indicate potential crop production in relative terms. Finally, climatic impact and the potential for abnormal food shortages are identified from these indices.

Agroclimatic/crop condition indices are based on derived climatic variables (e.g., soil moisture, plant water deficit and moisture stress) which directly determine the plant's response to environmental conditions, hence productivity. The selection of the regionally appropriate indices is in part determined through the use of episodic event data. For example, candidate indices are qualitatively compared to the historic occurrences of crop failure, food shortages, drought, flooding and other anomalous weather/non-weather events as determined from reports newspapers, and computerized data bases. Episodic data are also used to "calibrate" the index by establishing a critical threshold which is associated with crop failure and/or drought-related food shortages. The strengths and limitations of all available indices and models are collectively considered in the assessment process.

Documented food shortages determined by in-country field inspections reported by the UN/FAO and American Embassies during 1979-1983 confirmed that AISC assessments consistently provide a 3-6 month alert on potential food shortage situations. Typically, the initial AISC assessment on drought/food shortage conditions was issued 30-60 days before the beginning of the crop harvest. Actual indications of food shortage conditions were not evident until 2-4 months after harvest.

3 UPGRADED CLIMATIC IMPACT ASSESSMENTS FOR AFRICA

The NOAA/NESDIS Early Warning Program for Africa has been upgraded as the result of recent advances in the application of NOAA polar orbiting satellite data for agricultural assessment. Color-coded images and vegetation/biomass indexes are derived from the NOAA AVHRR sensor (Advanced Very High Resolution Radiometer) using daily four kilometer GAC data (i.e., Global Area Coverage). These satellite products have substantially upgraded AISC's ability to detect drought, determine its regional extent between sparse rainfall reporting stations, and assess relative vegetation/biomass conditions. Improved assessments of weather impacts on crops and rangelands are based on the integrated use of agroclimatic/rainfall index models and NOAA AVHRR satellite models. The integration of agroclimatic and satellite derived assessment information represents a classic opportunity for using a Geographic Information Systems (GIS) approach. In fact, AISC staff in Columbia, MO and Washington, D.C. are using a "light table" system to overlay and analyze various data for operational assessment of weather impacts on African agriculture.

The techniques presently used for processing information in interpretative visual analysis are time consuming endeavors which need to be automated, i.e., computerized. Drawbacks of the manual technique are unavoidable due to the sheer volume of spatial, temporal and thematic (special topic) data, along with the scale at which they are obtained. There is an urgent need to embrace a new generation of spatial analysis for climatic impact assessment, namely automated Geographical Information Systems (GIS).

To gain a better understanding of our diversified planet, scientists must gather, store, and analyze tremendous volumes of data on a wide range of subjects. Information needed in spatial analysis can range from bedrock structure, soils, weather observations, natural vegetation to population distribution, land use, political, economic and social structures. Understanding our world in terms of the physical and cultural landscape entails identifying and analyzing numerous relationships resulting from human and environmental interactions. The range, scope, and magnitude of information necessary for analyzing this world of the physical environment, socio-economic systems, and historical ties can become overwhelming.

In recent years government agencies have recognized the potential of Geographic Information Systems (GIS) as a technology suited for rapid data manipulation. GIS system, therefore, are viewed as an integral component in spatial problem analysis and assessment.

4 DEFINING GIS

A Geographic Information System is the overlaying of spatial data (i.e., population density, land cover types, land ownership, transportation network) for the same geographic space at a uniform scale. In a more sophisticated expression, a GIS is geocoded information in a spatial and/or tabular format with manipulation capabilities of georeferenced relationships.
Automated GIS are the results of methodological changes (i.e., digital data display in image format; superimposing of two georeferenced images) brought about by advances in technology. James and Martin (1981, p. 405) observe that:

. . . changes in the technology of observations and analysis not only provide spatial scientists (geographers) with more information than has even been available before, but also provide a means of storage and recall, and a way to carry out complex analysis.

The advent of automated or computerized data processing techniques in the 1950's and 1960's revolutionized the capabilities of creating and handling all types of information in problem solving procedures. With the aid of computers, vast amounts of data can be analyzed quickly and efficiently. Computers provided new means to analyzing complex problems through identifying relationships between variables never before thought possible. It was in the environment of the "quantitative revolution" that geographic data processing systems and GIS evolved.

An automated GIS is a system that has as its primary source of input a base composed of geographic referenced data coordinates, and the majority of processing is done digitally with a computer. This digital GIS can be viewed as a multilayered vertical structure of related georeferenced data sets designed for use in solving spatial problems in a computer environment (fig. 1). Further, GIS are capable of analyzing large volumes of data acquired from a variety of sources (Marble, Peuquet and Calkins, 1984).

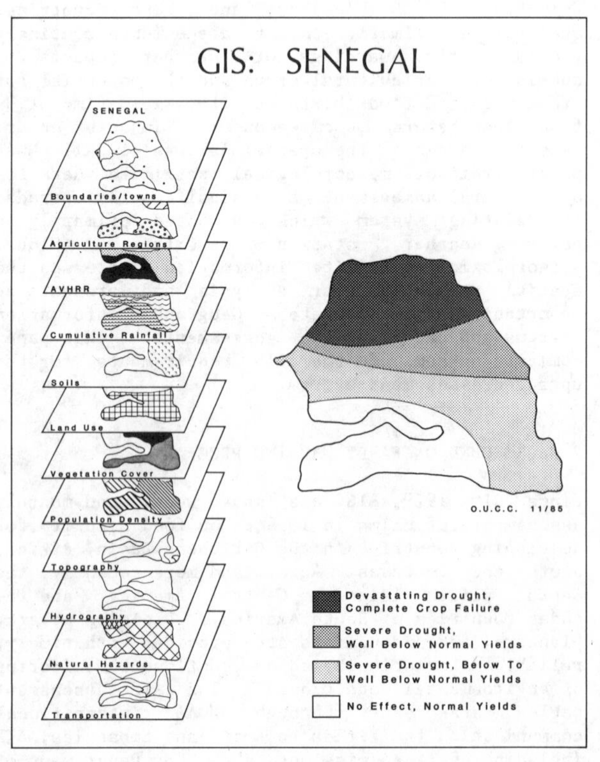

fig.1

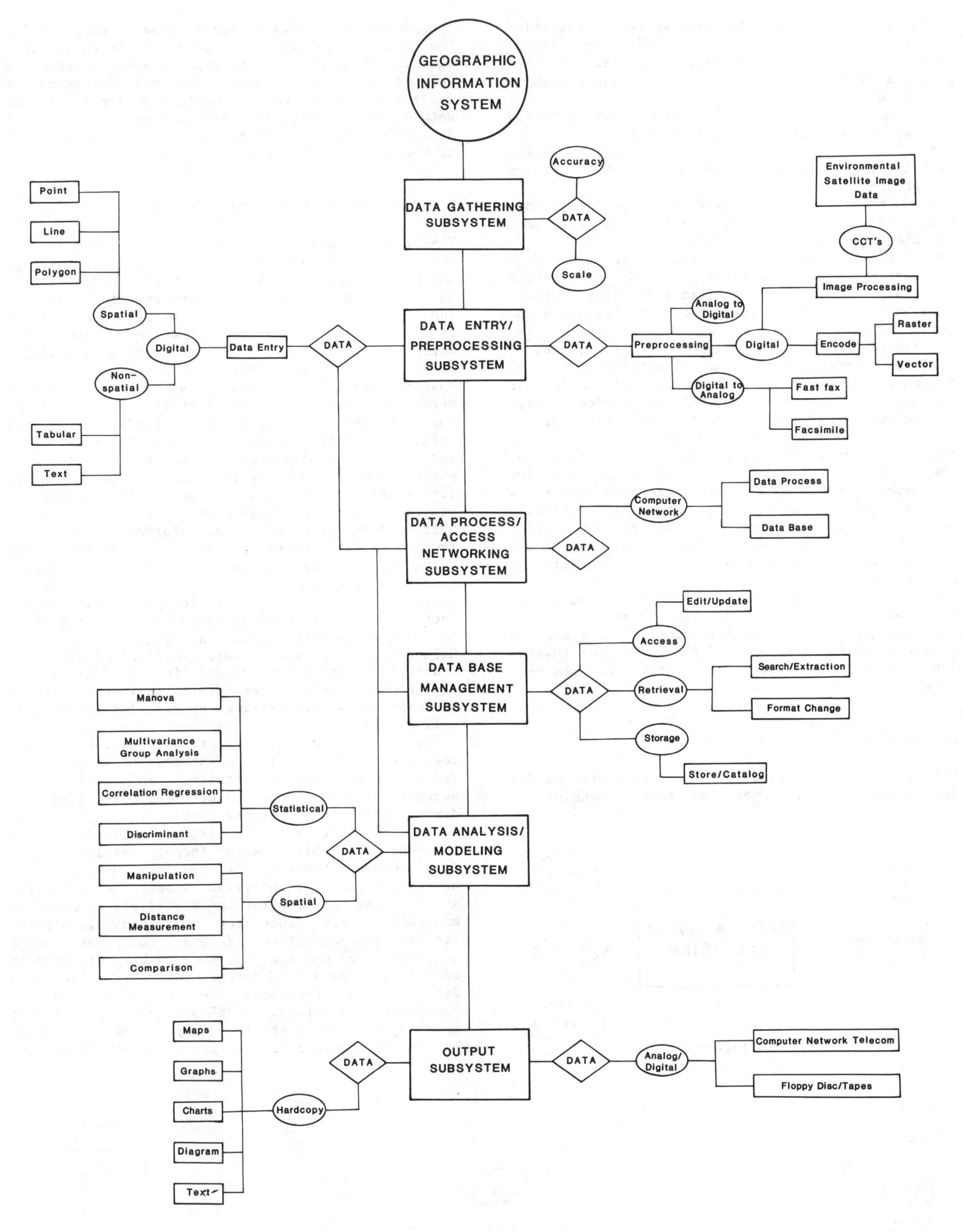

Figure 2

Dangermond (1984, p. 1-26) has identified a number of advantages of using Geographic Information Systems in spatial analysis, they include:

1) Data maintained in a physical compact form (e.g. the magnetic file).

2) Data can be maintained and extracted at a lower cost per unit of data handled (i.e., automated versus outmoded manual data retrieval).

3) Data can be retrieved with much greater speed.

4) Various computerized tools allow for a variety of types of manipulation including, map measurement, map overlay, transformation, graphic design and data base manipulation. (These operations would be cost and time prohibitive in a manual setting.)

5) Graphic and nongraphic (i.e. tabular and descriptive information) can be merged and manipulated simultaneously in a "related" manner.

6)Rapid and repeated analytic testing of

conceptual models involving geographic relationships can be performed (i.e. land suitability/capability). This facilitates both scientific investigation and policy analysis criteria over large areas in short periods of time.

7) Change analysis can be efficiently performed for two or more different time periods.

8) Interactive graphic design and automated drafting tools can be applied to cartographic design and production.

9) Certain forms of analysis can be performed cost effectively that simply could not be done efficiently if performed manually.

10) There is a resultant tendency to integrate data collection, spatial analysis, and decision-making processes into a common information flow context. This has great advantage in terms of efficiency and accountability.

Evaluating the capability of the techniques, as well as understanding the limitations of the systems, will permit the spatial scientist (assessor) to determine if a GIS can provide answers to spatial assessment questions. Knowledge of the basic data elements will help in the design of future GIS and avoid duplication or "reinvention of the wheel." The capabilities of a GIS for geographic problem solving are tremendous if systems are designed properly and output is optimally utilized. Uses of Geographical Information Systems are then only limited to the articulation of the problem to be studied and the availability of reliable data.

Many GIS have six primary functions in common: 1) data entry, 2) encoding, 3) preprocessing, 4) data base management, 5) spatial/statistical analysis and modeling and 6) statistical/graphic output (Myers, 1985, Brumfield, 1983, 1985) (fig. 2; see next page).

5 DATA GATHERING

Data gathering (i.e., selection) procedures of GIS development are the most important methodological consideration for GIS design outside of establishing the overall information needs of the user. This step is vital, because the type of data included and the scale of data resolution will determine the quality and usefulness of the information generated. Data accuracy must also be considered in terms of data gathering procedures, for accuracy is relative to the scale of data compilation (fig. 3).

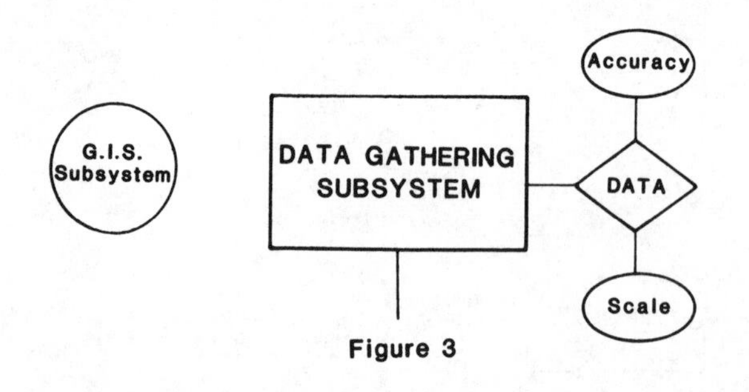

Figure 3

6 DATA ENTRY/ENCODING/PREPROCESSING

The largest portion of time and energy invested in developing a GIS system is in obtaining, entering, converting, and storing data. Useful data exist in various forms such as tabular, graphical, digital and remotely sensed video/digital. A primary problem of a GIS system, therefore, becomes the integration of these various forms into a single computer compatible format (fig. 4).

Spatial data are referred to as layers, fields or variables. Representation of these spatial entities must retain two basic characteristics of the original when the information is to be utilized in automated GIS processing: 1) the actual variable or characteristic, such as its name or value, and 2) its spatial location, or where it resides in geographical space (Dangermond, 1984).

Spatial data layers exist as information in one of four possible states: points, lines, surfaces, and polygons. These layers have their geographic components encoded by one of two basic techniques: vector/polygon format or raster/grid cell format. Vector form is created by assigning x,y coordinates to various points along a line or polygon. Cartographic entries into digital format are translated point for point and line for line (Marble and Peuquet, 1983). Raster form is located by its position on a predetermined spatial lattice or grid cell system.

These two forms of data transformation or geo-coding each have their advantages and limitations. Vector format depicts spatial information more accurately since it is point referenced data in N dimensional math coordinate space (e.g. x,y) and is more compatible with statistical analysis (Marble and Peuquet, 1983). Vector format results in less data volume, however, it requires more computer processing time for analysis. Raster format results in a loss of geographic specificity, however, manipulation and processing efficiency is higher. Vector representation refers to points whose positions are defined by the number of axis in coordinate space. Raster format represents a defined area, including the point position in coordinate space (e.g., AVHRR pixels). Spatial data modeling procedures are less complicated with raster because it provides more accurate registration, yet,

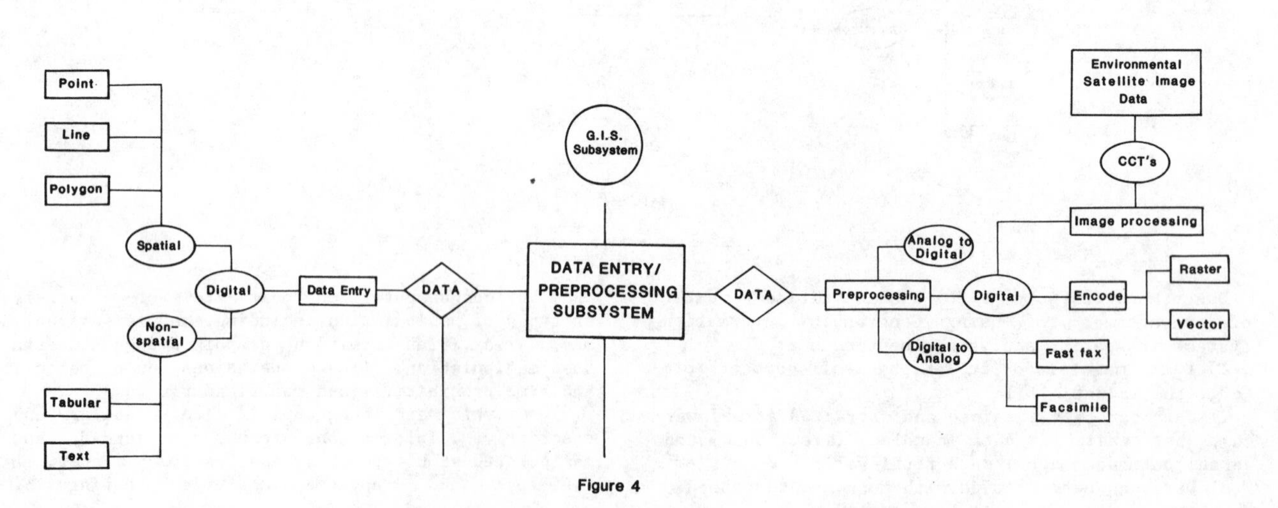

Figure 4

is positionally less accurate. The strengths of both formats (grid cell and x,y coordinates) can be utilized if the GIS is capable of dealing with both types of data and providing a means of conversion from one structure to the other (Myers, 1983, 1985; Tomlinson, 1983). In an applied situation, such as AISC, vector format is recommended for data storage and some data array manipulations. These include multivariate analysis of precipitation, temperature, soil moisture, terrain and ground cover data. The raster format includes AVHRR type data for overlay and display. For spatial analysis, vector data formats can be readily converted to raster in modeling procedures.

7 DATA BASE MANAGEMENT

A GIS must have the capabilities of supporting numerous data bases that are equally accessible by multiple users. An effective data base management system enables efficient data storage, retrieval and update (fig. 5).

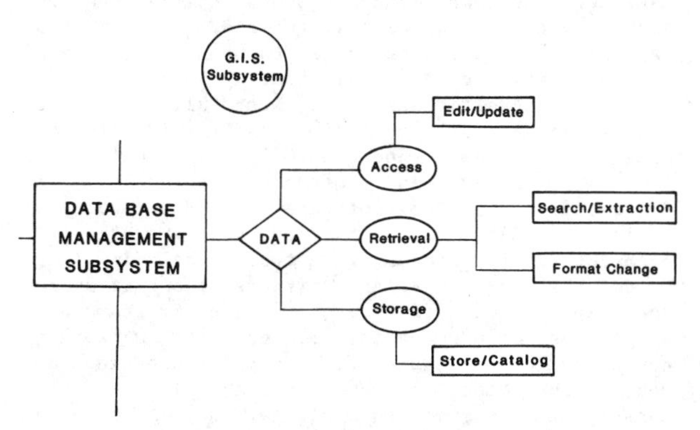

Figure 5

7.1 Data Accessibility and Telecommunication

Pertinent data collection and processing capabilities may not always reside within the particular agency interested in performing a specific task. In reality, information and specific processing capabilities are located at various national and international institutions. Further, not all institutions incorporate similar standards in terms of data format, data information content, or computing hardware and software. Telecommunication links and associated communication software enables data format transformation, transfer and processing to take place thus strengthening the capabilities of the single agency. The concept of telecommunication linkage brings together information and data processing capabilities needed in complex problem solving (fig. 6).

7.2 Data Analysis

Data analysis procedures enable the GIS user to statistically manipulate data sets so the specific information can be generated for particular interpretation. The procedures range from simple overlaying of data sets to complex statistical functions integrated into sophisticated spatial models. For example, canonical analysis can be applied to maximize discrimination of spectral grouping in AVHRR data as well as data reduction in feature extraction. The result is a means to better understand spatial relationships through examining a particular phenomenon and the relationships between phenomena (fig. 7).

Analysis of data from multiple spatial data layers requires processing techniques (algorithms) and procedures (processing sequences of algorithms)

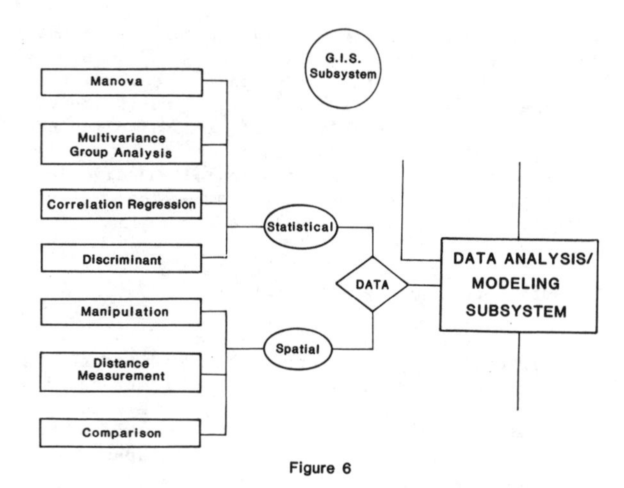

Figure 6

suitable for manipulation of either grid cell or x,y coordinate structured data. The core of a GIS is the computer's capabilities to process spatial data to the specification of the user. Automated GIS capabilities must include:

1) search, identification and extraction of spatial and non-spatial items in a data file, 2) format changing, 3) image data manipulation, 4) measurement and 5) comparison (Calkins and Tomlinson, 1977).

Computer based geographic information would be of little use if it could not be identified and accessed. Data search and retrieval operations select user specified data for direct application or subsequent manipulation. These extraction procedures may involve entire scenes, portions of scenes, or specified geographic entities. Search and extraction functions should include browsing capabilities, through both graphic and non-graphic data, windowing capabilities, query window generation capabilities (i.e., calling up specific geographic entities) and Boolean capabilities (i.e., specifying extraction of data based on non-graphic data querying) (Dangermond, 1984).

Manipulative functions are designed to change the data's physical quality to make it more useful or readable to the user and/or more amenable to further computer based spatial analysis (Calkins and Tomlinson, 1977). Manipulation functions include edge mapping, contouring, reclassification of polygons, map sheet manipulation (i.e., scale change and distortion removal), projection and coordinate change, coordinate rotation, and format change (i.e., vector to raster).

Measurement functions enable users to inventory spatial data. Spatial measurements can be end products in themselves, or they can be utilized in higher levels of statistical analysis. Measurement capabilities consist of determining distances, areas, and volumes. If temporal, multi-date information is incorporated, then rates and measurement changes can be determined.

Comparison functions allow users to detect, analyze and interpret relationships between different spatial data sets. These procedures

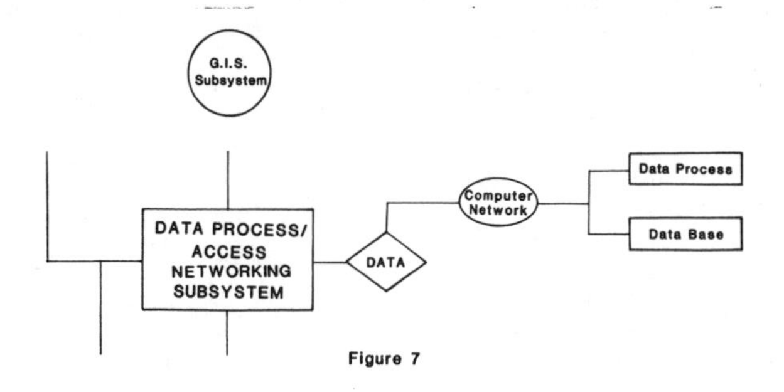

Figure 7

permit simple overlaying in grid and polygon format, which result in composite maps and dissolve functions which perform the opposite operation by creating individual maps out of complex multi-attribute maps. Comparison functions also include more involved statistical calculation of spatial data relation-ships. Operation include multivariate analysis, correlation analysis, time series analysis and canonical correlation analysis.

7.3 Data Output and Report Generation

Before any decision making process can be initiated, geographical information must be presented to the decision maker in a clear, concise and comprehensive format. A GIS must generate graphics and text to the exact specifications of the user. GIS output can take the form of images, charts, graphs, maps, facsimile data, or it can be generated as reports and tabular data (fig. 8).

8 RECOMMENDED GIS FOR NOAA's CLIMATIC IMPACT ASSESSMENT

The AISC GIS configuration must consist of an amalgamation of computer hardware, software, and peripheral entry and display instruments along with the information that is to be analyzed. All systems must function in unison to achieve the specified goals of AISC in natural disaster monitoring, impact assessment, and forecasting.

8.1 Delineating Data Variables

The following listing of data variables identifies inputs that must be integrated into a GIS for environmental monitoring, environmental and socio-economic impact assessment, and environmental and socio-economic modeling, forecasting and advanced disaster warning/preparation.

Weather Station Data
- precipitation
- temperature
- cloud cover (satellite)
- precipitation estimates (satellite)

Environmental Observations in Cartographic Form
soils (map)
topography (map)
hydrography (map)
natural vegetation cover (map/satellite)
vegetation response (AVHRR satellite)

Cultural Observations
political units (towns, borders, etc.) (map)
transportation (map)
population density/distribution/character (map)
land cover/use (map/AVHRR satellite)
- agriculture
  - crop suitability index
  - existing crop regions
  - production/yield observation
  - inputs (fertilizer, irrigation, mechanization, labor)
  - crop calendars
  - grazing
- livestock density
  - migration routes
  - water holes
- basic need support
  - markets
  - hospitals
  - good storage depots
  - refugee camps

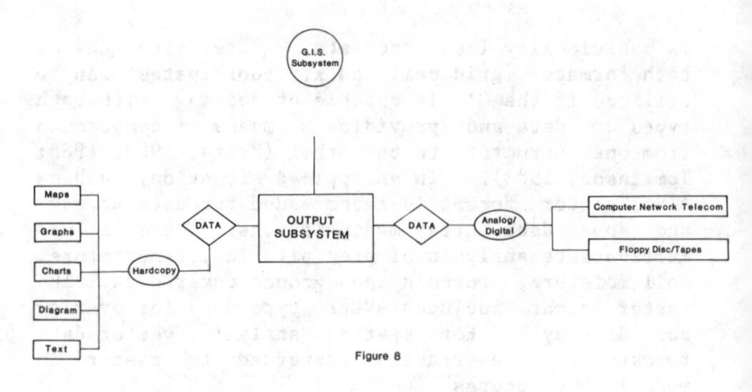

Figure 8

8.2 Software Needs

An automated GIS must incorporate a variety of software capabilities for data encoding, computer network communication, data entry, data base management, data analysis and data output. A complete GIS configuration must include software that is capable of encoding both spatial and non-spatial data, along with image data. A GIS must be capable of processing data through the use of spatial/statistical analytical procedures for modeling. Additionally, it must allow for data storage in an easily accessible and retrievable data management system. Such a GIS must include the following integrated software subsystems: 1) Geographic Subsystem software for data entry of analog to digitally converted data and data file creation; 2) Preprocessing software, including image processing and encoding software to format system compatible files in vector and/or raster data structures for analysis; 3) computer networking communication software for data processing and access on other network nodes; 4) Data base management subsystems to facilitate data file access (editing/update), retrieval (search/extraction/formatting) and storage/cataloging; 5) Data analysis software subsystem with spatial/statistical analysis capability for scenario modeling; 6) Output software subsystem for peripheral devices, e.g., photocopy, line plotter, CRT, floppy disk and computer tape.

Particular to AISC needs is that a GIS must be capable of assessing the impact of an event on an afflicted area in terms of the physical, social and economic indicators (i.e. vegetation response (NVI), crop production/yield, food prices, land use changes). The functions required in this process would include: combining data sets (overlaying) for visual interpretation, statistical analysis in determining variable relationships (i.e. impact), and various forms of correlation analysis for determining data redundancy in modeling.

GIS must finally be capable of producing modeling scenarios for forecasting and determining impact abating strategies. Examples would include, specifying in advance areas of impact and intensity thereof, forecasting crop production/yield, estimating food shortages, determining disaster assistant needs, improving policy and food security management decisions, relocating existing stocks of food or bumper crops in neighboring regions and recommending alternate crops capable of growing in expected weather conditions (Steyaert, 1984).

A complete GIS configuration must incorporate a variety of software capabilities to display facsimile data and digital images, outputting information to a CRT for animation and image graphics. Hardcopy output should include graphic representation of maps in grid cell and polygon format such as contour, choropleth, isopleth, and dot maps, as well as other graphic output including pie charts, bar and line graphs.

A GIS configuration must incorporate software capable of providing computer network linkage or telecommunication to other data sources. Further, a GIS must include software to communicate with different computer operating systems. Communication linkages can also provide an alternative method for early warning communication through facsimile data transmission.

The GIS software must be as versatile and comprehensive since the data sets are expensive. It must incorporate all functions previously defined, and must be operable on the above specified system. Very few GIS systems surveyed integrate GIS, image analysis, data base management and statistical functions required for the particular needs of AISC. Those identified and recommended include:

1) Land Analysis System (LAS). Of the specified requirements, this system is the most versatile and complete package to accomplish the goals specified and identified above. LAS was designed by NASA/GSFC with remote sensing data processing, GIS and data base management in mind. The software is designed to run on a VAX configuration.

2) System 600. The expanded version of System 600 is basically capable to do the same as LAS. However, it has limited statistics and imaging processing capability and designed to run on a VAX 8600/VMS version 4.2. The software cost ranges from $20,000 (limited version) to $150,000 for the expanded version. Other systems considered under this heading include Autogis, Geographics, ELAS/ARC-INFO, Champion, Deltamap, ELAS/SAGIS and ERDAS. However, none of these meets <u>all the requirements</u> set forth to AISC's suggested GIS requirements. Two software packages extracted from the survey have been identified for the workstation. They are:

1) ERDAS
2) Champion

8.3 Hardware Needs

In order to accomplish the data handling tasks specified in the preceding section, vast amounts of data must be processed in a proper computer environment. To exemplify hardware requirements, an analysis scenario has been created for Senegal. To maximize the efficiency of the computer system, an appropriate resolution must first be selected as the minimum spatial input. One square kilometer AVHRR data is suggested as the minimum resolution size. Therefore, in developing an assessment and modelling scenario for Senegal, 197,000 grid cells would be required for each data layer.

Using the currently applied five variables, standard for AISC, require a minimum of 985 Kilobytes of RAM. To make one manipulation of this data sets' five variables would require an additional 985 Kilobytes of storage. Several logical manipulations, such as simple overlay operations would quickly exceed the capacity of a micro computer. If multiple discriminate analysis were employed, a requirement of 32 bit word representation for real numbers would increase the RAM storage to 3.94 Megabytes per manipulation. These two factors of 32 bit word requirements and large RAM storage capabilities quickly exclude the use of any standard micro computer system. Only a super-micro computer system, such as the MICRO VAX II would be configured to meet the above needs. Further expansion of the current model to include five additional variables increases the minimum storage requirement by a factor of two.

Natural disaster analysis procedures for Senegal would necessitate the use of at least eleven data sets. On the basis of 197,000 data cells per variable, a minimum of 2.2 million bytes of RAM storage will be used by the data sets. To make one manipulation of this data set, involving all variables, would require at least one additional file of 2.2 million bytes of storage. Several manipulations, even for simple overlay operations in a GIS, can overwhelm the capacity of microcomputers. A further consideration in multiple discriminate analysis, is that real numbers are generated which require 32 bit representation (Brumfield, 1983). The internal bus structure and registers must therefore efficiently handle 32 bit words. Any form of multi-variable analysis for a study area the size of Senegal will unavoidably process megabytes of information. Computer processing requirements in GIS analysis are far beyond the capabilities of most micro-systems. Large 32 bit word size mini-mainframe or super-micro computer systems are needed for processing data in a GIS format to specifications required in AISC natural hazard monitoring, impact assessment and forecasting procedures.

8.4 Hardware Systems

The logical candidate to meet the hardware needs for such an extensive GIS would be a mini or mainframe computer system. Digital Equipment Corporation (DEC) provides a VAX series of mini computers which are compatible with the current NOAA VAX systems. The micro VAX II by DEC could provide an expanded environment with the forthcoming VAX cluster controller which would give NOAA/AISC an expandable processing system to handle the additional computational burden of the extensive GIS recommended in this paper. Further, as the global distribution of these data becomes involved in the development of regional network nodes, micro VAX II clusters would be cost-effective candidates at ninety percent of the processing capacity of a VAX 11/780, equally configured, for that development. The fact that the micro VAX II and the VAX 11/780 computer series are software compatible is also significant.

In addition to the essential computer configuration, the workstation would also include peripheral devices for all phases between data input and final analysis and display. This would consist of a digitizer table, 1600 and 6250 BPI tape drives, two 450 mega-byte Winchester disc drives, high resolution color display, four to eight DEC VT 220-type terminals, a matrix plotter printer, a laser quality printer and an IBM/PC computer with Polygon communication software for potentially needed workstation software development.

The above scenario would meet the requirements for an automated GIS to suit the needs of an organization such as NOAA/AISC at this time. The state of the art and the technology associated with the development of GIS changes rapidly. These developments are encouraging for the scientists who are forever searching for better ways to help alleviate some of the world's problems through objective means.

REFERENCES

Brumfield, J. & Brown W. 1983. Selected remotely sensed geodata bases consideration for GIS integration and modeling for energy resources management applications. Presented to the American Association for the Advancement of Science Annual Meeting, Detroit, Michigan.

Brumfield, J, Miller, A., Robinson, V., & Boyd R. 1985. Experiments in the spatially distributed processing of land resource data. Presented at the International Conference on Integration of Remotely Sensed Data in Geographic Information Systems for Processing of Global Resource Information. Washington, D.C.

Calkins, H. & Tomlinson, R. 1977. Geographic information systems, methods and equipment of land use planning. Reston, Virginia: U.S. Geological Survey.

Dangermond, J. 1984. A classification of software components commonly used in geographic information systems. In Basic readings in geographic information systems. Williamsville, New York: SPAD Systems.
James, P. & Martin, G. 1981. All possible worlds, A history of geographic ideas. 2nd ed. New York, New York: John Wiley and Sons.
Marble, D. & Peuquet, D. 1983. Geographic information systems and remote sensing. In Manual of remote sensing. Fall Church, Virginia: Sheridan Press.
Marble, D., Calkins, H., & Peuquet, D. (eds.) 1984. Basic readings in geographic information systems. Williamsville, New York: SPAD Systems.
Meyers, W. 1983. Strategies for interfacing geographic information systems and digital image analysis systems. In Proceedings of National Conference on Resource Management Applications: Energy and Environment. San Francisco, California: CERMA.
Meyers, W. 1985. Designing spectral, spatial, temporal analysis and modeling systems for the future: An academic perspective. In Advanced technology for monitoring and processing global environmental data: Proceedings of the International Conference of the Remote Sensing Society and the Center for Earth Resources Management Application. London, England: The University of London.
Steyaert, L. 1984. Climate impact assessment: Foreign countries. NOAA Professional Paper, Washington, D.C.: NOAA.
Tomlinson, R., Calkins, H., & Marble, D. 1976. Computer handling of geographic data. Paris: The UNESCO Press.

Symposium on Remote Sensing for Resources Development and Environmental Management / Enschede / August 1986

Comparison between interpretations of images of different nature

G.Bollettinari
Morfogeo s.a.s., Ferrara, Italy

F.Mantovani
Ferrara University, Italy

ABSTRACT: Data obtained analaizing images of different nature, are described to interprete from the neotectonic point of view some geomorphological feactures of the central and northern area of Perù. The purpose of this paper is the research of the relations ship between photointerpretation, obtained data and their reliability. Particularly merits and shortages,interpretative limits, and the fitness use of each type of images for the over mentioned work, will be described.

RIASSUNTO: Vengono descritti i dati ottenuti analizzando e confrontando immagini di differente natura, con lo scopo di interpretare in chiave neotettonica le caratteristiche geomorfologiche dell'area centro settentrionale del Perù. Scopo della presente nota è la ricerca dei rapporti intercorrenti fra fotointerpretazione, dati ricavati e loro attendibilità. In particolare vengono sottolineati i pregi e le carenze riscontrati, per ciascuna immagine, al fine di tracciare i limiti del dettaglio interpretativo e la destinazione d'uso più idonea.

1 INTRODUCTION

The following note is in reference to the preliminary, photointerpretative phase of a multi-faceted study aimed at identifying the principle neotectonic features of a high plain area in the Peruvian Andes.

In particular is here analyzed a photointerpretation process applied to different types of images and the morphoneotectonics classification also used. At the end, the various advantages and disadvantages of each type of image is underlined, indicating any limitation in interpretative detail and, based on the results, the use to which it is most suited.

2 PHOTOINTERPRETATION PROCESS

The area studied is in the Peruvian Andes N 10° Lat S within a 100-Km-wide area extending approximately 400 Km NNW-SSE and including the Andean high plain. The following documents were used:

a- LANDSAT images, average scale 1:250,000, spectral band 7;

b- LANDSAT F.C.C.(false colour composite) images obtained by automatic procedure of spectral bands 4, 5 and 7, average scale 1:250,000;

c- SLAR black and white photomosaic images, average scale 1:100,000;

d- Panchromatic black and white photomosaic aerial photographs, average scale 1:100,000.

The procedure followed two successive phases of the photographic documents available. The first phase consisted of a field assessment of those photoalignments of certain natural origin to which a tectonic meaning could be attributed; that is association with faults or fractures. The second phase consisted of a classification of the natural, tectonically significant photoalignments in the above-mentioned field: those which were not fully developed were eliminated unless fully aligned or associated with other alignments.

2.1 Phase 1. Assessment and distribution of total field

For assessment and distribution of the tectonically significant natural photoalignments of the total field the procedure was as follows:

a- Interpretation of the black and white LANDSAT images;

b- Interpretation of the F.C.C. LANDSAT images;

c- Interpretation of the SLAR images;

d- Interpretation of the panchromatic black and white photomosaic.

In particular the basically straight alignments as well as circular areas were separately identified for each type of image from their photographic and/or morphological expressions. The former were seen as variations in tone, texture, structure, contrast, glossiness, etc. of the object. The latter define various alignment patterns: lithological (contact between different lithotypes), structural (dip and strike, faults, fractures, etc.), hydrographic and orographic (water courses, valleys, watersheds, escarpments, etc.).

2.2 Phase 2. Classification of photoalignments from the total field

After assessment of the photoalignments had been completed they were classified on the following basis:

a- continuous photoalignment planimetry;

b- correspondence of the photoalignments in the different images used.

All those photoalignments which, during assessment, presented limited planimetric continuity were excluded in accordance with the photointerpretative limits established for a regional study. However, this rule did have exceptions both in the subsequent classification phase (correspondence of photoalignments to other images) and when the photoalignment belonged to a unit developing on a regional scale. Furthermore, any photoalignment identified in one type of image was sought on the others in order to establish to what degree interpretation was correct.

3 MORPHONEOTECTONIC CLASSIFICATION

All the tectonically significant natural photoalignments thus selected were classified according to any geomorphological features wich might indicate neotectonic activity. This classification was based on quality, quantity, congruence, evidence and meaning of the morphoneotectonic features, Panizza & al (1978), Panizza & Piacente (1978).

This method makes it possible to determine any modifications on the earth's surface produced by tectonic movement. In fact, the more recent the movements are the more marked and evident the modifications are. Therefore, one can infer these movements by means of a geomorphological analysis identifying the modifications.

There are numerous morphological elements making the elaboration a strictly morphogenetic profile possible and they may indicate recent deformation. The geomorphological modifications may be broken down in to two types: direct modifications such as escarpments, landslides, fissuring, recurring or aligned erosional forms, etc.; indirect modifications such as particular forms of slopes, ridges, peaks and of the hydrographic network. For example, the development of a rectilinear ridge may result from a fault escarpment or it may represent a summit or indicate regional uplifting. Any discontinuity in altimetric ridges may correspond to transverse faults. A simple escarpment may be linked to a fault having a vertical component or to area uplifting. Breaks-in-slope and landslide areas, as well as particular forms of erosion, may be linked to faults. Valleys with simple or double elbows may reveal the presence of fault with a horizontal component. Movement of this type may produce hooked or countercurrent fluvial confluence. Irregularities such as suspended confluence or truncated valley for example, may be associated with faults having different movements. Strong erosion or sedimentation areas could, respectively indicate relative uplifting or falling. Thus these and other morphological sculpturing are the indications of recent movements and, once they have been identified, verified, studied in depth and qualified, they then become proof of fact.

All the above geomorphological features were assessed and "lines" with a hypothetical neotectonic meaning (lineaments) were associated to them taking into account any comparison with the other images. The document which formed the basis for study was the band 7 LANDSAT. Herebelow, a description is given of the various photointerpretative phases characterizing the identification of linear and circular elements and leading to their classification as "lineaments" with likely neotectonic meaning. Thereafter, the advantages and disadvantages arising from the use of the different types of images are described. The attention of those interested in an interpretational method for attributing morphoneotectonic meaning to the obtained data is called to Panizza & Castaldini (1985).

4 INTERPRETATION

Band 7 LANDSAT images

The main advantages to using this type of image stem from the fact that with this type of regional study it is possible to follow continuity and recognize the aerial development of morphoneotectonic elements. A part from any implications due to the resolution provided by the type of sensor and solar azimut (the latter with negligible area effects), the main tectonic structures (folds, inverse and direct faults) are highly evident, surrounding the Andean high plain to the NE and SW.

In the present study the limitations of this type of image basically stem from the lack of details provided by the sensor: often signs are difficult to define and, at times controversial or contradictory. The clearest morphological evidence appears linked to highly developed polychronological forms although they do not, consequently, indicate recent movement. Vast, highly reflective areas within band 7 further impeded the geomorphological study.

F.C.C. LANDSAT images

In comparison to the previous images, those in F.C.C. made it possible to recognize a greater number of features such as, for example, highly eroded areas or those characterized by accentuated sedimentation, elements linked to the hydrographic network, etc.. This is so thanks to the different colours characterizing the various lithomorphological, vegetational and hydrographic entities as well as to a greater contrast between areas being eroded and the surrounding areas, higher interpretative resolution in those areas presenting elevated reflection in the band 7 LANDSAT images and, finally, to a clearer identification of morphotectonic features. Moreover, these same images have integrated and brought together some characteristic of the arising photoalignments such as: planimetric development, frequency and, in extrapolating regional development, bands of isooriented photoalignment.

SLAR photomosaic images

The SLAR images proved indispensable, especially in the high plain area since high sensor ground resolution made it possible for the microrelief conformation to stand out. It was, thus possible to identify numerous morphological features along the previously identified photoalignments requiring more precise neotectonic qualification. They likewise made it possible to eliminate some photoalignments which were essentially linked to the structural order. They proved equally useful in those areas where cloud or plant cover had partially obstructed the LANDSAT images.

To the E and W of the high plain, in areas where reliefs were quite strong, the use of this type of image was limited by the presence of extensive areas of shadow.

Panchromatic black and white photomosaic

This photographic support, which was to integrate the data drawn from the LANDSAT images, proved only partially useful. In fact, in the high plain area, it did not provide any more data than did the SLAR images. It only provided useful integrating images for neotectonic qualification of the photoalignments in the Cordigliera areas.

5 CONCLUSION

The black and white band 7 LANDSAT images proved lacking in information due to their inability to identify features which could, with any certainty, be attributed to recent tectonic movement. In comparison, the F.C.C. LANDSAT images provided more morphotectonic elements although these were not sufficient to neotectonically qualify the photoalignments. The SLAR images proved most useful in detailed study. The scale factor and improved ground resolution led to the identification of new basic patterns as well as elements accompanying the main trends which had emerged through the LANDSAT images.

All the above contributed to better defining the nature of the features, especially in regard to neotectonic. This thanks to the identification of the morphology and associated lesser structural elements as well as the unmistakable identification of layers and lithological limits.

On the other hand, the photomosaic proved most useful in the Cordigliera areas. It would have been possible to glean further elements by stereoscopically studying each individual photogram. This is especially true for the identification of morphological elements linked to a lack of altimetric continuity.

In conclusion it may be asserted that a regional study of this type requires the analysis of both "small scale" or limited resolution and "larger scale" or higher resolution images. This is so because no neotectonic conclusion can be draw from an initial wide observation and identification of photoalignments. Nor can they stem from a subsequent check and classification. They must, rather, be derived from the observation of local morphoneotectonic elements which only high resolution images can provide.

REFERENCES

Panizza, M. & al. 1978. Esempi di morfoneotettonica nelle Dolomiti occidentali e nell'Appennino modenese. Torino. Geogr. Fis. Dinam. Quat. 1:28-54.

Panizza, M. & Piacente, S. 1978. Rapporti fra Geomorfologia e Neotettonica. Messa a punto concettuale. Torino. Geograf. Fis. Dinam. Quat. 1:138-140.

Panizza, M. & Castaldini, D. 1985. Morphoneotectonics Analysis for applied studies. Atti Meeting I.G.U. Working Groups Morphotectonics Geomorphological Survey and Mapping. Czechoslovakia.

Global distributive computer processing systems for environmental monitoring, analysis and trend modeling in early warning and natural disaster mitigation

J.O.Brumfield
Marshall University, Huntington, W.Va., USA
H.H.L.Bloemer
Ohio University, Athens, USA

ABSTRACT: Hierarchial levels of environmental data availability and processing capabilities care discussed and illustrated for the following global configuration of components based on experimental design results: 1. National Analysis and Early Warning Workstations; 2. International/Regional Processing Nodes; 3. International Data and Processing Facilities; 4. Global Telecommunication Networking for Computers and Early Warning Systems. In as much as the components of computer hardware/software systems and source data are spatially distributed on a global scale, workstations and nodes must be able to communicate effectively (analog and/or digital) to maximize information exchange and decision making potential on a national or regional basis. Further information/data from global data bases are often necessary to supplement or augment locally or regionally derived information (e.g. climate data from NOAA and NASA in the USA and WMO in Switzerland and AVHRR and/or GOES data from NOAA in the USA). Also through global networking various national and international laboratories/centers and universities are accessible for scientific and technical expertise in problem solving in resource monitoring, management, and disaster mitigation. Furthermore, analog telecommunication links, such as facsimile transmission capabilities, can provide an alternative, effective early warning system/emergency 38.4 kbit/sec equivalent transmission rate for disaster mitigation.

Introduction

Global distributive computer processing systems for environmental monitoring and analysis must be capable of providing modeling senarios and telecommunications among systems components for early warning and national disaster mitigation. An example of modeling senarios might include: specifying in advance areas of impact and intensity thereof, predicting crop productivity/yield, estimating food shortages, determining disaster assistance needs, improving policy and food scarcity management decisions, relocating existing stocks of food or bumper crops in neighboring regions and recommending alternate crops capable of growing under expected weather conditions (Steyaert, 1984).

In recent years government agencies have recognized the potential of Geographic Information Systems (GIS) as a technology suited for rapid data analysis. GIS can be viewed as an integral necessary component in spatial problem analysis and assessment.

In order to decrease the devastating effects of severe environmental episodes, advanced data handling and analysis systems (ie. GIS) may be utilized to respond in a more timely fashion. A GIS that will meet the requirements of regional/global environmental modeling must be capable of automated data collection, data storage, data retrieval, and data manipulation, along with spatial/statistical analysis, modeling, and display capabilities for spatial georeferenced data (Brumfield 1983). To facilitate natural hazard analysis at a global scale, a GIS must include telecommunication linkage capabilities to worldwide computer information and processing systems to monitor, assess and predict trends for disaster mitigation (Boyd, 1983; Brumfield, 1985).

The International/National (Regional) Processing Nodes & Workstation

The global distributive computer processing systems, including software, must be capable of encoding, inputting and processing both nonspatial and digital graphics along with image data (digital/analog) and provide efficient data management through manipulative and analytic modeling operations. Also, a global system should have telecommunication/networking capability providing an adaptable tool to different users and for different applications. Figure 1 represents a international/regional processing node and workstation configuration we suggest incorporated into natural disaster trend assessment and forecasting systems. A complete hardware/software modeling system, in addition to a VMS or UNIX type operating system, with compilers and utilities might include:

1) Real time or near real time data link capability to GOES or polar orbiter type satellite weather data.
2) The ability to enhance and register multiple data sets to each other and animate sequential temporal data sets.
3) The ability to transmit/receive facsimile data by telephone, microwave, or satellite telecommunication for 38.4 kbit/sec. data transmission backup.
4) Software for image processing/geographic information systems.
5) Data base management software.
6) Statistical analysis software.
7) Data reformatting and package linkage capabilities among software systems.
8) Data capture systems and input/ouput peripherals, e.g. digitizers, tape drives, disc drives, high resolution video camera, with analog to digital and digital to analog conversion capability: video/facsimile, transmission/reception capability to facilitate 38.4 kbit/sec. equivalent transmission rate for standard telephone service with particular application in developing countries.
9) Spatial data/information file retrieval/transfer systems directory of available location, type and command codes; system software processing accessibility between computer network station nodes e.g. EARTHNET/Bitnet.
10) Off the shelf components and software system compatibility and reliability; minimum requirements for operational expertise and training; ease of system repair, component replacement and software upgrade with emphasis on meeting developing countries operating conditions.

The logical candidate to meet the hardware needs

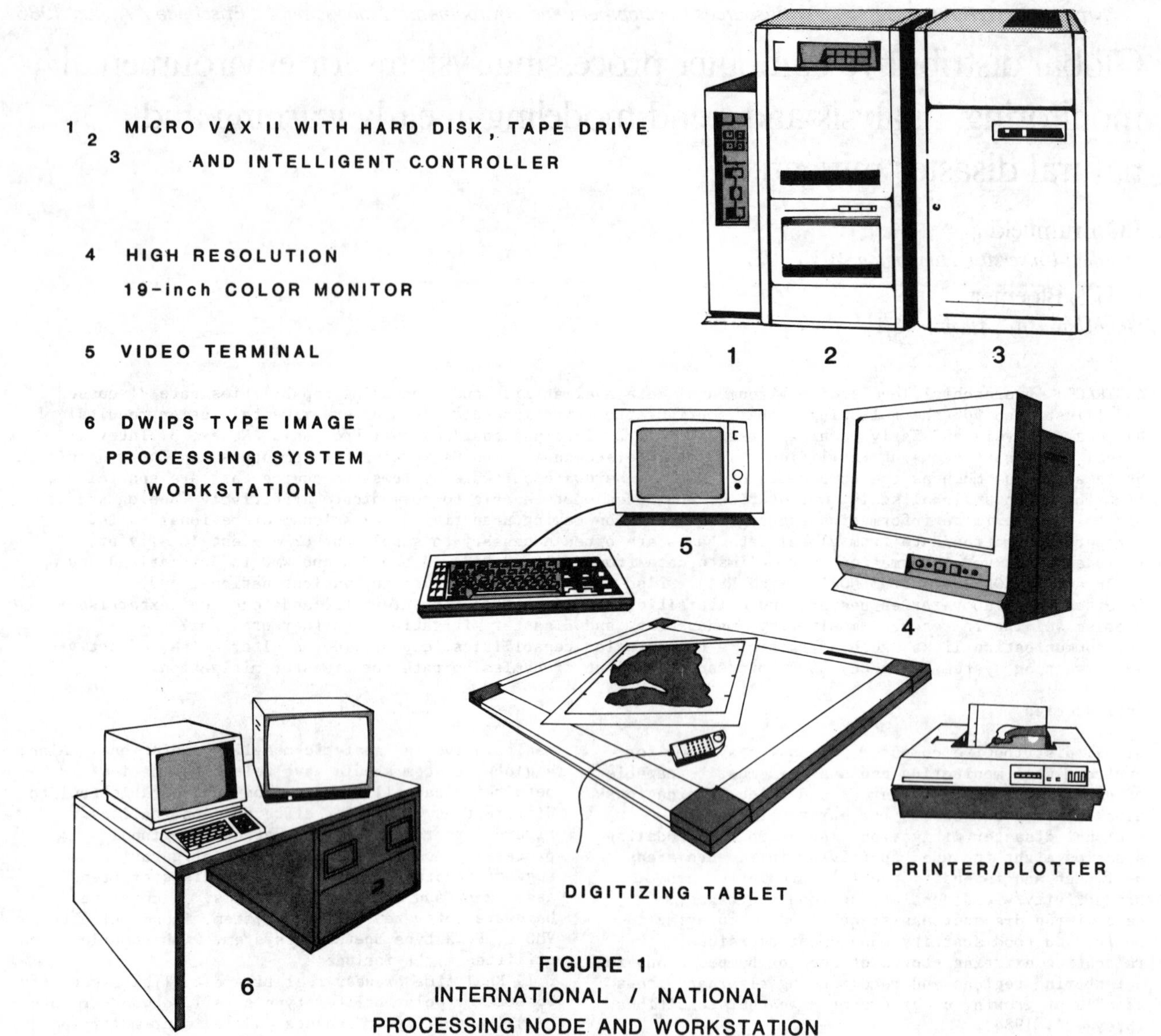

FIGURE 1
INTERNATIONAL / NATIONAL
PROCESSING NODE AND WORKSTATION
HARDWARE CONFIGURATION

would be a mainframe or mini-computer system. As an example, Digital Equipment Corporation provides an array of mini-computers systems (VAX). Based on the above arguments, the use of a MICRO VAX II type CPU could provide one plausable solution for the central processor. The MICRO VAX II can be configured at 90% processing capability of a VAX 11/780. Additionally, software compatibility exists among VAX computer systems operating under VMS. Particularly attractive of the VAX type system is its ability to expand as the needs of the user grows. For example, sometime this year, DEC will produce a cluster controller for the Micro VAX II with others in VAX series. Compatibility between systems is also a strength of VAX type computers. VAX type computers, located worldwide, are then capable of being linked together. This can be made network compatible with IBM micro mini and mainframe systems through polygon, JNET and RSCS communication software linkage capability permits distributed processing useful in large scale data analysis.

Beyond the basic computer processing system, extensive storage capabilities are required. Storage systems should include Winchester hard disk drive units capable of storing 450 plus megabytes of information, and two magnetic tapes (1600 and 6250 BPI) drive for providing backup and optional data entry capabilities, particularly for satellite data. Additionally, a micro-processor controler is necessary for coordinating activities between storage drivers and the computer. Peripheral devices needed in a complete node system configuration should include various data input and output system. Terminals, of course, are needed for interactive data input/output and software operation as well as color printer for graphics and text output, a high resolution color monitor for imge display and user interaction, and a digitizing tablet for spatial data input.

A further requirement for this particular configuration is a DWIPS (Digital Weather Image Processing System) type with a analog/digital interface for a Z80 CPU looping subsystem and an IBM PC controller subsystem, that inputs, processes, and displays AVHRR, RADAR, and GOES weather data. (Boice 1984) This system consists of a full resolution processor (FRP) and a digital image processing system for displaying animation loops. The full resolution processor can also accept a standard "GOES-Tap" signal, consisting of standard AM facsimile telephone circuit. The FRP digitizes and stores each transmitted picture as a full spatial resolution image data file. Transmission data is automatically and continually updated creating loops utilized in a digital weather image processing system.

DWIPS type imagery processing system, when

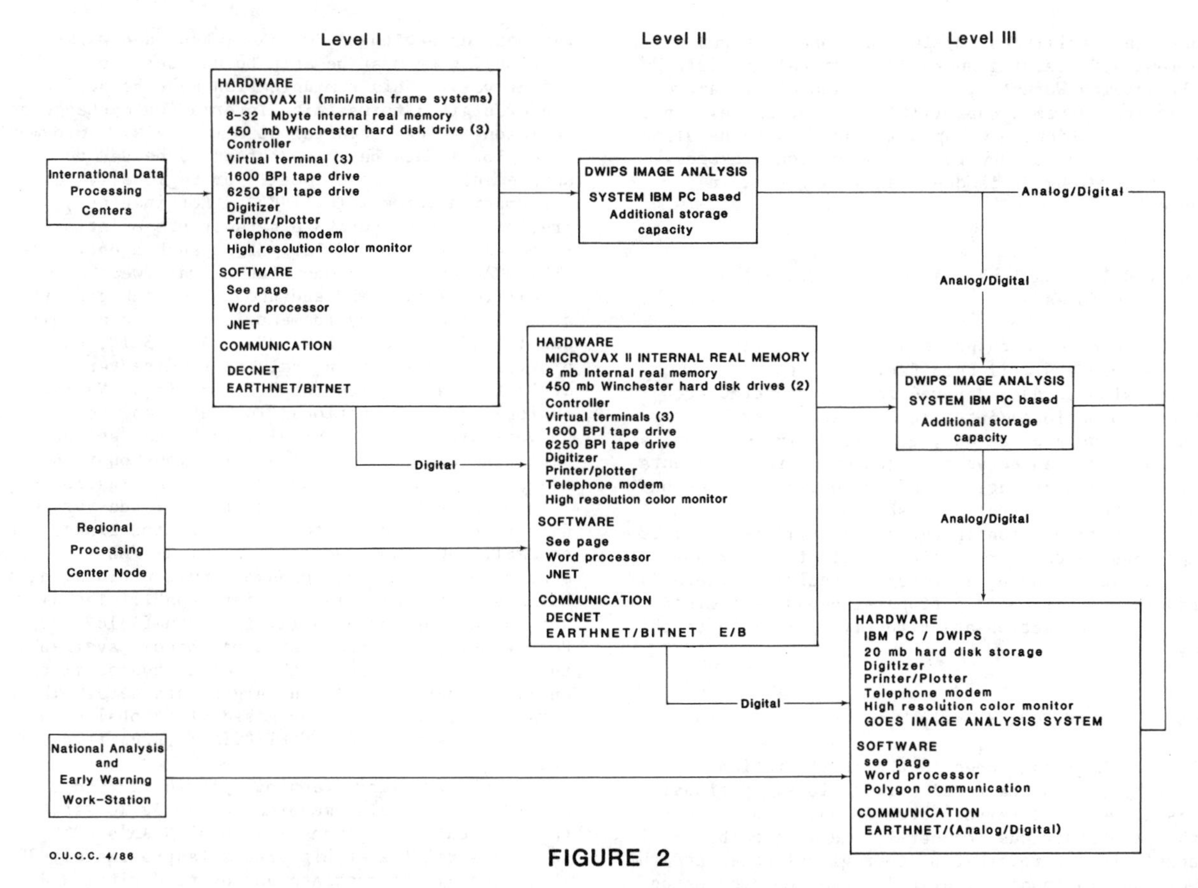

FIGURE 2

GLOBAL DISTRIBUTIVE COMPUTER PROCESSING SYSTEMS

integrated into the work station configuration, with ERDAS Image Processing/GIS Software can provide an automated means of determining temperature, precipitation, percent of cloud cover, and create image loops for analyzing temporal data as well as a simple image processing and GIS modeling system at the work station environment. Further, a DWIPS type image processing system is capable of accepting and outputting standard video/audio signals and digital data. These particular capabilities are useful as a backup telecommunication system for early disaster warning.

The above outlined computer configuration must be capable of expanding when the need becomes apparent. It is, therefore, necessary to stress the need to establish as a starting base, a 32 bit word processing unit for the nodes.

Suggested Global Distributed Computer Processing Systems

This section of the paper will outline a total systems approach which could be incorporated into natural disaster programs. Recommended for global distributed computer processing systems development is a hierarchical structure of information and data processing centers located at key positions. These nodes would include local/national centers, international/regional processing node centers, and instructional data and processing facilities. The global system should eventually encompass a global capability to facilitate and maximize numerous variable inputs not only for monitoring, assessing, trend modeling, forecasting, but as an early warning system in natural disaster mitigation. This global configuration must be flexible. Natural hazards represent a complex set of relationships between environmental and human socio-economic factors. Documenting and analyzing these relationships becomes a Herculean task, tor a global configuration must be capable of monitoring, assessing and forecasting these various types of phenomena and their impact. A global distributive computer processing system must be capable of accepting numerous types of data in diverse formats from a multitude of sources. A global system must be capable of accepting new software and hardware developments that would expand the powers of the system. A flexible system is capable of achieving the goal utilizing proper telecommunication and data input/output systems.

Global distributive computer processing systems configuration must be expandable. They must have expandable capacity to store and process more and larger data sets. Growth would be required with increases in data sets utilized in analysis and modeling procedures, spatial extent of surveillance areas, and temporal information expansion. As the sources expand, so must the ability to store and access these data. Finally a globally distributed configuration must be capable of expanding into regions of the world where natural disasters commonly occur, particularly less developed countries. Systems must be developed that are capable of operating in a third world environment of budget constraints, limited personnel, lack of system's maintenance, and perhaps even unreliable electrical sources.

The configuration must, therefore, be "user-friendly". That is, globally distributed computer processing systems must be technologically transferable, both in the capacity to operate for analysis and to interpret the output.

Taking into consideration the above needed approaches for establishing a globably distributed configuration, a step or phase approach can be incorporated in systems development and implementation. This approach should consist of three development levels taking place at three defined centers: 1) International/National Data and

Processing Facilities, 2) International/National (Regional) Processing Nodes, 3) National Analysis and Early Warning Workstations. Each stage will augment the entire system's capabilities by adding new areas of interpretation, new capabilities of data handling and processing, or new telecommunication linkages. Systems development is depicted as a matrix in Figure 2.

Suggested International/National (Regional) Processing Nodes

The expansion of data processing centers into regional centers will help facilitate the transmission of data derived in those particular regions to national/international agencies, organizations and nations, globally. In addition, a regional node can serve as a center of regional data collection and storage needed in analysis of regional events. This regional node should have the basic software-hardware configuration capabilities as more comprehensive data processing facilities, perhaps with not quite the data storage capacity. (Figure 1) Hardware, software and peripheral needs determined for the above section are directly applicable to this node.

National Analysis and Early Warning Workstation

Below the Regional nodes in the configuration hierarchy are the national workstations. National workstations will further provide data procurement within the individual countries and assist the transmission of information to regional nodes and the other internationally available data and processing centers. National workstations particularly developing countries could provide country-wide data for the overall system. The workstation however, will have limited capabilities of GIS, image analysis, statistical analysis, along with input and output capabilities, as a result of computer size and complexity. Basic environmental monitoring and impact assessment could be capable of being carried out at a national scale, however regional anlaysis will take place at regional and international processing centers. National monitoring capabilities would be further augmented through integrating a DWIPS Type image processing system along with the necessary transmission needs outlined below.

Computer processing capacity and capabilities might consist of an 8 or 16 bit micro-computer with 500 plus kbytes of internal real memory such as an IBM PC. Floppy disks and mini 20 Mbyte hard disk drives would be the median of data storage for this configuration. The software package for a national workstation must contain image analysis capabilities, basic GIS capabilities, such as overlaying and distance measurement functions, data management and report generation capabilities.

The workstation software packages do not include statistical analysis, extensive image processing and data base management. They have limited capabilities to be used as workstations in this context only. Devices, such as a high resolution monitor, printer/plotter, should also be incorporated as well as input devices such as digitizing tablet, and transmitting receivers for digital and analog data.

Global Telecommunication Networks and Early Warning Systems

Since the components of computer hardware/software systems and source data are distributed on a global scale (e.g. the node/workstations of the Sahel region potentially involving seven countries with workstations of IBM/PC quality with DWIPS type analog receive/transmission capabilities, and at least one regional node with MICRO VAX II type and workstation), they must be able to communicate effectively. This communication must be analog and/or digital to maximize information exchange and decision making potential at all levels. Information from global data bases may often be needed to supplement or augment locally or regionally derived information (Brumfield, 1985). For instance, precipitation/temperature data or pre-existing resource data are available from such agencies as WMO, NOAA and NASA along with climate/weather satellite data; UNEP and the World Bank can provide accessibility to environmental and socio-economic data; USDA, UNESCO, FAO, UNDRO AND USAID, can provide information, pertaining to disaster mitigation and agricultural production. Various national and international laboratories, centers and universities can also provide technical/scientific expertise and/or R & D in applications on resources monitoring, management and disaster mitigation.

In order to efficiently utilize all the various data bases, effective telecommunication links, such as JNET, RSCS (now available on MICRO VAX II, JNET) should be implemented to access multinational data bases and provide file transfer capabilities among the various proposed facilities (Brumfield 1985). For example, data processing may become available at the node for more in-depth modeling and statistical analysis, particularly on larger data sets. The data sets themselves may be accessed via global counter networking such as EARTHNET/BITNET (Brumfield 1985). (See Figure 3).

Analog (e.g. audio/facsimile) transmission can provide direct telecommunication among workstations to facilitate data transmission of NOAA's GOES weather satellite at high rates (approximately 38.4 kbits/sec.). Appropriate analog to digital and digital to analog conversion equipment such as DWIPS interfaced with the IBM/PC type workstation can provide a backup system to traditional computer telecommunication via standard telephone lines (Boice, 1984). The above system can be incorporated as an additional effective early disaster warning telecommunication link.

Summary and Conclusions

Hierarchial levels of environmental data availability and processing capabilities were discussed and presented for the following global configuration of components based on experimental design results:

1. National Analysis and Early Warning Workstations.
2. International/National (Regional) Processing Nodes.
3. International Data and Processing Facilities.
4. Global Telecommunication Networking for Computers and Early Warning Systems.

In as much as the components of computer hardware/software systems and source data are spatially distributed on a global scale, workstations and nodes must be able to communicate effectively (analog and/or digital) to maximize information exchange and decision making potential on a national or regional basis. Further information/data from global data bases are often necessary to supplement or augment locally or regionally derived information. Climate data from NOAA and NASA in the USA and WMO in Switzerland and AVHRR and/or GOES data from NOAA in the USA. and/or environmental data from UNEP, GENEVA; NOAA, USA; World Bank, USA; or NERC/Thematic Services, UK is only a partial list of examples of data bases available to a world of users. Also, through global networking, various national and international laboratories/centers and universities are accessible for scientific and technical expertise in problem solving in resource monitoring, management, and disaster mitigation.

Therefore, effective telecommunication in developed

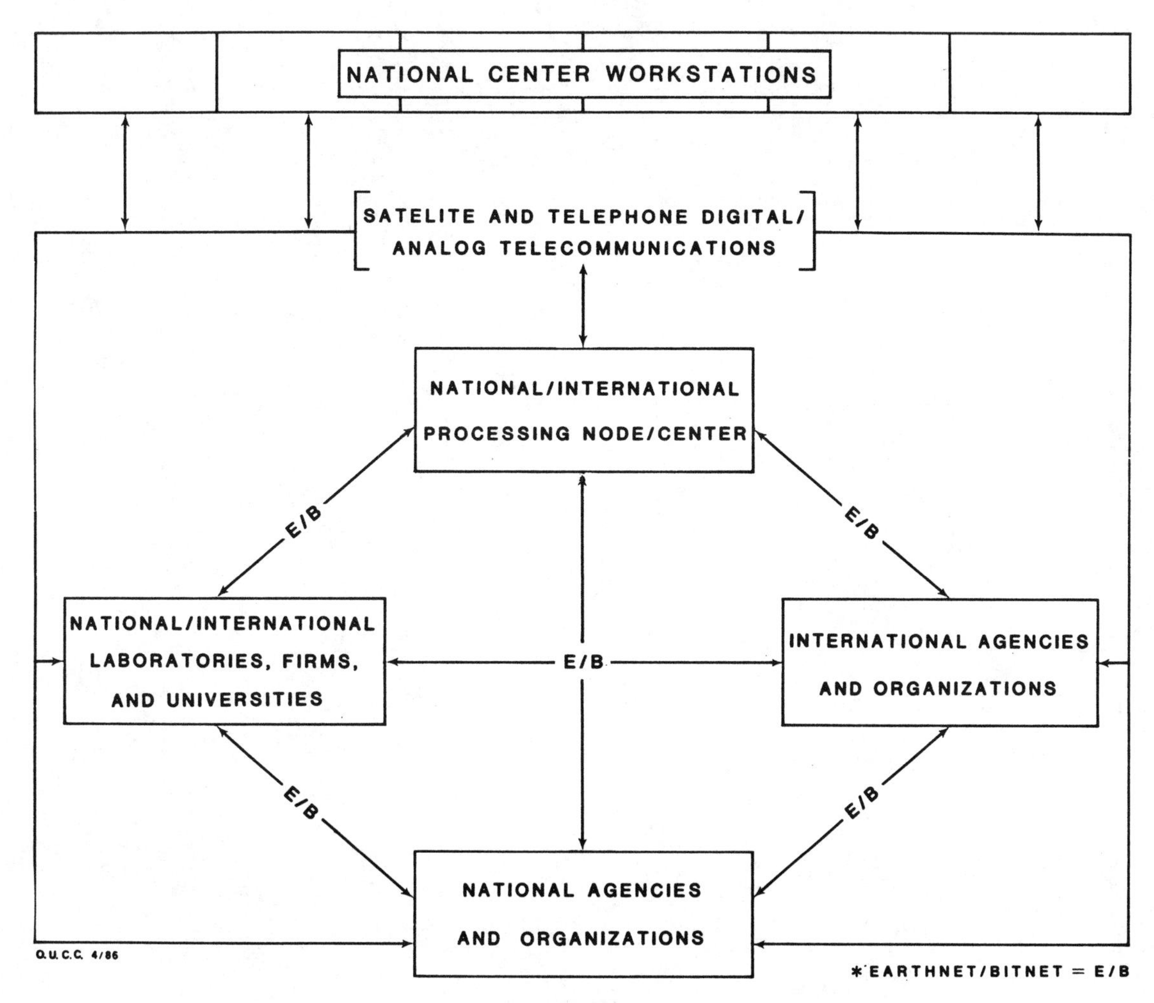

FIGURE 3
GLOBAL COMMUNICATION NETWORKING AND EARLY WARNING SYSTEM

and developing countries can be facilitated by nodal mini/mainframe type network links (e.g. JNET/RSCS on VAX to IBM 43XX or 30XX computers in a EARTHNET/BITNET computer networking environment) to multinational data/information bases as well as file transfer capabilities among workstations and nodes. Data processing is available on any node in an EARTHNET/BITNET by workstations for more in depth statistical/spatial analysis and modeling. This is necessary for larger data sets that may involve multimegabyte manipulations particularly from global data bases of regional analysis and modeling senarios derived from global computer networking such as EARTHNET/BITNET. Furthermore, analog telecommunication links such as facsimile transmission capabilities can provide an alternative effective early warning system/emergency 38.4 kbit/sec. equivalent transmission rate for disaster mitigation with around a 1% noise level depending on the telephone line quality.

Bibliography

Boice, Clarence, "Digital Weather Image Processing System", Image Processing Systems (IPS), Claremont, California, 1984.

Boyd, R., V. Robinson, J. Brumfield, "Data Communication and Processing Options for Global Resource Monitoring", invited paper presented at the National Conference on Resource Management Applications: Energy and Environment in San Francisco, California, August 23-27, 1983.

Brumfield, J.O. and W.R. Brown, "Selected Remotely Sensed and Georeferenced Data Base Considerations for GIS Integration and Modeling for Energy Resource Management Applications", invited paper presented at the American Association for the Advancement of Science Annual Meeting in Detroit, Michigan, on May 30, 1983. Abstract selected for publication by IEEE for the IEEE Transactions, Journal of the International Electrical and Electronic Engineers.

Brumfield, J.O., A. Miller, V.B. Robinson and A. Yost, "Experiments in the Spatially Distributed Processing of Global Environmental Information", paper presented at the International Conference on Advanced Technology for monitoring and Processing, global Environmental Information, Sept. 9-13, 1985, in London, UK and published in the proceedings.

Steyaert, L.T., "Climatic Impact Assessment Technology: Disaster Early Warning and Technical Assistance in the Developing World", U.S. Department of Commerce, NOAA, NESDIS/AISC, Washington, DC. 1984.

Symposium on Remote Sensing for Resources Development and Environmental Management / Enschede / August 1986

Geological analysis of the satellite lineaments of the Vistula Delta Plain, Żuławy Wiślane, Poland

Barbara Daniel Danielska & Stanisław Kibitlewski
Photogeological Department, Geological Institute, Warsaw, Poland

Andrzej Sadurski
Hydrotechnics Department, Technical University, Gdańsk, Poland

ABSTRACT: Regular pattern of Landsat lineaments is observed in the Vistula Delta Plain,North Poland. Geological interpretation of the lineaments was carried out using visual, analog and statistical methods. The results show that lineaments should be understood as traces of buried disjunctive structures of Upper Cretaceous brittle rocks underlying there the soft and loose Cainozoic sediments. The data obtained from remote sensing materials enable a re-interpretation of the geological model of the terrain and an improvement of the knowledge on tectonic character of peri-Baltic syneclize as well.

1 INTRODUCTION

The numerous lineaments of a different character might be observed in the Landsat satellite images in the area of the Vistula Delta Plain /Żuławy Wiślane/ i.e. in the mouth region of Vistula river, northern Poland /Fig.1/.

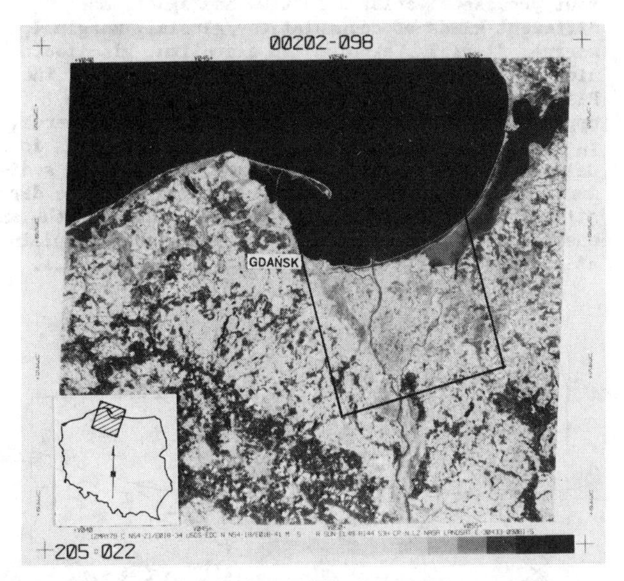

Figure 1. Scene: E-30433-09081 MSS-5,12 May 1979. Investigated area and image position are marked.

Some of the noticed lineaments are due to the human impact /canals,dams, embankments, roads, railways etc./ and were eliminated from the further discussion. The other ones, however, seem to be traces of the features of the inner structure, at least sub-Holocene ones. In contrary to the adjacent morainic plateaux the area of Vistula Delta Plain represents a flat, lithologically monotonous deltaic region with homogeneous relief, structure and lithology of Quaternary sediments and shows different vegetation cover. In the satellite images some regional lineaments might be noticed, which continuously and with no change of direction cut different geomorphological units as - from the west to the east - Kashubian morainic plateau, Vistula Delta Plain and Elbląg morainic plateau. It suggests therefore, that they reflect even deeper elements than those of Quaternary, or even Tertiary series, also in respect to the area of Żuławy themselves.

The hitherto recognition of inner structure of peri--Baltic syneclise, the unit containing Vistula Delta Plain, has been based only on a small number of deep boreholes. The near-surface geological structure, however, seems to be better recognized due to the numerous although not deep and dispersed boreholes done mainly to the hydrogeological purposes. Many geological problems remain unsolved, in this number - those of Quaternary deposits, tectonic pattern of Cretaceous and the whole Permian-Mesozoic series and Palaeozoic one.

The geological significance of the remote sensing materials interpretation has been recently more and more often confirmed by the results of the standard methods. It seems, therefore, reasonable to introduce this aspect of modern geological mapping to the geological recognition of the Vistula Delta Plain.

The photogeological interpretation applied to the area of Żuławy Wiślane is aimed at a characteristics of lineaments observed in Landsat images, as well as on explanation of their origin and geological significance.

2 METHODS USED

Photogeological interpretation based on Landsat images was done, namely scenes: E-2230-09063, E-02031-09021, E-02643-08513, E-21165-08355, E-30433-09081. They were analysed visually and with application of analog device /I^{2}S/, simultaneously by two persons /in aim to get the higher credibility of interpretation data/. An interpretation scetch was compared with the known surface elements of the geological structure /i.e. geological maps 1:200 000/. Since no relation between these elements and features under interpretation /as: tones, colours, structure and texture of the image/ has been found, the further work was concentrated on the recognition of the lineaments.

The lineaments display a distinct regularity, continuity and ordering /the last fact confirmed also by the statistical analysis results - Sadurski,Kibitlewski,Daniel-Danielska-in print/.

A map of lineaments pattern was compiled taking their visibility as the only the interpretation criterion into account. Such the pattern was geologically analysed and interpreted in relation to the existing geological materials including the results of the newest hydrogeological boreholes /Fig.3/. Most of them reach a top surface or uppermost strata of Cretaceous sediments.

Numerous geological cross-sections of the sediments building the delta plain unit were prepared. Interrelations between lithological and/or genetical boundaries and distribution of the lineaments along the cross-section lines were studied, as well.

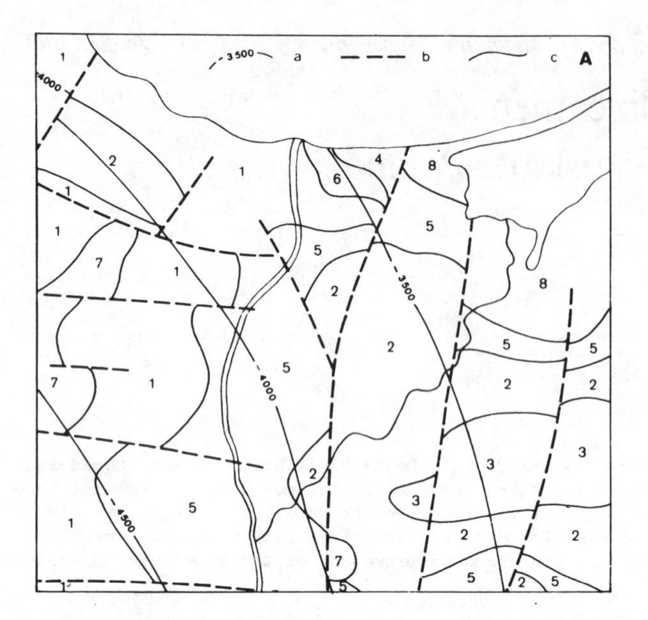

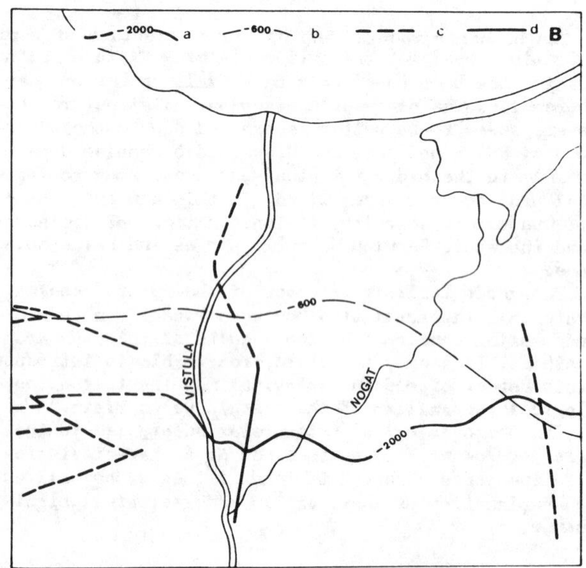

Figure 2. Geological structure of basement.
A - crystalline basement: a -contours of a top surface /in meters, after Kubicki, Ryka 1982/, b -faults, c -lithological boundaries. Lithology /after Karaczun, Kubicki, Ryka 1982/: 1 -granitoids, 2 -unsubdivided gneisses, 3 -unsubdivided gneisses and amphibolites, 4 -pyroxene gneisses and amphibolites, 5 -unsubdivided migmatites, 6 -pyroxene migmatites, 7 -anatectic and palingenetic granitoids, 8 - rapakivi like granitoids.
B - Zechstein-Mesozoic series /after Dadlez 1980/: a -contours of basal surface of series /in m/, b -contours of the base of Upper Cretaceous, c -faults penetrating lower part of the series /mainly Zechstein or Zechstein and Triassic/, d -faults cutting the whole series. Both structural surfaces inclined southwards.

3 GEOLOGICAL SETTING

In the Vistula Delta Plain area the crystalline basement of the peri-Baltic syneclize lies at depth of about 3500m /in the east/ and more than 4000m /in the west of the area/ - Karaczun, Kubicki, Ryka 1982.

The faulting character and wide-radius downwarps and uplifts of the top surface of the basement seem to influence the development of the later sedimentary cover of the platform. The top surface of Lower Palaeozoic cover occurs at the depth of about 1500m /in the northern part/ to about 2000m /in the southern part of the area/ - Fig.2A.

The Zechstein-Mesozoic series represents the following part of the geological section, the lower part of which displays some big faults /Dadlez 1980/ - Fig.2B. The upper part of the series is built up of calcareous and calcareous- siliceous deposits of Upper Cretaceous /Turonian to Maestrichtian in age/. These brittle sediments underlie, in turn, the loose and soft Cainozoic sedimentary cover the thickness of which varies from 50m /in Gdańsk vicinity/ up to 200m /east- and southwards from Gdańsk - in Tczew and Malbork region/.

Top surface of the Upper Cretaceous series occurs at the depth of approximately 90m dipping gradually eastwards and westwards reaching the morainic plateaux. It displays a distinct relief due to an intensive erosion, possibly facilitated by the faults -Fig.3. The local small faults are indentified in the numerous geotechnical works basing, however, only on analysis of denivelations of Cretaceous strata observed in boreholes /Mojski 1976/.

In the area of Żuławy Wiślane /Vistula Delta Plain/ the Tertiary series overlies the Upper Cretaceous sediments. This series has been locally strongly eroded too, due to the erosion and exaration which acted at Tertiary/Quaternary boundary as well as in Pleistocene, affecting - in the distinct area - even the youngest Upper Cretaceous strata /Fig.4/.

The sediments of three glaciation and the deltaic Holocene ones build the Quaternary cover. Pleistocene series seems to have very complicated inner structure what results from a combined influence of the different processes variated in time and space, as: different kinds of accumulation /glacial, marginal, marine, fluvial/, erosion and exaration, glacitectonics and possibly - tectonics. The thickness of the Pleistocene sediments equals in the Żuławy area approx. 70m. Their top inclines slightly northwards. In Żuławy area the relief of sub-Holocene series is in details insufficiently recognized. Pleistocene sediments surface is eroded and buried with Holocene deposits of thickness from 0 to 20m /increasing north- and eastwards/. These deposits compose the deltaic plain at the altitudes from about 1m b.s.l. to 1m a.s.l.

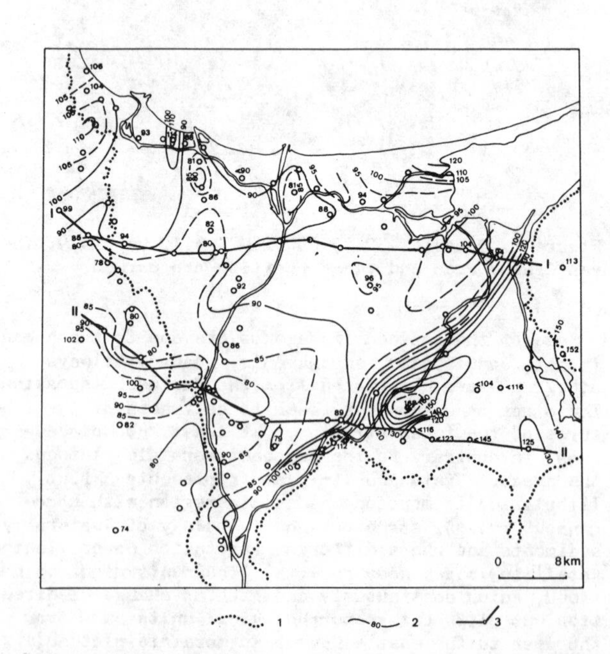

Figure 3. Top surface of Cretaceous series; 1 -boundary of morainic plateaux, 2 -contours of top surface of Cretaceous series /in m b.s.l./, 3 -boreholes and cross-section lines.

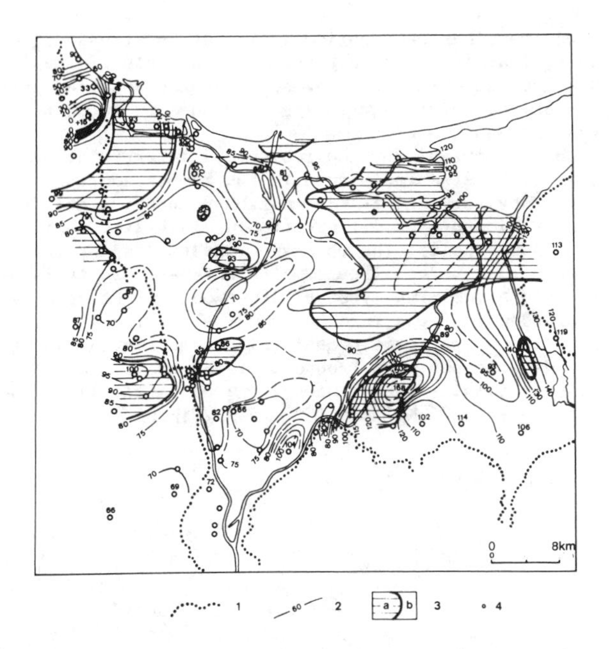

Figure 4. Substratum relief of Quaternary: 1-boundary of morainic plateau, 2 -contours of substratum surface /in m b.s.l./, 3 -lithologic units: a -Cretaceous, b -Tertiary; 4 -boreholes.

The present Vistula plain seems to be a result of the Holocene processes, mainly of deltaic accumulation /Mojski 1983/ and consists of moods, silts, sands, peats and river sands and gravels etc.

Its developments might have occured - at least - in two stages. The first one corresponds to the end of Tertiary. Due to the intensive Pliocene erosion affecting the Tertiary sediments or even Upper Cretaceous ones a wide depression was formed in this area which has survived till Pleistocene being significant for the development /accumulation and partially destruction/ of the glacial sediments following each other. The depression has survived even longer - till Holocene - despite the intensive Pleistocene infilling. In that last period glacial sediments have been partially eroded and removed especially in the time of the last glaciation by the Vistula river waters. The possible glaciisostatic movements and the oscillations of the level of world ocean basin caused the following succession of events in the region discussed: - lowering of the erosion basis /Yoldia period/, - the further uplift of the southern Baltic seashore zone /Ancylus period/, - marine transgression /in Litorina period - due to Scandinavian shield uplift/.

The last process caused - i.e. - the intensive fluvial accumulation characterised by changing directions of sedimentary transport as well as significant thickness of the deltaic sediments. The deltaic sediments were accumulated following the sea regression in a lagoon formed gradually due to the growth of a sand bar from the west. These different series of Quaternary sediments have been recognized in uneven degree.

It results from above that for the further discussion on character of the lineaments in the area examined such the elements seem to the most significant: Cretaceous sediments, their top surface relief and the nature of post-Cretaceous tectonic deformation sub-Quaternary surface relief, shape and preservation of the Quaternary sediments.

4 ANALYSIS OF GEOLOGICAL MATERIALS

The following maps:-of lithology, tectonics and relief of top surface of crystalline basement, -of tectonics of lower part of Zechstein-Mesozoic series /Fig.3,4/ as well as standard geological maps 1:200 000 show the occurence or suggest as a possibility of some faults in the area in question at different deep levels. The possible faults and fractures occuring in the brittle cretaceous rocks reactivated later due to younger and neotectonic movements could have influenced the Cainozoic plan of sedimentation as well as the distribution and circulation of waters in the whole Cretaceous-Cainozoic series of the region. The numerous authors assume the activity of many of such the faults in the period from post-Cretaceous till Quaternary inclusive and their influence on Żuławy development.

It might be, therefore, accepted that there exists a pattern of fractures and faults of a different rank still not recognized as a whole. It concerns the uppermost part of Zechstein-Mesozoic series. This suggestion seems to be confirmed also by the relief of Cretaceous top surface and the relief of sub-Quaternary surface and results from the comparison of the cited maps 1:200 000 as well.

As it has been already stressed, also in Polish literature /Bażyński 1982/ - in relation to the geologically different regions, the lineaments visible in the satellite images might represent a surface expression /projection/ of the inner faults and other tectonic elements. The fact has been observed in the regions with a thin Cainozoic cover as well as in those where such an overburden is much thicker than the thickness of Tertiary and Quaternary series in Żuławy Wiślane area.

As it results from the boreholes interpretation the Cretaceous top surface lies in the Żuławy region slightly higher than in surrounding morainic plateaux /Fig.5/.

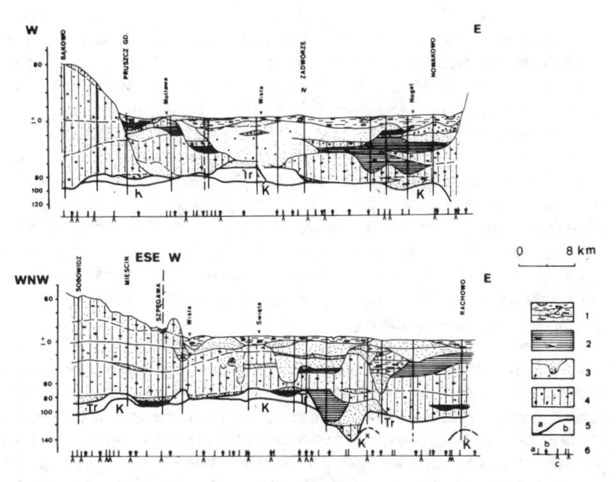

Figure 5. Geological cross-sections. Lithostratigraphic units: Holocene: 1 -organogenous deposits, varved clays, lake deposits, sediments of fluvial origin, marine and delta plain deposits; Pleistocene: 2 -limnoglacial deposits, 3 -glacifluvial and river deposits, 4 -glacial deposits, tills; 5 - sub-Quaternary rocks: a -Tertiary, b -Cretaceous; 6 - traces of lineaments on cross-sections: a -whole set of lineaments, b -lineaments of the most frequent directions, c -lineaments which create the regional the regional trends.

The relief of this surface displays some regular trends which may correspond to the ordered pattern of tectonic linears occuring within Cretaceous deposits or even seem to be characteristic for the whole Permian-Mesozoic series. The general relief pattern might be interpreted as the result of erosion facilitated by such faults. It might be suggested basing on a lithological character of sediments /brittle calcareous rocks/ and on a position of the studied area within the platform that the deformations discussed should occur mainly as not intensive entirely regular discontinuities of fracture type. These fractures could have been active in some periods resulting in development of the fault and block structures.

The sub-Quaternary surface seems to be varied too /Fig.4/. In the western part of Żuławy Wiślane there occurs a partly eroded Tertiary cover consisted of soft and loose /ductile/ sediments lithologically different from the Cretaceous ones. In the eastern part of this area as well as locally in the near-slope zone of the western plateau, Cretaceous sediments occur directly on the sub-Quaternary surface.

A distinct coincidence of the character of sub-Quaternary surface relief and that of Cretaceous one has been observed comparing a maps /Fig.3,4/. It might be suggested that pre-glacial and young Pleistocene erosion had the similar regional character as the post-Cretaceous processes. The general pattern of development of the post-Tertiary relief as well as that of post-Cretaceous seem to be similar in both cases - stronger erosion was marked in the eastern part of the area of Żuławy and locally - also - in the north-western part. The similarity concerns also the smaller relief forms of the Cretaceous top surface and the sub-Quaternary one as e.g. moderate broad culminations and depressions.

The coincides mentioned above might be, however, quite random - just due to the same number of boreholes and the same distribution of them i.e. to the same quantity of information in both the cases.

When one takes, however, different lithological structures of both the surfaces and the difference in factors conditioning erosion processes in pre-Tertiary, pre-glacial or Pleistocene periods into account, it seems to be possible that the relief /features/ in the surface of Cretaceous sediments have been influenced to a distinct degree by the tectonic disjunctive deformations of of these brittle - and strongly fissured - rocks. The features and dislocations of small amplitudes characteristic for the platform area and developed due to the relatively weak post-Cretaceous tectonic activity might have facilitated an erosion following the tectonic pattern of the Cretaceous rocks.

The relations of the post-Pliocene /pre-glacial/ erosion and the different Pleistocene processes /erosion, accumulation, exaration/ to the Cretaceous sediments - seem to be less distinct. Some deformations in Cretaceous rocks could have beem active, however, in this time under an influence of different processes i.e. activity of continental glacier, post-glacial processes, and finally - processes influencing the Baltic-sea development.

The phenomena characteristic for the area of Żuławy, as: changes in directions of palaeo-drainage channels in the different periods of Quaternary, positions of the fossil shore-line, stages, orientations and positions of a front of deltaic accumulation, as well as another Pleistocene and Holocene processes leading to variations of facies and thicknesses of the sediments seem to be connected with the reactivation of the tectonic deformations - from Cretaceous till recent.

5 GEOLOGICAL INTERPRETATION OF SATELLITE IMAGES

Vistula Delta Plain area represents a distinct morphogenetic units within surrounding uplands /morainic plateaux/ as it is seen in the satellite images /Fig.1/. The unit displays well visible boundaries due to the character of the tones different from the adjacent areas, as well as different structure and texture of the image. The difference mentioned seems to be mainly related to the soil and water conditions of the area of Żuławy i.e. different land-use character /arable fields, meadows etc./ and - in indirect manner - to the soil complexes /-mainly organic soils - of high water content, or a sandy ones - of alluvial origin/.

From the geological point of view the tonal pattern of Żuławy Wiślane seems to be weakly differentiated only with enhancement of tones corresponding to alluvial series in the river valleys, wide depression of the Druzno lake, seashore dune belt and the area of an increased lateral water supply /along the uplands margin/. No distinct relation of the tones and depression areas has been noticed. It must be stressed, however, that in the Żuławy area the analysis of the tones does not imply an discussion on the detailed geological structure expressing only some features of spatial management.

Most important for the geological interpretation seem to be lineaments seen in the remote sensing images. They form, as it was mentioned above, a net of linear elements of differentiated visibility, orientation and density. The observed linears display the length in interval from some to anywhere ten to twenty kilometres. Two types of lineaments might be distinguished: - long regional and - short local ones. The first type forms occasionally some regular longer trends with the most frequent direction of about 70°, less frequent - submeridional one and still less distinct ones: 20°-40° and 130°-150° /Fig.6A/.

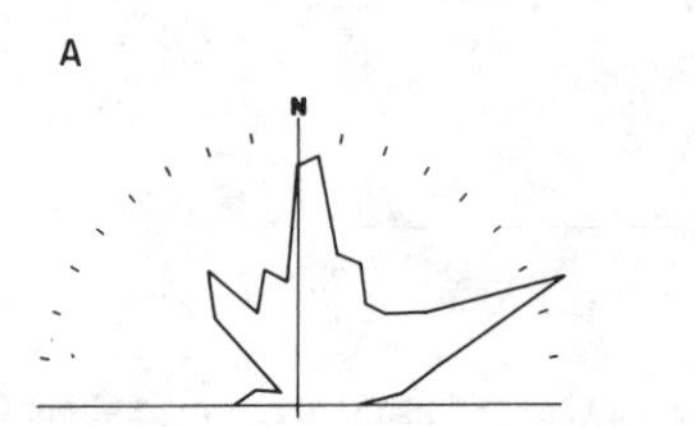

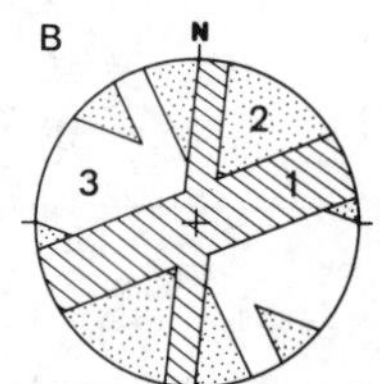

Figure 6. A - Azimuthal histogram of the lineaments for Vistula Delta Plain and adjacent morainic plateaux B - Structural diagram of lineaments /according to results of Vistelius method/: 1 -concentration zone, 2 -steady distribution zone, 3 -dispersion zone.

These lineaments occur in the more extense area - at least in the whole peri-Baltic syneclize. Numerous shorter and more dispersed as well as everywhere equally well visible lineaments display mostly the directions close to these regional trends. That is why the whole remote sensing pattern of Żuławy area has the regular i.e. ordered character despite the local domination of some directions or differentiation in the density of lineaments.

The statistical analysis has confirmed the results from above. The analysis itself included an estimation of the linear and surface densities of all the lineaments /even those less visible/, an evaluation of the correlation between their length and azimuths as well as an estimation of the statistically preferred orientations. It showed no correlation between the visibility of the lineaments and their length or azimuth in the area of Żuławy and/or adjacent plateaux. From the other side a high regularity of the lineament pattern, an existence of the preferred directions / 0-10° and 60°-80° - Fig.6B/ and a local increase in lineaments density /as e.g. in the central part of Żuławy/ have been stated. The zones of concentration of lineaments in trends remain in coincidence of about ±10° with the subclass of the best visible lines separated for statistical use.

The density of the lineaments displays local variations i.e. it is relatively low in the north: - in sea shore region, - in the belt along the Kashubian plateau margin and - at delta root, and locally in the center of Żuławy.

The more distinct changes might be noticed - in general - in the blocks marked by some trends of lineaments or their sets.

The high statistically proved regularity of the lineament pattern may be an evidence of the character of the factor which had caused their formation. It could be the same one in the area of Żuławy themselves as well as of the surrounding plateaux and in the whole peri-Baltic syneclize.

Since the genezis of the lineaments can not be referred to the Cainozoic sediments because of their specific, complex and irregular structure only the tectonics itself, especially - a fracture and fault

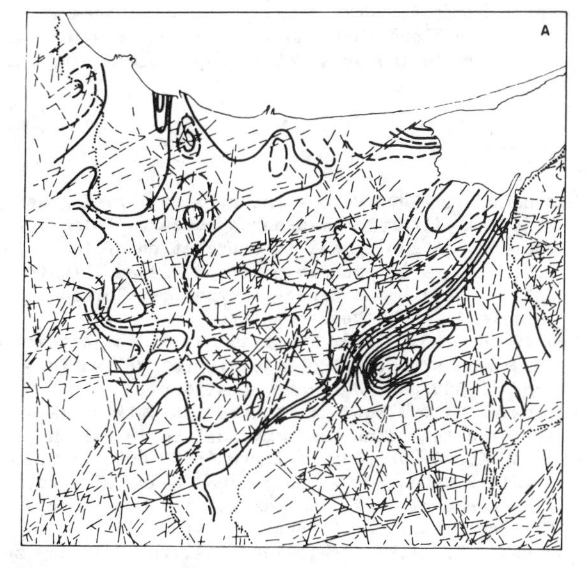

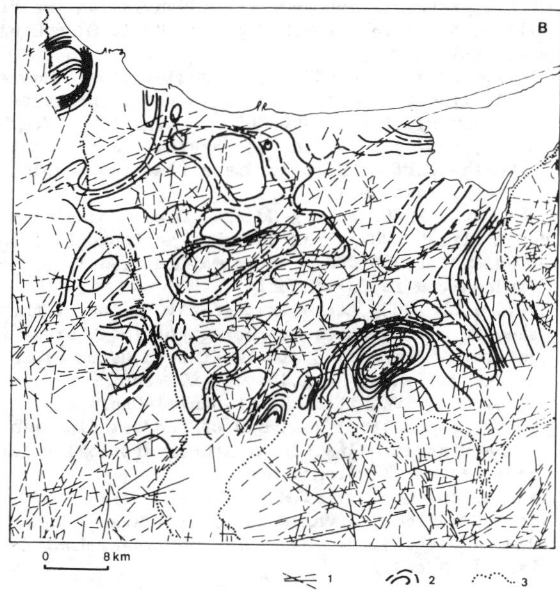

Figure 7. Landsat lineaments in relation to relief features in: A - top surface of Cretaceous series: 1 -lineaments, 2 -contours of surface /in m b.s.l./, 3 -boundary of morainic plateau; B - substratum relief of Quaternary: 1 -lineaments, 2 -contours of substratum surface /in m b.s.l./, 3 -boundary of morainic plateau.

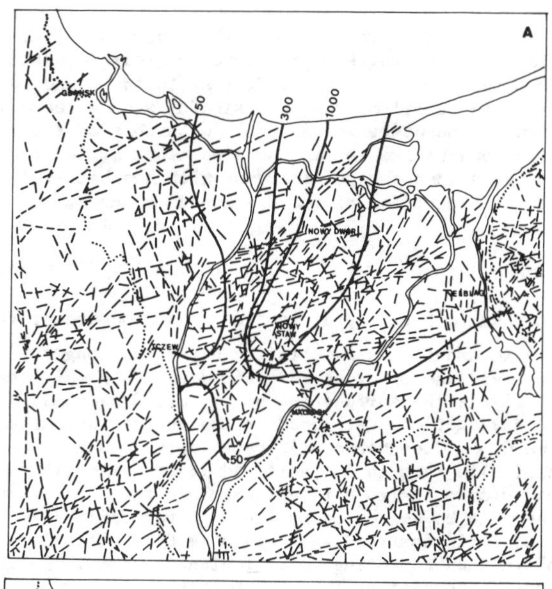

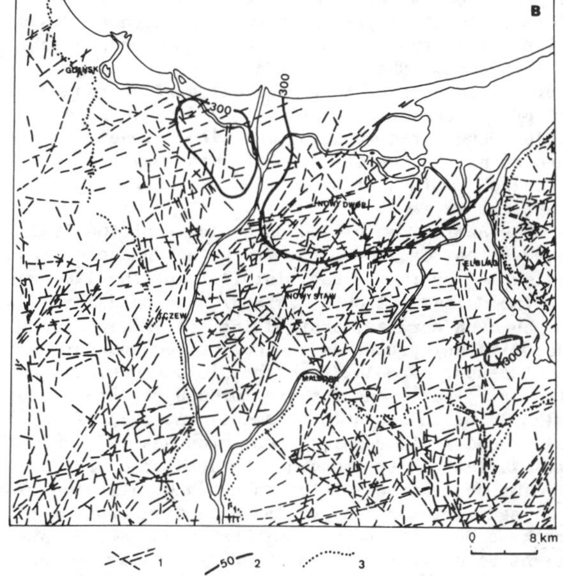

Figure 8. Landsat lineaments in relation to salinity of ground waters /after Kozerski and Kwaterkiewicz 1984/: A - in the Cretaceous aquifer, B - in Pleistocene aquifer; 1 -lineaments, 2 -contour lines of chlorine ion content in ppm, 3 -boundary of morainic plateau.

disintegration of the brittle carbonate series of Upper Cretaceous , seems to be responsible for the origin of the lineaments. This observed occurence of many small lineaments suggest that the fracture /of exactly this - not deeper - stratigraphical horizon/ have been projected onto the recent surface. The greater the depth - the smaller the chance of projection of small densely packed structures onto the surface.

There could occur, however, an influence of the deeper dislocations on the Upper Cretaceous tectonic pattern and on that of lineaments. Possibly - it refers to some few faults and lineaments corresponding to them, displaying the distinct trends and generating the block style of tectonics of the region. The blocking structure of the area corresponds both to the lineament pattern and to the geological data. It seems to be responsible e.g. for the different character of Cretaceous surface and sub-Quaternary surface reliefs of the areas east- and westwards from the submeridional lineament which cuts into a half the teritory between the rivers - Vistula and Nogat /Fig.7,2B/.

It is also important that both the zones of this lineament and its vicinity display the highest salinity of the underground waters in the Cretaceous carbonate sediments /Fig.8A/. The maximum salinity /above 1000mg/dcm^3 Cl^-/ corresponds at the same time to the area of maximum lineament density.

Some similar relations might be also observed as it concerns the salinity coefficient of the Pleistocene aquifer /Fig.8B/ - it falls also between two main rivers - Vistula and Nogat.

The sandy horizon of Cenomanian thins out eastwards in the same zone too, what together with the preceding observations suggest a tectonic significance of the discussed zone at least from Cretaceous, or possibly - from even earlier time.

The first order block structure seems to be accompanied with the blocking of the second order seen as moderate uplifts and downwarps, the big part of which might remain in coincidence with the main lineament trends.

The general relations described above are less distinct in some parts of the area possibly resulting from the worse and not sufficient tectonic and geological recognition. The precision of such a recognition is therefore incomparable with that of the remote sensing images /as for lineaments at least/ but the coincidence might increase with the better tectonic recognition of the Cretaceous bedrock.

6 CONCLUSIONS

It results from the disscusion presented above, that:

1 -Lineaments observed in the satellite images of Żuławy Wiślane might be understood with a great possibility as the surface traces of buried tectonic linear disjunctions developed in the brittle series of the Upper Cretaceous calcareous rocks and projected as faults through the soft and loose Cainozoic sediments onto the recent surface. Such the projection seems to be evidenced by the high regularity of the lineaments observed what would be difficult to explain when rejecting their tectonic origin. The geological conditions in not intensively disturbed platform sediments imply the regular pattern of disjunctive structures, its preservation and projection.
2 -Lineaments should be referred to the disjunctive structures closest to the recent surface, namely those of the Cretaceous series. The fractures in the calcareous series causing the block structure easily underwent the reactivation what is suggested by the detalled pattern of projection of inner structures onto the surface.
3 -The activation of tectonic blocks of the Cretaceous series occurred not only in the Holocene /what is evidenced by the present visibility of the lineaments on the delta sediments surface/ but could have taken place even earlier, controlling the erosional and depositional processes in the periods younger than Cretaceous, especially in the Pleistocene. During the time of glaciations the fractured Upper Cretaceous slab lying on unstable basement consisted of sands, clays and siltstones of Cenomanian was influenced due to glaciations by the changing load of overburden.
4 -The activity of the certain blocks affected also the sedimentation and/or the selective erosion of the rocks older than calcareous Cretaceous series what might be suggested by an extent the sandy member of Cenomanian. These sands occur in the block lying in the west from the line Tczew-Nowy Staw-Nowy Dwór Gd. Their extent remains in agreement with the run of two sets of lineaments, namely that parallel to the line Tczew-Nowy Staw-Elbląg, and that crossing them of submeridional /Nowy Staw-Nowy Dwór Gd./ orientation. The lineament pattern observed in the area of Żuławy Wiślane might be therefore characteristic not only for Cretaceous but also for the whole Jurassic-Cretaceous series.
5 -The hitherto presented facts prove the hypothesis concerning the significance of the lineament as a reflection of the fractures and dislocations occuring in the brittle rocks of the Upper Mesozoic. Since there exists simultaneously a high probability of an influence of these structures on sedimentation and erosional processes in the different periods of Cainozoic, the geological data there should be re-interpreted with regard to the lineament pattern. The lineaments themselves should be treated as the traces of possible disloca tion zones as well as of mobility of the basement. They shoul be taken into account when interpreting the geological cross-sections and constructing the geological maps. The authors propose to consider the lineaments pattern as an indicator of localization, orientation and continuity of the geological phenomena under prospection and recognition.

The presented above hypothesis on the possible role and significance of the lineaments in the area of Żuławy Wiślane should be practically verified as appear there newer geological data from the boreholes and other geological works /e.g. geophysical ones/.

Authors believe that the information from this area may be used in the areas of a comparable geological structure.

REFERENCES

Bażyński, J. 1980. Metody interpretacji geologicznej zdjęć satelitarnych wybranych obszarów Polski. /Methods of geological interpretation of the satellite images of choosen parts of Poland/. Instrukcje i Metody Badań Geologicznych, 44. Warszawa, Inst.Geol. - /In Polish only/.

Dadlez, R. /ed./ 1980. Tectonic map of the Zechstein--Mesozoic structural complex in the Polish lowlands 1:500 000. Warszawa, Inst.Geol.

Karaczun, K. & S.Kubicki & W.Ryka 1982. Lithological map of crystalline basement surface in Polish part of the East-European Platform 1:500 000. In S.Kubicki W.Ryka /eds./, Geological atlas of crystalline basement in Polish part of the East-European Platform, Table 2. Warszawa, Inst.Geol.

Kozerski, B. & A.Kwaterkiewicz 1984. The salinity zones of the ground waters and their dynamics on the Vistula Delta Plain. Arch.Hydrotech. Polit.Gdańskiej. - /In Polish only/.

Kubicki, S. & W.Ryka 1982. Structural-tectonic map of crystalline basement in Polish part of the East--European Platform 1:500 000. In S.Kubicki W.Ryka /eds./, Geological atlas of crystalline basement in Polish part of East-European Platform, Table 3. Warszawa, Inst.Geol.

Mojski, J.E. 1976. Arkusz Gdańsk. Mapa Geologiczna Polski 1:200 000 /Sheet Gdańsk. The Geological Map of Poland, 1:200 000/. Warszawa, Inst.Geol. - /Explanation in Polish only/.

Mojski, J.E. 1983. Lithostratigraphic units of the Holocene and surphace morphology of the bedrock in the northwestern part of the Vistula Delta Plain /Żuławy Wiślane/. Geol.Jb. A71:171-186.

Sadurski, A. & S.Kibitlewski & B.Daniel-Danielska /in print/. Statistical analysis of Landsat lineaments of the Vistula Delta Plain. Arch.Hydrotech. Polit. Gdańskiej. - /English summary/.

Sheets of the Geological Map of Poland 1:200 000: Sheet Grudziądz, A.Makowska 1972; Iława, A.Makowska 1976; Elbląg, A.Makowska 1977. Warszawa, Inst.Geol. /In POlish only/.

Symposium on Remote Sensing for Resources Development and Environmental Management / Enschede / August 1986

Analysis of lineaments and major fractures in Xichang-Dukou area, Sichuan province as interpreted from Landsat images

Lu Defu, Zhang Wenhua & Liu Bingguang
Institute of Geology, Academia Sinica, Beijing, China

Xu Ruisong & Jang Baolin
Institute of Geologic Technology, Academia Sinica, Guangzhou, China

ABSTRACT: The Xichang-Dukou area in Sichuan province of China was studied using Landsat imagery and existing geological maps.
A number of major and minor lineaments were identified which have been accounted for in available geological maps. N-S and ENE fault and lineament systems are present in the area. Rose diagrams were analyzed to assess direction and frequency of lineaments.
The fault zones are divided into two types: linear and arcuate.

INFORMATION AND METHOD

The area studied is located at the southern part of Sichuan province in China. Three Landsat images, scale 1:500,000, cover about 60,000 km² from north Minning-Yuexi in Sichuan province to south Yongren in Yinnan province. Geological maps of different type and scale are also available. Conventional visual interpretation methods were used, including modus of density slicing and local enhancement of lineaments for part of the area.

ANALYSIS OF LINEAMENTS

It is generally believed that fault zones and weak crustal zones are represented by lineaments on space imagery.

The lineaments on Landsat were interpreted by considering those lineaments which could be clearly determined as fault zone, fractured structure, linear river valley, linear tonal anomely zone etc,. The lineaments larger than 5km length were only drawn on this map scale.

On the basis of the directional distribution of the lineaments 4 groups have been distinguished, mainly in NS,NNW,EW,ENE direction, in which the NS (345°-5°) and ENE(65°-75°) trends are the strongest and the most distinct on the imagery. The lineaments in NS direction are principally present in the Anninghe valley and towards the west. They are characterised by their continuity and greater length. The EW lineaments occur in two areas: one from Dukou to Huapin and another one from Xichang to Muli. Their length is less than that of the NS lineaments.(Figure 1)

All linear features derived from the Landsat imagery and from existing geological maps were plotted in rose diagrams, as shown in Figure 2 and 3 respeclively.

The rose diagrams for length and frequency distribution are compiled in 5° interval classes.Fig.2 indicates that the lineaments longer than 5 km in length account for the dominant NNW, NS and ENE trend. Fig.3 shows a preferred orientation of the near NS and ENE major faults. A difference was found in preferred orientations between the lineaments as interpreted from the Landsat imagery and geological maps. The lineaments show two length distributions in the direction 345° and 75° on the Landsat imagery and in the direction 355° and 5° on the geological map. Directions 20°- 30° clearly present on the geological map are less exhibited on the Landsat.

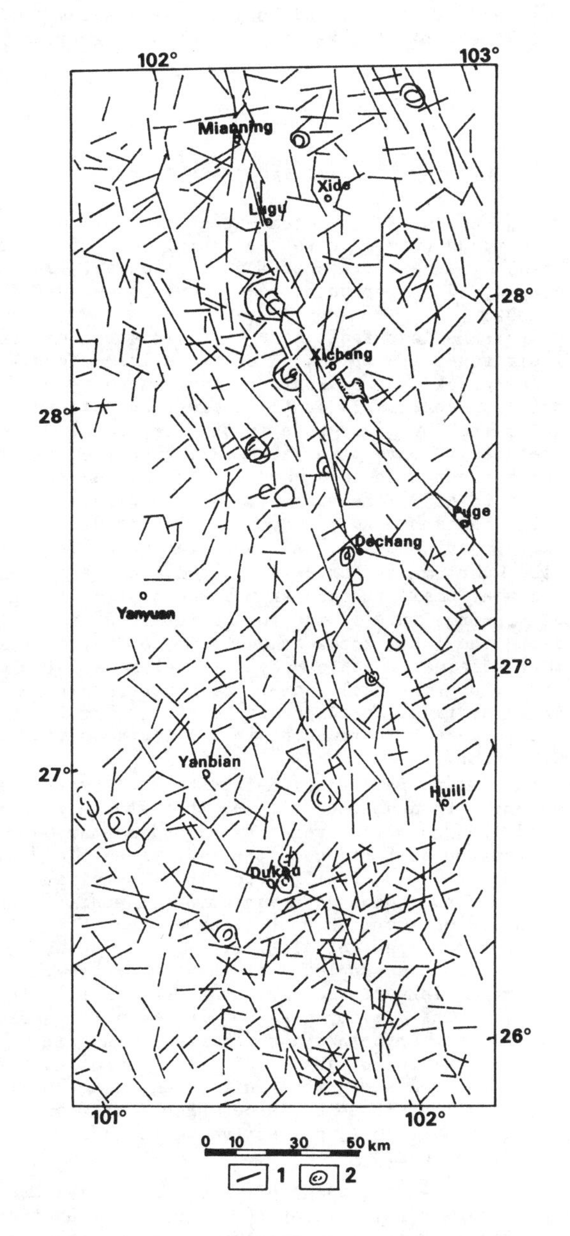

Figure 1. Major lineaments (>5 km) interpreted from Landsat imagery. 1. lineament 2. ring structure

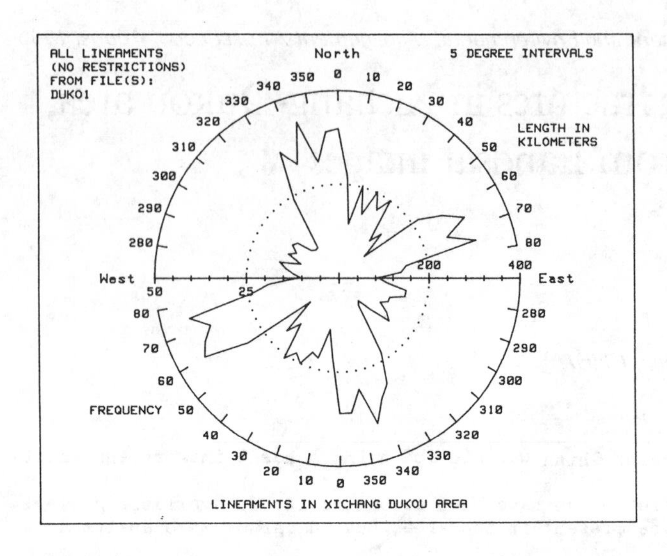

Figure 2. Frequency and length rose diagram for all lineaments (5 km) from Landsat imagery.

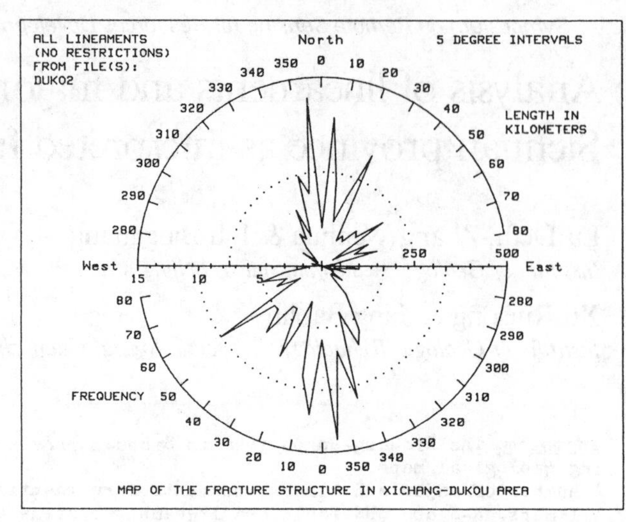

Figure 3. Frequency and length rose diagram for major fracture structures from geological map.

A DESCRIPTION OF MAJOR FAULT ZONES

On the basis of Fig.1 the map of the fracture structures (Fig.4) have been compiled of the study area. Lineaments interpreted from Landsat are here considered to represent major fault zones and fracteurs.

Anninghe deep fault zone (5): (location of the fault zones are indicated on Fig.4 in numbers between brackets) The Anninghe fault zone is not only easily recognizable on the imagery but has been determined by geologists in the field. The length measured is more than 350 km. It runs from north to south through the area and can be divided into two parts (Fig.5). The northern part to Dechang extends along the Anninghe valley near Xichang with the widest part about 15 km. There is series of elongated Quarternary basins developed along the valley, and many small rhombus shaped mountans to the north of Minning. This is due to tensional characterstics along the fault zone. In addition, offset of tributary streams indicate recent left-lateral displacement along the fault zone. The southern part of the Anninghe fault zone appears to be Zig-Zag in shape, but is genearlly trending in the north-southerly direction.
According to geological data, the fault zone has a long term developed history that strongly controlled the tectonic movements and sedimentary development on both sides of the fault zone (Cong Bai lin et.al,1973). A difference in alluvial and diluvial fans indicates a veriation in positive and negative movement on both sides of the valley in the northern part of Xichang.

Moreover the isopleths of the Bouguer gravity are extended along the NS direction from Shimain to Dechang. The gravity gradient in the valley area is higher than on both sides. Thirteen times in the recorded period of about 300 years a violent earthquake occured along the fault zone (Huang Zuzhi, 1979). All of these evidences show that The Anninghe deep fault zone is a active tectonic zone in recent periods.

Mopanshan-Luzhijiang deep fault zone (4): The Mopanshan-Luzijiang deep fault zone is a NS trending fault zone. It is situated towards the west of the Anninghe fault zone and appoximately parallel to it. The Mopanshan fault and the Xigeda fault are the principal fault in the zone. They can be clearly seen on the Landsat image. In many parts of the fault zone en echelon NNE directed fold structures can be observed. The structures are indicative for an ESE-WNW compressive stress resulting in a left-lateral transcurrent displacement along the fault zone. Many earthquackes are reported from Yuzao and Xigeda area. The basic-ultrabasic rocks of the last Caledonian stage to the early Hercynian stage and the Permian Emershan basalts are distributed widely in Hongge and Panzhihua area. This indicates a long term activity and large depth of faulting.

Ningnan-Huili fault zone (7): Although the NE lineaments are easily recognizable on the imagery, this fault zone is not important on the geological map. On the north-western side of the zone diorite and quartz-diorite of the Jinning stage are present, as well as diorite-gabbro of the Hercynian stage and the Emershan basalts of the Permian period. These rocks occur less on the southeastern side. Further a big sedimentary basin with Xigeda formation (Neogene) is developed mainly on the north western side.

Zemuhe fault zone (6): The general trend of the fault is towerds the NNW. The zemuhe fault and the Anninghe fault zone converge at the northwestern end of the area. The linearity of the river valley is clearly seen on the Landsat image. A quarternary basin is developed along the Zemuhe and Heishuihe valley. It is possible that the lake of Qionghai was formed as result of subsidence. Several earthquakes occurred in the area,illustrating present fault activity. The fault zone forms an important stratigraphic boundary.

Jinhe-Qinghe fault zone (2): It is a striking arcute type of fault zone. It converges upon the Anninghe fault zone at the northern end near Shimian. At the southwestern end,in the Yunnan province,it is cut off by the Yuanjiang fault. The trend of the fault zone varies from NNE in the Shumian-Lizhong-Zhoujapain area to N-S form Zhoujapain to Kuangshanliangzi, and NE trend from Kuangshan-liangzi to Qinghe (Fig.6).

There are related structures consisting of a series of closed folds occuring along the fault

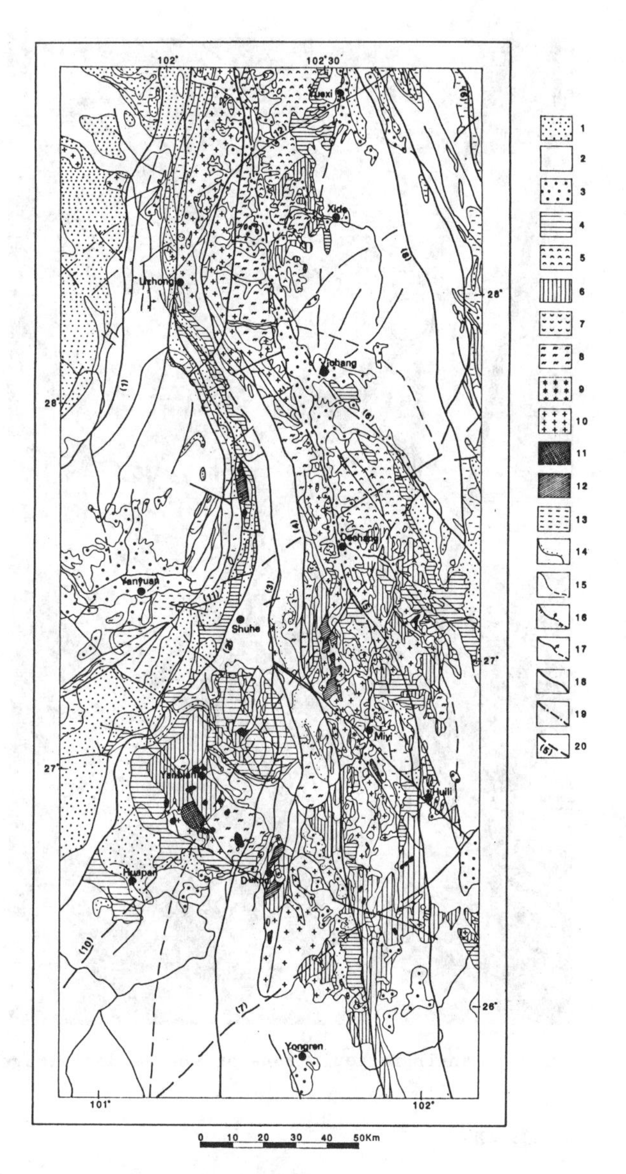

Figure 4. Landsat image and fracture structure map interpreted from Landsat imagery in Xichang-Dukou area. 1: Cenozoic; 2: Mesozoic; 3: Palaeozoic; 4: Sinian system; 5: Acid volcanic rocks system of Sinian period; 6: Proterozoic; 7: Unknown age metamorphic rocks; 8: Intermediate-acid-alkaline rocks of Jinning stage; 9: Intermediate-acid-alkaline rocks of Hercynian stage; 10: Intermediate-acid-alkaline rocks of Yanshan stage; 11: Ultrabasic and basic rocks of Jinning stage; 12: Ultrabasic and basic rocks of Hercynian stage; 13: Permian basalts; 14: Unconformity boundary; 15: Stratigraphic boundary; 16: Thrust fault; 17: Normal fault; 18: Fault zone; 19: Blind fault zone; 20: Conjecture fault zone.
Name of major fault zone: (1) Xiaojinhe fault zone; (2) Jinhe-Qinghe fault zone; (3) Lizhong-Dukou fault; (4) Mopanshan-Luzhijiang fault zone; (5) Anninghe fault zone; (6) Zemuhe fault zone; (7) Ningnan-Huili fault zone; (8) Xiluohe fault zone; (9) Puxionghe fault zone; (10) Huapan-Puge fault zone; (11) Xinping-Dechang fault zone; (12) Mianning-Yuexi fault zone; (13) Miyi-Huili fault zone.

trend, and thrust faults with thrust movement from NW to SE. The fault zone forms also an important geological boundary, controlling the sedimentary history on both sides of the fault zone since Palaeozoic. A few hundred meters of Emershan basalts occur along the fault zone (Yan Xianfu, 1981).

According to geophysical data, a negative gravity anomaly is formed along the fault zone, indicative for a large crustal thickness.

Most lineaments represent linear fault zones, except Jinghe-Qinghe fault zone and the Xijianhe fault. The latter two are arcuate fault zones which are considered to be formed under compressional stress conditions since a long history.

Figure 1 shows the distance between lineaments (linear fault and arcuate fault zone) in each direction to be approximately equal. For example, the NNE and NNW lineaments consitute a net-like structure. Their distribution is not limited by boundary lines of geotectonic units. In general, the close lineament belts reflect possible fault zones and /or deep fractures, confirmed by other geological evidences as gravity anomalies, seismic data, and sedimentary basin development, indicating their long geological history. Recent active faults are represented as clear lineaments on the Landsat image.

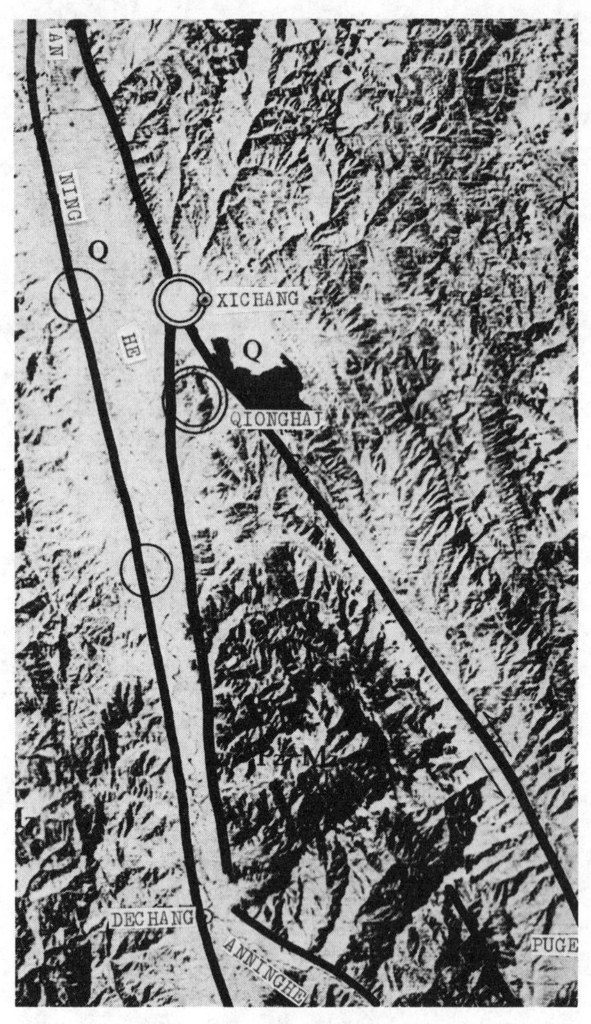

Figure 5. Anninghe fault zone on the Landsat image.

Figure 6. Jinhe-Qinghe fault zone on Landsat image.

CONCLUSIONS

Lineaments are well displayed on Landsat imagery. Considerable differences occur between lineament rose diagrams from Landsat and the geological map. The peak directions are present although they show differences in their distribution. Using Landsat image as complementary to existing data, a general limeament analysis can be made. Geological features such as faults/fractures are visible as full lineaments on the Landsat images. Several major lineaments systems are identified. To derive to a well-founded analysis, however, a very good knowledge of the geological structure of the area is required.

ACKNOWLEDGEMENTS

First and formost, to Dr. Bas N Koopmans for help with correction of the test; to N.H.W.Donker for digital processing of the lineaments; to some staffs of ITC for drawing of the figures and the typing.

REFERENCES

1. Cong Bailin, Zhao Dashang, Zhang Wenhua, Zhang Zhaozong and Yanmeie, 1973. Geol.Sec. IGAS, no.3 (in chinese).
2. Yan Xianfu, 1981, Geol.Journal, no.1(in chinese).
3. Huang Zuzhi, 1979, Acompilation seismic geological survey of the strongly seismic area in Sichuan-Yinnan region. p.144-161,(in chinese) Seismic Press, Beijing.

Symposium on Remote Sensing for Resources Development and Environmental Management / Enschede / August 1986

Application of remote sensing in the field of experimental tectonics

J.Dehandschutter
Royal Museum of Central Africa, Tervuren, Belgium

ABSTRACT : Geological documents of often display systematic arrangements of linear structural elements. The basic geometric configuration is a rhomb divided by its short diagonal. Various classes of geological structures repeatedly occupy the same position inside the basic rhomb. Examples are seen on remote sensing documents of both brittle and ductile tectonic domains in the Andes and in Central Africa. Photoelastic stress analyses implemented on homogeneous plates and on others cut by vertical discontinuities the elastic properties of which are different from those in the interior of the plate, suggest that several factors determine the geometric position of lineaments and structures. The rhombic pattern of intersecting discontinuities is one factor of lower order. Directions of vertical faulting inside the blocks and oblique-slip rotational strain inside the lineaments are predicted.

1 INTRODUCTION

The results presented here stem from conventional analyses of Landsat MSS imagery. Emphasis was put on the possible geological and economic significance of lineaments (Dehandschutter & Lavreau 1985). The latter are defined according to Hobbs' classic definition (Hobbs 1911).

Frequency diagrams issued by students of linears and lineaments from various parts of the world often show a remarkable similarity in the mutual angular relationships between several groups of lineaments and in their trends. Two sets generally occupy sub-latitudinal positions, while two others strike in a sub-meridional sense. A regional representative example selected from a survey over Shaba (Zaire) and northern Zambia is shown in fig. 1a. Figure 1b depicts the very comparable situation in the Eastern Cordillera of Colombia, South America. Dividing each of the regions in distinct areas results in diagrams revealing a strong positive correlation between one meridional and one latitudinal set. We may separate one eastern conjugate pair from a western one (fig.1a,c).On the scale of the area does in most cases one pair of sets largely subdue the other pair.

In plan view is this basic geometrical configuration translated into the picture of a rhomb divided by its short diagonal (fig. 1d). There is only limited three-dimensional control on the attitudes of the sensed structures and lineaments. It is tentatively assumed that the straight lineaments which are independent of topography, extent sub-vertically. The basic rhomb offers an hypothetical framework in which some large geological structures, i.e. rifts, plateau-uplifts, subsiding areas, oblique-slip sedimentary basins, fit in some characteristic spatial relationship. This relationship is recognised on various continents and underlying causative processes might have been operative during widely separated orogenic cycles.

2 SPATIAL RELATIONSHIPS

Geological structures can, for all practical purposes, be subdivided in three distinct classes according to the particular stress regime that prevails at the time of creation of the structure. It is furthermore convenient to limit the discussion to the major groups of structures, i.e. structures respectively related to vertical or horizontal compression taking along the third class, oblique compression, within any one of the former major classes.

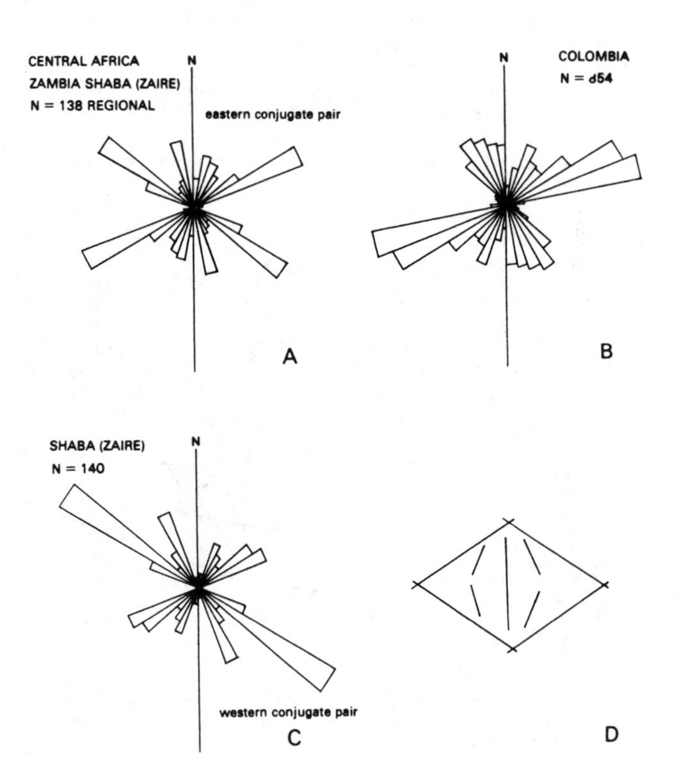

Figure 1. Representative examples of frequency diagrams (A,B,C); basic rhombic configuration (D).

2.1 Structures related to vertical compression

Rifts, normal faults, dilatancy

The conjugate couple of (sub)meridional-(sub)latitudinal lineaments appears clearly in downfaulted areas and/or areas under extensional strain.

Examples abound in the Central- and East-African rift belt (fig. 2a). The long axis of Lake Tanganyika is W from north close to and influenced by the NW-SE Precambrian Ubende trend. Both sets of lineaments constitute a western conjugate pair. Close to the NE-SW Precambrian Kibara lineament trend, the long axis projects east from north : the eastern couple. Observations of this kind are made on the Karroo rifts of Zambia (Luangwa, where the well known

Mwembeshi dislocation zone, De Swardt et al. 1965, on the one hand imposes a NE azimuth to the rift axis, the Ubende chain, on the other, deturns the axis towards the NW) and lots of downfaulted areas as Lake Mweru (eastern couple) and its southern extension in the Luapula River (western couple).

Surprisingly similar pictures arise from the analysis of the northern Andes (Colombia, fig. 2b). The Magdalena River flows from the south to the north linking a string of true grabens, half-grabens and faulted synclines, each segment of which in turn is limited by ENE and/or SE lineaments and faults. The Magdalena rifts have another particular characteristic in common : the individual segments are disposed in typical en-échelon fashion and some of the individual échelons are linked by tails protruding as either ENE or SE lineaments.

The latter features are easily recognised in Central Africa too (fig. 2a) : the geometry and shape of Lake Mweru Wantipa is one of a sigmoid rhomb, the main body of which strikes NE. The northern and southern tails extend as ENE lineaments. This eastern couple of lineamental directions is repeated in downfaulted (?) swampy areas (Luwala and Bwela swamps).

Quartz dykes, tens of kms long, have the same relationships to the confining ENE and SE lineaments as the NNE and NNW Phanerozoic rifts have (fig. 2a). The absence of regional compression in that part of northern Zambia (Mporokoso village) rules out a compressive stress field-related origin of the dykes. They are here interpreted as the products of dilatancy (Jaeger & Cook 1976) on the megascopic scale related to the formation of tensional joints and lineaments parallel to the axis of maximum horizontal stress.

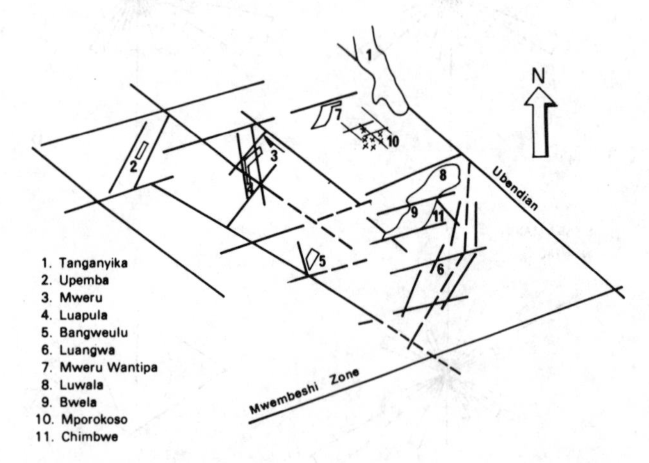

Figure 2a. Central Africa : schematical representation of spatial relationship between basic rhomb and vertical tectonic elements.

The examples suggest that rifts and vertical faults are linked to transverse lineaments. They are therefore easily recognisable on satellite imagery. The similarities in the geometry of all mentioned structures chosen in geologically (cratonic against mobile belt) different terrains are interpreted as a hint helping to understand some genetic aspects they have in common.

Pull-apart basins

Crust that is subject to an overall compressive stress field is amenable to local stretching and extension in the zones between the overlapping strike-slip master faults (Crowell 1974). These processes lead towards S- and Z-shaped rhombic pull-apart basins (Mann et al. 1983).

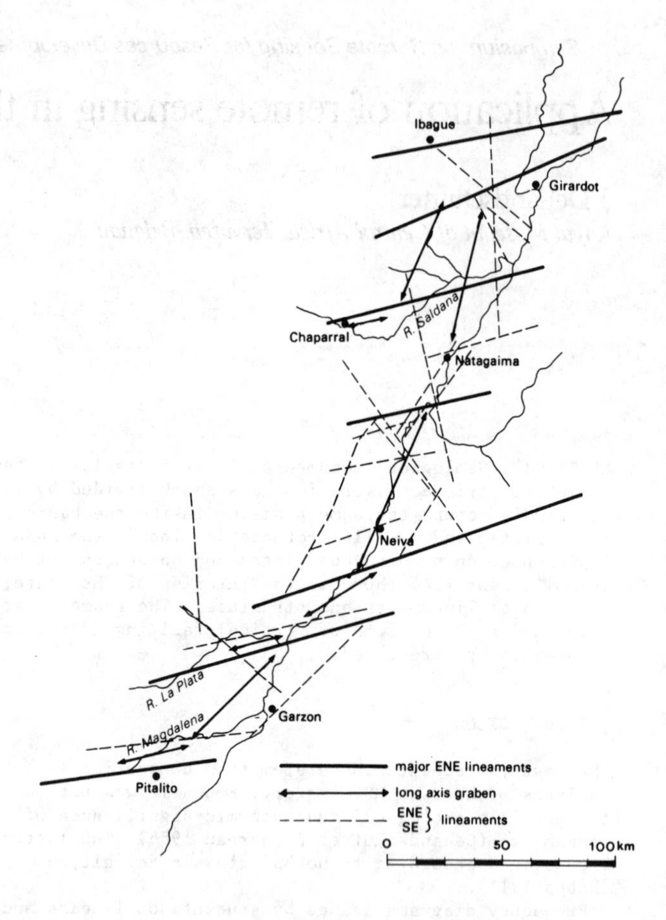

Figure 2b. En-échelon disposition segments Magdalena River, Colombia.

Figure 3 shows the photogeological interpretation of the Pre-Cambrian Chimbwe synform or S-structure, northern Zambia (see fig. 2a for localisation). The S-structure is found in a corner between major ENE (Kibara) and SE (Ubende) lineaments. The structure is interpreted as the outcome of the synsedimentary process of transform of slip-motion from one slip-line towards another. The southern half is composed of the synform or depocentre proper and is formed and deformed by left-lateral slip along ENE lineaments. The group of SE lineaments may represent crustal discontinuities or faults that accomodated the extension. The northern half, of less evident morphology than the southern one, has a weak Z-shape and was, in the present interpretation, pulled apart by dextral slip on the SE lineaments.

It is remarkable again that images obtained over the northern Andes, give evidence of the same fundamental geometric relationships between synforms and lineaments. Here the sediments are of Oligo-Pliocene age. It is obvious from figure 4 that the southern depocentres are controlled by an eastern couple of lineaments, the northern synforme strike to the NNW, their axes being influenced by strong NW-SE transverse lineaments (western pair). We may interprete the origin of these sigmoid structures in the same way we did for their African Precambrian counterparts. Other processes may however be invoked in explaining the origin of the depocentres. Another conceivable scenario, e.g., is represented in fig. 5b which illustrates the genesis of synforms comparable to the one under discussion. It can even be argued that the sigmoid shape is the result of compressive dextral simple -shear on the ENE lineaments after deposition of the sediments. This variety of possibilities requires a variety of slip-senses on the ENE and SE lineamnets. The paramount observation however, i.e. manifest strike-slip strain, cannot seriously be challenged. The region of concern is

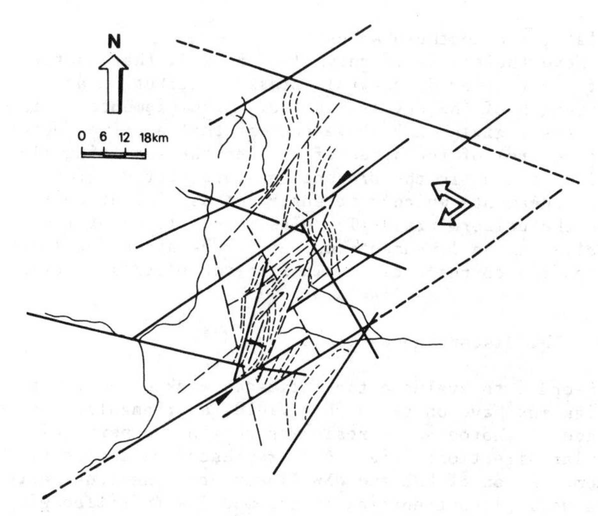

Figure 3. Photogeological interpretation of the Chimbwe-synform S-structure, Zambia. Depocentre created by sinistral slip on left-stepping oblique-slip faults.

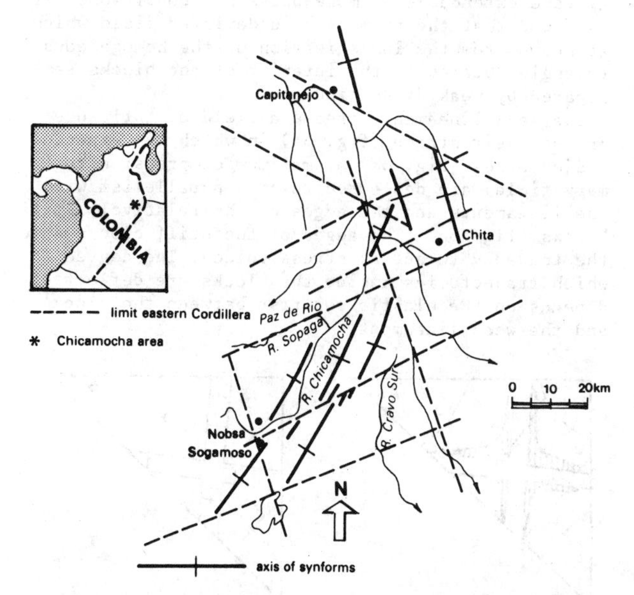

Figure 4. Photogeological interpretation of the Rio Chicamocha area, Boyaca, Colombia. The axes of synforms are strongly reminiscent of pull-apart basin geometries (S- and Z-basins).

limited to the east by the arm-pit of the Cordillera where recent uplift attained its highest levels (inset to fig. 4).

2.2 Structures related to horizontal compression

Figure 5a shows a combination of Landsat interpretation and published geological mapping (Raasveldt 1956) of the area around Girardot, Colombia. The rather chaotic image is caused by coeval shortening around two almost mutually perpendicular axes. Shortening is not equally distributed over the area. It is limited to well determined linear zones appearing as lineaments in which uncompetent strata are folded en-échelon around horizontal or plunging axes. More competent beds are merely uplifted or drape folded (Stearns 1978) over the linear zones of thickening. Figure 5b schematically represents the contours of these intersecting linear uplifts. In the field, they indeed appear as narrow and faulted anticlines. During the uplift of the anticlines, rocks inside the rhombic area outlined by them, is stretched and passively faulted. These

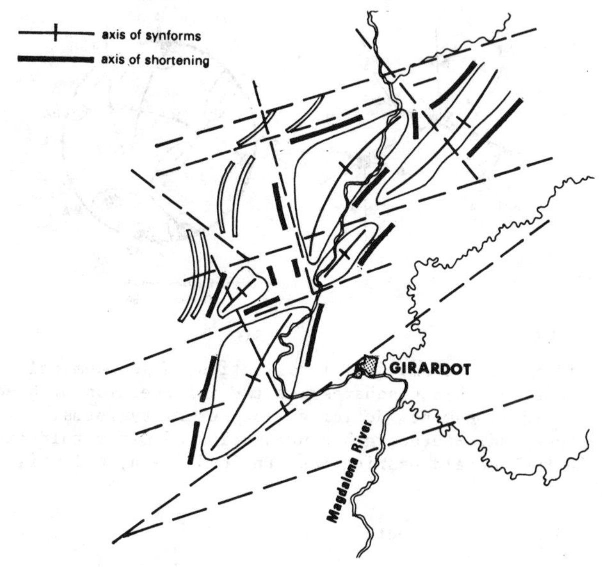

Figure 5a. Landsat interpretation of published geological map Girardot area.

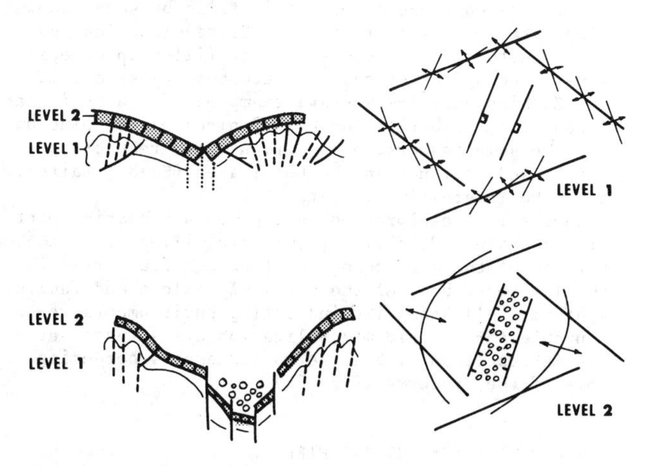

Figure 5b. Interpretation Girardot area.
Structural level 1 : active shortening and uplift in lineaments bordering a central stretched area;
Structural level 2 : passive drape folding and formation of synforms, syntectonic sedimentation.

areas appear as large-wavelength synforms. At the apogee of the border uplift, a central graben is created which links two corners of the rhomb. The graben is filled with molassic sediments.

This interpretation is corrobated by partial observations in profiles where shortening is evident and expressed in minor fold axes accompanied by axial plane and shear cleavage. Here, in the emerald bearing part of the same Eastern Cordillera, detail field mapping reveals that synforms are contained within a framework of crossing ENE and SE sub-latitudinal lineaments (fig. 5c). The lineaments are the topographic expression of lines of brecciated and metasomatically altered country rock. Some measured axes of minor folds group inside a statistical plane which is parallel to each of the sub-latitudinal lineaments (Ea and Wa in fig. 5d). More fold axes project within a plane comprising the sub-meridional lineament direction which is conjugate to the first sub-latitudinal member of respectively the eastern and the western pair (Wb and Eb in fig. 5d). On the megascopic outcrop scale, one sees two groups of narrow anticlines, each group with different axial strike and adjacent to the corresponding transverse brecciated lineaments. The interposed synform is terminated by the intersecting lineaments. The divergent strike of the axes of shortening or anticlines imposes important thinning in the lowermost portion of the rhombic synform (fig. 5b).

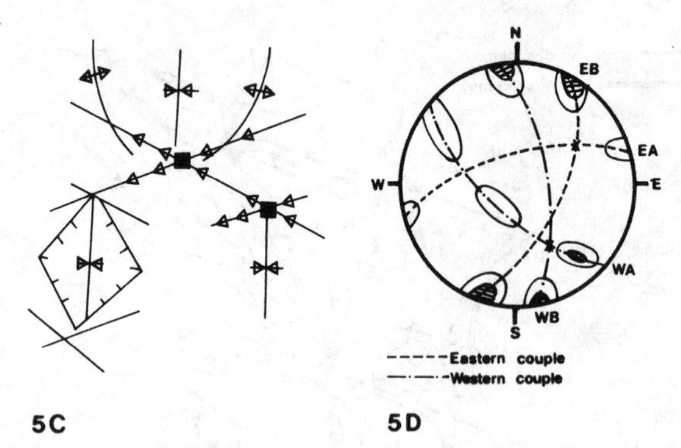

Figure 5c. Schematical localisation of geochemical anomalies (black squares) at the intersection of brecciated lineaments defining closures of synforms.
Figure 5d. Stereographic projection of minor fold axes in the emerald bearing Eastern Cordillera, Colombia.

2.3. Economic aspects

The deviatoric part of the horizontal stress field seems to split the crust in the direction of maximum horizontal compression before it fails by shear faulting. Fractures with this appropriate direction and relation to shear fractures appear filled up by epigenetic solutions of possible economic interest. If the Zambian quartz-dykes are examples of a non-mineralized fill, kimberlite dykes and pipes outcropping under the same favourable conditions with respect to strike and attitude in the Lofoi area (Shaba, Zaire), are more interesting targets.

Geochemical exploration in the emerald bearing part of the Eastern Cordillera, Colombia, illustrates the effects of the stretching and thorough fracturing in the lowermost beds of the passively folded and faulted synforms. All anomalies indicating environments favourable for emerald mineralization are situated at the intersection of brecciated lineaments bordering some of the synforms (fig. 5c).

3 THE CONTINENTAL STRESS FIELD

It is currently accepted that, except for uplifted and isostatically compensated regions where the stress field would be tensional (Artyushkov 1973), most of the continental plates are slightly compressional (i.g. Fleitout & Froidevaux 1983). The lithosphere act as a guide for this deviatoric field and may be treated as a continuum. The repetitive pattern of trends of lineaments and their spatial relationship with geological structures indeed suggest that the former are the effects of a uniform strain emanating from continuous and homogeneous processes. As far as the distribution of the stress field within the Earth's lithosphere and crust is concerned, the classical theory of a continuous elastic plate may be applied. Much of the observed strain in the crust however is of ductile nature and results from plastic processes. The applicability of elastic theory to geological processes in nature is thus restricted to the time period between the onset of loading of the planar continuum to the moment the elastic material fails by brittle fracturing or deforms in a plactic way.

3.1 The homogeneous field

The stress field in the plate interior depends largely on the strength of the applied forces, on the geometric outline of the loaded plate and on the boundary conditions. Pure shear does not rotate the strain axes and stress and strain remain co-axial. A study implemented on a triangular body (Dehandschutter in prep.) shows a good gross correlation of the obtained stress trajectories and lines of elastic slip with the observed strain in the triangular corner of the South America Plate, the northern Andes.

Nevertheless is it very clear as well that factors of lower order do control stress distribution and strength of the deviatoric field. The elements of the analysis given in 2 above suggest that a prime factor in control of the stress field and the resulting finite strain, is the presence of intersecting groups of lineaments which apparently act as discontinuities in the uniform crust. The elastic constants of the solid in the lineaments may either be higher or lower than the corresponding values in the plate's interior.

3.2 The discontinuous field

In order to evaluate the influence weak discontinuities may have on the stress field, a triangular plate made of photoelastic resin was cut in two perpendicular directions (figs. 6,7) representing a possible combination of ENE and NNW lineaments. The lineaments or weak discontinuities in the model were filled with another resin the modulus of elasticity of which is 5 times less stiff than the material in the interior of the plate. The photoelastic analysis was seconded by a numeric finite elements control (fig. 7).

Some results of the experiments done on the triangle are extendable to more universal conditions. It is found that the primary or undeviated field which is known from the investigation of the homogeneous triangle obtains in the interior of the blocks separated by weak discontinuities.

The weak lineaments create a field of influence around their strike (fig. 6a) in which the expected trajectories (fig. 6a) of maximum compression (primary field) are deflected towards parallelism with the lineaments and the edges of the relatively stiff blocks (fig. 6c). The adges of the stiff blocks have the tendency to act as stress guides. The degree to which trajectories inside the blocks are deflected depends on the elastic contrast between the block and the weak lineament.

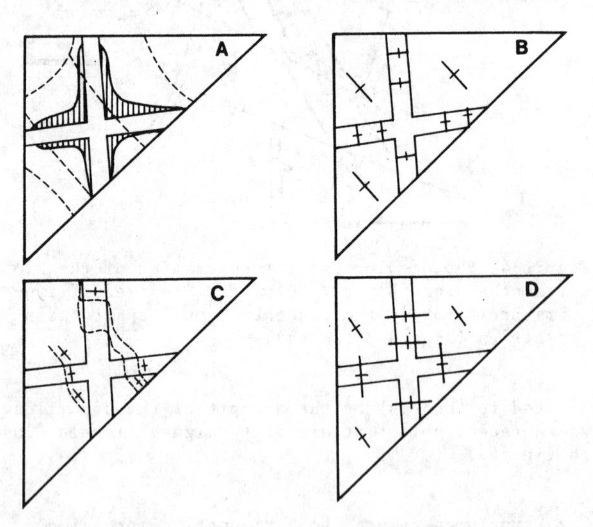

Figure 6. Elastic stress distribution in a triangle cut by weak discontinuities. A. Fields of influence of weak zones and sigma 1-trajectories in homogeneous triangle; B. Azimuth and magnitude of axes of maximum and minimum compression, thickness in lineaments = thickness in blocks; C. Refracted and reflected trajectories; D. Axes of max. and min. compression, thickness lineaments = thickness blocks/2.

Trajectories entering or leaving the weak lineaments are strongly refracted (fig.6c). The axes of minimum compression are re-oriented inside the zones where they strike parallel to the trend of the lineaments. The axes of maximum compression correspondingly turn towards perpendicularity with the respective lineaments. The angle of incidence of the

Figure 7. Elastic stress fields in a discontinuous body. A. Photoelastic analysis, only the axes of maximum compression are shown.

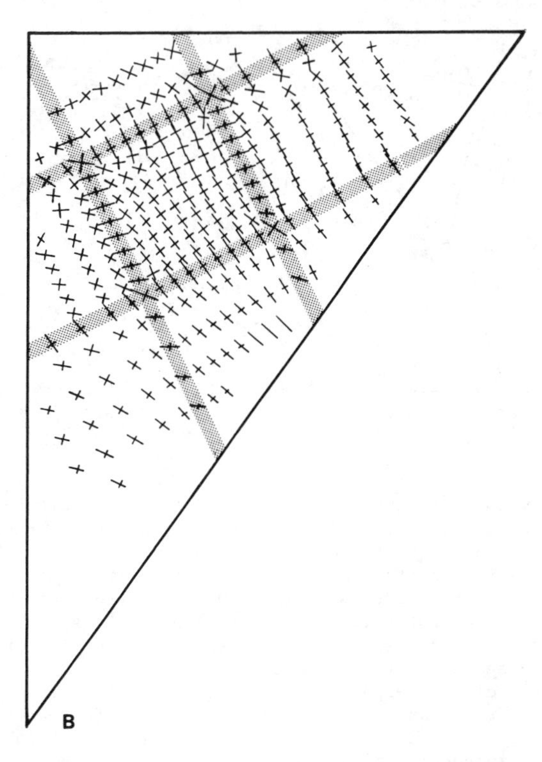

Figure 7. Elastic stress fields in a discontinuous body. B. FEM-analysis, axes of maximum and minimum compression.

trajectories entering the lineaments is determinative with respect to the consequences of the refraction : the axis of minimum compression of normally indicent trajectories is always parallel to the weak zone, inside as well as outside of it, and the trajectory may cross the weak zone freely.

Diminishing the thickness of the lineaments as compared to the fixed thickness of the blocks, greatly enhances the deviatoric field in the former (fig. 6d).

Finally do isotropic points of zero deviatoric stress which are loci of sudden changes in the directions of the trajectories, appear in two out of the four corners of the blocks surrounded by weak discontinuities. The exact location of the isotropic points within their corners depends on the boundary conditions of the loaded plate (fig. 8).

4 DISCUSSION

Accepting the possibility that lineaments are somehow related to crossing discontinuities within the elastic crust which are weaker than the more homogeneous crustal solids and which strongly influence the trajectorues of the primary stress field, provides possible, though not unique, clues in understanding the fixed spatial relationship to lineaments in which many geological structures appear.

The deviatoric field in the weak zones is always higher than the field in the block interior. The latter, with high mean stress, is often close to lithostatic.

Small increments of stress may bring the interior of the blocks to dilatation and vertical splitting in the direction of maximum horizontal compression. These directions are visualised by stress trajectories which appear to systematically contour the blocks. The direction of spreading of the trajectories over the block depends on the position of the isotropic points within the block. Figure 8 shows that, depending on the loading conditions, trajectories and resultant brittle tensional failure may be deviated from a diagonal position in the centre, towards a less inclined position with respect to the weak zones in the vicinity of the latter. This result of the investigation may explain the systematicity in the directions of geological structures and their relation to confining lineaments as weak zones which was discussed in 2.1. It is quite clear that the above discussion is limited in time to the first phases of rift formation. True extensional rifting will be accomodated by only 1 out of all possible sets of transverse structures which eventually evolved into a transform fault.

The weak zones proper, here modelled as an elastic medium, might in fact conform better with true elastic-plastic bodies or contain a visco-elastic element. A strong deviatoric field in these zones, which is absent from the interior of the blocks, may eventually lead to permanent strain, shortening and uplift. The compression from within the lineaments onto the rhombic blocks finally thickens the edges of the blocks. Stresses resulting from the consequential bending of the homogeneous rhombic block will then substitute the stresses emanating from the boundary forces. The block interior fails by vertical rupturing or is bent into

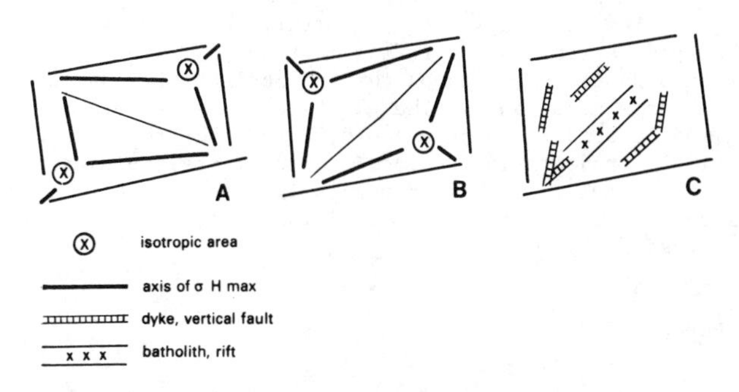

Figure 8. The position of isotropic areas determines the trajectories of maximum horizontal compression (A,B) and the directions of the geological structures this compression creates (C).

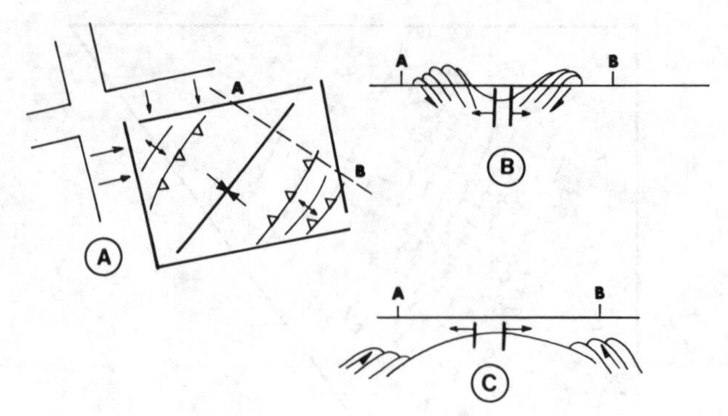

Figure 9. Thickening of the edges of a block by over-(C) or underthrusting (B) due to compression from within the weak zones (A).

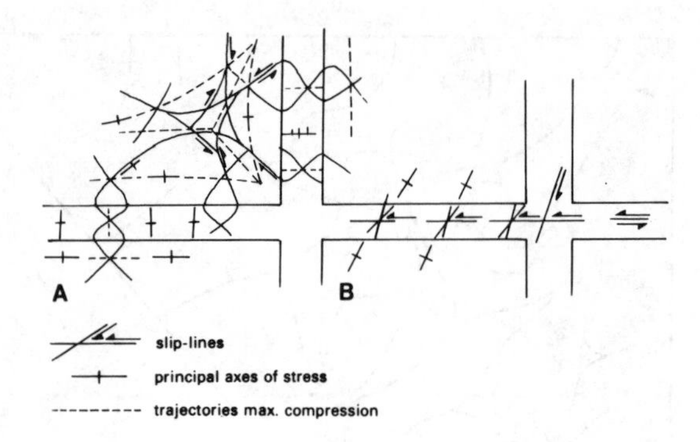

Figure 10. Possibilities of strike-slip faulting in lineaments. A : impossible; B : favoured.

a synform (fig. 9). This scenario offers an alternative to the first one although both may concur and joinedly affect the block interior.

The re-orientation of the axes of maximum compression towards perpendicularity with the trend of the weak lineaments impedes strike-slip faulting within the lineaments if the edge of the blocks are shear free and the principal axes of stress respectively parallel and perpendicular to the edge (fig. 10a). If however the axes are not refracted into a normal position, they are inclined to the edges of the block. The lineaments, filled with material of low shear strength, then offer ideal pathways for strike-slip faulting (fig. 10b). This fully confirms the observations in 2.2 which testified of oblique-slip strain inside the lineaments.

SUMMARY

Geological structures with precise tectonic significances and indicative to various degrees of the orientation of the paleo-stress field in which they were created, can be circumscribed within a framework of transverse lineaments visible on satellite imagery. Structures of this kind are tensional joints, vertical faults like rift boundary faults, quartz dykes, sedimentary basins of pull-apart origin. Even structures of compressive nature seem to relate to the same framework. The latter though might be the sequel of the presence of discontinuities like synsedimentary faults which were later reactivated during compression. During the opening phase the basin marginal faults were seemingly related to the framework. The are responsible for two directions of shortening (B normal to B) operative during the same orogeny.

The good positive correlation between the theoretical stress field within some homogeneous elastic triangle with observed large-scale tectonic features in the northern Andes confirm the validity of the approach. A second order factor in control of the stress field in the framework of intersecting weak discontinuities. The weak zones reflect stress trajectories in a way which is compatible with the observed variability of rift directions. This effect depends on the one hand on the physical conditions in the weak zones and in the blocks and on the boundary loading conditions on the other hand. The refraction of the trajectories inside the lineaments explains how several azimuthal groups of lineaments can give way in coeval strike-slip faulting under the same stress field.

REFERENCES

Artyushkov, E.V. 1973. Stresses in the lithosphere caused by crustal thickness inhomogeneities. Journ. Geophys. Res. 78 : 7675-7708.

Crowell, J.C. 1974. Origin of late Cenozoic basins in southern California. In Dott, R.H. & Shaver, R.H. (eds.), Modern and ancient geosynclinal sedimentation. Spec. Bull. Soc. Econ. Paleont. Miner. 19 : 292-303.

Dehandschutter, J. in prep. Lineaments in the northern Andes and their bearing on the geodynamic evolution in the leading corner of the South America Plate.

Dehandschutter, J. & Lavreau, J. 1985. Lineaments and extensional tectonics : examples from Shaba (Zaire) and NE Zambia. Bull. Soc. Belg. Géol. 94 : 209-221.

De Swardt, A.M.J., Garrard, P. & Simpson, J.G. 1965. Major zones of transcurrent dislocation and superposition of orogenic belts in part of Central Africa.Geol. Soc. Am. Bull. 76 : 89-102.

Fleitout, L. & Froidevaux, C. 1983. The state of stress in the lithosphere. Tectonics 2 : 315-324.

Hobbs, W.H. 1911. Repeating patterns in the relief and in the structure of the land. Geol. Soc. Am. Bull. 22 : 123-176.

Jaeger, J.C. & Cook, N.G.W. 1976. Fundamentals of rock mechanics. London : Chapman & Hall.

Mann, P., Hempton, M.R., Bradley, D.C. & Burke, K. 1983. Development of pull-apart basins. Journ. Geol. 91 : 529-554.

Raasveldt, H.C. 1956. Mapa geologico de la Rep. de Colombia, Plancha L9, Girardot. Bogota : Ingeominas.

Stearns, D.W. 1978. Faulting and forced folding in the Rocky Mountains foreland. In Matthews III, V. (ed.), Laramide folding associated with basement block faulting in the Western United States. Geol. Soc. Am. Memoir 151 : 1-37.

Thematic mapping from aerial photographs for Kandi Watershed and Area Development Project, Punjab (India)

B.Didar Singh & Kanwarjit Singh
Planning & Design St.Directorate (IB), Punjab, Chandigarh, India

Abstract: The authors have utilised the photo-interpretation techniques for the thematic mapping for geomorphological, hydrological, geological - hydrogeological, soil system and forestry studies of the Kandi Area in the State of Punjab under the World Bank aided Kandi Watershed & Area Development Project. The Kandi area lying at the foot hill of Siwaliks in Hoshiarpur and Ropar districts is in an under developed state on account of increasing population pressure, farming and raising livestock on erodible slopes, widespread felling of trees and fuel and fodder, poor irrigation and other infrastructural facilities, coupled with illiteracy, economic and social backwardness has inflicted serious damage by bringing about an increasing rate of top soil erosion. In order to reverse this ecological degradation and to protect as well as develop the agricultural and other infrastructural facilities in the area, the Govt. of Punjab with the assistance of British firms of Sir William Halcrow and Sir Murdoch Macdonald, prepared a detailed feasibility report for this project. The photo-interpretation techniques used in the present study proved to be very useful in exploration, monitoring and management aspects required for the developmental planning of an area.

1 INTRODUCTION

The Government of Punjab have long been aware of the growing damage being done to the land of Siwaliks in Punjab State, which borders the foothills of the Himalayan range. The Siwaliks in Punjab are mainly represented by the Upper Siwaliks formations of pleistocene to recent age and mainly composed of incoherent material like boulders, pebbles, cobbles, sands, silt and clays and outcropped in the NW-SE direction (Pascoe 1964). A large number of ephemeral streams locally termed as Choes flow down from the Sub-mountainous Zone of upper Siwaliks and immediately after a rainfall in the catchment area during the monsoon period, these streams swell into a flood. As the slope of the bed is steep, the velocity of flow becomes high and causes severe erosion of the soil cover. The flowing water charged with the heavy sediment load of the incoherent material of the catchment area, has a narrow course in the hills but as it reaches the plains, it spreads over a large area, deposits its sediment load to make the valuable fertile lands unproductive. Besides this the increasing pressure of a growing population, farming and raising livestock on erodible slopes, has led to the extensive destruction of protective vegetation. Together with widespread felling of trees, over-intensive use of the land has brought about an increasing rate of top soil erosion, which is also inflicting serious and permanent damage to the agriculture of the area lying at the foot-hill of Siwaliks in Punjab State, locally termed as the Kandi Area.

In order to alleviate these problems the Government of Punjab selected the Kandi area of Punjab State as the site of a World Bank aided project to base an extensive programme of development under a detailed and comprehensive plan called Kandi Watershed and Area Development Project.

2 PRINCIPAL OBJECTIVES OF THE PROJECT

The overall project plan is for the eventual improvement and development of a total of about 1,50,000 hectares of the Kandi tract lying in the State of Punjab.

In the first phase work has been concentrated on eleven representative watersheds covering an area of about 30,000 hectares in Hoshiarpur and Ropar districts of Punjab. Individual watershed has been taken as the unit of development in the Kandi tract and has been broadly split into two sub-units i.e. upper catchments and Kandi plains. The potential of each of the two subunits has been studied in details for the following development programmes.

2.1 Upper catchments

1 Afforestation & reforestation
2 Soil conservation measures-land levelling and terracing
3 Flood attenuation measures
4 Water storage & erosion control works
5 Minor irrigation schemes
6 Fish culture-where possible
7 Overall improvements to agriculture,horticulture and livestock of the area.
8 Improvement of area's infrastructure like improved water supply and all weather roads.

2.2 Plains of the Kandi Tract

1 Surface irrigation schemes
2 Groundwater irrigation schemes
3 Flood protection measures and stream diversion works
4 Canalisation of water-courses
5 Groundwater recharge measures
6 Overall improvements to agriculture & water management
7 Improvement of area's infrastructure like improved water supply and all weather roads.

3 PROJECT ORGANISATION

In order to give advice and assistance for the multi-developmental aspects of the project, it was decided by the Punjab Government to recruit a team of overseas consultants who would also supplement the expertise of the local professional staff of the Punjab Government. In the year 1980, the British firms of Sir William Halcrow and Partners and Sir Murdoch Macdonald & Partners against their joint proposal,

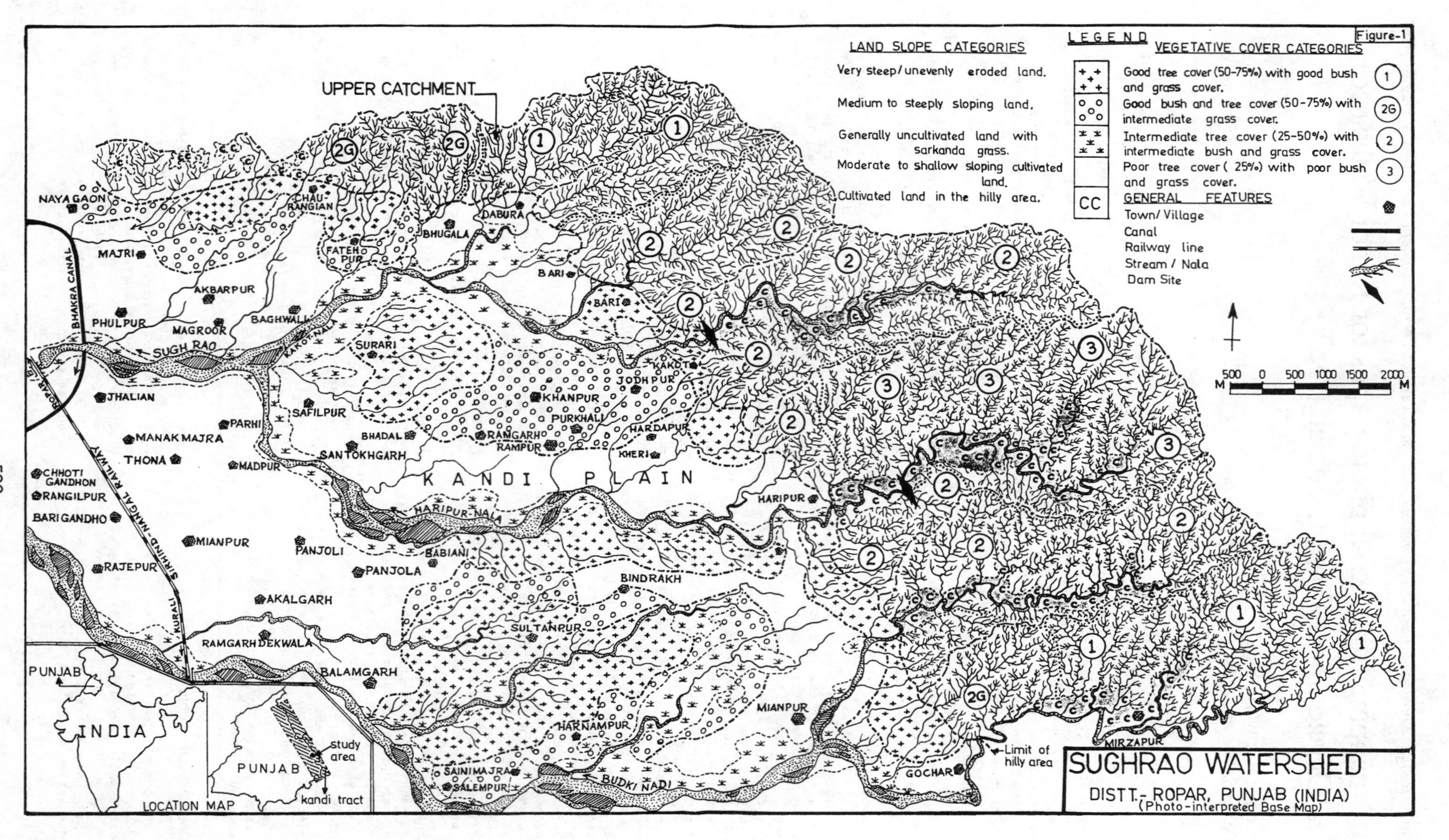
Figure-1
LEGEND
LAND SLOPE CATEGORIES
Very steep/unevenly eroded land.
Medium to steeply sloping land.
Generally uncultivated land with sarkanda grass.
Moderate to shallow sloping cultivated land.
Cultivated land in the hilly area.
CC
VEGETATIVE COVER CATEGORIES
Good tree cover (50-75%) with good bush and grass cover.
1
Good bush and tree cover (50-75%) with intermediate grass cover.
2G
Intermediate tree cover (25-50%) with intermediate bush and grass cover.
2
Poor tree cover (25%) with poor bush and grass cover.
3
GENERAL FEATURES
Town/Village
Canal
Railway line
Stream / Nala
Dam Site
500 0 500 1000 1500 2000
M M
UPPER CATCHMENT
NAYAGAON
CHAU-RANGIAN
MAJRI
FATEH PUR
BHUGALA
DABURA
BARI
AKBARPUR
PHULPUR
MAGROOR
BAGHWALI
SUGH RAO
BHAKRA CANAL
KAKOT NALA
SURARI
KAKOT
JODHPUR
KHANPUR
PURKHALI
HARDAPUR
JHALIAN
SAFILPUR
PARHI
MANAKMAJRA
BHADAL
RANGARHO
RAMPUR
KHERI
THONA
SANTOKHGARH
MADPUR
KANDI PLAIN
CHHOTI GANDHON
RANGILPUR
BARIGANDHO
HARIPUR
HARIPUR-NALA
SIRHIND-NANGAL RAILWAY
MIANPUR
PANJOLI
BABIANI
PANJOLA
RAJEPUR
BINDRAKH
AKALGARH
KURALI
SULTANPUR
RAMGARH DEKWALA
BALAMGARH
PUNJAB
INDIA
PUNJAB
study area
kandi tract
LOCATION MAP
MIANPUR
HARNAMPUR
SAINIMAJRA
SALEMPUR
BUDKI NADI
GOCHAR
Limit of hilly area
MIRZAPUR
SUGHRAO WATERSHED
DISTT.- ROPAR, PUNJAB (INDIA)
(Photo-interpreted Base Map)

were engaged by the Punjab Govt. to provide the necessary consultancy services. A resident team of 3 experts was made available to the Punjab Government, for two years. The team selected ten representative catchments of the Kandi tract for the detailed study and preparation of Feasibility Reports for each catchment during their stay of two years in the state of Punjab. The resident team was assisted by the local professional staff of Engineers, Hydrologists, Hydrogeologists & Economists of the Planning & Design Studies Directorate (I.B.) of Punjab Government.

4 BASE MAP REQUIREMENT FOR DEVELOPMENTAL PLANNING

An accurate and detailed base map depicting the upper catchments and the Kandi plains under each watershed taken up for the preparation of Feasibility Reports for their development was the prime necessity to carry out the detailed study. Thus it was decided to have a base map of each watershed on 1:20,000 scale,which was also adopted as the standard scale for all the thematic maps required under the project. It was further decided that the base map should provide the following minimum information for each watershed to serve for the detailed hydrological, hydrogeological, soils system and forestry planning programme of the project.

4.1 Upper catchments

1 Demarcation of overall catchment with sub-catchments
2 Demarcation of all major and minor drainage system
3 Demarcation of cultivated areas whether irrigated or under rainfed conditions
4 The sites of severe bank-cuttings along the major streams
5 Locations of suitable dam sites on major streams and small dam sites on their tributaries
6 Stream behaviour-perennial or non-perennial
7 Extent of baseflow in the major stream beds
8 Vegetative cover categorisation to evaluate the extent of erosion in the Siwaliks under the following norms:

Category 1 - Good tree cover (50-75%) with good bush and grass cover
Category 2G - Intermediate tree cover (25%-50%) with good bush and grass cover
Category 2 - Intermediate tree cover (25%-50%) with intermediate bush and grass cover
Category 3 - Poor tree cover (25%) with poor bush and grass cover.

9 Demarcation of habitation sites if ,any.

4.2 Kandi plains

1 Mapping of entire length of all the streams passing through the area
2 Demarcation of active flood plains which are generally devoid of any grass cover
3 Demarcation of old flood plains which are flooded only during the high flood period and are generally covered with Sarkanda grass
4 Demarcation of abandoned flood plains and palaeochannels where a fair distribution of tree cover and Sarkanda grass generally occurs
5 Sites of bank erosion along the stream length where flood protection measures like bunding, embankments, spurs or diversions if any are required
6 Landslope categorisation to plan for the overall development of Kandi plains for irrigation, soil conservation;horticulture programme under the following norms:

-Very steep unevenly eroded land
-Medium to steeply sloping land
-Generally uncultivated land with Sarkanda grass
-Moderate to shallow sloping cultivated land
-Cultivated land in the hilly area.

7 Demarcation of habitation sites if ,any.

As the existing topographic maps of the area were found unsuitable for such a detailed study, it was decided to have the base maps prepared from aerial photographs. The aerial photographs on 1:50,000 scale of the project area were procured from Government of India in the early stages of the Project.

The resultant base maps of each watershed were originally prepared on 1:50,000 scale and then blown up to the required scale of 1:20,000. Transfer of surface contours with a contour interval of 20 metres in the plains was done on photo-interpreted maps to serve as full fledged base maps. The infrastructure and expertise for aerial photo-interpretation existed in the Planning and Design Studies Dte,(I.B.), Punjab and it was carried out by the authors along with the requisite cartographic works with the help of drawing staff made available to the project.

5 UTILISATION OF BASE MAP FOR THEMATIC MAPPING

The base map prepared from photographs has been utilised for the preparation of a number of final thematic maps utilised in the Feasibility Reports of representative watersheds under the Kandi Watershed and Area Development Project, by combining the data derived from aerial photographs and the ground based data.The final maps prepared for each watershed are as under:

1 Map showing sub-surface lithology & location of deep tubewells
2 Water level elevation contour & depth to water contour map
3 Groundwater quality map
4 Present land use map
5 Land capability & soil classification map
6 Problem map of soils
7 Layout of dam based irrigation scheme
8 Map showing layout of ground water based irrigation scheme
9 Map showing flood plain and flood protection measures in the Kandi plains.

6 EXAMPLE OF PHOTO-INTERPRETED MAPPING

Fig.1 represents the sketch-photo-interpretted base map of Sughrao watershed, prepared from aerial photographs of 1:50,000 scale. Sughrao is a tributary of Budki Nadi which meets the main Satluj river downstream of Ropar town. Its two branches Kakot Nala and Haripur Nala merge to form Sughrao near village Baghwali at a distance of about 7 to 8 kms downstream of the hilly tract which later on flows in a northeasterly direction.

The upper catchment of Sughrao Choe system covers an area of 45.50 sq.kms. The entire drainage net work has been mapped along with the area under cultivation and the habitation sites. The vegetative cover categorisation of the upper catchment has been done to judge the extent of erosion and to plan the suitable remedial measures, as discussed earlier. The location of possible main dam sites and small dam sites has been marked to work out the hydrological studies of each site.

The area lying in the Kandi plains has also been subjected to detailed mapping. The main streams along with their net work of tributaries and all the habitation sites has been demarcated. It has also been subjected to land slope categorisation as discussed earlier to plan for the irrigation, soil-conservation, agriculture & horticulture development of the area. The main canals, roads, railway lines surface level countours (not shown in the present map) has also been mapped on the base map making it as a full fledged tool for thematic mapping. The present procedure adopted for thematiic mapping has resulted in the successful completion of ten number feasibility reports in a short span of two years.

CONCLUSION

It can be safely concluded that the wealth of details permanently recorded on aerial photographs when properly processed, interpreted and used with ground based data from other conventional sources can become a more viable tool for assessment, exploration, monitoring and management aspects required for developmental planning of an area.

REFERENCES

Leuder, D.R. 1959. Aerial photographic interpretation, principles and applications. McGraw Hill Book Co. Inc. N.Y.

Meijerink, A.M.J. 1970. Photo-interpretation in Hydrology, A geomorphic approach. ITC Publication.

Symposium on Remote Sensing for Resources Development and Environmental Management / Enschede / August 1986

Assessment of desertification in the lower Nile Valley (Egypt) by an interpretation of Landsat MSS colour composites and aerial photographs

A.Gad & L.Daels
Laboratory for Regional Geography, State University of Ghent, Belgium

ABSTRACT : The study area is situated in the lower part of the Nile Valley (Egypt) from the Giza- to the El-Menia province. The most important geomorphogenetic factors forming a threaten for the cultivated area are : wind action, especially in the Western Desert and the combined fluvial and wind action in the eastern desert. A visual interpretation of two satellite MSS colour composites was used to assess the desert encroachment. The interpretation has been verified by field work. Interpretation of Landsat MSS-images revealed the eolian deposits as bright coloured bands interupted by dark coloured patches of coarser erosional deposits coming from the mountains and hills in western desert. Aerial photo-interpretation had to be used to reveal the different orders of the hydrographic network and connection of drainage patterns and also the identification of the different types of erosional and depositional landforms. The combined interpretation of Landsat MSS images and aerial photographs made it possible to identify the landforms which are considered to be good indicators of desertification and so it was possible to recognize the endangered areas along the Nile Valley.

1 INTRODUCTION

Desertification, the extension of desert like conditions, is a severe problem in the arid and semi-arid regions. It happens by different processes (e.g. wind erosion, water erosion, vegetation degradation, salinization, compaction and crust formation.) The United Nations environment program (UNEP) stated that 20 million square kilometers have recently reverted to desert or desert like conditions.

Desertification in Egypt is not a new phenomenon. Closer to the Nile Valley, farmers have been fighting a lost battle with the sand for centuries. The dunes submerge the roads to the oases and encroach upon fields and complete villages.

The Eastern side of the valley is also not free from damage. The dense old drainage pattern in the Eastern desert accumulate a great amount of alterated material. Once a thunderstorm happens, it provides an enourmous amount of debris material down the slopes threatening the cultivated land and the villages.

The problem of desert encroachement is best understood by locating the associated features, studying the rates and direction of their movements. The purpose of the current investigation is to make a maximal use of remote sensing data to assess desertification in the study area. It was possible to minimize the need of ground information by analysing stereo coverages of some sample areas on aerial photo-interpretation were extrapolated upon enlarged landsat images.

2 LOCALIZATION AND ENVIRONMENT OF THE STUDY AREA

The investigated area is situated in the Northern part of the Nile Valley (Egypt). It lies approximately between latitudes 27°52' and 30°6' N and longitudes 29°46' and 31°41' E.(Fig.1)
It includes the governorates of Giza, Beni-Suef and El-Menia and the surrounding desert fringes. The three governorates represent one sixth of the cultivated area in Egypt.

The Western desert is a northern dipping plain of sedimentary rocks , composed mainly of sand stone in the South and limestone in the North (Said 1962). It occupies 681,000 Km^2, or more than two thirds of the area of Egypt. This flat region is interrupted by low-situated oases, and by granitic mountains in the South-West intersection between Egypt, Libya and the Soudan. Numerous sand dunes and sand sheet belts are crossing the rocky plateau of the desert.

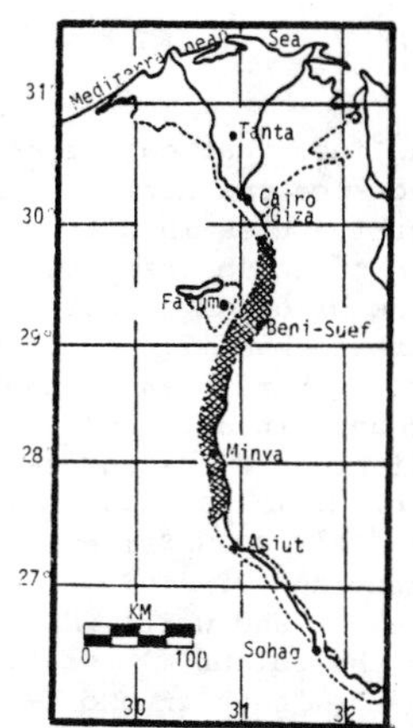

Fig. 1 - Localisation of the investigated area

The Eastern desert borders the Nile Valley eastward. It extends from the Nile Valley to the Suez Canal and the Red Sea, and has an area of about 223,000 Km^2. The Eastern desert differs markedly from the western one. It is intensely dissected by valleys and ravines. Murray(1951)and Butzer (1959) stated that although desert climatic conditions are prevailling over Egypt, the mountainous areas of the Eastern desert have received a higher amount of precipitations.

The flood plain is built up by alluvial deposits, formed from the sedimentation of mud which was carried by the annual floods during the most recent geological period. The mud is the product of igneous rocks forming the Ethiopian plateau.

The Western desert fringes of the study area are dominated by sand sheets and sand dune belts and varnished pebbles. El-Baz (1978) found, by using the Apollo-Soyuz photographs, that some longtudinal dunes are encroaching on fertile lands of the Nile Valley to the south. He confirmed that this process is active in several places along the western borders of the Nile Valley, south of Cairo. Kishk (1977) warned for the increase of dune encroachment on fertile land along the Western borders of the Nile Valley, especially after the completion of the High Dam and hence the absence of Nile alluvium. Monir et al (1984) refered also to the failure of reclamation projects, situated on the Western border of the Nile Valley, (El-Menia governorate).

On the Eastern side of the valley, the cultivated strip is exposed to the danger of occasional thunderstorms. The United Nation report (1980-1982) reffers that thunderstorms became more frequent during the last years. The El-Menia province was exposed to the destructive effects of the thunderstorms between the years 1965 and 1975. Studies are now performed to protect the New El-Menia town against the run-off caused by the thunderstorms (Salem et al 1982).

3 INTERPRETATION

3.1 Multistage analysis

The main idea of a multistage analysis is to make a maximum use of the different remote sensing documents, in view of organizing but also of minimizing the field observations. This approach is referred by Lillesand and Kiefer (1979) for geology mapping. It has been recommended for the assessment of desertification by Rapp (1974) and was successfully used by El-Hag (1984).

Landsat MSS imagery, conventional aerial photographs and ground truth have been used in the current study. Also, different tools have been employed to zoom on different features (e.g. over head projector and transparant screen, mirror stereoscope and zoom-transfer scope). It resulted progressively in more accurate information for corresponding smaller units of the study area. The detailed informations, obtained from the aerial photographs and from the field observations have been fed in the small scale interpretation of the MSS landsat imagery. That allowed us to establish a map of the soil conditions by comparing the image characteristics with the ground conditions.

3.2 Diazo processing technique

This technique is based upon the idea that each feature has its own specific reflectance characteristics. Thus, the grey tone of different features in different bands will differ according to these characteristics. It was noticed from the original negative images that the vegtation is dark in band 7 and light in band 5, whereas water is dark in band 4 and 5 and light in band 7. Light sand is dark in all bands. Consequently vegetation and light sand are indistinguishable on band 7 but different in band 5, whereas vegetation and water can only be distinguished by examining both band 5 and band 7. This comparison between two or more black and white images is inconvenient. Producing the colour composites by superimposing two or three bands, each in a different colour, is a more convenient method. The information contained in different spectral bands are combined into a single colour composite.

The preliminary documents used were negative black and white frames of three spectral bands (4, 5 and 7) on a scale 1:1,000,000. A photographic enchancement was performed by using Agfa-Gevaert 081P-Graphic Gevalith ortho film. It resulted in two different enchanced images for each band, one has the maximal contrast within the desert area and the other within the cultivated Nile Valley. The obtained images were then colour-coded using the colour diazo technique (Gad and Daels 1985).

For the desert area the colour composite was formed by magenta (band 7), cyan (band 5) and green (band 4). For the cultivated area the following combination was used : cyan (band 7), magenta (band 5) and green (band 4).

3.3 Visual interpretation

The different landscape elements are represented by the image characteristics. The interpretation of the landsat images was performed first for the colour composites on Macro and micro photomorphic unit levels. The used image characteristics were : the colour classes, the texture, the quality of the unit boundaries, homogeneity of shape and size and relative size. Some features have been choosen to be followed in more details by using a 10 times magnification of band 7. Some representative photomorphic units have been choosen to be studied in detail on aerial photographs. The aerial photographs were studied with a mirror stereoscope and the zoom-transfer scope.

4. RESULTS AND DISCUSSION

The delineation of photomorphic units in the desert area, supplemented with field observations and the existing base maps, reveals the different physiographic regions.

4.1 Western desert

It was possible to distinguish the different erosional and depositional features from the different documents. It was obvious that the wind erosion process is dominant obscuring the drainage pattern. Multistage interpretation enabled us to localize the belts of sand sheet, sand dunes and the eroded limestone plateau.

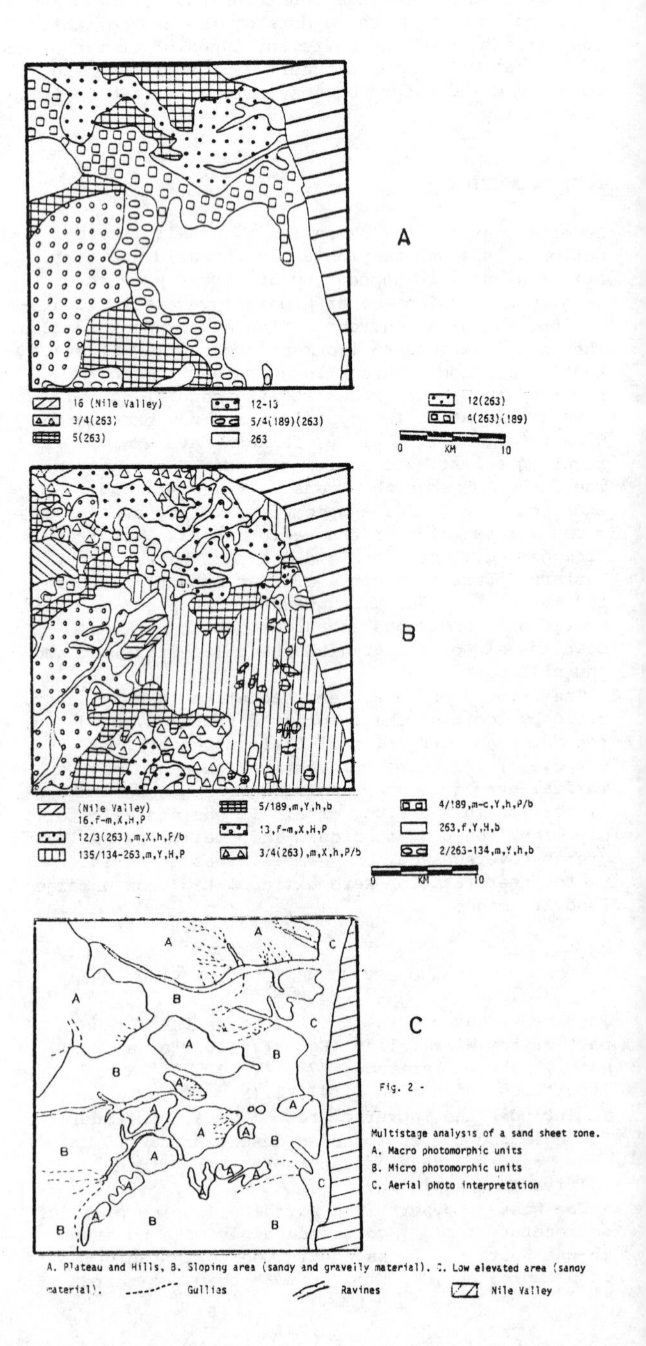

Fig. 2 - Multistage analysis of a sand sheet zone.
A. Macro photomorphic units
B. Micro photomorphic units
C. Aerial photo interpretation

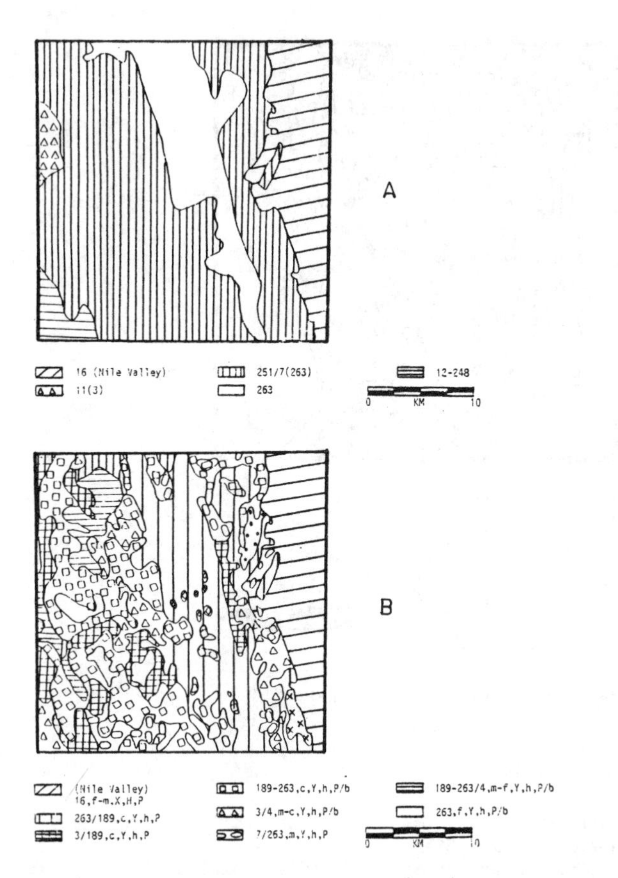

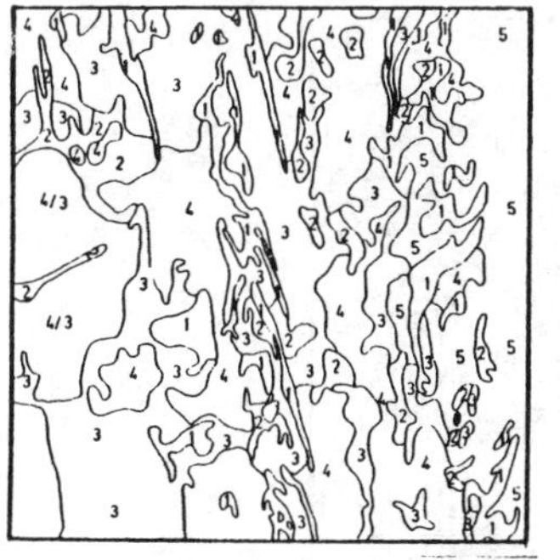

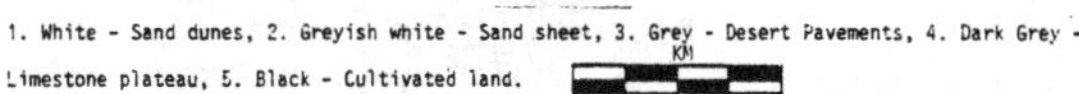

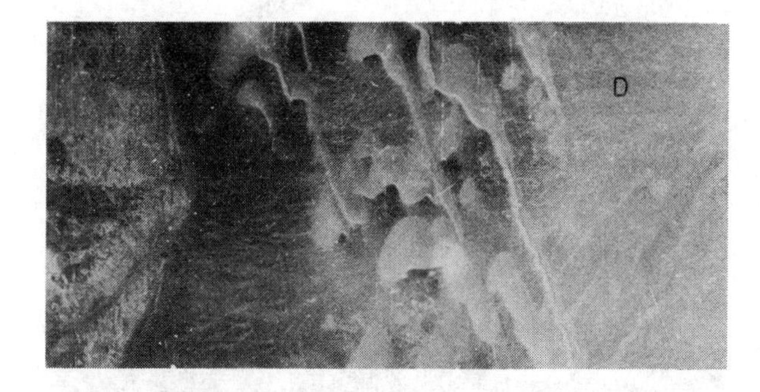

Fig. 3 - Multistage analysis of a sand dune zone
A. Macro photomorphic units
B. Micro photomorphic units
C. Aerial photo interpretation
D. An aerial photograph of a sand dune belt (scale 1:40,000)

4.1.1 Sand sheet area

This landform is visible on the colour composites, just west of the Nile Delta and along the northern part of the Nile Valley. The Macro-photomorphic units (fig. 2A) are characterized by bright coloured wide patches. The field investigation indicates that some parts, within this area, are covered by gravel and pebble deposits rather than by sand. It was found also that the deposits are characterized by the alternation of layers (1-2 cm) of medium to fine sand and gravel. The subtle colour differences, appeared in the micro-photomorphic level(fig. 2B), may represent differences in size, spacing and composition of pebble deposits (El-Baz and Maxwell 1979).

The aerial photo-interpretation of a representative area of sand sheet (fig. 2C) made it possible to see that it is characterized by an undulating topography. It was also possible to follow the weak drainage pattern in the rocky surface, which is controlled by NW SE direction. That direction corresponds to the NW SE direction of the fractures and the faults(Said 1962).

4.1.2 Sand dune belts

This physiographic unit was observed in the macrophotomorphic unit level interpretation on the colour composite (fig.3A) as a light coloured area (e.g. bluish white, mixed with light purplish pink). The microphotomorphic level subdivided this area into smaller units of different colour shades (fig. 3B). The interpretation of that area on a scale 1:100,000 (fig. 3C) revealed the individual longtudinal dunes and the interdunes, and made it possible to estimate its morphometry and to see that the main direction of the moving dunes is S-N, which corresponds to the main wind direction. Although the resolution of the landsat images is too small to detect the barghan dunes, it was noticed that the linear pattern of the longtudinal dunes takes a rounded shape, with a light colour in the lee-ward of some individual dunes. The detailed study of aerial photographs (fig. 3D) revealed the existence of the barghans among the longtudinal dunes

It was also possible to see that the barghans are moving in a North-South direction with some deflation, having a West-East direction.

4.1.3 The plateau

The Western desert, in general, is a north-dipping plateau of sedimentary rocks. That rocky terrain appears on the landsat images as dark coloured macro-photomorphic units (fig. 4A). The interpretation on micro-photomorphic level subdivided the macro-photomorphic units into different shades (fig. 4B). These shades could be attributed to the importance of local bedrock outcrops, and stages of erosion. Multistage - interpretation defined the dark coloured areas as individual patterns and as composite patchy patterns. The patchy pattern is controlled by the existence of the light coloured eolian deposits. The individual dark patterns occur where a single hill or mountain exists. It was also possible to detect the very faint main drainage patterns on the landsat images. On the aerial photographs the small tributaries could be seen (fig 3D). The drainage pattern is dendritic and has a lighter colour compared with the surrounding hills. It was believed that these drainage patterns indicate the fluvial processes during the pleistocene, which

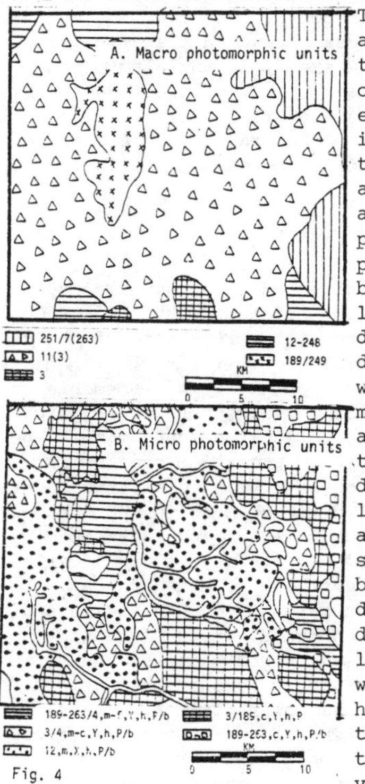

Fig. 4
Multistage analysis of a plateau area.

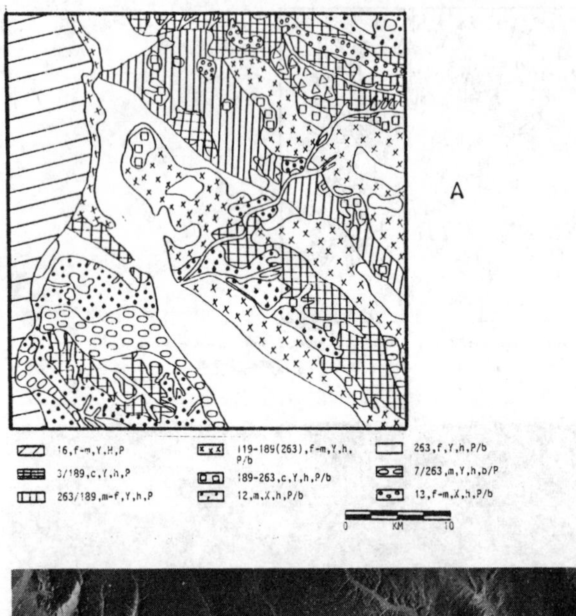

Fig. 5 - Multistage analysis of a part of the eastern desert, near to the Nile Valley
A. Micro photomorphic units
B. An aerial photograph (scale 1:40,000)

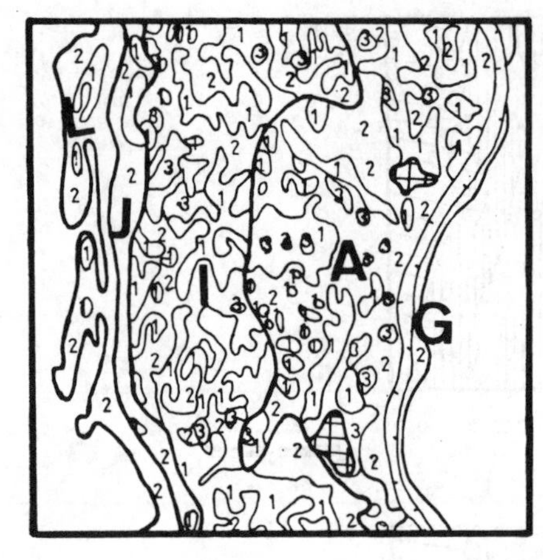

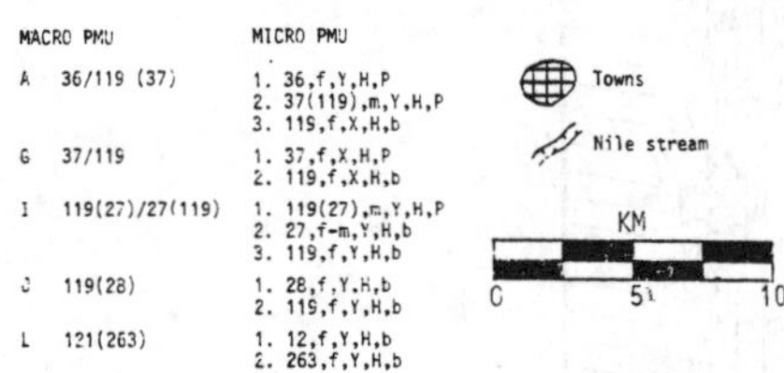

Fig. 6 - Photomorphic units for a part of the Nile Valley.

was responsible for the erosion of debris and its transportation northward in a manner similar to that of the present day River Nile system (El-Baz 1979).

The field observation confirmed that composite dark patches indicate the desert pavements surrounded by the light coloured sand sheet. It was also possible to conclude that the dark coloured linear patterns in the sand dune area are gravelly surface corridors The desert pavements and the gravelly dune corridors are protecting the underlaying sandy material.

4.2 Eastern desert

The colour composites revealed the rugged landscape of the Eastern desert in different dark shades. Comparison between the image interpretation and the geologic map made it possible to attribute to the different colours a geologic meaning. It was possible to map the dense network of wadies and ravines, which are characterized by a distincet bright colour. The photomorphic units on the macro and micro levels made the delineation possible of high altitude areas with a dense dendritic drainage pattern and the down slope area of debris accumulation, characterized by a light colour and a poor drainage network.

The study of aerial photographs made it possible to follow the shape and path way of the wadies and its tributaries by using the elements of tonality, texture and relief. It was found that the dendritic drainage pattern is the basic type in the Eastern desert, especially in the upper plateau.

The bright colour patches found in the colour composites of the adjacent part of the Nile Valley and in the wady bottoms (fig. 5A) were studied in detail on aerial photographs. It corresponds to the modified dendritic (e.g. Dichotomic) and to the braided river pattern (fig. 5B).

These patterns are indicative for depositional processes and are associated with a coarse soil material.

The following landscape forms could be distinguished;
1. The existence of a cradle shaped valleytype in the broad gently sloping shallow valleys in the down slope area. This valleytype indicates a low rate of erosional force or a large rate of debris accumulation.
2. The existence of a V-shaped valleytype in the low order tributaries of the upper plateau. This valleytype indicate a strong vertical erosion.
3. The U-shaped valleytype is occuring in the main valleys. That type of valley might be developed after a strong vertical erosion, or where the pathway followed a fracture or reached a hard rock.

We may conclude that the Eastern desert is characterized by a dominant fluvial erosion, especially during the thunderstorms. The eroded material is deposited in the old wadies and ravines. The debris material is transported and accumulated in the adjacent part of the cultivated Nile Valley.

The interpretation of enchanced landsat images of the Nile Valley resulted in eleven PMU's different in the dominant colour, texture, unit boundaries, homogeneity and relative size. Each of these PMU's could be subdivided into a number of microphotomorphic units (fig. 6).

The dominant colour gives some information about the vegetation cover. Where no red or green is visible, the vegetation cover is regarded as less than 1%, faint green-brown or green-brown 20-40%, a faint brown-red 40-60%, a pronounced brown-red or magenta 60-80% (Mitchell 1981). It is obvious that the microphotomorphic units characterized by the red colour are dominating inside the valley, whereas they become less important toward the desert fringes. A photomorphic unit of almost faint green colour and fine texture is bordering the cultivated Nile Valley on the West.

Field work confirmed that the desert fringes are mostly covered by grasses and dead vegetation resulting in the faint green colour on the colour composites. Inside the valley, the crops are strong and have an active photosynthesis. Some areas inside the valley are reflecting strongly in the near infra red radiation of the E.M.S.-resulting in a continuous magenta, which indicates a healthy vegetation. These observations could be made easely on the enlarged colour composites (scale 1:100,000).

Band 7 was used to delineate the areas in the desert fringes, which are still resisting the desertification and these which have been covered by the sand sheet. Sample areas were located upon aerial photographs and studied in detail. The steroscopic view revealed the encroachement of the eolian deposits along the Nile Valley and covering already some fields (fig. 7).

Fig. 7 - An aerial photograph of the interference zone between the Nile valley and the western desert (scale 1 : 40,000).

It was also possible to detect and map the damaged irrigation network by sand deflation, the remnants of the damaged villages were also visible.

This study is concerned with the assessment of desertification in the areas bordering the Nile Valley, although different photomorphic units could be distinguished, in the Nile Valley they were not taken into account.

5. FIELD OBSERVATION AND LABORATORY-ANALYSIS

An intensive purposive field observation was performed. Linear traverses scheme , edited by Justice and Townshened (1981) has been followed. It resulted in a detailed description of the terrain, necessary for the understanding of the previous image interpretation. 70 soils samples were collected occuring in the different physiographic units. Particle size distribution was determined for all the samples. Frequency histograms of particle size distribution show in most of the sand sheet and interdune samples (fig 8A) a wide range including an amount of fine material. The frequency distribution of sand dune samples (fig 8B) revails a narrow range of grading without a marked amount of fine material. The interference zone samples show a higher amount of the coarse fraction at the surface of the cultivated area (fig.8C) than in the sub-surface. The contrary was found in the samples of the non-cultivated "desertified" areas (fig.8D). That might indicate the continuous contribution of eolian sediments in this zone.

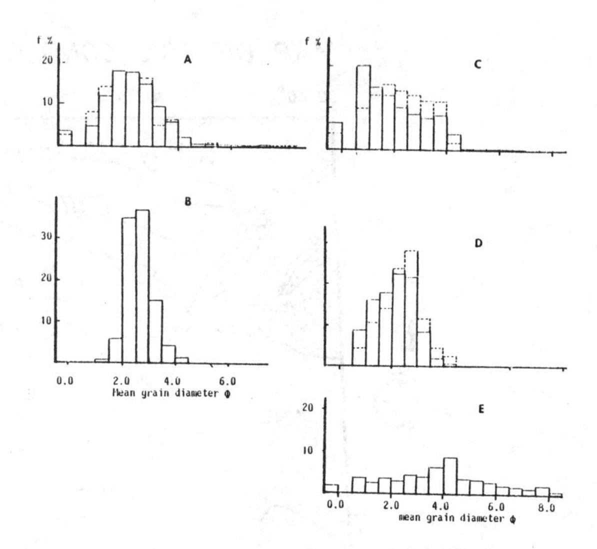

Fig. 8 - Histograms of the particle size distribution of different physiographic areas.

A. An interdune (———) and a sand sheet sample (-----)

B. Sand dune sample

C. Non-cultivated area of the interference zone

D. Cultivated area of the interference zone

——— surface sample
------- subsurface

E. Sample of the regular cultivated area in the Nile Valley.

The histogram of the particle size distribution of the regular cultivated Nile Valley (fig. 8E) shows a wider range of grading and the absence of marked modal frequency in most of the samples. Some samples, in the western borders, have a naarow range of grading which could be attributed to an encroachment of desert sediments inside the valley. A relative increase of the coarse constituents was found in the sites adjacent to the Eastern desert cliffs. That might reflect the influence of the Eastern desert.

6. MAPPING OF SOIL CONDITIONS

All the results from the landsat and aerial photo-interpretation, field observation and laboratory analysis have been fed back in the photomorphic unit maps. A map of soil conditions (fig. 9)was derived on the base of a system modified from El-Shazly et al (1978). The study area has been divided, according to the potential land use, to arable and non arable. The arable area is further subdivided, according to the priority in agricultural development into the following classes : Grade I ; soils of the river Nile flood plain, Grade II ; soils of mixed Nile alluvium and eolian sediments. Inside this group a sub-divisious is possible in cultivated land and desertified land. The size of these units is so small that they could not be mapped on such a small scale. The non arable areas are divided into following grades, Grade III ; soils of sand sheet belts, Grade IV ; soils of sand dune belts and Grade V the desert pavements.
Grade VI ; the slightly eroded plateau, Grade VII ; highly sloping eroded plateau, Grade VIII ; wady bottom and debris accumulation zone with coarse soil material.

7. CONCLUSIONS

The results of remote sensing application in the assessment of desertification could be considerably improved by using a multiple view approach ; multistage, multiscale and multispectral.

7.1 Multiscale sensing

The multiscale approach indicates the analysis of satellite data in conjunction with the aerial photographs. The synoptic view, provided by the satellite

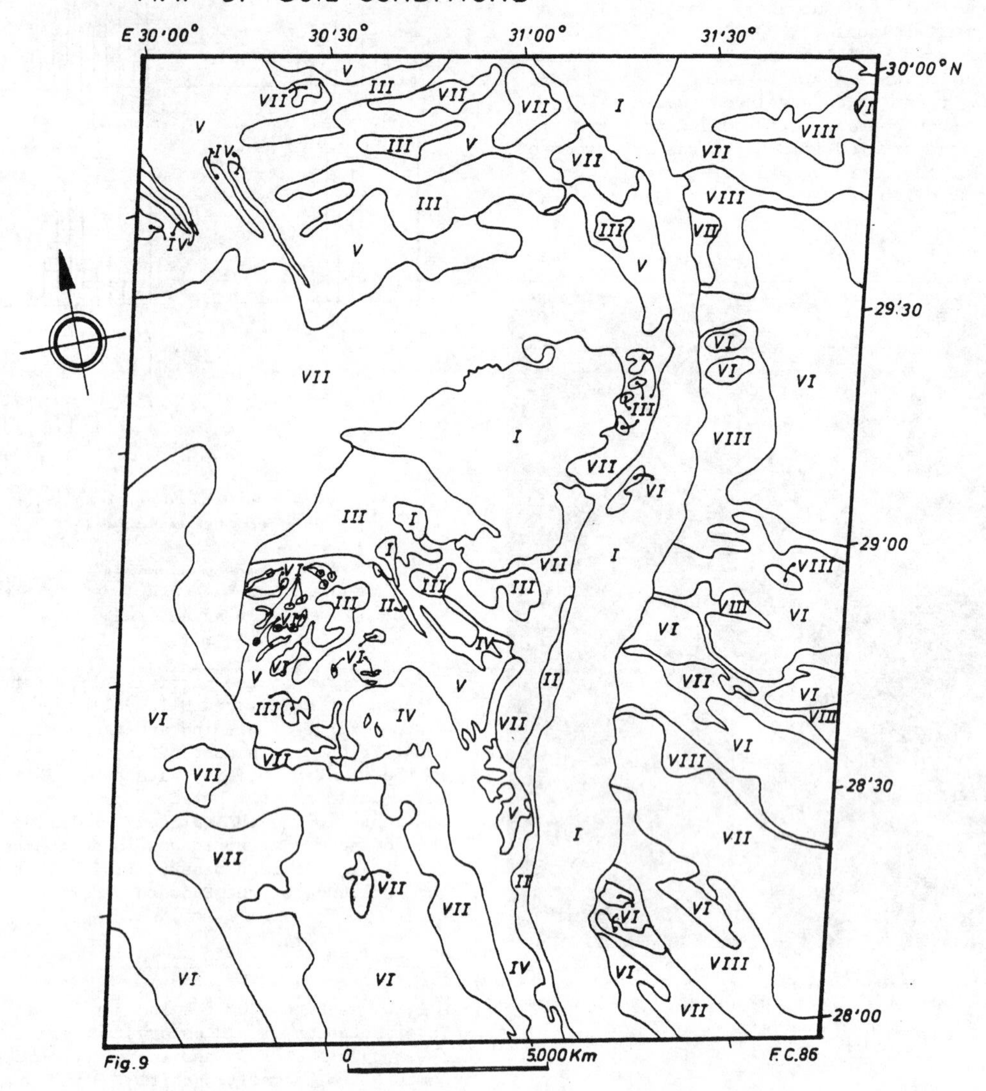

images is the only possible way to understand the relationship between specific landscape elements and the dominant processes in large areas. Enlarging the landsat images made the mapping easier. Aerial photographs, with a higher resolution, are necessary for understanding and observing the morphologic characteristics of smaller landforms. Information obtained at a lower level of observation may then be extrapolated to higher levels of observation.

7.2 Multispectral sensing

Multispectral imagery provides more information than data collected in a broad spectral band. Different phenomena ; the condition of the vegetation and differences in soil conditions show a specific spectral reflectance only discernable on multi-spectral images. Field work and laboratory analysis are necessary for understanding the relation between certain phenomena and their recorded reflection.

With a visual or an analog interpretation a general mapping could be performed on a scale of 1:250,000 and for some areas upon a scale of 1:100,000 with a reasonable accuracy. This method is most suitable for the assessment of the factors leading to desertification.

8. APPENDIX

Delineation of the photomorphic units (PMU)

1.- Macrophotomorphic units

The macro PMU's were delineated according to the dominating colour. The colour scale of the "Manual of colour photography 1968" was used for the colour description. The mentioned numbers on the maps correspond to the following colours :

scale n°	colour	scale n°	colour	scale n°	colour
2	S. pink	27	deep Y pink	134	V.L.Y.G.
3	deep pink	28	L.Y pink	135	L.Y.G.
4	L. pink	29	m Y pink	136	m Y.G.
5	m. pink	35	S. ro	153	g white
7	P. pink	36	deep ro	180	V.L.B.
11	v. red	37	m. ro	184	v. p. B.
12	S. Red	40	S.rBr	189	b white
13	deep Red	115	v.y.G.	248	deep p pK
16	d. Red	116	brill Y G	249	L.p.pK
25	v.y. pink	119	L.y.G.	251	d.pK.
26	S.y.pink	121	p.y.G.	263	white

* S = strong ; L = light ; m = medium ; v = very ; R = Red; B = blue ; G = green ; p = purple ; y = yellow ; ro = orange.

Whenever more than one colour exist, a combined code is used, e.g. :
R(y) = red with inclusions of yellow
R/y = both colours exist but red is dominant
R-y = both colours exist at the same proportion.

2.- Micro PMU's were delineated by using the criteria indicated in the following descriptive formula :

colour, texture, quality of boundaries, homogeneity of shape and size and the relative size.
1.) colour : the macro PMU are subdivided into smaller units of different colours.
2.) texture : textural elements are grouped into three categories : fine(f), medium(M) and coarse(c).
3.) quality of boundry :
X = very distinct, sharp and high contrast.
Y = not distinct, not sharp and low contrast.
4.) homogeneity of shape and size within the PMU.
H = homogeneous size and shape.
h = heterogeneous size and shape
5. relative size : p. = relative large size ;
b. = relative small size.

9. REFERENCES

Butzer, K.W. 1959. Environment and human ecology in Egypt during predynastic times. Bull.Soc.Geogr. Egypt 32:42-82.
El-Baz, F. 1978. The meaning of desert colour in earth orbital photographs. Phtogramm. Eng.Remote Sens. 44(1):7-11.
El-Baz, F. 1979. Monitoring the desert environment from space. In A. Beshay & W.G.Mc ginnies (eds.), Advances in desert and arid land technology and development, 1,p. 383-397.
El-Baz, F. & T.A. Maxwell 1979. Eolian streaks in South Western Egypt and similar features in Ceberus region of Mars. Proc. 10th. Lunar Planet.Sci.conf., p. 3017-3030.
El-Hag, M.M. 1984. Study of desertification upon landsat imagery (North Kordrfan-Sudan). Rijksuniversiteit, Thesis Ph.D.; Ghent, Belgium.
El-Shazly, E.M., M.A. Abdel-Hady, M.M. El-Shazly, M.A. El-Ghawaby, S.M. Kawasik, A.A. Haraga, S. Sanad & S.H. Attia 1978. Application of landsat imagery in the geological and soil investigation in the Central western desert, Egypt. Twelfth International symposium on remote sensing of environment, April 1978, Manilla, Philippiens.
Gad, A. & L. Daels 1985. Assessment of wind and fluvial action by using landsat MSS colour composits in the lower Nile Valley(Egypt). International symposium on parameterization of land-surface characteristics, use of satellite data in climate studies and first results of ISLSCP. Rome, Italy, 2-6 December 1985.
Justice, C.O. & J.R.G. Townshend 1981. Intergrating ground data with remote sensing. In J.R.G.Townshend(ed.), Terrain analysis and remote sensing, p. 38-58. London : George Allen & Unwin.
Kishk, N.A. 1977. A warning against the forth coming damage of Egytian soils. Proceedings of scientific symposium on effects of the High Dam on soil fertility and water quality in Egypt, April 16-1, 1977. Assiut, Egypt, 1 : 39-52.
Lillesand, T.M. & R.W. Kiefer 1979. Remote sensing and image interpretation, p. 94-184. New York : John Wiley Mitchell, C.W. 1981. Reconnaisance land resource surveys in arid and semi-arid lands. In J.R.G. Townshend (ed.), Terrain Analysi and Remote sensing, p. 169-183. London : George Allen & Unwin.
Monir, M.M.,R.F. Misak & A.M. Ahmed 1984. Wind-Blown deposits and sand dunes stabilization in Egypt. Desert Research Institute. Cairo, Egypt.
Murray, G.W. 1951. The Egyptian climate a historical outline. Geogr. J. 117: 422÷424.
Rapp, A. 1974. A review of desertification in Africa-water, vegetation and man. P. 77. Lund : Lund universitets geografiska institution.(Report N° 39).
Said, R. 1962. The Geology of Egypt. Amsterdam, New York and London : Elsevier.
Salem, M.H., H.T.Dorah & A. Fateh Elbab 1982. Design of the New El-Menia town-Thunderstorms study. Mins. Cons. (ARE), 1982, (In Arabic).
UNEP, 1983. Desertification Control, p.51. Nairobi : UNEP.

Symposium on Remote Sensing for Resources Development and Environmental Management / Enschede / August 1986

Spring mound and aioun mapping from Landsat TM imagery in south-central Tunisia

Arwyn Rhys Jones & Andrew Millington
University of Reading, UK

ABSTRACT: Groundwater upwellings on two playas in south-central Tunisia form rare geomorphological features known as spring mounds and aioun. They form clusters in certain playa facies. Lineament analysis of Landsat TM imagery of the playas was carried out on single bands, FCC's and convolved imagery. Lineament directions indicated strong underlying hydrogeological controls on these features related to Alpine folding and associated faulting and jointing.

RESUME: Eaux de la terre montant lentement sur deux playas en Tunisie sud-centrale forme les phenomenes geomorpholgiques connus sous le nom de monticules aux sources et aioun. Ils forment des amas sur certines parties du playas. Traitement d'image lineament sur l'image de Landsat TM etait entrepris utilisant les FCC's et les filtres directionals. Les directions des lineaments indiquaient la forte maitrise hydrogeologique sur les monticules aux sources et aioun ayant rapport a des plissment alpin associe a des faillies et des joints.

1. INTRODUCTION

Spring mounds and aioun are both geomorphological expressions of groundwater upwelling found on playas. Spring mounds are formed when deep groundwater escapes to the playa surface as a spring and there is either chemical deposition of groundwater salts or aeolian and lacustrine deposition of playa surface debris. The height of the mounds depends on the piezometric and discharge properties of the groundwater (Reeves, 1968). In Tunisia they commonly range from 10 to 15 m and cover areas of up to 7.5 Km^2, although more commonly the maximum is about 2.5 Km^2.

Figure 1. A spring mound found on the Chott el Fedjadj Tunisia, note its vegetated centre clearly visible on TM imagery

They have been recorded in Australian, North American (Mabbutt, 1977) and North African (Cooke and Warren 1973) playas. An ain (pl. aioun) may be a related feature, but their origins are far from clear. It is thought that they may be related to solution subsidence in areas of rising groundwater (Cooke and Warren, 1973); however whether this is due to shallow or deep groundwater movements is unclear. If they are deep groundwater features they are undoubtedly related to spring mounds but if they are shallow groundwater features the factors controlling their distribution should be entirely different. Coque (1962) describes typical aioun on the Chott el Djerid as being about 4m deep and 5m in diameter. Over the last three years the same aioun have, however, shown very little surface relief (only up to about 10cm); this may be due to infilling of the collapse structures with aeolian or lacustrine sediments. Their most striking features are the occurrence of concentric circular patterns of salts on playa surfaces. They have been recorded in North African and Iranian playas (Cooke and Warren, 1973; Coque, 1962).

Spring mounds and aioun occur in clusters which we have termed spring mound and aioun fields respectively. The former are more common in south-central Tunisia. Two major spring mound fields are found - the largest is on the Chott el Djerid to the south of the Djebel Tebaga and a second, smaller field is found in the south-west Chott el Fedjadj. One large aioun field is also found in the north-east of the Chott el Djerid. These latter two fields are examined in this paper in more detail.

2. HYDROGEOLOGY OF SOUTH-CENTRAL TUNISIA

South of the North African Atlas Mountains, at the junction of the Saharan Platform and the folded Atlas Block are a series of playas occupying a zone of subsidence. Known locally as chotts they stretch from the Sebkhet el Hamma (about 20km west of Gabes in southern Tunisia) to Chott el Melhrir (in central Algeria) occupying a zone of subsidence. The largest of these playas is the Chott el Djerid which has an elongated north eastern arm in the core of an eroded anticline, known as the Chott el Fedjadj.

The hydrological regimes of both chotts are dominated by groundwater seepage from the south in a surface quifer - the Continentale Terminale - and a deeper aquifer - the Continentale Intercalaire. In addition there is groundwater seepage from alluvial fans flanking the mountain ranges to the north of the chotts in the winter wet season. Although annually evaporation exceeds precipitation, for most of the year there is winter surface runoff in the region and the water levels on some parts of the chotts are dominated by this runoff. These areas do not, however, have spring mound or aioun fields.

As spring mounds are formed by upward groundwater seepage they are frequently associated with specific underlying lithologies and structures, particularly well developed faulting and jointing Reeves (1968). The overall structure in this area consists of a series of sedimentary horizons, with known aquifers, dipping to the north. This structure itself is not conducive to groundwater upwellings despite the fact that groundwater is present. The Alpine folding and faulting of the Atlas Mountains affects the sedimentary strata on the northern margin of the Saharan Platform. Folding of these sediments on the Saharan Platform led to the formation of the Chott el Fedjadj anticline. The Djebel Tebaga is the prominent southern limb of this anticline and it plunges westward under the Chott el Djerid. The area still suffers from infrequent tectonic activity. It can be hypothesised that the occurrence of spring mounds are related to faulting and jointing associated with this folding. This can be tested to examining the directions of the alignments of these isolated features on Landsat TM imagery by lineament analysis.

In addition the distribution of aioun can also be examined and related to the structural trends. This should enable the origins of aioun to ascribe to either deep or shallow groundwater sources.

3. IDENTIFICATION OF SPRING MOUNDS AND AIOUN ON SATELLITE IMAGERY

Difficulties exist in attempting to identify and map spring mounds and aioun by ground survey Geomorphological problems are encountered during the identification and mapping of aioun because of their lack of relief and the difficulty in mapping subtle variations in salt concentrations on flat playa surfaces. Spring mounds are far easier to identify in the field because of their relief and the presence of vegetation. Vegetation is commonly found on the less saline margins of playas. It mainly takes the form of halophytic shrubs and grasses and the centre of the spring mounds are also vegetated, often with less salt tolerant species, in response to spring-flow. The logistical problems of field work on playas are immense; they mainly concern difficulties of access, heat, salt glare and the location of ground control points. As a consequence most topographic maps mark playas as featureless voids!

Remotely sensed imagery is therefore an extremely important tool in the identification and mapping of playa features such as spring mounds and aioun. The advantages of remotely sensed imagery in such studies not only lie in their ability to overcome the geomorphological and logistical problems but also because of the shape and spectral characteristics of the features themselves.

A very important identification criterion for spring mounds and aioun is their circular shape; this makes them readily identifiable on remotely sensed imagery (Figs, 2 and 3). However it can be seen on Fig. 3 that the spring mounds develop long tails to the SW. These tails are deposits of gypsiferous sands deposited in the lee of the spring mounds in the prevailing wind directions.

In addition there are significant variations in spectral responses between the features and the adjacent playa surfaces, Table 1. Spring mounds are often vegetated with either a natural vegetation community of tamarisk and date palms or, more commonly in this area, they form the nucleus of irrigated oases with date palms and an understory of smaller trees and ground crops. As they are

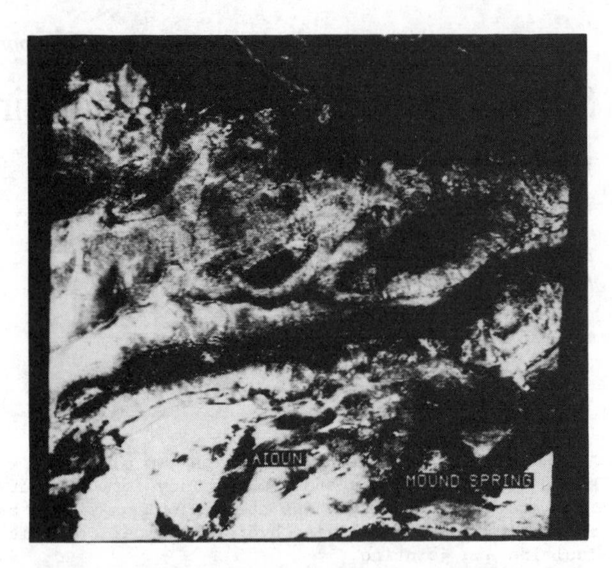

Figure 2. Band 3,4,5, false colour composite of the Chotts el Djerid and el Fedjadj showing the playa facies (from Mitchell, 1982), the aioun and spring mound fields and the Chott el Fedjadj plunging anticline.

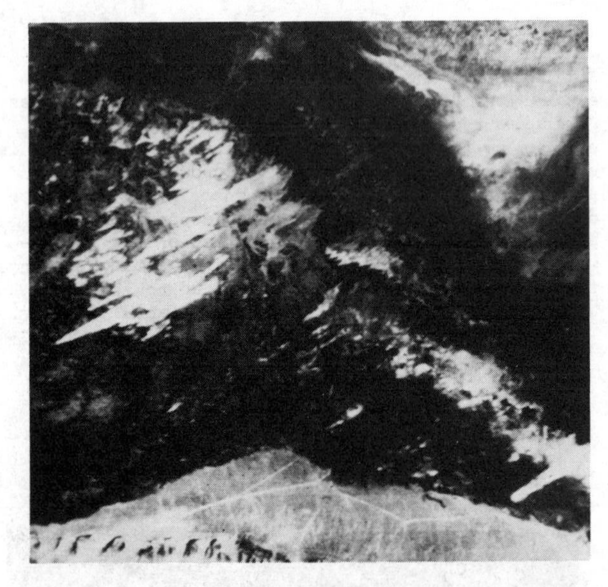

Figure 3. Band 3 image of Seftimi showing spring mounds with sand tails. The resistant limestone beds forming the southern limb of the Chott el Fedjadj anticline can be seen at the bottom of the image.

located in marginal playa facies they can be surrounded by unvegetated surfaces or halophytic plant communities. Where they are surrounded by bare surfaces identification is easy because of the differences in absorption between vegetation and bare soils; particularly in TM Bands 3 (0.63-0.69μm) and 4 (0.79-0.90μm). It is also relatively easy to spectrally separate spring mound vegetation from the surrounding halophytic vegetation. As spring mounds have adequate water supplies all the year around their vegetation is rarely under stress and maintains a low reflectance in TM Bands 3 and 4 throughout the year. Halophytic vegetation does however exhibit some seasonality and partially dies back in the summer; stressed halophytic vegetation is seen on TM imagery in this area. In addition the

Table 1. Profiles of digital values across spring mounds, aioun and associated sand tails.

SAND TAILS	177	169	171	176	178	179	202	224	222	221	212	204	197	181	171	161	159	164
	174	180	180	180	179	178	176	189	205	211	207	193	182	180	178	174	176	174
SPRING MOUNDS	203	201	206	204	203	210	0	218	229	196	0	98	140	124	111	118	0	118
	133	120	59	0	48	94	110	0	131	110	221	196	181	0	48	43	203	222
AIOUN	90	93	91	77	89	78	75	70	71	70	70	70	68	70	73	68	83	94
	100	101	103	98	101	100	82	86	72	71	72	75	77	74	73	71	68	83

halophyte communities have lower plant densities than the mound spring communities and consist of single storey shrubby or herbaceous plants. As a consequence of the differences in community structure and density single pixel spectral responses in the halophyte communities are a mixture of salt-tolerant vegetation and bare playa surfaces. Furthermore, the dominant types of vegetation are often grey-green or reddish green, compared to the deeper greens of date palms and other irrigated crops.

Aioun only occur in the central playa facies in the Chott el Djerid where there is no vegetation. However there are concentric circles of salt efflorésences, indicative of variations in surface salt concentrations. These create a very distinctive circular pattern of variations in reflectance at all visible and infra-red wavelengths and so can be detected in all TM bands.

4. DIGITAL IMAGE PROCESSING AND PLAYA GEOMORPHOLOGY

Digitally processed MSS and TM imagery has been used to map playa facies of the Chott el Djerid (Jones, 1986a,b; Mitchell, 1982; Munday, 1985) and the Chott el Fedjadj (Jones, 1986a,b). From Fig. 2 it can be seen that the distribution of spring mounds is confined to marginal playa facies and that aioun are restricted to a more central facies, (Mitchell, 1982).

This study builds upon previous work by examining in more detail the distribution of spring mounds and aioun within playa facies. All of the image processing reported on in this paper was carried out on a TM image (Path 192; Row 36) taken on 29 January, 1983.

The detailed image processing was carried out on a subscene of this larger image. This was located so that it encompasses the spring mound field on the Chott el Fedjadj. It was apparent from a visual inspection of different single band images of the area that the data showed significant intercorrelation between bands, although some bands, particularly 5 (1.55-1.75μm) and 7 (2.08-2.35μm) contained more geomorphological information than others. Table 2 shows the correlation matrix derived from the six TM reflective bands in the spring mound test area. High correlations were found between all bands (for all correlations r= +0.688) with very high correlations between the three visible bands, the two middle infra-red bands and Band 4 and all other bands. It is well known that the TM was designed primarily for vegetation discrimination with bands selected to take advantage of the spectral response of vegetation (Salmonson et al., 1980). The implication for geomorphological investigations of bare surfaces, such as playas, is that after studying an infra-red or FCC image that additional single band images provide little additional information unless a narrow pixel value range is utilised.

After comparing the images it was concluded that Band 3 (0.63-00.69μm) was the most useful image for analysis.

This was due to the clear depiction of vegetation (by chlorophyll absorption) on spring mounds and playa surfaces, which allowed the easy identification of spring mounds. The playa facies with spring mounds

Table 2 TM Bands correlation matrix for spring mound area

1	2	3	4	5	6
1.000					
.957	1.000				
.925	.983	1.000			
.869	.932	.936	1.000		
.734	.840	.878	.872	1.000	
.608	.706	.738	.746	.944	1.000

was delimited and a contrast stretch was applied to this area to enhance the contrast between features (Fig. 2). All further image processing was carried out on this contrast stretched image.

One of the main aims of the study was to see if linear patterns could be detected in the distribution of spring mounds that could be related to faulting and jointing patterns in the underlying rocks. The alignment of isolated circular geological features to indicate underlying geological phenomena has been attempted in volcanic terrain but we believe this is the first attempt to use such analytical techniques in folded sedimentary strata; although lineament analysis of faults, lithological boundaries and fold axes is well known in similar terrain.

To highlight the edges in the image prior to lineament analysis a series of directional edge enhancements were applied to the single band image. Edge enhancement operates by passing a digital filter or kernal, in the form of a matrix, over the data. It has been successfully used in many studies of geological lineaments (Bailey et al., 1982). Four directional filters were selected (N, NW, W and SW) for this study.

Ratioing of spectral bands was also examined initially as it is known to reduce topographic noise and enhance spectral differences between surface features. However, to effectively use such methods, the relationship between the surface materials and the spectral responses must be understood. Whilst a partial knowledge of the reflectance properties of playa salts it known the actual ratio images produced poor results and were not used in the later analysis.

4.1 Lineament analysis of spring mound distribution

Lineament analysis of spring mound distribution were carried out on five images; a Band 3,4,5 FCC, four single Band 3 images with different directional filters (N, NW, W and SW), (Fig. 4). In addition lineament analysis of the known lithological boundaries in the area was carried out on a Band 3 image.

Software for the lineament analysis was written by D. Greenbaum (British Geological Survey, Keyworth, Nottingham, UK) for the IIS image processing system. The technique involves the visual interpretation and drawing of lineaments on the VDU. Lineaments were defined as straight lines joining three or more spring mounds. The spring mound area is roughly rectangular (Figs. 2 and 3) and this means that there is a greater probability of longer lineaments, connecting greater numbers of mound springs, in

Table 3. Summary statistics of lineament analysis on spring mound distributions with different directional filters.

Filter direction	No. of lineaments identified	Lineament distance (m) min.	max.	mean	Dominant directions	Minor directions
None	42	0.90	8.00	2.89	NE; E-SE	N-NNE
N	48	0.49	7.99	2.52	E-SE	ENE
NW	26	0.89	5.83	2.86	ESE; SE-SSE	-
W	36	0.56	7.68	2.31	E-ESE; SE	N-NNE
SW	32	0.30	5.80	1.91	SE-SSE	NNE; ENE; ESE
None*	12	0.72	7.19	3.07	ESE	

* Only known rock outcrops analysed.

directions parallel to the long axis of the spring mound area than across it. This will become apparent in the results to a certain extent. After all lineaments have been identified full statistics (starting and finishing pixel coordinates, distance in Km and directions) are tabulated; in addition summary statistics are displayed as a semi-circular rose diagram.

The statistics summarising the results of the lineament analyses with the different directional filters are shown above (Table 3). The number of lineaments detected with different treatments ranged from 26 to 48. The length statistics of the lineaments was quite similar under all filters except the SW filter. The smallest lineament was always less than 1 Km; ranging from 300m in the SW filter image to 900m in the unfiltered image. The longest lineament detected under different filters fell into two internally consistent groups. The NW and SW filtered images had longest lineaments of 5.83 and 8.8 Km respectively. The other group consisted of the N and W filtered and unfiltered images; here the longest lineament ranged from 7.68 Km (W filtered image) to 8.00 Km (unfiltered image). The directional data on the lineaments was more useful than the number of lineaments and their lengths in assessing the effects of the different filters and for geological interpretation. The directional data are tabulated (Table 3) and summarised in semi-circular rose diagrams, (Fig. 5)

The effect of directional filtering on the Band 3 imagery seems to have a marked effect on the lineaments in the N to NNE sectors, when compared to the unfiltered image. In the latter image NE was a dominant lineament direction, but in the filtered images all of the lineaments with directions between N and ENE were minor when compared to those with directions between E and SSE. Generally however it can be seen from Table 3 and Fig. 5 that the dominant lineament directions fell between E-W and SE-NW and that a secondary direction - NNE-SSW - perpendicular to the dominant direction can be identified. The other directions, N-S; ENE-WSW and SSE-NNW, are unimportant. This information is summarised in Figure 6.

5. DISCUSSION

It was suggested earlier that the distribution of spring mounds and aioun in the Chotts el Djerid and el Fedjadj might be related to Alpine folding and faulting of sedimentary strata on the Saharan Platform. This can be examined by comparing the directions of the lineaments to the structural trends in the region.

5.1. Structural controls on spring mound distribution

If underlying geological structure controls the distribution on spring mounds on the Chott el Fedjadj the distribution of lineaments should reflect the stress patterns associated with the folding of the Chott el Fedjadj anticline. Reconstructed folding of this anticline (Fig. 7a) suggests that two parallel E-W fold axes were present during folding. Stresses in the brittle strata of the region associated with the folding would probably have created a series of faults parallel to the fold axes, i.e. approximately E-W. The present-day geological structure of the Chott el Fedjadj is similar to the dome and basin structural association suggested by Hobbs et al., (1976) which would fit the folding and faulting patterns. This can partly be seen in Fig. 7b. This shows that the Chott el Fedjadj anticline has been breached as a result of two major faults parallel to the fold axis which have led to the downfaulting of the sediments in the crest of the southernmost fold. Undoubtedly other minor parallel faults exist within the structural association, particularly tensional faulting in the upper strata of the fold where the hinge angles were far less acute than in the deeper strata. Faulting parallel to the fold axes probably therefore partially accounts for the dominant direction of alignments of spring mounds in the area. However in the core of the anticline, under Quaternary playa sediments, sedimentary strata, undercrop with a strike parallel to the fold axes as well. Some of these horizons particularly the sandstones, are known aquifers. It is likely therefore that some of the linear alignments relate to spring lines along junctions between aquifers and aquicludes under the playa sediments. The strata subcrop is also probably responsible for the extent of the spring mound field in the south-west Chott el Fedjadj. Particularly the fact that it is found only in the southern part of the chott and that there is no equivalent to the north.

The secondary lineament direction is orthogonal to the main direction. It is a well known physical phenomena that jointing and faulting patterns related to stress release occur in brittle materials perpendicular to the main stress directions (Park, 1983; Whalley, 1976). If this is the case in this area on a large scale then the secondary lineament direction can also be explained in terms of the folding of the Chott el Fedjadj anticline. There is evidence to support this hypothesis along the southern limb of the anticline. Here the resistant dolomitic limestones which form a number of cuestas, the highest of which is the Djebel Tebaga, display a regular series of wind gaps orientated perpendicular to the fold axes and parallel to the secondary lineament direction (Fig.3). These can be seen as continuations of the lineaments on the imagery and therefore are related to the hypothesised faulting and jointing beneath the playa sediments.

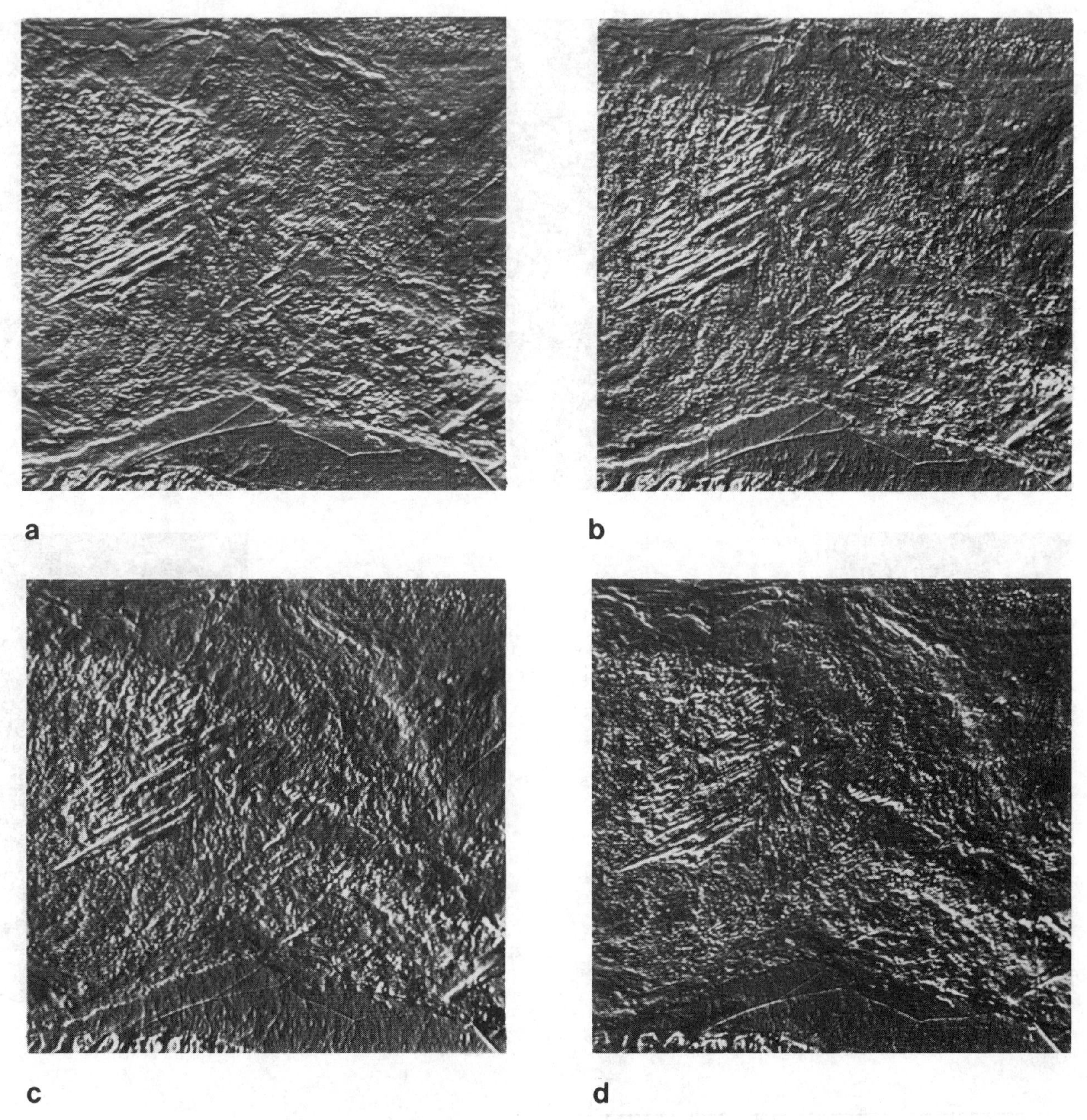

Figure 4. Band 3 images (cf. Fig.3) after directional filtering: a) North b)North-West c) West d) South-West

5.2. Aioun distribution and structural control

It is not possible to determine alignment directions of aioun in the central Chott el Djerid. This was because of the lower number of aioun when compared to spring mounds. However the entire aioun field appears as a large lineament orientated NNE-SSW.

If aioun are related to deep groundwater upwelling it should be possible to relate the orientation of the aioun field to the Alpine structural trends in the region. If there is little relationship between them then it might suggest that aioun are controlled by shallow, rather than deep, groundwater movements. The evidence relating the aioun field orientation to structural trends is at the present time somewhat circumstantial. The aioun field appears to be located at the end of the Chott el Fedjadj anticline. This anticline plunges to the west and is overlain totally by playa sediments of the Chott el Djerid about 8km before the eastern edge of the aioun field. Geological strata in the Djebel el Asker, to the north of the Chott el Djerid, curve southwards in the direction of the aioun field (Fig. 2). If the southerly structural trend of the Djebel el Asker rocks is continued the aioun field falls along its course. Appealing though this relationship may appear it still does not explain why there is deep groundwater coming up to the surface along this trend line. It may be that the entire structure is related to the buried nose of the Chott el Fedjadj anticline but without an adequate sub-playa reconstruction this can only be a matter of conjecture at the present time.

6. CONCLUSIONS

It has been shown that the detection of spring mounds and aioun on playas from remotely sensed imagery has many advantages over conventional ground survey both in terms of overcoming logistical and geomorphological aspects; and that it enables advantage to be taken of the shape and spectral characteristics of these features. A further major advantage of mapping the

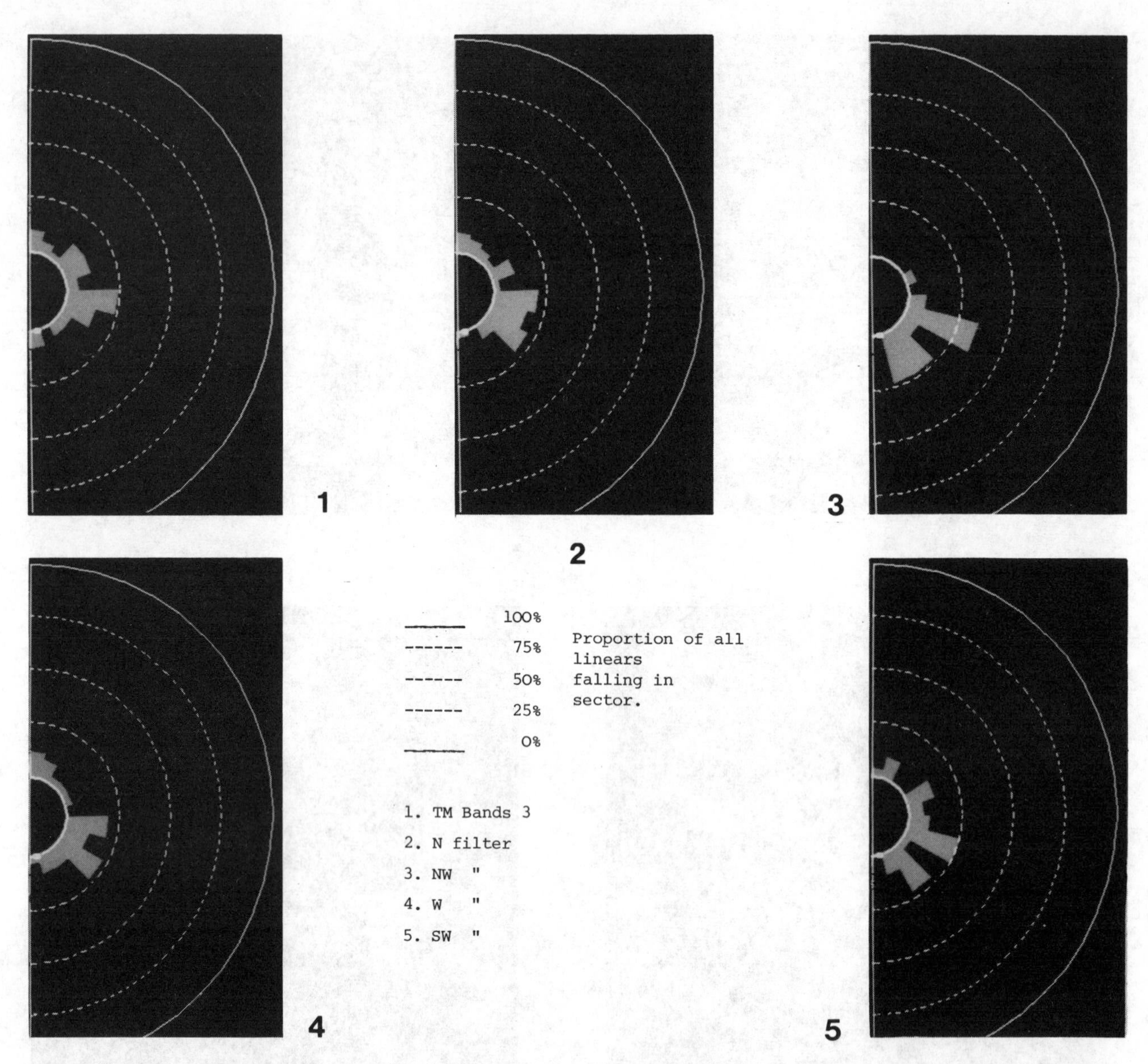

Figure 5. Rose diagrams of the lineaments identified of the different processed imagery for the spring mound field.

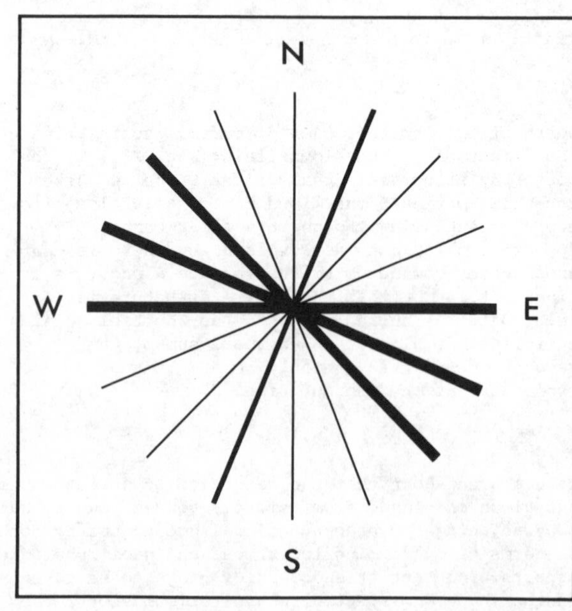

Figure 6. Summary diagram of lineament directions.

distribution of such features from satellite imagery is that directional trends can readily be identified and summarised.

On the Chott el Fedjadj spring mound lineaments occurred in one main direction, E-W to SE-NW, associated with rock subcrops and faulting parallel to the fold axes of the Chott el Fedjadj anticline. A secondary direction, NNE-SSW, was identified which is related to faulting and jointing perpendicular to the main fold axis. Wind gaps along the Djebel Tebaga are probably associated with this faulting and jointing. The aioun field in the Chott el Djerid is orientated in a NNE-SSW direction and appears to be related to the undercrop of aquifers or large scale faulting at the buried nose of the Chott el Fedjadj anticline.

ACKNOWLEDGEMENTS

Part of this work has been carried out as a NERC Studentship GT4/83/GS/87 to one of the authors, Arwyn Rhys Jones. We would like to thank Geoff Wadge for his constructive comments and to Chris Holland, Sheila Dance and Erika Meller for the technical preparation of the paper.

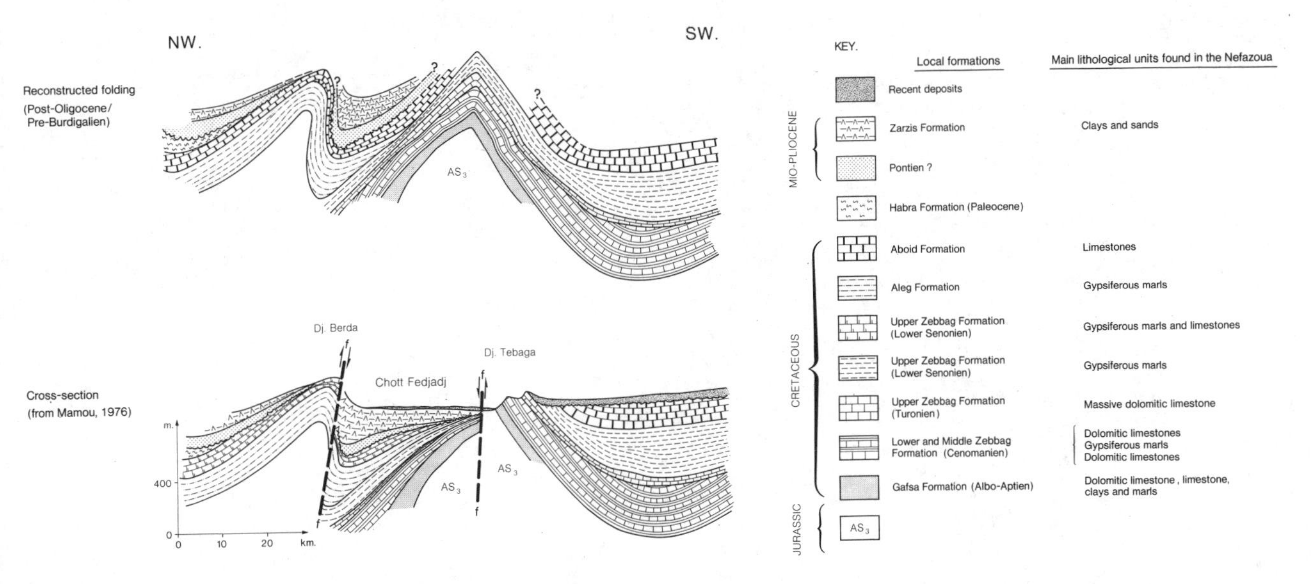

Figure 7. Cross sections of the Chott el Fedjadj anticline (a) before and (b) after breaching (after Mamou, 1976)

REFERENCES

Bailey, G, Dwyer, J. and Francica, R. 1982 Evaluation of image processing of Landsat data for geological interpretation. Proc. 2nd Thematic Conf., Remote Sensing for Exploration Geology, Forth Worth, 555-577.

Bellair, P. 1957 Sur les sols ploygonaux du Chott Djerid (Tunisie) C.R. Hebdom. Acad. Sci. Paris, 224, 101-3.

Cooke, R.U. and Warren, A. 1973 Geomorphology in Deserts, Batsford, London.

Coque, R. 1962 Le Tunisie presaharienne Armand Colin, Paris.

Hobbs, B.E., Means, W.D. and Williams, P.F. 1976 An outline of structural geology, J. Wiley and Sons: New York.

Jones, A.R. 1986a Use of Thematic Mapper Imagery for Geomorphological Mapping in arid areas (This proceedings)

Jones, A.R. 1986b An evaluation of satellite Thematic Mapper imagery for geomorphological mapping in arid and semi-arid environments Proc. 1st Int. Geomorphology Conf., Manchester, U.K. Sept. 1985 (in press)

Mabbut, J. 1977, Desert Landforms, MIT Press, Boston Massachusetts.

Mamou, A. 1976 Contribution a l'etude hydrogeologique de la presqu'ile de Kebili, Unpub. Ph.D. Thesis, L'Universitie Pierre et Marie Curie, Paris

Meckelein, W. 1977 Zur geomorphologie des Chott Djerid Stuttgarter Geog. Stud. 91 (Unterschingen am Nordland der tuneischien Sahara, ed. W. Meckelein), 247-301

Mitchell, C.W. 1983 The soils of the Sahara with special reference to the Mahgreb The Mahgreb Review, 8, 29-37

Munday, T.J. 1985 Multispectral remote sensing of surficial materials in an arid environment. Unpub. Ph.D. thesis, University of Reading

Park, R.G. 1983 Foundations of Structural Geology Blackie, Glasgow

Reeves, C.C. 1968 Introduction to Paleolimnology Elsevier, Amsterdam

Salmonson, V., P. Smith, A. Pink, W. Webb and T. Lynch 1980 An overview of progress in the design and implementation of Landsat-D systems, IEEE Trans. Geosci. Remote Sensing 18.

Whalley, W.B. 1976 Properties of material and geomorphological explanation, Oxford U.P. Oxford

Symposium on Remote Sensing for Resources Development and Environmental Management / Enschede / August 1986

Application of MEIS-II multispectral airborne data and CIR photography for the mapping of surficial geology and geomorphology in the Chatham area, Southwest Ontario, Canada

A.B.Kesik
Department of Geography, University of Waterloo, Ontario, Canada
H.George & M.M.Dusseault
Department of Earth Science, University of Waterloo, Ontario, Canada

ABSTRACT: The surficial geology of the Chatham area in southwestern Ontario has been mapped using B&W panchromatic, CIR air photographs and MEIS-II airborne, multispectral scanner imagery. The study shows that the application of multisensor and multispectral data offers significant advantages for the enhanced discrimination of surficial materials geomorphological features and aggregate exploration.

1 INTRODUCTION

The mapping of surficial geology was prompted by a shortage of aggregates in the rapidly developing Windsor-Toronto corridor in Southwestern Ontario. A multisensor approach was adopted for effective mapping in view of difficulties experienced by previous investigators in determining geologic unit boundaries with the aid of only black and white panchromatic air photographs (E. Sado, pers. comm., 1984). Figure 1.

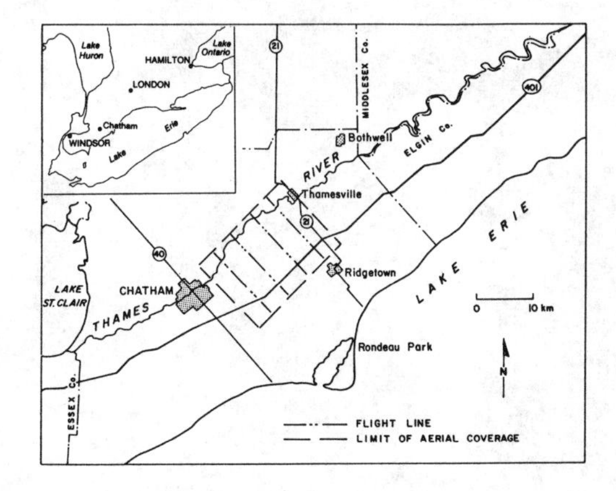

Figure 1. Location of the study Area in southwestern Ontario, Canada

The objectives of the study were:
 1. To compare photo-geological interpretations of panchromatic and CIR air photographs with that obtained from digital analysis of multispectral data (MEIS-II).
 2. To explore different models for Quaternary sedimentation likely to lead to the explanation of the origin of coarse aggregates within the study area based on a revised interpretation of the surficial geology.
 The fundamental hypothesis was that multisensor remote sensing and related image analysis could provide an improved means by which scattered coarse fluvioglacial sediments, known to occur at shallow depths below the ground surface, could be detected and delineated.

2 STUDY AREA

The study area is 5 km east of Chatham, in SW Ontario and comprises 160 sq km of the physiographic unit known as Chatham Flats (Chapman and Putnam, 1984). Bedrock consists of Upper Devonian shale and occurs at a depth of approximately 23 m. The Quaternary sediments overlying bedrock are associated with continental glaciation - till, proglacial sands and gravels and glacio-lacustrine sediments, mostly silts and clays. Deltaic sands are usually present as a discontinuous blanket at the ground surface.
 Potentially usable fluvioglacial sediments are masked by fine textured glacio-lacustrine and deltaic sediments making aggregate exploration difficult.

3 DATA ACQUISITION

The remote sensing data used for this study included the following:
 1. Black and white panchromatic air photographs, 1:15,840 taken during late summer. 1978.
 2. Colour Infrared (CIR) air photographs, 1:26,000 taken during Spring (May 85).
 3. Multispectral scanner data from MEIS-II (Till et al. 1983) taken in May 1985. Spectral ranges of the channels are as follows:

Table 1. MEIS-II imagery specifications

Band Identifers and spectral ranges of filters (nanometers)	CH00: 522-735 CH01: 793-893 CH02: 626-703 CH03: 508-601	CH04: 542-605 CH05: 456-518 CH06: 751-787 CH07: 613-687
Nominal ground resolution	2.77 metres	
Date and Time of Imagery Acquisition	May 22, 1985 10:43-11:20 AM	

Soil sampling for textural and moisture content analysis were collected during the time of acquisition of imagery in May 1985.

4 METHODOLOGY OF RESEARCH

4.1 Visual analysis

The B&W, panchromatic and CIR photographs have been subjected to conventional air photo analysis using an Old Delf Scanning Sterescope, and a Zoom Stereoscope. Analysis was supported by the ancillary information (maps, reports, relevant literature), and by field studies of surficial geology and geomorphology.

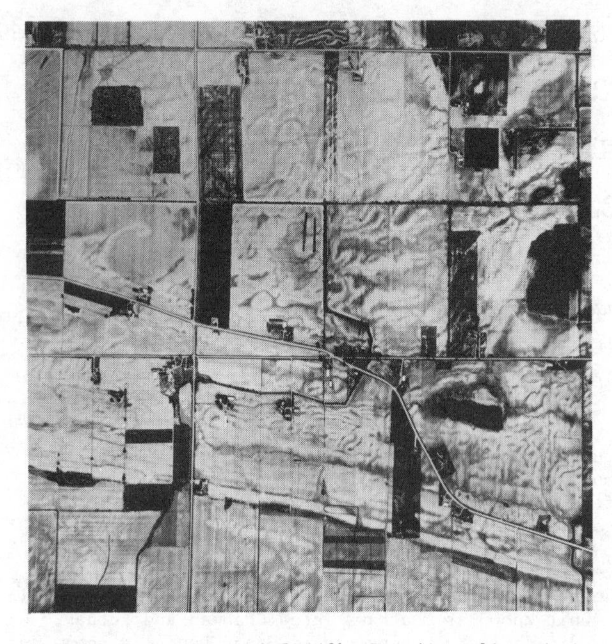

A

B

C

D

Photographs: Selected photomorphic features presented on colour IR air photographs. Original scale 1:26,000

A. Linear features representing paleocoastal environment of proglacial lake: beaches, sand bars, spits, troughs
B. Convoluted features representing ice-collapse, glacio-lacustrine sediments.
C. Elongated to irregularly shaped paleo-dunes
D. Contemporary floodplain of the Thames River.

4.2 Digital Analysis

Airborne multispectral scanner data (MEIS-II) have been subjected to image analysis procedures using a Dipix, (Aries-II) System. Data processing and enhancement operations included optimal band selection, contrast stretches, spatial filtering and principal component transformation, prior to visual interpretation.

5 RESULTS

5.1 Black & white panchromatic air photographs

Panchromatic, B&W air photographs, taken in late summer provide limited information for surficial geological mapping due to the masking effect of vegetation and subdued relief. Only broad delineation of physiographic units likely to reflect geological units are possible.

5.2 Colour infrared air photographs

The colour infrared photographs depict terrain conditions during early spring, prior to the emergence of crops. Soil conditions are therefore better exposed than in the panchromatic air photos. The six major photomorphic units which were not clearly defined or identified in previous studies include: (Photo 1):

a. Linear features representing deposits and landforms associated with the paleocoastal environment of proglacial lakes: viz. beaches, sand bars, spits, troughs.

b. Convoluted features representing ice-collapse glacio-lacustrine sediments.

c. Elongate to irregularly shaped paleo-dunes.

d. Contemporary flood-plain of the Thames River.

e. Moraine ridges.

f. Hummocky moraine pitted with kettles.

5.3 Digital Imagery

a. Selection of optimal band triplet

The selection of three optimal bands was achieved using a ranking technique suggested by Sheffield, (1985). The top ranked 3 bands are CH05 (456.5-517 nm), CH00 (522-735 nm) and CH01 (793-893 nm). Imagery from these bands were subjected to contrast enhancement and spatial filtering prior to the visual interpretation of composite images.

b. Contrast enhancements

Histogram-equalization stretch enhancement provided the best results for visual analysis of displays of colour composites formed using the optimal band triplet.

c. Spatial filtering

Best results were obtained using high-boost filtering. The enhanced image was particularly useful for accentuating the convoluted textures of ice-collapse glaciolacustrine deposits and the boundaries of paleo-dunes.

d. Principal component transformation

A principal component transformation (Jenson and Waltz, 1979), provided discrimination of surficial material and geomorphology equal to that of the optimal three band colour composite.

e. Band ratioing

After elimination of the four highest correlated bands, the four remaining bands (CH00, CH01, CH02, CH05) were used to construct six unique band-ratio images. Best results for discrimination of surficial material, soil patterns and drainage features were obtained using band ratio CH02 (green)/ CH05 (blue).

Enhanced digital images particularly the histogram-equalisation stretch and band ratioing contributed to better identification and image expression of photomorphic units and their boundaries. They have been particularly useful for verification of information derived from the analysis of CIR photographs.

The increased detail present in enhanced images does not always contribute positively to the recognition of surficial material. Local features of conditions such as variations in moisture content, humus content, soil erosion, micro-relief and plowing, can cause additional information noise after enhancement, leading to large number of photomorphic units. This must be taken into account during geological interpretation. An increase in the apparent number of photomorphic units does not necessarily represent the number of truly separable geologic units.

In summary, the best analytical approach for surficial geology mapping and aggregate exploration was obtained through the analysis of colour infrared photographs complemented by MEIS-II multispectral data. Timing of the imagery acquisition, scale of the imagery and availability of ancillary data were important factors in subsequent data analysis. Such factors must be carefully examined when advanced remote sensing systems are considered as technical aids in similar studies.

REFERENCES

Chapman, L. J. & D. F. Putnam 1984. The physiography of southern Ontario. Ontario Geological Survey. Special, vol. 2, pg. 270.

Jenson, S. K. & F. A. Waltz 1979. Principal component analysis and canonical analysis in remote sensing. Proc. of the American Society of Photogrammetry. 45th Annual Meeting. March 18-24. Washington. D C., pg. 337-348.

Till, S. M., McColl, W. D. and Neville, R. A., 1983. Development, field performance and evaluation of the MEIS-II multi-detector electro-optical imaging scanner. Proceedings, Seventh International Symposium on Remote Sensing of the Environment. Ann Arbor, Michigan, preprint.

Sheffield, C. 1985. Selecting band combinations from multispectral data. Photogrammetric Engineering and Remote Sensing. 51:681-687.

Symposium on Remote Sensing for Resources Development and Environmental Management / Enschede / August 1986

Remote sensing methods in geological research of the Lublin coal basin, SE Poland

Stanisław Kibitlewski & Barbara Daniel Danielska
Photogeological Department, Geological Institute, Warsaw, Poland

ABSTRACT: Some connections were found comparing the patterns of lineaments obtained from Landsat images and side-looking radar /Toros/ ones with faults in Lublin coal basin area /LCB/, SE Poland. It concerns the satellite lineaments pattern and the radar one corresponding to the Upper Palaeozoic- and Meso-Cainozoic faults, respectively. The conclusions suggest possibility of projection onto the recent surface the deep linear structures through the thick sedimentary cover which consists of soft and loose deposits - in LCB region.

1 INTRODUCTION

In the papers on photogeology there exist many evidences of a tectonic character of lineaments observed on the different remote sensing images of the structurally controlled areas exposed in a geological sense. Many geologists treat sceptically such opinions when they concern the regions covered with a thick layer of soft and loose sediments especially in relation to lineaments in a local or subregional scale. Big lineaments - of regional or transcontinental quality - seem to be, however, accepted as the traces of features of tectonic origin despite of the nature of their host rock.

The great part of territory of Poland has been covered with soft and loose thick Cainozoic sediments which bury older structures and render these structures recognition difficult using standar methods. That is why it seems to be of special significance here to apply the photogeological data as possible indicators of tectonic phenomena /i.e. faults/.

The paper presents the results of a test done in aim to estimate the significance of an appication of the satellite Landsat images and airborne side-looking radar images /of Soviet system Toros/ to examination of the geological structure of coal-bearing region near Lublin, SE Poland.

2 TEST DESCRIPTION

To check out a coincidence between remote sensing data and the geological ones the test has been done in an entirely small area chosen in respect to geological recognition. The area tested covered all the characteristic structural units of the Lublin coal basin in aim to extrapolate the testing results to the whole basin - in case of positive resuts of the test. In that very case it would be also possible to introduce photogeological data to the construction and gradual improvement of the structural model still existing.

The Lublin coal basin area is covered by the thick Cainozoic sediments and displays either block-faulting or block-folding character of the inner structure /Żelichowski 1972, Bojkowski and Porzycki 1980/. It is geologically well recognized at the several depths due to a black-coal mines development.

3 GEOLOGICAL CHARACTER OF THE AREA

Lublin coal basin is situated in the south-west marginal part of East-European platform in the region between Vistula and Bug rivers /Fig.1/. The terrigenic sediments of Upper Proterozoic platform /zonally preserved/ lie on the crystalline complex of Lower Proterozoic, top of which plunges from about 1000m b.s.l. in NE part, to 7000m in SW part, Fig.2A/. The following overlying strata of Lower Palaeozoic /of total thickness of 950-2500m/ are represented by the terrigenic sediments of Lower and Middle Cambrian and/with depositional gaps/ those of Ordovician and Silurian. The sediments are folded only in the zone of Caledonian movements adjacent from SW to the platform margin /outside the area tested/ - Fig.3.

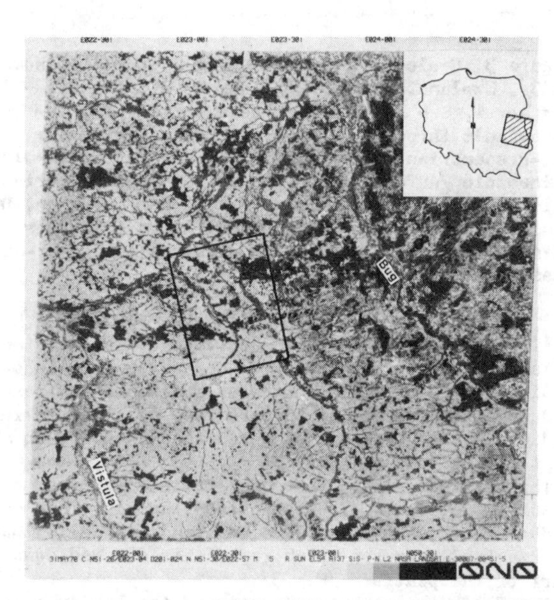

Figure 1. Scene: E-30087-08451 MSS-5,31 May 1978. Investigated area and image position are marked.

The platform conditions occur in the whole territory discussed since Devonian. The Lower Devonian sediments are terrigenic, while Middle and Upper Devonian ones are succesively those of lacustrine-marine and terrigenic, partially carbonate /i.e.dolomitic/ as well. The Devonian sediments thickness varies from 1500 to 6300m. Their top surface occurs recently from 800m b.s.l. in the NE part of the studied area down to 2600m b.s.l. in the SW part of it.

Due to the Bretonian movements distinct system of horst and grabens with NW-SE and SW-NE orientation and high amplitude has been developed. The NW-SE orientation seems to prevail in SW part of the area under investigation while that SW-NE - in NE part. Those structures have been intensively eroded prior to Carboniferous sedimentation.

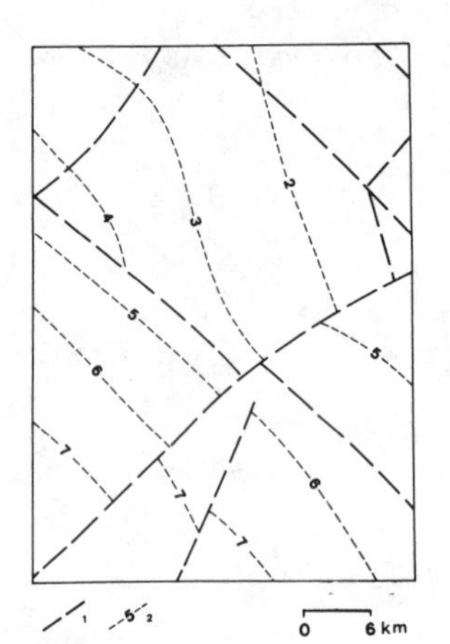

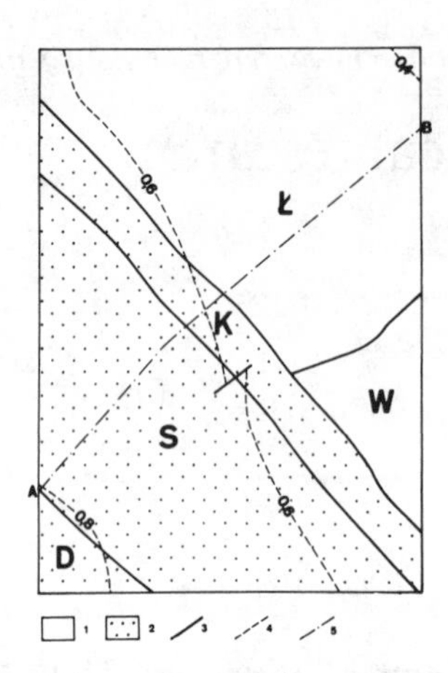

Figure 2. A - The top surface of crystalline basement /after Ryka 1983/: 1 - faults, 2 -isohypses /in kilometres b.s.l./. B - The top surface of Palaeozoic: 1 -elevated part of platform, 2 -downwarped part, 3 - main faults, 4 -isohypses /in km/, 5 -cross-section line /symbols as Fig.3/.

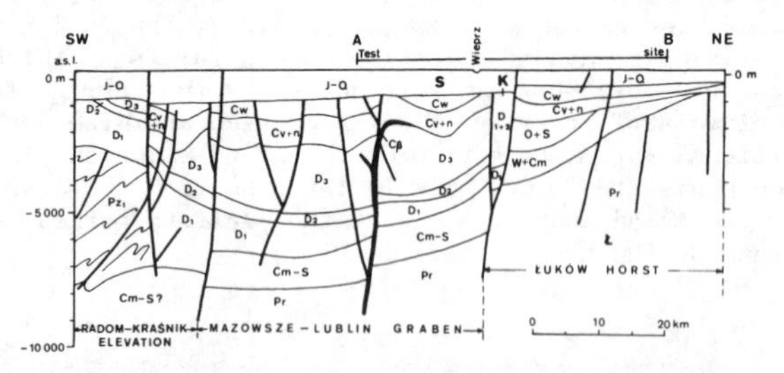

Figure 3. Geological cross-section /after Żelichowski 1983/. Explanation: K - Kock horst anticline, Ł - Łuków horst, S - Stoczek-Dorohucza depression, W - Włodawa fault through, D - Dęblin-Krasnystaw swell; Pr -Precambrian, W+Cm -Wend and Cambrian, Cm-S -old Palaeozoic /unfolded/, O+S -Ordovician and Silurian, Pz1 -old Palaeozoic /folded/, D1 -Lower Devonian, D2 - Middle Devonian, D3 -Upper Devonian, Cβ -diabases, Cv+n -Visean and Namurian, Cw -Westphalian, J-Q -Jurassic to Quaternary.

After the strong volcanic activity of Tournaisian /diabases and tuffites/ there occured the continuous sedimentation from Visean to Westphalian which resulted in formation of the thick series of coal-bearing sediments the thickness of which reaches 2000m in SW part of the basin.

The initial shape of the coal basin was formed by Asturian movements of post-Westphalian age. They caused a development of NW-SE elongated unit called Mazowsze-Lublin graben which represents a moved down part of the platform /Fig.3/.

The following structural units of the Lublin coal basin district occur in the area tested /Fig.2B/: - Kock horst anticline, Łuków horst and Włodawa fault through /which all belong to the elevated part of the platform/ and - Stoczek-Dorohucza syncline /with Abramów-Żyrzyn horst/ - a part of Dęblin-Krasnystaw swell.

The numerous faults forming the zone of well developed set of NE-SW direction /transversal faults -Bretonian in age/ and less distinct NW-SE set /longitudinal -Asturian in age/ occur to the north-east of Kock horst anticline. Such the net of faults implies a relatively simple block tectonic pattern of Palaeozoic sediments in that part of the region.

The area to the south-west seems to be tectonically more complicated. It consists of numerous elongated anticlines and synclines stretching NW-SE and complicated by transwersal and longitudinal faults.

The top surface of carboniferous sediments in Lublin coal basin has erosional character and is inclined SW /Fig.2B/ from about 400m b.s.l. to about 800 m b.s.l.

The thick Mesozoic series of Middle Jurassic to Upper Cretaceous sediments lies on the Carboniferous deposits. Jurassic has sandy-clayish and carbonate character. In Lower Cretaceous, after erosion due to Young Cimmerian movements the sedimentation of carbonates has begun starting from Albian or Cenomanian and persisted till Coniacian. The sediments of Upper Cretaceous display an increasing content of clay material. The sedimentation of this type prevailed up to Palaeocene. The general thickness of Mesozoic in Lublin coal basin district increases from about 200m in NE to more than 1000m in Mazowsze-Lublin graben.

In the top of Mesozoic there occur numerous disjunctive deformations of a discussed origin when it concerns their relation to the Palaeozoic deformations. According to the opinion of some authors a Mesozoic cover has been dislocated to some extent separately from the structures of the older basement /after decollement/. It is evident, however, /i.a. Henkiel 1983/ that some of the existing dislocations or their systems show a distinct relation to the tectonic elements of Palaeozoic, as e.g. Mesozoic lineament Kock--Łęczna corresponding to the Palaeozoic Kock horst anticline. The character of Mesozoic faults in both the side-areas of this horst changes similarly to the palaeozoic pattern. In the area to NE from the horst the regular pattern of Meso- and Cainozoic faults corresponds /to some extent/ to that of Palaeozoic ones. To SW, however, the total Mesozoic disjunctive pattern seems to be less distinct and displays an increasing number of fractures and faults. The directions there are subordinated to the general stress field orientation connected with a strike-slip character of the faults in the marginal zone of Mazowsze-Lublin graben. There exist also some tectonic zones in Mesozoic sediments which show no relation to the known structures of the older basement.

Tertiary in the area of Lublin coal basin is represented by the strongly eroded and dissected sediments /marls, opocas, gaises, glauconite sands, clays, sands and siltstones/ of locally varied thickness from 60m in Mazowsze-Lublin graben to 200m farther to NE.

The Quaternary cover lies either on the Tertiary sediments or directly on the Cretaceous strata. It is built of the following sediments: loose and soft glacial sediments /sands, tills and loesses/ of Pleistocene age and total thickness 0 - 100m, of Holocene lake deposits /thickness 0 - 60m/, as well as - of river and eolian deposits.

When discussing the tectonic processes in the Meso- - and Cainozoic sediments of the area examined the most complete information has been presented by Henkiel /1983/- Fig.5. He distinguishes several phases of tectonic movements influencing the faults and fractures development. The oldest disjunctive structures /faults and small grabens/ with NE - SW orientation are believed to be of Eocene age while some W-E and NE-SW faults and grabens with N-S orientation are defined as post-Sarmatian ones. Finally - some long - latitudinally oriented fault structures as well as several sets of differently oriented faults are believed to be due to the last phase of Alpine movements.

4 REMOTE SENSING MATERIALS AND THEIR INTERPRETATION

Such the remote sensing materials have been used in the paper: - Landsat images, namely scenes: E-2244-08442, E-2946-08211, E-2298-08434, E-2892-08240, E-2928-08221, E-0087-08451, E-21090-08142, E-2155-08512 and E-0448-08513. They have been visually interpreted in the form of false colour compositions /FCC/ and photos of different bands of MSS /black and white, papers/ enlarged to a scale 1:500 000, as well as colour compositions obtained by the means of Additive Color Viewer /I²S/; - airborne side-looking radar images of the Soviet system Toros. They have been used in a form of diapositive films and interpreted with a help of Zoom Transfer Scope - Bausch and Lomb.

In aim to compare the interpretation results of both

the methods discussed above the lineament patterns obtained from Landsat materials as well as those of airborne radar images has been converted to the same scale of 1:200 000 adopting the criterion of repeatibility of the elements observed in the different materials and by the different authors.

The comparison of the both patterns resulted in observation of distinct differences in abundance, continuity, size and density of satellite and radar lineaments. The differences themselves although partly due to the different components of the techniques of the images applied /active and passive systems, distance from the sensor to the object etc./ might offer some information on different geological features, too.

5 GEOLOGICAL MATERIALS

The following geological materials have been used to analyse the relations between different lineament patterns and geologeal structure of the area discussed: map of the surface deposits /Mojski 1968, Malinowski and Mojski 1978/ and its generalized form showing morphogenetic /physiographic/ regions /Fig.4/, map of sub-Quaternary surface relief, pattern of faults, faults zones, disjunctions and dilatations of different kinds and other weakness zones in the Mesozoic and Lower Tertiary rocks /Fig.5/, structural maps of a top surface of a Palaeozoic series with a fault patterns /Fig.6,7/ and structural map of crystalline basement /Fig.2A/.

6 GEOLOGICAL INTERPRETATION AND DISCUSSION

6.1 Geological interpretation of the radar data

From the spatial and directional points of view the

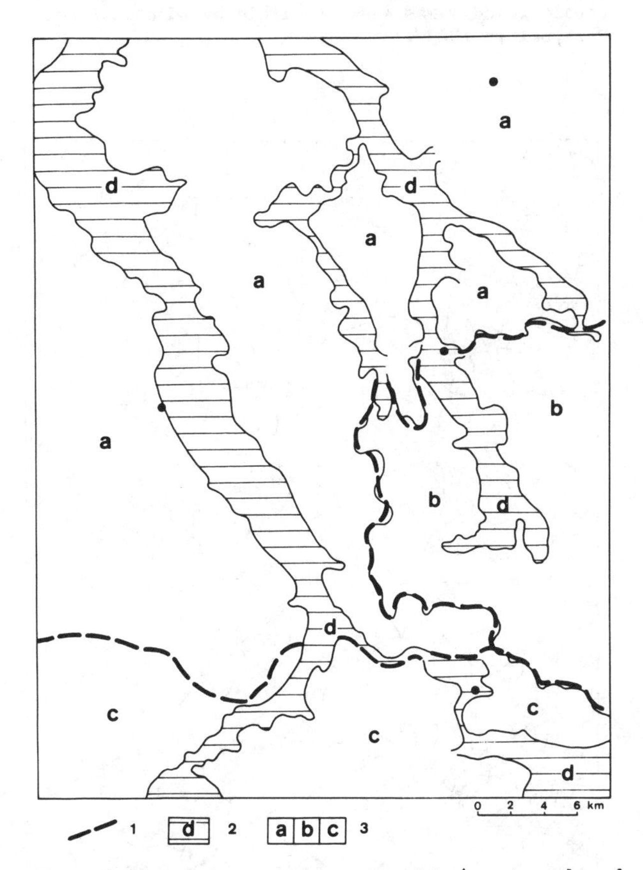

Figure 4. Morphogenetic /physiographic/ regions based on surface deposits map: 1 -boundaries, 2 - valley areas /d/, 3a -morainic plateax, 3b -depression with prevailing lake accumulation deposits, 3c -upland covered with loess.

radar lineaments seem to be rather irregular. One can tell from the comparison of their pattern and morphogenetic map of the area /Fig.8,4/ that the differences in the density of these lineaments /Fig.9/ correspond to the morphogenetic units. The maximum density of the lineaments might be observed in the southern upland covered with loesses, in adjacent depression with a prevailing lake accumulation as well. Distinct density might be also noticed in the area corresponding to the bottom and some slopes parts of the Wieprz river valley. The minimum density of the radar lineament pattern refers to the morainic plateau. Radar lineaments observed are generally not long /3km in average, maximum 6km/ only occasionally forming longer trends. Although in general they display a distinct dispersion of orientations, some intervals of an increased frequency might be distinguished, namely: WNW to NW and NE to ENE /main azimuths 30^{0}, 110^{0}, 130^{0} - compare Fig.10/. The directions of the lineaments discussed are mostly dispersed in the area of the morainic plateau while those on the plain with lake accumulation, as well as in loess upland seem to be more regular.

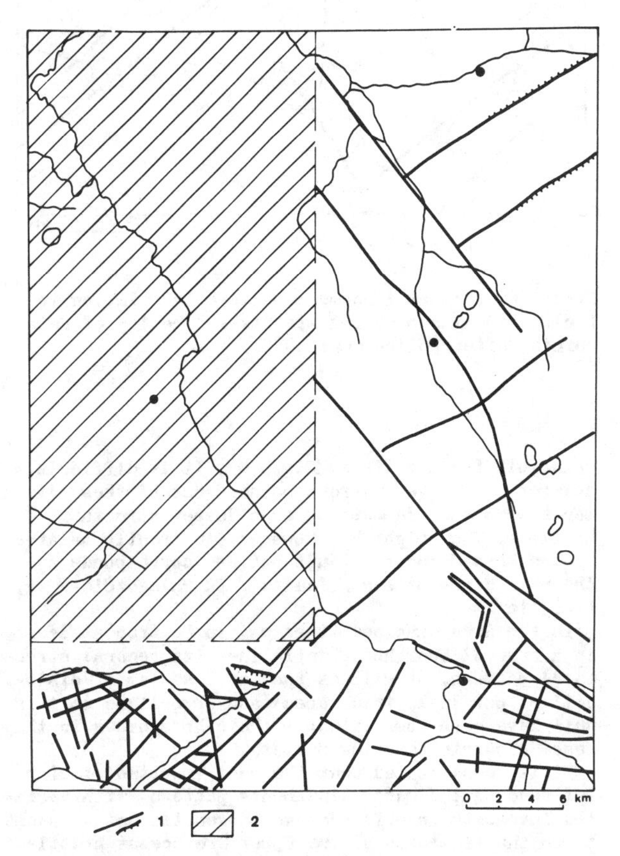

Figure 5. Faults and other disjunctive structures in Meso-Cainozoic series /1/, no information /2/; after Henkiel 1983.

Resumming the facts from the comparison above it seems that the radar lineaments are more distinct, frequent and they display more regular net in the areas covered with a relatively homogeneous surface deposits /lake sediments and loesses/ than in those of morainic plateaux lithologically more complex and covered with different glacial deposits.

Some spatial relations may be noticed comparing the map of the sub-Quaternary surface relief and that of morphogenetic units. There exists also coincidence between radar lineament pattern and the details of sub-Quaternary surface in individual morphogenetic units.

In many places, especially in the loess area, the radar lineament patterns seem to be close to Meso- and

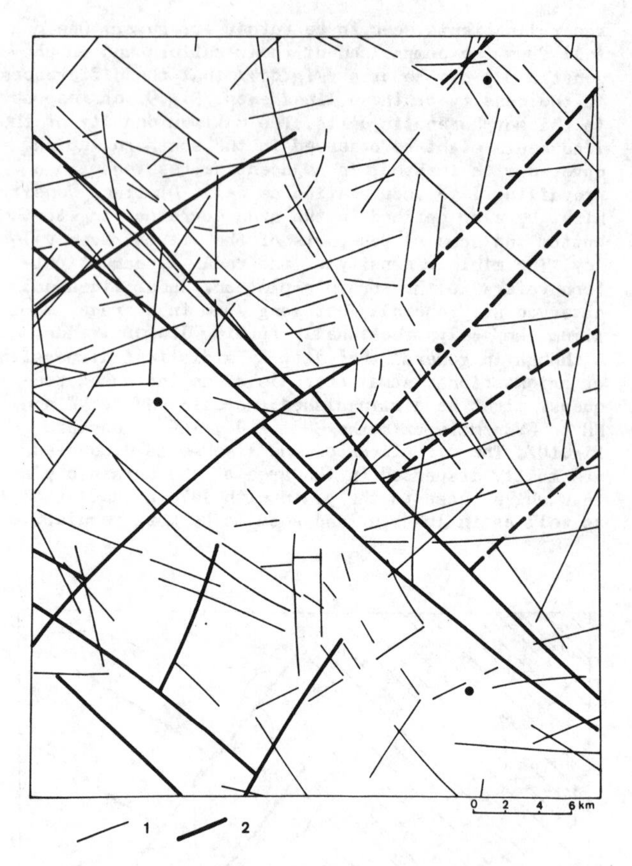

Figure 6. Landsat lineament pattern in relation to faults: 1 -lineaments, 2 -faults in the top of Palaeozoic /after Żelichowski 1972/.

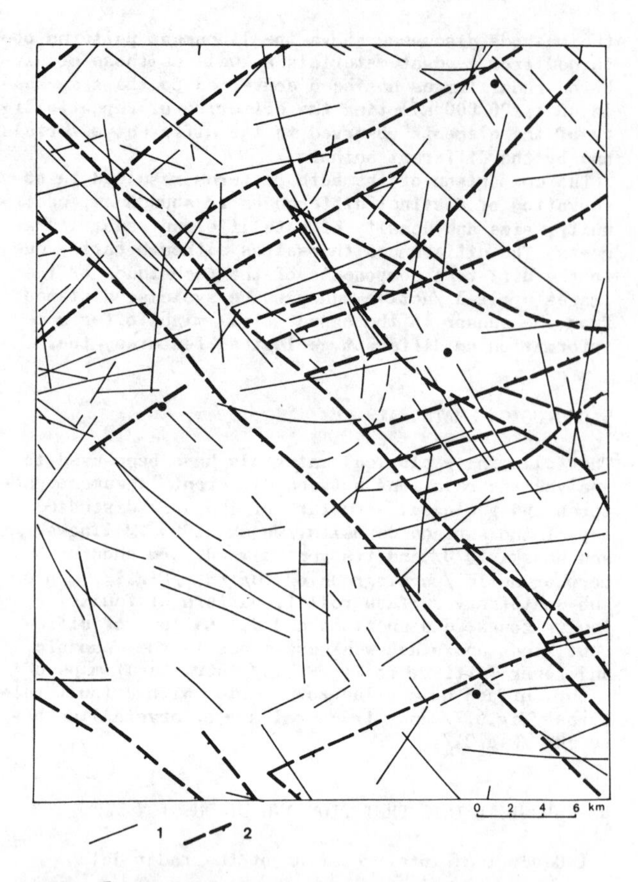

Figure 7. Landsat lineament pattern in relation to faults: 1 -lineaments, 2 -faults in the top of Palaeozoic /kindly was made available by Zdanowski and Porzycki in 1983/.

Cainozoic fault pattern /Fig.5,8/. It is difficult, however, to prove a direct coincidence of these lineaments with the elements of even deeper structural horizons. They might be, however, indirectly related if the Meso-Cainozoic fault pattern corresponds to the main dislocations / faults / in Palaeozoic /Fig. 11a, 11c/.

In the area examined it seems to be true at least in a case of Kock horst anticline. Its general structural pattern, as well as that of a system of dislocations parallel, transversal and/or oblique to this unit, have - to some extent - their homologues in the Meso-Cainozoic faulting pattern.

It is to be stated that the most important factor influencing the radar lineaments pattern /with variated intensity in different morphogenetic units/ seems to be the tectonics of the Upper Cretaceous brittle rocks i.e. directly - by simple projection of the faults on surface, or indirectly - by the influence of sub-Quaternary relief on later sedimentation. This first case of influence seems to be the most distinct i.e. easily observable in the remote sensing images in the areas of the lithologically homogeneous surface cover /e.g. loess/, less distinct in the lithologically differentiated surface deposits /e.g. glacial sediments /Fig,10, 11a/.

6.2 Geological interpretation of the Landsat data

The pattern of Landsat lineaments displays regular uncomplicated distribution in the region in question /Fig.12/. The maximum density of the lineaments occurs in the following areas: the south-western part of Kock horst anticline and north-eastern region adjacent to it - the area corresponding to the axial part of Stoczek - Dorohucza syncline /Fig.12,2B/.

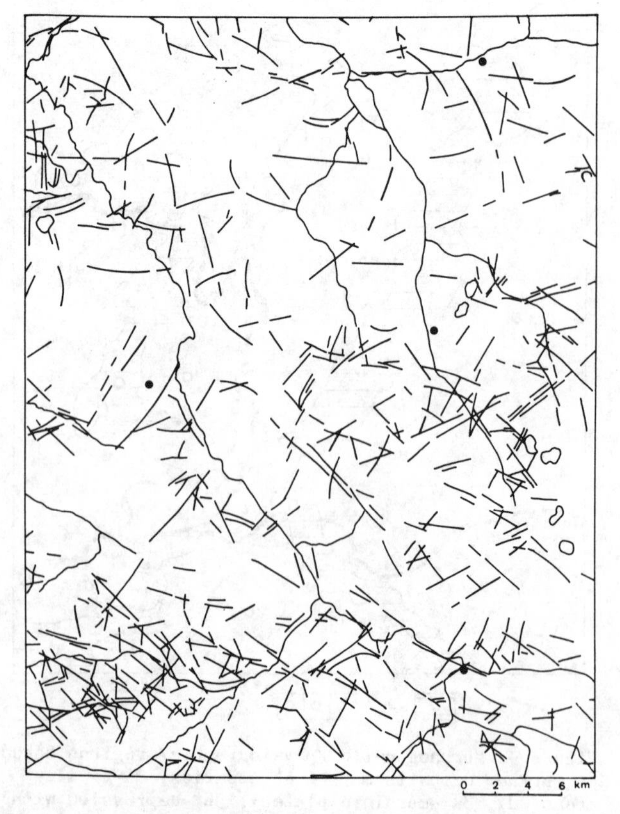

Figure 8. Radar lineament pattern

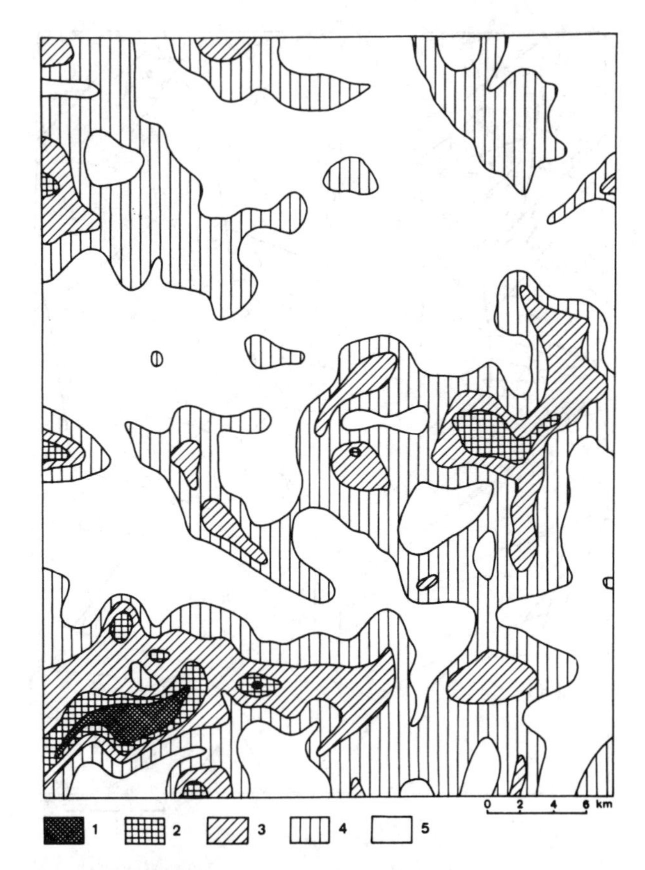

Figure 9. The pattern of radar lineaments density /in km/km^2/: 1 ->2,0; 2 - 2,0÷1,5; 3 - 1,5÷1,0; 4 - 1,0÷0,5; 5 -<0,5.

The satellite lineaments are the continuous lines which extend 8km in average /maximum 16km/ and locally form the longer trends /up to 50km/. The lineaments of NW-SE /130°-150°/ directions, as well as those of ENE-WSW /70°-80°/ and submeridional /300°-10°/ prevail.

The first group of lineaments mentioned above /NW-SE orientation/ corresponds to the main structural direction delimited by the edge of the platform i.e. in this area - by Kock horst anticline and Żyrzyn-Abramów-Świdnik horst separated by Stoczek-Dorohucza depression /Fig.7,2B/. The second group corresponds to the directions of the faults /oblique and/or perpendicular to the Kock horst anticline/, especially well developed in the platform area adjacent from NE to this horst.

Basing on the comparison of the remote sensing interpretation with two stages of succesive geological recognition /Fig.6,7/ it might be stated that the more

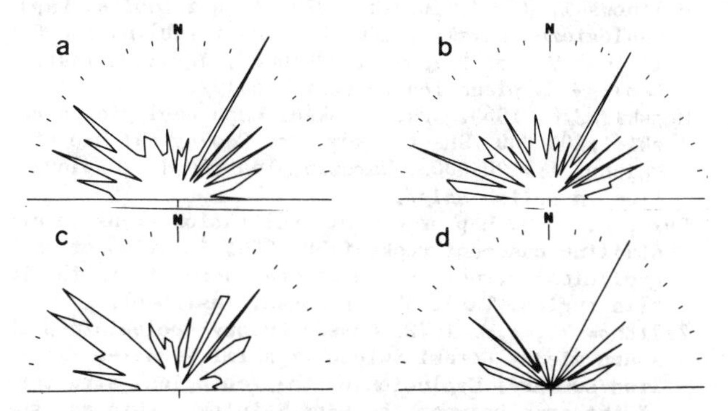

Figure 10. Statistical diagrams of radar lineaments: a -for the whole area, b -for morainic plateaux, c -for the area covered with loess, d -for depression with lake sediments.

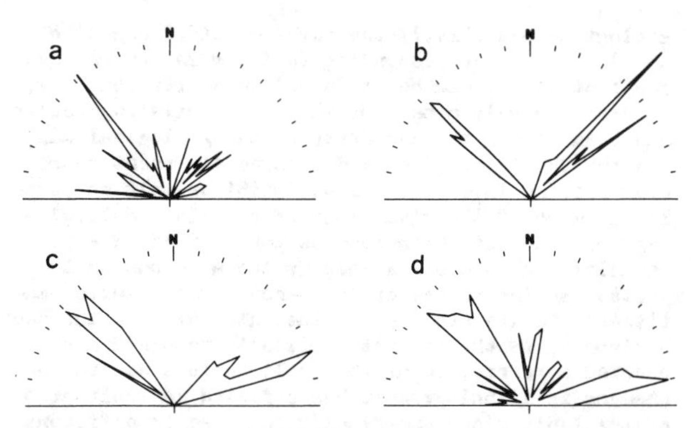

Figure 11. Statistical diagrams showing: a -faults in Meso-Cainozoic sediments, b -faults in Palaeozoic sediments /state from 1972/, c -faults in Palaeozoic sediments /state from 1983/, d -Landsat lineaments.

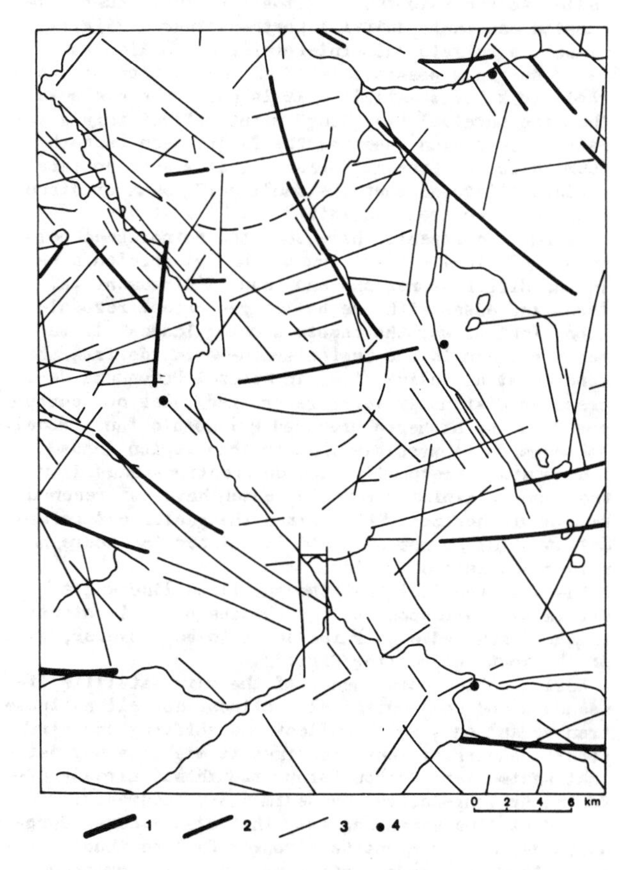

Figure 12. Landsat lineament pattern: 1 -best visible, 2 -less visible, 3 -weakly visible, 4 -main settlements /for comparison only/.

complete geological data - the better coincidence of Landsat lineaments and geological linear elements /faults/. It means, therefore, that Landsat lineaments display weak similarity against the earliest fault pattern interpretation /in the Palaeozoic level/ remaining however, in a distinct coincidence with the further one based on the more detailed geological recognition of the area. Landsat lineament pattern seems also to give more information, as it concerns the direction of the faults oblique and perpendicular to the Kock horst anticline in the area extended towards north-east.

It is evident that Landsat lineaments display azimuthal relationship with the faults in Palaeozoic top /compare also with the statistical diagrams - Fig.11b, 11c and 11d/ and might represent their surface projection.

Still comparing remote sensing elements with the

geological map /Fig.7/ the parallel displacement of the lineaments corresponding faults might be observed. The fact itself can be explained in a different way, namely: 1 - only a part of the really existing faults has been recognized and drawn on the geological map, 2 - the faults could have displayed different character i.e. an interpreted fault might be only an idealized image of the wider zone of parallel fault planes, only one of which reaches the surface, 3 -several parallel faults could accompany the main one on the surface as the traces of the second order faults, antithetic faults etc., 4 -in the interval from the deep horizon up to the present surface there could have occured a refraction of the fault surface due to the changes in lithology what has affected an inclination of the fault plane and finally resulted in different location of the traces of this plane in the different stratigraphic horizons and recent surface as well.

In the NE side of the Kock horst anticline there occurs a distinct trend of the Landsat lineaments parallel to the main fault of the structure under discussion and their shifting north-eastwards /Fig.7/. Since the seismic data interpretation /Żelichowski 1972/ suggests possibility of NE inclination of the whole Kock horst anticline it is not to be excluded that the parallel "shifting" might reflect this asymmetry. Discussion concerns the faults seen beneath 800m below the present surface. Taking such an interpretation into account the fault surface under discussion would be inclined /dipped 60°SW/.

It must be repeated here that the coincidence between geological recognition of the interpreted area in the different regions and Landsat lineament patterns increases with the better geological recognition. That is why the remote sensing Landsat lineaments pattern in the region south-westwards from the Kock horst anticline /i.e. in Stoczek-Dorohucza depression distinctly worse recognized/ does not correspond to the hitherto proposed Palaeozoic fault model. It seems to be possible that in this region Landsat lineaments correspond to the deformations caused by the younger Alpine movements, which have not reached Palaeozoic horizon. Still since the geological information might be not complete the discussion there has rather academic character.

The observed fact that the satellite lineaments distinctly correspond to the changes of coal-potential in the discussed depression seems to be, however, one of the most interesting/Fig.13/.

Basing on the coincidence of the main satellite lineaments and coal-potential isolines, as well as those transversal ones which reflect the shifting and directional changes of these isolines it might be suggested that sedimentary Carboniferous trough was already affected by long-active, synsedimentary movements.

Landsat lineaments suggest that also Stoczek-Dorohucza depression might be stronger faulted than it results from the present structural map. The proposed interpretation of the remote sensing data applied to the coal basin recognition demands, however, to be checked in the further geological recognition of coal deposits.

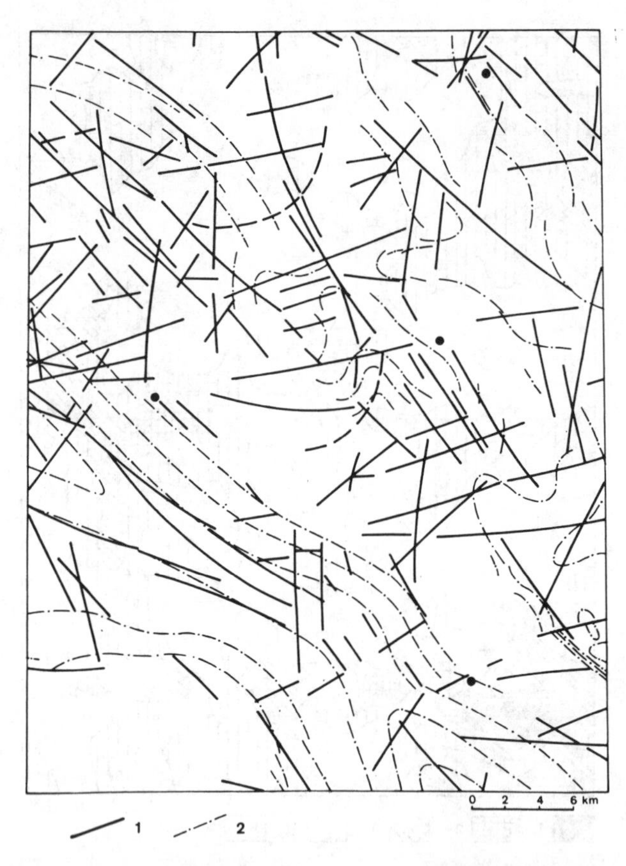

Figure 13. Landsat lineament pattern and coal-potential changes: 1 -lineaments, 2 -coal-potential isolines/kindly was made available by Zdanowski and Porzycki in 1983/.

7 FINAL REMARKS

As it seen from the facts above a distinct coincidence between Toros radar lineaments, satellite Landsat ones and the geological structure of LCB occurs. Radar lineaments seem to be connected with not deep-lying structural horizon /the top surface of brittle Mesozoic and Lower Tertiary sediments/ and might have some significance for the further geological recognition of the cap rocks of the coal-bearing sediments.

From the other side Landsat lineaments correspond to the geological structure of the Palaeozoic series which contains the coal-bearing sediments. Such the relation is so distinct that basing on the standard geological methods and the remote sensing techniques a creation of a certain model of the structure of the area might be postulated. Such the model might be - in particular - in case of Lublin coal basin quickly checked and verified due to the progressive development of the black-coal deposit.

In general - the model itself - can be applied to improve the geological recognition of the other coal basins with a comparable geological setting.

REFERENCES

Bojkowski, K. & J.Porzycki 1980. Geological problems of coal basins in Poland. Warszawa, Wyd.Geol.

Henkiel, A. 1983. Tektonika./Tectonics. In Cainozoic of Lublin Coal Basin.Symposium in Lublin, 9-11.IX/. Kenozoik Lubelskiego Zagłębia Węglowego. Lublin, UMCS, p.41-64. - /In Polish only/.

Malinowski, J. & J.E.Mojski 1978. Arkusz Lublin. Mapa Geologiczna Polski 1:200 000 /Sheet Lublin. The Geological Map of Poland, 1:200 000/. Warszawa,Inst. Geol. - /Explanation in Polish only/.

Mojski, J.E. 1968. Arkusz Łuków. Mapa Geologiczna Polski 1:200 000 /Sheet Łuków. The Geological Map of Poland, 1: 200 000/. Warszawa Inst.Geol. /Explanation in Polish only/.

Ryka, W. 1983. Map of ore mineralization signs in crystalline basement rocks 1:500 000. In Atlas of the geological structure and mineral deposits in the Lublin region, Table 39. Warszawa, Inst.Geol.

Żelichowski, A.M. 1972. Rozwój budowy geologicznej obszaru między Górami Świętokrzyskimi i Bugiem /English summary: Evolution of the geological structure of the area between the Góry Świętokrzyskie and the river Bug/. Bull.IG, 263. Warszawa, Inst.Geol.

Żelichowski, A.M. 1983. Tectonic map 1:500 000. In Atlas of the geological structure and mineral deposits in the Lublin region. Table 34. Warszawa, Inst.Geol.

Symposium on Remote Sensing for Resources Development and Environmental Management / Enschede / August 1986

Photo-interpretation of landforms and the hydrogeologic bearing in highly deformed areas, NW of the gulf of Suez, Egypt

E.A.Korany
Qatar University, Doha

L.L.Iskandar
Secondary Schools, Cairo, Egypt

ABSTRACT: West of the gulf of Suez, the area is highly deformed. The Precambrian Shield was developed into an elongated fault block. It is overlapped on both eastern and western margins by thick sequences of younger sedimentary rocks. They are all broken by five sets of faults trending N-S, NE-SW, NW-SE, E-W, and WNW-ESE.

The surface is developed into a higher Plateau occupies the Red Sea Range of the Basement rocks. The coastal strip is covered by Miocene and younger sediments and occupies the outlets of the dissects drainage arteries shedding from the higher plateau.

An image interpretation of landforms is carried out on the bases of, qualitative and quantitative studies of relief criteria, the drainage network and the hydrographic features.

More attentions are focused on the hydrogeologic bearings, the control upon water flow either on the surface or in subsurface, and the impacts upon the groundwater conditions in the Miocene aquifer.

The study approach is based on the stereoscopic examination of aerial photographs, the study of photo-mosaics and landsat imageries, and the field check and measurements.

LOCATION OF THE AREA

The area of study is located in the Eastern Desert of Egypt along the coast of the gulf of Suez. It is bounded by latitudes 27° 45` and 28° 25` N, and longitudes 32° 45` and 33° 45` E (Fig. 1).

GEOLOGIC CONDITIONS

The gulf of Suez lies within the stable belt of Egypt. It runs in a NW-SE direction forming an elongated depression seperating the massive central Sinai fron those of the Eastern Desert. It is being regarded as a complicated rift type graben structure initiated during Oligocene time and controlled by NW-SE normal faults(Shalom,1954; Said,1962; Youssef, 1968; Said ,1969; and Abdeine,1981).

West of the gulf of Suez, the area is developed into NW-SE parallel fault blocks forming successive grabens and horsts. The horsts have the form of anticlines and the grabens have the form of synclines (El-Tarabili,1970, and Iskandar, in preparation).

Five systems of faults are recognized in the area. They are trending N-S, NE-SW, NW-SE, E-W, and WNW-ESE. They are responsible for the development of certain geomorphologic landforms and the hydrographic features.

The surface is built of the Basement rocks in the higher Plateau, while it is formed of younger sedimentary rocks along the coast. The Miocene section is the thickest single stratigraphic unit. The record culminated with thick evaporites of Middle Miocene age.

The Rudeis Formation represents the wider distributed unit in the Middle Miocene section in the area. It is built of sands and sandstones at base, and shales at top. It maintains the main aquifer in the area. It yields brackish water having higher concentrations of Ca, Na+K, Cl, SO_4, and HCO_3 (Iskandar,in preparation).

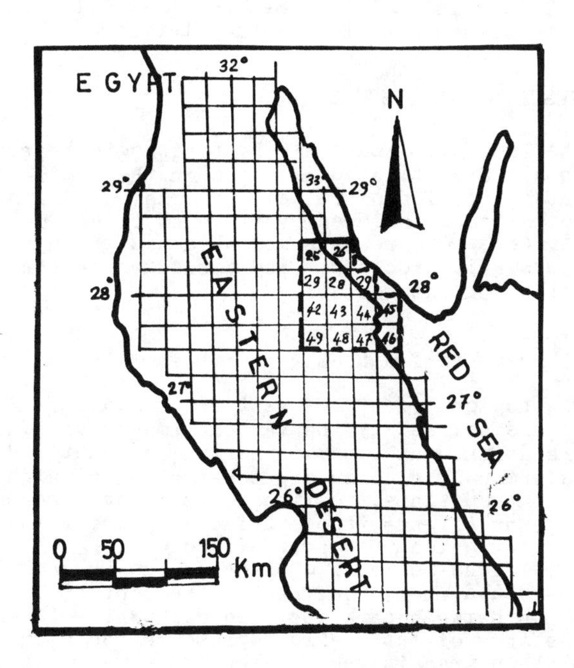

Figure 1. Location of the studied area, with an index of aerial photos and mosaics.

CLIMATIC CONDITIONS

The area is characterised by arid conditions. It has very low rainfall intensities (average of 12.7 mm/ year) and high intensities of evaporation and evapotranspiration (average of $8.1(10)^3$ mm/year).

Nevertheless, the occassional heavy showers during winter months along the higher Plateau and slopes represent possible routs for seasonal contribution to the water budget either on the surface or in subsurface (Korany,1980).

STUDY APPROACH AND TECHNIQUES

The main task of the present work is the interpretation of the landforms in the area of interest and their hydrogeologic bearings. The control upon water flow either on the surface or in the subsurface and the impacts upon groundwater conditions in the Miocene aquifer are the objects.

The study approach depends on both the qualitative and quantitative techniques either in the lab or in the field. Among these techniques are;

1. Stereoscopic examination of 249 aerial photographs of scale 1:40,000,
2. Assembling the photo-mosaics covered the area (about 13 photo-mosaics of scale 1:50000)
3. Assembling the landsat imageries which include the area of interest. Three of the ERTS imageries of bands 5,6,and 7 are selected to represent the conditions of January and June of the year 1976,
4. Infiltration tests by using double rings in selected sites representing the different surface deposites,
5. Sampling of rock, soil, and water bodies. The collected samples are studied petrographically or chemicaly,
6. Field survey and measurements. Distances, angles and levels are obtained,
7. Hand templet and sloted templet methods for map compilation are applied, and
8. Statistical treatment of the collected data either manually or computerized.

RESULTS AND DISCUSSION

The landforms and their hydrogeologic bearings are investigated as based on the photo-interpretation of relief criteria and drainage patterns followed by field check and measurements. The results are obtained through systematic procedures and discussed in the following categories:

1. Regional features

The regional features west of the gulf of Suez are defined by the landsat imageries. They are defined as based on normal and subtle differences in color and tones along straight or curved lines,the lineaments in drainage and the alignments of surface features. The following features are defined (Fig.2):

1.1. The regional outline of the Precambrian sheild,where defined by the contact bounaries between dark grey and dark white tones. The area of the sheild represents the higher plateau area in the Eastern Desert of Egypt. It slopes in general to the east and the west.

1.2. The shape of the coast line of the gulf of Suez, where it runs into zigzag line in both sides and trending NW-SE. It is highly controlled by the regional trends of faulting in the gulf region. The N-S and WNW-ESE faults play the pioneering role in this manner.

1.3. The regional patterns of drainage basins, where they run either to the east or the west of the basement sheild. The higher plateau maintains the water divide and the up-stream parts of the drainage basins.

1.4. The local bodies of salt and brackish water on the surface, where they occupy the lowlying areas along the coast. They represent the marsh and swamp areas. They are developed mainly due to subsurface intrusion of salt water from the gulf and seeps from the groundwater in the land.

1.5. The structural ridges , where they de-

Figure 2. A combined mosaic of the ERTS imageries indicating the regional features west of the gulf of Suez.

veloped parallel to the coast line and mainly attributed to the influence of NW-SE faults.

1.6. The evaporite exposures along the western coast of the gulf, where they occupy the areas of whitish tone. They represent the upper unit of Middle Miocene section in the gulf of Suez region.

2. Drainage and hydrographic patterns

When dealing with the drainage and hydrographic patterns the photo-mosaics are examined. The drainage basins are delineated and the boundaries are defined. They are defined by the name of the main valley. The drainage patterns represent the main channels and tributaries in each basin or sub-basin. A combined map of the area is compiled after the examined mosaics (Fig. 3).

Almostly, the main channels have a trend from west to east in the area west of the gulf of Suez. This reflects the influence of the initial slope trend and the structural framework.

Certain patterns of drainage are defined qualitatively and quantitatively in the present drainage sub-basins. They are differentiated into several types as based on some criteria such as, the degree of integration, the density, the degree of uniformity, the orientation, the degree of control, the angularity, and the angle of juncture (Ray,1960; Thornbury,1962; Howard,1967; and El-Etr & Yousif,1978).

About 13 of photo-mosaics are examined (Nos.,25, 49, inclusive). Nine sub-basins are defined from north to south of the area respectively. They all shared together with the outlets along the coastal plain. The main channels issuing from the high basement plateau.

The following features are concerned when dealing with the drainage sub-basins in the studied area(Fig. 3 and Table, 1) :

2.1. The surface area of the drainage sub-basins varies between 125.5 sq. Km. and 962 sq. Km.

2.2. The total length of the main stream varies between 27 Km and 82 Km. It increases from north to south of the area. The main stream changes into meander one in the southern sub-basins.

2.3. The stream order assumes the fourth to the fifth order in all the drainage sub-basins. This reflects the same slope trend and magnitude and lithologic characters along the present sub-basins.

2.4. The drainage density has an average value changing from 5.7 to 8.7 one per sq.Km. This reflects moderate density in the high plateau. While in the coastal plain the drainage density assumes the minimum value and ranges between 2 and 5 one per sq.Km. This indicates more or less homogeneous lithology and slopes, where the porous and permeable rocks are exposed on the surface.

2.5. The dominant types of drainage patterns in each of the nine sub-basins vary greatly. The common types along the high plateau are the dendritic and the sub-parallel types of drainage. While the dominant types in the coastal plain area are the braided and dichotomic types. This variation reflects the changing in surface facies and local slopes in the same sub-basin from west to east and between the nine sub-basins. The dominant structure plays an important role in the development of certain assortment of drainage patterns in the area. The trillis, the pinnate and the barbed types of drainage patterns reflect the influence of the dominant trends of faulting and folding in the area.

The drainage network which developed in the area of study is an outcome product of the last fluvial periods of the Pleistocene and Recent times. The present dry vallies and tributaries were already engraved during that fluvial periods. They are now filled with surficial deposits of gravels, sands and clays having high infiltration capacities(0.5 to 1.5 mm/ sec.). The occassional showers during winter times at present along the high plateau and slopes maintain short-period floods where great amount of water infiltrate dowenward at the foots of the plateau and in the coastal plain to contribute the groundwater in the Miocene aquifer. The drainage sub-basins shedding water almostly to the east, while other local trends of northeast are present.

3. Geomorphologic units

The area of study is built of three geomorphic units. These units are distinguished as based on the relief criteria and drainage patterns. The system proposed by Verstappen (1977) is considered. The structural deformation of the area plays an important role in the development of these units. While the lithologic variation along the surface reflects the chief events of the geomorphologic history of the area(Figs. 4 and 5).

3.1. The coastal plain

It occupies the eastern part of the area. It forms a longitudinal strip of land running parallel to the gulf of Suez in the NW direction. It has a width varies between 2.5 Km and 9.5 Km. It increases in width from north to south. The surface elevation ranges between zero level and 70 m above sea level. It has gentle slope which assumes 5.9 m/ Km averagely(slope angle of less than 1/2 degree).

It includes lowlying marsh lands (El-Malahat) which occupies a longitudinal area parallel to the coast.

The surface of the coastal plain is built of Middle Miocene facies, exposed in parts and covered in others by Recent surficial deposits(Fig. 5). It is built of a great syncline running in the NW-SE direction and developed along the graben parallel to the gulf of Suez.

It represents the area of the water collector west of the gulf of Suez, where it recieves water either by surface runoff through t the drainage basins or by subsurface flow in Miocene aquifer.

3.2. The pediment

It occupies the gentle inclined plain at the foot of the higher plateau and represents the transitional zone between the coastal plain and the high plateau. It has a width ranges between 18 and 24 Km. The surface elevation attains the maximum of about 300 m above sea level. It has the gradient of about 10m/ Km averagely.

The surface of the pediment is built of the Middle Miocene facies. It is dissected by dry vallies and tributaries issued from the high plateau in the direction of the coastal plain. The channels are filled by surficial deposities of high infiltration capacities.

The pediment plain is formed by degradation and retreat of the mountain front in arid or semiarid regions across rocks of varying lithology (Easterbrook,1969). A break in slope

Table 1. Drainage analysis and hydrographic network of nine sub-basins in the studied area.

Ser.no.	Name	Area(sq.Km)	Length(Km)	Order	Drainage density(one/sq.Km)			Mosaics no.
					min	max	aver	
I.	Northern	296.0	27.0	4 th	4	17	8.7	25,26
II.	Wadi Abu-Haad	296.5	40.5	4 th	5	9	6.7	25,26,29
III.	W. El-Darb	125.5	33.0	4 th	4	11	7.3	25,26,29
IV.	W. Khreim	188.5	46.0	4 th	2	14	7.3	25,26,28
V.	W. Um-Yasar	188.0	30.0	4 th	2	11	5.7	28,29
VI.	W. Khurm ElUyun	333.5	42.7	4 th	5	11	7.7	28,29
V.	W. Dara	672.5	51.5	5 th	4	7	5.7	28,29,42,43
VI.	W. Dip	539.0	53.5	4 th	4	12	7.3	42,43,44,48
VII.	W. Abu-Had	962.0	82.0	5 th	4	11	8.7	42,43,44,48,49

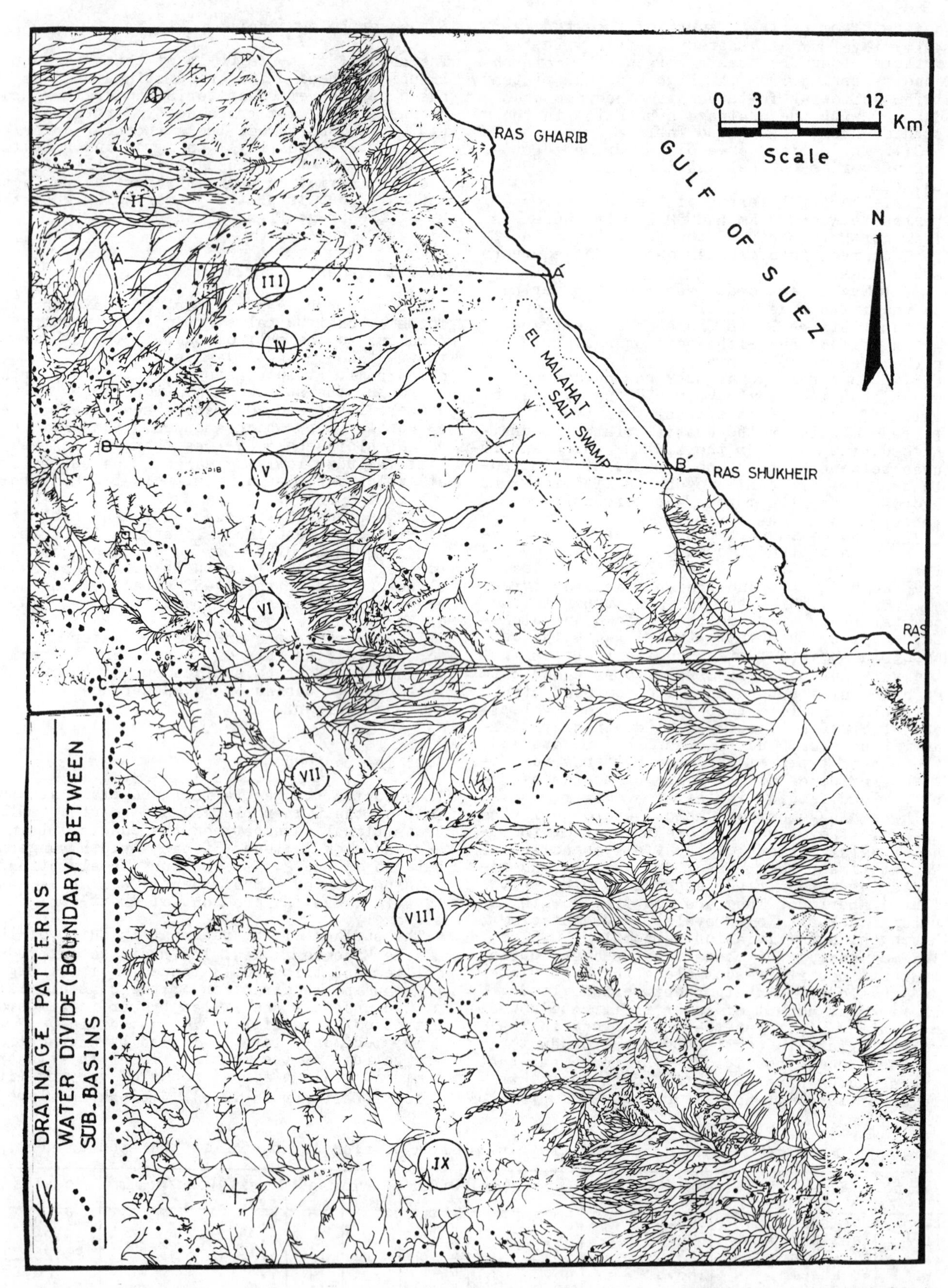

Figure 3. The compiled map from photo-mosaics covered the area. It indicates Nine sub-basins running from west to east. They are defined from north to south respectively as following: I. The northern sub-basin, II. Wadi Abu-Haad sub-basin, III. Wadi El-Darb sub-basin, IV. Wad i Khreim sub-basin, V. Wadi Um-Yasar sub-basin, VI. Wadi Khurm El-Uyun -Gharib sub-basin, VII. Wadi Dara sub-basin, VIII. Wadi Dip sub-basin, IX. Wadi Abu-Had sub-basin.

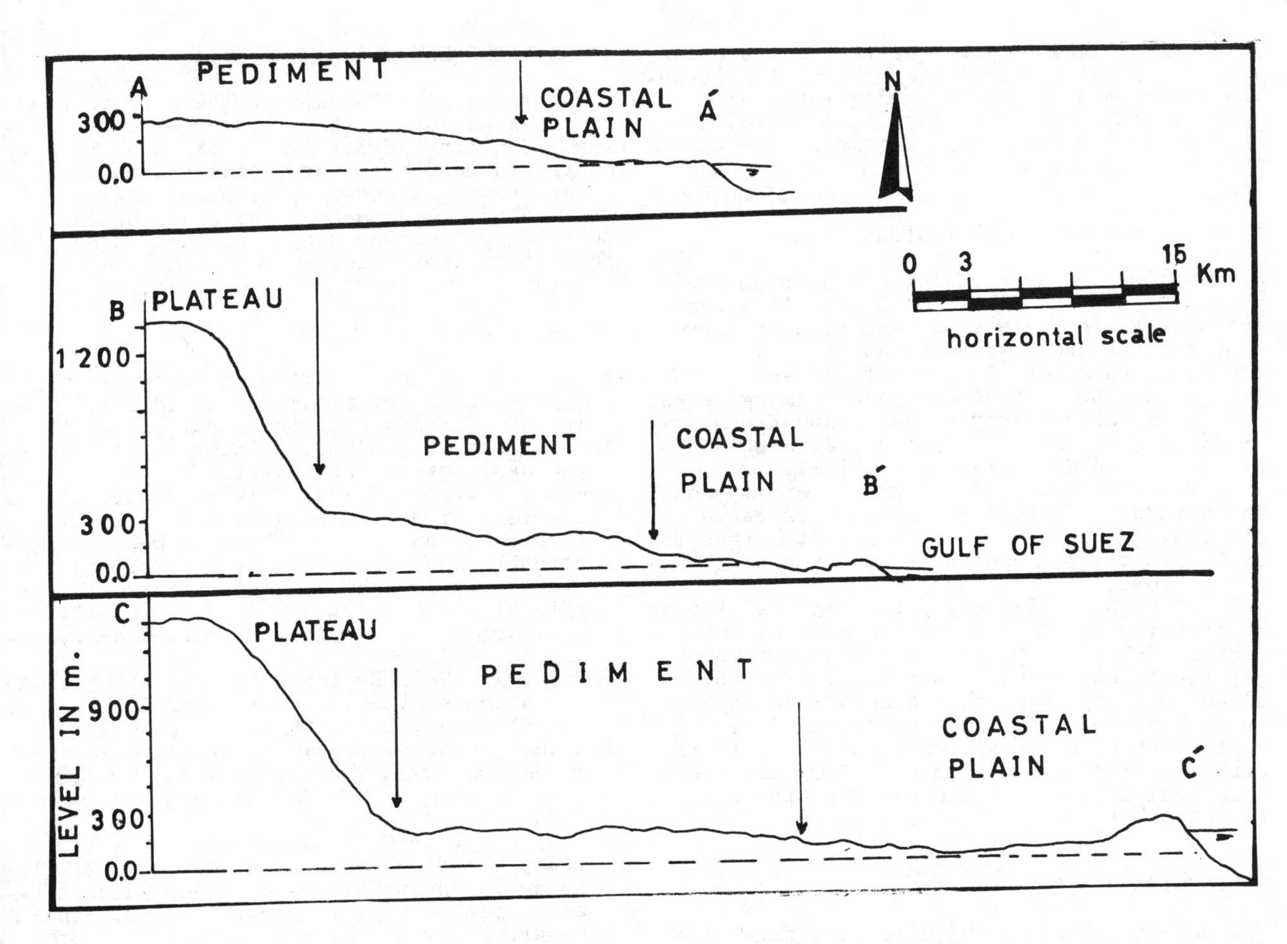

Figure 4. West-east traversed profiles along the area west of the gulf of Suez(refr.Fig. 3).

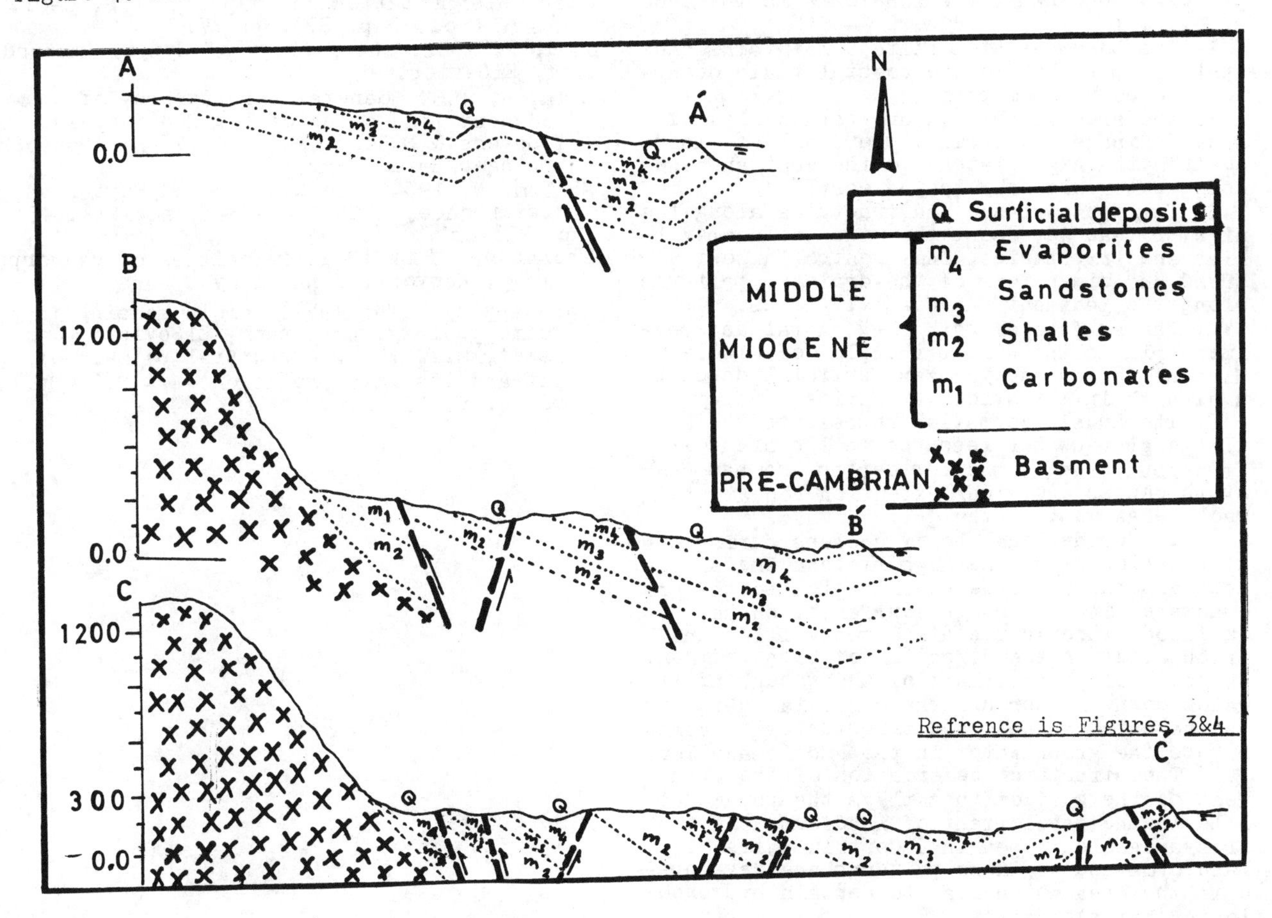

Figure 5. West-east projected geologic cross-sections along the area west of the gulf of Suez.

is distinguished between the pediment and the mountain front (Foot of the high plateau), while no break with the coastal plain.

The pediment plain is highly deformed. It is affected by faulting and folding trends in the gulf region.

3.3. The basement high plateau

It occupies the area of the Precambrian sheild in the Eastern Desert of Egypt. It possesses highly steep elevated land with an elevation reaches the maximum of about 1751m above the mean sea level. It is built of series of high mountains of basement rocks. Among these are, from north to south; Gebel Abu-Khashaba (+1461 m), G. Gharib (+1751 m), G. Riseis (+949 m), and G. Ghuweirb (+1359 m).

It represents the topest part where the water divide is delineated and th upstreams of all drainage basins are occured. It merges to east with the pediment plateau but with a break in slope.

It is highly affected by the regional trends of faulting and folding in the gulf of Suez region. The fault planes and fractures which represent weak planes, especially of NE-SW direction, have been turned into wide and deep vallies (maximum width reaches 2 Km).

It receives an occassionaly heavy rainfalls during winter period (Korany, 1980) and shedding them either to the coastal plain or to the Nile Valley.

CONCLUSION

The results obtained and discussed give rise to the following conclusion about the hydrogeologic bearing of the landforms in the area of study :

1. The basement high plateau maintains the upstream part, while the coastal plain occupies the downstream part.

2. The area is built,hydrographically, of nine drainage sub-basins, where all are issued from the high plateau to the west and shedding water almostly to the east.

3. The fault planes and fractures along the high plateau represent the wide and deep vallies and tributaries. They control also the trend and dimensions of the drainage patterns along the pediment and coastal plains.

4. The surface of both the coastal and pediment plains is built principley of Middle Miocene facies and younger surficial deposits of high infiltration capacities.

5. The Rudeis Formation represents the available groundwater resource in the area. It is encountered at shallow depths and has great thickness and wider extention beneath the whole area west of the gulf of Suez.

6. The occassional heavy showers during winter period along the high plateau and slopes maintain the possible contribution to the groundwater in the Rudeis aquifer.It causes peak floods through the dissected vallies and tributaries in the direction of both pediment and coastal plains. Part of the water infiltrates downward through the surficial deposits in the main channels and tributaries to contribute the groundwater in the Rudeis aquifer.

7. The structural deformation of the area west of the gulf of Suez plays the paramount role in the development of landforms and their hydrogeologic bearing. Certain units of landforms and drainage patterns are developed in the area which lead to certain hydrogeologic characteristics.

ACKNOWLEDGEMENT

The authors wish to express their gratitudes to the director of the Egyptian Geologic Survey and Mining Authority for the sincere help in providing the aerial photos and mosaics.

Thanks are extended to head and staff of the Geology Department of AinShams University Cairo,Egypt for the kindly help and providing laboratory facilities.

REFERENCES

Abdine, A.S. 1981. Egypt's petroleum geology, good grounds for optimism. World oil, Circle 109 in reader service card, p.99-112.

Easterbrook, D.J. 1969. Principles of geomorphology. Newyork,McGraw-Hill.

El-Etr, H.A. & M.S.Yousif 1978. Systematic analysis of drainage pattern of the Qift-Quseir region, central Eastern Desert, Egypt.Bull. Soc. Geogr. Egypte, p.25.

El-Tarabili, E. 1970. Contribution to the origin of Red Sea depression, Origin of its northern part. 7th Arab Petrol. Congr.,Kuwait.63, B-3.

Howard, A. 1967. Drainage analysis in geologic interpretation, a summation. Amer. Asso. Petrol. Geol. Bull. 24. 1, p. 2246-2259.

Iskandar, L.L. In preparation. Hydrogeology of Shagar area, Eastern Desert, Red Sea governerate, ARE. M.Sc. Thesis, AinShams Univ.

Korany, E.A. 1980. Peak-runoff calculations and preventing the risk of occassional flooding in Sannur drainage basin, Eastern Desert, Beni-Suef governerate, Egypt. 5th Congr. Stat. Cairo, p.505-534.

Ray, R.G. 1960. Aerial photographs in geologic interpretation and mapping. U.S. Geol. Surv. Prof. Pap. 373, p.227.

Said, R. 1962. The geology of Egypt. Amsterdam, ElSevier.

Said, R. 1969. General stratigraphy of the adjacent land areas of the Red Sea. In T.Degens & D.Ross (eds.), p.71-81. Newyork, Springer-Heidelberg.

Shalom, N. 1954. The Red sea and Erythrean disturbance. 19th Int. Geol. Algiers. 15, p. 223-231.

Thornbury, W.D. 1962. Principles of geomorphology. Newyork, John Wiley & Sons.

Verstappen, H.Th. 1977. Remote sensing in geomorphology. Amsterdam, ElSevier.

Youssef, M.I. 1968. Structural pattern of Egypt and its interpretation. A.A.P.G. Bull. 52, 4, p. 601-614.

Symposium on Remote Sensing for Resources Development and Environmental Management / Enschede / August 1986

Monitoring geomorphological processes in desert marginal environments using multitemporal satellite imagery

A.C.Millington & A.R.Jones
University of Reading, UK

N.Quarmby & J.R.G.Townshend
NERC Unit for Thematic Information Services, Reading, UK

ABSTRACT: Methods for geomorphological process monitoring and change detection using digitally processed multitemporal Landsat TM & MSS imagery are evaluated in south-central Tunisia. Three categories of changes are detected - sub-sampling unit, seasonal and long-term changes. Hydrological and geomorphological changes in Tunisian playas are examined within these categories. Both surface water and groundwater are important in determining salt and sediment budgets on these playas

RESUME: Les methodes geomorphologiques des processus de mesurement et de detection des changements en utilisant le traitement des images multitemporales Landsat TM and MSS sont evaluees au centre-sud de la Tunisie. Trois changements de categories sont detectes - unite sous-echantillon, saisoniere et changements a long terme Les changements hydrologiques et geomorphologiques dans les playas tunisiennes sont examines dans les memes categories. Les eaux de surface ainsi que celles sous terre sont importantes pour determiner la balance entre sel et sediment des playas

1 INTRODUCTION

The monitoring of sediment transfer processes in arid and semi-arid environments has many important applications for engineering problems (e.g. Cooke et al., 1982; Doornkamp et al., 1980) and land evaluation procedures (Mitchell, 1982; Purdie 1984). A closer examination of these problems focusses attention on the difficulty of monitoring sediment transfer processes to obtain information for resource assessment and environmental planning.

Geomorphological processes on the desert margin are characterised by:-

1. high-magnitude and low-frequency events,
2. a strong seasonality,
3. occurrences as spatially discrete and, often, uncorrelated events.

These characteristics make semi-arid geomorphological processes difficult to monitor using conventional ground-based instrumentation because there is a very low probability of measuring any geomorphological event in a specific locality using site-specific instrumentation. However the probability is increased if either the time or spatial dimensions are increased. The time dimension is, for all practical monitoring purposes, inflexible. However events can be monitored in a relatively short time period if a suitably large area is examined which will increase the number of sites where processes are active. The synoptic capability of satellite data provides large enough areas for this and makes the monitoring of sediment transfer processes, utilising change detection algorithms on multidate imagery, a distinct possibility.

This methodology is currently being evaluated in south-central Tunisia using Landsat MSS and TM data acquired between 1981 and 1985. Attention is being focussed on geomorphological change in three major process-domains - alluvial fans, braided river systems and playas. This paper presents the first observations from this project and concentrates on playa environments.

2 BACKGROUND TO THE STUDY AREA

2.1. Geomorphological Processes

Water plays a central role in the transfer of sedimentary materials in south-central Tunisia although wind becomes an increasingly important agent further south. The main components of the fluvial system on the desert margin are: i) Catchment slopes and pediments ii) River channels iii) Alluvial fans iv) Enclosed depressions containing playas (known locally as Chotts).

Over time materials are transferred from mountain slopes by slope erosion, river channels and alluvial fans. Finer textured material is deposited in the playas, whilst the coarse material is left in the channels and fans and on the slope as lag deposits. It is clear that this transfer of sediment is a far from continuous process and episodic erosion and deposition prevail often with long periods during which materials are stored in the components of the fluvial system.

Aeolian activity is important in the dry season when gypsiferous sands are deflated from the playa surfaces, and redistributed on the surrounding landforms. The wind-blown material deposited in channels is then flushed through the system during the next spate of fluvial activity. Sandy materials are also blown into the area from the south in the late winter and early spring.

2.2. Geology

The study area is dominated by Cretaceous and Tertiary sediments which have been subjected to Alpine folding, (Burollet, 1967; Coque and Jauzien 1967). The area is dominated by a series of eroded anticlines and synclines which give rise to many resistant dolomite and limestone cuestas with intervening valleys formed by the differential weathering and erosion of marls and clays which are now filled with late Tertiary and Quaternary sediment.

2.3. Climatology

Rainfall in the region is concentrated in the autumn, winter and spring seasons. Both temperatures and evaporation increase markedly in May and remain high until late September. Consequently runoff events are restricted to the wet season and streams are dry throughout the summer as there is no base flow contribution and soil moisture deficits are high. The climatological pattern is slightly different in the south of the study area, where rainfall totals are lower and consequently runoff events are rarer, and in the north where it is slightly wetter.

3. SATELLITE DATA AND DESERT GEOMORPHOLOGY

3.1. Previous work

Satellite data have been used for geomorphological investigations in arid and semi-arid areas by several workers and most applications have involved landform mapping (eg. Mitchell et al., 1982; Sunha & Venkatachalam, 1982 and Van Steen, 1982) and surficial material survey (eg. Asem et al., 1982; Bird et al., 1982; Davis et al., 1982; Gladwell, 1982; Hamza et al., 1982; McCord et al., 1982; Sunha & Venkatachalam, 1982 and Townshend & Hancock, 1981). Few workers have attempted to monitor geomorphological change although Graetz & Pech (1982) and Klemas & Abdel-Kader (1982) have measured river channel changes and flooding in arid and semi-arid environments.

Three problems are apparent in these previous studies:

1. They have been limited by the relatively coarse spatial resolution of MSS sensors.
2. They have been limited by the restricted spectral resolution of MSS data. The inclusion of middle IR bands (1550-1750 and 2080-2850nm) on the TM has greatly enhanced the possibility of discriminating between surficial materials, (Bodechetel, 1983; Gladwell, 1982; Hunt, 1980; Kahle, 1984). This enhanced power of surficial material discrimination is crucial to any interpretation of sediment dynamics.
3. The few geomorphological monitoring studies that have been undertaken using Landsat data have been severely restricted by image availability. Archival material has been compared with current imagery (Klemas & Abdel-Kadar, 1982) but change detection utilising concurrent image interpretation and ground verification has been far less satisfactory (Graetz & Pech, 1982) because of the costs involved in data availability, acquisition, ground station receiving policies and atmospheric conditions.

3.2. Change detection

Jones (1986a, b) has thoroughly evaluated the potential of digitally processed TM imagery for geomorphological mapping in this area of Tunisia. Whilst this research shows the applications that can be made using single date imagery, monitoring geomorphological change using digital imagery requires the use of a multidate imagery and different change-detection algorithms.

In this project image data was supplied as CCT's and analysed digitally using a I2S Model 75 image processor. Change detection procedures involve either a multidate or a post-classification comparison approach. A multidate approach combines the two unprocessed images to produce one output data set. A post-classification comparison approach involves an initial supervised or unsupervised classification of the two images. In this study the multitemporal approach to change detection was preferred because quantitative comparisons between the two techniques have shown it to be more accurate (Singh, 1984).

Any change detection study involves scene-to-scene image registration to ensure that pixels correspond to the same ground locations in each image. Six ground control points were used to co-register the two TM quadrant images used in this study with an average RMS erros of ±0.37 pixels, compared to twenty seven ground control points used to co-register the anniversary 512 x 512 pixel MSS images. This confirms the expected improved geometric fidelity of the TM compared to the MSS. The ground control points chosen in the co-registration procedure were permanent features, in the landscape such as road junctions and road/rail crossings.

Any misregistration of the imagery will produce errors in the change detection output images since boundary pixels corresponding to one surface cover type may be compared with boundary pixels of the adjacent cover type, resulting in spurious changes being detected later. In order to remove these possible edge effects a median filter, with a 3 x 3 pixel square kernel, was passed over all the images used in this study before the change detection algorithms were applied to the data.

Image differencing, image ratioing and principle components analysis were found to be the most meaningful change detection algorithms. Vegetation indices (Howarth and Boasson, 1983; Singh, 1984) and the ratio differences technique were unsuccessful in detecting change. Possibly vegetation changes in semi-arid environments are too subtle to be detected on the imagery despite the fact that in some environments, for instance on playa margins, there are geomorphologically significant vegetation changes. The most useful spectral bands were MSS Band 7 (800-1100nm) and TM Bands 3 (630-690nm) and 7 (2080-2250nm).

Differenced images were produced by subtracting the median filtered image for the first date from that for the second date, and adding a constant to ensure that the output values were positive. Ratio images were produced by dividing the image for the first date by that for the second date. For the principal component analysis, the two images being compared were treated as one date set. In the analysis, Bands 4 (500-600nm) 5 (600-700nm) and 7 (800-1100nm) were used for each MSS image, and Bands 4 (760-900nm), 5 (1550-1750nm) and 7 (2080-2250nm) were used for each TM image. Previous research into the use of principal component analysis in change detection indicates that gross differences due to overall radiation and atmospheric changes are contained in principal component 1, and that statistically minor changes associated with local changes in land cover appear in the minor component images (Byrne et al., 1980; Lodwick et al., 1979; Richardson and Milne, 1983). Consequently principal components 2 and 3 were used to detect geomorphological change in this study.

Thresholds, chosen on the basis of previous research (Nelson, 1983; Singh, 1984), were applied to all change detection output images at $\pm 1\sigma$ from the mean value. FCC images were produced by assigning the changes which corresponded to pixels with values of $<-1\sigma$ to the red gun of a colour monitor, those $>+1\sigma$ to the blue gun, and either the MSS Band 7 or TM Band 7 image to the green gun in order to preserve spatial detail.

4. GEOMORPHOLOGICAL MONITORING AND CHANGE DETECTION IN PLAYA ENVIRONMENTS

4.1. Environmental setting of study playas.

Three playas have been intensively studied using satellite imagery and ground observations in this project.

The Chotts el Djerid and el Fedjadj occur to the south of the Atlas Mountains in a zone of subsidence and they form part of a series of playas stretching from southern Tunisia to central Algeria . Chott el Djerid is the largest of these playas (5360km^2), it has an elongated north eastern arm, the Chott el Fedjadj, which continues eastwards into the Sebkhet el Hamma (770km^2). Their geomorphology and hydrology have been the subject of a number of investigations (Coque, 1962; Coque & Jauzien, 1967).

Chott el Guettar (75km^2 in area) is situated in an enclosed basin about 20km SE of Gafsa. It is bounded to the north by the Djebel Orbata and the south by the Djebel Berba; both of these mountain ranges have active fans encroaching on the chott. Low ground and smaller mountains and hills are found to the east and west of the chott.

4.2. Geomorphological changes on playas

Geomorphological changes have been detected on these playas from satellite imagery and ground survey, and they can be attributed to the seasonality of the hydrological regimes, surface salt concentrations, interactions with adjacent landforms and aeolian activity. These factors support the existance of seasonal patterns of climatically-controlled geomorphological change. Other geomorphological phenomena respond to fluctuations with either shorter or longer periodicities than the seasonal patterns and are not necessarily related to climatic fluctuations. If the temporal and spatial parameters of the geomorphological changes on playas are combined they can then be classified into 3 groups (Table 1). This subdivision is of fundamental importance because it recognises that specific types of change can be observed with different types of satellite data and different intervals between image acquisition as distinct from those that can only be observed by other techniques.

Seasonal changes are readily detected by change detection algorithms applied to satellite imagery from the different seasons. Imagery corresponding to wet and dry conditions have been compared for all three playas in this study and four types of geomorphological changes in parameters affecting the geomorphology of the playas have been identified -

1. - surface moisture (standing water and subsurface moisture)
2. - surface texture and composition
3. - vegetation cover
4. - aeolian activity

Seasonal patterns related to surface moisture and vegetation are best illustrated by examining change detection images between winter 1983 and summer 1985 TM imagery of the Chott el Guettar (Fig. 1). The greatest changes are those related to the fan delta at the end of the eastern fork of the Gafsa Fan and the adjacent chott. The fan delta is relatively well vegetated with Crassula spp, Limoniastium gyonianum and Limonium spp, and slopes very gently towards the chott without any obvious distal channel. Subdued sparsely vegetated topographic depressions act as surface water collecting areas which drain into rills, these feed into channels of about 2m depth and 50-100m width which drain into the chott. The individual channels and their restriction to the distal end of the fan delta can be clearly seen (Fig. 1). The adjacent chott surface is moist throughout the year and has a similar, but less dense, vegetation than fan delta.

The change detection imagery (Fig. 1) indicates higher middle IR absorption on the fan in winter than summer. This can be explained by a combination of higher levels of soil and foliar moisture. The channels and adjacent chott areas are characterised by higher reflectance in the summer than the winter. The high summer middle IR absorption is related to higher soil moisture levels found in these areas, a response to the relatively high groundwater seepage through the Gafsa Fan due to irrigated agriculture. The higher winter reflectance values appear paradoxical considering the increased surface flow but are probably related to the fact that as flow is intermittent, water evaporates between flow events leaving surface salt effloresences. Combined field and satellite data suggests that surface water gathers on the fan delta and flows into the eastern part of the chott in winter and that in the summer the area is dominated by high levels of groundwater seepage. The area influenced by winter surface water and high summer moisture forms a roughly semi-circular area adjacent to the fan with a lobe to the south. Similar patterns in winter surface water contributions have been seen on MSS change detection imagery.

The other areas contributing surface water appear to be far less important. Increased winter surface or subsurface flow along the Oued el Rahr and Oued es Sedd systems to the east and northeast of the chott respectively is visible (Fig. 1). However, the chott area affected by water from the east is far less extensive than the west and according to the imagery is more variable in extent. Water flowing onto, or seeping through, the alluvial fans to the north shows little seasonal variation on the imagery. This is probably due to a lack of surface flow in the years examined, the dominance of a relatively constant groundwater seepage or the 'filtering' effect of the El Guettar oasis. Most of the fans to the south of the chott also show a similar effect despite the lack of a similar vegetation 'filter'. However substantial changes in moisture levels in the south-east chott were seen on 1981 change detection imagery (Fig. 2) which can clearly be linked to surface water flow in fan-channels.

As there is little evidence of any kind of aeolian activity groundwater seepage and surface water control the transport of salts and sediments onto chott and their redistribution within the chott. The main seasonal changes in surface water and near-surface moisture are related to winter runoff from the eastern fan delta, and to a lesser, extent the channels draining Guettaria. A less frequent source of surface water is from the southern alluvial fans. Surface runoff acts as the main transport mechanism for sediment movement into the chott. This is particularly important in the areas adjacent to the southern fans. In addition there is an area of active gullying on fine-grained old chott sediments to the south-west of the chott. These gullys drain into the chott and act as a further sediment source area. They can be distinguished on the imagery as a cluster of small patches of high winter middle IR reflectance (Fig. 1). The sediment moving into the chott from the south forms a south-central depositional wedge which is recognisable in all change detection images as an area of lower summer Band 7 reflectance (Figs. 1 & 2). This is due to the increased water holding capacity of the sediments resulting in high soil and foliar moisture levels in winter. Salt transport, unlike sediment transport, is related to both surface water and groundwater fluxes and it is likely to mirror the hydrological regimes more strongly than the sedimentation patterns.

Seasonal changes in surface texture and composition are most noticeable on imagery from the Chott el Djerid. These are particularly well developed in the north-central Chott el Djerid where the Fatnassa-Deajache road traverses it (Fig 3). Mitchell (1982) has divided the Chott el Djerid into 7 surface detail classes and 4 of these occur on the area of the chott traversed by the road:-

1. Wind sculpted, hummocky terrain
2. Thick salt crusts
3. Aioun (standing water and salt efflorescences)
4. Blistered thin crust with polygonal ridging

Seasonal change is greatest in the area of blistered thin crust with polygonal ridging. In winter much of the area of covered by shallow standing water which by late spring exhibits a relatively featureless thin crust. As evaporation increases through the summer polygonal ridging and blistering develop. The areas of thick salt crusts show far less variation and the main differences relate to water on the surface in winter and the thrusting of the thick salt crusts during the summer. The area of aioun shows little variation as well and the circular patterns of groundwater associated with it (Jones & Millington, 1986) can be seen in both the wet and dry seasons. All three of the above areas however show one major change between the wet and dry seasons. In the winter wet season many of the areas exhibit standing water and upon drying out a relatively clean salt crust develops. However, in the summer, dry season much silty and sandy material is blown onto the salt crusts. This gives them a thin veneer of brown material and affects the amount of reflectance in all bands. The wind sculpted and hummocky facies show the least variation between seasons of all of the zones. The only noticeable difference which will affect reflectance patterns is the slight increase in surface moisture during the winter.

Longer term changes can be identified from anniversary change detection images which compare images taken at similar times but in different years. Satellite orbits and problems of image acquisition means that it is almost impossible for imagery of the same day and week to be acquired. Imagery acquired during the same month is feasible especially in the dry season at the desert margin but it is more difficult to obtain in the wet season because of increased cloud cover.

Changes detected on anniversary images should ideally indicate longer-term changes as well as areas of little change indicating areas of long-term stability. However there are two problems that need to be considered in interpretating geomorphological change in anniversary images in desert margin environments. Firstly, the erratic nature of wet season rainfall and runoff changes in fluvial and lacustrine geomorphological phenomena may be attributable to rainfall and runoff fluctuations within a season. Consequently the changes may be seasonal rather than long-term. Secondly, intrinsic geomorphological changes may be detected which although representing a long-term change in the system do not indicate long-term instability as would be the case with extrinsically-controlled changes.

Examples of longer term changes on the Chotts el Djerid and el Fedjadj can be divided into two groups. Firstly, processes which operate each year creating an annual incremental adjustment of the system. These can be subdivided into chott marginal processes, strongly influenced by geomorphological and hydrological processes in the areas adjacent to the chotts, and processes operating in the centre of the chotts, which are mainly artefacts of internal adjustment of the water-salt-sediment balance. The second group of processes occur less frequently and fall into the high magnitude - low frequency category. These are most likely to be linked to catastrophic events such as storms with very long return periods and tectonic activity. Erosion and sedimentation are responsible for long-term geomorphological changes on the chott margin. Chott margin erosion occurs in two ways. Firstly, by cliff retreat at the junction of the playa and fan sediments (eg. the fan delta and Chott el Guettar) or at the junction of old remants or terraces of playa sediments and the contemporaneous playa surface. Secondly, between the bare playa surface and the vegetated chott margins where runoff from adjacent slopes erodes into the vegetated margins forming a series of runnels which enlarge into embayments.

Other marginal facies are dominated by sedimentation related to either the influx of sheetwashed material from actively eroding adjacent areas or debris transported out of alluvial fans. The first situation is commonly found on the Chott el Fedjadj, and in places occurs in association with marginal erosion. Deposition of alluvial fan material is more commonly found in the southern Chott el Guettar and the northern Chott el Djerid (Fig. 4).

Marginal processes operating each wet season, result in small rates of annual geomorphological change. Consequently, detection of the retreating vegetated-bare playa boundary and chott marginal cliffs is unlikely to be seen on remotely sensed imagery unless the interval between image acquisition is relatively large.

Sedimentation from alluvial fans onto the chott is dependent on discharge events with greater return periods than sheetwash. Consequently, they fall into the high magnitude - low frequency event category and the resultant sedimentation can easily be seen on remotely sensed imagery (Fig. 4)

In the more central facies of the larger chotts in Tunisia geomorphological redistribution of material on the chotts related to surface water movement and groundwater seepage can be identified on remotely sensed imagery. Examples can be seen in the Chotts el Fedjadj and el Djerid. (Fig 5 & 6,)

Winter runoff gathers on the Sebkhet el Hamma each wet season and then flows westwards into the Chott el Fedjadj along very low gradients. Well developed geomorphological features related to these flow patterns are found in areas where the playa is narrow. In Sebkhet el Hamma an inland delta to the north and a large, mobile sand body to the south constrict the playa (Fig. 5). Here it is about 2.5km wide and, for the most part, sparsely vegetated. The central Chott el Fedjadj is constricted by a 2m high cliff to the north and the gypsiferous Djebel Klikr, to the south. Here the playa is less than 2km wide and mainly unvegetated. Field observations made in May 1986 found evidence of recent water flow such as eroded vegetation hummocks, scoured channels and ripple marks. These flow features can be seen on the imagery (Fig 5) and represent zones of relatively fast water flow for a playa environment and are areas of actively eroding playa sediments.

On the Chott el Djerid, flow- and splay-like features are evident in thin salt crust areas. The trends evident in the salt crusts suggest they were formed by salt-rich water flowing into a topographic depression or an area with a lower water table. In some of the flow lines, which are up to 0.5km wide, darker channels can be seen suggesting more than one flow event. Further evidence to support a multiple flow hypothesis can be seen at the splayed ends of the features where there is evidence of the uppermost crusts overlapping other crusts, or salt-rich zones, beneath them.

These flow features start to the south of the embanked road (Fig. 6) and measurements of the water table carried out in 1984 showed that this embankment had a marked affect on the water table up to 2km away. It is likely that these features represent flow of surface water which has dammed up to the north of the road and then reappears after flowing under the embankment. It then continues to flow southwards to the depression or low water table area.

5. CONCLUSIONS

Evidence from southern Tunisia has shown that seasonal and long-term changes in playa geomorphological phenomena can be detected using digitally processed Landsat and SPOT imagery. However, when examining such changes it is important to establish a framework of the temporal and spatial dimensions of change for comparison with changes detected on different types of imagery.

Three classes of geomorphological changes were identified with reference to satellite data - subsampling unit changes, seasonal changes and long-term

Table 1. Geomorphological change detection categories on Tunisian chotts using remotely sensed imagery

Sub-sampling unit changes (cannot be detected on imagery as less than spatial and/or temporal resolution)

Seasonal changes (can be detected on multidate imagery corresponding to different sensors)

- surface moisture
- surface texture
- surface chemistry
- vegetation cover
- aeolian activity

Longer-term changes (can be detected on anniversary imagery)

- in marginal facies
  - erosion
  - sedimentation
- in central facies
  - flow features

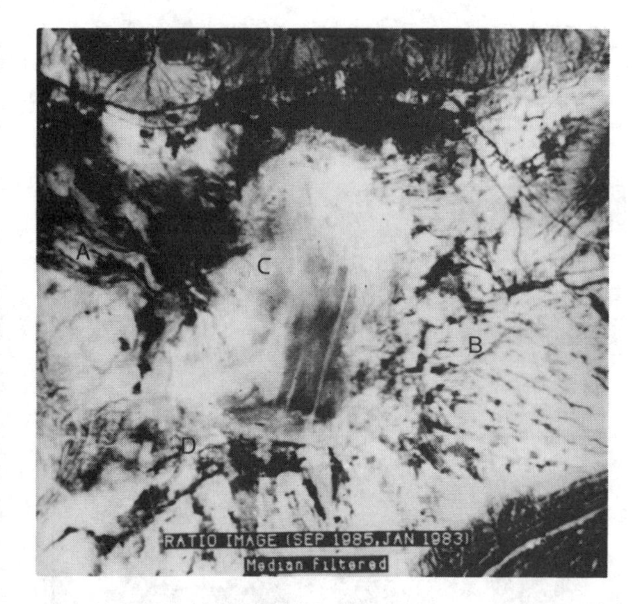

Figure 1. Change detection image of Chott el Guettar produced by ratioing TM Band 7 imagery of January 1983 with September 1985. El Guettar oasis is indicated (EG). The areas of highest surface moisture change can be readily identified - the areas of winter surface runoff on the fan delta (a) and from Guettaria (b) and the area of high summer moisture levels on the chott related to groundwater discharge from the Gafsa Fan (c). The gullied areas from the southwest can also be seen (d).

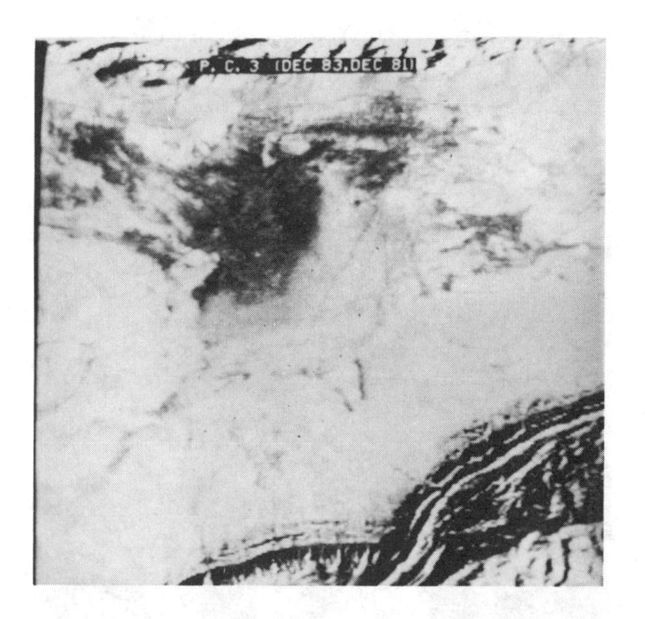

Figure 2. A third principal component image of Chott El Guettar MSS data from the wet and dry seasons of 1981. Areas with high levels of moisture change can be seen in one of the fans to the south of the chott and south-central chott. This area of change is related to a runoff event on the fans which also eroded sediments from the fans and deposited them on the chott.

changes. The latter two types of changes can be seen on both MSS and TM imagery. This suggests that the changes are large enough to be detected even with coarser spatial resolution MSS imagery. However the spectral resolving power provided by TM imagery does enable changes in salt and sediment patterns and moisture levels to be more readily recognised. Whereas on MSS imagery many of the changes that can be seen are related to large variations in overall reflectance and vegetation.

The important role of surface water in redistributing sediment and salts on Tunisian playas is evident in this study. In many facies it appears to be equally if not more important than groundwater. Furthermore, interactions with adjacent landforms and the role of aeolian activity are also locally important. The extent and magnitude of the contribution of these factors in the geomorphological evolution of playas needs to be monitored using remotely-sensed data backed up by ground observations wherever possible.

ACKNOWLEDGEMENTS

This work has been carried out as part of a NASA Principal Investigatorship under the Scientific Applications of TM Program to two of the authors (ACM & JRGT). Part of the work was also carried out as part of a NERC Studentship (GT4/83/GS/87) to another of the authors (ARJ).

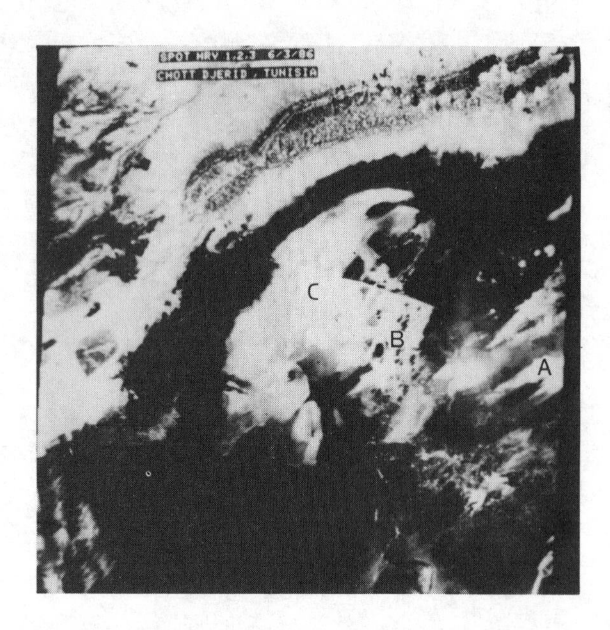

Figure 3. SPOT-HRV FCC of the northern Chott el Djerid, the road can be seen traversing the chott. The areas of aeolian activity (a), aioun (b) and salt crust (c) are easily identifiable.

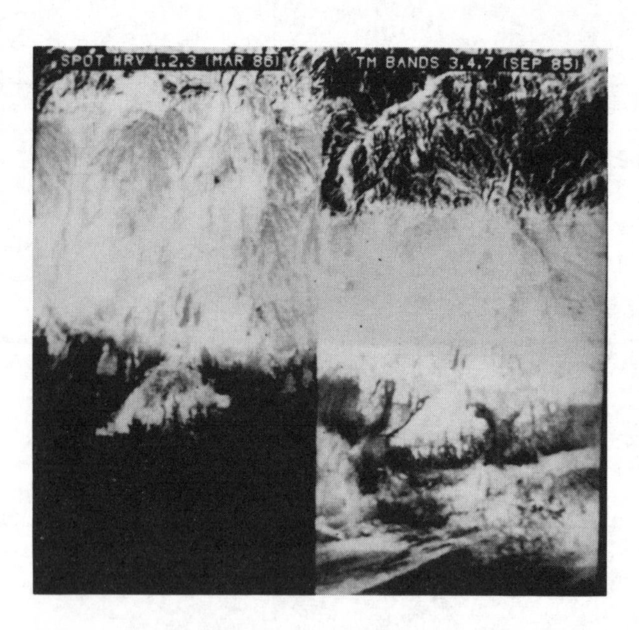

Figure 4. Landsat TM (right) and SPOT-HRV (left) FCC's from 1983 and 1985 respectively showing sedimentation of alluvial fan material onto the northern chott el Djerid from a fan flanking the Djebel el Asker. There has been little change in the small sediment fan in the time between image acquisition.

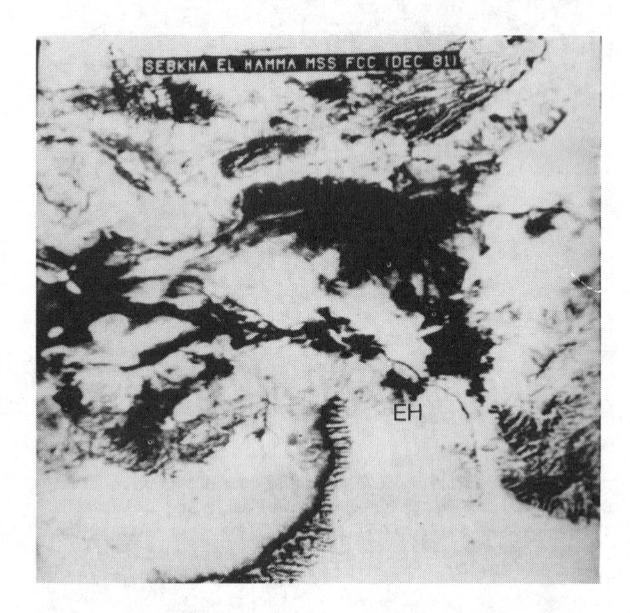

Figure 5. Landsat MSS FCC of the Sebkhet el Hamma from 1981. The detail in the chott has been increased by contrast stretching the chott at the expense of the surrounding areas but El Hamma (EA) can still be seen. Vegetation and salt patterns relating to surface water flow from the surrounding uplands is seen funnelling through the neck of the sebkhet as it enters the Chott El Fedjadj to the west.

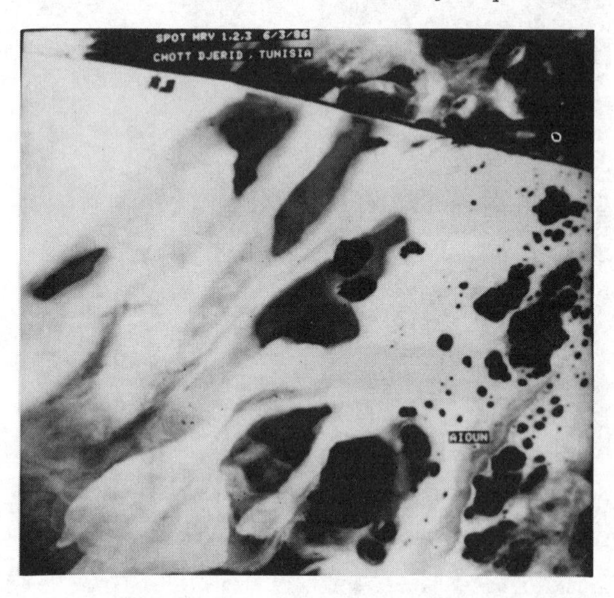

Figure 6. SPOT-HRV FCC image of part of the Chott el Djerid, March 1986. Two features are noticeable (i) the variable size of the aioun in the east of the image and (ii) the multiple flow-like features in the thin salt crusts, with cross-cutting relationships in the south-west, due to surface water flow.

REFERENCES

Asem, A., Khalaf, F., Attusi, S. & Palou, F. 1982 Classification of surface sediments in Kuwait using Landsat data. Proc. 1st Thematic Conf. of Remote Sensing of Arid and Semi Arid Lands Cairo, 1057-64

Bird, A.C., Williams, T.H., Barrett, M.E., Munday, T.J. and Townshend, J.R.G. 1982 The Imperial College multi-channel electronic image classifier and its applications to the classification of surface types by multispectral analysis Proc. 1st Thematic Conf. on Remote Sensing of Arid and Lands, Cairo, 665-675

Bodechtel, J. 1983 Requirements for spaceborne remote sensing in geology. Remote Sensing: New satellite systems and potential applications, ESA SP-205, Albach: Austria, 53-57

Burrollet P.1972 General Geology of Tunisia in L. Martin (ed) Guidebook to the Geology and History of Tunisia, Soc. of Libyan Petroleum Geologists, Handbook 3, Tripoli, 51-58

Byrne, G.F., Crapper, P.F. and Mayo, K.K. 1980 . Monitoring land cover by Principal Components Analysis of Multitemporal Landsat Data. Remote Sensing Env. 10, 175-184.

Cooke, R.U.; Doornkamp, J.C.; Brunsden, D. & Jones, D.K.C. 1982 Urban Geomorphology in Drylands, Oxford Univ/UN Univ. Presses, Oxford.

Coque, R. 1969 La Tunisie Pre-Sharienne: Etude Geomorphologique Colin, Paris.

Davies, P.A., Grolier, M.J., Schultejann, P.A. & Eliason, P.T. 1982 Discrimination of Phosphate, Gypsum, Limestone, Halide and Quartz-sand deposits in south-central Tunisia by cluster analysis of Landsat Multispectral data Proc. 1st Thematic Conf. of Remote Sensing in Arid and Semi Arid Lands, Cairo, 337-360

Doornkamp, J.C. et al. 1980 Geology, Geomorphology and Pedology of Bahrain Geobooks, Norwich

Gladwell, D.R. 1982 Application of reflectance spectrometry to clay mineral determination in geological materials using portable radiometers Proc. 2nd Thematic Conf. of Remote Sensing for Exploration Geology, Fort Worth, 29-38

Graetz, R.D. and Pech 1982 The utility of Landsat for monitoring the ephemeral water and herbage resources of arid lands: An example of rangeland management in the Channel Country of Australia Proc. 1st Thematic Conf. of Remote Sensing of Arid and Semi-Arid Lands, Egypt, 1031-1046

Hamza A., Mami, A. & Sadowski, F. 1982 Land use Mapping from Landsat Imagery applied to central Tunisia Proc. 1st Thematic Conf. of Remote Sensing of Arid and Semi-Arid Lands Cairo, 1099-1112

Howarth, P.J. and Boasson, E. 1983 Landsat digital enhancements for change detection in urban environment. Remote Sensing Env. 13., 149-160

Hunt, G.R. 1980 Electromagnetic Radiation: The Communication Link in Remote Sensing In Remote Sensing in Geology (Eds, B.S. Siegal & A.R. Gillespie) John Wiley New York.

Kahle, A. 1984 Measuring spectra of arid lands in F. El-Baz (ed) Deserts and Arid Lands Martinus Nijhoff, The Hague.

Klemas, V. and Abdel-Kader, A.M.F. 1982 Remote Sensing of Coastal Processes with emphasis on the Nile Delta Proc. 1st Thematic Conf. of Remote Sensing of Arid and Semi-Arid Lands Cairo 389-416

Lodwick, G.D. 1979 Measuring ecological changes in multitemporal Landsat data using principal components. Proc. 13th Int. Symp. Remote Sensing Env., Ann Arbor, Michigan 1131-1142

Jones, A.R. 1986a An evaluation of satellite Thematic Mapper Imagery for Geomorphological Mapping in Arid and Semi-Arid Environments Proc. 1st Geom. Conference (Manchester) John Wiley (in press)

Jones, A.R. 1986b The Use of Thematic Mapper Imagery for Geomorphological Mapping in Arid and Semi-Arid Environments (This proceedings)

Jones, A.R. & Millington, A.C. 1986 Spring mound and aioun mapping from Landsat TM imagery in south-central Tunisia (This proceedings)

McCord, T.B., Clark, R.N., Melroy, A. Singer, R.B., Adams, J.B. and El-Baz, F. 1982 An example of the application of a procedure for determining the extent of erosional and depositional features, and rock and soil units in the Kharga Oasis Region, Egypt, using Remote Sensing Proc. 1st Thematic Conf. of Remote Sensing of Arid and Semi-Arid Lands, Cairo 909-920.

Millington A.C. & Townshend, J.R.G. 1986 The Potential of Satellite Remote Sensing data for Geomorphological Investigations: an overview. In Proc. 1st Int. Geomorphological Conference, Manchester, England Sept. 1985 (in press)

Mitchell, C.W. 1983 The soils of the Sahara with special reference to the Mahgreb Mahgreb Review 8, 29-37

Mitchell, C.W., Howard, J.A. and Mainguet, M.M. 1982 Soil degradation mapping from Landsat in North Africa and the Middle East Proc. 1st Thematic Conf. of Remote Sensing for Arid and Semi-Arid Lands, Cairo, 899-908

Nelson, R.F. (1983) Detecting forest canopy change due to insect activity using Landsat MSS. Photog. Eng. & Rem Sensing, 49, 1303-1314.

Purdie, R. 1984 Land Systems of the Simpson Desert Region CSIRO, Nat. Resources Ser. 2.

Richardson, J.A. and Milne, A.K. 1983 Mapping fire burns and vegetation regeneration using principal components analysis. Proc. IGARSS, San Francisco, California 51-56

Singh, A. 1984 Tropical forest monitoring using digital Landsat data in northeastern India Unpub. Ph.D thesis, Univ. of Reading.

Sinha, A.K. & Venkatachalam, P. 1982 Landsat spectral signatures studies with soil association and vegetation Proc. 1st. Thematic Conf. of Remote Sensing of Arid and Semi-Arid Lands, Cairo, 813-822

Townshend, J.R.G. and Hancock, P. 1981 The use of remote sensing in mapping surficial materials. In: Terrain Analysis and Remote Sensing, (ed) Townshend, J.R.G. George Allen & Unwin, London

Van Steen, L.A. 1982 Landsat Data in the Sahel: Their use and accuracy for small-scale soil surveys and their time and efficiency Proc. 1st Thematic Conf., of Arid and Semi-Arid Lands, Cairo, 91-110

Symposium on Remote Sensing for Resources Development and Environmental Management / Enschede / August 1986

Remote sensing assessment of environmental impacts caused by phosphat industry destructive influence

S.C.Mularz
University of Mining and Metallurgy, Poland

ABSTRACT: A number of remote sensing techniques, such as colour aerial photography, black and white aerial photography and thermal imaging have been used to detect adverse environmental impacts associated with the location of the phosphogypsum dump area. As an effect of processing and interpreting remotly sensed data, assessments of water pollution, solid waste disposal impacts on toe-failure deformations zones and the degree of vegetation cover damages, have been specified. Results of experiments indicate that remote sensing methods are an irreplaceable tool to solve environmental monitoring, as well as planning problems. Wide-area repetitive coverage by remote sensors provides information which is not readily available by conventional measuring techniques.

1 INTRODUCTION

In a wet-process of phosphoric acid production from apatite and phosphorite, phosphogypsum is generated as the major by-product. In such technology large quantities of waste are produced (as a rule, 5.5 tons of phosphogypsum are produced for each ton of phosphoric acid produced).

Phosphogypsum is composed mainly of gypsum ($CaSO_4 . 2H_2O$), which makes 93-95% of total waste mass. The remaining 5-7% is composed od different admixtures like: noncombined molecules of phosphoric acid and phosphates (together 0.4 - 1.5% of P_2O_5), siliceous acid, fluorine compounds, organic matter and also some radioactive substances. Phosphogypsum is very burdensome substance to the natural environment, and causes its degradation because of harmful contamination. Therefore, it is necessary to study and to monitor environmental changes occuring in phosphogypsum dumping areas.

Remote sensing techniques are very useful for this purpose, as they make it possible to record the complex data, which is necessary for a thourough interpretation of natural conditions. They also make a qualitative and quantitative analysis of phenomenas and processes that happen on dumping and adjacent regions, possible.

The results of investigations, based on remote sensing techniques, of a phosphogypsum dumping area of a chemical factory, near Szczecin, are presented in the paper. Presently, 3.5 miliion tons of waste are acumullated here annually, which accounts for 70% of the phosphogypsum generated in Poland.

The main purpose of the investigation was to asses the usefulness of remote sensing techniques for:
- the creation of a complex environmental monitoring system for the dumping area,
- periodically mapping the progress in dumping operations,
- the assessement of the hazard for staff, heaping machines and constructions, which are located close to the dump area.

The objective of the research reported here was to also examine the environmental damages in the dump influence zone.

2 STUDY AREA, PROBLEM DEFINITION AND DATA

Geographically, the dumping area is located near the Odra river estuary, north of Szczecin (Figure 1).

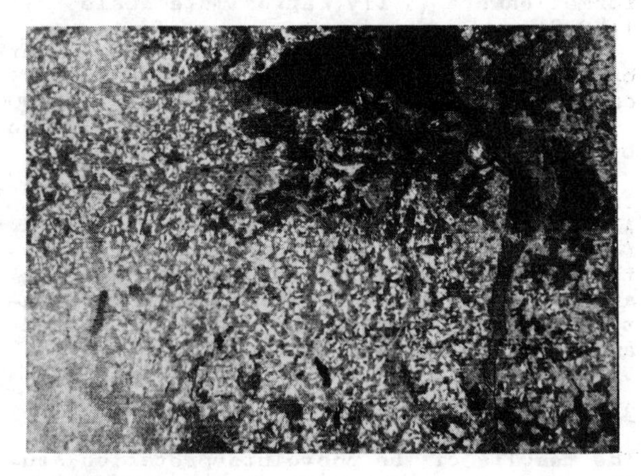

Figure 1. Location of the study area on Landsat MSS (Band 4) image portion.

Morphologically, it lays on the flooding terrace of the Odra river. Originally the terrain was flat, with an average height of 1 m below sea level. Considering the geological conditions, it was found that the dump subgrade is composed of holocene deposits, which are organic in the upper part (peats, organic outwashes) and, in the lower part, consist of sand and gravel layers. They are underlaid by older pleistocene clays, sometimes interbeded with sands. The holocene deposits are 8-11 meters thick, locally up to 15 meters. The pleistocene deposits are several tenths of a meter in thickness.

From a geotechnical point of view the upper organic layer is very weak, practically unstable, easy to deformate under loading. This feature of the base was confirmed by toe-failure which occurs on a large scale during the dump formation. As a result of the toe-failure process, a wide zone of upliftings was formed around the dump. This zonecovers not only the close surroundings of the dump body but also a far forefield up to 350 m.

As a result of toe-failure, the layer of organic soils is destroyed and phosphogypsum can have direct contact with the water-containing sands and gravels layers. As a consequence, the contaminations can easily penetrate to underground water causing its pollution by toxic and radioactive components. The contaminations can also freely migrate with rainfall-water to the nearest channels and the Odra river. Water-soluble components and suspended matter, which flow down with surface water run-off are the main degradating factors to the open-water environment, even far from the dumping area. It must be stressed that degradating factors are active for along time, have a wide-range of influence, and their impact cannot be controlled. That is why, the use of remote sensing techniques for an environmental inventory was justified.
The following remote sensing experimantal images were taken:

1. Aerophotogrammetric black and white photos (6 series, approximate scale vary from 1:2500 to 1:4000) ;
2. Infrared photos - taken by a small format camera (2 series, approximate scale 1:10 000, 1:8000);
3. Colour aerial photos - taken by a small format camera (1 fly, aproximate scale 1:8000);
4. Multispectral photos - taken by a four-band NAC camera and MSK-4 Carl Zeiss Jena camera (3 series, approximate scale 1:10 000).
5. Thermal infrared aerial images - taken by an AGA-780 with magnetic recording;
6. Thermal infrared terrestial images - taken by an AGA-780 Thermovision System.

All the materials listed above were used for qualitative study. Photogrammetric photos were also used for mapping the dumping area, tracing the terrain and dump deformation, and calculating the volume of heaped and uplifted masses.

3 RESULTS

The results of the photointerpretation studies have shown that the phosphogypsum dumping process should be considered within the two following aspects:

1. How engineering-geological conditions affect the technology of dump formation, and
2. What impact it has on the natural environment.

The impact of both elements depends mainly on the dump base instability. The shifting of the dump front is the cause of the propagation of the deformation process in distant parts of the forefield and can provoke a hazard for objects and constructions situated close to the dumping area, as well as danger for heaping-machine operators.
In the uplifted zone the intensive mass movement can be seen. As an effect of it complicated folding, forms are created which can reach up to 10 m in height above the original terrain surface (Figure 2).
The photogrammetric measurement has shown that the mass displacement within the toe-failure zone varies from 10 to 30 cm/day.
On the aerial photos it is easy to recognize all the structural elements which exist in a forefield (uplifted zone contour, folding forms) as well as within the dump body (cracks, fissures, dump body dislocations, and toe-failure origin slides, which may appear in the front zones of the dump, see Figure 3).

Figure 2. The toe-failure zone in the dump forefield.

Figure 3. The typical dump body deformations.

It was also stated, on the basis of thorough analysis of all remote sensing date, that the toe-failure process is not a continuous one, but is a cyclic one due to the continuous movement of the dump front.

This conclusion was drawn on the basis of an analysis of the shape and situation of a toe-failure zone border, as well as, the directions of uplifted mass dislocations, which were registered on the succesive series of photos.

A photointerpretation made it possible to distinquish at least the two following pecu-

liar stages of mass movement:
- a preliminary stege, and
- a stage of intensive movement.

The preliminary stage is characterized mainly by upward-lifting movements, which occur within the already existing deformation zone. In the second stage, a rapid growth of the toe-failure zone, toward the forefield, is noticed. This is accompanied by toe-failure landslides at the front part of the dump body.

The cyclic character of the toe-failure process is also proved by the comparison of the volumes of the heaped and uplifted masses. This data was extracted from the digital terrain models.

To characterize the toe-failure process, the ratios of the increase of the dump volume (P_z) to a volume of uplifted masses (P_w), were calculated for succesive stages (see Table 1).

Table 1.

Period	P_z / P_w
May 78 - July 78	1.176
July 78 - April 79	0.942
April 79 - August 79	1.646
August 79 - June 80	1.235

In the data listed above, which refers to a particullar period of time, generally, the volume of dumped masses exceeds (from 18% to 65 % average 25 %) the volume of uplifted masses (except stage 3, April 79, where the data reflects the consolidation process of the dumped material and the fluffing process of the uplifted ground).

The changebility of P_z / P_w ratio shows that the toe-failure process can be cyclic.

The interpretation of remote sensing data also shows that the range of the toe-failure process depends on the height of the dump layer and the engineering-geological conditions existing on the forefield.

The influence of the engineering-geological condition is represented by different shapes of toe-failure borders for the comparable loads of dumped masses.

We can also expect that for soil which has the highest moisture content, the shear strength can be significantly lower. Therefore, for such regions, we can expect that the toe-failure process will be more intensive. The regions with the highest moisture content can easily be recognized on infrared and multispectral photos (see Figure 7).

However, we can state generally, that the increase of the toe-failure zone is mainly caused by the increase of the dump height.

Interpretation of remotely sensed data has also shown the evidently negative influence of phosphogypsum dump on the natural environment. This refers mainly to the open surface water and vegetation cover.

It was observed, that contaminations flowing down from the dump area can easily penetrate into the surface water.

The water circulation is facilitated by the flooding (for high water level) or the draining (for low and normal level) of the Odra river. The water run-off is accelerated by the existance of the dump body together with the upward-lifting zone which creates morphological elevation.

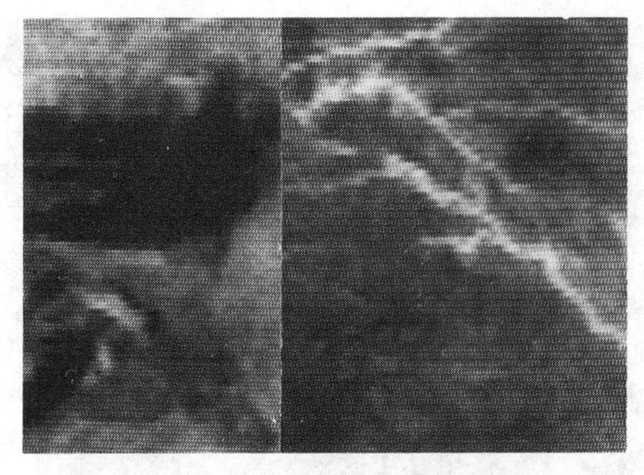

Figure 4. Contaminants propagation on the dump forefield give a specific thermal pattern (thermal infrared aerial image produced by an AGA-780 Thermovision System).

Figure 5. Thermal infrared ground image of the dump body. (Black and white photo is for comparison).

The new drainage pattern is easy to recognize in all kinds of aeriel photos, especially on the infrared and thermal infrared images. On such images, the flooding parts and places where polluted water flows into the river water can be precisely determined. The run-off of polluted water very often gives a clear thermal effect (see Figure 4).

The largest amount of suspended matter penetrates into river from the recent front part od the dump. This occurs because the phosphogypsum, in this part of dump is not yet consolidated. The consolidation process and other changes of the dump body can be traced on thermal images.

In Figure 5 zones of radiation temperatu-

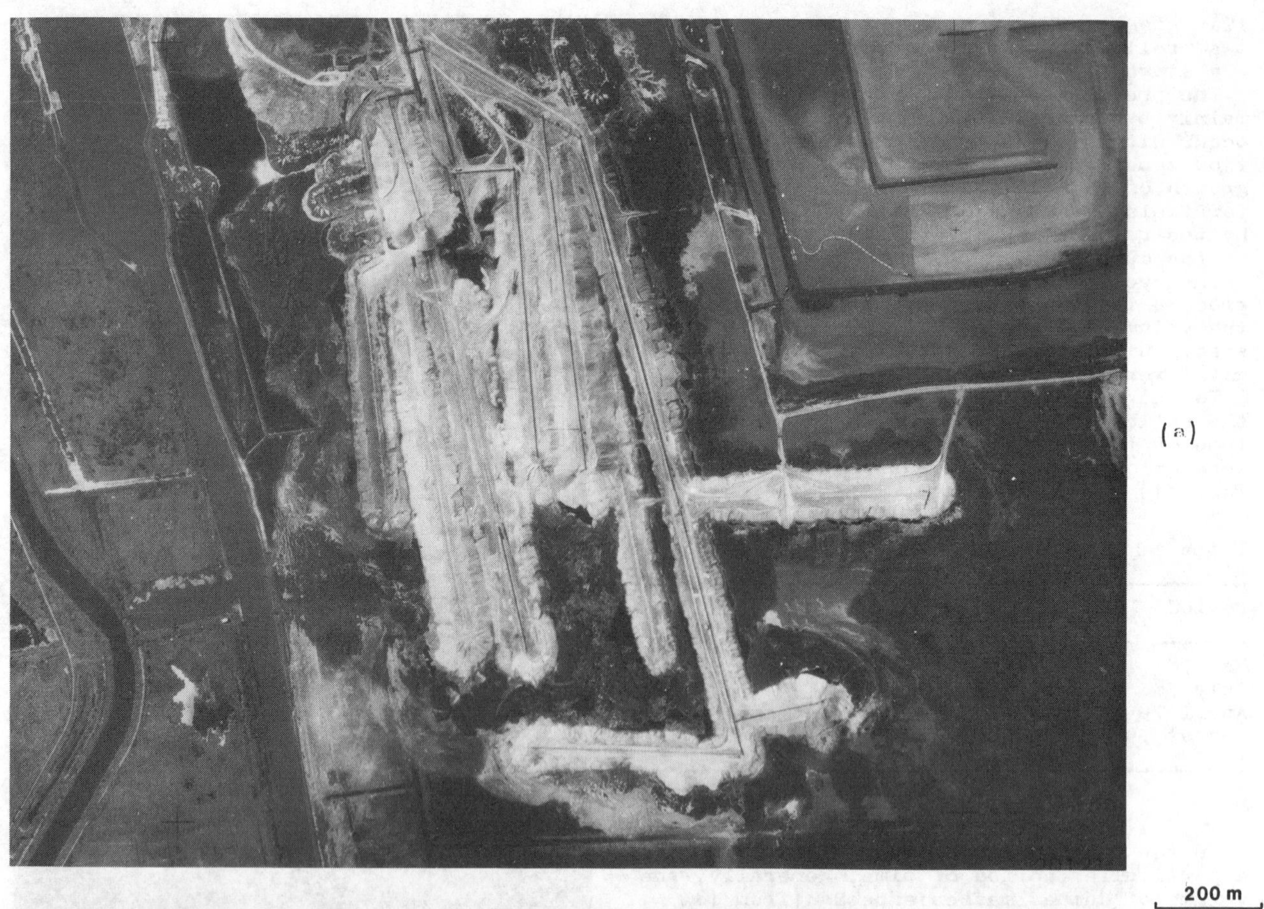

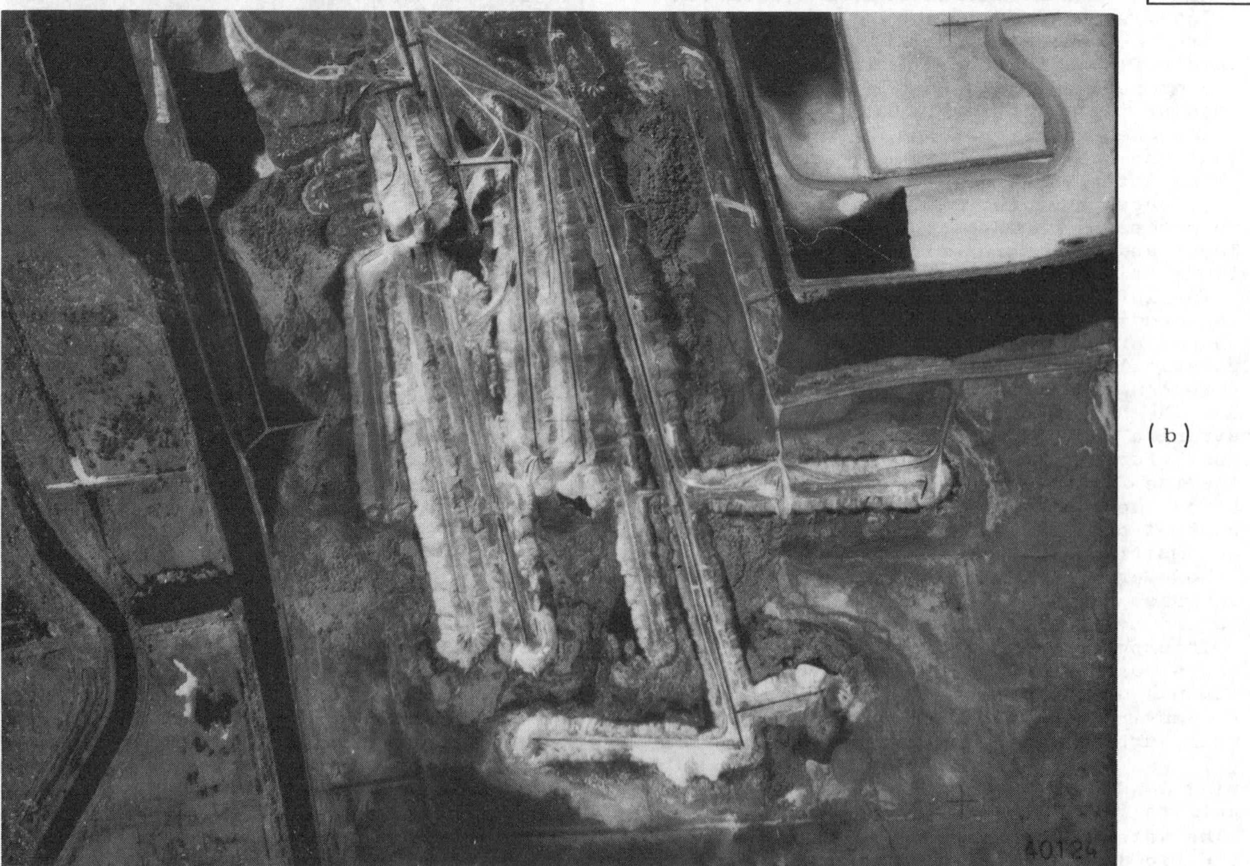

Figure 6. Multispectral photography of the dump area in blue (a) and green (b) bands (Kodak 2402 aerofilm ; Carl Zeiss Jena MSK-4 multispectral camera) .

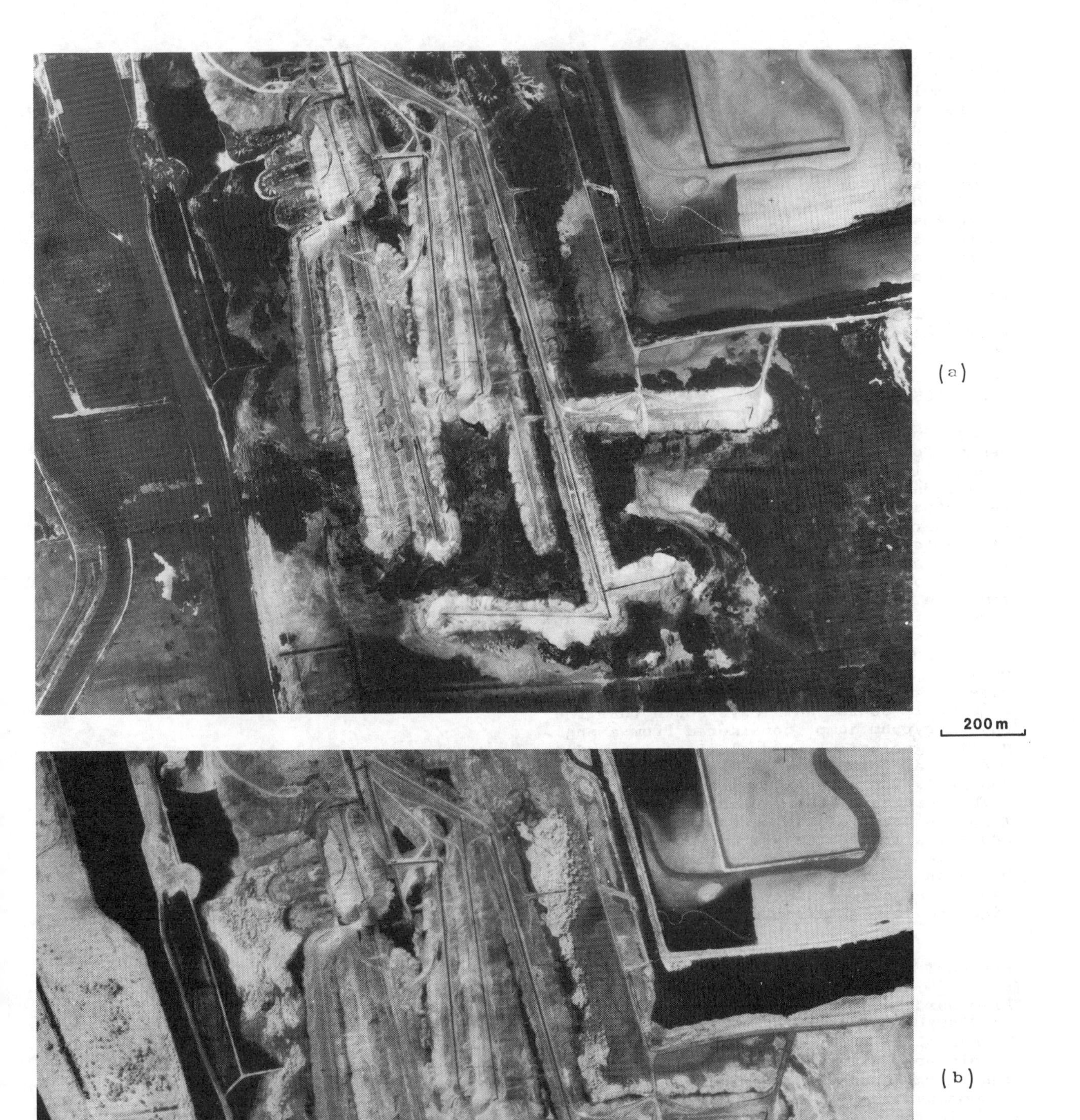

Figure 7. Multispectral photography of the dump area in red (a) and infrared (b) bands (Kodak 2402 aerofilm and Kodak 2424 infrared film; Carl Zeiss Jena MSK-4 multispectral camera).

re distribution can be seen, which correspond to the succesive stages of dumping.

The multispectral technique is very useful for the detection of the degree of water pollution, as well as, the determination of the ranges and directions of polution propagation.

The evaluation of the damage to the vegetation cover in the dump surroundings can be succesfully made by use of multispectral photos see (Figures 6 and 7). The colour composits of green, red and infrared bads make it possible to distinquuish different types of vegetation, as well as differentiating lifeless vegetation from stiil living vegetation.

4 CONCLUSIONS

The remote sensing techniques are very useful for the assesment of environmental changes in the phosphogypsum waste dumping areas. Based on the analog and digital data processing of a series of successively taken aerophothos, the quantitative nature of the toe-failure process was given. Dump forming on an unstable ground base can cause, not only technological problems connected with the method of dumping but, what is also very important, have a significant destructive impact on the natural environment

The surface and underground waters, as well as, vegetation are under the greatest stress. The results of the interpretation of remotly sensed data have shown that the phosphogypsum dump, considered from a engineering-geological, hydrological and hydrogeological point of view, can cause serious damage to the natural environment.

The results of the conducted research might be used for the creation of an environmental monitoring system for such areas, which can be effective and complete only if based on remote sensing techniques. This system can also satisfy many different design and documentation needs.

REFERENCES

Blanchard M.B., Greeley R., Goettelman R., 1974. Use of visible near infrared, and thermal infrared sensing to study soil moisture.NASA Tech. Mem. X-62, 343:8.

Idso,S.B.,Jacson,R.D.,Reginato,R.J.,1976. Compensating for environmental variability in the thermal inertia approach to remote sensing of soil moisture. J.Appl. Met.,15:811-817.

Mularz,S.C.,1973. Toe-failure processes on an open-cast mine waste-dump. Studia Geotech.4 (1):23-34.

Mularz,S.C.,Rybicki,S.,1977. Subgrade and dump deformations caused by dumping of mine waste. Engineering Geology,11:189-200.

Mularz,S.C.,1985. Thermovisional observation of the slopes in a strip mine. Fotointerpretacja w Geografii, T.8 (18):104-112.

Symposium on Remote Sensing for Resources Development and Environmental Management / Enschede / August 1986

Remote sensing for survey of material resources of highway engineering projects in developing countries

R.L.Nanda
Nigerian Building & Road Research Institute, Lagos

ABSTRACT: The third world countries are embarking on big projects of road construction so as to meet the needs of developing economy. But the conventionally used hard stones are invariably not available at or near the site of construction. These have to be transported from long distances, thereby increasing the cost of construction. With a view to achieving economy, emphasis is being laid on the judicious utilization of locally available materials such as soil, gravel, laterite and calcrete (calcareous aggregates). These low grade aggregates though not as hard as conventionally used hard-stones, still have adequate mechanical strength to be made use of in the bases and sub-bases of the pavement.

For optimum utilization of these regional deposits, inventory of material resources is a pre-requisite, which can be prepared expeditiously only by the application of remote sensing technique, even if the region is inaccessible.

Using this scientific technique, attempts have been made to conduct survey of soil for engineering purposes and identification and location of sub-surface calcrete, commonly occuring in the arid and semi-arid regions of India and Nigeria, covering an area of about 13,000 sq. km. and 18,000 sq. km. respectively. The timely information about the location of locally available materials will go a long way in economising the cost of road construction in future development programmes.

The paper discusses the remote sensing techniques adopted during the experimental surveys, their relative usefulness and limitations, particularly for the developing countries.

1. INTRODUCTION

Large scale road construction programme is being embarked in developing countries so as to meet the needs of developing economy. The known sources of aggregates are getting progressively depleted due to the expanding construction activity. Furthermore, there are certain areas where construction of network of roads is in progress but the conventionally used hardstone is not available at or near the site of construction. It has to be transported from long distances thereby increasing the cost of construction. With a view to achieving economy in the road development programme, emphasis is being laid on the judicious utilisation of locally available materials such as soil, gravel, laterite and calcrete (calcareous aggregate). These regional deposits though not as hard as conventionally used hard stone still have adequate mechanical strength to be made use of in the bases and sub-bases of the pavement.

In order to utilize these local deposits to the maximum possible extent, an inventory of material resources is a prerequisite for obtaining information regarding type, location and extent. This type of inventory can be prepared either by ground survey or by the application of scientific technique of remote sensing. With the conventional method of ground survey the procedure is not only time consuming, but also laborious as the whole area is to be traversed and numerous bore-holes are to be made. Moreover the progress of survey is liable to the handicapped by inaccessibility to land due to lack of roads, paths or existence of dense vegetation and high sand dunes. Such is not the case when the technique of remote sensing is applied as the land is brought to the laboratory and studied overthere without any hinderance.

2. SCOPE AND LIMITATIONS OF REMOTE SENSING IN HIGHWAY ENGINEERING

2.1 Scope

The technique of remote sensing is of special interest and of great assistance in the field of highway engineering. Some of the important aspects are as follows:

(i) Materials survey
(ii) Route location
(iii) Field performance of highway pavement.

It is the first aspect which has greater significance in a developing country where the road system is to be developed at a fast rate in order to meet the needs of expanding economy. It is essential to systematise the knowledge of road making materials, including soils available in the country. The investigation is likely to reveal the presence of large quantity of low grade aggregates which can be judiciously utilised either as such or after scientifically processing. This amounts to considerable saving in the cost of material as well as transportation. The materials survey with the help of remote sensing can be conducted not only with less manpower but also much faster. The timely information about the location of locally available construction materials will go a long way in economising the cost of road construction in future development plans.

2.2 Limitation

Remote sensing is a composite term which includes many types of sensing. But for the purpose of engineering soil and material survey conducted in India and Nigeria the techniques have been limited to the following three types:

(i) Aerial photographs (black and white)
(ii) SLAR imagery (side looking airborne radar)
(iii) Landsat imagery (photographic)

(i) Aerial photographs (black and white)

The panchromatic (black and white) aerial photographs (23cm X 23cm) have continued to gain wide acceptance both in India and Nigeria. This is mainly due to high resolution, large scale and stereoscopic capability which is essential for the interpretation of sub-surface calcrete. For engineering and soil material survey a scale of 1:25,000 has been found to be quite satisfactory.
As the scale is large the land area covered on a single photograph is relatively small, the reconnaisance survey of the whole terrain is difficult. However a format of large area can still be studied by arranging the air photo mosaic. Although the system is passive and depends upon solar illumination, nevertheless it provides the best source of qualitative data extraction for rapid survey.

(ii) SLAR imagery (side looking airborne radar)

It is an active system and independent of solar energy. This it is suitable for mapping areas obscured by cloud as is the condition prevailing in Southern Nigeria. The image strip of small scale (1:250,000) currently available in Nigeria, can cover large areas on a single photograph, thereby presenting a synoptic view of the terrain. But it is devoid of stereo coverage which is necessary for the identification of subsurface calcrete land forms. Though general view of the terrain can be studied but delineation of soil boundaries for engineering interpretation and classification is not practicable as the scale is 10 times smaller than the aerial photographs.

(iiia) Landsat imagery (photographic)

A false colour composite landsat imagery has been studied to ascertain the potentiality of landsat imagery to the survey of calcrete in N.E. Nigeria. The imagery used is a small scale photograph and has colour reflections in different bands. The tonal variation between calcrete-bearing soil and other soil types is a reflection of the colour contrasts, which is not quite obvious in this area. Generally, the imageries have low colour contrast. The colour reflection of sandy soil is light brown with little bias towards the calcrete bearing areas, particularly where vegetation is fairly pronounced and along erosional channels or long depression. They are only capable of depicting regional linearments like the sand dune ridges, but without stereoscopic vision. Hence it has poor potential of interpretation relating to relief and sub surface calcrete. Further probe for the location of calcrete has to be followed by the interpretation of aerial photographs. The lack of stereo image and the corresponding small scale are inherent problems associated with the interpretation of Landsat imagery. Though it has a special advantage of exploiting differences in tone signature of terrain objects image in various bands, but being passive system offers no solution to the acquisition of imagery through the dense cloud cover of the southern parts of Nigeria.

(iiib) Digital landsat imagery

Digital landsat imageries have been used in developed countries. Their processing and classification in this way can greatly enhance value of remotely sensed data. But the newer digital techniques require heavy investment in computing equipment and output devices. This is a stumbling block for those working in technologically less advanced countries and also for users with low budget and lack of duly trained technical manpower. Infact, much can be achieved on relatively standard equipment so as to avoid adding cost of interpretation.

3. ENGINEERING SOIL SURVEY

Soil is the cheapest locally available construction material. This can be used either as much, or after scientifically processing in various layers of the pavement. For any highway project , reconnaisance and detailed soil survey are required to be conducted for assessing the suitability of soil as sub-grade, sub-base and base course. Information about the characteristics of soils can be had expeditiously using remote sensing technique, even without access to the site.

3.1 Classification of soils in alluvial plains of India

As a pilot study (Nanda 1969) an area of about 650 sq. kilometre around Ambala (Haryana) in the alluvial plains of north-western part of India had been interpreted from the aerial photographs, for the identification of various soil types. Some of

SOIL MAP OF AMBALA AREA INDIA
(BASED ON AIR PHOTO INTERPRETATION TECHNIQUE)

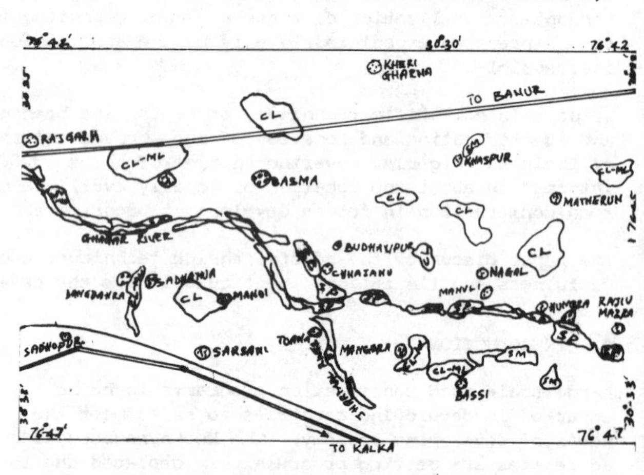

LEGEND

SM - SILTY SAND
SC - CLAYEY SAND
MH - CLAYEY SILT
CL - SILT CLAY
ML - SANDY SILT
SP SANDY
ROAD
RLY LINE
VILLAGE & TOWN
RIVER

FIG 1

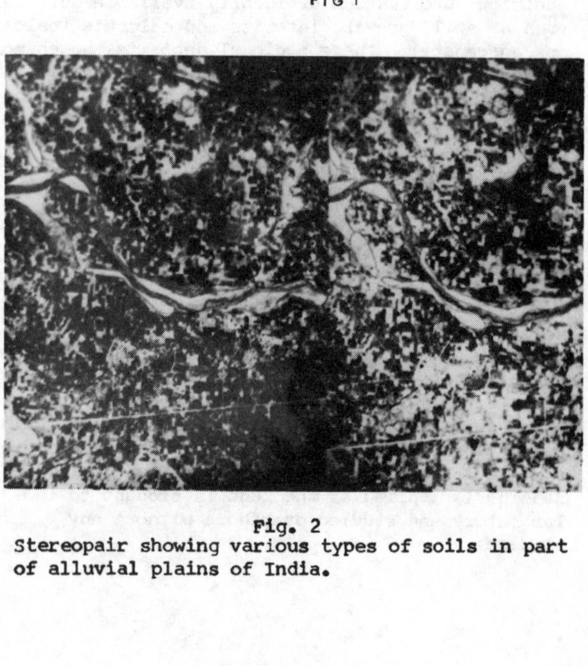

Fig. 2
Stereopair showing various types of soils in part of alluvial plains of India.

Table 1 Air photo interpretation and verification of soils around Ambala (Haryana) India

Site Location SL.	Soil identification by Airphoto Interpretation	Unified Soil Classification as per lab. testing	Remarks
1. Jathsar Village	Silty Clay (CL)	CL	
2. Luthar Majra Village	Silty Sand (SM)	CL	Kalar area (Sulphate infested)
3. River bed opp. to Arjana Kaund	Sand (SP)	SP	
4. Umla Nala Trough	Silty Sand (SM)	SM	
5. Near R. D. 200 right side of Canal (Jhensa to Jansol)	Clayey Silt (ML)	CL-ML	
6. Thaskali Village	Clayey Sand (SC)	SC	
7. Adopur Village	Silty Sand (SM)	SM	
8. Khwaspur Village Pond	Clay (CH)	CH	
9. 8/7 right side Ambala Chandigarh Road	Silty Clay (CL)	CL	
10. Ghagar River bed	Sand (SP)	SP	
11. Ghhajnu Village	Clayey Sand(SC)	SC	
12. Near R.D.205 Umla	Clayey Silt (ML)	CL-ML	

the soil thus delineated were silty clay (CL), silty sand (SM), sand (SP), clayey silt (ML), clayey sand (SC) and clay (CH). A part of the soil map thus produced and a stereo pair are given in Fig. 1 and 2 respectively. On ground verification, it has been revealed that it has been possible to interpret engineering properties of soil from the aerial photographs. The correlation of some of the interpreted sites is given in Table 1.

3.2 Delineation of soil boundaries in the sandy terrain of Nigeria

Similarly, soils occuring in the sandy terrain of North-East Nigeria (part of Borno and Kano States) covering an area of about 18000 km^2 have also been interpreted from the aerial photographs and classified according to Unified Soil Classification System. Predominantly the soils identified are silty sand (SM) clayey sand (SC). Their photo tones vary from light grey to dark grey and sometimes with whitish tinge due to the occurence of sulphate and calcareous deposit. The other soil types identified and met with in certain areas are sandy silt (ML), clayey silt (MH), silty clay (CL) and clay (CH). A stereo pair and a typical map are given in fig. 3 and 4 respectively.

4. SURVEY OF CALCRETE

Calcrete is an English term used for calcareous aggregates commonly met with in the alluvial plains and desert terrain of India and under aeolin deposits in North-East Nigeria. Locally these aggregates are called 'Kankar' in India and Jiglin in Northern Nigeria. Generally they are not visible at the surface and occur at shallow depth as a semi continuous horizon under an overburden of 0.5 -2m of soil as shown in Fig. 5. They are of different shapes, forms and sizes. They may be granular or in the form of clod. The colour is usually light grey. A typical hard variety of calcrete is shown in Fig. 6.

The deposits of calcrete in tropical countries are said to be formed by the process of calcification. Carbonated rain water dissolves calcium carbonate (CaCo3) present in the local soil, converting it to

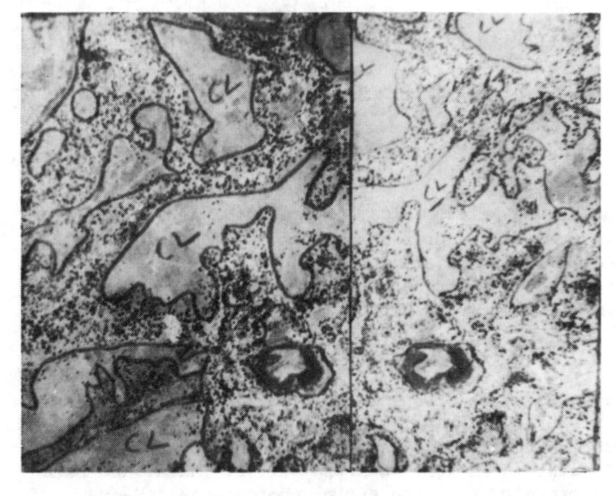

Fig. 3
Stereopair showing silty clay (CL) in part of North East Nigeria.

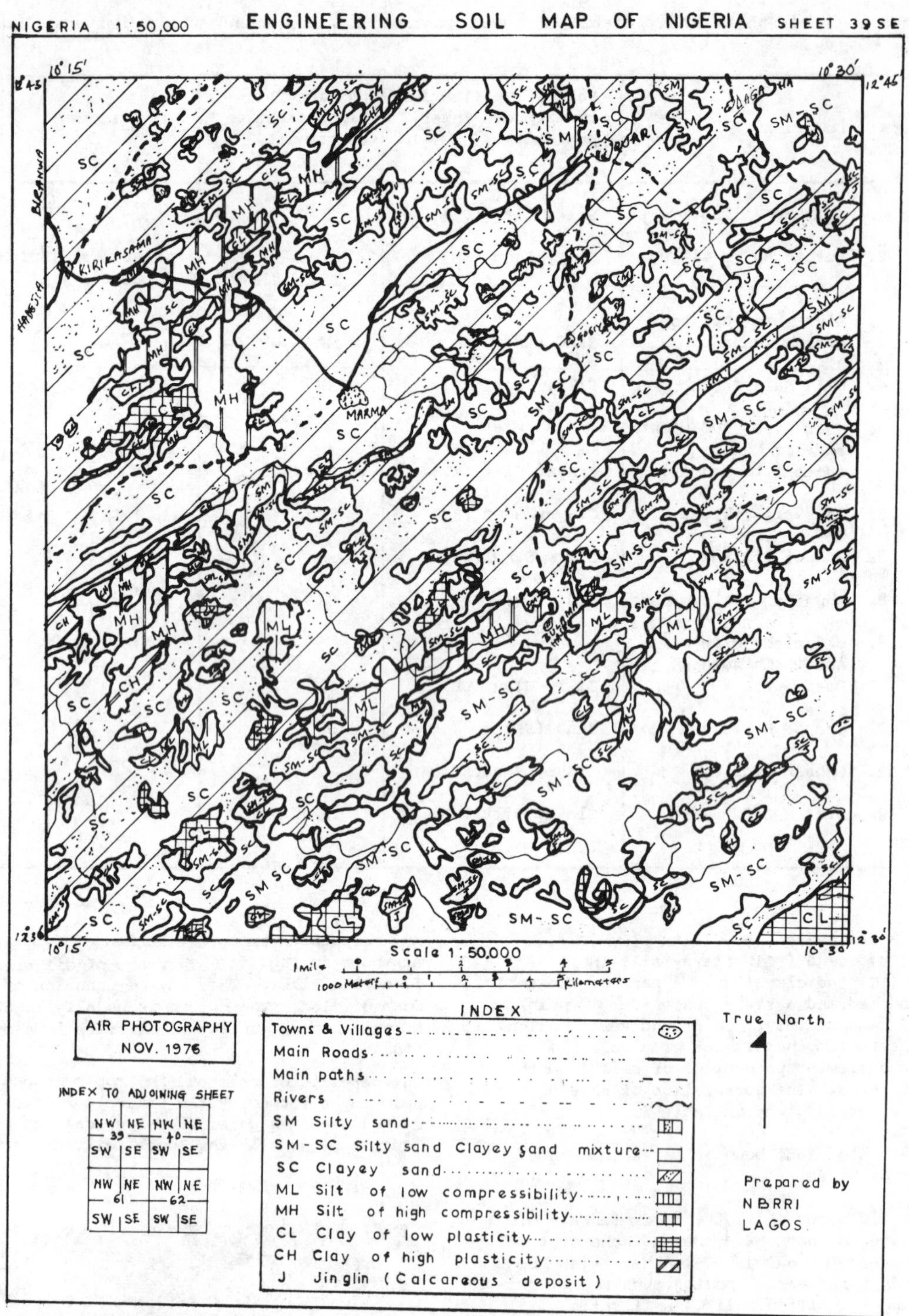

FIG. 4

a soluble calcium bicarbonate $Ca(HCO_3)_2$. The later is then washed down the slopes of depressions and hollows of undulating topography to form calcium carbonate, after water and carbon dioxide have been evaporated, as illustrated in the equation below:-

$$CaCO_3 + H_2O + CO_2 \rightleftharpoons Ca(HCO_3)_2.$$

4.1 Location of calcrete in Indian desert

The calcrete bearing areas in desert terrain of Rajaisthan in North-Western part of India, had been interpreted from the aerial photographs. (Nanda 1970). They are interdunal sandy plain or shallow depressions of light grey tone, showing signs of wind erosion and with scattered bushes. 37 such sites had been identified in an area of 13,000 km^2. Field verification revealed that 75% of the site confirm air-photo interpretation. A typical stereopair is shown in Fig. 7.

4.2 Prospecting of calcrete in sandy terrain of Northern Nigeria.

Similarly, air photo interpretation study for the location of calcrete has been carried out in the sandy terrain of North-East Nigeria (Borno and Kano States), covering an area of 18000 km^2. The calcrete has predominantly been interpreted to be in the interdunal sandy plain and shallow depressions with light grey tone, showing signs of wind erosion and without remarkable vegetation, as it has

Fig. 5
A typical cut section showing horizon of calcrete in Indian desert.

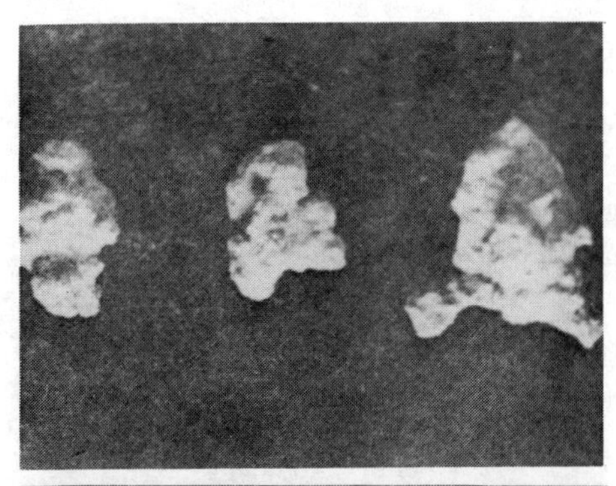

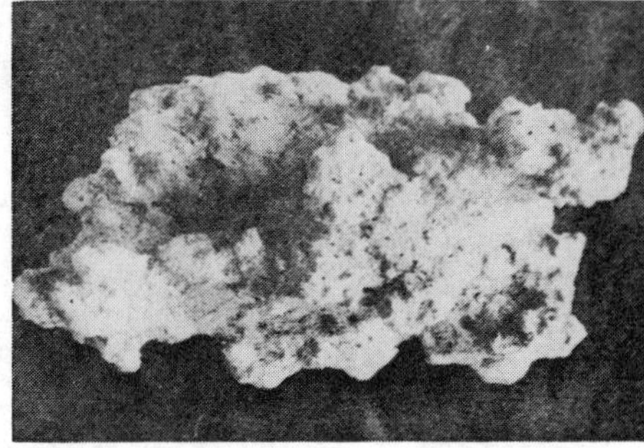

Fig. 6
Specimens of hard variety of calcrete

been the case in India. However, another prominent landform encountered in the region is interdunal hollow. These are characterised by whitish grey tone without land use aspect. A typical stereopair is shown in Fig. 8. On ground verification, the occuracy of interpretation has been found to be of the order of 90%.

5. CONCLUSIONS

(i) Air photo interpretation provides an accurate and expeditious means of conducting survey of highway material resources, even if the region is inaccessible.

Fig. 7
Stereopair showing interdunal sandy plain of calcrete bearing areas in Indian desert.

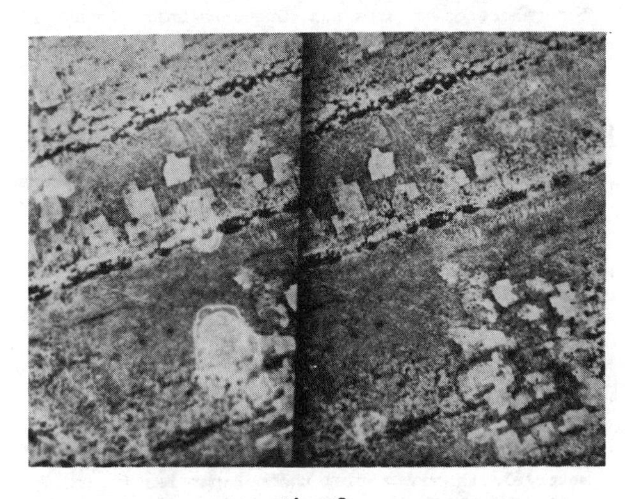

Fig. 8
Stereopair showing interdunal hollow depressions having calcrete deposits in part of North East Nigeria.

(ii) Though the location of calcrete and delineation & classification of soils for engineering purposes can be carried out through the aerial photographs, certain amount of field verification is desirable to avoid any discripancy or anomaly.

(iii) The air photo patterns developed for the identification and location of calcrete bearing areas have their applicability and future scope not only in the terrains studied but also in other developing countries where calcareous deposits exist in similar topographical and physiographical environment.

(iv) Both SLAR and Landsat imageries studied have depicted synoptic view of large areas of sand dune ridges and long narrow depressions, thereby providing some clue about the calcrete bearing landform. However, they lack 3 - Dimensional view of the terrain which is essential for the interpretation of sub-surface calcrete.

ACKNOWLEDGEMENT

The paper is presented with the permission of Dr. A. O. Madedor, Director Nigerian Building & Road Research Institute (N. B. R. R. I.) Lagos. The Author is thankful to him for his keen interest in the study and valuable discussions.

Some of the data presented is extracted from the interpretation work carried out at N.B.R.R.I. in connection with the project, "Preperation of engineering geological map of Nigeria" and other is of my earlier work at Central Road Research Institute, New Delhi (India), thankfully acknowledged.

BIBLIOGRAPHY

Allen, P. E. T. 1979. The use of side looking airborne radar. (SLAR) imagery for the production of land use and vegetation study of Nigeria UN/FAO training seminar on remote sensing of earth resources, Ibadan (Nigeria).

Beaumont, T. E. 1979. Remote sensing for the location and mapping of engineering construction materials in developing countries. Q. J. Engineering Geology. 12 : 147 - 158.

Lawrence, C. J. 1984. Remote sensing in the search for calcretes in Botswana, Overseas unit, Transport and Road Research Laboratory, U. K. Proc. eighth regional conference for Africa, Soil mechanics and foundation engineering, Harare (Zimbabwe).

Madedor, A. O., Nanda, R. L., Akinyede, J. O. 1984. Engineering soil and material survey with remote sensing technique. Proc. eighth regional conference for Africa, Soil mechanics and foundation engineering, Harare, (Zimbabwe). 1:23-32.

Nanda, R. L. 1969. Application of air photo interpretation for engineering soil survey, Indian Roads Congress, Bull 13, 62 - 82.

Nanda, R. L. 1970. Survey of Road Construction Materials in Rajasthan through aerial photographs, Jr. Indian Roads Congress, 33 part - 3

Nanda, R. L. 1982. Survey of road construction materials through aerial photographs for future road development programme in India. International seminar on impact of aerial survey on National Development, held at Regional centre for training in aerial surveys, Ile - Ife (Nigeria).

Nanda, R. L. and Akinyede, J. O. 1985. Application of remote sensing for the survey of calcrete in semi-arid region of Nigeria. Proc. eleventh International conference on soil mechanics and foundation engineering, Sanfrancisco (U. S. A.), 4 : 2447 - 2452.

Natterberg, R. 1978. Prospecting for calcrete road materials in South and South West Africa. Jr. Civil engineer, South Africa, JAN (Die Siviele inginieur in Suid Afrika).

Nigerian Building and Road Research Institute. 1983. Engineering soil and materials surveys in parts of N. E. Nigeria, Annual Report.

Nigerian Building and Road Research Institute. 1985. Identification and location of road building materials in North East Kano State, consultancy report, B.R.R.I./TE/I.

RIB, H. T. 1982. Examples of remote sensing application to highway engineering. Proc. second national workshop on engineering application of remote sensing, Canada centre for remote sensing, Ontario, 99 - 104.

Uppal, H. L. and Nanda, R. L. 1968, Survey of hidden calcrete in alluvial plains of India by air photo interpretation. Proc. Forth conf. Australian Road Research Board, 4: 1677 - 1688.

Symposium on Remote Sensing for Resources Development and Environmental Management / Enschede / August 1986

Remote Sensing applications in the Eastern Bolivia Mineral Exploration Project (Proyecto Precambrico): Techniques and prospects

E.O'Connor & J.P.Berrange
British Geological Survey, Nottingham, UK

ABSTRACT: LANDSAT imagery in various modes, and conventional aerial photography, played an indispensible role in the planning, construction of topographic basemaps, field surveying and identification of exploration targets for Proyecto Precambrico. The project was sponsored by the British Government's Overseas Development Administration and carried out the first ever systematic geological mapping and mineral exploration in a Precambrian shield area of ca 220,000 km^2 largely covered with tropical forest.

No basemaps existed for the northern half of the project area and a series of eight 1:250,000 topographic basemaps and one at 1:500,000 scale were prepared using enhanced false-colour imagery, panchromatic aerial photography and limited ground data.

LANDSAT hardcopy prints in black and white and false colour at various scales from 1:100,000 to 1:1,000,000, and panchromatic aerial photography in stereoscope mode was used routinely for photogeological interpretation and mapping of the region as a whole.

Interactive computer processing of LANDSAT CCT's of selected areas included colour-compositing, ratioing, ratio compositing , edge and contrast stretch techniques. The processing was carried out at intervals in different laboratories e.g. Purdue University (LARS-PDP system); NCRS, Farnborough, (IDP system), Silsoe (Gems system), Nottingham University (PDP 11/34 base system) and BGS, Keyworth (I^2S system).

The resulting transparencies and colour-prints provided important additional data for geological mapping and mineral prospecting purposes. This included the detection of forest covered granitoid plutons with related fracture systems and the recognition of topographic and/or vegetational anomalies related to important mineral discoveries in the region. These are the Cerro Manomo carbonatitic complex with P, RE, U & Th mineralisation; silica caprocks with rich underlying garnieritic Ni deposits formed on ultramafic rocks of the layered Rincon del Tigre Igneous Complex; and banded Fe and Mn formation (BIMF) in the SE sector of the shield.

Detecting and mapping of different volcanic stages and other geomorphic features by Landsat images in 'Katakekaumene', Western Turkey

F.Sancar Ozaner
MTA Jeologi Etüdleri Dairesi, Ankara, Turkey

ABSTRACT: Kula and its surroundings which is located in Western Turkey has been chosen for the study. Some geological and geomorphological units which were differentiated by this study have already been mapped by different researchers. The aim of the study is to prove that, some of the units which were mapped by long and expensive field works already, could also differantiated on the Landsat images. The boundries of three eruption stages of basaltic rocks on the southern part, and andesitic rocks on the northern part were mapped. Shifting of the eruption sites from north to south, and, changing of the volcanics from andesitic type to basaltic,have been detected on the Landsat images. Thus, new targets related to further earthquake studies become evident.

1. INTRODUCTION

The study area have already been worked by numbers of researches. A well known historian, Strabon, visited the area before birth of Christ and named Kula basalts as "Katakekaumane". Later on, (Philippson 1913), (Canet and Jaoul 1946), (Beekman 1964), (Bergo 1964), (Erinç 1970), (Şenol and Karabıyıkoğlu 1977), (Ercan et al 1977,1980), (Bircan et al 1982), (Ozaner 1984), worked and mapped the area. Some of these researches focused their attention especially on mapping of Kula basalts, while the others also mapped sedimentological, geological and geomorphological features of the area. The auther first used Landsat images of the area, involving with the project, (Bircan et al 1982), and realized that most of the features which had already been mapped are also detectible,– in some cases more accurately mappeble– on Landsat imagery. In Addition, synoptic view advantage of Landsat images has created fruitful conclusions for the tectonics of the area which are not contradict to global plate tectonic results of Western Turkey. The author helped and partly quided to the study of Hakim (1983) at this part of the country.

2. INTERPRETATION

Landsat image interpretation has been made on the image which was recorded on 16 September 1975. 1: 500.000 scale of Band 5, Band 7, and colour composite image, and 1: 35.000 scale aerial photographs of the area were used during interpretation. Colour compasite image was not printed for abstaining from extra expenses of publication.

As above mentioned, volcanic terrain units are the most striking features of the images. For this reason, interpretation will start with these units. In the north of the imagery, two giant andesiticriolitic heaps, namely Asi Tepe and Yağcı Dağ, are very distinct forms by its circular shape, relief impression and radial drainage patterns. Due to fact that andesitic lavas are less fluid than basaltic lavas, it gives a mass morphology. The visual impressions of the ruggedness was formed by differences of illumunation of the slopes which were transformed into different grey tones. There is not an absolute age for this rocks. However it is estimated to be of Early Upper Miocene (Ercan et al. 1977).

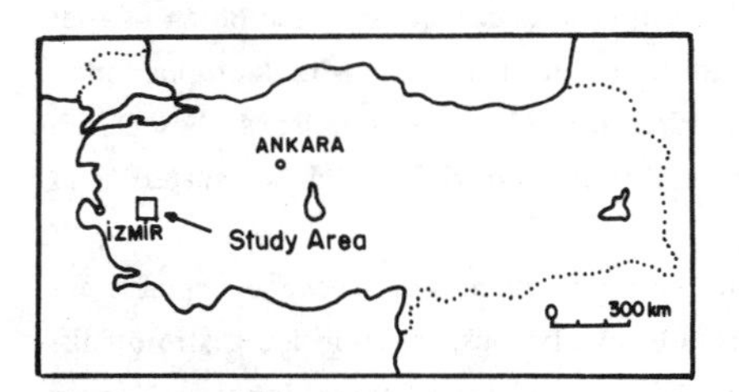

Figure 1. Location map

Approximately, at the centre of Landsat image (in other words: southern part of andesitic heaps), a well known "Katakekaumane" is located. Here, three main basaltic eruption phases could be mapped by typical image characteristics, such as grey tone differences, site and shape.

The first stage basaltic lavas take place at the southern part of older basalts. This intermediate lavas do not show circular shape, because they have not been severely dissected yet. Thus, they don't show morphological impression. However, they do show a distinct dark grey tone which starts at close to watersheds and are bounded with Gediz River. In Band 7, and in coloured composite image, this site effect assisted to the interpreter for coming the conclusion that they had been formed by flows of fluid lavas. The boundries of this stage are seen clearly in Band 7 and in colour composite image. In Band 5, very dark grey tone at point A, reflects vegetetation on limestones and confuses the boundry. Vegetation appears red in color compasite image and clears the problem. In aerial photos, under steoroscopic view, these two volcanic phases are easily discriminated by their different altitudes, which represent total erosion between two eruption times. Most of craters and cones of this stage can be recognised on Landsat imagery. This lava is estimated to be of Late Upper Miocene (Ercan et al, 1985). First and second lavas appears with naphta colour in colour composite image.

Recent basaltic flows appear with three distinct black patches in Landsat images. In Band 7 and 5 and in colour composite image there is not any difference between lake of dam and third stage of lavas, from the point of colour. Site effect which has the same characteristic as outlined for the second stage, assisted interpreter for discriminating these lava flows from the dam lake. Whitish areas in the northern part of lake which indicate the presence of accumulations of very recent sediments is another criteria for discriminations. Most of the craters of the youngest stage are visible in Landsat images.

According to radiometric data, this lava is about 25.000 years old (Ercan et al 1985).

Mapping of the volcanic rocks on Landsat imagery shows approximately the same accuracy as from 1: 35.000 scale aerial photographs of the area.

Metamorphic rocks appears with distinctive morphological impresssion which is determined by watersheds and drainage patterns. Ruggedness of the unit is much apperant in Band 7.

Soft Neogene formations are mostly seen in whitish tones and could be mapped by the help of Band 5, and colour composite image. White tones represent bare slopes consisting of sandy silty, tuffaous materials which reflect much of the electromagnetic energy in dry season.

Limestone plateaus which have been formed in Neogene basin rocks are also distinct terrain features, occur in the north of first stage basalts. They are easily detectable units with flat topography, faint signs of internal drainage and site characteristics. These terrain features have been isolated by scarp zone which is transferred to badlands.

Badlands occur in the areas where soft Neogene deposits are capped by limestone or basalt plateous. They appears as stronly dissected fine textural drainage network in Landsat imagery. Narrow straight ravines cause the relief impression. Badland boundries were not mapped but typical areas were indicated.

The boundry between Neogene soft formations and Ouaternary deposits of Gediz Graben is very clear in Band 5, and especially in colour composite image. Graben buttom appears in darker tones in Band 5 and bright red in colour composite, which reflects parcels of various crops and orchards and makes a distinct contrast with the white Neogene sediments.

Capital letter M in Band 7, points faint meandering belts of Gediz River.

2.1 Analysis of lineaments

The following linear features have been analysed on the Landsat image:

Certain faults; distinct on Landsat imagery and proved by earlier researches in the field.

Possible faults; clear on Landsat imagery as distinct lineaments and have been decided possible faults in the field.

Faults; derived from the rows of craters and cones which are visible in Landsat.

Most of the faults can be seen on Landsat images clearly. In addition, synoptic view advantage of Landsat imagery has created fruitful conclusions for the tectonics of the area. It can easily be recognized that the oldest andesitic eruptions occured in the northern part, while relatively younger three basaltic eruptions took place in the middle and southern part of Landsat image. Even these three basaltic stages have been arranged in rows which are continuously shift towards south.

According to the rows of cones, the first stage basaltic lavas have been erupted from roughly E–W trending faults or cracks. The

Figure 2. M.S.S. Landsat Band 5

Figure 3. M.S.S. Landsat Band 7

rows of the second stage craters which are situated south of first stage, imply that they came to the surface by WNW–ESE and SW–NE trending faults an cracks. The youngest lavas also erupted by the same trending weak zones as happened in the second stage, but shifts south, relative to second stage.

At the upper most south of image, a young depression can be seen. This NW–SE trending graben is recent active formation of south shifting tectonic activity. In the light of such tectonic pattern, one can understand that further tectonic studies and earthquake researches have to be targeted at the southern part of the area.

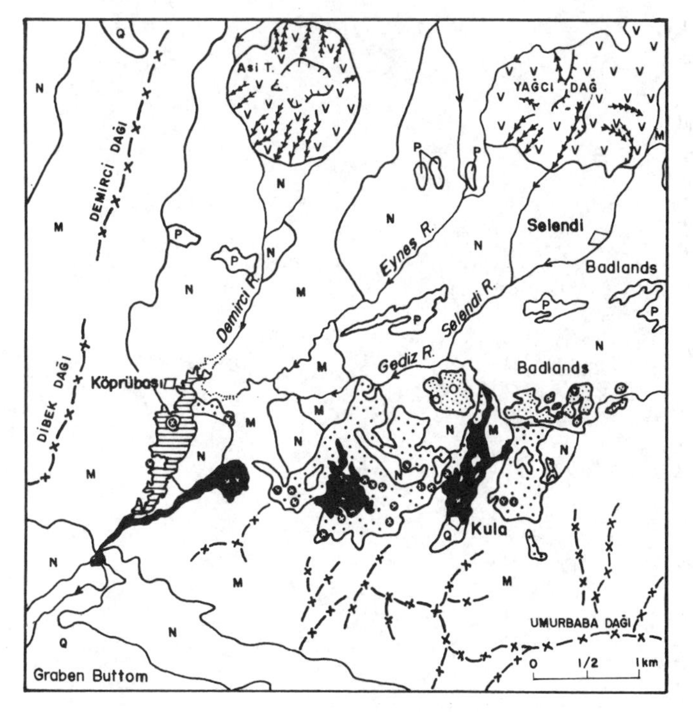

Figure 4. Landsat geological–geomorphological map

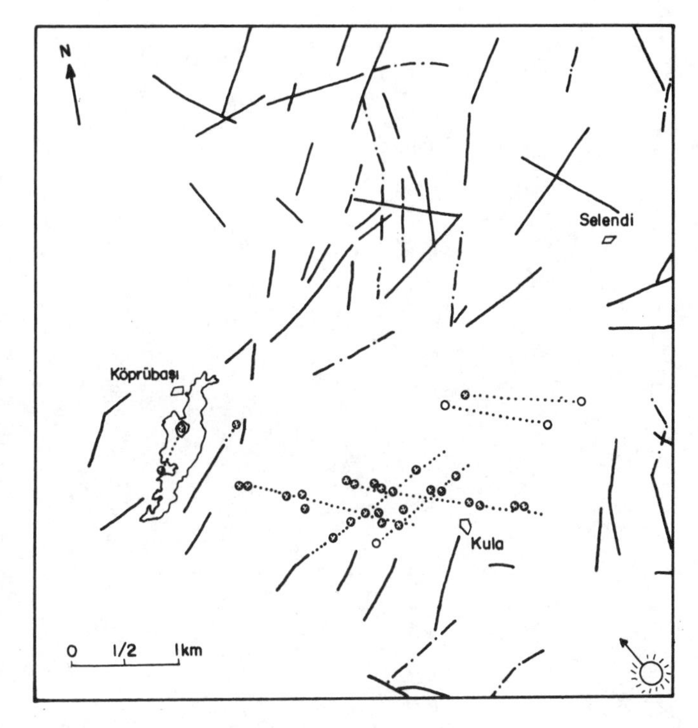

Figure 5. Landsat lineament map

REFERENCES

Beekman, H.P. 1964. Geological investigatins near Kula and Borlu: MTA report, Ankara.

Bergo, G. 1964. Kula yöresinin volkanizması: MTA raporu, Ankara.

Bircan, A. et al. 1982. Gediz Graben Sisteminin Jeomorfolojisi ve Genç Tektoniği: MTA Raporu, Ankara.

Canet, J. and P. Jaoul. 1946.Geologie de la Region de Manisa–Aydın–Kula–Gördes: Rapport de MTA, Ankara.

Ercan, T. et al. 1977. Uşak yöresinin jeolojisi ve volkanitlerinin petrolojisi: MTA Raporu, Ankara.

Ercan, T. et al. 1980. Kula–Selendi yörelerinin jeolojisi ve volkanitlerinin petrolojisi: MTA Raporu, Ankara.

Erinç, S. 1970. Kula ve Adala arasında genç volkan reliyefi: İst. Üniv. Coğ. Ens. Dergisi c. 9, s.17: 7–22.

Philippson, A. 1913. Das vulkangebiet von Kula in Lydien, die Katakekaummene der Alten: Pet. Georg. Mitt. c.2: 237–241.

Şenol, M. and M.Karabıyıkoğlu. 1977. Kula ve Köprübaşı yöresinde uranyum içeren Neojen dolgularının sedimentolojisi: MTA raporu, Ankara.

Hakim H.G. 1983. Geomorphological reconnaissance survey and mapping in WSW Anatolia–Turkey using remote sensing technigues I.T.C. thesis, The Netherlands.

Ercan et al. 1985. Batı Anadolu'daki volkanik kayaçlarda yeni yapılan kimyasal analizlerin 87 Sr/86 Sr ölçümlerinin ve radyometrik yaş belirlenmelerinin yorumu Türkiye Jeoloji Kurultayı bildiri özetleri, Ankara.

Ozaner, F.S. 1964. Kula–Selendi yörelerinin jeomorfolojisi ve morfotektoniği: MTA raporu, Ankara.

Symposium on Remote Sensing for Resources Development and Environmental Management / Enschede / August 1986

A remote sensing methodological approach for applied geomorphology mapping in plain areas

Eliseo Popolizio
Centro de Geociencas, Universidad Nacional del Nordeste, Argentina

Carlos Canoba
Instituto de Fisiograffia y Geologia, Universidad Nacional de Rosario, Argentina

ABSTRACT: Methodological problems faced along for more than a decade in applied geomorphological mapping of extense areas of the Argentine plains,allows the formulation of a methodological criteria based on remote sensing. According to plain singularities and scales perception levels , relations between each study objectives and final cartographic scales is discussed. Relations objective and scale are analyzed for three taxonomic levels: regional, zonal and local as well as what it is perceived in the imagery and what can be plotted using an adequate symbology.

INTRODUCTION

Present work objective is the description of methodological problems that have been faced during a decade and more for applied geomorphological mapping of large plain areas, using basically remote sensing support.

On the base of such experience methodological criteria were elaborated for terrain mapping with several objectives, in different perception levels and in corresponding scales.

For that matter it seems convenient to start from the systems theory conception of geomorphology, which was reached after successive approximations (Fig.1) (Popolizio 1980).

This presupposes the existence of four principal subsystems (interoperated between them under the parametric controls of three universes). They are: lithostructural, biotic, hydric and edaphic.

Forms and processes are system outputs, the replies external parametric controls, that is to say, they are dynamic and they have memory. On the other hand, as Callieux-Tricart (1965) say, the size of the geomorphological unit being analyzed (taxonomic level) decisively influences on the relative importance of each subsystem. All this is seen on the landscape and obviously it is reflected on the remote sensing imagery.

Therefore it is necessary to be clear on the fact that rather past is seen than present in a temporal cut, within an evolutionary tendency where variables and subsystems are differently perceived according to the perception level.

When work objective is applied cartography, a new problematic is introduced which is entailed to work objective and representation scale,which can coincide or not with that of the image.

So before starting a work it is imprescindible to ask oneself the following questions: a)Which is work objective?. b) Which is the reality observed?. c) What is reflected from this reality on the available imagery? d) Which scale is to be used in the presentation of work result?

Due to the huge extension of large plains they basically require a totalizing and synthetic vision of landscape, so as not to risk falling into partiality; nevertheless detailed work is also needed so as not to fall into generality. Few methods allow both aspects to be managed efficiently, rapidly and cheaply; remote sensing constitutes and advantageous technique, particularly when applied geomorphological works are performed.

GATHERED EXPERIENCE

Exposed concepts are the result of a large experience in the use of photointerpretation translated in numerous works. It has been performed at detail and semidetail level covering thousands of square kilometers which represented a permanent synthesis effort through succesive aproximations which still continue.

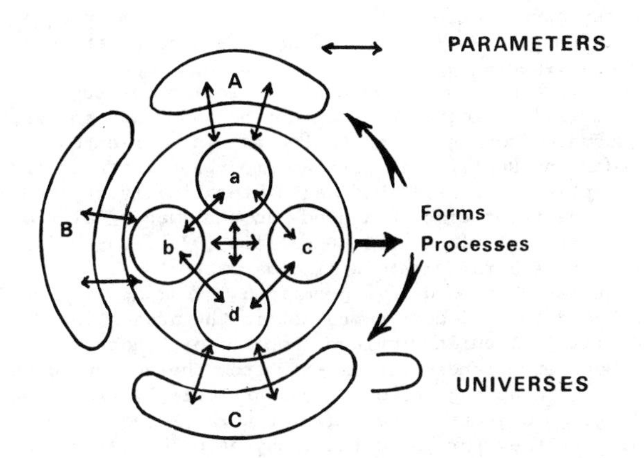

Figure 1. Geomorphological System. A.Climatic; B Anthropic; C Geodynamic; a Hydrologic; b Edaphic; c Lithostructural; d Biotic

We should not forget to mention thath the former works which systematically employed remote sensors for argentine large plains interpretation were performed by P.Pasotti (1966) and, on the other hand, were always performed as the base for Applied Geomorphology for engineering matters.

From the beginning of investigations the authors were confronted with the questions to already stated problems especially in what concerns to a valid and operative relationship between the landscape, the work objective and its cartography.

At those times satellite imagery were not available and even now proffesionals do not easily obtain them or they are not willing to employ them. Nevertheless authors permanent thoughts to seek for the adequate level perception to permit synthesis obtention; that also implies the need of scale changes. This is nowadays much more easily to solution than in those days.

This problem which appeared to be negative in former times resulted as the best profit in our methodo= logical experiences, due to the fact that it allowed to look for ways to find synthesis with available resources and not to leave aside any document or perception level (as Verstappen 1983 has refered in multiphase method). This point is not frequently taken in consideration by those investigators at their initial stage of application of the technique.

Each document original or generated (p.e.cartography, photoindex, photomosaics, etc) has its own different value but not less important than the others, except in objective function and work scale. There since most of the success depends on knowing how to choose and giving priority to the use of remote sensors documents, since, summing up, reality can be considered as the interaction of partial aspects reflected in each document.

Pattern mapping of Paraná and Paraguay abandoned river valleys was one of the former works performed. (Canoba, Popolizio 1968). It was basically reflected by vegetation physiognomy; changes of scale perception allowed the detection of these courses displacement dynamics which were of capital improtance for engineering projects in this zone. Lately Popolizio (1973) elaborated a work on genesis and evolution of Paraná river valley from Corrientes to Esquina by means of a combined analysis using Geographic Militar Institute cartography and photoindexes from Corrientes Province. That was the base for the studies of the "Plan Maestro de Defensa de la ciudad de Goya", stability of river banks and protection studies (Popolizio 1978, 1985)

In these works the cartographic problems were perceived, particularly when the considered area was relatively large, running the risk of loosing outstanding aspects of dynamic character which could be inferred in those original documents.

Since 1967 and following Ab'Saber experiences (San Pablo, Brasil) Popolizio has elaborated a paper on geomorphological cartography by which he tried the first aproximation towards morphogenetic, morphographic, morphometric, morphodynamic and morphocronological aspects representation which was adapted for the large plains cartography, employing the support of aerial photographs. Such method was shown in post graduated courses which took place in Corrientes, Posadas and Resistencia with an important interest and receptivity. This method was sistematically applied for the first time on a study about "Cuenca del Río y Estero del Riachuelo" (Popolizio 1973), for hydric resources arrangement and floods control.

It was also used in a geomorphological applied study related to road design project in the area "Dique el Cadillal",Tucumán Province (Popolizio, Canoba 1970).

Both experiences let us know from the beginning the need of adapting cartography, particularly the symbology, to objective and scale of work. Moreover the difficulties for the cartography of processes dynamic and its tendencies, fundamental factors for engineering projects and natural resources management, could clearly be perceived.

These problems were also the means of preocupation of different geomorphologists and Canoba was the one who had the opportunity of knowing and training himself under the guide of Prof. Verstappen and Dr. Van Zuidam in 1974 in the use of the named ITC System for Geomorphological Survey (1968) which also tackles the mentioned problems and constitutes an excellent aproximation for its resolution.

A highly important step due to work magnitude and the theoretical generation that it caused, was the Geomorphological Study of Bajos Submeridionales (Popolizio et al 1976)

The first fundamental aspect of this study is the clear perception of the importance of paleopatterns related to climatic conditions different from nowadays ones, that even when it compelled the mapping rerun of the third part of the area, was the starting point of a conceptual and methodological review. On the other hand space division in taxonomic geomorphological units was applied in a systematic and rigorous way.

To fulfill this task it was necessary to establish first the criteria for classification and consequently to determine the priority order of each geomorphological subsystem in its type and patterns. Second the level perception of geomorphology, hydrology and phytogeography for each unit had to be also determined.

Then it was necessary to study the whole conceptual and methodological tool intended to elaborate legend and cartography adapted to previously mentioned aspects

The result of what was already exposed was the clasification, typification and corresponding legend of the surface run-off systems (Popolizio 1973a) and also of the surface run-off network typification (Popolizio 1973b) by means of a combined matricial scheme.

By this way a method that allowed to state dynamic and tendency of natural and anthropic processes through typology, patterns and legend, especially in what concerns to surface run-off systems was obtained.

This fact has not been casual but it responds to intrinsic factors of geomorphological plains system where surface run-off systems and corresponding nets constitute a circulatory system analogy and so its detailed analysis allows the understanding of the major part of the geomorphic system global dynamic.

Finally, the recently shown evaluation has originated an immediate consecuence: restatement of geomorphologic concepts from the point of view of the general system theory and including logically the interaction between the anthropic and natural universes.

By this way the whole perspective of the problem was modified and it was considered as multivariable in space and time. But moreover, criteria introduced a new problem: how to interpret through static imagery a dynamic reality which has memory and which subsystems present different times of responses, so that imagery are not a present but a reflection of different past upon which present acts and both are projecting the future.

The return to the origin of the problem allows the reinforcement of the idea about the employment of all remote sensing resources, generated imagery and produced cartography, improved reality perception, since they are optative shapes of reality that we intend to aprehend and moreover, as many as the shapes perceived are, as nearer we are to the essence.

From this point of view we should remark that while Popolizio and his collaborators in Centro de Geociencias started from the analysis towards the synthesis, Pasotti, Canoba and their colleages from Instituto de de Fisiografía y Geología started from the synthesis to the analisis in adjacent areas which constitute a geomorphic unit of first magnitude.

Results convergence let the establishement of differently used imagery potentialities and its complement for the study of large plains and also for practical application.

METHODOLOGICAL CRITERIA

Work Objective

According to the questions stated in the introduction (a) which is work objective? Which is the reality observed?, What is reflected from this reality on the available imagery? Which scale is to be used in the presentation of work results?, the first problem to be solved is the clear visualization of work objective to be performed.This idea is directly related to the involved geomorphological unit size (fig. 2)as it was considered in a former work (Popolizio 1981) the importance of bioclimatic conditions, lithoestructural conditions and anthropic influence is different in relation to the relief size considered, particularly on plains.

Actually, it is not really the specifical objective the one to be considered especially, but the space dimension engaged by the objective and the intrinsic imagery capability matching to these scale and dimension, so as to reflect by means of the patterns, the aspects of reality which influence upon objective.

That is to say, it has been stated that this objective spatial aspect automatically circumscribes types of document and the aspects that can be detected through these documents.

At the regional level, plain dominant factors are lithoestructure and large vegetation units which frequently respond to major climatic controls.

Satellite imagery, photoindex or photomosaics assemblies reduced or perceived at the same scale of those, constitute the basic work element and allow the obtention of space division into subunits through detailed analysis of models, limits and transitions.

This first division must be immediately correlated with all other available landscape information (topography, vegetation, climate, soils, space organization, etc.)which has similar perception level; particular especial emphasis must be put on spatial correlations of models and changes in limit zones and transitions.

With a work like this, a first approximation can be obtained although very valuable about functional aspects of the geomorphological system,included those introduced by anthropic action and which can be altering natural tendency. All this will influence on the concrete possibilities of reaching the proposed objective and will make evident retroalimentation processes related between that and the geomorphologic system tendency (natural or anthropic).

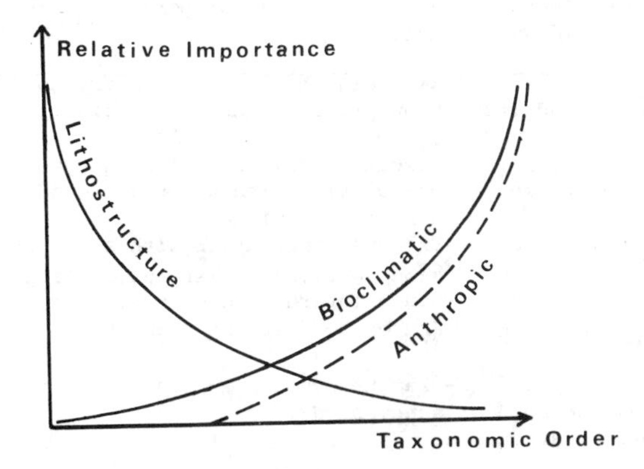

Figure 2: Scheme about forms size and its relation with processes and lithostructure influence.

At zonal level, any study complexity is the maximum as it can be deduced from fig.2 since determinant factors have the same or similar importance rank, making difficult the causalistic analysis. This is also seen in the election of remote sensors documents by which it is difficult to know which of them presents the best posibilities. In spite of what was said, these factors (lithoestructural, bioclimatic and anthropic) have their own dynamic and consequently they have different ways of expression which are concretely traduced into patterns imagery.

For that reason detailed pattern study particularly those corresponding to drainage net, surface runoff system and vegetation physiognomy , constitute the problematic fundamental aspect. But also it is imprescindible the detailed study of other patterns (edaphic, anthropic, etc), basically to correlate them with those already named. From this correlation the the interaction between different geomorphic subsystems can be deduced and from these to the anthropic universe. Such correlations reflect the natural tendency and the alteration being originated and lastly represent the basic determinants for the objective concretion.

In local level dominant factors are decidedly the bioclimatic and anthropic, but frequently it is important they were perceived at least as a reference level, into a zonal one which allows its integration in a spatial and functional whole.

Documents to be selected are strongly conditioned to the objective and it should be done carefully due to its high cost. On the other hand the photointerpreter adapted to this work scale has to be selected, as the requirements can be extremely specific (p.e. how to distinguish vegetal species into a phisiognomic unit).

In the geomorphologic aspect and specially in plain areas , patterns and nets of laminar and transitional systems (Popolizio 1973a b) which they are not frequently taken in consideration, constitute the interpretation key.

Different patterns correlations reach a most high specificity that requires indefectively a detailed field work.

To finish with this first point referred to the objective it seems convenient to emphasize that independently from the level where one work it is always indispensable to manage the other levels as reference ones, but of course with less intensity.

Observed reality

The second question of our interest about how the reality observed is, constitutes the most difficult problem to solve since we are able to aproximate just to what the essence is through the forms by which it shows itself.

These facultative forms or shapes of showing itself in direct observation (landscape) as well as to indirect (imagery) or generated (cartography) are the perception base of reality and its behaviour.

There we find the explanation why we cannot separate field work (direct observation)with imagery interpretation (indirect) and cartography, since all of them represent the same thing which we want to know and represent, so as to reach our established objective.

In the regional level plains are directly associated to the sedimentary basin dynamic of platforms reliefs and consequently controlled by crustal dynamic and subdivided by geotectonic domains.

In this scale, climatic zonality and mild typologic transitions impose apparent landscape monotony, even when there is a huge susceptibility in the plain to minimum topographic variation and anthropic action. Originated differences are perceived on the images through an indirect mechanism with retroalimentation, which is initiated with surface run-off and which is traduced in soils and vegetation patterns.

For the zonal level it is necessary to reiterate that it corresponds to the major complexity. In this level tectonic looses significancy and predominance of morphophysiologic conditions over the lithoestratigrafics and to eolic and fluvial paleopatterns.

Breaks of slope so important in plain, must be carefully studied since, even when they can be of tectonic origin, they can also come from erosion processes or from paleoclimatic origin. Surface run--off nets and systems together with vegetal physiognomies constitute the basical aspects observable which define geomorphologic subunits.

Patternspresented by anthropic universe, correlated with the previous ones admit the perception of natural tendency and provoked alterations due to detected anomalies.

In local level the smallest elements which form the landscape are considered; individual processes and microforms are perceived, which make the perception of organization patterns very difficult.

That is why this level must be simultaneously taken into account according to zonal level which constitutes the local level conditions context, thus facilitating its adequate perception.

Imagery and reality

The third question: how that reality is reflected on the imagery?. It can have different answers according to sensorial method used, to its resolution and scale.

Nevertheless, the important thing to be remarked is that whichever imagery were ,they just constitute a reality representation that turns to be known according to photointerpreter ability,to understand and relate the elements of imagery and reality.

Atregional level the difference between the analyzed geomorphological unit size and the relief forms and those daily perceived is enormous, so that except an austronaut, a common photointerpreter has no direct experience of this perception level and then he requires the maximum abstraction level for imagery interpretation.

On this point there is a clear relationship with

Table 1

	REGIONAL	ZONAL	LOCAL
OBJECTIVE	-State work objective clearly and the space involved. -Verify objective compatibility with size of the geomorphic unit. -Compile available information about topography, vegetation, climate, soils, space organization, etc. of similar perception level. -Use and documents selection: satellite imagery, reduced photomosaics and photoindexes. -Information evaluation, emphasizing specific correlations between models (patterns), and changes in limits and their transitions. -Try a global primary correlation; verify correlations.	-State work objective clearly and the space involved. -Verify objective compatibility with size of geomorphic subunits. -Compile available information about topography, vegetation, climate, soils, space organization, etc. of similar perception level including important field work data. -Use and documents selection: medium scale airphotos, enlarged satellite imagery. Stereoscopy is important. -Information evaluation, emphasizing particular aspects of models and nets (natural and anthropic). Special care in causes study. -Perform intrinsic correlation and also with the other two perception levels. Correlation with field work data in first approximation.	-State work objective clearly and the space involved. -Verify objective compatibility with forms and processes that can be perceived in the image. -Compile available information about topography, vegetation, climate, soils, space organization, etc. of similar perception level including very important field work data. -Use and documents selection: large scale airphotos. Relief study and stereoscopy are indispensable. -Information evaluation, supported by photointerpretation highly specialized, strong correlation with field work data; processes study emphasis. -Perform intrinsic correlation and also with superior perception levels.
REALITY	-With all the information (documents) try to find control parameters, climatic, geodynamics and anthropic. -Detection of regional breaks of slope (neat and transitional limits) and of anthropic models (zones) breaks. -With available information try to find cohesion factors of each zone. -Execute drainage net and surface runoff system analysis (former points correlation) and limits adjustments. -Execute soils and vegetation analysis on the base of tone and texture variations; correlation with former points. -Execute anthropic models analysis verify its correlation with former points. -Study interference degree of systems and subsystems. -Study how the system runs.	-With all the information try to know geomorphic subsystems and its dynamics; first correlations. -Execute models detailed analysis; correlation with constituent elements and processes also. -Execute paleoforms and paleomodels detailed study. -Execute correlation study between models of analyzed subsystems, including paleosystems. -Execute analysis of anomalies and lacks of correlation between models; determine structural (organization) and functional causes. -Execute dynamic approach and tendency study.	-With all the information try to know processes and tendencies. -Execute models detailed analysis; correlation with processes and constituent elements. -Execute correlation with sup. perception level, anomalies and deviations. -Study internal model variations and/or constituent elements. -Execute dynamic approach and tendency study.
IMAGERY	-Select images to be employed, those appropriate for units determination at this level. -Establish models and its correlation with reality, (large units, relief forms assemblies, lineaments and geotectonic models, major surface runoff systems, large vegetation and soils units, anthropic model organization of space. -Establish dynamic aspects detectable on images (e.g. headwaters retreat, eolian activity, etc.). -Elaborate global problem interpretation reflected on the imagery (static, dynamic and the tendency).	-Select images which offer best contrast of patterns and identification of constituent elements. -Establish specific models of analyzed systems and subsystems, (surface runoff nets, vegetation physiognomy and soils, anthropic space organization) and paleosystems as well. -Establish elements and organization net which constitutes each model; define limits types in detail. -Establish dynamic aspects detectable on the images and relation between elements and organization nets; feed back mechanisms relation. -Elaborate global problem interpretation reflected on the imagery (static, dynamic and the tendency).	-Select images that allows identification of elements and patterns, direct correlation with reality. -Establish elements and its associations and/or its spatial and morphologic relations. -Elaborate networks and constituent elements observed details. -Establish dynamic aspects that can be detected on the images and its tendency. -Elaborate punctual problem interpretation that have to be correlated with previous perception level.
CARTOGRAPHY	-Establish representation scale ≥1:100.000 -Establish representation priorities .Perceived geomorphic units and subunits. (taxonomic superior orders). .Surface runoff nets and systems. .Surface runoff and geomorphic processes dynamic and tendency. .Anthropic aspects and tendency (according to the objective). -Elaborate legend. .Homogeneous units designation on the base of geomorphology, climate, vegetation and soils. -Elaborate symbology.	-Establish representation scale 1:100.000 to 1:25.000 -Establish representation priorities .Models and limits. .Surface runoff nets and systems in detail. .Surface runoff and geomorphic processes dynamic and tendency. .Difference between actual models and paleomodels. .Anthropic aspects and tendency (according to the objective). -Elaborate legend. .Taxonomic and homogeneous designation of models. -Elaborate symbology.	-Establish representation scale ≤1:10.000 -Establish representation priorities .Elements and models. .Variations between elements and organization models. .Surface runoff systems and networks in detail. .Geomorphic and anthropic processes dynamic and tendency. -Elaborate legend. .Taxonomic and homogeneous designation of elements and models. -Elaborate symbology.

cartographic experiences, as a consequence obtained when small scale maps are carried out; it is clearly perceived what can be seen of reality.

Logically, in that perception level large geomorphological units or relief forms assemblies and structural, lithological or bioclimatic important control effects can be perceived.

By the way , according to the relation between the relief form size and its evolution time, paleomodels influence is far too strong particularly in the relief forms and the surface run-off networks. That is why, special care should be taken in the stablished correlations between the different geomorphic subsystem engaged, due to the fact that they represent different pasts.

Care must be taken in small details observations or any kind of anomaly. What it is of interest in this level is organization of models and nets as well as limits and/or transitions which the associated elements can indicate and not the isolated element itself.

Consequently, we could say that what it is reflected in this level, whichever the considered aspect were, is the maximum degree of landscape space-time organization and so multi-interdisciplinary vision of imagery interpretation is required.

At zonal level perception experience is not excesively far away from the everyday reality and it is near to one observed from an airplane by which interpretation is easier. Nevertheless as it was suggested in fig.2 reality perception reaches the maximum of complexity since, as it was already stated, the intervenient factors influence the same or similar range of importance.

Frequently, and as it was already seen, some element arrangements can be interpreted as provoqued by factors different from those which gave them origin. For example a clear depressions alignment which could be interpreted as of tectonic origin, may be of karstic, pseudo-karstic, eolic, etc. origin.

For the same reason and in spite of that plain breaks of slope are so important, they are not always tectonic and when they are so, they can be displaced of their original position by means of erosion or paleoclimatic processes.

Summing up we coul say that what it is reflected in this level is maximum correlation and the space-time interference between geomorphic and anthropic processes. From this point of view we could say that this perception level is the key for morphodynamic interpretation an consequently for the system natural tendency and the anthropic influence.

At local level perception is almost identical to the habitual one, and for the same reason the photo-interpreter could get the ability to see "what he must see", that is to say, a great degree of specialization.

This is the optimum level to recognize individual processes, microforms and smallest elements which integrate the landscape.

Nevertheless it should be remarked that so as to finish this point, different perception levels are complementary, since at local level perception from organization space-time correlation and interference degree is minimum.

Scale representation

The last question formulated:Which scale is to be used in the presentation for work result? It leads us to one of the principal point of the paper.

It can occur that interpretation were performed at one scale and mapping carried on in a similar one, or that a totally different scale were used for cartographic representation. This makes us reiterate that the existence of an important analogy between imagery and cartography. Both are realities "representations" and then what we have stated for different perception level imagery is valid for the cartography.

That is to say, we cannot expect to obtain from the mapping elaboration process more than what reality can be reflected at that scale. This suggests the importance of the correct scale election and the problems than can be derived when image and map have not the same or similar scale.

Another very important aspect of cartography is that it should represent by means of symbols or legends those geomorphologic basic aspects.: Morphometry, morphogenesis,morphocronology, morphotaxonomy and morphophysiology. In spite of that the taxonomic or perception leveldetermines the major and minor importance of these aspects and the corresponding elements of each of them. It should be remembered that the purpose of the study also plays its role.

It is also necessary to remark, since it is not as much considered, the importance that correct and strict limit of geomorphologic elements represented, since correct model perception results from mentioned accuracy. All what was stated reaches more importance as far as work scale becomes greater. From that it can be deduced that detailed geomorphologic survey performed is important in every level, even when it implies the use of adequate symbology and appropriate generalization at each of them.

Taking in consideration that reality perception can be performed just as much as dimension scale, limits, forms and background and figure contrast are known. Then we can understand that when we speak about cartography, we want to point out that those aspects have to be correctly represented,so as not to risk the relations between models with non correct causes.

Experience gathered from all performed plain works indicates how immensely important election is and/or geomorphologic symbology and legend preparation and more still when work is done of applied character.

Difficult task constituted by thematic cartography causes numerous questions. The first of them is whether all those geomorphological aspects above mentioned will be represented by symbols, legends or both combined.

The second is that the individual symbol or that associated in a model must in any way be related to habitual perception of geomorphological event or element to be represented.

The third question stated is that the symbols or its association must reflect, immediately, the relative importance of what we want to represent.

The fourth treates about linear elements must be represented with lineal symbols and in the same way the areolar ones.

The fifth question is about legend which organization must allow the comprehension of geomorphic processes, dynamic and tendency thus pointing out dominant process or agent, its temporality and the corresponding form.

Not intending to have an exhaustive enumeration, effective possibilities of printing elaborated cartography and the related problems have to be considered. To that respect we lack possibilities for color printing as well as the support of experienced cartographers for this task.

About the cartographic elaboration, Verstappen points out: "The basic problem of the cartographic elaboration of geomorphological maps is the great variety in types of information that one could include. Restraint in this respect is an absolute necessity; otherwise the maps produce will be complex, costly in printing and difficult for the user to read. The limited range of cartographic means of expression restricts the tendencies to undue perfectionism, although great care is always required to not overload the maps. Simplicity by emphasizing the most essential information and generalizing or omitting the less important should be the guiding priciple" Part of the information required may be included in the description of the geomorphological phenomena given in the legend and not in the map itself".

The above mentioned questions were gradually solved according to these large plains geomorphological characteristics which make the cartographic concept elaboration more difficult, particularly at semidetailed level (zonal level, the more complex). Due to its importance plains surface run-off system was studied a classification elaborated. The classification llows a systematic study reordering the existent terminology, moreover the conceptual content of each term was reformulated according to its dynamic. (Popolizio 1973a)

Linear symbols and letters were used for the cartography. It could be pointed out that the transfluence from one basin to the other is particularly indicated, since it is a frequent phenomena in Argentine plains and very important from the applied geomorphology point of view. This work made the photointerpreter task and its corresponding cartography easier.

At the same time and considering that different types of surface run-off frequently tend to reach an ordered spatial association, though not so well definned as in the major amplitude relief, surface runoff network must be specially taken into account (Popolizio 1973b)

Fourteen elemental network models were defined and also characterized into integrated or desintegrated, whether they are divergent or convergent. After these models we can pass to more complex patterns. The last ones are the most frequent all the more as larger the analyzed network is and climatic, edaphic, vegetational, geological, or geomorphological variance of the area where it is developed. The advantage of considering elementary models, consists in that any variance from one towards the other, indicates certain conditioning by any of the above mentioned factors, without leaving aside such a widespread influence as the anthropic. Evidently these model changes can require an adequate observation scale and so it is not advisable to manage only one.

Using a matrix in which diagonal appears pure elements, possible models have been indicated and supposed elements combinations in twoes, the composed network.

By means of an adequate descriptive formula the type of network is defined, which allows also to indicate the existence of two models interjoined in a sole net, a paleonet presence and the sequential trend towards a determined net. In this way dynamic aspects can be established to the use of formulas that have been incorporated to the legend. (Popolizio 1973b).

On these fundamental elements, connect to the correct interpretation of plains geomorphological characteristics, the thematic cartography elaboration was carried on.

For a geomorphological map at semidetailed level, scale 1:50.000 (La Escondida) an areal symbology was designed for corresponding mapping units. They belong to forms of structural origin, forms of fluvial origin, paleoforms of eolic origin and corresponding symbology to morphometry, transport and urban infraestructures, also.

It should be pointed out that lithology is not included in these maps, appart from corresponding printing difficulties. From one hand the relative homogeneity of quaternary sediments must be considered and moreover, if its variation has an important geomorphologic value, it is indicated on legend.

Due to originated importance or vegetation influence to define geomorphic units, the same as from the applied point of view, phytogeographic chart at equal scale is elaborated at a physognomy level. Corresponding limits are determined and accompaning letters are shown and defined in the legend. Geomorphologic maps at scale 1:250.000 were also elaborated where structural elements are indicated (lineaments) and the units of second order limit, as well as the corresponding subunits of third and fourth order. They were accompanied by numerical symbols related to decimal classification system, which were expressed in the legend by unit definition and its corresponding vegetal association.

On the same areas and at equal scale synthesis maps about surface runoff dynamic were elaborated. There, basins, river courses and its features, transfluence, headwater retreat, etc. were defined.

In every basin identified by a letter the surface run-off system is described in the legend.

From the applied geomorphology point of view and the water resources management, this thematic map is essential and very useful.

On the pampean plain sector which corresponds to the south of Santa Fe Province and its adjacent zone belonging Paraná river valley and its delta, a geomorphological mapping at small scale 1:1.000.000 was performed on the base of Landsat imagery.

On these areas members of the Instituto de Fisiografía y Geología have carried on previous studies along two decades and more, using aerophotographies and topographic maps at different scales and including field works.

Structural lineaments are indicated in cartography to which the limits of second order units are associated and third order subunit are considered as forms of structural origin, hydrographic networks of major hierarchy and lakes were included using conventional symbol. On the base of detectable elements on Landsat imagery; fourth order subunits were detected and indicated separately using conventional symbology also (Canoba 1982)

The geomorphological mapping at 1: 1.000.000 scale of one adjacent sector of Paraná river delta was a very valuable experience where the ITC system for geomorphological surveys were used. Geomorphological units and subunits determinations could be used due to the fact that geoforms can be perceived in the small scales images thus allowing its interpretation. This occurs for the reason that Paraná river Delta has an unusual dimension compared to those corresponding to other deltas in the world.

It should be pointed out that previous studies of some of these areas with terrain classification purpose on the bases of their geomorphological characteristics were performed at 1:100.000 scale. Geomorphological surveys at 1:20.000 scale was carried on in key sectors. (Pasotti et al 1976).

Aerophotography and cartography at different scales utilization and particularly Landsat images from different periods of the hydrologic year, allows the improvement of some aspects of fluvial dynamics interpretation.

CONCLUSIONS

As conclusion of our experience about geomorphological studies on Argentine plains, using remote sensing, we realize that frequently the same method corresponding to the zonal level is used (been the one where the most of the experiences were carried on).

Nevertheless it is considered that it should not be so and is very particularly reiterated for plains studies.

Each survey level implies a corresponding perception level and a particular methodological adjustment (Table 1).

We can also define three types of photointerpretation (image interpretation according to the above mentioned levels: (the generalist, the integrator and the especialist). Even when they are complementary, each of them has a great importance either from the interpretation point of view or from applied geomorphology. The regional level surveys are used for the sake of taking decisions, those of zonal levels are for planning and those of local level are for specific projects or works. That is why synthesis stated on Table 1 is presented as a contribution, being supported by our experiences and the expressed concepts. So we can consider it as a useful base for work performance as much as for photointerpreters formation.

REFERENCES

Canoba, C. 1982. Geomorphological mapping using Landsat imagery: a case study in Argentina. ITC Jour. 1982-3, p.324-329. Enschede.

Canoba, C. & Popolizio, E. 1968. Estudio aerofotográfico de paleopotamología en un sector ribereño del río Paraná-Chaco. Inst.Fis.y Geol. Notas N°2 UNR

Pasotti, P. & A. Castellanos 1966.Rasgos geomorfológicos generales de la llanura pampeana. Conf.Reg. Latinoam. Vol.III.UGI, México.
Pasotti, P., Canoba, C. & W. Catalani 1976. Aerofotointerpretación del Delta Entrerriano. Inst.Fisiog. y Geol. Pub. LX, UNR. Rosario.
Pasotti, P. & C. Canoba 1979. Tectonic lineaments in a sector of the Argentine plains (pampa). In Podwysocki M. & J. Earle (eds.) Proc. 2nd. Int. Conf. on Basement Tectonic. p. 435-443. Delaware.
Popolizio, E. & C. Canoba 1970. Informe Geológico-Geomorfológico para el proyecto vial circuito turístico Dique El Cadillal. Dirección Provincial de Vialidad. Tucumán.(inédito).
Popolizio, E. 1972. La carta geomorfológica. Lab. de Geol., Fac. Ing. y Fac. Humanid. UNNE. Serie A. N°9. Resistencia.
Popolizio, E. 1973a. Los sistemas de escurrimiento. Cent. Geociencias Apl. UNNE. Serie C. Inv. T2 N°2 Resistencia.
Popolizio, E. 1973b. Las redes de escurrimiento. Cent. Geociencias Apl. UNNE. Serie C. Inv. T2 N°3. Resistencia.
Popolizio, E. 1973c. Contribución a la Geomorfología de Corrientes. Ins. Fisiog. y Geol. Notas N°6 y 7. UNR. Rosario.
Popolizio, E. 1973d. Informe de Geomorfología. Estudio de recuperación de áreas inundadas, Estero y Río Riachuelo. Corrientes. T. 1 y 2. Min. de Defensa. DIGID.
Popolizio, E. 1976. Geomorfología de los Bajos Submeridionales en el área de Chaco y Santa Fe. Convenio CFI-UNNE (inédito).
Popolizio, e. 1978. Fotointerpretación aplicada al estudio de la cuenca del río Negro. Chaco. Cent. Geociencias Aplic. Serie C Inv. T14. UNNE. Resistencia.
Popolizio, E. 1981. La teledetección como apoyo a la neotectónica del nordeste argentino. Exposed at 26° International Congres Geol. 1980. París. Abst. p.829.
Popolizio, E. 1983. Teoría de sistemas aplicada a la Geomorfología. Cent. Geociencias Apl. UNNE. Serie C N°9 . Resistencia.
Subsecretaría de Recursos Hídricos de Corrientes. 1985. Plan Maestro de Defensa contra inundaciones de la ciudad de Goya . Corrientes. Informe interno.
Tricart, J. 1965. Principes et methodes de la Geomorphologie. Masson. París.
Verstappen, H. & R. Van Zuidam 1968. ITC System of Geomorphological Survey. ITC text books VII. Delft.
Verstappen, H. 1977. Remote Sensing in Geomorphology. Amsterdam. Elsevier.
Verstappen, H. 1983. Applied Geomorphology. Amsterdam Elsevier.

Symposium on Remote Sensing for Resources Development and Environmental Management / Enschede / August 1986

Use of (stereo-) orthophotography prepared from aerial and terrestrial photographs for engineering geological maps and plans

Niek Rengers
International Institute for Aerospace Survey and Earth Sciences (ITC), Enschede, Netherlands

ABSTRACT: Orthophotography, prepared by optical mechanical correction of aerial or terrestrial photography, is imagery with a uniform scale and without relief displcement. Together with so called "stereomates", prepared from overlapping imagery, stereovision can be obtained. Examples of stereo-orthophotography are shown with their engineering geological interpretation. A comparison is made of the interpretability of stereo-orthophotography at various scales as compared with normal stereo photography for various terrain details which are used for delineation of engineering soil and rock units as well as for slope instability phenomena.

ORTHOPHOTOGRAPHY

Stereo aerial photography and stereo terrestrial photography are important tools for the engineering geologist. Interpretation of the stereo imagery under the mirror stereoscope is very helpful during the early stages of preparation of an engineering geological map or plan for a civil engineering project (IAEG, 1981).

One of the main problems encountered when using photography is that due to the central projection the image shows important variations in scale and radial relief displacements when the terrain is not completely flat. Thus the transfer of information from a photo overlay to a topographical map (a vertical parallel projection of the terrain on the horizontal datum plane) causes a large amount of extra work and an important source of errors.

Orthophotography offers an elegant way out of these problems. An orthophotograph is made from an ordinary aerial photograph by optical-mechanical correction of the radial relief displacement and scale differences with help of an orthophoto projector. With modern equipment orthophotography imagery of excellent quality can be produced. The photographic quality is not notably affected by the projection process. The topographic quality (positional accuracy) depends to a large degree on the accuracy of the available digital terrain model but in normal circumstances errors in position should not exceed the order of magnitude of 5-10 m on orthophotography at a scale of 1:10,000 ($\leqslant$ 1 mm in the orthophoto). At these scales such an accuracy for an engineering geological map is quite acceptable. Orthophotography can be printed on the (uniform) scale of the map on which the engineering geological information must be plotted. The transfer of data from a photo to its orthophoto is still extra work but the sources of error have decreased strongly as features on the photo and orthophoto can be correlated much easier with each other than features from photo and map.

Orthophotography can be made from aerial photography, but also from terrestrial photography if a metric camera was used to make the terrestrial photographs (Grabmaier, 1983).

Figure 1 shows part of an aerial orthophotograph at a scale of 1:5,000 onto which the contour lines of a topographic map with contour line interval of 10 m are printed. This orthophotograph has been prepared from original negatives at a scale of 1:22,000.

STEREO ORTHOPHOTOGRAPHY

With help of the orthophoto projector a "stereomate" can be produced of photography overlapping the photography from which the orthophotograph was prepared. Into the stereomates an artificial relief displacement in x direction is introduced which causes x parallaxes between orthophotograph and stereomate which equal the x parallaxes of the original stereo model.

When orthophotographs and stereomate are viewed stereoscopically with a mirror stereoscope a three dimensional image of the terrain is visible which is not distinguishable from the original stereomodel.

When stereo orthophotography are prepared exactly at the scale of the map to be prepared the overlay on which the photo interpretation results are plotted can be used directly as a map after introduction of the contourlines from a topographic base map.

INTERPRETATION OF (STEREO) ORTHOPHOTOGRAPHS

Aerial photography.

The interpretation of aerial stereo orthophotography follows exactly the same lines as the interpretation of normal aerial photography. Engineering soil and rock units, slope instability phenomena and erosion features can be recognised and outlined on the photo-overlay (Soeters and Rengers, 1981). A comparison of the quality of the engineering geological interpretation between stereomodels of aerial photographs and their orthophotographs and stereomates has shown that orthophotography gives comparable and in some cases even better results (Sissakian, Soeters and Rengers, 1983).

Terrestrial photography.

In cases where vertical outcrops have to be mapped in detail for engineering purposes, the transfer of information obtained from stereoscopic interpretation of overlapping terrestrial photography to a base map is particularly difficult. The reference plane should be a vertical plane on which the interpretation results must be plotted with parallel projection in a direction normal to the reference plant.

In this case orthophotography (fig. 3) can be particularly useful and the digital terrain model obtained photogrammetrically from the metric photography can be used to plot as well an equidistance

Figure 1. Orthophotograph at a scale of 1:5,000 prepared from aerial photography at a scale of approximately 1:22,000. Contourlines have been plotted from the original photography and were introduced photographically.

Figure 3. Orthophotography of part of a rock outcrop. The upper part of the orthophotography is disturbed due to the fact that the Bench is not completely represented in the original photographs.

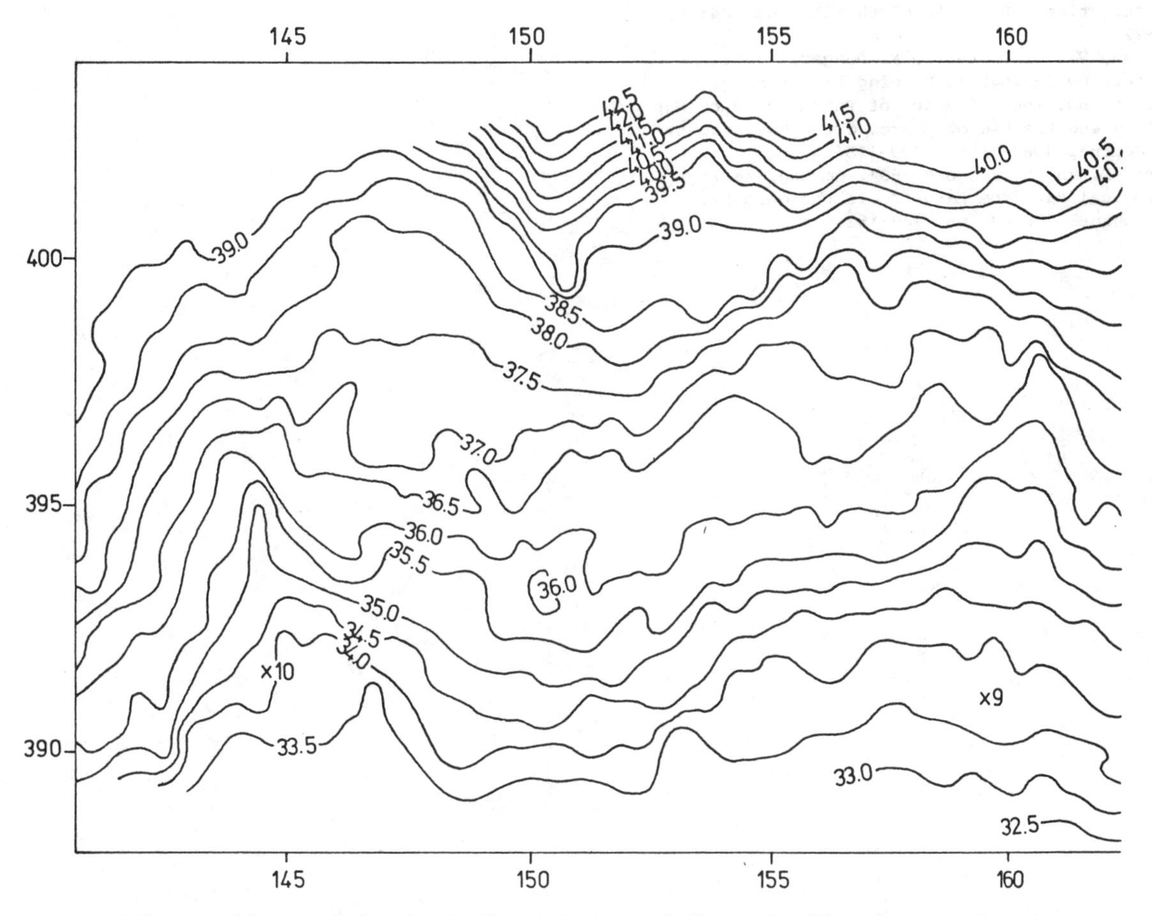

Figure 4. Equidistance line map belonging to the orthophoto of figure 3. All number are given in m. Equidistance indications with reference to the camera base line.

	1:20 000		1:10 000		1:5 000		1:3 000	
	AP	OP	AP	OP	AP	OP	AP	OP
Differentiation of rock and soil units	-/o	-/o	o/+	o/+	++	+++	++	++
Recognition of slope instability phenomena	o	o	+	++	+	++	++	+++
Recognition of erosion phenomena	o	o	+	+	++	++	++	++

- = inadequate; o = of limited use; + = useful; ++ = very useful:
+++ = extremely useful

Figure 2. Comparison of the quality of interpretation using aerial photography (AP) and orthophotography with stereomates (OP), at various scales prepared by enlargement from original photography at a negative scale of 1:22,000.

line map (fig. 4) on which the photo interpretation results can be plotted.

REFERENCES

Grabmaier, K. 1983. Production of proper stereo orthophotographs. ITC Journal 1983-2: pp. 119 to 122.

IAG, 1981. Report of the IAEG Commission on Site Investigations, Bulletin of the IAEG No. 24: pp. 185-226.

Sissakian, V., R. Soeters & N. Rengers, 1983. Engineering Geological Mapping from Aerial Photographs: the influence of Photo Scale on Map Quality and the Use of Stereo orthophotographs. ITC Journal 1983-2: pp. 109-118.

Soeters, R. and N. Rengers 1981. An Engineering Geological Map from Large Scale Photography. ITC Journal 1981-2: pp. 140-152.

Symposium on Remote Sensing for Resources Development and Environmental Management / Enschede / August 1986

Small scale erosion hazard mapping using landsat information in the northwest of Argentina

Jose Manuel Sayago
San Miguel de Tucumán, Argentina

ABSTRACT: A methodology for the erosion hazard mapping at small scales is described through a survey carry out in an extensive region of the northwest of Argentine with climatic condition fluctuating from semi arid to wet sub-tropical. Interpretation of multi-temporal landsat images and conventional aerial photo-interpretation permit delineation of mapping units characterized by the recurrence of landform and land use pattern. Field measurement of the USLE factors-rainfall erosivity, soil erodibility, slope steepnes and lenght and coverage-allow to stimate the erosion hazard expressed in potential rates of soil loss following the criteria suggested by S.A. El-Swaify (1977).

INTRODUCTION

In developing countries small scale erosion maps might be an important tool in order of the regional development. They contribute to detect areas seriously affected by erosion processes where application of conservation measures is prioritary. In areas of natural vegetation ready to be transform in cultivated lands through deforestation they allow to stablish the potential risk of degradation. Finally, they identify those regions that for the complexity of its erosive features and the variety of the environmental influences must be studied with more detail.

The region under study, representative of the northwest of Argentine for its climatic and geomorphological conditions, appears severely affected by differents erosive processes, although there are areas still not disturbed. The objectif of this work is to carry out an erosion hazard survey in order to stablish methodological criteria for the mapping of extensive regions at scales ranging between 1:250.000 to 1:1.000.000.

For the purpose of this work the geomorphological units defined by the recurrence of internal landscape characteristic represent the basic information for the mapping of erosion.

In that sense, the use of satelite information assures acceptable planimetric base, a multidisciplinary perspective and an objective pursuit of the environmental changes. For the evaluation of erosion by water the application of parametric criteria contributes more objectivity particulary through the use of methods as for example the Universal Soil Loss Equation.

THE GEOMORPHOLOGICAL SECTORIZATION

Based on the ideas of Verstappen H. Th.(1983), Zonneveld J.I.S.(1983) and Tricart J.(1982), a relief classification for the study areas has been established. In this classification relief and climate appear as independent factors acting as the "substratum" for the other landscape factors (soil, vegetation, hidrology, fauna, etc.). The classification includes five levels of complexity or scale: geomorphological province, geomorphological region, geomorphological association, geomorphological unit and relief element. Each category correspond to a determined scale interval that represents a hierarchical level of relief sectorization. At each hierarchical level, information on landscape factors such as for example hydrology, vegetation, soils or fauna, with similar levels of complexity are to be integrated (Sayago J.M.1982). At the scale of this work seven geomorphological associations, (FIGURE 1) following partly the concepts of Van Zuidam (1983), were characterized on the basis of the regional meso-climate conditions and the geomorphological features of major recurrence.

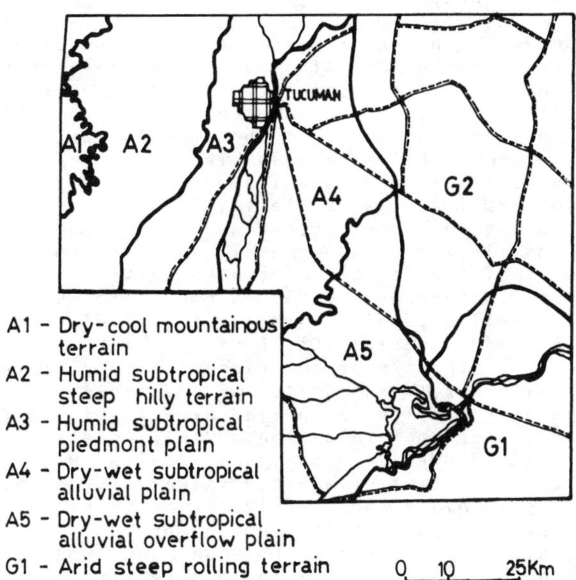

FIGURE 1 Geomorphological associations characterized by a meso-climate and a typical relief pattern.

THE EVALUATION OF EROSION HAZARD

Erosion hazard can be described as the chance that acelerated erosion will start in the near future. In cases where accelerated erosion is already in progress the erosion hazard would be the degree of further erosion that can be expected in the near future. Erosion hazard can decrease when effective

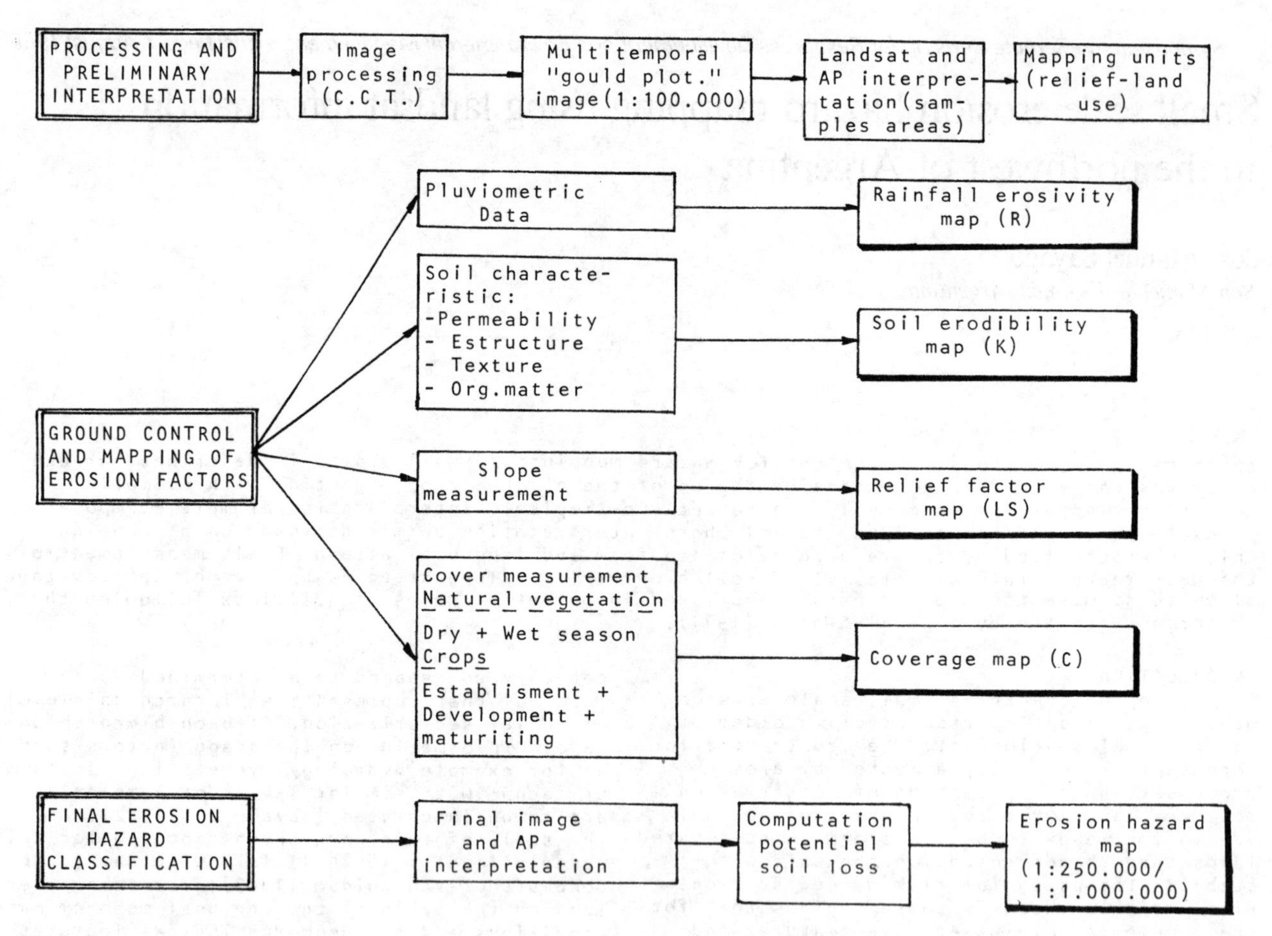

FIGURE 2Scheme of the main stages for processing and sampling the information related to the erosion hazard mapping

conservation measures are taken in the area. In this context erosion hazard is the combined effect of all erosion factors: climate, relief, soil, land use and management (Bergsma, 1981).

For the evaluation of the erosion hazard each geomorphological association was subdivided into mapping units defined on "gould plotter" images at scale 1:100.000 by a distinctive land use pattern and minor land forms. In every mapping unit field measurements of the factors of the Universal Soil Loss Equation were made: rainfall erosivity (R), soil erodibility (K), relief influence (LS), and coverage (C). Finally, ranks of soil loss in tn/ha/year corresponding to different erosion hazard categories were establish following the criteria applied by S.A. El Swaify (1977) for the evaluation of soil loss in Hawaii (FIGURE 2). The innacuracy derived from the computation of average values of the USLE not disminishe the usefullness of this approach for the erosion mapping at regional level. With respect to this Janssen(1983:120) said, "in many areas a rough potential erosion map based on the USLE and covering large areas might be useful to organizations which intend to undertake erosion control measures to improve agricultural production".

THE LANDSAT INFORMATION

The use of Landsat images gave an acceptable planimetric base considering the poor reliability of the regular cartography at small scales. In the same way the "gould plotter" enlargements permitted the clear definition through visual interpretation of mapping units integrated by typical landforms (f.e.alluvial fans, fluvial valleys, etc.) generally covered by uniforms land use pattern.

The coverage evaluation from multi-seasonal images reflecting different biomass densities was obtanined both by digital processing and visual interpretation. The use of those approaches depends both on the availability of adquate equipment and the cost-benefit of their application taking into account the extension of the studied area and the requiered mass of field information. According with Townshend (1981) a visual interpretation can prove to be much more cost effective for many task than computer implemented methods. The amount of data in a single Landsat frame is enormous and consequently the computational time for even a large computer system can be very high. If small scale map production is required, then pixel by pixel classification may be unnecesarily detailed.

RAINFALL EROSIVITY

The rainfall erosivity indexes are parameters which derive from the characteristics of the rain; for their direct correlation with various erosive processes (splash, sheet, gully, etc.), they are used in the predictions of the soil loss. The one which is best know has been developped by Wischmeier and Smith(1978) who based their investigation on the relationships between soil loss and the characteristics of the rainfall (quantity, intensity, impact and drop moment). The index expresses the product of the cynetic energy and the maxima 30 minutes of intensity of a rainfall

(EI_{30}). Nevertheless, its application shows limitations in areas lacking pluviographical records, a fact which is frequent in most of the developping countries. To avoid this difficulty an index obtained by Arnoldus(1978) was used, i.e. $\sum_{i=1}^{12} \frac{p^2}{P}$, where p=monthly rainfall and P=annual rainfall. This index has the adventage of employing simple meteorological datas and good correlation with the values of "R" given by the USLE. To achieve such a correlation, the following general equation was established: $R = a \times \sum_{i=1}^{12} \frac{p^2}{P} + b$, where "a" and "b" are constants derived from regional climate conditions. This equation was tested in the United States and in Africa where a high correlation with the pluviographical values of "R" was obtained(Arnoldus,1978). Following this methodology, iso-erosivity curves were calculated for the studied area (FIGURE 3). Its analysis revealed that in the eastern part of the studied area the rainfall erosivity is relatively weak and increases toward the west in parallel with the increase of the orographic precipitations. On the intermediate high slope of the eastern mountainous area the values fall abruptly, thus responding to the diminution of the rainfall. In summary, a high correlation between pluviosity and rainfall erosivity has been verified.

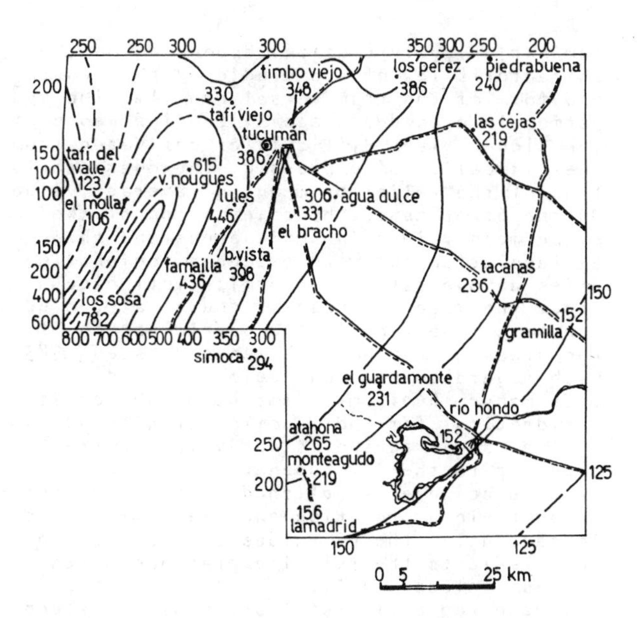

FIGURE 3 Iso-erodent map(factor R of USLE) of the study area

SOIL ERODIBILITY

The sense of the term "soil erodibility" is different from the one of the concept "soil erosion": The mass of soil loss caused by erosion can be influenced at a higher degree by the slope, the coverage or the management than by the intrinsic proprieties of the soil. Nevertheless, some soils get more eroded than others even when all the other factors are similar. This difference, due to internal proprieties of the soil is called "erodibility" (Wischmeier and Smith,1978). Among the methods applied for its determination, the most practical one is the well know nomograph included in the Universal Soil Loss Equation. For the calculation of the soil erodibility of the area, the most representative types have been sampled and their erodibility calculated on the basis of the information required by the mentioned nomogram(percentage of silt and very fine sand, percentage of sand, percentage of organic material, structure and permeability).

The values of the soil erodibility in the area are shown in FIGURE 4; there is an obvius relationship with the distribution of the original materials and the regional climate. Thus, it can be seen that in the extreme west the semi-arid conditions and the presence of siltstones of the Tertiary determine a moderate erodibility. On the eastern slope of the Aconquija and Cumbres Calchaquíes, under wet sub-tropical climate, the decrease of the erodibility reflects the presence of the mountainous wood and its positive influence on the structural stability of the soils. On the opposite, eastwards, the values increase in parallel with the presence of loess and inversely with the rainfall diminution. In summary, the relatively high values of erodibility given by the soils of the area must be attributed mainly to the constant presence of loessoid material among their original materials.

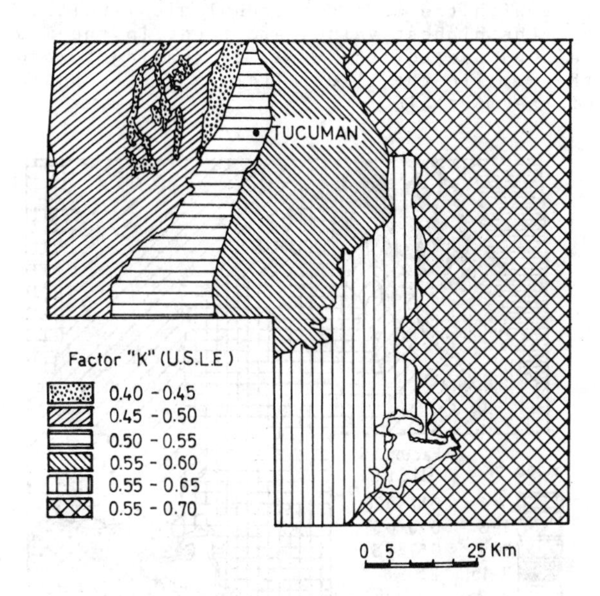

FIGURE 4 Regional distribution of soil erodibility average values

THE SLOPE INFLUENCE

The two aspects of the slope which have the greatest influence on the erosion processes are its steepnes and its length. The steepness has an influence in the sense that the steeper the slope stronger is the direct effect of the rainfall(splash erosion). In the same way, the infiltration time will be shorter because of the higher runoff speed which on turn increases its erosion capacity. On the other hand, the longer a slope is, the greater are its possibilities to receive rain. The kind of influence of the slope shape on erosional processes is not exactly known, because the complexity given by the changeable influence of the type of soil and vegetal coverage. Nevertheless, it is assumed that the concave slopes tend to concentrate the runoff and facilitate the gully erosion, while the convex slopes disperse the runoff, facilitating the laminar or the rill erosion. On the straight slopes predominate either laminar or gully erosion, depending on the characteristics of the gradient, the length and form of the transversal profile. The evaluation in extended areas of the influence of the slope arises difficulties derivated from the irreality of expressing the characteristics of the relief by means of average values. On the other hand, the calculation of the slope mean values in small areas and their extrapolation for more extended regions implies an overestimation of its influence, since the more extended is the area studied, the more important is the participation of the areas with minor gradients(Janssen,1983). With regard to studied region, considering that many landforms seems be morphometrically identical for their common origin and evolution, some ranks of slope values corresponding to the morphogenetic units of major recurrence were established by means of aerial photo-interpretation and field measuring. Consequently, the LS values of each mapping unit reflects the relative presence of the shapes included in it.

In the regional distribution of the steepness and slope length values, it can be seen (FIGURE 5) that these values generally coincide with the main geomorphological units. So, the highest values are found in the mountainous area of the west, decreasing toward the east in the alluvial plain at the bottom of the mountain.

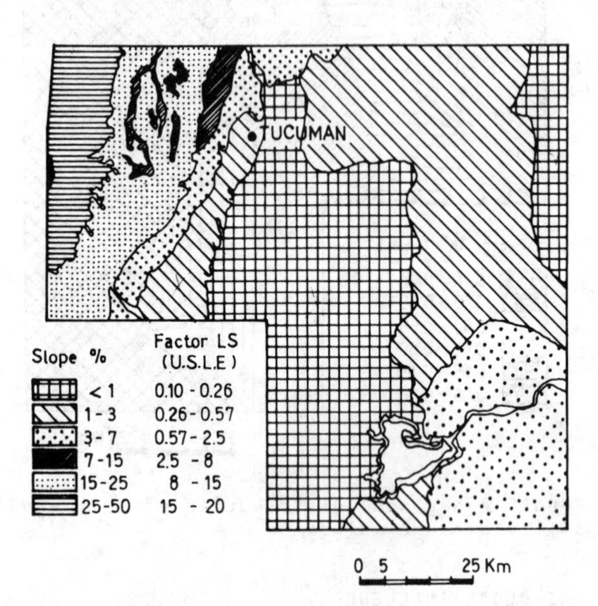

FIGURE 5 Regional slope influence

THE COVERAGE FACTOR

It is well know that the vegetal coverage protects the soil from the direct impact of the rainfall, facilitates the storage of the water and the biological activity, and reduces the dripping. Nevertheless, its importance increases in the tropical and sub-tropical regions because of the aggressiveness of the climate factors. In the studied area, the evaluation of the vegetal coverage as a factor of erosion risk is complicated by the marked seasonal climate contrasts and the wide extension and varieties of physiognomies and crops types. The criteria applied to obtain the coverage values (factor "C" of the USLE) were the following ones:

a) Determination of the area occupied by each culture type and vegetal physiognomy within the area.
b) Setting of the extrems of the natural vegetation development (dry and wet season).
c) Definition of the calendar of the different vegetative cycle stages of the main crops in the area.
d) Representative coverage measurements of the different types of natural vegetation growing during the dry and the wet season.
e) Representative coverage measurements of different crops in three stages of its vegetative cycle (establishment, development, maturing).
f) Calculation of the factor "C" of the USLE for each type of crop and natural vegetation.
g) Calculation of weighted averages of "C" values, corresponding to different coverage types, in each mapping unit.

The evaluation registred by of the area occupied by each culture was made on the basis of historical records given by technical agencies. The determination of the percentage of land occupied by the diverse coverage types in each mapping unit was made through the interpretation of Landsat pictures taken during the dry and the wet seasons, aerial photos, and ground controls. The coverage measurements and calculation of the "C" values were carried out after the criteria recommended by the Manual 357 of the United States Departament forAgriculture.

On the map of FIGURE 6 it can be seen that there is a high correlation between the soil occupation, the climate and the coverage values. The eastern slope of the mountainous region, covered with subtropical wood, possesses the greatest coverage ($C<0.1$). Toward the west, the presence of deciduous forest and summit grass, which can be considered as a indication for minor precipitation and temperature, show up a slightly coverage diminution (0.11-0.15). In the central and eastern part of the area, the Chaco forest, though affected by degradation processes, shows moderate coverage values (0.16-0.20) in comparison with the ones of the main crops.

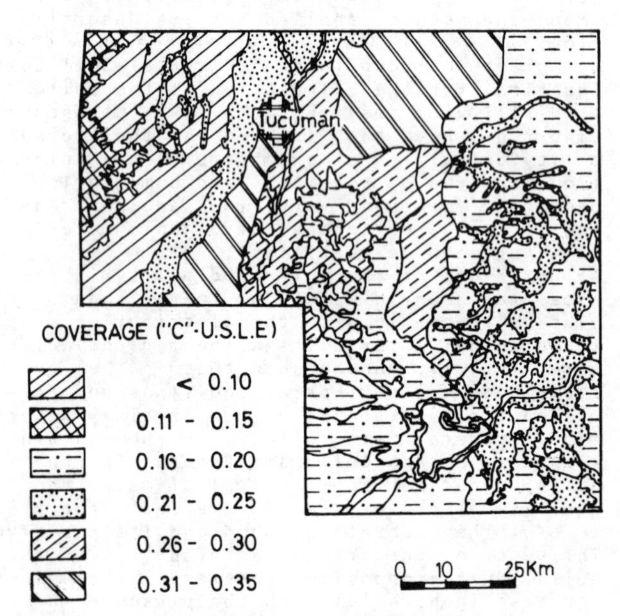

FIGURE 6 Coverage values of the main crops and natural vegetation units in the region

In this sense, the citrus plantations in the piedmont plain shows lower coverage values (0.21-0.25) despite of lesser exploitation, which are similar to the ones of the sorghum in the ondulated eastern plain. Finally, in

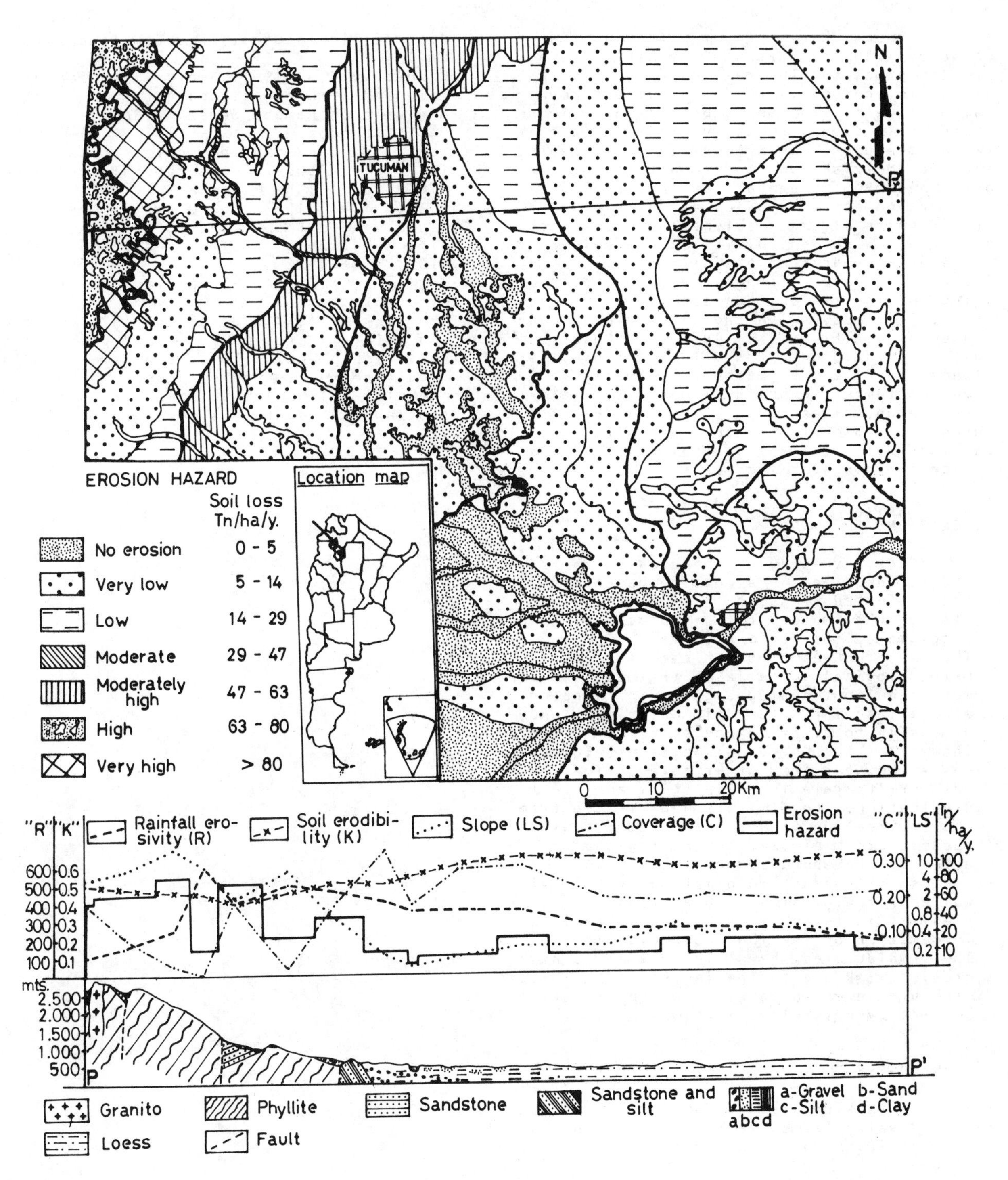

FIGURE 7 Schematic erosion hazard map of the study area with a profil (PP') showing the regional distribution of erosion factors and soil loss

the alluvial plain, the sugar cane and cereals cultures are the ones which show the lowest coverage values (0.26-0.35), thus reflecting the prolonged period of their vegetative cycle during which they remain with little coverage.

THE EROSION HAZARD

The schematic map of FIGURE 7 shows the erosion hazard classes in the area expressed in different rates of potential soil loss, as it can be seen in the attached TABLE 1 and in the profil also included in figure 7, the erosion factors in the extended studied area show a great variation which is reflected in the wide range of values obtained for the potential soil loss.

TABLE 1 Variation extremes of erosion factors

Erosion factors	Values interval
Rainfall erosivity (R)	110-600
Soil erodibility (K)	0.30-0.65
Slope (LS)	0.15-16
Coverage (C)	0.005-0.34
Potential soil loss	1.7-90 tn/ha/y

In many cases the values of potential risk coincide with the present intensity of the erosion processes. The maximun erosion risk arises in the cultivated areas of the mountainous region with wet sub-tropical climate where it is urgent to apply practical conservation measures. The areas still not cleared shows a low risk, on condition that the present management is preserved. In the plain areas the potential erosion values are moderate, despite of low gradient, because of the high erosion susceptibility of the soil and the aggressiveness of the climate.

The coincidence of the extreme of soil loss values of the region, characterized by a subtropical climate, with the categories established by S.A.El Swaify (1977) for Hawai must be pointed out: the values for Hawai are higher than the ones measured normally in regions of temperate humid climate. The relatively high values of soil loss can be attributed the regional climate aggressiveness, the instability of the vegetal coverage and the predominance of loessic soils.

CONCLUSIONS

- The quick inventory of the erosion hazard in extended regions allows to detect the most affected areas, a fact which contributes to the implementations of conservation programs at the regional level.
- The use of a geomorphological basis, the methodological criteria of the USLE and remote sensing techniques gives the mapping at small scales scientific objetivity and allows it to be rapid and economic.
- The use of the Universal Soil Loss Equation has advantages from other parametrical methods for its relative simplicity and great divulgation. The limitation derivated from the application of averages of the erosion factors are minimized by the regional character of the appraisement.
- The use of Landsat information by both visual and/or digital interpretation gives planimetric reliability and multi-temporal information of the landscape dynamics.
- The studied area shows, in general, a high erosion risk due to the increasing anthropic pressure on extended areas which are still covered by natural vegetation.

ACKNOWLEDGEMENT

I wish to tank Dr. Van Zuidam for his assistance and valuables comments during my permanence in ITC and also in the field work period in Argentine.

REFERENCES

Arnoldus, H.M.J. 1978. An approximation of the rainfall faotor in the universal soil loss equation. In assesment of Erosion. M. de Boodt and D. Gabriels (eds.) John Willey and Son, New York.

Bergsma, E. 1982. Aerial Photo-interpretation surveys. ITC lectures notes part.II and III.

El-Swaify S.A.1977. Susceptibility of certain tropical soils to erosion by water. From Soil conservation and management in the humid tropics. John Wiley and Sons.

Janssen, M. 1983. Land erosion by water in different climate. Upsala University, Depto. of Physical Geography. Report N°57.

Sayago, J.M. 1982. Las unidades geomorfológicas como base para la evolución integrada del paisaje natural. Acta Geológica Lilloana 15-4.

Townshed, J.R.G. 1981. Regionalization of terrain and remotely sensed data. In Terrain Analysis on Remote Sensing. George Allen and Unwin, London.

Tricart, J. 1982. Taxonomical aspects of the integrated study of the natural environment. ITC Journal 1982-3:344-348.

Van Zuidam, R. 1983. Guide to geomorphology aerial photographic interpretation and mapping. Section of Geology and Geomorphology ITC, Enschede.

Verstappen, H.Th. 1983. Applied Geomorphology. Elsevier.

Wischmeier, W.H. & D.D. Smith. 1978. Predicting rainfall erosion losses. A guide to conservation planning. U.S. Department of Agriculture, Agriculture Handbook N°537.

Zonneveld, J.I.S. 1983. Some basic notions in geographical synthesis, Geo-Journal 7.2:121-129.

Symposium on Remote Sensing for Resources Development and Environmental Management / Enschede / August 1986

The study of mass movement from aerial photographs

Varoujan Kh.Sissakian
Geological Survey and Mineral Investigation, Baghdad, Iraq

ABSTRACT: The applicability of large scale (1:5 000 and larger) aerial photographs in studying of mass movement is described. Their interpretation for different aspects, dealing with the mass movement like classification, prediction, recognition, dating and activity, is discussed and highly recommended in all projects, in which the land is involved.

INTRODUCTION

Mass movement is a term used for all kinds of movements which take place in rock and/or in soils on slopes due to one or more reasons. The movement could be of different shape, size, origin, and type along one or more shear plane (then it is called landslide) or within thick zone consisting of a system of partial sliding planes (Zaruba & Mencl 1969).

Mass movements and landslides may cause huge damages to engineering works, properties, lives and human activities, due to these reasons many authors tried to classify the movements to different types. Among them are Heim 1882, Howe 1909, Almagia 1910, Terzaghi 1952, Ladd 1935, Shape 1938, Emelyanova 1952, Varnes 1958, Eckel 1958, Ter-Stepainan 1966, Zaruba & Mencl 1969, and Nemcok et al. 1978 .

These authors come to many classifications depending on different aspects like

1. Form of sliding surface
2. Kind of material moved
3. Age of rate of movement
4. Stage of development

But still some difficulties occur in identifying the type of movement, because the type of movement does not depend only on the material in which it takes place, but also on the scale of the movement and also because "only in a few cases the movements occur in pure form" (Yague, 1978).

Aerial photographs, especially of large scale 1:10 000 and larger can be used in studying different aspect of mass movements like their type, origin, causes, potentially critical areas, activity, and the prediction of the movement. From all of these, it is very neccessary to study aerial photographs of large scales before doing any engineering work or other activities in which the land is involved. Due to this it is always recommended, in large projects, to take inconsideration the use of aerial photographs. Their use is recommended by (Zaruba & Mencl 1969, Harold & Taliang 1978, Nossin 1978, Yague 1978, Varnes 1976).

1 CLASSIFICATION OF MASS MOVEMENT FROM AERIAL PHOTOGRAPHS

In this study the classification of the mass movements is based on that followed by Nemcok et. al. 1978, these are:

1.1 Slide

Within this type are inclosed all those movements of coherent masses along one or more well défined shear surfaces. Two main kindes of slides can be recognized from aerial photographs, these are slides in rocks and slides in soils. These can be differentiated by recognizing:

1. Where the scarpe is, i.e. in rocks or in soils.
2. Whether the moved mass is rock or soil.
3. The surrounding materials of the phenomenon.

Obviously the two kinds are very easily distinguishable on aerial photographs.

1.2 Creep

This type includes long-term movement of non-increasing velocity without a well defined sliding surface. Zaruba & Mencl 1969 confined this too.

It is neccessary to mention that the creep process may lead to different types of mass movements if the creep is accelerated due to any reason. The movement stops after reaching equilibrium. Hence any recognized creep on old aerial photograph may be find as another phenomenon during field check.

Usually creep occur in soils, the toe area of which is associated mainly with small sliding. Creep occurs in rocks too, but their recognition is very difficult.

1.3 Flow

This includes mass movements in rocks and soils analogous to the movements in liquids. Movements take place due to liquification due to any reason. According to Varnes 1978, some authors use the term creep for indicating the flow. Zaruba & Mencl 1969 use different terms for the flow like earth flow and debris flow. The most common movement of this type is the mud flow.

1.4 Fall

This is a sudden mass movement. The mooving mass looses its coherence and for a short time also its contact with the ground. The most common movement of this type is the rock fall. The size of the fallen blocks may reach up to tens of metres, whereas the involved area may reach up to hundreds of metres. "Usually rock falls are small but they occur at extremely high velocity and thus are capable of damage or to generate fairly large moves" (Schuster, 1978).

2 RECOGNITION OF MASS MOVEMENT FROM AERIAL PHOTOGRAPHS

Without any doubt the interpretation of aerial photographs is one of the most promissing methods in recognition the mass movements, especially the large scale aerial photographs.

Many mass movement phenomena can not be recognized in the field without the help of aerial photographs, specially when they are old, inactive and big enough that the limits can not be seen.

The main procedure for recognition of mass movement from aerial photographs is to look for the three main parts of it (fig.1), which are :

1. The root area or the crown (scarp area).
2. The tongue area (displaced materials).
3. The toe.

The recognition becomes more difficult when these parts are already removed or flattened by erosion processes. On the other hand the recent and active phenomena are easily recognizable, because the morphology often reflects the occurance of mass movement phenomena by an irregular slope form which is not in harmony with the surroundings.

Another difficulty arises in recognition of mass movement phenomena when the moved mass is very large and old that it might be missed and explained as another geological process such as faults.

The clues which could be observed from aerial photographs in mass movement recognition are:

1. Existance of cracks on steep slopes (when big enough).
2. Hummocky slopes.
3. Erosion front in the foot of a steep slope which faces stream.
4. Existance of bulges in the foot of slopes.
5. Existance of steep scarps on a slope.
6. Existance of concave or spoon-shaped slope.
7. Existance of accumulated mass at the bottom of a steep slope or cliff.
8. Steep slope having large mass(es) of loose soil and rock (Varnes, 1978).
9. Steep break(s) in a slope giving it a steeped shape.
10. Existance of ponds on slopes.
11. Narrowing of a valley which faces a steep slope and has no apparent relation with underlying bedrock.
12. Changes in the direction of a valley in arch shape, facing the instable slope with a clear widening of the valley, both below and above the arched section.
13. Assymetrical valley with active erosion on the steeper side.
14. Internal drainage on slope.
15. Existance of highly saturated areas which show a different tone on slopes (Nossin, 1973).
16. Seepage zones (Harold & Taliang, 1978).
17. Sudden change in valley gradient.
18. Accumulation of scree on slopes.

The existance of only one of the above mentioned clues can not be taken as indication for recognition of movement because it may indicate another process. Usually few of the clues have to exist for confirming a kind of movement, which is finally to be proved by adequate field check. (The amount of the movements also can be estimated from aerial photographs by observing the offset of any linear feature, Zaruba & Mencl, 1969).

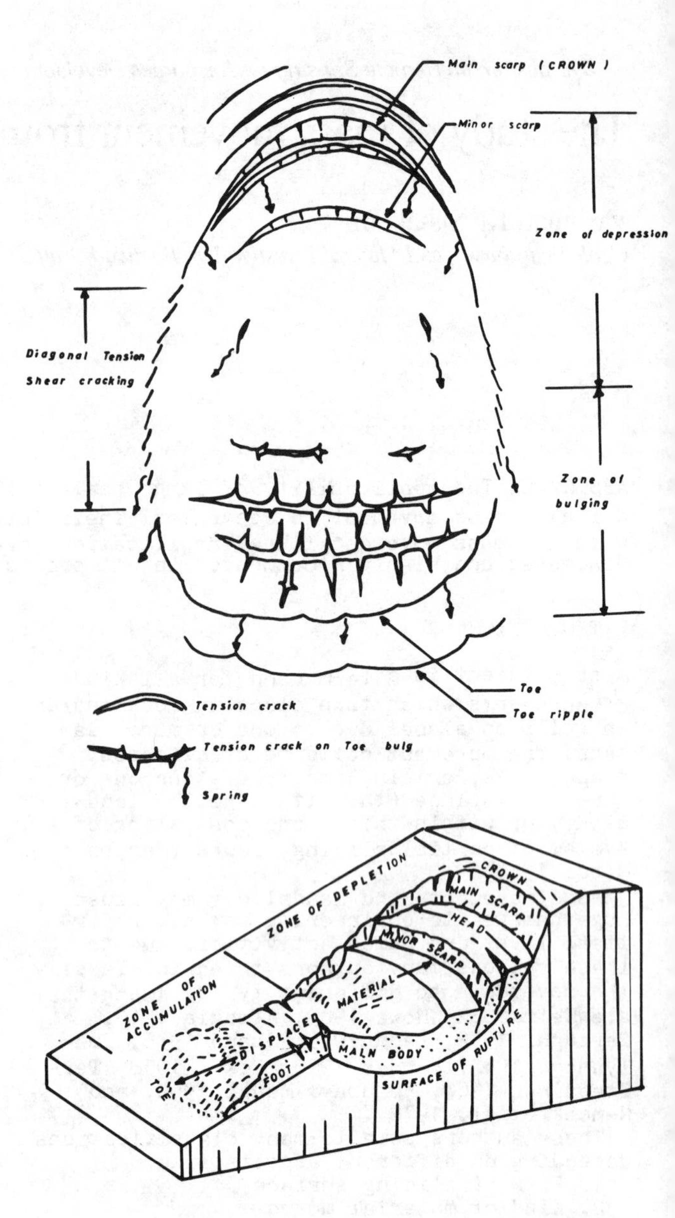

Figure 1. Showing details of mass movement (after George F. Sowers and David L. Royster).

3 CAUSES OF MASS MOVEMENT

Although there are many causes for the mass movements, only those which can be recognized from aerial photographs are listed below:

3.1 Geological conditions:

Peletic rocks when are overlain by thick competent rocks and are situated on steep slopes tend to behave as a lubricating surface below the competent rocks.

Fine grained clastic rocks, poorly consolidated soils and pelitic soils interbedded with hard rocks (flysch) are extremely likely to give rise to mass movements (Terzaghi, 1950).

Such combination of rocks can be easily recognized from aerial photographs, hence the cause can be known.

3.2 Weathering:

Rocks loose their cohesion due to weathering, specially mechanical weathering which increases the possibilities of water penetration, this leads to an increase of the pore pressure and consequantly decreases the cohesion and internal friction angle. This will trigger the movements on slopes.

Weathering phenomena are visible in aerial photographs of large scales only.

3.3 Vegetation:

It is well known fact that vegitation increases the stability of slopes, as plant cover provides protection against surficial slides and erosion (Cotechia,1978). However (Prandini et.al. 1977) declear that some authors believe that deforestation helps to stabilise slopes. Others believe that the angle of repose of a slope covered by vegitation is 60 but it decreases to 36 when is barren from vegetation.

The roots of plants play role in keeping the stability of slopes, on the other hand they play role in desintegration of materials during penetration. The weight of large trees also might disturb the equilibrium due to overloading of slopes.

Changes in vegetation can be easily observed from aerial photographs, but their relationship with mass movements has to be determined in the field.

3.4 Human activities:

Some mass movements occur due to human activities like quarrying excavations, terracing of slope for agricultural purposes, deforestation, blasting etc., such factors can be observed on aerial photographs if the photography is made after ward.

3.5 Tectonic activities:

Any sudden change of slopes due to tectonic activity like fault may lead to mass movement, especially when the slope is in critical equilibrium and can be triggered due to any reason.

Obviously faults can be easily recognized from aerial photographs, hence the cause of any mass movement phenomenon which occur in faulted area may be explained due to triggering of the fault.

4. PREDICTION OF MASS MOVEMENT

Aerial photographs help in the prediction of different types of mass movement. Their interpretation before starting any engineering work is highly reccomended. Construction activities for engineering works can trigger the potential areas and as soon as movement has started they can be stopped only with great difficulties and high costs.

Prediction of mass movement from aerial photographs can be done by observing the following clues (1-9), but it can't be estimated when they will start to move, since they may never move:

1. Cracks on steep slopes, when big enough to be visible on the photographs.
2. Thick colluvial soil on steep slopes. (Gray et. al. 1977).
3. Clear bedding planes dipping towards the slope. (Barton, 1977).
4. Highly saturated areas and seepage zones.
5. Old mass movement areas.
6. Active erosion at the foot of slopes.
7. Areas showing disturbed vegitation.
8. Hummocky surfaces.
9. Small depressions on slopes.

The author believes that most of the above mentioned clues are recognizable from aerial photographs, but it is neccessary to remined that the existance of one of the mentioned cues must not be considered as a certain indication for prediction of mass movement.

5 DATING AND ACTIVITY OF MASS MOVEMENT

The age of the mass movement can be estimated from aerial photographs; approximately by recognizing few aspects which are mentioned below (E_1-E_4), whereas the activity of the movement i.e. wether it is active or not, recent or old can be known adequately, specially in large scale photographs (1:3 000 and larger):

5.1 Vegetation

The age of trees or shrubs can be estimated from their heights, which can be measured from aerial photographs, hence the age of the tree which occur on the movement area indicates, roughly the age of the movement. Comparision between the age of the trees, which occur on the movement area and the surroundings, also indicates the age of the movement.

5.2 Erosion

Erosion processes indicate the activity of the movement only. These easily can be observed from aerial photographs, hence the activity of the phenomenon can be known, which is very important in planning different human activities.

5.3 Human activities

Recent and active mass movement areas are usually abandoned from human activities, Recognition of abandoned houses, roads, farms, quarries etc. may indicate and lead to a rough estimation of the movement. Obviously such recognitions are easily done from aerial photographs, especially recently photographed photographs.

5.4 Recognition of sedimentation processes

If any movement is covered or taken place by/in terrace, alluvial fans, colluvial deposits etc. then the estimation of the age of the sedimentation processes indicates, roughly the age of the movement. But the estimation of the age must be done very carefully, and to be confirmed in the field, otherwise it will be missleading.

The disturbance of recent sedimentation processes also may indicate the acitivity of the mass movement, if other clues, mentioned in 2 & 4, exist which indicates the existance of the movement.

6 REPRESENTATION OF MASS MOVEMENT

After interpretation of mass movement from aerial photographs and checking them in the field, it is neccessary to represent them on engineering geological maps. Different aspects of mass movement which can be seen from aerial photographs are represented on maps such as the scarp "crown area", moved mass, toe, scar area, ripples, cracks, saturated areas and potential area. All these can be differentiated in recent and/or active, old and/or nonactive "still" forms, from aerial photographs and has to be represented by different colours, usually red for recent and/or active, black for old and/or nonactive "still".

Such interpreted engineering geological maps are very important in most engineering works, although such maps are not common all over the world (Pulinova et. al. 1977). Depending on such maps all the potential areas, in which movement could be triggered, has to be avoided as much as possible.

7 CONCLUSIONS

It is very neccessary and recommended to study aerial photographs of large scale 1:5 000 and larger to indicate all the mass movement and potential areas, in which movements can be triggered due to any human activities.

Because without avoiding the indicated potential areas, large damages in lives and properties may take place, keeping in mind that such potential areas can not be seen or observed easily in the field, without using aerial photographs, especially those which are recently photographed.

It is clear from all, above mentioned subjects that all the interpreted informations, from aerial photographs has to be checked and confirmed in the field. Otherwise the interpreted information can not be taken in consideration adequately.

REFERENCES

Barton M.E. 1977. Landslide along bedding plane Bull. IAEG No. 16.

Cotechia V. 1978. Systematic reconnaissance mapping & registration of slope movements. Bull. IAEG No. 17 (5-37).

Gray R.E. & Gardner G.D. 1977. Process of colluvial slope development of McMecken, West Virginia. Bull IAEG No. 16 (29-32).

Nemcok A., Pasek J. & Rybar J. (1972). Classification of landslides and other movements. Rock mechanics Vol. 4/2 (71-78).

Nossin J.J. 1973. Use of airphotos in studies of slope stability in Crati basin(Calabria, Italy), Geologia, Applicate e Idrogeologia, Vo. 8, part I.

Prandini L., Guidicini G., Bottura J.A., Pocano W.L. & Santor A.R., 1977. Vegitation in slope stability. A critical review. Bull. IAEG No. 16 (51-55).

Schuster R.L. 1979. Reservoir induced landslides. Bull. IAEG No. 20 (8-15).

Sissakian V. 1982. Applicability of aerial photographs and orthophotographs at various scale for engineering geological mapping. M.Sc. Thesis submited to International Institute For Aerial Survey and Earth Sciences (I.T.C.), Enschede The Netherlands.

Terzaghi K. 1950. Mechanism of landslides in application of geology to engineering practice (Berkey Vol.). Geol. Soc. America (5-7).

Varnes D.J. 1976. Landslides-causes and effects. Bull. IAEG No. 14 (205-214).

Varnes D.J., Harold T.R. & Taliang 1978. Landslides-analysis and control. National Academy of Sciences. Washington D.C.

Yague A.G. 1978. Modern methods used in study of mass movement Bull. IAEG. No. 17 (65-71).

Zaruba Q. & Mencl V. 1969. Landslides and their control. Czechoslovak Academy of Sciences, Prague. Elsevier.

Symposium on Remote Sensing for Resources Development and Environmental Management / Enschede / August 1986

An evaluation of potential uranium deposit area by Landsat data analysis in Officer basin, South-Western part of Australia

H.Wada & K.Koide
Power Reactor and Nuclear Fuel Development Corporation, Tokyo, Japan

Y.Maruyama & M.Nasu
Asia Air Survey Co. Ltd, Tokyo, Japan

ABSTRACT: An evaluation of potential uranium deposition area was carried out by the Landsat MSS data analysis. As a denomirator land coverage, depth of basement rock and density of vegetation are chosen. The area well evaluated in Officer basin area lie on the mining or the prospecting area at present time.

1 INTRODUCTION

Economic uranium mine is identified into some typical type by geological condition, (1) sedimentary deposit (2) related with unconformity (3) hydro thermal type (4) volcanic origin, etc. The purpouse of this study is to evaluate the possible area of sedimentarty uranium deposit with using Landsat imagery.

In general, the ultimate sources of uranium are granite, granitic detritus, and silicic volcanic ash and flows. Uranium was leached from the source rocks and transported in solution by oxygen-rich ground water and then migrated into porous sandstone or conglomerate beds. Within these eventual host rocks the migrating water encountered reducing conditions caused by the presence of organic material, natural gas or hydrogen sulfide,etc. The environment of chemical change from oxidizing to reducing caused the uranium to precipitate as oxide minerals, primarily uraninite, which coat sand grains and fill pore spaces in the host rock. According to these processes of development the environment of uranium deposition can be estimated to have following conditions.

A. The area is surrounded by the mountains composed of the ultimate source of uranium such as granitic rocks.

B. The depositional condition

(1) The area of sedimentary rocks distribution is extensive.

(2) The thickness of sedimentary rocks is high (possiblity of ground water).

(3) Existing the high porocity in the bed (host rock).

(4) Succesion of high porocity zone (paleo channel).

(5) The reducing agent : organic material or natural gas in the host rocks.

The sedimentary uranium deposit can be realized under two conditions, one is the area suppling uranium to oxdized ground water and the other is the reducing zone to precipitate it in the sedimentary rocks.

All of the condition for uranium deposition, which mentioned above is not possible to be analysed only by Landsat imagery. Thus, the some of those condition for uranium deposition are analysed by Landsat MSS data in this paper. And evaluation of the possible area was taken place from the view points of the thickness and extent of sediments and succession of high porocity zone (paleo channel).

2 AREA STUDIED

2.1 Location

The area studied is located in the southern part of Officer basin, South-east Australia. There is among Long. 122°30' ~ 125°00' E and Lat. 28°00' ~31°00' N (Fig. 1).

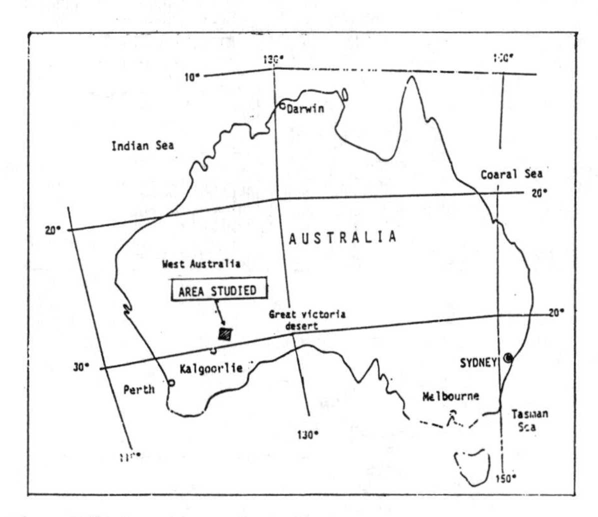

Figure 1. Location of studied area

2.2 Topography and climate

It is on the south-western part of Great victoria desert. The basin generally is flat or gently undulated, of which is mainly sand dune.

Climatic condition is arid ~ semi arid. As annual amount of presipitation is 180 ~ 200 mm and evaporation is more than 2000 mm, the river is dried up in the most of year.

Average temperature is 30 ~ 35°C in summer and is 18°C in winter. Area is covered by barren vegetation such as spiniphex, marry, marblegum, etc.

2.3 Geology

Geological structure of West Australia is composed of pre-Cambrian western shield and some of the sedimentary basin on it.

Officer basin extending on the eastern margin of Yilgarn block is composed of granitic and metamorphic rock. Sedimentation was taken place during lower Paleozoic to Tertiary under the neritic and continental environment. Lower part of sedimentary rocks is called Patterson formation which is clay dominant Permian system. This formation was eroded to make the continuous valley in it. Sedimentation in Cretaceous and Tertiary covered these valley unconformably. They are composed of conglomerates, sandstone, silt and clay.

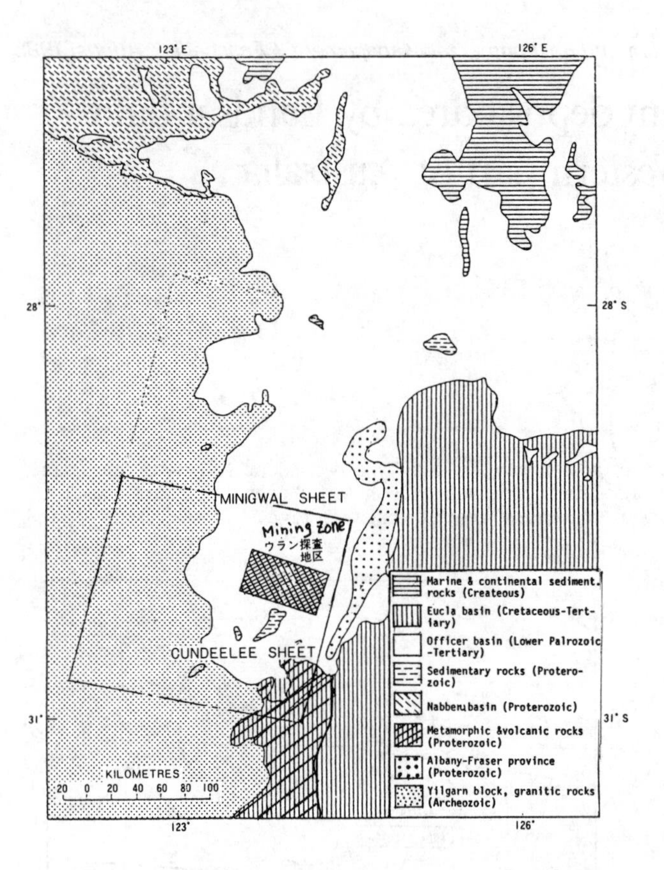

Figure 2. An outline of geology

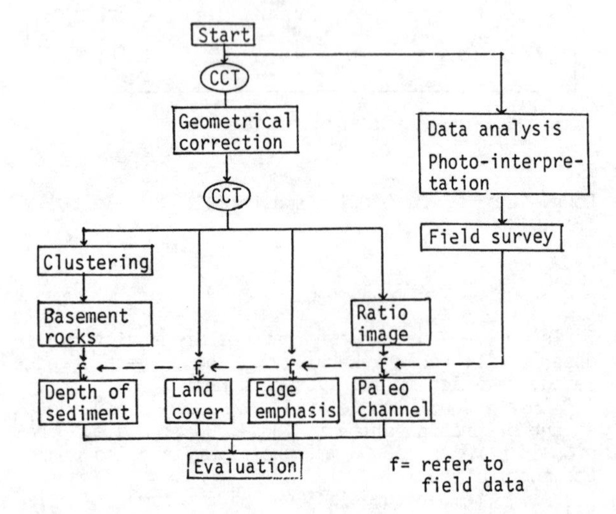

Figure 3. Process of evaluation

3 METHOD

3.1 Data

Landsat MSS (CCT) path 115 row 81
DATE 6th February 1973

3.2 Process of evaluation

After several trial, depth estimation of sedimentary rocks, land coverage, ratio images are employed for the evaluation. Process of study is shown on the figure 3.

4 DEPTH ESTIMATION OF THE SEDIMENTARY ROCKS

4.1 Hypothesis for estimation

Basement rocks mainly are affected by mechanical erosion agency under arid and semi arid climate. This hypothesis was presended by Büdel (1957) as the conception of "double surface of leveling" (see figure 4) According to his conception the level of the sediment surface is changed by flood and climatic condition, (a) Wash plain of seasonal flooding is as much as 100 m above the weathering front. Pediments fringe the wash plain. (b) Wash plain is lowered by rejuvenation or climatic change. Inselberg and marginal pediments are exhumed or regraded to the lowered wash plain. This geomorphological process indicate that the wash plain being more extensive area or longer distance among inselbergs, have relatively deeper sedimentary deposits. Using this hypothesis, depth estimation was done with Landsat imagery data.

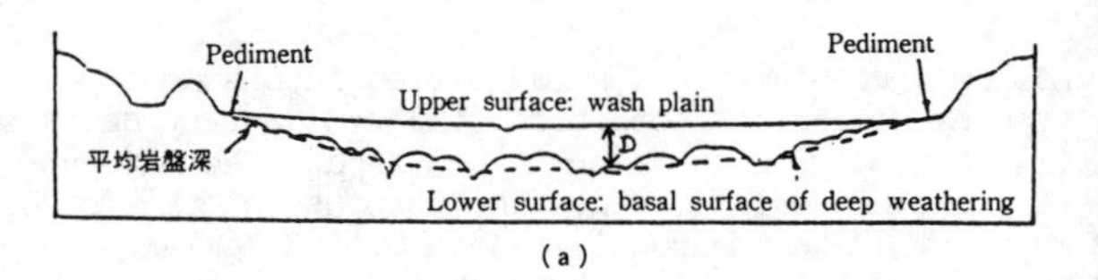

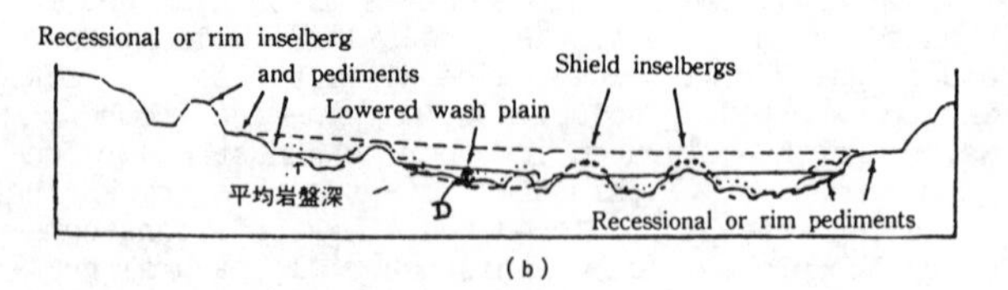

Figure 4. Schematic explanation of hypothesis of "double surface leveling"

4.2 Depth estimation map

The map is prepared with the following processes.

(1) Basement rocks distribution

Because of the heterogeneous nature of the study area, the unsupervised classification method is used. The analytical process involved the clustering method using an algorithum to divide the four-channel Landsat data into groups of points having homogeneous spectral characteristics.

Coverage type was classified into sixteen items by the clustering. Then, three of them are interpreted that basement rocks are comming on or near the ground surface, by means of the ground survey and aerial photo-analysis. This pixel map is called "Basement rock distribution map".

(2) Density of basement rocks

Number of pixel classified to basement rock were counted in each mesh unit, 1 Km x 1Km. The maximum number of it is 100, and minimum is 0. The original distribution map is shown on Fig. 5.a. The result was, however, too noisy to understand the trend of its distribution. Then, smoothing technique was employed to solve this problem. The original data was replaced by the average number of basement rock pixels of the surrounding mesh units. The smoothing window size was examined by 5 Km x 5Km, 7 Km x 7Km, and 9 Km x 9 Km. In this case window size, 9 Km x 9 Km was the most reliable for the purpouse (Fig. 5.b).

(3) Depth estimation

Basically trend of basement rock density must be

corresponding to the thickness of younger sedimentary rocks. Thus, the patlern of the basement rock density must be indicating the thickness or the relative depth to the basement of sedimentary rocks.

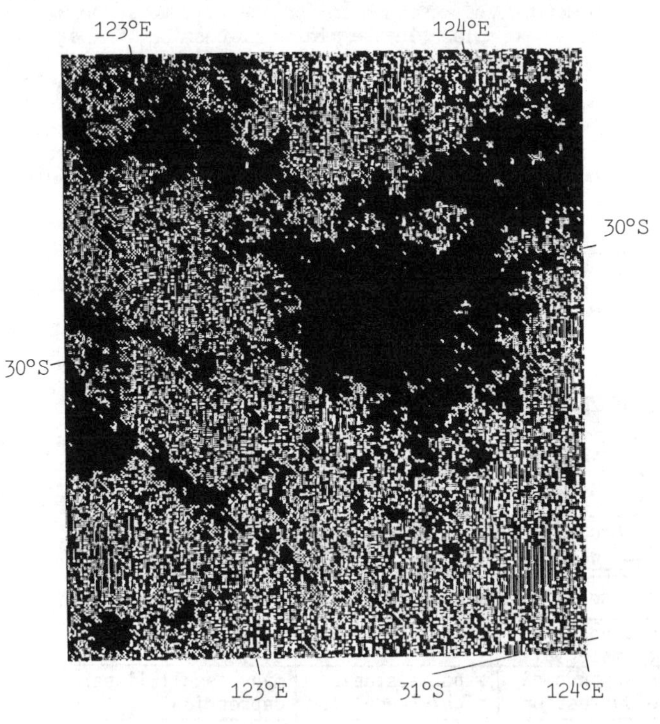

(a) Original distibution

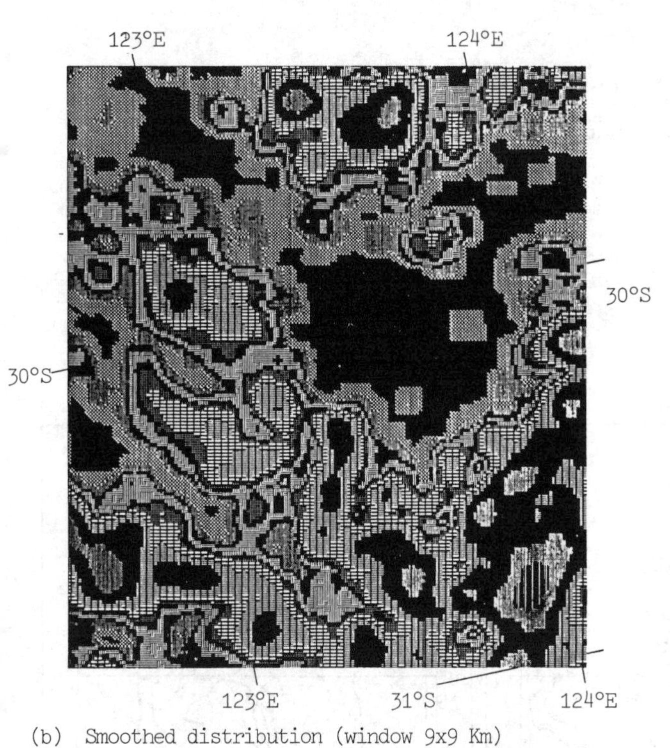

(b) Smoothed distribution (window 9x9 Km)

A B C D E F G H I J K L M

0 2 4 6 8 10 12 15 20 30 50 70 100

frequency

Figure 5 Basement rock distrbution map

4.3 Verification of the result

It is necessary to compare the real depth with the other data such as boring test. However, it is not possible to receive real depth in such vast area. Thus, we compared it the gravity map. Depth estimation map in figure 6 is analysed by double fourier series with two harmonics in both directions to get the effective trend surface.

There is more or less the same trend of relative depth in both method, Landsat data analysis and an airbone gravity. According to such verification, this method and the result can be trustworthy for the pourpose.

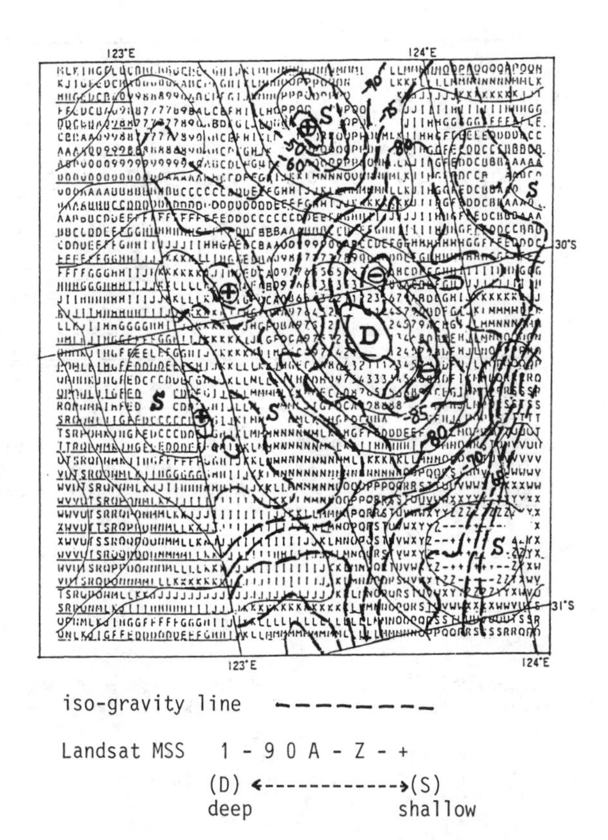

Figure 6. Comparison of estimated depth between by Landsat MSS and by airbone gravity survey

5 PALEO CHANNEL DETECTION

5.1 Relationship of paleo channel and present topography

Paleo channel mean the old river which had been covered by the new sediments. When there was the river channel the climatic condition was more rainy than present time. The river transported more and coaser sand/gravels. Climatic condition was gradually changed to dry and hot. Thus, total amount of sediments was relatively small. And also there was not influenced so much by the crustal movement in the recent term. The river channel could be remained its characteristics as the paleo channel under those two geological environment.

The paleo channel is composed of coaser materials and it makes permeable zone for the ground water.

By means of aerial photo interpretation and field observation, the relations between topography, geology and vegetation which is summarized on the figure 7.

The highly undulated area is sand dune, where have barren vegetation. And depressional area have flat bottom, where is covered by dense vegetation with short trees and grass. There is composed of fine materials on the top and located over the shallow ground water table. These depression area still supply the water to underground during rainy season. Because of these phenomena, density of vegetation was employed for a denominator of paleo channel detection.

5.2 Method for vegetation density survey

Discussion to find the suitable method for vegetation density was done between MSS band ratio 7 / 5, 7 / 4 and streched-ratioed colour composite image. Finally MSS ratio 7 / 4 was chosen as the most favourable technique to emphasize the density of vegetation in this area.

most appropriate weight was given, which was, land coverage : depth : paleo channel = 3 : 10 : 10.

The result of evaluation is shown on figure 9. The mesh units which evaluated the highest potential are shown in black and the second potential area is in dots. The research and prospection for uranium are taken place in the mining concession, the square in the figure. And some of good result of the prospection were reported from that area. Most of the highly scored mesh units are distributed in the mining square. From such view point it can be summerized that the result of this evaluation suggest the area where have favorable environment for uranium deposit beneath there.

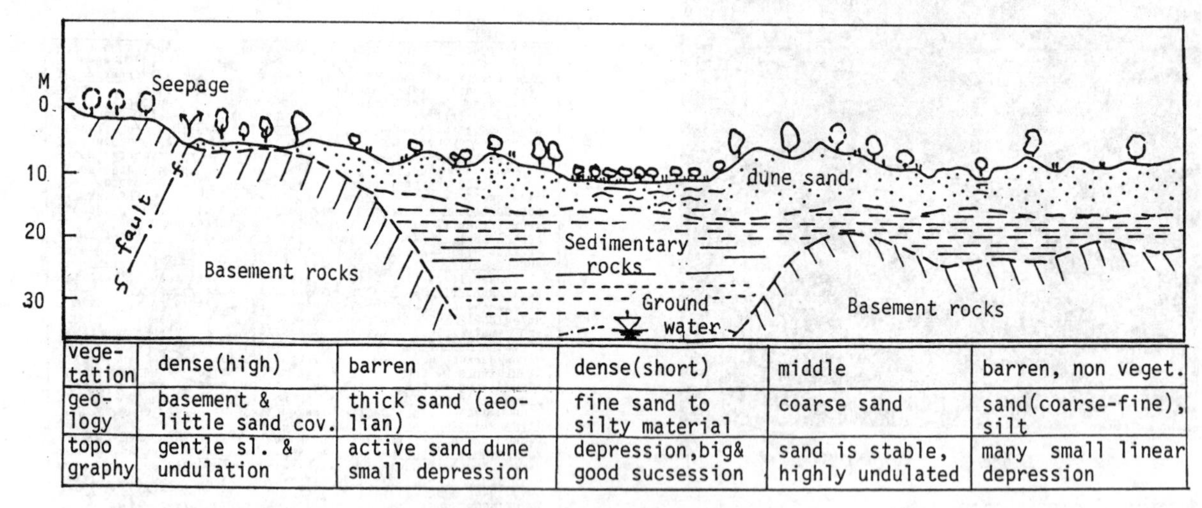

vegetation	dense(high)	barren	dense(short)	middle	barren, non veget.
geology	basement & little sand cov.	thick sand (aeolian)	fine sand to silty material	coarse sand	sand(coarse-fine), silt
topography	gentle sl. & undulation	active sand dune small depression	depression,big& good sucsession	sand is stable, highly undulated	many small linear depression

Figure 7. The relation between vegetation, geology and topography

5.3 Relation between Vegetation and paleo channel distribution

The distribution of dense vegetation zone is shown on figure 8 (a), of which pattern is like of the river channel. However, there are not developed any rivers on the surface at present time. With using the borehole data the depth of the basement rock was analysed, which is shown on figure 8 (b).

The basement rocks form like valley shape and the paleo channel is developed in that valley. There are very similar pattern of development between vegetation and paleo channel, which can be observed in map (a), (b).

Generally the depressional area is densely vegetated. Occasionally the lineaments and the sinkholes are developed in there, which seem to supply ground water to the paleo channel even at present time.

6 EVALUATION

Land coverage, depth thickness distribution of sedimentary rocks and the paleo channel analysed by Landsat imagery are employed for the factors to evaluate the area where are the suitable environment of the uranium deposit.

For the evaluation, we scored these items with certain weight in each mesh, 1 Km x 1 Km. After several time of try and error the

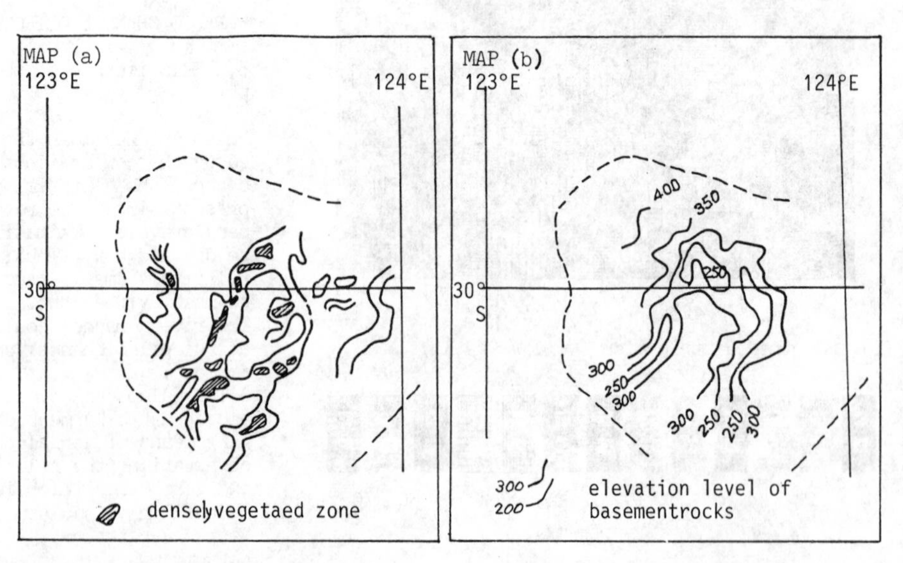

Figure 8. Comparison between pattern of vegetation distribution (a) and (b) morphology of basement rock

7 CONCLUSION

Landsat data analysis can be used for the possible area evaluation of sedimentary uranium deposit. However, still some technical problems are remained on the each procedure. For instance, some part of laterite was not identified from the basement rock, the physical relation between surface phenomena and geological structure is not completely made clear yet, and so on. These ploblems can be solved in the furture to study in comaparison with other data, such as field survey, geophysical prospecting,boring test, etc.

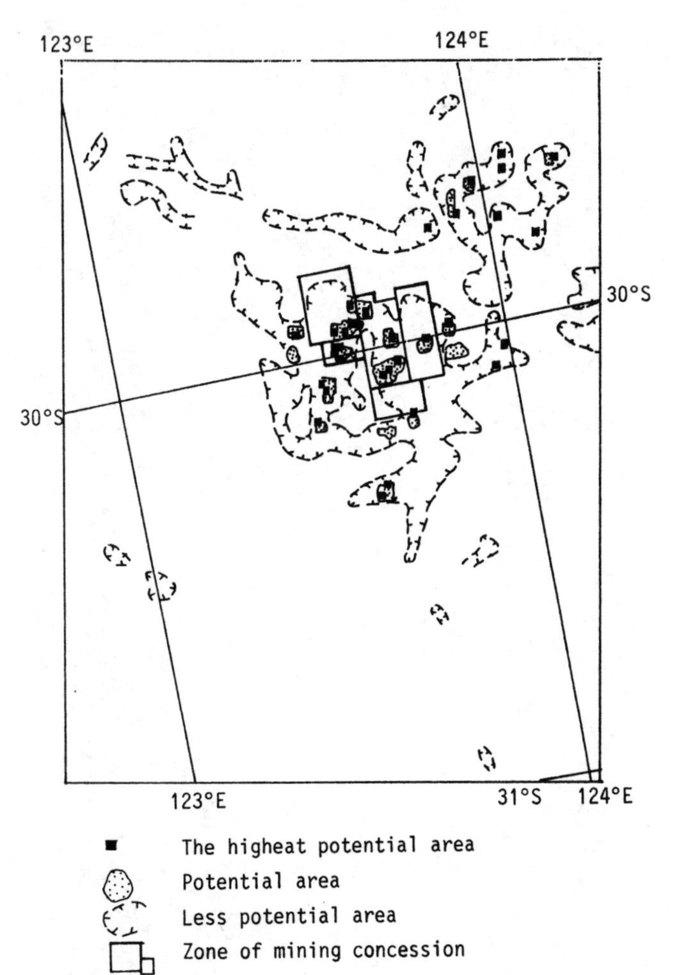

Figure 9. Evaluation map of potensial zone of uranium deposit

REFERENCES

Childers, M.O. 1974. Uranium occurence in upper Cretaceous and Tertiary strata of Wyoming and northern Colorado. Mountain geology Vol. 11, p. 131-147.

Floyd, F.S. 1978. Remote sensing, Principles and pretation W.H.Freeman and company, San Francisco.

Missallati, A. , Prelat, A.E. & Lyon, R.J.P. 1979. Simatlenious use of geological, geophysical, and Landsat digital data in uranium exploration. Remote sensing of environment 8 : 189-210.

Smith, A.F. 1977. Interactive digital image processing of Landsat data for geologic analysis. Proceedings of the international symposium on image processing, Interactions with photogrametry and remote sensing, Graz, p. 197-212.

Vincent, R.K. 1977. Uranium exploration with computer -processed Landsat data. Geophysics Vol. 42, No. 3.

Symposium on Remote Sensing for Resources Development and Environmental Management / Enschede / August 1986

Digital analysis of stereo pairs for the detection of anomalous signatures in geothermal fields

E.Zilioli, P.A.Brivio, M.A.Gomarasca & R.Tomasoni
Istituto per la Geofisica della Litosfera, CNR, Milan, Italy

ABSTRACT: A campaign of infrared aerial surveys for mapping heat flow anomalies was performed in the geothermal field of Travale (Italy). Direct methods of quantitative analysis of thermographies and deduction of secondary physical quantities were applied: three spots of possible geothermal interest have been individuated. Further investigation was developed on the basis of spectral signature of vegetation; a third flight was accomplished in springtime to provide infrared false colour stereo-pairs at 1/12,500 scale. Color transparencies have been scanned by a drum microdensitometer and a quantitative computation was carried out. The suspected areas so identified have been compared to other spots where ground and depth conditions are known.

1 PREVIOUS WORK

This paper deals with the remote sensing activity in the frame of the Energy Project sponsored by the National Research Council of Italy.

Previous work is here summarized:

Data Acquisition. Campaigns included 2 thermal IR surveys by a dual-channel analogue scanner DS-1230 realized at the steady-state conditions of the diurnal thermal transitory, in wintertime. During the flight operations meteorological stations were arranged for corrections and calibrations. A stereo IR False Colour coverage was acquainted on the same area in springtime.

Thermal Data Processing. The black-body referenced thermographies were digitized and a correlation between the calibration temperature levels and the Digital Numbers (DN) was obtained (Bolzan et al. 1982).

Table 1. Summary of spectral bands utilized .

Means	Wavelength (μm)	Band
	0.5-0.6	Green
Photography	0.6-0.7	Red
	0.7-0.9	near IR
	1.0-2.0	near IR
Scanner	4.5-5.5	middle IR
	9.5-11	thermal IR

By means of calibrated thermographies we obtained 3 new derived quantities we called Function of Apparent Thermal Inertia (FATI), Entropy Variation (EV) and Sensitivity (S) respectively (Zilioli et al. 1985). The common solutions of these equations individuated 3 spots on the ground of possible geothermal interest.

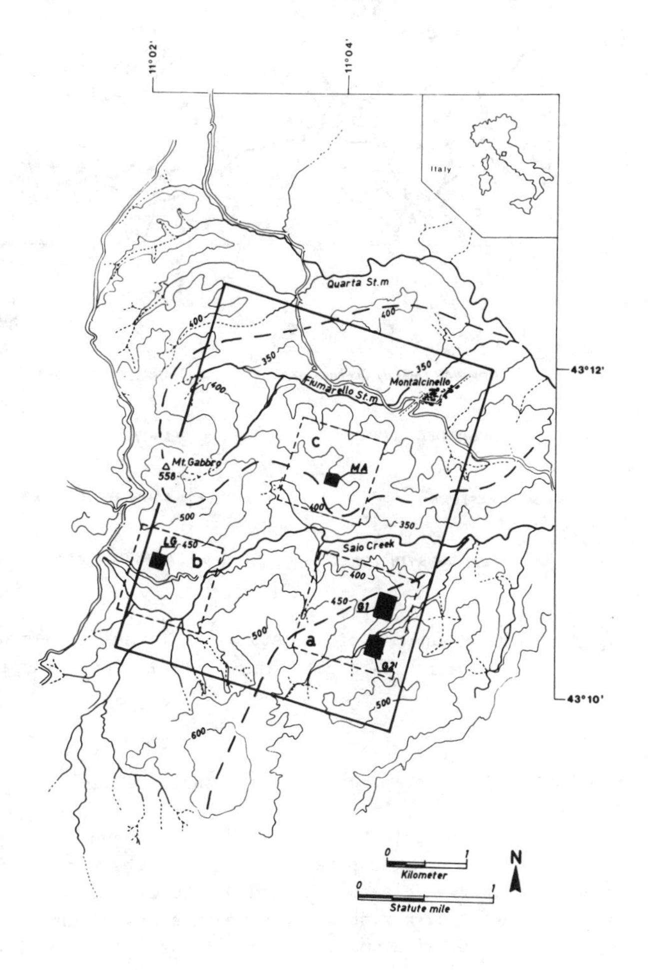

Figure 1. Location of the study area. Frames and subframes are pointed out.

2 DESCRIPTION OF THE SUSPECTED AREAS

Location of the suspected areas is shown within the frames outlined in fig. 1. According to the classification by Pavari (1942), the study area is included in the phytoclimatic belt of Italy: Castanetum, subzone cold with Quercus spp deciduous woods and defined by a hilly zone ranging between 300 and 800 meter. Land use is about 75% forested and the remaining 25% pasture and sown, almost distributed on the right and on the left slope of the Saio creek basin, respectively.

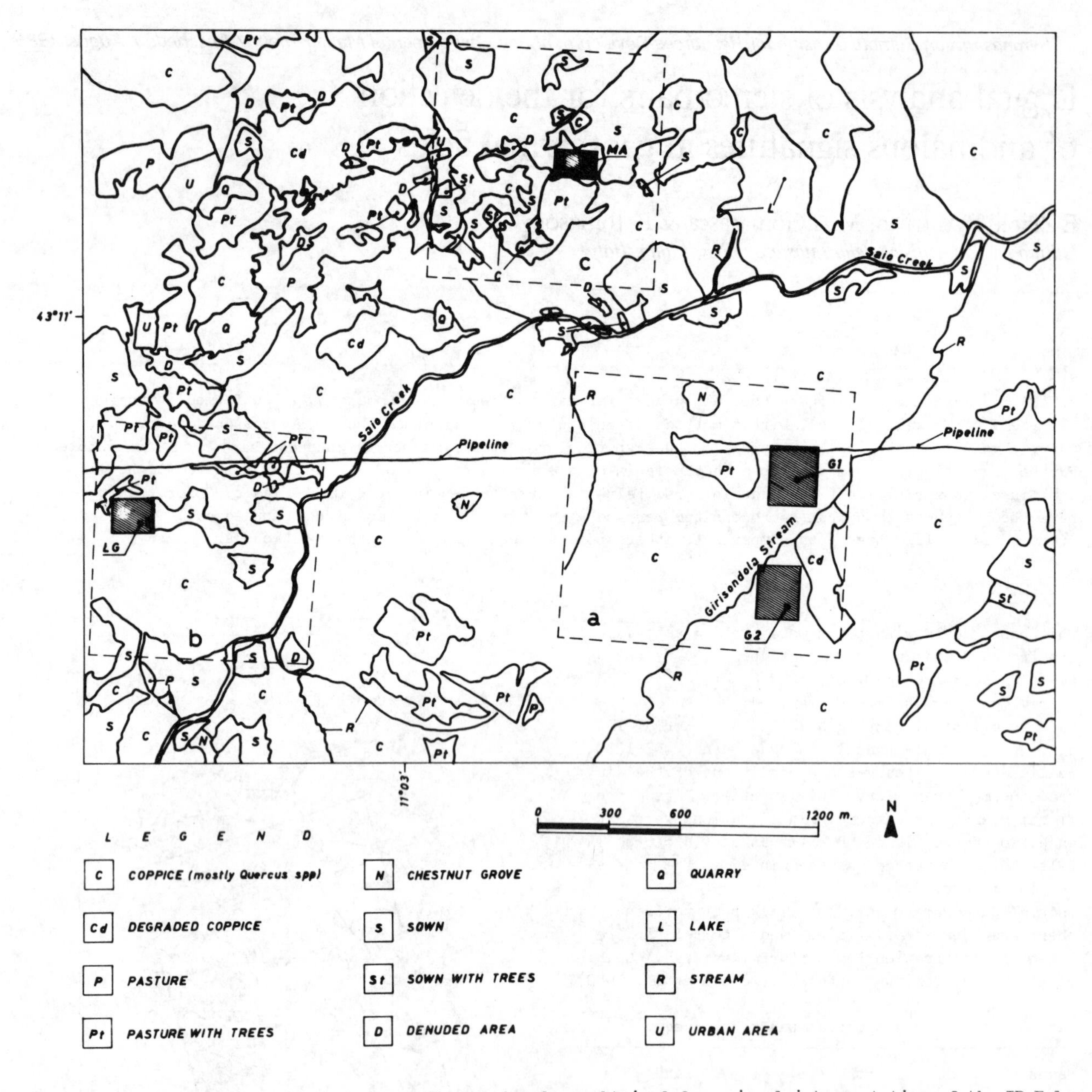

Figure 2 . Land use map of the study area. The map has been obtained from visual interpretation of the IR False Colour stereo pairs. Take date: May 1, 1981. The 3 suspected areas and subframes are outlined.

Brief separate descriptions of the 3 suspected zones follow.

a. Fosso Girisondola. Two sub-frames have been cut out where the highest textural homogeneity was preserved; they have been labelled G1 and G2 respectively. They concern with densely and mostly uniform vegetated spots, with small foliage of the individual tree.

b. Lagoni. It coincides with the old exploited geothermal field. The sub-frame LG corresponds to an irregular deciduous wood due to different growing stages. Dimensions of trees are greater than in the other considered sites.

c. Poggio Mauriccia. The subframe MA corresponds to a woody area where trees show a more uniform structure than in the previous two cases.

2.1 Geothermal outline

Recent studies (Batini et al. 1985) have produced maps where distributions of physical parameters of the reservoir are plotted.

Table 2. Presumed values of depth, temperature and pressure of the top of the reservoir, at 4 sites under investigation.

	Depth from the surface (m)	Temperature (°C)	Pressure (kg/cm^2)
G1	1150	100	35
G2	800	80	30
LG	120	150-200	30
MA	1100	150-200	35

Data concerning G1 and G2 sub-frames are estimated and extrapolated since no drill has ever been tried.

The top of the reservoir is rather shallow at site LG which can be referenced as surely affected by geothermal influence. The following cross section, after Calore et al. (1979), illustrates the geothermal situation schematically.

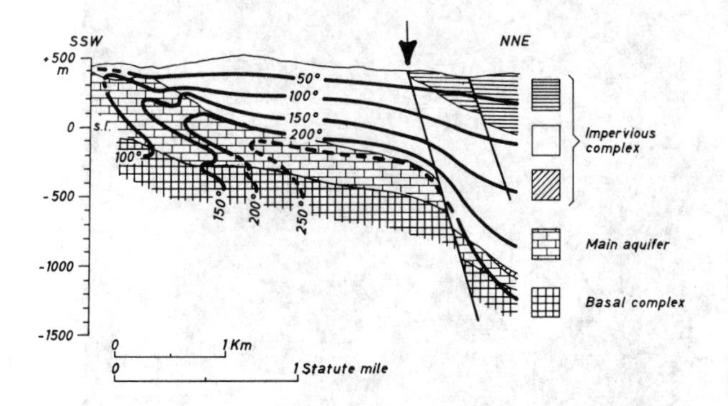

Figure 3. Geothermal cross section of the Travale field. Arrow outlines location of the sub-frames G1, G2 and MA.

The top of the reservoir dips north-eastward, beneath the cover of an impervious complex. The sub-frames G1,G2 and MA are sited in correspondance of a tectonic feature -the graben of Travale- where associated fractures are the easy vehicle of the geothermal fluid. This structural remark confirms the geothermal potentiality of the suspected zones.

3 DIGITAL ANALYSIS OF THE I.R. AERIAL PHOTOGRAPHS

The investigation has continued on the basis of spectral signature of vegetation, analysed on the IR False Colour stereo pairs.

3.1 Image processing

In order to avoid faults due to aberration of the camera lenses used and to the film developing processes we selected the frames so that each test area occupies the central portion of them. Frames no. 035, 077 and 101 resulted to be in good agreement with the requirements.

The 4 sub-frames previously identified were scanned, at a 100 μm resolution, and digitized by means of a drum microdensitometer Optronics C-4500.

Table 3. Dimensions of the windows digitized from the original transparencies.

Site	Frame no.	Dimensions (no. pixels)*	Area (ha)
G1	077	212 x 173	5.7
G2	077	188 x 164	4.8
LG	035	119 x 137	2.5
MA	101	98 x 135	2.0

* 1 pixel = 1.25 m x 1.25 m

Each window was scanned under separate reading of the 3 colour components of the film layers Blue, Green and Red which correspond to the reflected energy Green, Red and Infrared.

Moreover, since the object of our investigation is to outline fine anomalies within visually homogeneous vegetated zones, we also have applied to the band ratios IR/Red, IR/Green and the so called Normalized Difference Vegetation Index (NDVI) where:

$$NDVI = \frac{IR - Red}{IR + Red}$$

On the basis of frequency distribution of DN present in the NDVI image and in agreement with the actual values of NDVI, we have operated a suitable level slicing as coded as follows.

Table 4. Correlation between DN intervals and actual values of NDVI represented on the image.

DN intervals	Grey Level Code	NDVI Values
0 - 126	0	< 0.0
127 - 141	30	0.00 - 0.14
142 - 156	70	0.15 - 0.29
157 - 171	110	0.30 - 0.44
172 - 186	150	0.45 - 0.59
187 - 201	190	0.60 - 0.74
202 - 216	230	0.75 - 0.89
217 - 255	255	> 0.9

The band ratios IR/Red and IR/Green correlate the highest values of reflectance to the lowest and to the Green peak of vegetation, respectively.
NDVI enhances the different stages of the vegetation vigour.

The additive view of the 3 ratios also has been attempted by different coding for visual interpretation. The most useful colour combination has resulted: IR/Red → Blue, Green/IR → Green and IR → Red.

The results obtained by the digitization of IR component and their elaborations have been imaged in a plate (see fig. 4).

3.2 Output analysis

Data deriving from the image processing have been represented under the form of images and statistical parameters.

Since the sub-frames correspond to small and well confined areas, the statistical approach has seemed to be useful and more suitable than others.

In particular, the arithmetic mean m, the standard deviation σ and the coefficient of skewness s have been considered.

Values of standard deviations confirm that the sub-frame MA and G1 are the most uniform units.

The values of σ for IR/Green and NDVI images appear to be smaller than in the other cases.

The skewness of ratios remains almost negative; that means the baricenters of the histograms are shifted to high DN of representation in relation to the primitive images.

Mean values have been plotted and correlated.

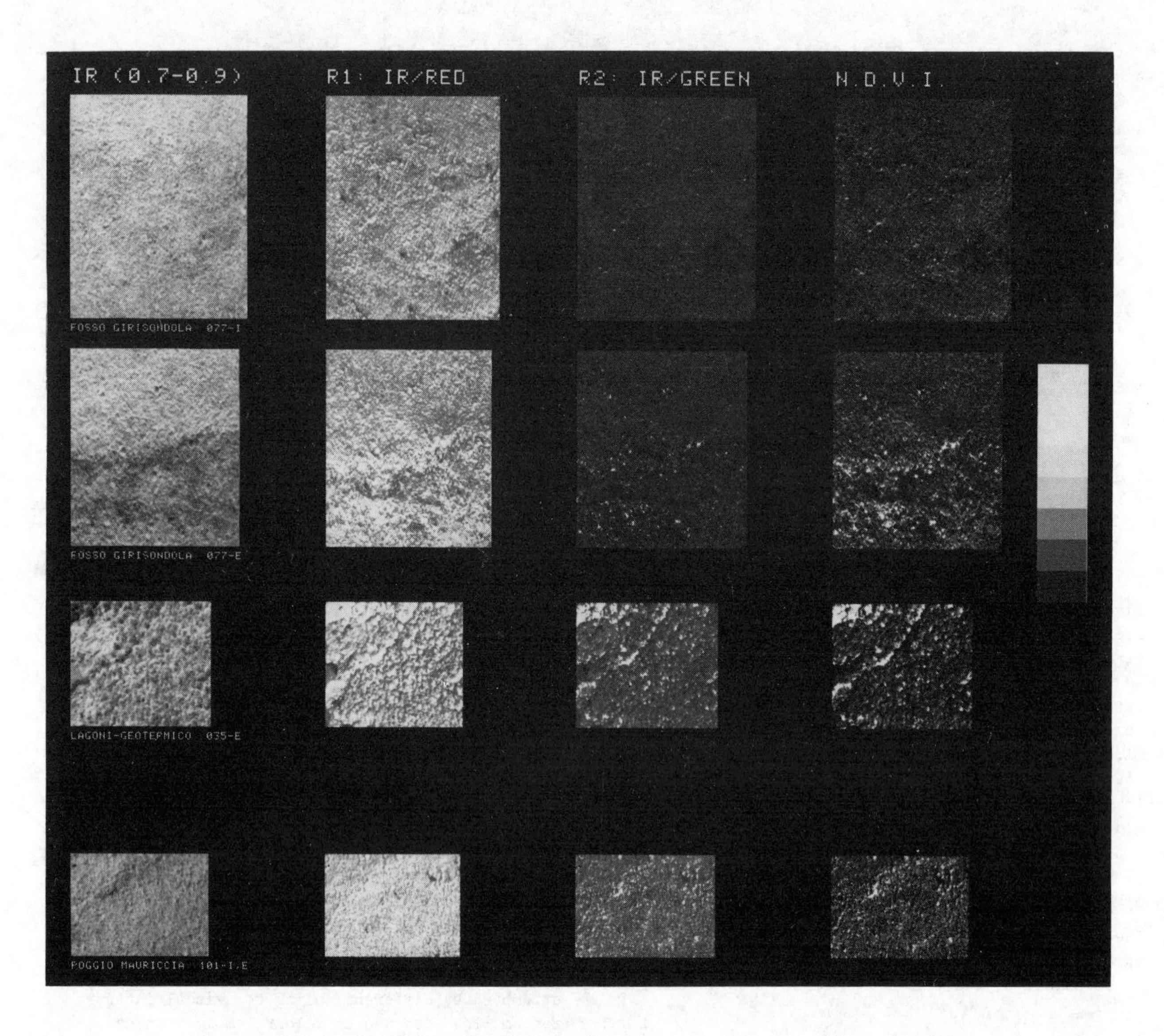

Figure 4. Image representation of the 4 sub-frames analysed in relation to the band Infrared and respective band ratios. The discrete grey scale at 8 levels on the right is the reference scale for NDVI images.

Table 5. Values of statistical parameters concerning to the base colour components Green, Red and Infrared and to band ratios of the 4 sub-frame set. Arithm. Mean = m, Standard Deviation = σ and Skewness Coeff. = s.

Sites	Statistical parameters	Green	Red	IR	IR/Red	IR/Green	NDVI
GIRISONDOLA 1	m	149	108	165	157	77	148
077-I G1	σ	21	24	16	26	8	8
	s	0.02	0.24	-0.08	-0.27	-0.35	-0.16
GIRISONDOLA 2	m	130	86	147	177	80	154
077-E G2	σ	28	29	24	37	15	13
	s	0.51	0.36	0.72	2.09	-0.31	-0.6
LAGONI	m	124	85	148	180	126	157
035-E LG	σ	36	34	24	43	34	18
	s	0.37	0.03	0.19	1.73	3.7	-0.54
POGGIO MAURICCIA	m	109	73	141	197	132	160
101-E,I MA	σ	22	22	14	36	23	12
	s	0.41	0.26	0.33	1.6	-0.52	-0.49

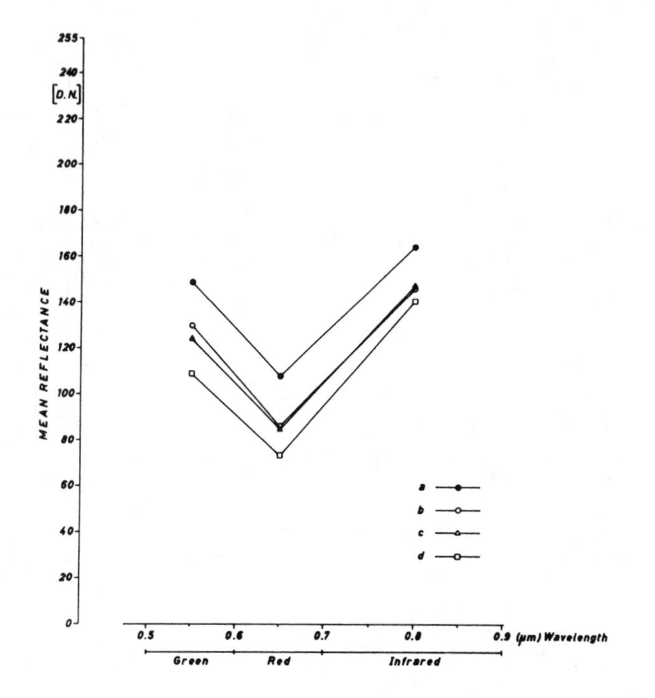

Figure 5. Diagrams of the mean reflectance derived from the mean values of the base colour components. a) G1, b) G2, c)LG and d) MA. DN stands for Digital Numbers.

We remark a good matching between the diagrams of the subframes G2 and LG; we also observe a decreasing of the mean reflectance from the sub-frame G1 up to MA.

This fact would mean a high correlation between the response in reflectance of the surely geothermal zone LG and the zone G2; in both cases thermal IR surveys indicated zones where EV was very small, almost zero.

Moreover, the sub-frame MA was the most suspected area, since it was the only spot indicated at the same time both by FATI and by EV function solutions.

On the other hand, the sub-frame G1 shows the highest reflectance values and that could fully be correlated to the highest values of FATI previously checked.

The arithmetic means of each band ratio have been correlated to their standard deviations (see fig.6).

We remark here that arithmetic means of NDVI subframes keep almost the same values. That confirms the spots selected show similar vegetational vigour and the mean values of actual NDVI are ranging between 0.15 and 0.44 which correspond to the middle levels in the grey scale of fig. 4.

Moreover, a noticeable difference appears in the mean values of the IR/Green ratio between G2 and LG; on the contrary, their respective reflectance features shown in figure 5, resulted to be almost the same. That means the ratio IR/Green has a more discriminating power than the ratio IR/Red and than the spectral signature alone.

In particular, the site MA shows the highest values in both ratios and, at the same time, it also presents the lowest reflectance mean values in according to the basic concept of a general decay when vegetation is submitted to the presence of geothermal fluids.

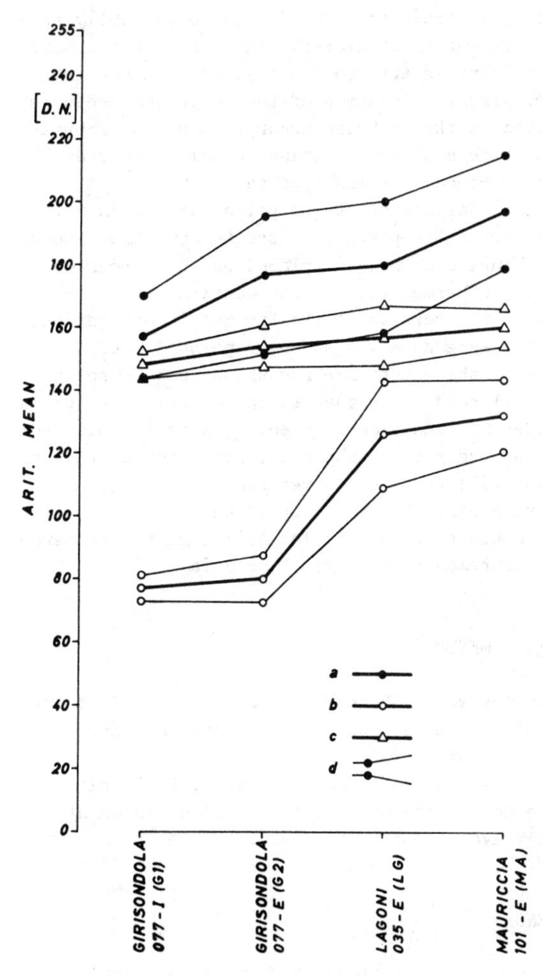

Figure 6. Diagrams of the mean values of band ratios at different sites. a) IR/Red, b)IR/Green, c) NDVI and d) standard deviation amplitude.

These arguments authorize to take the suspected zone MA Poggio Mauriccia into a higher consideration than the other sub-frames.

Scattergrams of data also have been analysed, although no particular indication has outcome beside the high correlation that has been found between the Green and Red images, almost in all 4 cases. Scattergrams of band ratios IR/Green and IR/Red confirmed the lowest correlation is found for the subframe MA.

4 CONCLUSIONS

The application of the Derived Thermal Quantities method in the Travale geothermal field has given some unexpected results.

At first, neither the old field of Lagoni nor the new one to the north have been particularly pointed out in relation to other surrounding areas. That gives rise to some perplexity on the validity of the method.

Nevertheless, some other interesting indications have been provided. Although the checked sub-frames have been interpreted as uniform, they actually present different spectral reflectance and furthermore, band ratioing is able to distinguish them one from another. Also, the location of the suspected areas is connected to the tectonic escarpment of the Travale graben where geothermal fluids actually can reach the sub-surface more easily than elsewhere.

Digital analysis of stereo pairs has concentrated and supported the previous suspects onto the site of Poggio Mauriccia which is placed on the border of the new geothermal field. That confirms also the goodness of the results of the thermal investigation where the same area was strongly pointed out.

However, the experience has shown that these surveys still need to be checked in several other cases at different conditions; moreover, a noticeable knowledge of the geo-environmental situation on the ground is still required as reference.

Nevertheless, the next step of this research will be the achievement of shallow drillings for temperature measurements at Poggio Mauriccia.

ACKNOWLEDGMENTS

The authors would thank Mr A. Colli and Mr G. Bolzan for their technical assistance in drawings and production of outputs.

Particular thanks also to Dr Baldi, Enel-Unità Nazionale Geotermica of Pisa for precious information on the Travale geothermal field.

REFERENCES

Batini, F., P. Castellucci & G. Neri 1985. The Travale geothermal field. J.Geothermics. 14:623-636.

Bolzan, G., P.A. Brivio & E. Zilioli 1982. The thermal infrared prospecting in geothermal exploration. Proc. Int. Sym. ERIM, 2nd Th.Conf., Fort Worth, 1: 331-340.

Calore, C., R. Celati, P. Squarci & L. Taffi 1979. Studio termico dell'area di Travale. Proc. 1st Sym. S.P. Geotermia, Rome, 259-269.

Pavari, A. 1942. Dispense di selvicoltura: classificazione fitoclimatica. Ed. Univ., Florence

Zilioli, E., P.A. Brivio & R. Tomasoni 1985. The infrared aerial surveys for heat flow anomalies mapping in the geothermal fields. Proc. Int. Sym.ERIM, 4th Th.Conf., San Francisco, 2:417-426.

6 Hydrology: Surface water, oceanography, coastal zone, ice and snow

Chairman: K.A.Ulbricht
Co-chairman: Mikio Takagi
Liaison: R.Spanhoff

A methodology for integrating satellite imagery and field observations for hydrological regionalisation in Alpine catchments

R.Allewijn
Department of Hydrogeology and Geographical Hydrology, Institute of Earth Sciences, Free University, Amsterdam, Netherlands

ABSTRACT : Years of intensive fieldwork in the N-Italian Dolomites have demonstrated that in complex Alpine environments predictions of stream runoff and sediment yield for ungauged watersheds require a semi-distributed, physically-based regionalisation model. The physically-based character of the model has to be guaranteed by detailed field observations, while Landsat remote sensing data can be quite valuable in quantifying the distributed nature of the model.

A hierarchical regionalisation procedure is presently being applied to the N-Italian Dolomits :

(1) A broad physiographic zone of Permo-Triassic-Liassic rocks has been delineated with Landsat MSS images (1:200.000).

(2) Vegetation and landuse units are identified by a supervised digital classification of Landsat data. Other patterns of landsurface-physical features are delineated by the visual interpretation of Landsat MSS and TM imagery (1:100.000-1:25.000).

(3) After an insight and quantification is gained of the relation between the landsurface-physical variables and the hydrological character of the Permo-Triassic-Liassic rocks, by comparison with hydrological field survey data, surface water systems and related groundwater flow systems are identified.

(4) The spatial characteristics of the hydrological units are stored in a Geographic Information System, which serves as a data bank for semi-distributed water and sediment yield models.

(5) Landsat data can further be used to correlate reflectance indices with specific field-data-based model parameters.

This procedure is being developed for a reference area and will be tested for a control area during a later stage of the investigation.

In conclusion, if one knows how to use remotely-sensed information, this data source could be an important additional tool in solving the hydrological regionalisation problem.

1 INTRODUCTION

1.1 Problem-oriented regionalisation approach

In Alpine environments, where a detailed gauging network is often missing, reliable predictions of stream runoff and sediment yield are most needed. The extrapolation of records to ungauged catchments can be performed by several regionalisation methods (Figure 1). The choice for a specific approach depends on the nature of problems to be solved, the scale at which a solution is required and the complexity of the research area (Simmers, 1984).

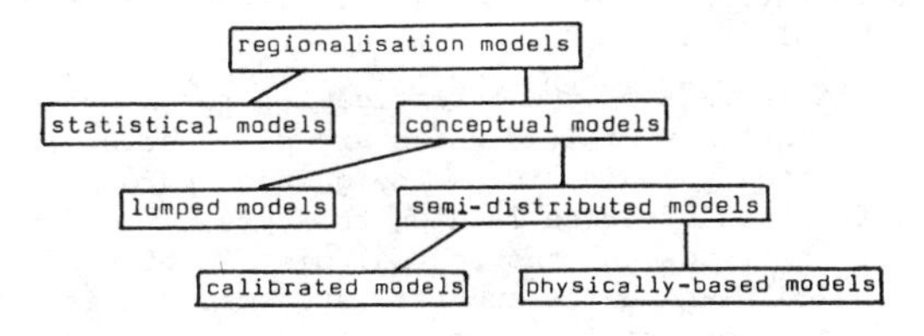

Figure 1. Regionalisation models.

In (Alpine) regions, where a large number of factors controls the hydrological regime, it seems to be very difficult to predict the runoff of ungauged catchments by purely statistical methods (Mosley, 1981; Ebisemiju, 1979). In such areas a conceptual model is more appropriate, in which the types of hydrological processes and their interrelationship with landsurface-physical variables are considered.

Lumped models, such as the Stanford Watershed Model IV (Crawford and Linsley, 1966) and the SCS curve Number Model (Soil Conservation Service, 1972) cannot be applied to catchments with a complex spatial distribution of hydrological units. A more realistic way to deal with the problem of heterogeneity is to divide the catchment into a number of homogeneous subareas and to model the hydrological processes for each subarea separately (semi-distributed models such as the USDAHL-74 Model; Holtan et al., 1975).

Calibrated catchment models, in which the parameters have no unique relation to field measurements of hydrological phenomena or to landsurface-physical variables, are not suited for the extrapolation of the results of gauged watersheds to ungauged watersheds. Consequently, in complex Alpine regions, a physically-based model with a (semi)distributed character is recommended.

1.2 Use of satellite imagery in catchment modelling

Until now, the evaluation of the applicability of remote sensing techniques to infer basic model parameters has been focused on existing catchment models (Peck et al., 1981; Engman, 1982; Rango, 1985). However, these models do not have a significant potential for using remotely-sensed information (Peck et al, 1982).

The incorporation of Landsat data in catchment response modelling has been described, among others, by Ragan and Jackson (1980 : SCS Curve Number Model), Groves and Ragan (1981 : Modified Stanford Watershed Model) and Fisher and Ormsby (1982 : USDAHL-74 Model). As the Landsat bands can give only indirect estimates of hydrological parameters, one has to determine how landsurface-physical features are related to hydrological characteristics of the area. Field observations are indispensable for this purpose (Figure 2).

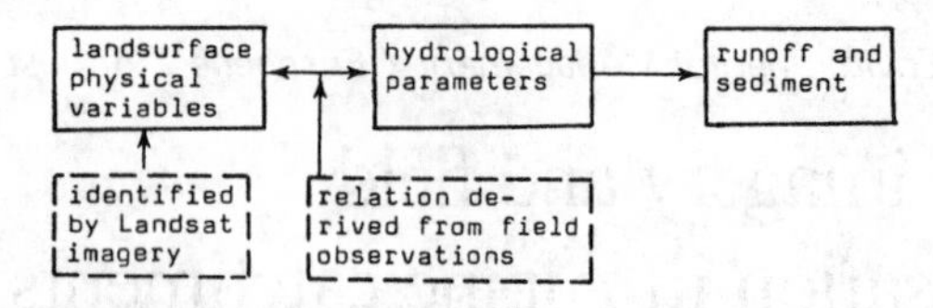

Figure 2. Use of Landsat imagery in catchment modelling.

In most models which use Landsat data the spatial variability of the hydrological processes is ignored. However, in catchments containing complex patterns of recharge and discharge areas, Landsat imagery can be of great use in quantifying the field heterogeneity in a semi-distributed model. This is especially true for inaccessible mountainous areas.

2. METHODOLOGY

A phased methodology is suggested for the regionalisation and runoff prediction in heterogeneous Alpine areas (Figure 3).

I Reconnaissance stage: Identification of broad physiographic zones (maps and Landsat MSS imagery 1:200.000)
II Mapping, field surveys and modelling in a reference area 1. Identification of patterns of land units - visual interpretation of Landsat MSS and TM 1:100.000 - 1:25.000) - digital classification of Landsat MSS and TM 2. Hydrological quantification of land units, resulting in hydrological land units - field observations 3. Synthesis of hydrological land units in flow model - systems analysis
III Simulation and verification of the model in a control area

Figure 3. Methodology.

2.1 Reconnaissance stage : Identification of broad physiographic zones

In the reconnaissance stage the study area is restricted to one broad physiographic zone of a few thousand to tens of thousands of square kilometres. No hydrological information transfer is allowed from one zone to another.

Within a climatic zone these physiographic zones are mainly distinguished on basis of their lithology and structural geology. The general morphology is used for further differentiation (Meyerink, 1976).

Apart from information from small scale maps (1:200.000), the visual interpretation of Landsat MSS imagery is an important data source.

2.2 Mapping, field surveys and modelling in a reference area

In the next stage a reference area is selected which is expected to contain all the hydrological diversity within that particular physiographic zone. A hierarchical mapping and modelling procedure is performed in this reference area (Figure 3).

On the lowest level patterns of landsurface-physical features (land units) are identified. For the delineation of vegetation and landuse units a digital classification of Landsat MSS and TM data is pursued. The identification of geomorphological features, fault patterns, etc., is performed on basis of the visual interpretation of Landsat MSS and TM imagery (1:100.000-1:25.000).

Land units derived from imagery are checked in the field and quantified hydrologically, based on hydrological characteristics like the generation of peak discharge or baseflow, sediment yield, soil moisture content, etc. Thus, the land units are converted to hydrological land units.

These two-dimensional hydrological units are composed by a systems analysis into interconnected flow systems of surface waters, soil waters and groundwaters. Groundwater flow systems are e.g. subdivided in deep regional, deep to intermediate subregional and shallow local systems (Engelen, 1984).

For the identification of regional groundwater flow systems, which cross main surface water divides, Landsat MSS imagery is of great use (scale 1:100.000). On this system level regional fault patterns are analysed. Subregional and local flow systems, within the major subcatchments, are mainly delineated on basis of geomorphological features and vegetation and land use patterns. Landsat MSS and TM images (scale 1:50.000-1:25.000) are an important tool for identifying these features.

Generally, for high Alpine environments with fractured or karstic rocks the recharge takes place over broad areas, while the outflow of the systems is limited to concentrated spring areas. The relatively poor resolution of Landsat MSS and TM is a severe limitation for the identification of concentrated spring areas. Furthermore, spring areas are mostly not uniquely related to a specific vegetation type. Thus one has to rely on field observations and aerial photographs (scale 1:25.000-1:10.000).

The hydrological character of a catchment is not only modelled by the conventional lumped parameters and variables, but in this semi-distributed flow model, the spatial distribution of the hydrological land units, like distance to outlet, area, mean height and aspect, is also quantified. This spatial information, derived from a combination of Landsat imagery maps and field surveys, is stored in a Geographic Information System (G.I.S.). Although the desired grid scale of this data bank depends on the variable and the type of region to be modelled, for most input data a grid scale of 30 m (resolution TM), or even 80 m (resolution MSS), seems sufficient (Hendriks, in prep. : study in East-Luxembourg).

The response time and the initial state of water contents of the various flow systems are obtained by hydrograph analysis, analysis of spring discharges and soil moisture measurements.

Some model parameters of the hydrological land units, like the interception storage capacity, have to be measured in small field plots. The field-data-based parameters may be correlated with reflectance indices of Landsat MSS and TM.

2.3 Simulation and verification of the model in a control area

In the last stage of the investigation for a control area the patterns of land units are delineated from Landsat imagery. On basis of the gained hydrological quantification of these land units a prediction is made of the types and the spatial distribution of hydrological land units and flow systems. The stream runoff and (potential) sediment yield of the control area is predicted by the developed semi-distributed flow model. The mapping and modelling results are verified by fieldwork and aerial photography (1:25.000).

3. CASE STUDY

The three-phased methodology is presently being applied to the complex Alpine environment of the N-Italian Dolomites (Figure 4). The study forms part of an extensive research and graduate training program on mountain hydrology in the Dolomites, carried out since 1966 by the Department of Hydrogeology and Geographical Hydrology, Institute of Earth Sciences,

Amsterdam Free University, the Netherlands (Engelen, 1974; Seyhan et al., 1985).

The broad physiographic zone of the Permo-Triassic-Liassic rocks of the Central-Dolomites has been selected as the research area. In the summer of 1985 and the spring and summer of 1986 a detailed mapping and discharge measuring program has been executed in the Upper-Boite reference area (Figure 4). The control areas are also indicated in Figure 4.

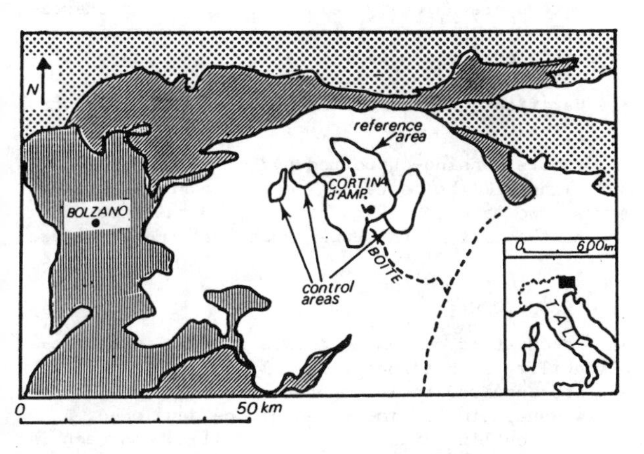

Legend:

- Gneisses and schists of the Central Alps
- Quartz-phyllite basement of the Dolomites
- Quartz-phorphyries of the Bolzano region
- Permo-Triassic-Liassic rocks of the Dolomites

Figure 4. Broad physiographic zones.

3.1 Reconnaissance stage - Broad physiographic zone

The heterogeneous character of the Permo-Triassic-Liassic rocks of the Dolomites can be explained by three main factors (Engelen, 1963; 1974) :

1. Lithology : A wide range of rock types are present in this region. Apart from shales, gypsum, sandstones and limestones, huge dolomitic reef masses are found, which interfinger with a complex of marls volcanic ashes and lavas. Due to a differential erosion, the dolomite rock presently stands out 1000 à 1500 m above the tuffaceous marl formations.

2. Structural geology : The dolomite (and limestone) formations are strongly influenced by gravity-tectonics, which resulted in a large number of faults (Engelen, 1963). The marine-volcanic formations between the reef dolomites and the underlying plastic shales and gypsiferous marls are squeezed upward diapirically by the subsiding dolomite reef masses. This resulted in an intricate outcrop pattern of rock types.

3. Quaternary morphology : Glacial erosion features, glacial and fluvioglacial deposits, Holocene talus cones of coarse dolomite rubble and recent mass-movements in the weathered marly and tuffaceous rocks gave this physiographic zone its final shape. Most geological and morphological phenomena can be identified with Landsat imagery.

The mean annual precipitation at a mean height of 1600 m is approximately 1000 mm, with a maximum in the three summer months of 410 mm (Fliri, 1975). The Penman potential evaporation at 1600 m is estimated to be 620 mm/year and 290 mm in the summer months (based on climatological data of Fliri, 1975).

3.2 Mapping, field surveys and modelling in the Upper-Boite reference area

In the summer of 1985 the Upper-Boite catchment (210 km²) was selected as a reference area. The discharges of about 20 subcatchments have been measured in the surroundings of the main town in this area, Cortina d'Ampezzo. The reaction of the hydrological units to snowmelt was observed during the spring of 1986.

A comprehensive description of all kinds of hydrological land units and flow systems is beyond the scope of this paper. However, to illustrate the mapping procedure two complexes of interrelated flow systems, with distinctive hydrological characteristics are selected (Figure 5 and 6) :

1) The fractured dolomite rock in combination with the coarse dolomite rubbel.

2) The deeply weathered and locally slumped marly tuffs.

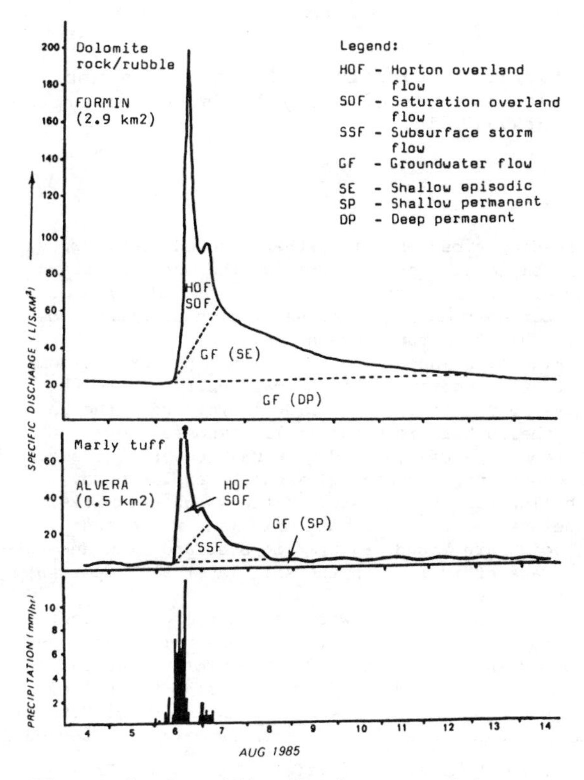

Figure 5. Specific discharges of two main complexes of hydrological flow systems with an indication of the types of flow systems.

The hydrological groundwater flow systems which can be distinguished with the field observation-based systems analysis are sketched in Figure 6. A provisional separation of the hydrographs is also made on basis of these field observations (Figure 5).

The higher parts of the dolomite rock recharge area (of the deep permanent groundwater flow system) are completely bare, while the lower parts are generally covered with dense pine forests on thick soils of (semi)impermeable moraine material. In a transition zone one finds open pine forests and shrubs on thin soils. The dolomite rubble, with a shallow, episodic flow system on top of the permanent flow system in

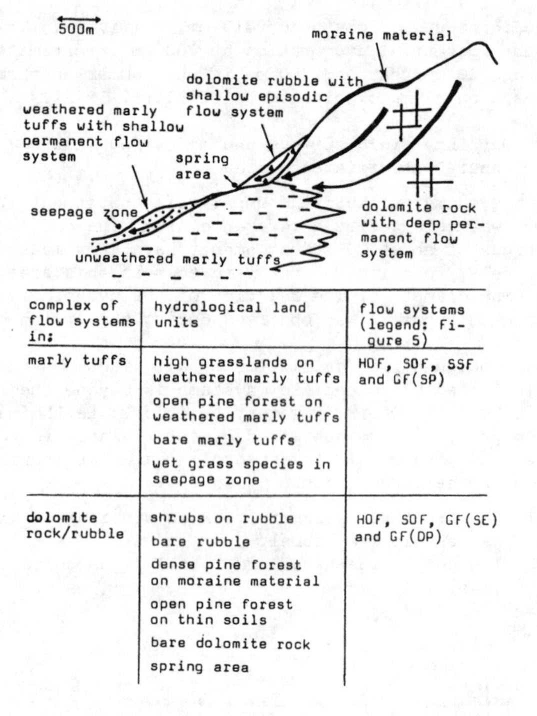

complex of flow systems in:	hydrological land units	flow systems (legend: Figure 5)
marly tuffs	high grasslands on weathered marly tuffs	HOF, SOF, SSF and GF(SP)
	open pine forest on weathered marly tuffs	
	bare marly tuffs	
	wet grass species in seepage zone	
dolomite rock/rubble	shrubs on rubble	HOF, SOF, GF(SE) and GF(DP)
	bare rubble	
	dense pine forest on moraine material	
	open pine forest on thin soils	
	bare dolomite rock	
	spring area	

Figure 6. Hydrological land units and flow systems of marly tuffs and dolomite rock/rubble.

the deeper bedrock, is either covered with shrubs, or completely bare. Generally, the water discharges as structurally controlled concentrated springs. Horton overland flow and saturation overland flow are the storm runoff mechanisms.

High grasslands with some open pine forests dominate the vegetation of the marl unit. Due to the mass-movements in the residual overburden the unweathered bare rock is locally exposed. Here a Horton type of overland flow may occur. The small amount of groundwater flow can be explained by a shallow permanent system in the weathered marls. In the seepage zones of this unit many wet grass species are found. Saturation overland flow and subsurface storm flow in the soil layer are important runoff processes.

The first effort towards a digital classification of the main vegetation and landuse units was performed in March/April 1986. Therefore a Landsat TM tape of August 1984 and two Landsat MSS tapes of August and October 1984 have been analysed. Several supervised classification methods have been tested :

1. Two band box classifier (MSS 7-5 and TM 4-3, 4-5 and 5-3).
2. Maximum likelihood classifier using all TM bands.
3. To eliminate the shadowing effect a fast parallelepiped two-features classification has been executed, using :
   *combinations of TM ratios 4/5, 3/4 and 3/5.
   *"greenness" and "yellowness factor" (leaving out the "brightness factor") of a principal component transform of Landsat TM bands.
4. Multitemporal classification using the near infrared MSS bands of August and October 1984.

A final evaluation of the classification results is scheduled for the end of 1986. The results will be compared to a map, aerial photograph (1:25.000) and field data-based classification.

Hydrograph analysis of the 1985 and 1986 discharges of the subcatchment of the Upper-Boite area will provide the parameters of the groundwater flow systems. During the spring and summer of 1986 the field plot measurements of interception and soil moisture storage capacity, maximum percolation rate, etc. will be sampled for each hydrological land unit.

Although it is premature to give definite reflectance indices which may be correlated to hydrological parameters, at this stage combinations of ratios of MSS 5/7 and TM 3/4, 4/5, 3/5, "greenness", "yellowness" and "brightness factors" and the thermal infrared values seem to be promising as hydrologically important features. It should be realized that only for physically-based models in which the hydrological parameters are uniquely related to landcover characteristics, such a correlation with Landsat reflectance indices is possible.

3.3 Verification of the flow model in several control areas

The remote sensing-supported hydrological mapping and modelling will be tested in several control catchments. Two catchments are situated to the west and one to the east of the Upper-Boite catchment (Figure 4).

4. CONCLUSIONS

As the existing catchment models are of limited applicability in heterogeneous (Alpine) environments and are unsuited to incorporate remotely-sensed data, a new generation of models should be developed. A possible outline of such a new modelling approach is described in this paper. The methodology is focused on two items :

1. The identification of the spatial distribution of hydrological units and processes is strongly supported by satellite imagery.
2. The basic input parameters are physically-based which makes it possible to relate field data-based parameters to Landsat reflectance indices.

In the near future this remote sensing-supported semi-distributed model will be applied to several catchments in the N-Italian Dolomites.

REFERENCES

Crawford, N.H. and R.K. Linsley. 1966. Digital simulation in Hydrology : Stanford Watershed Model IV. Technical Report no. 39, Department of Civil Engineering, Stanford Univ., Standford, CA.

Ebisemiju, F.S. 1979. An objective criterion for selection of representative basins. Water Resources Research, 15(1) : 148-158.

Engelen, G.B. 1963. Gravity tectonics in the N.W. Dolomites. Geologica Ultrajectuna No. 13, Rijksuniversiteit Utrecht.

Engelen, G.B. 1974. Hydrogeology of the Sasso Lungo Group. A Dolomitic Reef Stock in the Alpine Dolomites of North Italy. J. Hydrol. 21 : 111-130.

Engelen, G.B. 1984. Hydrological systems analysis. A regional case study. Report OS 84-20. Institute of Applied Geoscience TNO-DGV, Delft.

Engman, E.T. 1982. Remote sensing application in watershed modeling. In Applied modelin in catchment hydrology, Proc. of the Int. Symp. on rainfall-runoff modeling. Water Resources Publ. : 473-494.

Fliri, F. 1975. Das Klima der Alpen in Raume von Tirol. Monographiën zur Landeskunde Tirols. Universitätsverlag Wagner, Innsbruck-München.

Fischer, G.T. and J.P. Ormsby. 1982. The application of remotely sensed observations to hydrologic models. In D.N. Body (ed.), Application of results from representative and experimental basins : 409-428.

Groves, J.R. and R.M. Ragan. 1983. Development of a remote sensing based continuous streamflow model. In Proc. of the 17th International Symposium on remote sensing of environment : 447-456. Ann Arbor, Michigan.

Hendriks, M.R. in prep. Regionalisation of hydrological data. Ph.D. thesis, Inst. of Earth Sci., Free Univ., Amsterdam.
Holtan, H.N., G.J. Stiltner, W.H. Henson and N.C. Lopez. 1975. USDAHL-74 revised model of watershed hydrology. Technical Bulletin no. 1518, United States Department of Agriculture, Washington, D.C.
Meyerink, A.M.J. 1976. A hydrological reconnaissance survey of the Serayu River basin. Nuffic Project ITC/GUA/VU/1, Final Report. Vol. 2 : 25-64.
Mosley, M.P. 1981. Delimitation of New Zealand hydrologic regions. J. Hydrol. 49 : 173-192.
Peck, E.L., T.N. Keefer and E.R. Johnson. 1981. Strategies for using remotely sensed data in hydrologic models. NASA-CR-66729, Goddard Space Flight Center, Greenbelt, Md.
Peck, E.L., T.N. Keefer and E.R. Johnson. 1982. Suitability of remote sensing capabilities for use in hydrologic models. In Proc. Int. Symp. on Hydrometeorology : 59-63. AWRA, Denver, Colo.
Ragan, R.M. and T.J. Jackson. 1981. Runoff Synthesis using Landsat and SCS Model. In Journal of the Hydraulics Division, Proceedings, ASCE, Paper 15387, 106 : 667-678.
Rango, A. 1985. Assessment of remote sensing input to hydrologic models. Water Resources Bulletin. Vol. 21, no. 3 : 423-432.
Seyhan, E., A.A. van de Griend and G.B. Engelen. 1985. Multivariate analysis and interpretation of the hydrochemistry of a dolomitic reef aquifer, Northern Italy. Water Resources Research. Vol. 21, no. 7 : 1010-1024.
Simmers, I. 1984. A systematic problem-oriented approach to hydrolgoical data regionalisation. J. Hydrol. 73 : 71-87.
Soil Conservation Service. 1972. National Engineering Handbook Section 4 : Hydrology. U.S. Department of Agriculture, Washington, D.C.

Symposium on Remote Sensing for Resources Development and Environmental Management / Enschede / August 1986

The JRC program for marine coastal monitoring

J.A.Bekkering
Joint Research Centre, Ispra (Varese), Italy

ABSTRACT: An overview is given of the result of the JRC activity "Coastal Transport of Pollution", since its start in 1980. It covers essentially RS from space, in-situ measurements and modeling relative to its main test site, the Northern Adriatic Sea. The activity is strongly based on a collaboration with institutes and organizations of the EC member-states. For the near-future the CTP envisages to extend its collaboration, in particular concerning the modeling effort.

INTRODUCTION

Since 1976 the JRC , together with many national institutions, is engaged in an effort to investigate the possibilities to monitor the marine ambience from space. The effort started in the wake of the Barcelona Convention of 1975 ,which was organized by the UNO to favor an international collaboration to limit and reduce the disastrously increasing pollution in the Mediterranean basin.

The first major enterprise of the activity was the EURASEP OCS Experiment, held in the North Sea in 1977, and essentially meant as a simulation experiment for the CZCS, to be launched in 1978.

In 1980, in the frame of the new multiannual research program of the JRC, the activity has been baptised "Coastal Transport of Pollutant" (CTP) and its scope has been redefined. For practical reasons, mainly based on meteorological conditions and on its vicinity to the JRC ,the Northern Adriatic sea has been selected as the main test site.

Eversince the objective of this Communitarian activity has been to collaborate on the development of an operative system, based mainly on RS from space, capable of monitoring and forecasting major marine events in coastal area's, like plankton production, algae blooms, transport, transformation and degradation of pollutants and nutrients, sedimentation and resuspension of suspended matter, etc.

The system is meant as an aid to the management of marine resources and environment. Presently the activity is stil confined to the Adriatic sea and up to now only spaceborne sensors operating in the visible and IR are evaluated.

The CTP activity is engaged in a continuous effort to get as many national institutions as possible collaborating or participating in the project.

The collaboration can have many different forms, but roughly 3 main forms can be distinguished:

- concerted action, in which the action has been agreed between the partners, but each is responsible for his own part, technically and financially,
- shared cost, in which the JRC bears a part of the cost sustained by one of the partners, essentially meant as an encouragement to start the action or to adapt it to a common goal,
- contract, in which the JRC bears the full cost and just acquires the service or the product.

The activity has 3 major branches :

- Remote Sensing from space,
- Continous and discontinuous in-situ measurements,
- A computer run model

The JRC personnel involved in the CTP is around 13 man.year/year and the activity is roughly subdivided as follows :

- Development of procedures and algorithms for the processing of spaceborn sensor data : B.Sturm, S.Tassan
- Biological and optical in-situ measurements, data elaboration and evaluation : P.Schlittenhardt, M.Ooms
- Atmospheric mearurements and evaluation : G.Maracci
- Chemical measurements and evaluation : G.Ferrari
- Management of the activity CTP : J.A.Bekkering

The image processing software has been developed in a joint effort with the activities involved in land application of RS , and special reference is made here to

- System management of the image processing facility: B.Dorpema
- Coordination of application software development: W.Mehl

A part of the work has been executed by persons or institutions outside the JRC; whenever the case they will be quoted together with the description of the work

SPACE BORN SENSORS

-Nimbus 7 - CZCS

For completeness the most salient characteristics of the CZCS are listed in table 1, and the spectral ranges are indicated in figure 1.

A fundamental problem with marine application of RS is the fact that so few radiation is reflected out of the water.

Even on a clear bright day only some 20% of the radiation arriving at the sensor originates from the

Table 1
The most salient characteristics of used sensors

Sensor	CZCS	AVHRR(4)	AVHRR(5)	TM
Satellite	Nimbus 7	NOAA 6	NOAA 9	Landsat 5
height	950	833	870	709 km
inclination	99.24	98.74	98.90	98.25 dgr
equat time	11.50 (n)	7.30 (s)	14.30 (n)	9.45 (s)
pixel leng.	800	1100	1100	30 m
width	800	800	800	30 m
swath	1400	2400	2400	180 km
daily cov.	.5	.8	.8	.07
launch	24.10.78	27.6.79	8.11.84	5.3.82

(n)= North bound

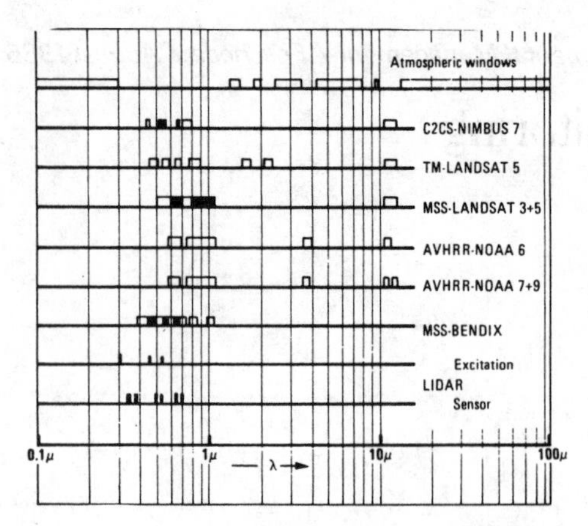

Fig 1. Spectral range of the channels of the sensors used during the "Adria 84"

water, this making soffisticated methods a must to finish up with an acceptable precision.

The processing of the CZCS data consists essentially of the following phases:

- Sensor sensitivity and callibration. The CZCS has an internal calibration system, which has never indicated an appreciable deterioration of the sensors (except for CH 6, the thermal channel). Nevertheless there were clear indications of a deterioration, obviously an optical surface outside the calibration circuit. Deterioration was negligible for CH 4, but increased substantially with decreasing wavelength for the other channels, a phenomenon indicating (cosmic) dust deposit as the most probable cause. Sensitivity loss coefficients have been established for the channels 1,2 and 3 relative to CH 4, using the upwelling radiation of the clear-water pixels of all CZCS scenes ever elaborated at the JRC.
- Atmospheric correction. The fact that the sea reflects so few radiation makes the atmospheric correction a must. Although the basis of the correction is still the one proposed by Gordon, quite some refinements have been added eversince. Three effects are of importance :
  - absorption on the illumination source welling down into the sea and on the reflected radiation welling up to the sensor.
  - Rayleigh scattering or molecular scattering
  - aerosol scattering

  The first 2 effects can be calculated straight forward, the 3d effect is an everchanging quantity depending on place and time.
- Chlorophyll and Total Suspended Substance (TSS) determination. An emperical relationship has been developed, based on in-situ measurements and the RS signature, to establish CHL and TSS concentrations.

The CZCS spectral resolution does not allow to distinguish between CHL and Yellow Substance. In coastal regions with no fixed relationship between CHL and YS, this might lead to substantial errors. Part of this work has been done by a visiting scientist, M.Viollier.

- Geometric correction. For comparison of different images, they must be brought on a common geometrical footing, in our case the Mercator projection (projection cylinder going thru the equator). The classical geometrical correction method fails when not sufficient well distributed reference point are available, as is regularly the case over sea area's. NASA provides for each image so-called anchor points, synthetic reference points of which the location is given, derived from sensor relative data.

Experience however has learned that the anchor points are not sufficiently precise and can give rise to errors of some tens of kilometres.

A procedure has been developed, essentially based on sensor relative data, only corrected with a minimum of ground control points, which need not to be equally distributed over the scene. The program has been developed under contract (M.Langemann, A.Popella) under the leader ship of prof. Ph. Hartl, of the Technical University of Berlin.

- Final product. The elaborated images are hard copied for subsequent visual interpretation, or once again geometrically corrected to fit the grid used for the hydrodynamic model of the Adriatic Sea.

-NOAA AVHRR

The sea has a much more uniform temperature compared to land, reason why sea temperature, when to be significant, must be established with much greater precision.

The thermal channels of the AVHRR operate with an equivalent precision of around .1 K (at 293 K), but due to the atmospheric presence, and in particular to the water vapor, errors of several degrees are possible, the measured value always being lower than the real one.

The water vapor quantity can be established in two ways :

- with the split window technique, e.i. derived from the difference between CH 4 and CH 5. Only NOAA 9 is equipped with a split window, NOAA 6 has only 4 channels.
- by means of the HIRS (High resolution IR Sounder), installed on both NOAA satellites, but for this scope considered less precise. A procedure has been developed, based on the split window technique by a visiting scientist G.Dalu (Istituto di Fisica dell' Atmosfera, CNR, Roma).

The procedure consists eesentially of the following phases:

- calibration.
- determination of brightness temperature for CH 4 and CH 5
- determination of the quantity of water vapor
- determination of the SST (Sea Surface Temperature) from CH 4 and the water vapor quantity.

The objective was to arrive at a precision of .3 K. Up to now published SST maps refer essentially to scenes with relative large temperature differences, e.i. the gulf stream eddies, or the strait of Messina, but the Adriatic Sea, especially during the summer has very small and slow temperature differences, which makes high precision necessary.

The SST derived from surface radiation refers only to the utmost top layer of the water body (20 microns) and may be more than 1 K below the water bulk temperature (at 30 cm depth), depending on water evaporating conditions and water turbulence. Actually no procedure for geometric correction of AVHRR images is available. The processing program is implemented on the Amdahl mainframe.

-Landsat TM

With the probable shutdown of the CZCS in near future, the marine community will be deprived from a specific Ocean Color Sensor for an unknown stretch of time. The TM might constitute at least a partial substitute for the CZCS.

The major problems with the TM are :

- less spectral resolution, broader channels, thus more difficulties to distinguish CHL and TSS.
- small swath, thus low daily coverage
- early morning pass, thus low illumination, specially in winter,
- very expensive on a covered surface basis.

The JRC will participate in the North Sea experiment in May 1986 during which the TM capabilities as an ocean color sensor will be investigated.

-New sensors

When no risk exist of over stretching the activity, new sensors will be investigated, as substitute for

actual operative sensors or to enlarge the number of marine parameters detectable.
Specific reference of course is made to the active microwave sensors on board of N-ROSS, TOPEX and ERS1, which promise to open quite new horizons.

-Image Processing Facility

For completeness a very short summary of the processing facility is given.

Main facility (general image processing)

computer	main mem.	mass mem.	display	hardcopy
VAX 785	4 Mbyte	900 Mbyte	TRIM	
PDP 11/24			2 TRIADE	MATRIX
PDP 11/24				VIZIR
PDP 11/23			HEIMANN	

Satellite facility (marine image processing)

computer	main mem.	mass mem.	display	hardcopy
VAX 750	2 Mbyte	120 Mbyte	TRIADE	MATRIX

The satellite facility is housed in an other building and connected to the main facility with a fast link.
The application software is essentially developed in-house, except the spider package. It would be really a tremendous advantage whenever the European Image Processing Community, or atleast the Communitarian part of it could find each other and arrive at such a level of normalization to render software exchange (and user exchange) possible or at least a bit less frustrating as it is today.

IN-SITU MEASUREMENTS AND CAMPAIGNS

While the JRC is situated rather distant from the sea and does not dispose of proper means like ships or airplanes, in-situ measurements are only effectuated during campaigns, which are organized atleast once a year and last regularly one or two weeks.
The campaigns comprise normally the following measurements:

- biological measurements
- in-situ optical measurements, underw. and over w.
- physical measurements
- chemical measurements
- in-situ atmospherical measurements
- airborn sensor data acquisition

Often the campaigns are organized by the JRC, but sometimes the JRC participate in campaigns organized by others.
The most massive campaign organized by the JRC up to now was in 1984 (28.8 till 7.9)
The participating institutions are listed in table 2.

The JRC organized campaigns are essentially based on a concerted action or shared cost, with the JRC bearing only the organizational cost and occasionally the rent of an airplane or special instrumentation, and where the JRC scientists find hospitality on the ships or fixed marine platforms of other institutions. Every institution is responsible for its own means and its own data acquisition and elaboration, although a substantial effort is made by all to bring the measurements on a common footing. Essentially all data is afterwards exchanged between the participants. A continous effort is done to get more institutions involved, but up to now we have only succeeded to interest Italian and German institutions, probably due to the distance.
The JRC organized parties were always centered on the North Adriatic Sea, but several times the JRC has participated in campaigns outside its regular test site, like:

- 1980 (29.1-21.2) Benguela current , South Africa
- 1983 (10.1-25.1) Atlantic Ocean , Senegal
- 1986 (5.5-17.5) North Sea , Germany

TABLE 2
Institutions participating in the ADRIA 84 Campaign

Institution		platf.	main measurements
- DFVLR, Oberpfaffenh.	D	DO-28	Bendix,OCR,SCR,PRT5
- Univ. Oldenburg	D		LIDAR
- EREO, Firenze	I		IR scanner, LLtv cam
- Univ. Firenze	I		Bio/optical
- Regione Emilia-R.nga	I	Daphne	Current, bio/chem.
- ISDGM, Venezia	I	Litus	Bio., STD, Radiom.
- IBM, Venezia	I	d'Ancona	Bio, STD, Radiom.
- Univ. Venezia	I		Bio/chem.
- OGS, Trieste	I		Current
- Scient. Data Padova	I		Atmosph.
- Inst.Meer. Hamburg	D		Part. size
- Regione Friuli-V.zia	I	Cessna	
- Univ. Regensburg	D		Bio/optical
- JRC, Ispra	I	DO-28	Bio/chem/opt., atm.

OCR =Ocean Color Radiometer,
SCR =Six Channel Radiometer
PTR5=Precision Radio Thermometer(Barnes),
STD =Salinity-Temperature-Depth.

A specific study has established how far fluorescence spectrography can be used to determine Yellow substance concentrations (Ph.D thesis, M.C.Russo, London)

CHEMICAL MEASUREMENTS

A substantial part of the major pollutants, like heavy metals, are not detectable from distance, however when it would be possible to establish a relationship between detectable substances, like CHL, TSS, sediment and temperature and the adherence of heavy metals when present to the particles, it would be possible with few in-situ measurements and RS maps to establish the spacial distribution of heavy metals over large area's. As regularly the case, once this investigation was started, the problem resulted more complicated as anticipated, but the concept looks still promising.

AIRBORN SENSORS

Airborn sensors are only flown during campaigns and solely considered as an aid to space born sensor interpretation and as an addition to in-situ measurements, or for sensor development.
A specific effort is dedicated to the understanding of the atmopheric influence, for which in the past the MSS Bendix scanner was and in future the Deadalus scanner will be flown on different altitudes.
The associated analitical work together with the development of the image processing software is done in collaboration with R.Guzzi, Zibordi the Istituto per lo Studio delle Metodologie Geofisiche Ambientale, CNR.

THE ADRIATIC SEA MODEL

Already in the early stages of the CTP activity the need was felt for the development of a model as a complement to RS data and in-situ data.
Essentially the model should respond to the following needs:
- Short term (days or weeks). To predict propagation of actual pollutants
- Medium term (months). To predict propagation of nutrients and their secondary effects (f.e. eutrophication).
- Long term (years). To predict propagation,transformation and sedimentation of pollutants and nutrients and their possible resuspension.

For the model development a contract was made with the university of Liege, J.C.J.Nihoul, F.Clement.
The model development was envisaged in the following phases:

- A 3 dimensional hydro dynamic model with passive constituents (decay, sedimentation),
- A model describing active constituents (plankton/ algae growth and decay, resuspension of sedimented matter, etc),
- A residual current model,
- The creation of a library of model-reconstructed cases, defined by their RS signature.

Actually we are around half way the final goal.

The hydrodynamic model is a 3D model, actually consisting of a coarse grid (11.445 x 5.723 km) model covering the entire Adriatic down to the strait of Otranto, and a fine grid (2.861 x 2.861 km) model covering solely the Northern Adriatic Sea , north of the line Pula-Rimini. The depth is divided in 25 steps. The fine model derives its sea-boundery conditions from the coarse model.

The initial conditions and the boundary conditions of the model runs will be based as far as possible on RS data from space, integrated with meteorological data and in-situ data. The model is presently implemented on the Amdahl mainframe.

THE ADRIATIC MODEL CALIBRATION CAMPAIGN

Up to now the JRC modeling effort was essentially by contract and and we would be glad to enlarge its basis to a concerted action together with several institutions or a shared cost action. Furthur on the calibration of the actual JRC model is hardly possible with RS data and requires extensive in-situ measurements, essentially of a type hardly acquired up to now (current).

It is in this context that the JRC tries to organize a collaboration to arrive at:
- a redifinition of the model to make it a common objective of more collaborating institutions
- the organisation of a (or more) massive campaign in the Adriatic basin to collect all data needed for the calibration of the model and for a better understanding of the marine ambience

The first phase is envisaged to start this fall.

Workshops are envisaged to investigate the merits and drawbacks of the various modeling approaches and to arrive at a common modeling concept for the Adriatic Sea, supported by several institutions, who will engage in further development and implementation.

REFERENCES

Camagni, P., Galli de Paratesi,S., Gillot,R., Maracci,G., Omenetto,N., Pedrini,A., Rossi,G., Schlittenhardt,P., Mehl,W. Sturm,B., Tassone,G., Toselli,F., 1983, "Marine remote sensing activities of JRC-Ispra". EARSEL General Assembly & Symp., 26-29 April 1983, Bruxelles (Belgium)

Clement F.,Nihoul J.C.J., 1984, "Three dimensional hydrodynamic models for stratified semi enclosed seas. Aspects and computation". Applied Numerical Modeling Nat. Cheng Kung University, Taiwan, 27-29 December.

Ferrari, G., Schlittenhardt,P., 1983, "Correlation study between water-derived elements and total suspended matter derived from CZCS-satellite data". Intern. Conf. on Heavy Metals in the Environm.,6-9 Sept. Heidelberg(FRG).

Guzzi G., Maracci G.C., Rizzi R., Siccardi, A., 1985, "Spectroradiometer for ground-based atmospheric measurements related to remote sensing in the visible from a satellite". Applied Optics, Vol 24, no 17, 1 Sept 1985.

Langemann, M., Popella,A.,1983,"Geometric Correction of CZCS Data". Contr. No. 2017-82-12 EP ISP/D (Final Report).

Maracci, G., Mostert,S.A., Schlittenhardt, P., Shannon,L.V., Sturm,B., 1983, "Results of the Nimbus-7 CZCS Experiment in the Benguela Current, February 1980". Intern.Symp. on the most important upwelling areas off Western Africa, 21-25 November 1983, Barcelona(Spain).

Nykjaer, L., Schlittenhardt,P.,Sturm,B., 1984, "Qualitative and quantitative interpretation of ocean color-Nimbus 7-CZCS imagery of the northern Adriatic sea from May to September 1982". JRC Report No. S.A./1.05.E2.84.05.

Ooms, M., Maracci,G., Schlittenhardt.P., 1983, "Underwater irradiance measurements at 4 wavelength(440,520,550,670 nm) in the northern Adriatic sea". Intern. Coll. on Spectral Signatures of Objects in Rem. Sens. 13-16 October 1983, Bordeaux(France).

Russo, M.C. 1983, "Evaluation of fluorescence spectroscopy as a method for measuring 'Gelbstoff' in aquatic environments". PhD Thesis submitted to the University of London.

Shannon,L.V., Schlittenhardt,P, Mostert, S.,A., 1984, "A Nimbus-7 CZCS experiment in the Benguela current region off southern Africa, February 1980." Journ. of Geophys. Research. 89.

Schlittenhardt, P. 1980, "Preliminary report on comparision of chlorophyll-a analysis in some laboratories". EURASEP Sea-truth working group meeting,8-11 April 1980, CCR Ispra ORA 19620.

Schlittenhardt,P., editor , 1983 "Workshop on remote sensing of coastal transport in the northern Adriatic sea", held 11-12 Oct. 1983 at JRC Ispra. Proceedings, JRC report S.A. 1.05.E2.85.03 (1985).

Sturm, B., 1980, "Atmospheric correction of satellite and aircraft remotely sensed data in the visible spectral range and the quantitative determination of suspended matter in the surface layer of water-bodies from remotely sensed upwelling spectral radiance". Summer School on Rem. Sens. in Meteorology, Oceanography and Hydrology, EARSEL, Councel of Europe, 1-20 Sept. 1980, Univ. of Dundee, ORA 19792.

Sturm, B., Maracci,G., Schlittenhardt,P., Ferrari,G., Alberotanza, L., 1981, "Chlorophyll and total suspended matter concentration in the North-Adriatic sea determined from Nimbus-7 CZCS". ICES Statutory meeting, 6-10 Oct. 1981, Woods-Hole(USA), ORA 30235.

Sturm, B.,Tassan,S., 1983, "An algorithm for sediment retrieval from CZCS data, with low sensitivity to the atmospheric correction uncertainty". XVIII General Assembly of Intern. Union of Geodesy and Geophysics (IUGG), 15-17 August 1983, Hamburg(FRG).

Sturm, B., Schlittenhardt,P., 1983, "Coastal Zone Color Scanner (CZCS) imagery from the West African upwelling". International Symp. on the most important upwelling areas off Western Africa, 21-25 Nov. 1983, Barcelona (Spain).

Sturm B., 1985, "CZCS Sensitivity Decay Study" NET Meeting GSFC Greenbelt, 8-9 May.

Tassan, S., 1983, "A method for the retrieval of phytoplankton and sediment contents from remote measurements of sea color in the coastal zone. Results of validation tests". Intern. Symp. on Remote Sens.(?), 13-17 Sept. 1982 Toulouse (France).

Tassan S., 1985, "Evaluation of the potential of the Thematic Mapper for Chlorophyll and Suspended Matter determination in Sea Water". JRC Internal Report.

Symposium on Remote Sensing for Resources Development and Environmental Management / Enschede / August 1986

Shape and variability of the absorption spectrum of aquatic humus

H.Buiteveld* & F.de Jong
Delft University of Technology, Netherlands
* *Present address: Rijkswaterstaat, DBW/RIZA, Lelystad, Netherlands*

R.Spanhoff
Rijkswaterstaat, DGW, The Hague, Netherlands

M.Donze
Kema laboratories, Arnhem, Netherlands

ABSTRACT: Shape and variability of the absorption spectrum of aquatic humus is investigated. The exponential description of the shape is not accurate enough for remote sensing applications. Considerable improvement in the accuracy of the interpretation of airborne reflection measurements may be expected when actually measured absorption spectra of aquatic humus (part of the optical seatruth) are used as input for the deconvolution algorithm.

1 GENERAL INTRODUCTION

In numerous studies it has been demonstrated that airborne passive remote sensing of surface water in the optical window may yield a wealth of synoptical information. This information consists of convoluted data of the effect on the light-field of several physical, chemical and biological compounds of interest. Deconvolution is done using 'algorithms'; the results of these calculations are calibrated by a statistical technique against some seatruth measurements on the compounds of interest. The calibrated values are subsequently used for interpolation and extrapolation to construct distribution maps of these compounds.

These maps are as yet of limited use due to the noise and variability in space and time observed in the 'constants' yielded by the calibration procedure. This problem increases in importance going from the open ocean to estuaries; in freshwater bodies it may even be greater.

A number of factors may contribute to this undesirable situation:

1. Lack of measurements; instrumental noise.
2. Natural noise in the environment.
3. Natural phenomena that are not recognized in the measurements nor covered by theory.
4. Inadequacy of theory; especially nonlinearity in the relationships between concentrations and optical results cannot be treated in a statistically satisfying way, given the amount of noise in the data.

When we develop the instrumental and theoretical apparatus to distinguish such factors we can reach a position from which reliability of remote sensing observations in dependence on the quality of instrumentation, seatruth observations and local conditions, can be judged.

As a first step it may be expected that 'optical seatruth', consisting of optical measurements in situ and spectroscopy of water samples, will be much more usefull to develop precision in remote sensing than attempts to directly calculate chemical concentrations from airborne measurements. This translation can be separately done with the spectroscopic data.

Purpose of our research is to contribute to a program as sketched above.

1.1 Aquatic humus

Humus (or yellow substance or gelbstoff) is a general name covering dissolved organic compounds of large molecular weight. Its definition actually consists in the methods of isolation (primarily pore size of the filterpaper) and measurement; such as total organic compounds, extinction and fluorescence, or any more elaborate set of properties.

Pure water has fixed optical properties. Humus has a variable concentration and variable optical properties (Zepp and Schlotzhauer 1981; Bricaud et al. 1981). Together these two determine the optical background in which the contribution from particulate material must be studied. The optical properties of water were reviewed by Smith and Baker (1981). In the present paper the absorption spectrum of humus is discussed.

The shape of the absorption spectrum of humus can in first approximation be described by an exponential function (Kalle 1966); absorption decreases strongly with wavelength in a monotonous fashion.

$$a(\lambda) = A\, e^{d(\lambda-\lambda_0)} \qquad (1)$$

with $a(\lambda)$ = measured absorption coefficient in m^{-1}
λ = wavelength in nm
λ_0 = arbitrary constant in nm
A,d = calculated by least squares

This exponential form is usually applied in marine optics (Prieur and Sathyendranath, 1981). The accuracy of this description was studied by Bricaud et al. (1981) and Zepp and Schlotzhauer (1981).

In this model A may be roughly equated with the concentration of humus, while d roughly describes the shape of the spectrum. In fact both parameters depend on the choice of λ_0, the measured wavelength range and individual deviation from the model. Zepp and Schlotzhauer (1981) observed that values for A do correlate with total organic carbon, with an uncertainty of a factor of about 2.

Bricaud et al. (1981) determined the constant d using a linear regression fit to equation 1 in the range 375-500 nm. The value of d varied from -.02 to -.01 nm^{-1}, with a mean of -.014 nm^{-1}. Zepp and Schlotzhauer (1981) found, in the case of freshwater humus, d values between -.0116 and .0175 nm^{-1}, with a mean value of -.0145 nm^{-1}, using the wavelength range 300-500 nm.
The exponential function (1) with d = -.014 nm^{-1} is often used as model for the humus absorption. But it appears that the variability of the absorption spectrum of humus in nature and the deviation from the exponential function, with fixed d, are considerable.

2 MATERIALS AND METHODS

Samples of surface water were collected at 10 different locations in The Netherlands.

In the field the water was filtered over Whatman GF/C filters, followed within 1 day by a filtration over .2 μm membrane filters. Organic compounds passing this filter are here defined as aquatic humus.

Absorption spectra of humus were measured with a Perkin-Elmer 551S double-beam spectrophotometer, using cuvettes of .01 or .1 m. Reflection spectra, of the compound system humus and water, were measured with a multispectral scanner, built from a grid-monochromator and a photodiode array. Signals from both instruments were handled by a Hewlett-Packard 86B computer.

An aquarium was used to measure reflection spectra. A parallel beam from a slide-projector at an angle of 8° with the vertical was used as a light source. This beam passed the layer of water of .345 m and was diffusely reflected by a white bottom plate. Radiance emitted from the aquarium was measured.

In underwater optics usually the attenuation coefficient for a particular wavelength is used, implying both light absorption and scattering. For the present samples light scattering was insignificant. In the spectrophotometric measurement this could be checked using an integrating sphere; in the reflection geometry the measured radiance was dominated by the light reflected by the white plate, given the small depth of the water column. So in this work we use the spectroscopic definition of extinction coefficient, as applied to clear solutions.

The wavelength range considered here is 300-800 nm. The short-wavelength cut-off is chosen such as to avoid contributions to the absorption from inorganic compounds; the upper limit followed from instrumental restrictions.

3 RESULTS

The absorption spectrum of freshwater humus deviated significantly from the exponential funtion with $d = -.014\ nm^{-1}$. The extinction, on a logarithmic scale, of one sample is given in figure 1. In the same figure the model result, with $d = -.014\ nm^{-1}$ and $\lambda_0 = 440$ nm, is given.

In the spectrum shown in figure 1 the difference between the measured extinction and the exponential function is about 50% at 600 nm. This difference is of the same size as the water extinction at that wavelength, and is therefore not negligible.

Another illustration of the error introduced by application of the exponential function (1) is the

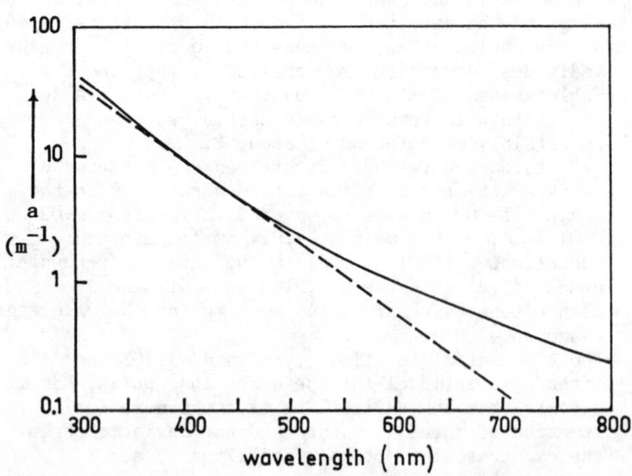

Figure 1. Difference between measured absorption spectrum and exponential model. The extinction scale is logarithmic. The dashed line is the exponential, with $d = -.014\ nm^{-1}$ and $\lambda_0 = 440$ nm, normalized to the measured extinction at 440 nm.

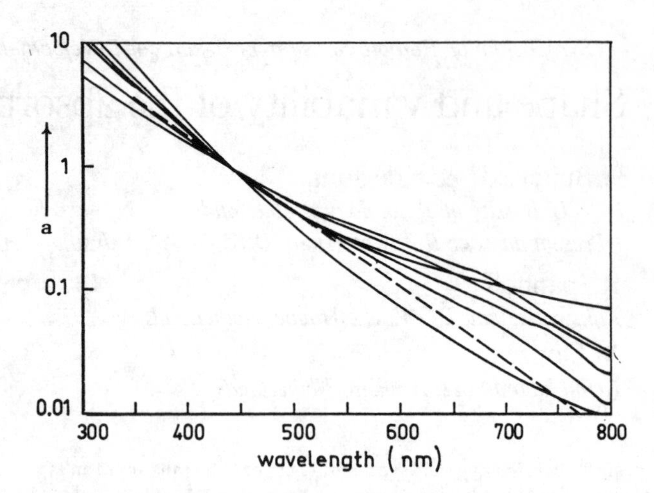

Figure 2. Illustration of the variability of the absorption spectrum, on a logarithmic scale, of freshwater humus from different locations in The Netherlands. The spectra were normalized at 440 nm. The dashed line is the exponential model, with $d = -.014\ nm^{-1}$ and $\lambda_0 = 440$ nm.

following, simplified, example of a depth measurement in the aquarium from remotely sensed data. Reflection spectra from a dilution series of humus with demi-water, in the aquarium, were measured at a water depth of .345 m. It was found that the reflected flux decreased exponentially with the sum of extinctions by water and the variable amount of humus. This indicates that in this case the scattering of the water sample is insignificant.

From this exponential dependence the bottom depth can be calculated, using the reflection at two arbitrary wavelengths, 500 nm and 630 nm. Other inputs for the calculation were: a constant and known bottom reflection, water absorption at 500 nm and 630 nm according to Smith and Baker (1981), and the ratio of the humus extinction at these two wavelengths. Two ratios were used: the ratio of the measured extinction of the humus sample shown in figure 1 and the ratio based on the exponential model, with $d = -.014\ nm^{-1}$.

Table 1 contains the result of the depth calculation for the dilution serie. As a parameter for the humus dilution the humus extinction at 500 nm was used. The calculated depths, using the measured humus extinction and the exponential function, and the relative error, when the exponential function was used, are given.

Table 1. Error introduced by the exponential humus absorption spectrum in an experimental depth measurement. Depth was calculated from reflection data, using directly measured extinction values and values derived from the exponential function, for the humus dilution serie. The dilution of the humus is represented by a(500), the extinction at 500 nm. Relative errors are shown in the last column.

a(500)	depth		rel.error
	measured ratio	model ratio	
m^{-1}	m	m	%
3.55	.353	.545	58
2.67	.344	.492	43
1.32	.354	.435	26
0.62	.340	.382	11
0.10	.348	.362	5

The calculated depth using the ratio of measured humus extinction corresponds well with the depth of .345 m. The calculated depth based on the exponential absorption is systematically higher than the real depth. The relative error increases with the humus concentration.

The choice of the two wavelengths was arbitrary. A pair of wavelengths that introduce a smaller relative depth error is possible. The shape of the humus extinction spectrum determines the wavelength range where the exponential can be used. This range is different for humus from an other place, as can be seen in figure 2.

The variability of the absorption spectrum of humus from different locations in the Netherlands is shown in figure 2. The spectra are normalized at 440 nm. The extinction at 440 nm varied from 1.06-9.33 m^{-1}. The dashed line is the humus spectrum according the exponential function, with d = -.014 nm^{-1} and λ_0 = 440 nm.

It turns out that the differences among the samples are even greater than the deviation from the exponential model in the example shown before.

4 DISCUSSION

The results, obtained with spectrophotometric measurements, demonstate that the shape of the humus absorption spectrum cannot be described by an exponential function. It also appears that the variability of the humus absorption spectrum is considerable. The simplified depth experiment illustrates that substantial errors are made, when the exponential function, with a fixed value for d, is used.

Considerable improvement in the accuracy of the interpretation of airborne reflection measurements may be expected when actually measured absorption spectra of aquatic humus (part of the optical seatruth) are used as input for the deconvolution algorithm.

Futher progress in the mathematical description of the absorption spectrum of humus can only be made by the introduction of models more complicated than an exponential curve. As a first attempt we applied conventional factor analysis to the data-set (figure 2). Preliminary results indicate that four components are sufficient to cover the observed variability of shape (Krijgsman and Buiteveld, unpublished results). It will likely be possible to obtain more accurate correlation between optical and chemical data of humus than was possible with the exponential model.

In further studies factor analysis will be applied to introduce error calculation and noise-analysis in our experimental procedures and in the analysis of reflection spectra of more complicated and realistic situations, e.g. by the introduction of silt and algae in the system.

REFERENCES

Bricaud, A, A. Morel and L. Prieur 1981. Absorption by dissolved organic matter of the sea (yellow substance) in the UV and visible domains. Limnol. Oceanogr. 26:43-53.

Kalle, K. 1966. The problem of Gelbstoff in the sea. Mar. Biol. Ann. Rev. 4:203-218

Prieur, L. and S. Sathyendranath 1981, An optical classification of coastal and oceanic waters based on the specific absorption curves of phytoplankton pigments, dissolved organic matter and other particulate materials. Limnol. Oceanogr. 26:671-689.

Smith, R.C. and K.S. Baker 1981. Optical properties of the clearest natural waters (200-800 nm). Applied Optics 20:177-184.

Zepp, R.G. and P.F. Schlotzhauer 1981. Comparison of photochemical behavior of various humic substances in water: III Spectroscopic properties of humic substances. Chemosphere 10:479-486.

Rainlog and preslog: Novel tools for in-situ measurements

D.de Hoop
International Institute for Aerospace Survey and Earth Sciences (ITC), Enschede, Netherlands

1. INTRODUCTION

The rainlog and preslog electronic water level loggers were developed at the International Institute for Aerospace Survey and Earth Sciences (ITC), Enschede, and after some redesigning, are produced by Siemens Netherlands, The Hague.

The instruments measure and record water pressure over long periods (for example, at sixty minutes intervals for a period of 12 months) without requiring maintenance or changing of batteries. Expensive test stations are not required. The loggers, together with the data-processing facilities, provide a new scientific approach to field measurements.

The pressure recordings are controlled by a microprocessor and are stored in a removable internal memory block (the "Eprom"). The capacity of the Eprom is 10240 records. The records are read by a personal computer (Epson, type PX-8) with an "Eprom reader". Depending on the software chosen the results can be presented raw or in calculated form.

Two types of loggers are available; both have the same electronics. The difference is based on the type of sensor used. Rainlog has a relative pressure sensor; preslog has an absolute pressure sensor. The standard range for rainlog is 2 meters. The standard range for preslog is 0 to 4 bar, which at sea level is equivelent to a range of 30 meters.

The precision of the measurements is approximately 0.2 percent for preslog and 0.5 percent for rainlog. The resolution is 2000 for the rainlog and 4000 for the 4 bar-type preslog. The measuring interval can be set between 30 seconds and 60 minutes (60 minutes is standard). A built-in audio signal provides an operational check.

The loggers were designed for both accuracy and durability. Durability requires permanent attachment to the internal panel of all parts except the removable memory block (Eprom) and also very firm placement of the panel in the steel cylinder. Opening the cylinder to replace the Eprom or to re-set the recording interval requires a vise, which is usually not available at a test site. We therefore recommend transporting the entire logger to a field office where the necessary equipment can be kept and where there is less danger of damage to the Eprom.

1.1 Preslog applications

River level
To record river levels, a logger must be anchored in the riverbed. After installation, it is left for a maximum period of 12 months. The instrument functions without disturbance, despite shifting bottom sand and gravel. After the operating period, the logger must be retrieved and opened; the memory block is removed and read by the personel computer. An additional "dry" preslog is required for barometric correction. One dry logger covers an extensive area and thus can be combined with several wet loggers.

Groundwater level
Preslog can also function as a groundwater level logger. Its three inch diameter allows placement in waterpipes for pump-test. Data from several tests can be recorded in sequence by placing a "marker" in the recorded data to indicate the end of one test and the beginning of the next.

1.2 Rainlog applications

To record rainfall, a rainlog is connected to the bottom of a rain collection container. Only increases in pressure are recorded. Decreases in pressure caused by evaporation or emptying the container are ignored. A simple program later shows the results of a year's data collection.

Symposium on Remote Sensing for Resources Development and Environmental Management / Enschede / August 1986

Simple classifiers of satellite data for hydrologic modelling

R.S.Drayton
University College, Cardiff, UK

T.R.E.Chidley & W.C.Collins
University of Aston, Birmingham, UK

ABSTRACT: Hydrological models are normally based on parameters extracted from conventional maps such as drainage density and land cover. In this paper we examine the use of parameters derived from thematic classifications of satellite imagery. The advantage of this technique is that the satellite image becomes an up-to-date, easily managed source of data for monitoring of water resources. The methods of classification range from simple density slicing to supervised classification. A model was constructed in which run-off characteristics were regressed on thematic classifications. The problem of achieving sufficient variance in the hydrological parameters within one satellite image are discussed, and recommendations for distributed modelling are made.

1 INTRODUCTION

One of the principal aims of applied hydrology is to make estimates of streamflow. Frequently, the hydrologist is asked to mak- such estimates in the absence of observed, historic streamflows, and is required to use his skills to construct mathematical models which can predict future flows. These fall into two broad categories: statistical models which provide estimates of river flows having a specified probability, or process models which provide estimates of streamflow resulting from certain specified rainfall sequences. In this paper we investigate the ways in which satellite data can be employed in such models. The advantage of this approach is that the rectified and classified image becomes an up-to-date, easily manageable source of data for monitoring water resources.

The obvious but rather uninspired way to use satellite data is to construct the conventional maps from which existing hydrological models draw their information. The benefits of timelines and global coverage have been claimed many times, while the limitations of satellite sensors in evaluating hydrological variables have been explored by Rango et al, 1983 (thematic features) and by Drayton and Chidley, 1985 (geomorphic features). The conclusion appears to be that satellite sensors may be used to evaluate conventional parameters to a precision suitable for regional models.

A far more promising approach is to construct new models which are appropriate to the view of the world offered by the satellites. The fundamental difference in this view is that it is holistic; it brings together the various aspects of the earth's surface rather than separating them, as happens with maps. Thus, the reflectance of a patch of ground depends not only on its vegetal cover, its soil type and its moisture content, but also on the interactions which occur between them, and with climate. All of this information is also relevant to its hydrological response, and is conveniently integrated or compressed into this single characteristic, namely its spectral signature. Thus the judicious use of parameters based on spectral characteristics is likely to yield hydrological parameters which are far more powerful than any parameters derived from maps.

Thus, our aim was to produce interpretations of our study catchments, based on their spectral properties, to build hydrological models around them, and to compare their efficiency with existing models using conventional classifications. This is different from the approach by Gurnell et al (1985) in which it was heathland vegetation which was considered to be the mechanism which brought the hydrological information together, and the satellite systems were used only to monitor the heathland.

The benefits of using satellite data are not confined to thematic considerations alone. There is an enormous benefit to be gained from the structure and organisation of the data, which is distributed uniformly over the scene, and is organised in a regular grid cell structure. Distributed hydrologic models do exist but their usefulness has been diminished by the logistical problems of estimating values at many node points. Satellite data is clearly well provided with the means to overcome such problems.

Our current work is concerned with ways in which the holistic nature of satellite data can be preserved in hydrological models. As a first step we have examined its use in "lumped" models, so that the thematic benefits can be clearly distinguished from those of distribution. Work is in hand to examine the benefits of data structure, in relation to finite difference groundwater models, and will be reported later.

2 BACKGROUND TO MODEL

The study reported here was based on a group of river basins in the South Wales area, for which hydrologic data were available. (For details see Chidley and Drayton, 1985). Although lying within the same region the basins show a good variety of vegetal, geologic and climatic characteristics.

Our aim was to evaluate the usefulness of the thematic classifications which are unique to satellite data. So, as a first step we considered a lumped model in which we could eliminate the benefits derived from distributed data. We considered a generalised lumped rainfall-runoff model, which would provide estimates of annual runoff (catchment yield), based on the inputs of annual rainfall and a set of characteristics evaluated from the satellite data.

We assumed that any one year's runoff could be considered as being composed of the sum of runoff from n different runoff types, and that each type has a response which depends on the magnitude of rainfall. Then total runoff,

$$RO = \sum_{n} f(RF)_n A_n$$

In this exploratory model we assume that the distribution of run off types is constant from year to year. Then the model reduces to the simple statement

BAND 4/BAND 6

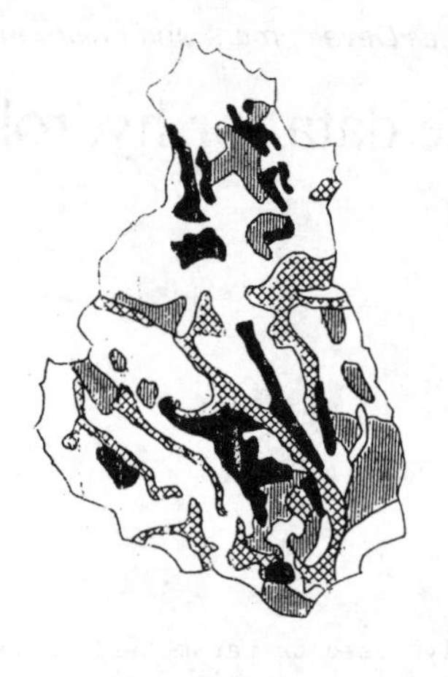

LAND USE CLASSIFICATION

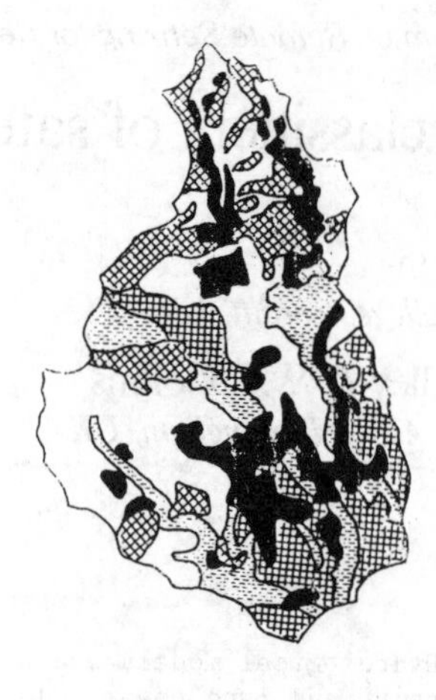

PRINCIPAL COMPONENTS

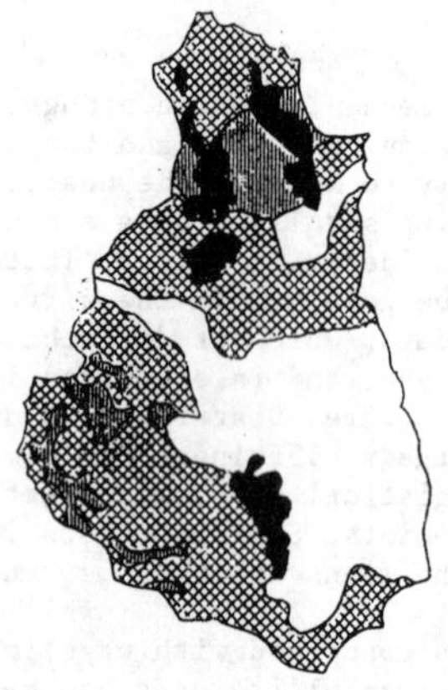

BIOMASS INDEX

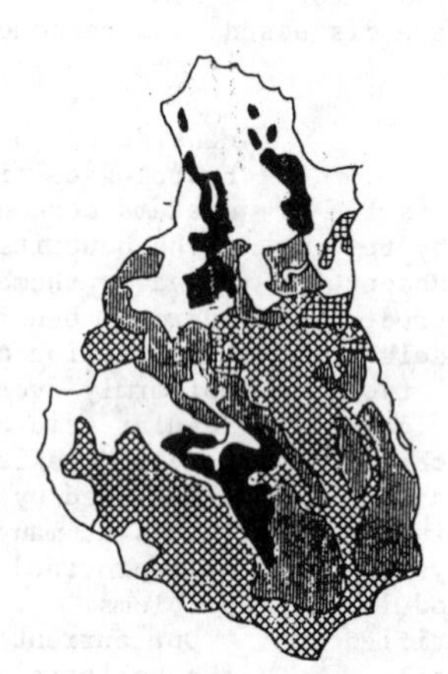

3 BAND COMPOSITE

BAND 7

Figure 1. Thematic partitions of the Taff River basin

$$RO = f(RF)_c A_c$$

where the suffix c refers to the whole catchment.

We examined hydrological records for this area and found that the runoff function was a simple linear relationship, and hence the yield in millimetres over the catchment,

$$\text{Yield} = \frac{RO}{A_c} = a\,RF - b \qquad \text{(mm)}$$

Therefore, at the simplest level, our satellite classifications will be useful if they can be used to predict the values of proportional runoff, a , and losses, b , for any catchment.

Our approach was to derive parameter values based on the conventional image processing techniques of density slicing and classification. Thus we could slice a Near Infra Red image and estimate the total area in the catchment in each of say 5 levels. The same process could be applied to Biomass Index, Principal Component or any band ratio. Alternatively a classification could be made, and the areas falling within various classifications measured. This applied to supervised or unsupervised classifications and any colour composites.

Thus we could derive a set of variables such as LEVELA1, LEVELA2...LEVELA5, LEVELB1...LEVELB5, CLASSA1, CLASSA2,.....CLASSA5, etc., where LEVELA refers to a density slice in NIR, CLASSA refers to an unsupervised classification etc. The values of constants a and b could then be evaluated by multiple regression.

Clearly a very large correlation analysis could ensue, so some preparatory work was needed. Only those parameters which exhibited a large variance and showed a strong correlation with streamflow, and showed little correlation with other parameters were considered. Once this weeding was done, then a multiple regression could be carried out.

3 METHOD

A suitable set of 'satellite' catchment characteristics was chosen. Parameter values were extracted for the study catchments. Annual rainfall and runoff values were acquired, and for each catchment the constants a and b were calculated for the equation

$$RO = a\,RF - b$$

The constants a and b were regressed in turn on the set of catchment characteristics, and the optimal regression equations were found. The equations were tested by estimating values of annual runoff not previously used.

4 DISCUSSION AND RESULTS

The first impression was that the various slices and classifications enabled many different partitionings of the catchments to be made e.g. Fig. 1. However, when the proportions in each classification were calculated, the difference between the methods of classification became less obvious (Table 1) i.e. as soon as the data were lumped the statistics became similar, even though the meanings were different.

Secondly, it was observed that many strong correlations occurred between variables. These appeared as negative correlations within groups e.g. a high proportion of LEVELA1 corresponding to a low proportion of LEVELA5, and also appeared as positive correlations across groups e.g. a high proportion of FOREST corresponding to a high proportion of LEVELA1.

The constants a and b were regressed on this reduced set of variables, but not significant correlations were found. Although this is a sad result, it is useful in that it clarifies some issues and points the way to possibly happier conclusions.

The underlying problem is that of continuity of the satellite data. It was necessary for the integrity of the regression that each set of parameter values e.g. CLASSA, LEVELA should be statistically homogeneous. The simplest way of ensuring this is to evaluate their values from only one image, in which case the characteristics of radiation, the calibration of sensors etc. will be consistent between each catchment. This requires that the catchments studied should lie within an area of 185km x 185 km for Landsat MSS, or a smaller area for Landsat TM or SPOT. This is such a small geographic region that there is inevitably a high degree of hydrological and environmental similarity between the catchments. Thus, the dependent variable does not have sufficient variance to make a good regression possible.

To achieve a higher variance we should need to include in the regression catchments which are much more dissimilar in their hydrology. To achieve dissimilarity we should include catchments which are remote from one another, but this would mean interpreting more than one image, and different images will inevitably be of different dates and seasons, and different thematic properties. Thus, it appears that we should reconsider our spectral characteristics, and use only those which are absolute and consistent in range and scale from place to place and time to time. The logical conclusion of this argument is that we cannot achieve what we set out to: either we have insufficient variance for a regression model, or we have non-homogeneous data.

Apart from the problems with regression, we have clearly lost much useful information by lumping the catchment characteristics. In an attempt to separate the two main benefits which we saw in satellite data we have lost both.

Although the regression on constants a and b was unsuccessful, useful regression equations were found for average yield from the catchments and for mean annual flood. Of course, these are very crude summary statistics, but the fact that useful regressions were found using 'satellite' parameters in the place of conventional characteristics is encouraging, and supports our view that satellite data can be useful if the right model is found.

5 CONCLUSIONS

The hydrological characteristics of an area were seen to depend on the same features as the spectral characteristics. Thus an attempt was made to construct an hydrologic model which employed spectral characteristics as its parameters. The attempt failed because the variance of the hydrological characteristics from the area contained in one Landsat frame was too small. To achieve greater variance data would have to be drawn from more than one image. The problems of achieving homogeneity in data from different images and different dates appears to be insuperable.

Table 1. Breakdown of classifications - proportion of total area in each class

	Classification					
	a	b	c	d	e	f
Class 1	0.62	0.38	0.44	0.57	0.45	0.35
2	0.22	0.20	0.23	0.16	0.36	0.29
3	0.15	0.16	0.17	0.14	0.11	0.23
4	0.007	0.15	0.16	0.13	0.08	0.10
5	-	0.12	-	-	-	0.03

If satellite data is to prove useful then different models will be required. Crude hydrological statistics can be estimated satisfactorily, using regression models, but to achieve greater sophistication it appears that we shall have to turn to the benefits of distributed models.

REFERENCES

Chidley, T.R.E. and Drayton, R.S. (1985) Visual interpretation of standard satellite images for the design of water resources schemes. Proc.Int. Workshop on Hydrologic Applications of Space Technology. International Association of Hydrologic Sciences. Florida 1985. In press.

Drayton, R.S. and Chidley, T.R.E. (1985) Hydrologic modelling using satellite imagery. Proc.Int.Conf. on Advanced Technology for Monitoring and Processing Global Environmental Data, Remote Sensing Society, London. 219-225.

Gurnell, A.M., Gregory, K.J., Hollis, S. and Hill, C.T. (1985) Detrended correspondence analysis of heathland vegetations: the identification of runoff contributing areas. Earth surface Processes and Landforms, 10(4), 343-351.

Rango, A., Feldman, A., George, T.S. and Ragan, R.M. (1983) Effective use of Landsat data in hydrological models. Water Res. Bull., 19(2), 165-174.

Symposium on Remote Sensing for Resources Development and Environmental Management / Enschede / August 1986

The delineation and classification of inland wetlands utilizing fcir stereo imagery

Stephen A.Estrin
Mahopac, New York, USA

ABSTRACT: The fastest and one of the most accurate methods of classifying and delineating inland wetlands and their adjacent upland is through the interpretation of medium-to-large-scale (1:12,000 and larger) fcir stereo imagery. This is accomplished by manual photo interpretation of those readily recognizable properties of inland wetlands and uplands; vegetation types, presence of standing water, soil water content and topography-elevation and slope. Vegetation is one of the best indicators of water quantity, quality and permanence. Specifically, in wetlands, vegetation is one of the principal factors causing differences in spectral reflectance; therefore, noticeable differences in vegetation, vis-a-vis their spectral reflectance, depicted on fcir imagery are directly related to the presence of water, both standing and soil water content. Consequently, the delineation of the boundary between a wetland and its adjacent uplands can clearly be ascertained based upon these differences in spectral reflectance. Classification and delineation of the wetland-upland complex was found most accurately determined in the northeastern and central United States from spring fcir stereo imagery flown at an altitude of 6,000 feet AMT (1:12,000). In comparison, the modern methods of field ecology survey were found to be too time consuming, difficult and costly. In particularly fragile ecosystems, comprehensive biological field studies tended to cause considerable physical environmental damage, while the necessary ground truth program covering only a small portion of the area had no adverse impact. Naturally, the subsequent imagery analysis causes no adverse environmental impact.

1 INTRODUCTION

Wetlands are an important element of the natural environment and the increased public concern with environmental issues has lead to the enactment of federal, state and local legislation protecting inland wetlands.

Due to the inadequacy and inaccuracies of the majority of existing maps and wetland inventories as well as their heterogenity, areal extent, time and money constraints, many governmental agencies at the federal and state levels have utilized remote sensing as the primary method of delineating and classifying wetlands. Remote sensing affords both a practical and economic means for their accurate delineation and classification. It differs from conventional wetland data collection in that the recording methodology is not in direct contact with the ground. It is airborne. In the majority of instances, the choice of a remote sensing technique is a function of a series of interrelated factors; size of the project area, accessibility, time of year, cost, data to be obtained, accuracy, and level of detail required.

In 1977, one of the most extensive studies of inland wetlands was undertaken by Carter, Garrett, Shima & Gannon of the U.S.G.S., The Great Dismal Swamp located in Virginia and North Carolina. These scientists/photogrammetrists believed that the use of conventional wetland data collection methods would be too expensive, time consuming, and too difficult to interpret for such a large, diverse and inaccessible area. They, therefore, decided on the use of seasonal low- and high-altitude color infrared photography utilizing manual interpretation techniques. The wetland maps were prepared at a scale of 1:24,000.

These analysts found that imagery obtained in the spring, during dormancy, allowed the identification of wetland boundaries, areas covered by water, the drainage pattern, the location of coniferous vegetation and its classification, and the classification of the understory vegetation. Photographs obtained during the summer were used to classify deciduous vegetation.

1.1 A real example - Longridge Corporate Park, New York

In the case of the Longridge Corporate Park, the determination of type of remote sensing was never a factor because the site had been disturbed to such an extent from October, 1982, that only a vague and inaccurate reconstruction of previous conditions would be possible. Therefore, historical remote sensing data, predating October, 1982, was the only practical and accurate means available for the delineation and classification of the wetland.

In March, 1973, and again in April, 1974, Stephen A. Estrin, Inc., as part of its contract with the County of Putnam for the design of a Comprehensive Land Development Plan and Sewer Study for the Town of Southeast, had its Division of Photogrammetry contract with Grumman Ecosystems for an aerial photomapping mission, to its specifications, which produced stereo color cartographic mapping photography and false color infrared imagery at the scales of 1"=2000' and 1"=1000'.

It is the April 20, 1974 fcir, 1"-1000', imagery that forms the basis of the delineation and classification of the Longridge Corporate Park Wetland. The interpretation of this imagery was by manual means utilizing imagery interpretation and transfer equipment, ground truth and a wetland image analysis key developed by Stephen A. Estrin, Inc. in 1972, updated in 1976. In addition, the color cartographic photography was analyzed to determine the hydrogeology of the site and in conjunction with a Balplex 760 High Precision Stereoplotter, verification of the topography shown on the survey of the site was made.

Finally, historical black and white serial photography on file in the photogrammetric library of Stephen A. Estrin, Inc., dated 1968 and 1970, with a scale of 1"=2000', provided the means of synoptically analyzing the site over a critical six year period; critical due to the interdiction of this once continuous and extensive wetland by the construction of I-84. Essential to wetland delineation and classification is the impact of man-made structures.

2 REMOTE SENSING IMAGERY ANALYSIS

2.1 General

Most remote sensing imagery analysis requires the photogrammetrist to have at his command data other than that directly obtained from the imagery being examined. This is referred to as Ground Truth.

The development of a County Land Development Plan and the Southeast Sewer Study generated a considerable amount of geophysical data relating to this site. The photogrammetrist, therefore, had available Ground Truth on soils, geology, topography, groundwater systems, drainage, vegetation, wetlands and existing land use. The sources of this data were both in-house field surveys and county agencies, most notably the County Soil and Water Conservation District. Additionally, the State of New York, Department of Geology has published Bedrock Geology Maps, and the Kensico Water Shed Maps prepared photogrammetrically for the New York City Department of Water were also utilized.

2.2 Interpretation Techniques

Although the field of Remote Sensing has developed rapidly since the early 1970's, most applications of Remote Sensing for Inland Wetlands have, and still use, aerial photography and imagery in conjunction with manual interpretation rather than digital imagery analysis techniques of multi-spectral sensed data. The reason for this is due to the heterogeneity of wetlands, the varying vegetation species and their locations within an environment of varying moisture content.

Under such conditions, "Spectral Signatures" are often overlapping, and therefore, confusing. The Spectral Signature of a feature is a set of values for the reflectance or the radiance of that feature, where each value corresponds to the reflectance or radiance averaged over a different, well-defined wavelength interval. It is within these different wavelengths of the electromagnetic spectrum that the overlap occurs.

An "Imagery Signature", on the other hand, refers to a distinguishing characteristic only associated with a single feature or group of similar features defined by six parameters. These parameters include:

1. Color - as expressed as a hue and tone, i.e., gray/green, red, magenta-violet-red
2. Height - as expressed as a vertical distance from the ground, in feet
3. Texture - as expressed by a word picture, i.e., coarse, coarse (mottled), fine, smooth
4. Shape - as expressed as a geometric arrangement, i.e., rounded open crowns, arching limbs and long pendulous branchlets
5. Site - as expressed as a landuse type and terrain form, i.e., farmland, glacial till
6. Association - as expressed as the interrelation of natural features to one another, swamp-standing water-tree species - ferns - club mosses at the outer boundaries.

These six parameters of fcir imagery interpretation are of varying importance in terms of wetlands delineation and classification. Color, however, is undoubtedly the most important, followed by Height. The remaining four are supportive of these two.

2.3 False Color Infrared Film

Kodak Aerochrome Infrared Type 2443 is the accepted standard fcir film used for wetlands delineation and classification by photogrammetrists.

When viewing this imagery, there is a "field-of'view" which defines the common picture element. This is known as a Pixel. It is the grouping of these individual Pixels that determines what is seen on the imagery and permits interpretation by the photogrammetrist. In the fcir used for wetlands delineation and classification, the Pixels are representative of reflected energy. Therefore, before consideration can be given to the specifics of wetland delineation and classification, it is necessary to understand how fcir imagery functions.

2.4 Light and Electromagnetic Spectrum

Light, as the human eye perceives it, can be referenced to the electromagnetic spectrum of which the micrometre (Um) is the basic unit of measure. It is divided into various wavelengths, representative of different types of reflected or emitted energy. In terms of the human eye and most photographic film, sensitivity is very narrow, between 0.4 and 0.7 Um. Fcir imagery extends that sensitivity to 0.9 Um. In both the human eye and fcir, the energy sensed is reflected.

In terms of target interactions, reflection is defined as the amount of energy at various wavelengths that is returned to the sensor (camera) from a given object. This energy return can be quantified as either a unique spectral or imagery signature which is similar to a fingerprint. In fcir imagery, this signature is defined by six parameters of which Color is the most important. Therefore, wavelength selectivity, the interaction of electromagnetic energy with objects (targets) within a given environment, is at the heart of fcir discrimination between these objects. The 2443 film has three emulsion layers; each sensitive to a different portion of the electromagnetic spectrum, the green and ultraviolet portions - yellow layer; magenta portion - green layer; and infrared portion - red layer.

3 FCIR IMAGERY - THE ADVANTAGES OF ITS USE

1. Vantage Point - unique vertical perspective of terrain
2. Resolution is greater than that of the human eye
3. Spectral sensitivity over a range of the electromagnetic spectrum about twice as broad as the human eye; 0.4 Um to 0.9 Um

4. High geometric accuracy
5. Permanence of records
6. Ability to synoptically analyze a given environment over a specific time span

However, it is the ability of fcir imagery to enhance the subtle differences in reflectance that are barely discernible to the human eye which gives it its greatest importance to wetland delineation and classification. This is particularly so in delineating the change from wetland, to transition zone to upland, where frequently, spectral differences between vegetative species, water and soil moisture content, which are essential in establishing competitive advantage, are so small that they can and often are overlooked in field surveys; particularly field surveys of sites that have been impacted heavily by human actions.

4 MANUAL IMAGERY ANALYSIS

With an understanding of how fcir imagery functions, a return to the techniques of manual interpretation is in order. As previously stated, there are six interpretation parameters: color, height, texture, shape, site and association. The methodology of how these parameters are employed to establish wetland boundaries and vegetation types is accomplished in the following manner.

First, it is necessary to hypothesize, to make an educated assumption, of what exists within the project area. In the matter of Longridge, a "Given" had been provided by the client, determination that a Red Maple Swamp of approximately eighteen acres, existed. At issue were the size, boundaries and classification.

The hypothesis to be tested, therefore, was one primarily concerned with delineation. Classification would essentially be used as confirmation. Manual imagery analysis, as applied to this hypothesis, is basically a deductive process in which features and their identifying signatures lead the photogrammetrist to his delineation and classification.

The next step in analyzing any terrain is to establish a point of beginning, a base line from which each element of the terrain is considered in a logical progression. In wetlands analysis, the photogrammetrist establishes a series of transect lines at right angles to the base line. In Longridge, the stream was chosen as the base line and transects were established at 100-foot intervals. These transects ran east to west, progressing outward and topographically upward from wetland to upland. Interpretation took place along these lines for classification; whereas delineation took place at those points in which a change in classification took place.

The specific delineation and classification of a wetland is the result of the convergence of evidence. Critical examination of this evidence clearly establishes that only one interpretation is correct. It is very important to understand that imagery analysis is not performed in a vacuum; it requires three distinct and different aids. It is the utilization of these three aids which forms the third step in the analysis process.

The first of these is Ground Truth, data obtained directly in the field either onsite or in a terrain similar to the one being analyzed. This ground truth can be generated in-house or can be obtained from published documents of local agencies such as a soil and water conservation district.

Image Analysis Keys are the second aid, and these keys can be totally in-house generated, but in most cases have incorporated some signatures and descriptions established by others working in the same field of wetlands delineation and classification. The key utilized for Longridge was developed in 1976, and expanded over a four-year period. It consists of signature characteristics and word descriptions and is organized on the basis of elimination. The following is an example from the key:

Spring Film 2443

Level I - Classification
Descriptive Image Characterists
Wetland Type
Live Deciduous Tree Swamp
Signature: standing water, trees 15 ft. or taller, trees have short trunks, many branches, rough texture, tight crowns

Level II - Classification
Signature:

Color	Height	Texture	Shape	Wetland Type
Green-Blue	15ft+	Coarse	Tight Crowns	Red Maple Swamp
Grey-Green	Low	Fine	Dense	Grassy Wet Meadow

The final aid available to the photogrammetrist is Imagery Interpretation and Transfer Equipment. Due to the scale of the imagery utilized, it is absolutely essential for interpretation that the photogrammetrist have available viewing equipment which permits him to view the site stereoscopically and enlarged. Additionally, he must be able to measure horizontally and vertically.

In the case of Longridge, a series of instruments were utilized as is the accepted practice by Stephen A. Estrin, Inc. First, for synoptically viewing the site and surrounding terrain, a Dietzen Mirror Stereoscope, 6x power was used with a parallax bar. The parallax bar permitted vertical measurement of heights of trees to be made utilizing the floating dot principle. This permitted an overall evaluation and allowed a Class I classification of the wetland. For detailed classification and delineation, a Bausch & Lomb Zoom Transfer Scope was utilized. The ZTS permitted the imagery to be enlarged fourteen times, and the transfer of delineated boundaries directly to a base map by scale matching. It also superimposed the imagery on the base map when the scale was matched, and thereby permitted the photogrammetrist to directly draw the wetland boundary as he was interpreting it. Finally, the ZTS permitted the photogrammetrist to isolate areas requiring more detailed examination, and therefore more accurate classification. This was achieved by maximum enlargement of fourteen times imagery scale and 360 degree imagery rotation.

5 CONCLUSION

Many photogrammetrists have demonstrated the applicability of remote sensing to wetland vegetation analysis in a variety of terrains. Seher and Tueller (1973) studied Nevada Wetlands utilizing color and fcir imagery. In

1974, Enslin and Sullivan produced maps of the Pointe Monillee Marsh, Michigan, from medium-scale fcir.

By 1975, U.S.G.S. was mapping wetlands at a scale of 1:24,000 using seasonal fcir as reported by Carter and Stewart. By 1979, U. S.G.S. produced wetland vegetation maps, 7.5 quad sheets of the Tennessee Valley area on which six vegetative classes and twelve sub-vegetative classes were shown. Also in 1979, Howland completed the wetland mapping of the Shelbourne Pond, Vermont area, in which he tested three types of imagery, conventional color, fcir, and color-enhanced multi-band imagery. He concluded that the fcir imagery was best after delineating and classifying sixteen separate and distinct wetland types

The modern methods of field ecology survey are often difficult, time consuming and too clostly to apply to wetland delineation and classification. In particularly fragile ecosystems, comprehensive biological field studies can cause far more physical damage than would an appropriate ground control program in a small portion of the area and subsequent imagery analysis.

REFERENCES

Anderson, P.H. 1977. Delineation of Deciduous Wetland Forests in Northeastern Connecticut. Connecticut: Univ. of Connecticut.

Brown, W.W. 1978. Wetland Mapping in New Jersey and New York. Virginia: American Society of Photogrammetry. 44:303-314.

Carter, V., D.L. Malone & J.H. Burbank. 1979. Wetland Classification and Mapping in Western Tennessee. Virginia: American Society of Photogrammetry. 45:273-284.

Cowardin, L.M. & V.I. Meyers. 1974. Remote Sensing for Identification and Classification of Wetland Vegetation. New Jersey: Lakewood Publications. p. 308-314

Ernst-Dottavio, C.L., Hoffer & R.M. Mroczynski. 1981. Spectral Characteristics of Wetland Habitats. Virginia: American Society of Photogrammetry. 47:223-227.

Helfgott, T., Kennard, W.C. & M.W. Lefor. 1976. Inland Wetlands Definitions. Connecticut: University of Connecticut.

Kennard, W.C. & M.W. Lefor. 1977. Evaluation of Freshwater Wetlands Definitions. Connecticut: University of Connecticut.

Lefor, M.W. 1975. Remote Sensing of Wetlands. Connecticut: University of Connecticut.

_____. 1975 Wetlands Mapping. Connecticut: Univeristy of Connecticut

Lillesan, T.M. 1976. Fundamentals of Electromagnetic Remote Sensing. New York: Syracuse University.

_____. Manual of Remote Sensing. Vol 1. Theory, Instruments and Techniques. 1983. Virginia: American Society of Photogrammetry.

_____. Manual of Remote Sensing Vol. 2. Interpretation and Applications. 1983. Virginia: American Society of Photogrammetry.

Steward, W.R., V. Carter & P.D. Brooks. 1980. Inland (Non-Tidal) Wetland Mapping. Virginia: American Society of Photogrammetry. 46:617-628.

Symposium on Remote Sensing for Resources Development and Environmental Management / Enschede / August 1986

A hydrological comparison of Landsat TM, Landsat MSS and black & white aerial photography

M.J.France & P.D.Hedges
Aston University, UK

ABSTRACT: Landsat TM is evaluated for its accuracy in delineating, mapping and measuring, water bodies, drainage networks, catchment areas and landcover. A comparison is made with results from Landsat MSS imagery and 1:50000 scale black & white aerial photography, for the same area in North Wales, U.K.

Landsat TM is found to have significant advantages over Landsat MSS for recording drainage network information. Lakes as small as o.6 hectares can be identified using Landsat TM imagery and the delineation of small streams aids in accurately defining catchment boundaries.

1 INTRODUCTION

Satellite remote sensing provides an attractive way of collecting hydrological data and is of particular significance in areas where traditional methods of hydrological data collection are inadequate or impractical. The data can be used as input to morphology or landcover -based hydrological models.

This research is directed at a comparison of Landsat Multispectral Scanner (Landsat MSS) imagery, Landsat Thematic Mapper (Landsat TM) imagery, 1:50000 scale Black & White aerial photography and the 1:50000 scale Ordnance Survey map of the area, using the same test area.

Earlier work on Landsat MSS, with its 80m resolution found encouraging results in areas of high relief (Rango et al., 1975; Killpack and McCoy, 1981) but limited success in most British situations . It was thought that the Landsat TM with its greater spatial and spectral resolution might prove more successful.

Table 1. Comparison of Landsat TM and MSS sensor characteristics.

	TM Data		MSS Data		
Band	Spectral range, μm	Spatial resoln., m	Band	Spectral range, μ m	Spatial resoln., m
1	0.45-0.52	30	4	0.5-0.6	80
2	0.52-0.60	30	5	0.6-0.7	80
3	0.63-0.69	30	6	0.7-0.8	80
4	0.76-0.90	30	7	0.8-1.1	80
5	1.55-1.75	30			
6	10.4-12.	120			
7	2.08-2.35	30			

Work concentrated particularly on the Landsat TM imagery as there is little relevant published material on this imagery. Specific aims were:

1. To ascertain the efficiency of drainage mapping using Landsat TM imagery.
2. To specify the most useful image enhancements for this purpose.
3. To compare the efficiency of interpretation of Landsat MSS imagery, Landsat TM imagery and Black & White aerial photography.

The area selected for this research, some 235km^2 in extent, is the Dolgellau and Coed-y-Brenin area of North Wales, U.K. (see Figure 1). The area is, in the main, an upland region of hills and mountains with a maximum elevation of 754 metres. A Lowland valley runs across the southern part of the region. Streams and rivers in the area are bordered by various lowland and upland vegetation types and frequently pass through forests or woodland , giving opportunity for the examination of several water/vegetation interfaces.

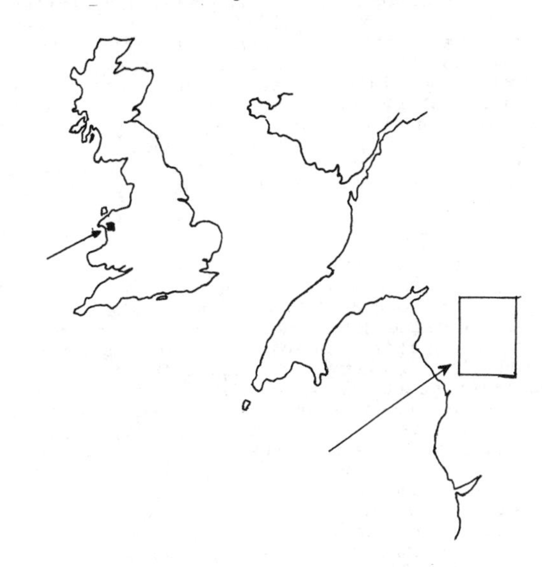

Figure 1. Location of study area.

2 METHOD

Four Landsat MSS images, with minimal or no cloud cover, were available for the area, the best of which is a scene from May 1977. One Landsat TM quarter scene from July 1984, with little cloud coverage was available from the U.K. National Remote Sensing Centre. These images were enhanced using the GEMS image processors at NRSC, Farnborough, and Silsoe College, Bedford and an I^{2}S system at NERC, Swindon.

Image enhancement was mainly directed at the Landsat TM data as little published work exists on the processing of this material. Individual bands, band combinations, principal components analysis and several edge enhancement techniques were evaluated for their enhancement of linear water bodies. Similar techniques were applied to the Landsat MSS imagery. Hard-copies of enhancements were taken using a matrix camera linked to the image processor, which recorded images on 35mm colour slide film.

The colour slides were projected onto transparent overlays using a 'Lamprey box' (see Figure 2). From these projections, details of drainage systems were drawn onto the transparent overlays and then input to a Hewlett Packard digitiser, coupled to a BBC microcomputer. A series of software programs were applied to calculate stream lengths and lake areas. Quantitative comparisons were made between various Landsat MSS and Landsat TM enhancements,the Black & White aerial photography and drainage details extracted from a 1:50000 Ordnance Survey map of the area.

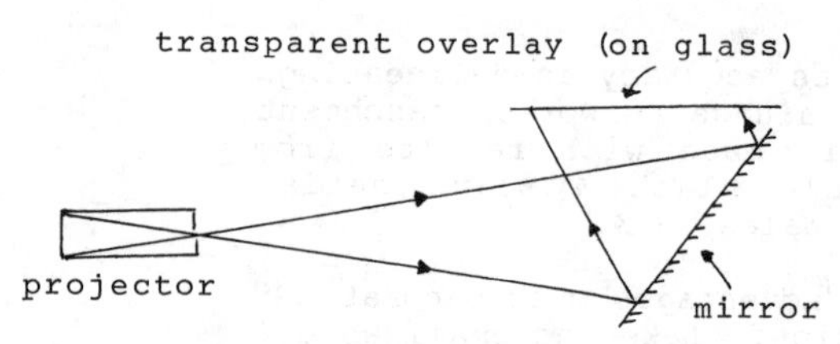

Figure 2. Lamprey box configuration.

The black & white aerial photographs were examined using a Wild ST4 with 8x magnification. Drainage details were recorded on acetate sheets and a mosaic of sheets created to cover the research area. Details from the mosaic were transferred to a single transparent overlay and input to the digitiser.

An initial field survey of the area was carried out from 9-11 March 1985. This survey concentrated on recording stream width measurements and a first look at the general landcover types in the area. The measurements were taken as "ground truth" for comparison with interpretations of each imagery type, for quantitative assessment of image capabilities. A further field survey was carried out from 24-26 June 1985, to coincide with the season of image of acquisition.

3 RESULTS AND DISCUSSION

In comparing Landsat MSS and Landsat TM imagery the most striking difference is the effect of the increased spatial resolution and spectral possibilities of Landsat TM, permitting near 'true colour' representations, not unlike small scale colour aerial photography. The increased spatial resolution results in substantially easier recognition of terrain features, this being aided by shadow effects which provide some idea of topography. Image interpretation is further aided by band combination possibilities which facilitate the creation of images with similar colour renditions to those experienced in the field. These effects are subjective ones affecting the visual processes and as such cannot be quantified, however they result in much easier image interpretation.

When examining individual Landsat TM bands, bands 4 & 5 revealed the most detail, though a three-band combination is more useful in discriminating streams and rivers from their surroundings by providing a 'true colur' contextual background. Band combinations 1,4,5 ; 2,4,5 and 3,4,5 assigned respectively to blue, green and red colour guns on the image processor, proved to be the best band combinations. There is little to choose between them though the author prefers the 1,4,5 combination. Two-part linear or manual stretches were applied to these band combinations. The use of principal components analysis was found to be less informative than simple band combinations. Edge enhancement was found to increase drainage detail, the Laplacian filter, Sobel operator and directional fitering (Seidel, Ade and Lichtenegger, 1983) using 3x3 kernels proving the most successful.

When enhancing the Landsat MSS imagery the best results were obtained using a standard 4,5,7 band combination assigned respectively to blue, green and red, with a gaussian stretch and Laplacian filtering of band 7, using a 3x3 kernel.

The drainage network was most easily discerned over moorland areas, less easily so in agricultural areas where there were small field sizes and with some difficulty in thickly wooded or forested areas, where the spectral contrast is small. Discrimination between roads and rivers which frequently proves problemmatic using Landsat MSS imagery is greatly reduced in the case of Landsat TM due to topographic and contextual effects.

A study of 32 small lakes in the study area and surrounding region revealed that the Landsat TM imagery could be used to identify water bodies as small as 0.3 hectares in areas of open moorland, however an overall figure of 0.6 hectares is more realistic due to difficulty in identifying small water bodies within forested areas. This compares with figures of 2.4 hectares for Landsat MSS and a suggested value of <0.5 hectares for SPOT in multispectral mode i.e. with a 20 metre spatial resolution (Chidley and Drayton, 1986). Using aerial photographs the spectral contrast between water bodies and other terrain features was sufficient to enable lakes of 0.2 hectares and smaller to be identified.

Table 2a Water body identification.

Imagery	Resolution	Minimum area of lake detected
Landsat TM	30 m	0.6 ha
Landsat MSS	80 m	2.4 ha
SPOT XS	20 m	<0.5 ha
B & W Photos	1:50000 scale	<0.2 ha

Landsat TM imagery permits much easier landcover identification than using Landsat MSS imagery. In addition to areas of open water, areas of forest, moorland, lowland agriculture and urban areas can be easily and accurately identified.Using aerial photography the various textures enable discrimination between these major landcover types. Relation of these landcover types to their hydrological characteristics is important for hydrological modelling.

Use of a 2x zoom on the Landsat TM imagery was found to provide the optimal imagery for interpretation. The time required to analyse this four-fold increase in quantity was more than compensated for by the much greater ease of interpretation. However, over large areas this method of analysis may prove prohibitively expensive. Little geometric distortion was found in the TM imagery and the minor necessary corrections were 'manually' applied.

Table 2.Comparison of channel lengths and drainage densities.

	MSS	TM	B&W Photos	Map
Total Channel Length (km)	25.21	88.75	121.50	272.24
Drainage Density (km/km^2)	0.11	0.38	0.52	1.15

Table 2 gives details of total channel length and drainage density identified on all three types of imagery as compared with details extracted from the1:50000 scale Ordnance Survey map. It should be borne in mind that the drainage details on the 1: 50000 map are taken from 1:10000 scale maps of the area. On the face of it these figures look very poor, however on closer examination of the Landsat TM details and comparison with field work readings the situation is much more promising (see Table 3). The reason for this improvement is due to over half the drainage network consisting of streams less than 3 metres in width.

Table 3. Details of streams identified from Landsat TM imagery.

	Number in field	No.on TM imagery	% Correctly identified
Streams over 5 metres wide	28	26	92
Streams over 3 metres wide	43	36	83

The main drainage channels were easily determined on the enhanced Landsat TM imagery (see Figure 3). The detail of minor streams enables greater accuracy in delineating the catchment boundaries, than is possible using Landsat MSS imagery. Many smaller streams, some only 0.5 metres wide could also be identified, though there was great inconsistency of results at these widths, identifiable streams frequently being associated with minor tributary valleys a few metres deep. Landsat MSS enhancements were only consistent in delineating stream widths of 10 metres or more (of which there are few in this area). Some streams as narrow as 3 or 4 metres appear sporadically.

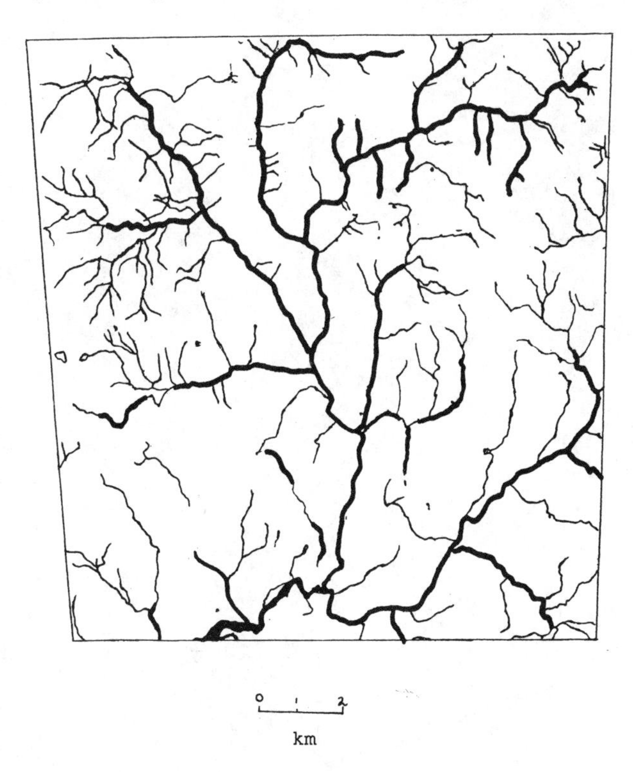

Figure 3. Comparison of stream delineation.

When streams were ordered according to Horton's scheme of stream ordering (Horton,1945), the following results were obtained:

Table 4. Stream orders from Landsat TM, aerial photos and 1:50000 map.

	Stream Order 1	2	3	4	5
1:50000 map	156	36	15	3	1
Landsat TM	33	6	2	1	
Aerial photos	72	16	5	1	

Table 4 shows a variation in number of stream orders between the 1:50000 map and the satellite and aerial imagery. The fifth order stream on the map corresponds to the fourth order stream on the satellite and aerial imagery. It is at the first order where the discrepancies occur and these have an effect on subsequent stream orders. The drainage network on the 1:50000 map is derived from 1:10000 maps and includes many first order streams only 0.5 or one metre wide. Some of these can be delineated from the aerial photography but few from the Landsat TM imagery.

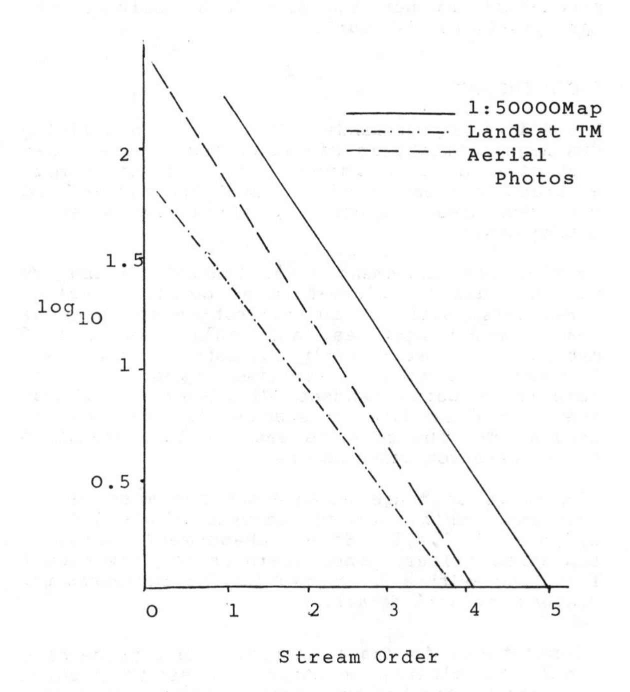

Figure 4. Number of streams v Stream order.

Under normal basin conditions the gradient of the graphs should fall within the range 3-5 (Smart,1972). All three graphs lie within this range, their values being 3.51 for the 1:50000 map, 3.16 for the Landsat TM imagery and 4.03 for the aerial photography.

Using the Landsat TM imagery catchment boundaries can be defined with reasonable accuracy, though other catchments need to be examined before this accuracy can be realistically quantified. Landsat MSS imagery cannot be used for delineation of catchment boundaries in this area, due its delineation of only the largest streams in the area. Over larger catchment areas it has proved more useful- see Drayton and Chidley, 1985.

One noticeable feature revealed during the period of this research was the variation in brightness on the colour slides when similar image enhancements had been carried out. This is due to one of four factors, namely the different image processors used, the level of illumination in the image processing area, the different matrix cameras used or the film processing. The film type and processing house were constant throughout.

4 COST COMPARISON

The following figures are intended to give some idea of the relative costs of imagery (at April 1986). The costs of U.K. photography were from an area of $900km^2$ and the costs for Zimbabwe are based on standard calculating figures.

Table 5. Cost comparisons.

	Landsat TM-whole scene	Landsat TM -1/4 scene	B&W aerial photos U.K. 1:10000	Zimbabwe 1:25000
Cost of imagery in £/km^2	0.06	0.12	16.55	6.20

These costings of satellite imagery are for the whole area which in most circumstances will not be fully required. The cost of digital image processing is not included (U.K. hire rates: April,1986, £35-80/hour).

5 CONCLUSIONS

The efficiency of Landsat TM imagery in defining drainage networks is significantly greater than that of Landsat MSS imagery. This results in more accurate delineation of catchment boundaries and catchment area estimation, particularly for small catchments.

Within the catchment area, Landsat TM imagery provides useful information on geomorphological parameters, with the identification of lakes as small as 0.6 hectares, and small streams (3-5 metres wide) being fairly consistently recorded. Several different landcover types can be identified using Landsat TM imagery. Although there is more data to examine in the case of Landsat TM the relative ease of interpretation compensates for this factor.

In terms of image enhancement the most useful waveband combinations for Landsat TM are 1,4,5 ; 2,4,5 and 3,4,5. Edge enhancement using a Laplacian filter, Sobel operator or directional filtering, with a 3x3 kernel further increases the drainage network detail.

Consistency in image recording and processing should be closely monitored so as to produce consistent results, thus enabling more meaningful comparisons.

ACKNOWLEDGMENTS

The author would like to thank the Overseas Development Administration, NRSC Farnborough and NERC, Swindon for their assistance and the use of image processing facilities.

REFERENCES

CHIDLEY, T.R.E. & DRAYTON, R.S., 1985. The use of SPOT simulated imagery in hydrologic mapping, Int. Jour. Rem. Sens., in press.

DRAYTON, R.S. & CHIDLEY, T.R.E., 1985. Hydrologic modelling using satellite imagery. Proc. Int. Conf.on Advanced Technology for monitoring and processing global environmental data, Remote Sensing Society/CERMA, 1985.

HORTON, R.E.,1945. Erosional development of streams and their drainage basins: hydrophysical approach to quantitative morphology. Bull.of the Geological Soc.of America 56,275-370.

KILLPACK, D.P. & McCOY, R.M., 1981. An application of Landsat derived data to a regional hydrologic model, Rem. Sens. Quart. 3(2),27-33.

SEIDEL,K.,ADE,F., LICHTENEGGER, J.,1983.Augmenting Landsat MSS data with Topographic Information for Enhanced Registration and Classification, IEEE Transactions on Geoscience and Remote Sensing, Vol. GE-21, No. 3, July 1983.

SMART, J.S., 1972.Channel networks. Advances in Hydroscience 8,305-346.

RANGO,A.,FOSTER,J.,SALOMONSON,V.V.,1975.Extraction and utilization of space acquired physiographic data for water resources development. Water Res. Bull. 11(6), 1245-1255.

Application of remote sensing in hydromorphology for third world development: A resource development study in parts of Haryana (India)

A.S.Jadhav
Dept. of Geography, Vidyanagar, Kolhapur, India

ABSTRACT: Remote sensing technique is nowadays being widely employed in various studies and exploratory work. In present study, remote sensing techniques both conventional and satellite, are used for collection of the information regarding distribution and quality of groundwater. The main aim of the study is to determine the nature of the distribution, depth and quality of groundwater. A special attention is given to delineate the saline water areas and fresh water pockets in the semiarid area in the parts of Haryana state of India. In present paper, the landscape method alongwith analysis of landforms for hydrological deciphering is applied. The study was carried out in three stages i.e. prefield interpretation, field work and post field work. Aerial photographs, imagery and toposheets were chief tools used in the study. It is found that potable water pockets can easily be located with help of remote sensing techniques. The occurrance, distribution and degree for salinity of groundwater is closely associated with geomorphological processes and landforms of the region.

1 INTRODUCTION

Whether for irrigation, power generation, drinking, industry or recreation, water (surface and groundwater) is one of the critical resources and it is a subject matter of hydrology. Remote sensing can be used in variety of ways to help monitoring the quality, quantity and geographic distribution of the water.

A precise geohydromorphological map gives a good deal of information on the groundwater potential of an area. The purpose of this study is to obtain information from aerial photographs and landsat imagery supplemented with fieldwork on quality, quantity and distribution of groundwater.

2 GEOGRAPHICAL SET-UP OF STUDY AREA

The study area is bounded by the latitude 27°35'2" and 28°1'10" N. and the longitude 76°58'27" and 77°24' E. and covers the part of Haryana, Rajasthan and Uttarpradesh states of India.

The topography is rugged and undulating with moderate to high hills. There are two main hill ranges in the region which run close and parallel to its west boundary in the north eastern direction. Beside these ranges, there are a few isolated hills scattered on the eastern side of Ajabgarh series. There are number of streams which do not meet any major stream or river and disappear beneath the permeable surface of the sandy plain and alluvial plain.

The region experiences semiarid climate characterised by extreme climatic conditions during summer and winter months with appreciable range in diurinal temperatures. The rainy season is much shorter in duration; the rains are higher and often irregular. During the greater part of the year, the weather is hot, dry and often dusty. The rains occur in the form of the sudden heavy thunder showers, intersparsed with short spells of fair weather. The essential feature of the climate is that potential evapotranspiration from soil and vegetation exceeds the average annual precipitation. As a result of the rainfall deficiency, there is no constant supply of groundwater. In such climatic zone, increasing aridity is marked by a gradual deterioration in the vegetal cover from poor grass land and scrub with some bare patches of rock or soil through a wide variety of semiarid plant community. The annual rainfall in the region varies from 500 mm. to 800 mm. and about 60 to 75% of the annual rainfall occurs during the rainy season. The annual mean temperature varies between 11°C and 35°C.

3 METHOD AND MATERIAL

The studies on groundwater resources management are carried out in laboratory by using aerial photographs (black & white) on scale 1:50,000, landsat imagery (FCC and band 5,6 and 7). In subsequent phase, the interpreted data were varified and confirmed in field. In addition, new data was collected by a few observations and checking of the critical sections. The information regarding groundwater condition and quality was collected from unpublished official records. The entire photo coverage of the area was examined in the laboratory during the post fieldwork and the information was revised in accordance with the field observation, the data collected during the field traverses and the data collected from governmental and non-governmental agencies. The maps and plans were finalized and redrawn.

4 WATER BEARING PROPERTIES

In the study area quartzite and quaternary deposits have different water bearing properties.

4.1 Alwar series

Rocks of Alwar series of Delhi system include massive, very thickly carbonacious phylite, garnateferous mica schist, schist and staurolite schist. Groundwater in these rocks occurs and moves through the joints, fractures, foliation plane and weathered zones. Well tapping these formations yields discharge ranging from 40 to 80 cubic metres per day.

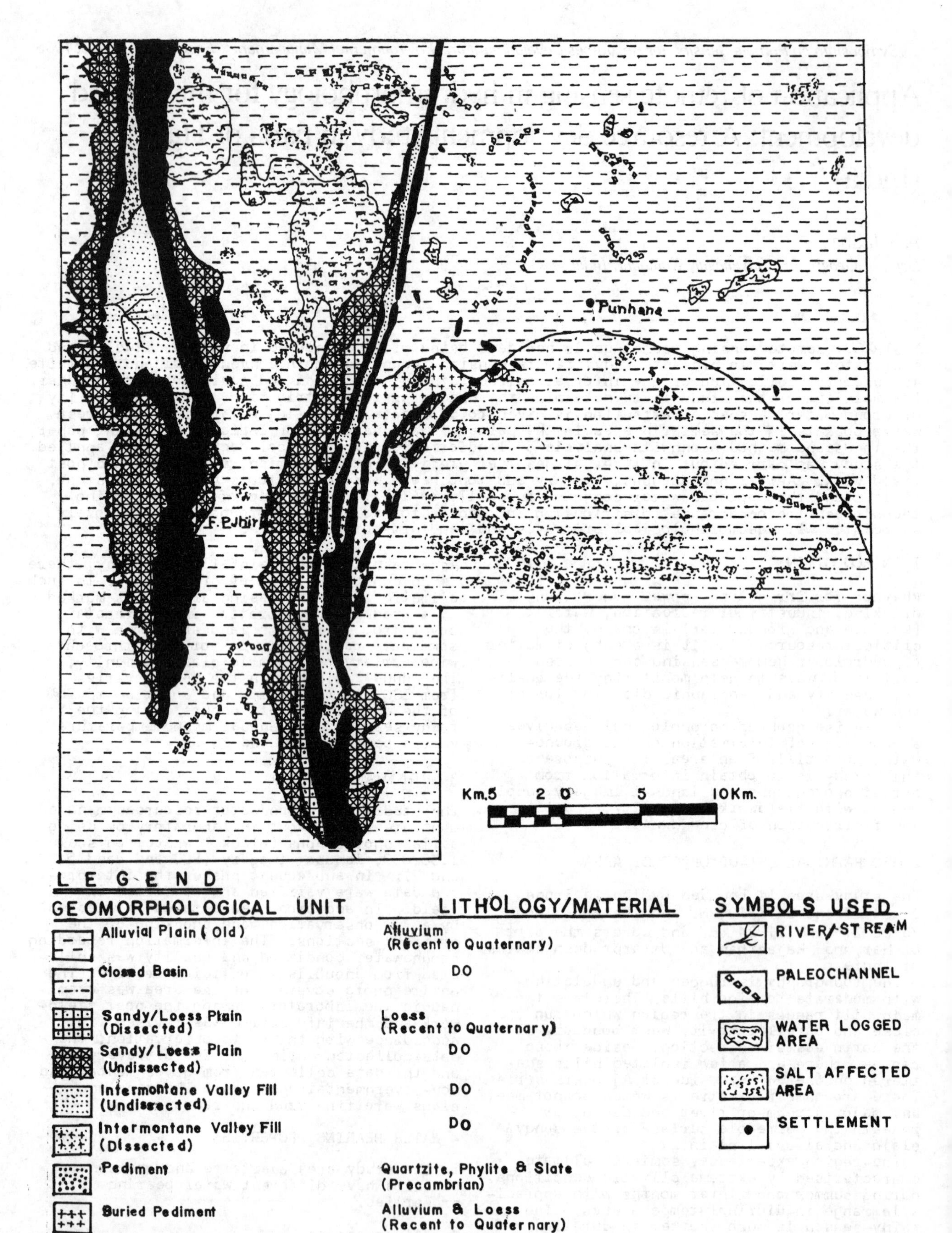

Figure 1. Geohydromorphological map (Based on Aerial Photographs & Landsat Imagery).

4.2 Ajabgarh series

The Ajabgarh series of rocks comprise of carbonacious phylite, slate, thinly bedded quartzite, chlorite, sericite schist and quartzitic schist. These are better aquifers than Alwar rocks. There is a lack of primary porosity and permeability due to metamorphism, but secondary porosity and permeability is found due to joints and fractures which facilitates harbouring of the groundwaters under watertable conditions mainly. Well in these rocks yield discharges ranging from 3.6 to 10 cubic metres per hour.

4.3 Quaternary deposits

Recent to Sub Recent deposits of Alluvium and loess with good primary porosity and permeability has high recharge potential. They form aquifers and contain potable to brackish water. Wells tapping these deposits are capable of yielding good supplies, ranging from 30 to 40 cubic metres per hours. The yield of well varies from place to place, but generally wells located in the vicinity of Yamuna river and Agra canal yield more than other wells.

5 GROUNDWATER POSSIBILITIES IN DIFFERENT GEOMORPHIC UNITS

5.1 Structural and Denudational hills

Structural hills of hard quartzite rocks form a runoff zone. However, some groundwater occurs and moves through joints, ractures and faults. Delhi quartzites with high relief and steep slope show high runoff zone and very poor recharge. No well has been dug in this unit due to poor aquifer characteristics. Near the Siva temple, Firozpur Jhirka, one dug well in Alwar quartzite is noticed. Depth to water is about 8 metres. It is also located in the faulted zone of the Siva temple. A perennial spring is also noticed close to dug well. The well might be recharged through groundwater collected in fracture zone, otherwise no dug well has been reported.

5.2 Pediment

Owing to hardness of area, slope and very limited weathering, pediment forms mainly runoff zone. It possibly provides nominal recharge surface for lower pediment zone. The good quality water is found in this unit.

5.3 Buried pediment

This unit consists of clay, silt and sand with nominally weathered, jointed rocks beneath. It is a recharge zone. Water is found under watertable conditions. The thickness of deposited materials varies from place to place. Depth to watertable varies from 8 to 20 metres. The quality of water ranges from potable to brackish.

5.4 Intermonatane valley fill

This unit consists of wind blown materials. The unit is located between two ridges, so it does not carry much groundwater weightage in present study area.

5.5 Obstacle sand dunes and sandy loess plain (dissected)

Most of the area is covered by sand. This area is barren and intensive gally network has been developed indicating a high runoff zone. The material is porous and permeable, still owing to steep slope, there is very poor recharge, and almost all precipitation contributes for runoff. There is no cultivation over this unit. Water is potable but at places slightly brackish.

5.6 Sandy loess plain (undissected)

Considerable area is covered by aeolian sand which is being cultivated by well irrigation and at some places rainfed cropping is also practiced. Sand is fine grained, well grounded with high porosity and permeability. Only local development of Kankar is found. Depth to watertable in area ranges from 8 to 20 metres.

5.7 Alluvial plain

This unit comprises of old flood plain and closed basin. The alluvial plain country is composed of the clay, silt and sand. The watertable varies from a few metres to 20 metres. In old alluvial plain salt affected areas occur and paleochannels fills are having shallow watertable and good recharge. The good quality of water is found in paleochannel. The quality of water ranges from potable to saline.

6 DEPTH TO WATERTABLE AND WATERTABLE EVALUATION

The depth to watertable, in general, lies within 8 metres below ground level in a large part of the area; the shallowest being 0.2 metre below ground level and deepest 32.85 metres. It is evident from figure 2 that the depth to watertable is more in the vicinity of hilly tracts. Shallow watertable is found in a large parts of the area east to the ridges, and west to Agra canal. The watertable contours generally follow the surface topography. The altitude of watertable ranges from 179.6 to 281.5 metres above mean sea level. In the eastern part, hydraulic gradients are moderately steep.

The watertable slopes southwards in the northeastern part of the area with an average hydraulic gradient of 0.6 metres per kilometres. In the south western part, in the valley between two ridges, groundwater movement is NNE with hydraulic gradient of the order of 1.45 metres per km. NNE - SSW trending ridges form a groundwater divide.

The fluctuation of watertable is of two types, the long term and seasonal. Broadly, the seasonal fluctuations are caused by seasonal variation in rainfall and groundwater draft whereas the secular variations are mainly a result of groundwater development. During the period of June 1974 to 78 a rise in watertable has been observed in most of the hydrographic stations of the order 0.41 to 2.23 metres and a recession of about 1 metre in two stations. The five years average seasonal fluctuations in the various blocks of the region varies from 1.2 to 2.5 metres. Considering the long term fluctuations between 1961-63 and 1975-76, it is seen that there is a rise in water level in most of the part of the region which ranges from 0.1 to 8.6 metres.

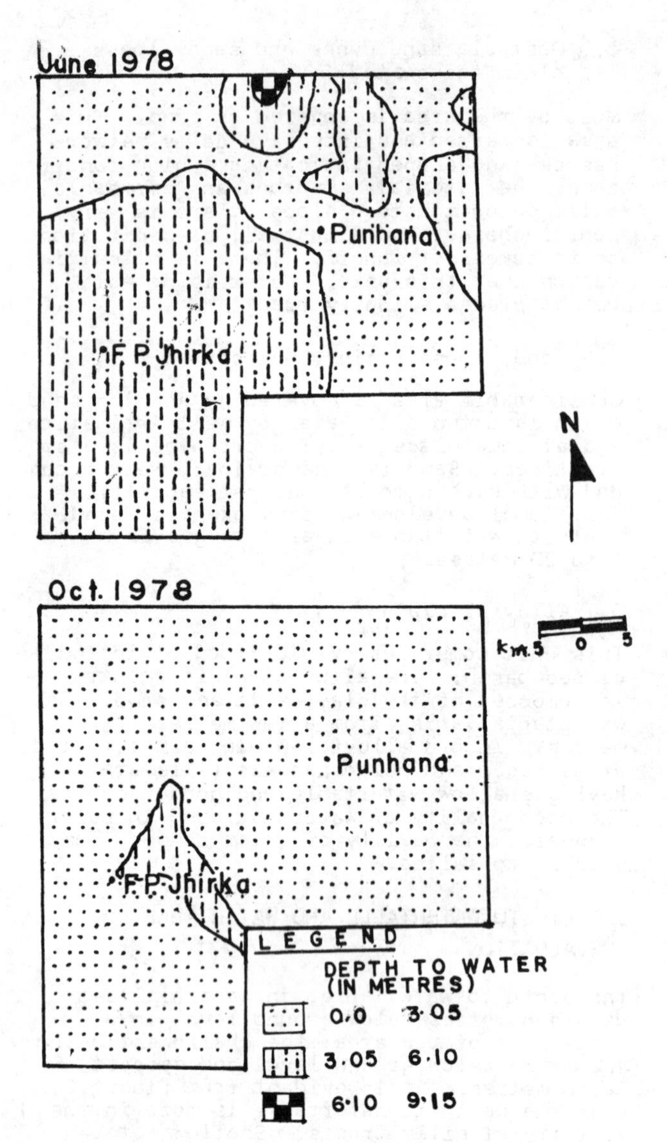

Figure 2. Maps showing depth to water; June and October 1978.

7 GROUNDWATER QUALITY

It is clear from figure 3 that shallow groundwater in the region shows great variation in chemical quality from place to place. Some of the shallow waters in area are saline and are not suitable for irrigation. The specific conductivity of shallow waters in the area ranges between 446 and 34160 micro hos/cm at 25°C. Ec values of more than 6000 micro hos/cm at 25°C i.e. areas of saline groundwaters, occur in the central part of the study area around Pataudi and Firozpur Jhirka. It is found that fresh water occurs all along the piedmont zone of Alwar formation and Ajabgarh formation on either sides. As one moves towards plain from piedmont zone salinity increases. Some patches of fresh water have been found in plain area surrounded by the saline water areas. These patches of fresh water areas are on paleochannel fills of the abandoned river courses of recent past times. Due to granular material deposits in paleochannels and better drainage as well as flushing conditions sweet water is found. Saline water mainly occurs due to the physiographic conditions of the area in depressions. There is no outlet in the study area. Rainfall occurred during rainy season inundates the area and makes water logging thereby endangering human life and property. Since there is no outlet in deadly flat area, there is no flushing which contributes to the increasing salinity.

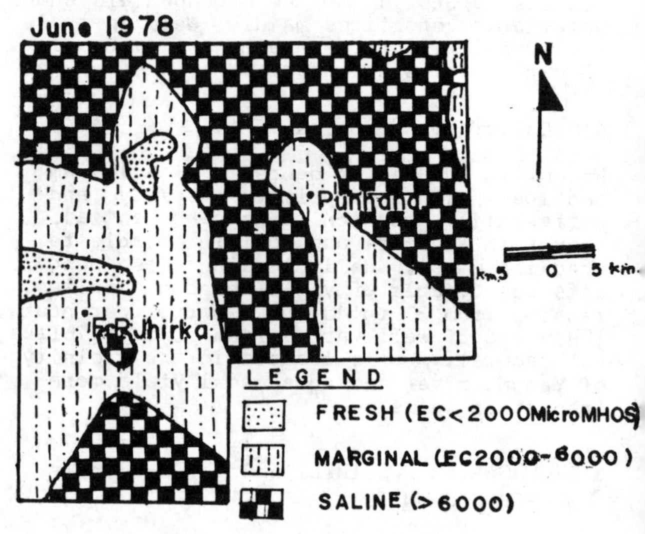

Figure 3. Water quality map.

Water quality map (figure 3) shows that salinity has negative correlation with depth and relief. As the depth increases, salinity decreases and as the relative relief increases, the salinity decreases.

Based on above, it can be said that area is having good potential of groundwater, but faces acute groundwater problems in respect of quality. The paleochannels can be used to extract good quality of potable water.

REFERENCES

Adyalkar, P.G. 1964. Paleography, sedimentalogical frame work and groundwater potential of the arid zone of Western India. Proc. Samp. Prof. Indian Arid Zone, New Delhi.

Heron, A.M. 1917. The Geology of the North Eastern Rajputana and Adjacent Districts, Mem. GSI., Vol.XLV, No.1.

Vertappen, H. Th., and R.A. Zuidam; System of Geomorphological Survey, ITC, Vol.7, No.2.

Symposium on Remote Sensing for Resources Development and Environmental Management / Enschede / August 1986

Remote sensing of flow characteristics of the strait of Öresund

L.Jönsson
Dept. of Water Resources Engineering, University of Lund, Sweden

ABSTRACT: The strait of Öresund is located in a densely populated region and subject to many activities. Thus, it is important to understand the flow characteristics of the strait properly. The paper presents results of the use of NOAA and Landsat imagery for exracting information on the large-scale flow pattern and the possibility of distinguishing certain flow phenomena. These findings are discussed in relation to known properties of the hydrography of Öresund. The remotely sensed flow information is also discussed as far as numerical modelling of the flow in Öresund is concerned.

RESUME: Le détroit d'Öresund est situé dans une region populeuse exposé a beaucoup d'activités. Ainsi c'est important de savoir les mouvements de l'eau profondement. Cet article present des resultats de l'utilisation de NOAA et Landsat data pour l'extraction d'information sur la circulation a grande échelle et sur la possibilité de découvrir certains phénomènes hydrodynamiques. Ces résultats sont discutés en relation avec la nature de l'hydrographie d'Öresund. L'information d'écoulement obtenu avec télédetéction est aussi discuté en relation avec des modèles numériques d'écoulement en Öresund.

1 INTRODUCTION

Knowledge of hydrodynamic processes and circulation patterns in coastal waters is important when conside-ring resource or quality aspects of the waters or in connection with technical/constructional measuers in the coastal zone. Data on large-scale water circula-tion by means of satellite imaging provide one type of information which could be used either directly or in connection with numerical flow models for in-creasing this knowledge. This paper discusses the use of NOAA and Landsat data from the strait of Öresund. The study has been made possible by the support of the Swedish Space Corporation, the Swedish Natural Science Research Council and the University of Lund.

2 BACKGROUND

The strait of Öresund forms one connection between the southern part of the Baltic Sea and Kattegatt (Fig. 1). The length is approximately 100 km and the width 10-20 km.

The flow in the strait is rather complex and at times one could distinguish three different water types - Baltic water with salinity S = 8-10 o/oo, Kattegatt surface water S = 18-24 o/oo and Kattegatt bottom water S = 30-34 o/oo. The surface flows to the north 66% and to the south 34% of the time mainly due to wind and pressure induced water level diffe-rences between Kattegatt and the Baltic. The most common overall flow could be described as a two-layer flow situation with northbound surface flow and southbound bottom flow. However, the bottom flow is normally blocked by the shallow-8-10 m - depth in the southern part of the strait. Thus one could think of Öresund as an estuary for the "river" Baltic most of the time (Harremoës et al 1966). The main types of surface flow patterns based on (rather scarce) field measurements are shown in Fig. 1.

Öresund is bordered by densely populated areas im-plying a heavy demand on the coastal water for a num-ber of sometimes conflicting uses. This is one rea-son, besides purely technical aspects, why detailed investigations of the consequences of interferences with the coastal waters are often needed. Thus a num-ber of studies have been performed involving knowled-ge of and consequences for the flow behaviour in more or less detail. In several cases numerical modelling of the flow has been used. In other cases the prob-lems have been approached by means of physical model-ling, prototype simulations or rather simplified cal-culations. Some examples are given below (ref. Fig. 2).

A - an intended railway tunnel on the bottom of the strait. Risk for blocking of salt, nutritious bottom water. Ecological consequences. Numerical models based on vertically or laterally integrated hydrody-namic equations.

B - planned, extensive land fillings which might affect water exchange, navigation, erosion. Studied by means of vertically integrated (shallow water) flow equations.

C - discharge of cooling water from nuclear power

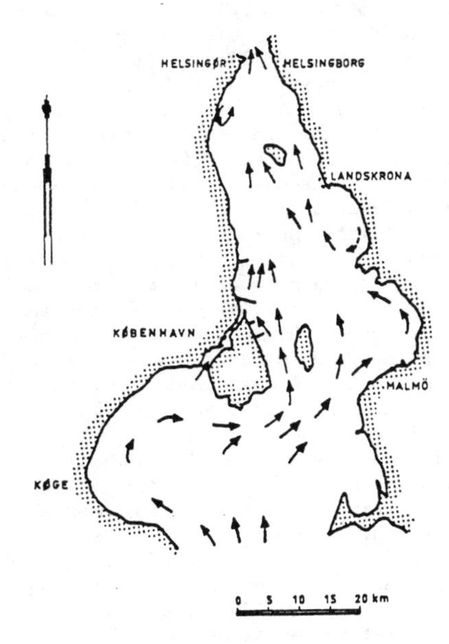

Figure 1. Large-scale flow pattern in Öresund. Northbound flow.

Figure 2. Map of Öresund

plant. Very limited studies.

D - parts of this shallow bay was planned to be filled up. The effect on the circulation in the bay was investigated using a physical model.

- location of second sea outfall for treated sewage water was determined on the basis of field simulations of discharges by means of tracers.

E - planned extension of small island was anticipated to influence water exchange at nearby beach. Field measurements of some representative flow patterns were used as a basis for assessment.

F - sand suction activities have been going on for a long time in Öresund, producing a number of holes on the bottom. In areas with weak circulation there is a risk for deteriorating water quality of the water in the holes.

Satellite derived data might be used for these kinds of problems in different respects

1) contributing a data base for the flow behaviour in Öresund

2) contributing in the development and use of flow models.

3 SATELLITE DATA

Digital NOAA data and Landsat TM-data have been studied - mainly contrast-stretched - on an EBBA-II image processing system. Water surface temperature and suspended mterial (pollen, silt) were used as indicators of water movements and identification of different water masses. As for NOAA far-infrared data the eight least significant bits were used because of the small water temperature differences involved. Fig. 3 shows two consecutive passages of NOAA-7:

CH 4: 1984-08-22 at 14.10 and 1984-08-23 at 13.58 The darker the water surface the colder is the water. In both cases (especially on the 23rd) the surface waters of Öresund seem to be colder than the water of Kattegatt or the Baltic. One possible explanation for these temperature conditions is the behaviour of Öresund as a stratified estuary. A surface flow directed northwards is mostly accompanied by a dense, cold bottom flow of high salinity to the south and this cold water is entrained into the surface water thus lowering its temperature. Another indication of this explanation is the fact that the cold water region seems to start in the Copenhagen-Malmö area where a sill is located.

The discharge of the cold Öresund water into Kattegatt takes place as a plume somewhat growing in the NW-direction during the 24h. According to field observations the flow was directed to the north during this period.

Another interesting detail of Fig. 3 is the seemingly developing eddy outside the southern coast - just

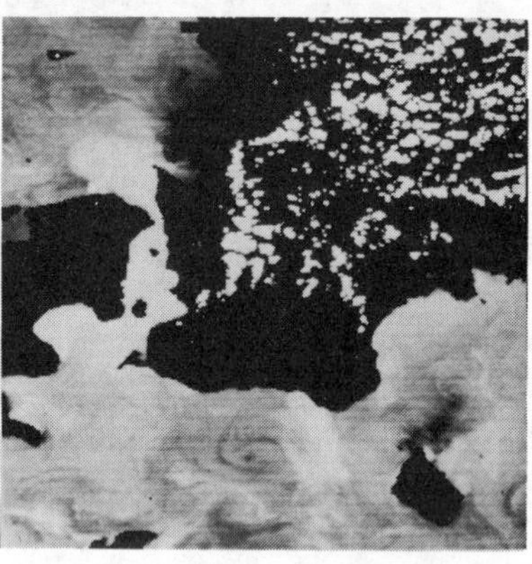

Figure 3. NOAA-7 1984-08-22 and 1984-08-23 CH 4. Öresund and the Baltic

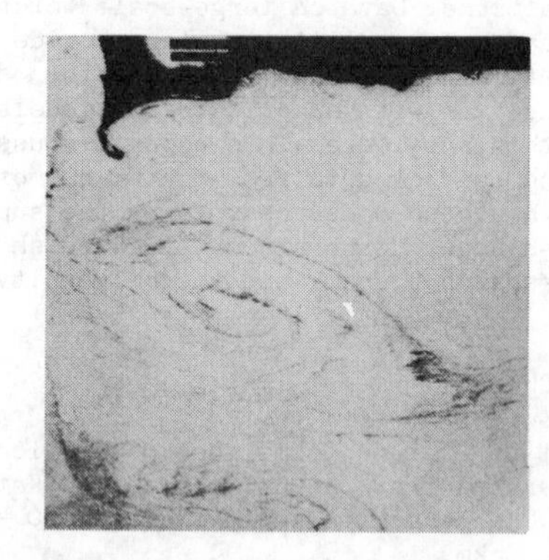

Figure 4. Landsat-2 1975-08-08 CH5. The waters outside Falsterbo - SW Sweden.

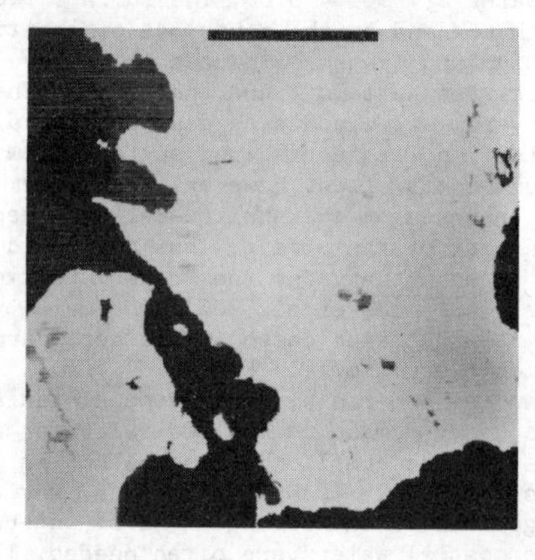

Figure 5. Landsat-5 1984-05-13 TM 6 Öresund and part of Kattegatt.

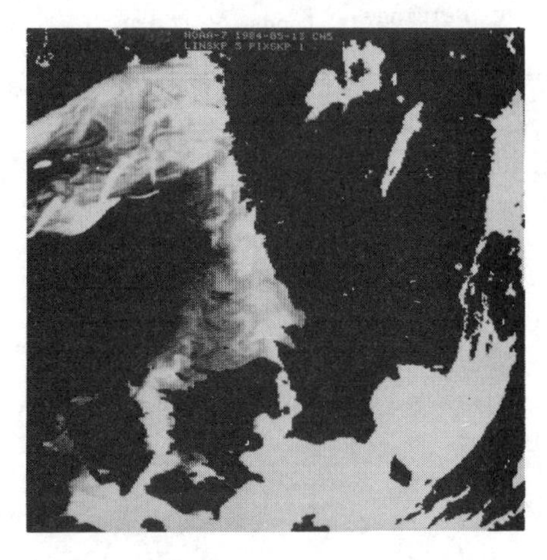

Figure 6. NOAA-7 1984-05-13 CH 5 Denmark and Southern Sweden.

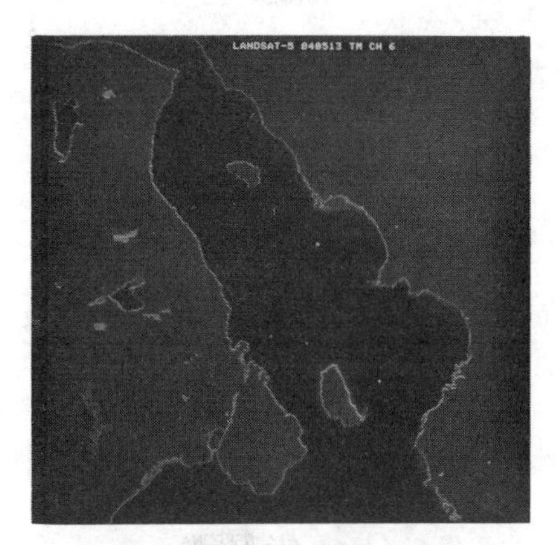

Figure 7. Landsat-5 1984-05-13 TM 6 Öresund

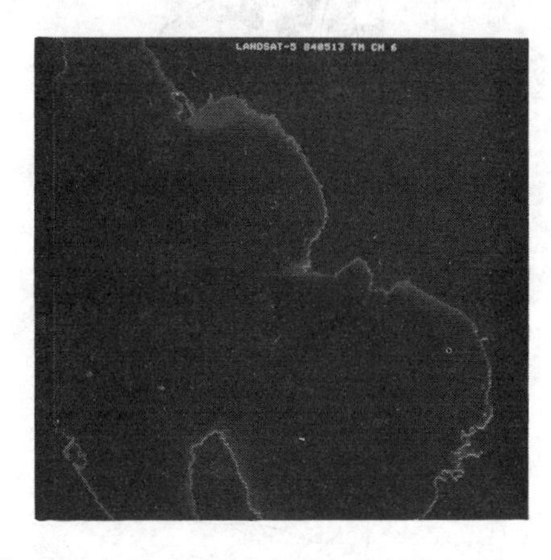

Figure 8. Landsat-5 1984-05-13 TM 6 Landskrona - Barsebäck area.

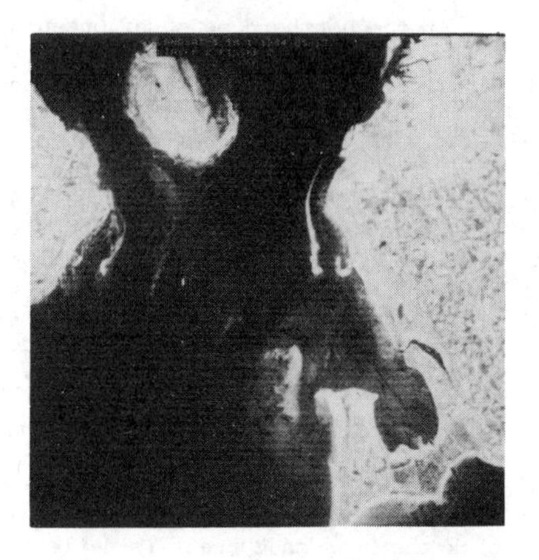

Figure 9. Landsat-5 1984-05-13 TM 1 Köpenhamn - Malmö area

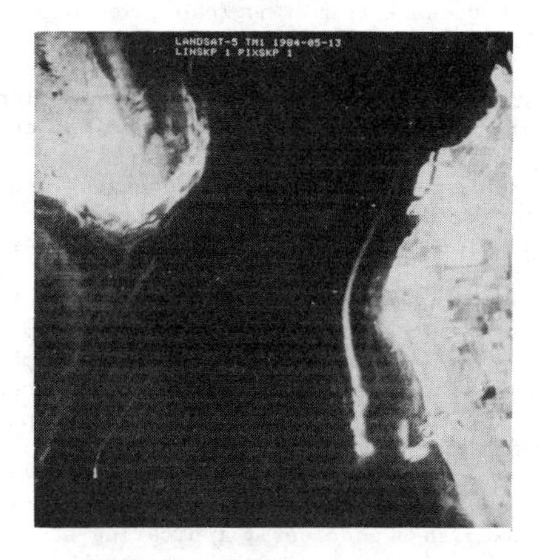

Figure 10. Landsat-5 1984-05-13 TM 1 Saltholm - Malmö area

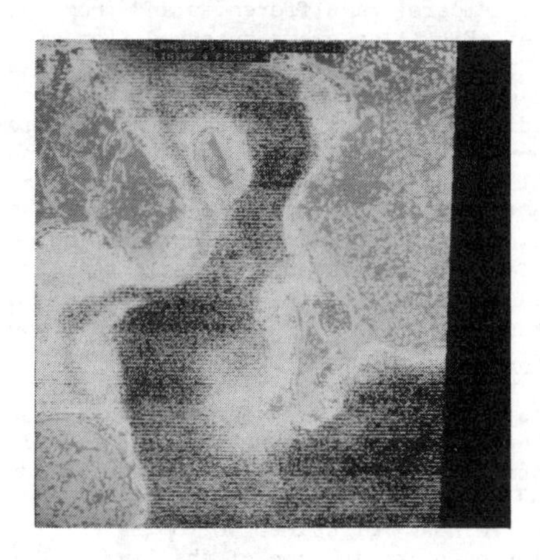

Figure 11. Landsat-5 1984-05-13. TM 1 + TM 6 southern Öresund

a dark dot on the 22nd and an eddy structure on the 23rd. The dot has also moved westwards during the 24h.

The westward moving large-scale eddy structures could also be seen at another occasion by Landsat-2 CH5 1975-08-08 probably due to pollen, Fig. 4. The eddy outside the entrance to Öresund is seemingly rotating anticlockwise. It is not known if the eddy enters the strait and complicates the flow pattern. It is, however, certain that the eddy is not a local phenomenon but has been generated far away in the Baltic (Jönsson, 1984).

As the resolution of NOAA is too small for resolving any details of the flow in Öresund one has to resort to Landsat imagery (especially Landsat-5) and the rest of the Öresund flow information is based on a Landsat-5 passage of 1984-05-13. Fig. 5 shows a TM 6, 360 m resolution image of the whole of Öresund (at 09.35 GMT) and for comparison a NOAA-7 CH 5 full resolution image from the same day (at 14.52 GMT) is also depicted, Fig. 6. For the Landsat-scene cold water is dark whereas the opposite is true for the

NOAA-scene. At the northern mouth of Öresund the discharge of cold water to Kattegatt along the Swedish side is clearly visible in both scenes. This tendency might be due to the Coriolis force which tries to move water to the right on the northern hemisphere . The cold water plume originating from Öresund seems to be more extended in the NOAA-scene than in the Landsat-scene. This could partly be explained by the difference in time for the two detections (field measurements indicated a northbound current). However, one could also expect that the plume looks smaller in the Landsat scene due to the smaller temperature sensitivity of Landsat.

Fig. 7 shows a full resolution TM 6 image of Öresund. Cold (dark blue) water is entering the southern part of Öresund in a band not occupying the whole distance between Malmö and Köpenhamn. It passes both sides of Saltholm, bypasses Lommabukten on the whole. At Barsebäck the Öresund current is deflected to the north, entirely avoiding Lundåkrabukten (between Landskrona and Barsebäck (se also Fig. 11). One could further notice two discharges of cooling water (bright blue) on the Swedish side - Fig. 8:

- at Barsebäck from a nuclear power plant directed almost entirely to the west
- at Landskrona from industrial processes. This warm water seems to flow southwards along the shore indicating the existance of an eddy in Lundåkrabukten.

During the passage of Landsat-5 dredging and dumping operations were going on in connection with a laying of a gas pipe between Köpenhamn and Malmö causing a lot of suspended inorganic material in the water visualizing streamlines. Fig. 9 shows a TM 1, 60 m resolution of the southern part of Öresund. Four sites can be seen along a line between Malmö and Köpenhamn acting as sources for suspended sediment. Traces of the sediment can be seen extending northwards on both sides of the island Saltholm. A dumping site outside Malmö harbour can also be seen clearly. Fig. 10 shows a TM 1, 30 m resolution, image of the Swedish side clearly indicating streamlines - notice especially the right one which could be of direct interest to problem E.

Because of the dredging operations there is a unique possibility of comparing flow information from water surface temperature differences and from suspended sediment. Fig. 11 shows a superposition of TM 1 and TM 6. The cold water is shown in dark blue whereas the suspended sediment is depicted as white yellow. The sediment trace on the Swedish side follows the boundary between the colder main current and warmer coastal water. These findings support the assumption that both temperature differences and suspensions are passive indicators of the actual flow.

The few satellite scenes investigated thus point to several details of the flow structure:

- flow discharge to Kattegatt where probably both wind effects and the Coriolis force are important
- a kind of "curving" flow in Öresund not occupying the whole width, i.e. far from channel-like flow
- the existance of large-scale eddies along the southern coast of Sweden and which might advect into Öresund
- entrainment of colder bottom water
- large eddies in some of the bays in Öresund
- streamlines in parts of Öresund

The satellite derived flow information could be compared with Fig. 1. Generally one could say that the former either supports the latter (streamlines in the southern part of Öresund and the tendency to an eddy in Lundåkrabukten) or complements it.

4 NUMERICAL FLOW MODELS

The effect on the circulation and salinity contents of the strait of the proposed tunnel on the bottom between Helsingborg and Helsingör was studied by means of a flow model based on the laterally integrated flow

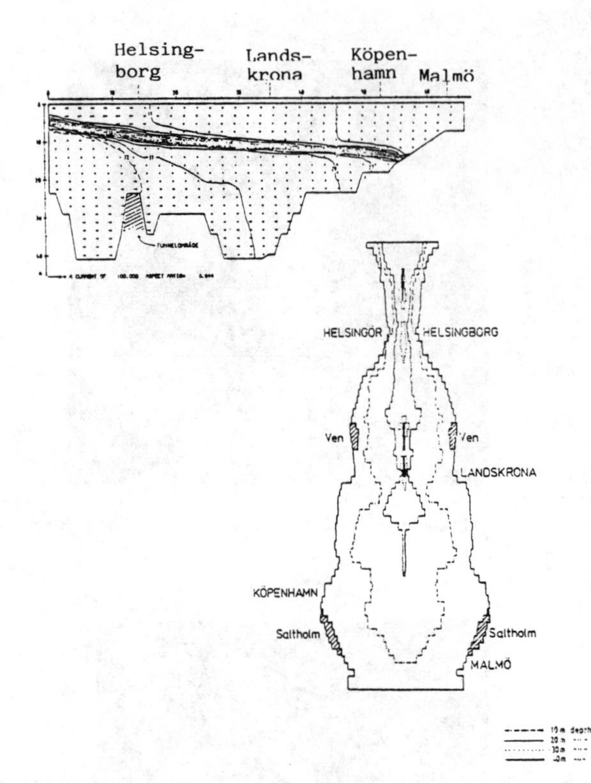

Figure 12. Representation of Öresund for a laterally integrated computational flow model (after Svensson 1978).

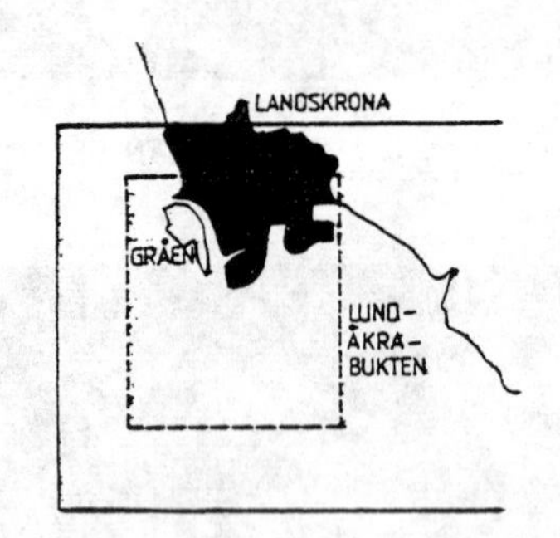

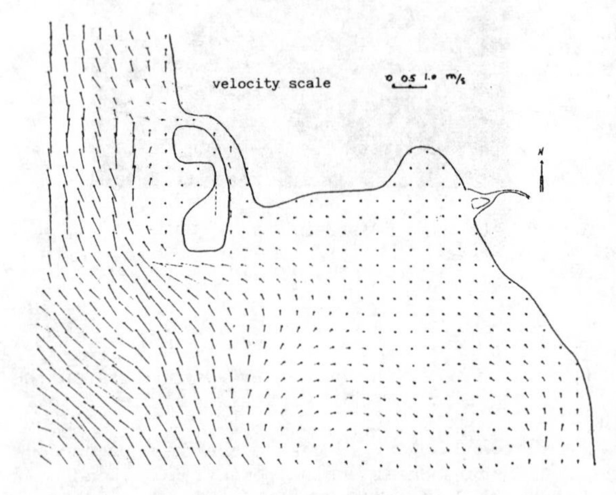

Figure 13. Coastal waters outside Landskrona. Continuous line-boundary for computational flow model. Dashed line-boundary for detailed computations (left).
Computed flow in the coastal waters of Landskrona. Northbound flow in Öresund.

equations - i.e. the strait was considered as a channel where the velocity in each cross section just depends on the depth (Svensson 1978). The representation of Öresund in the computational model is shown in Fig. 12. A longitudinal section is also shown in the same figure with inflow boundary at Malmö and outflow boundary north of Helsingör-Helsingborg. The tunnel is the shaded part. A calculated pycnocline is clearly visible.

The approximations of the flow inherent in the "channel" approach could be compared to satellite data. One could point at two discrepancies:

- according to Figs. 7, 8 the flow seems rather curving with significant accelerations perpendicular to the channel axis
- the flow is not homogeneous in a cross section - within the strait and at the southern boundary, Figs. 7, 11.

Another study concerned the effect of landfillings at Landskrona on the flow pattern from navigational, erosional and water quality point of views (Larsen 1975). The 2-D shallow water equations were used and boundary values for a small computational water area were partly obtained by computing the flow in the whole Öresund. Fig. 13 shows the present conditions with a small island. The continuous line shows the boundary for computations. Fig. 13 also shows the computed flow pattern for overall current to the north in the strait and for the present geometry. It is evident that correct boundary values of the flow are of utmost importance for predicting the flow within such a small area, especially when the boundary is cutting across an eddy, Fig. 8.

These two applications indicate the usefulness of satellite flow information in the context of numerical modelling:

- choosing an appropriate form of approximation of the hydrodynamic equations (1-D, 2-D vertically or horizontally)
- discerning important flow mechanisms that should be modelled (Coriolis effect)
- choosing suitable boundaries for detailed flow computations (avoiding eddies on the boundary)
- obtaining flow characteristics on the model boundaries (inflow to the southern part of Öresund)
- calibration and validation (use of detected streamlines)
- determining density and distribution of grid points (i.e. a dense net close to Barsebäck where the flow seems complex).

5. CONCLUSION

Satellite imaging has demonstrated its potential of providing flow information in Öresund. The findings have implications in several respects as to numerical flow modelling.

In order to fully exploit the possibilities of remote sensing in coastal water studies the imaging must have good spatial and temporal resolutions in suitable wave length bands (especially the far infrared). However, there is no satellite today combining these properties. Furthermore, atmospheric conditions often prohibit satellite based remote sensing in Sweden.

Thus methods which are more independent of weather conditions will be of interest in the future when studying dynamic processes such as water circulation. One such method is based on active micro-wave techniques in combination with Synthetic Aperture Radar. The mechanisms involved in imaging the water surface are, however, far from fully understood.

REFERENCES

Harremoës et al 1966. Report on the investigations of the Swede-Danish Committee on Pollution of the Sound 1959-64.

Jönsson, L. 1984. Remotely sensed surface temperatures and suspended material as sources of information on water circulation - a study on coastal waters in Kattegatt and Scania. Dept of Water Resources Engineering, University of Lund, Sweden.

Larsen, P. & Wittmiss, J. 1975. Landskrona-influence of proposed landfillings and dredgings. Hydro-technical investigation. Dept of Water Resources Engineering, University of Lund, Sweden.

Svensson, J. & Wilmot, W. 1978. A numerical model of the circulation in Öresund. Evaluation of the effect of a tunnel between Helsingör and Helsingborg, SMHI, Sweden, Nr RHO 15.

Symposium on Remote Sensing for Resources Development and Environmental Management / Enschede / August 1986

Present state, changes and quality of Sologne and Brenne, two French large wetlands, studied with the MSS and TM Landsat data

Michel Lenco
Ministry of Environment, Quality of Life Delegation, Neuilly s/Seine, France

Jean-Pierre Dedieu
Company SFERES-Télédétection, Montrouge, France

ABSTRACT : This work was undertaken to give original informations to local administrative and scientific authorities as an assistance to decision. At first, from Landsat TM and MSS data, the study yields cartographical (at 1/50.000 scale) and numerical informations in 23 classes on the present state and the late 1975/84 changes of the watershed land cover of two french large wetlands. Secondly, Landsat TM data processing in supervised classification have afforded mapped informations about depth, aquatic vegetation, turbidity and outer temperature of the ponds in ascertaining thresholds in channels 5, 1, 2, 3 and 6.

RESUME : Ce travail a été entrepris pour fournir des informations originales aux autorités administratives et scientifiques locales en tant qu'aide à la décision. En premier lieu, à partir des données Landsat TM et MSS, l'étude donne les informations cartographiques (à l'échelle de 1/50.000) et numériques en 23 classes sur l'état actuel et les changements récents 1975/84 de l'occupation biophysique du sol du bassin versant de deux grandes zones humides françaises. En second lieu, des traitements des données Landsat TM en classification supervisée ont apporté des informations cartographiques sur la profondeur, la végétation aquatique, la turbidité et la température de surface des étangs en procédant à des seuillages dans les bandes spectrales 5, 1, 2, 3 et 6.

1. INTRODUCTION

The aims of the study carried out in 1985 was to know the present state and the late changes of ecozones and land cover of the watershed and the water areas of Sologne and Brenne, which are two large wetlands of 130,000 and 145,000 ha extent respectively, situated in the center of France below the Loire river. There are several schemas both to bring out and protect these two natural media where the changes were important during the last ten years owing to hunting and fishing or maize cultivation development, and cattle breeding withdrawal. The remote sensing seemed a good tool to give cartographical, quantitative and qualitative data on these humid areas as an aid to the decision. The study was managed in collaboration with local administrative and scientific authorities.

2. METHOD

Two Landsat data recordings were used, a 7/27/1975 MSS tape and 6/17/1984 TM tape. An improved and restored false color image was elaborated for each date at 1/50.000 scale. After restoration, the ground resolution (pixel) was 50 x 50m in the first case, and 30 30m in the second one. The three wavelenghts choosen to establish the two restored images were 4, 5 and 7 for Landsat MSS, and 2, 4 and 5 for Landsat TM.

The second image was interpretated to study the land cover of the two watersheds with the support of topographical, geological and potential vegetation maps and a sampling of photographies drawn out of the last national coverage at 1/30.000 scale. It was possible to obtain a classification with 23 items (see hereafter results). Landsat TM data appeared interesting to detect easily meadows, cereals, conifers, water surfaces and wetlands with the level of the soil humidity in the classes. Some difficult points needed a control on the ground, during the mapping processing : classification of wet vegetation on humid areas near water surfaces.

The changes since 1975 were distinguished by comparison of interpretated TM data with corresponding MSS data called on a screen in a processing station.

Between the two dates a perceptible increasing of ponds in number and surface and a rise of the drainage to develop the maize cultivation were noticed. The interpretated map was treated on computer to get statistics in hectorage for each class and changes.
The channel 5 of 1984 Landsat TM data gives the possibility to draw out the water surfaces and the wavelengths 1, 2, 3 and 6 allow to study the characteristics of water surfaces (depth, nature of the bottom, turbidity, aquatic vegetation and temperature) in selecting thresholds* in supervised classification with a discriminant analysis, because some ponds are regularly analyzed and watched by local scientists and administrative authorities and were used as ground data. It was possible also to observe on the ground some ponds in june 1985 at the same period of the year that the catch of TM data. In the middle infra red, the channel 7 has brought nothing more than channel 5 to study the water surfaces.

* some ratios

3. RESULTS

3.1 Classifications for mapping at 1/50,000 scale

LAND COVER CLASSIFICATION

code	category
	1- Building areas
11	urban, dense
12	urban, mean and few dense
13	urban, diffuse
14	urban hold (industrial areas...)
15	bare soil, quarry, building yard
	2- Agricultural areas
21	arable land
22	fields in copse (arable land and grassland)
23	wet meadows
24	reaped grassland and dry grassland
25	orchards, vineyards, nurseries
26	draining areas and areas in changing
	3- Natural areas
31	hardwoods (oaks species,...)
32	softwoods (pines)
33	mixed forests
34	cuts
35	reafforestings (in conifers)

36 dry waste lands and heathlands (brooms and heathers)
37 degraded wooded areas (scattered oaks and birch trees on dry heathlands)
4- Wetlands
41 wooded vegetation of water banks (poplars, alder-trees)
42 wet waste lands and moors (reeds, mushes)
43 shrubby vegetation on the pond edges and on marshes (willows)
E water surfaces (rivers, ponds) and associated aquatic vegetation
T peat bogs

WETLANDS CLASSIFICATION (LANDSAT TM DATA)

code biophysical class

1 wooded vegetation of water banks (poplars, alder trees, ash trees)
2 shrubby vegetation of the pond and river banks (willows)
3 floating or immersing herbaceous vegetation of water areas (aquatic weeds, water lilies)
4 pond or reservoir without vegetation in open waters
5 drained pond or in laying out
6 permanent streams

CHARACTERISTICS DATA OF WATER AREAS (LANDSAT TM DATA)

Depth, 8 classes arranged in 3 items

IA 0-1m/1,2m
IB 1m/1,2m-2,0m
IC 2,0m and more

.Turbidity (and ground kind.), 8 classes arranged in 4 items

IIA very limpid pond (sandy ground)
IIB limpid pond (sandy and muddy ground)
IIC turbid pond, rich in organic substances (phytoplankton) (muddy ground)
IID turbid pond, rich in mineral substances (colloids) (clayish ground)

.Temperature (outer layer), 8 classes from 17 to 25°C arranged in 4 items

IIIA <20°C
IIIB 20-21°C
IIIC 22-23°C
IIID 24°C and more

3.2 Statistics on the land cover of Brenne watershed area (see tables 1 and 2)

The main noted facts are an important increasing during the 1975/84 period of draining areas or areas in changing on one hand, and of small ponds on the other hand

Table 1 : Evolution 1975/84 of the Brenne ponds higher than 1 ha

	number	surface (ha)	average surface by pond (ha)
1975	957	8,455	8.8
1984	1,014	9,025	8.9
evolution 1975/84	+ 57	+ 570	+ 0.1
evolution 1975/84 in %	+ 6%	+ 7%	+ 1%

source : Ministry of Environment

Table 2 : Land cover of the Brenne watershed in 1984 and 1975/84 changes (in ha)

1984 state		1975/84 changes	
code	surface (ha)	code	surface (ha)
11	90	26/21	1,550
12	755	26/22	230
13	755	26/24	180
14	125	22/26	310
15	120	24/26	275
s/total	(1,845)	36/26	1,550
21	19,350	42/26	20
22	56,990	E/26	25
23	2,705	31/34	60
24	10,225	32/34	15
25	330	34/35	5
26	2,300	E/43	25
s/total	(91,900)	s/total	(4,245)
31	20,430	22/E	45
32	3,395	24/E	110
33	7,810	26/E	5
34	200	33/E	30
35	5	34/E	10
36	7,120	36/E	205
s/total	(38,960)	42/E	210
41	275	43/E	5
42	5,355	s/total	(620)
43	1,095		
E (1)	10,480	TOTAL	4,865
s/total	(17,205)	(%)	(3,2%)
TOTAL	149,910		

(1) including 9,025 ha in ponds
source : Ministry of Environment

4. CONCLUSION

The study on the Brenne and Sologne wetlands has shown the high interest of the channels of Landsat TM data to know at 1/50,000 scale the state of the large wetlands land cover with details about the vegetation of humid media. It gives the possibility also to discriminate the characteristics of depth, aquatic vegetation, turbidity and temperature of water areas.

The observation of the land cover changes during the 1975/84 period in humid media between Landsat TM and MSS data was possible and did not present too many difficulties.

Symposium on Remote Sensing for Resources Development and Environmental Management / Enschede / August 1986

The quantification of floodplain inundation by the use of LANDSAT and Metric Camera information, Belize, Central America

S.T.Miller
University of Aston, Birmingham, UK

ABSTRACT: Investigations are made into the quantification of a flood event using a single post flood LANDSAT 3 image aided by basic topographic and hydrological information. Comparisons are made between LANDSAT, Metric Camera and ground survey data to assess the contribution that space borne remote sensing can make where limited ground-truth is availible. The results indicate that such images can provide detailed accounts of flood behavior and estimates of flood volumes in circumstances where ground surveys are impracticable.

I. BACKGROUND

The Belize river catchment covers an area of more than 8,000 km^2. in Central America and its floodplain is inundated every few years. Economic dislocation due to the flooding of the road link (the Western highway) between the capital Belmopan and the main seaport of Belize City, is a serious concequence. The floodplain area is largely uninhabited and has few other roads. Access within the area is difficult at all times and impossible at times of flooding. In such circumstances remotely sensed images provide an excellent opportunity for the quantification of flood extents. In this paper the flood of December I979 is investigated.

While the flood distribution is considered throughout the floodplain area, estimates of flood volumes are made only for the part above Davis Bank gauging station, the upper floodplain. In the case of the lower floodplain, comparisons are made with other information sources as to floodwater distribution and the locations of flooding and destruction of the Western highway. The LANDSAT scene 020/48, 29th. December I979 was aquired at least two weeks after the inception of flooding.

2. THE UPPER FLOODPLAIN

2.I. The hydrological data base

In common with many developing countries, hydrological and meteorological records in Belize are short, increasing the problems of flood volume assessment. The first stage of this investigation was to establish a base for return period peak flows that could be used in flood volume calculation.

Several options were considered before the 'Peak-Over-Threshold (P.O.T.) method was selected, for the following reasons(Flood Studies Report I972):

I. It is based on a simple exponential series, the validity of which is easily tested.

2. It is suitable for short periods of records.

3. It is adaptable for seasonal and non-seasonal variations.

4. It is adaptable to conditions where flow peaks do not adhere to the same distribution at all stages.

Five gauging stations, two outside the floodplain and three within it, were checked for conformity to an exponential distribution of high flows. The three floodplain stations were Big Falls Ranch, Bermudan Landing and Davis Bank having 3, 4 and 2 years record respectively, obtained between I968 and I972. It was noted that while the two stations outside the floodplain conformed to a consistant P.O.T. distribution at all stages, the three inside did not. Their distributions were found to have varying exponential gradients.

Since the P.O.T. distributions depend upon the threshhold value (qo) and the exponential gradient (b), the former was set to assure an exponential gradient and distribution of observed peak flows for return periods of five years or more. The number of peaks per year that exceed the threshold (qo) is termed λ. The equations that define return period flows are of the form $Q(T)= qo + b(\ln \lambda . \ln T)$ and are given below in table I for the floodplain stations. (Flood Studies Report I972:I89).

Table I. P.O.T. formulae for floodplain stations

Station	P.O.T. formulae
Big Falls Ranch	$Q(T)= 400 + 26.33(\ln 2 + \ln T)$
Bermudan Landing	$Q(T)= 400 + 20.20(\ln I.5 + \ln T)$
Davis Bank	$Q(T)= 280 + 28.00(\ln I + \ln T)$

where Q(T), the T year flood and qo are in cumecs.

In this way a flood peak return base was identified for later use, incorporating a slow growth in keeping the limited ability of the river channel within the floodplain to conduct flood discharge.

2.2 Synthesis of the I979 flood event

Having determined a suitable hydrological base for the determination of flood peaks, an assessment of the flood volume of the I979 flood event was necessary.

No hydrological records of the event were available. However past hydrological and meteorological records enabled the construction of unit hydrographs. Two previous events occuring in I969 and I97I were used and from their averages unit hydrographs for each station were obtained (Flood Studies Report I972:375-402).

The time period for the description of these unit hydrographs was four days, determined by the nature of the rainfall periods and daily rainfall records. They were converted to one day unit hydrographs for convenient application to the I979 flood (Flood Studies Report I972:397)

The reconstruction of the I979 flood flow hydrographs was completed using rainfall rcords from all available meteorological stations. Areal weighting was applied, as were estimates of probable base flow at the time of the I979 event, consideration of the I969 and I97I flow hydrographs indicated that a I5% increase in the derived hydrograph was appropriate in the case of each station.

Although the practical limitations of such methods are recognised (Flood Studies Report I972:379-38I), these proceedures provide the only flow estimates against which those obtained from LANDSAT imagery can be compared.

An estimate of the return period of the 1979 rainfall was made, as was an estimate of the flow peaks of the 1979 event, as obtained from the unit hydrographs. An interesting conformity was seen and is presented below in table 2.

Table 2. Rainfall and synthesised flow peak return periods

Station	Rainfall	Peak flow
Non-floodplain station I	II years	9.3 years
Non-floodplain station 2	9 years	8.5 years
Big Falls Ranch	9 years	3.7×10^2 years
Bermudan Landing	9 years	9.6×10^8 years
Davis Bank	9 years	9.2×10^4 years

In all cases, rainfall returns approximated to nine years as did the flows of the non-floodplain stations. In contrast, the floodplain stations' return periods were extremely high.

It was clear that by applying a nine years return limit to the hydrographs of the floodplain stations, a good estimate of their true maximum discharges for the 1979 event could be found. The difference between this and the synthesised flood flow would provide estimates of loss from the Belize river to the floodplain. Table 3 below presents these losses and also includes a value of the runoff contribution from the floodplain catchment (Fiddes 1977).

Table 3. Belize river losses to the floodplain

Station	Loss (m^3) to floodplain
Big Falls Ranch	0.83×10^8
Bermudan Landing	1.48×10^8
Davis Bank	2.02×10^8
Local catchment contribution	1.04×10^8
Total Flood Volume	5.37×10^8

While these values cannot be substantiated by other means and are obtained from a series of calculation methods, these methods are based on well tried hydrological principals. From the study of the flow hydrographs of the 1979 event it was apparent that the LANDSAT scene was taken between 14 and 16 days after the start of the flooding which first occured at or around Big Falls Ranch.

2.3. LANDSAT estimates of the 1979 flood

Preliminary investigations of the LANDSAT flood scene showed that slides of computer compatible tapes could provide the most information if projected at a scale of 1:50,000. Ground control point measurements were made to assess the distortional effects of the projections and indicated that while they were indeed present, they were not great. It was not possible to identify these distortional effects in terms of their vertical and horizontal components due to the distribution of the ground control points. The overall relationship of the distortions can be described by:
Map scale = 0.99898 + 0.00283 x Slide scale

From 1:50,000 scale topographic maps, spot heights were transposed onto the base map obtained from the LANDSAT flood scene. Sufficient were present at the flood margins to identify a maximum flood level of between six and seven metres above sea level. The overall extent was estimated as 265.6 square km. Not all the flood extent was observable due to the dense vegetation cover and the maximum level was extrapolated to identify flooded areas under dense vegetation.

Sufficient spot heights to provide average flood depths, were present only in the basinal areas close to permanent lagoons. This depth was three metres, obtained from cross-sections through the area but the consideration of the wider spread of spot heights indicated an average flood depth of two metres, or possibly less.

In these circumstances, a range of depth/volume estimates were made to provide a comparison with that obtained by the flow hydrograph study. They are given below in table 4.

Table 4. LANDSAT estimated flood volumes

Flood depths (metres)	Flood volumes (m^3)
1.5	3.98×10^8
2.0	5.31×10^8
2.5	6.63×10^8
3.0	7.97×10^8
hydrograph estimate	5.13×10^8

It can be seen that the estimates from both sources are close and while no direct ground measurement of the flood is available, the correlation is clear. It is possible that during the later stages of the flood process, that water could flow back to the Belize river. However, the observation of the floodwater distribution shows it concentrated in the area of the river still in active flood, the area providing 85% of the floodwater. Any return to the river is likely to have been relatively small.

3. THE LOWER FLOODPLAIN

3.1 LANDSAT investigations

The information of flood plain inundation can be summarised, as obtained from the LANDSAT scene, by figure 1. The identification of both source and distribution can be made, as can the areas of its influence upon the Western highway. In the lower floodplain river flows are not recorded, spot height information is very restricted and so flood estimates are not attempted. The distribution of the floodwaters and their effects on the Western highway is the most significant factor under study.

While local catchment and Belize river contributions to flooding have been identified, the lower floodplain is shared with the Sibun river, another source of floodwater. Figure 1 shows the areas of flooding as presented by false colour composite slides. The position of the flooding can be located to about ± 150 metres according to ground resolution and slide dist-

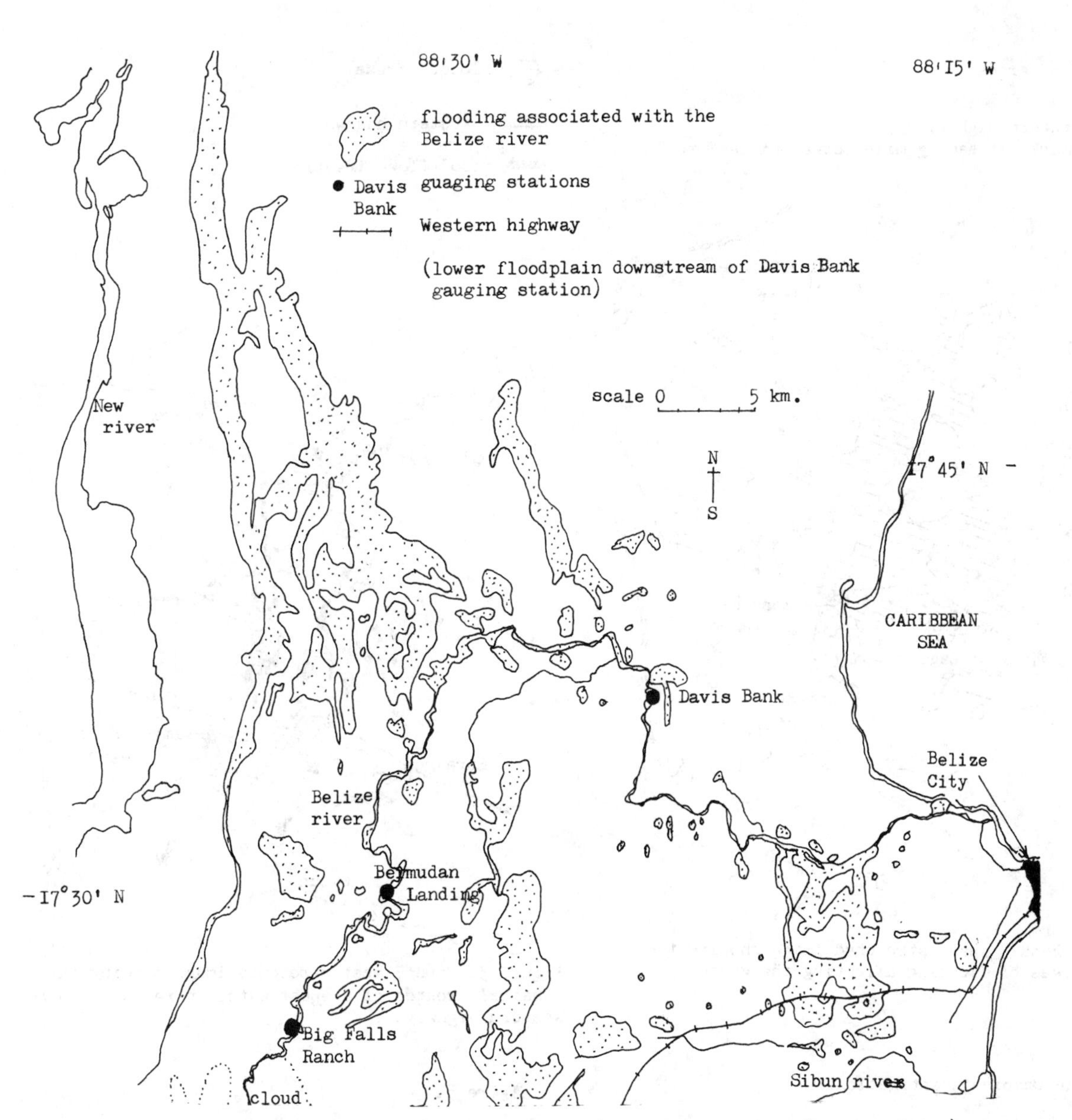

Figure I. Areas of visible flooding. LANDSAT scene 020/48 29th. Dec. 1979 (F.C.C. slides)

ortion constraints. The source of the floodwaters is not easily confirmed, since while that related to the Belize river is clearly identified, dense vegetation around the Sibun river and the dispersion of the local contribution prevent the tracing of all source locations. The area around the Straight and Almond Hill lagoons (approximately miles $9\frac{1}{2}$ to $12\frac{1}{2}$ on the road) is clearly a focal point of inundatery waters that spread south from the Belize river. They appear to spread as a broad front across the floodplain but image processing techniques identified two main channels, focal routes for floodwater passage.

Band 7, undecimated images of the area were examined and manipulated by the process of piece-wise stretching. A histogram of pixel brightness values was obtained and their distribution provided density slice boundaries. A series of stretches of increasing severity were applied to the extent of the erosion of lagoon margins. Two main channel features were seen within the broad front of the floodwaters, linear features about which the movement from the Belize river was concentrated.

Figures 2 and 3 show the least and most extreme stretches, respectively and may be compared to the distributions of figure I and to the more detailed figures 4, 5 and 6. Figure 3 indicates the probable routes taken by the floodwater, though it must be stated that any contribution from the Sibun river is hidden by dense vegetation.

The pattern of these floodwater channels can be seen to have a close correlation to the elevations shown by Metric Camera and ground survey information, most importantly in the area of the Western highway close the Straight and Almond Hill lagoons. The stretches also show that while the floodwaters are located around specific channels on the floodplain, they are derived from a broad margin of the Belize river and that relatively little flooding occurs to the north of the river. Table 5 gives stretch boundaries.

Table 5. Boundary brightness values for stretches

Stretch no.	Original b.v.	Redesigned b.v.
I	0	0
	20	200
	2I	20I
	55	255
2	0	0
	I0	I70
	II	I7I
	55	255

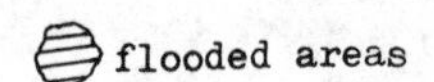

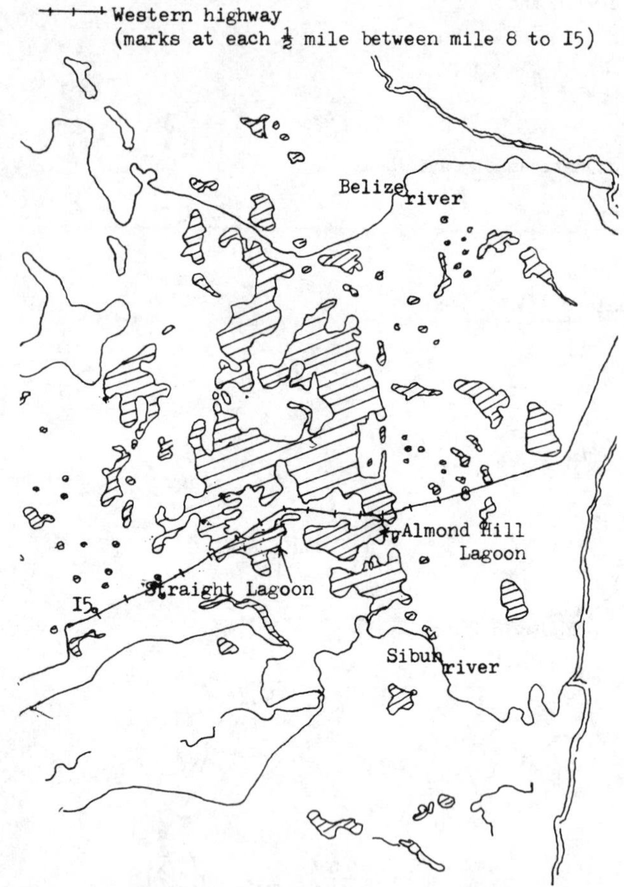

Figure 2. Band 7 least stretched image showing the general areas of flooding affecting the Western highway.

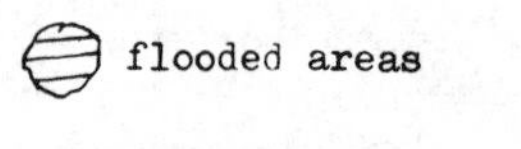

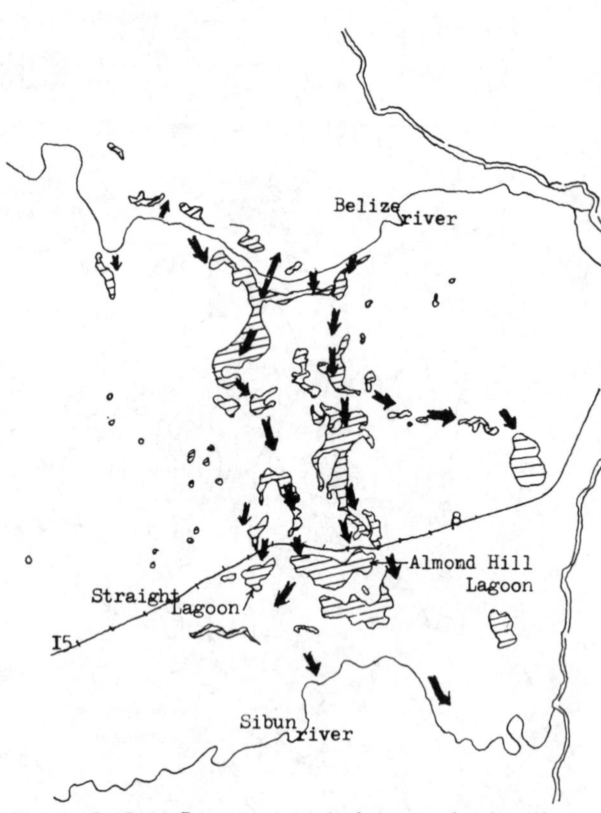

Figure 3. Band 7 most stretched image showing the 'channel' routes of deepest water in relation to the Western highway.

3.2. Metric Camera investigations

Metric Camera imagery of the lower floodplain was obtained from scene 090I with a nominal ground resolution of about 20m. (Schroeder I985). The stereoscopic possibilities of this information source were not suitable for an area of such low relief (Dowman I985) but enlargements at I:250,000 and I:50,000 scale provided durable prints despite some degradation of the image. Distortional effects of the enlargements were present and the relationship of the image to map scales were identified as being in the average form:
Map scale = I.0706 - 7.56 x Image scale
The interest of this imagery is not only to identify its ability in presenting the features of the floodplain but also the extent to which these features may be presented without the advantage of a flood event occurring. The Metric Camera imagery was found to give an accurate identification of vegetation distribution as an adjunct to LANDSAT data, where the spectral signatures of different, mixed land cover types may confuse such identification.

From the observation of Metric Camera imagery at and around the area miles $9\frac{1}{2}$ to $I2\frac{1}{2}$ on the Western highway the locations of relatively high and low land were seen. Differences in vegetation density, the presence of human activity (consistantly confined to the 'pine ridge' higher ground in this region), tracks and other associated features provided indicators. Features less than IO metres in width were seen by contrast with their surroundings.

The disposition of higher features indicated a concentration of floodwater around the Almond Hill and the Straight lagoon sections of the highway as can be seen in figure 5.

3.3. Ground survey investigations

The survey (Richards and Dumbleton I98I) was undertaken to identify the location of most severe flooding on the Western highway and propose remedial measures. It provides precise elevations between miles 0 to I5 on the highway but was less concerned with the internal drainage of the floodplain. The opinion of the report was that "..this section of the road (miles $9\frac{1}{2}$ to $I2\frac{1}{2}$) is believed to be affected by the backwater from the Sibun river.. and overflow from the Belize river" (Richards and Dumbleton I98I:84).

The elevations of the flood levels are not strictly defined by the report and are taken from anecdotal sources and estimated as being 2metres over this part of the highway (Richards and Dumbleton I98I:85).
The consideration of channel-full capacities of the Belize river led the authors of the report to draw the following conclusions:

I. That overspill from the Belize river around Davis Bank would be relatively small.

2. Upstream of Davis Bank, the attenuation of the flood peak by the channel was probably more important than loss to the floodplain, though overspill was not ruled out.

3. No correlation could be found between the three possible sources of floodwater.

4. Flooding appeared to be closely linked to high levels in the Sibun.

Figure 6 illustrates the survey's levelling of the flood prone part of the highway and may be used in

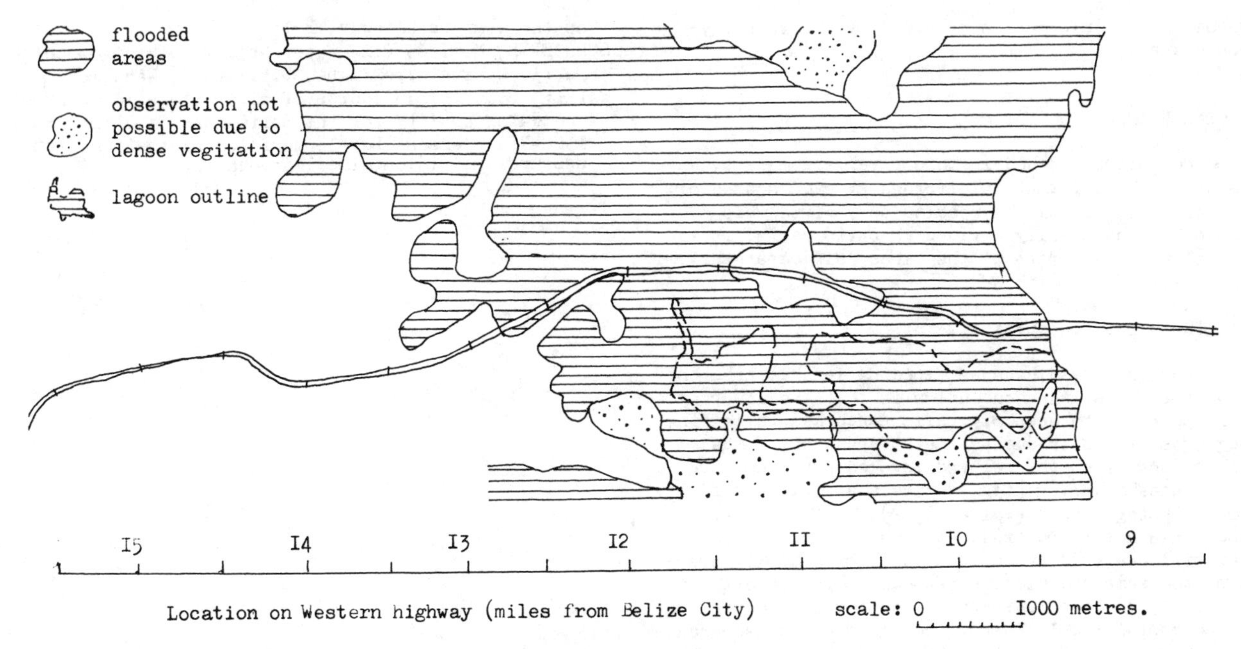

Figure 4. Location of flooding on the Western highway, LANDSAT 3 computer compatible tapes.

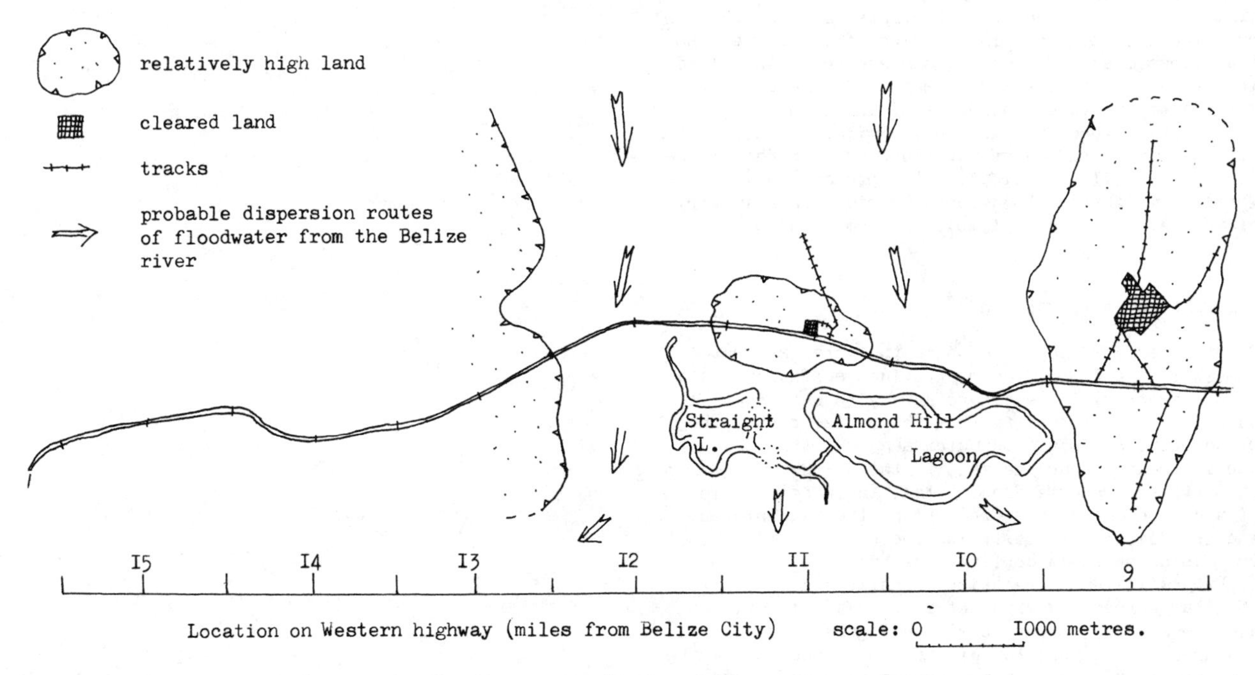

Figure 5. Focal points of probable flooding on the Western highway, Metric Camera.

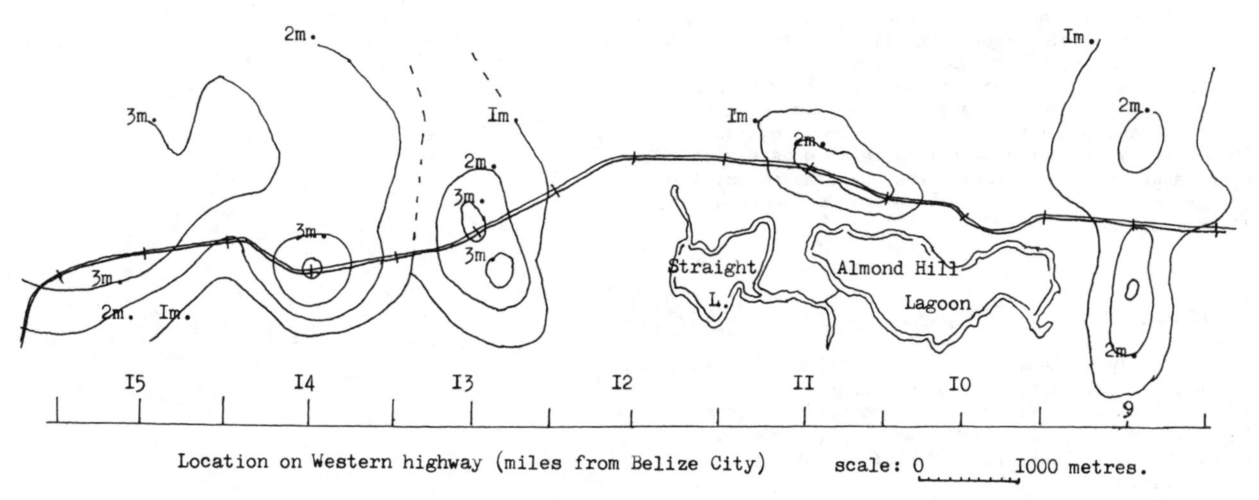

Figure 6. Elevations on the Western highway, ground survey report.

assessing the information provided by the remotely sensed imagery.

4. CONCLUSIONS

The conclusions reached in this paper are made by comparing the information afforded by each source of data and identifying the results they give. This section is necessarily divided into two parts, so as to deal with each part of the floodplain separately.

4.I. The upper floodplain

The ground survey discusses flooding in this area only briefly but its main conclusion is that the lowering of flood peaks between Big Falls Ranch and Davis Bank gauging stations is due to the attenuation of the peak by the river channel and invokes previous work that states "..no overflow channels are visible from aerial photographs." (Walker I972). LANDSAT imagery shows this not to be true. At least 5 locations of overspill are evident and can be seen on map figure I. Using the same calculation methods as the report (Fiddes I977), the contribution of the local catchment can account for only 26% of the visible flooded area - an expansion of more than 88 square kilometres over normal conditions.

The LANDSAT scene not only provides this information but also displays the distribution of floodwaters so that areas most affected and the source of the flood may be seen. With the aid of basic spot height data the imagery provides estimates of the flood volume of the I979 flood that are very similar to those obtained from studies of the flood hydrographs of the period, as well as indicating the probable maximum level to which the flood rose. Metric Camera imagery of this area is not available for comparison.

4.2. The lower floodplain

Quantative flood assessments in this area are not possible. The LANDSAT scene provides sufficient detail in agreement with the ground survey to show that flood water is present, moving from the Belize river. The broad front of this floodwater suggests that the volume is greater than the report states. The scene also supports the report's finding that the concentration of the flooding is centered around the lagoon area and in addition it identifies the channel routes that the flood takes. It confirms that the rest of the highway will remain relatively unaffected.

Similar agreements with Metric Camera imagery can also be seen by comparisons with figure 5. The subtle features of high and low ground can be seen to be the same as identified by the ground survey in figure 6, despite their differences of elevation - only two to three metres, the landform causing the bifurcation of floodwaters being approximately ½ square km.

It is evident that with multitemporal imagery and more sources of supplimentary data remotely sensed information can provide detailed insights to flood processes and provide estimates of flood volumes. This imagery gives its best results when used with other sources of information but where necessary it may be used by itself to provide unique insights into flood processes.

REFERENCES

DOWMAN, I. I985. Report on the Metric Camera workshop II-I3 February I985. Symposium paper.

FIDDES, D. I977. U.K. Transport and Road Research Laboratory. Supplimentary report 259, paper 5.

FLOOD STUDIES REPORT, I972, U.K. Institute of Hydrology, H.M.S.O.

RICHARDS and DUMBLETON INTERNATIONAL, I98I. Report on the Western highway, Belize.

SCHROEDER, M. I985 Flight performance of the Spacelab Metric Camera experiment. D.F.V.L.R., Wessling.

WALKER, S.H. I972. Records of river stage, discharge and water quality for the Belize and Sibun rivers I968-7I. Overseas Development Administration, Land Resources Division miscellaneous report I26.

Symposium on Remote Sensing for Resources Development and Environmental Management / Enschede / August 1986

Remote sensing as a tool for assessing environmental effects of hydroelectric development in a remote river basin

W.Murray Paterson & Stewart K.Sears
Ontario Hydro, Toronto, Canada

ABSTRACT: The development of new hydroelectric generation projects in Ontario (Canada) requires environmental studies to be carried out at various stages in the project life cycle. The feasibility of using remotely-sensed LANDSAT satellite data to assist in these studies is assessed based on a pilot project carried out in a remote northern Ontario river basin. Results suggest that remote sensing technology offers a potentially effective and economical means of collecting, interpreting and presenting environmental information for studies related to broad level river basin planning, conceptual assessments, project scoping, impact assessment, and post-development project follow-up and monitoring.

1 INTRODUCTION

In Ontario (Canada), existing legislation governing the planning and development of new hydroelectric projects requires environmental studies to be carried out at various stages in a project's life cycle. In the planning and design phase, studies are undertaken to assess the acceptability of selected development schemes and to gain approval under Ontario's Environmental Assessment Act. During the project construction and operation stages, studies are conducted to monitor construction activities, to verify predicted effects, to check the effectiveness of implemented mitigation measures, and to confirm the operating integrity of the facility. Completion of these studies requires the collection and analysis of large volumes of data over an extended period of time. The cost and effectiveness of such studies are strongly influenced by sampling methods used, and by the quality and vintage of available data.

Much of the remaining, undeveloped hydroelectric potential in Ontario is located in remote, northern regions of the province. Baseline data acquisition and long-term monitoring in these areas can be both time-consuming and expensive. Therefore, cost-effective means of collecting and maintaining "suitable data" are highly desirable.

Satellite remote sensing technology has been recommended as a useful and cost-effective tool for performing environmental studies (Ross and Singhroy 1983). In order to gain a generic assessment of the value of remote sensing, and to demonstrate ways the technology could augment or improve future environmental studies for hydroelectric projects, a pilot project was conducted by Ontario Hydro in 1985, in conjunction with the Ontario Centre for Remote Sensing (OCRS). The project involved mapping a remote study area - the 15,200 km^2 Little Jackfish River (LJR) basin - according to a variety of cover types, using multi-date LANDSAT MSS satellite data. A series of secondary tests were also run on a sub-area of the drainage basin to further evaluate the data extraction and interpretative capabilities of the technology.

1.1 Study Area

The LJR basin is located north of Lake Nipigon in Ontario (Canada) at about Latitude 50° 23'N,

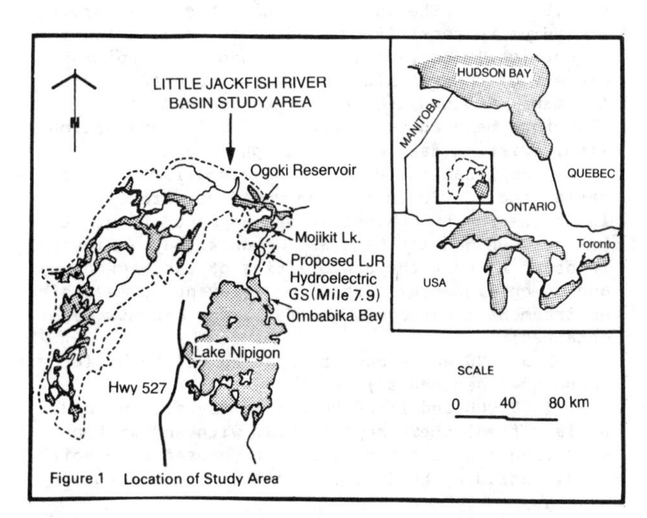

Figure 1 Location of Study Area

Longitude 88° 18'W (Figure 1). The basin is physiographically quite uniform and is typical of boreal forest environments on the Canadian Shield.

Ontario Hydro is proposing construction of hydroelectric generating facilities on the LJR to develop about 132 MW of available hydraulic potential. A tentative dam site has been identified at approximately 12 km upstream of the river mouth. Studies have now entered the detailed environmental assessment (EA) phase and station in-service is expected in 1993. Results from this remote sensing pilot project proved to be very timely for use in the current EA studies on the LJR. Their direct use by the LJR project team provided an opportunity for some hands-on evaluation of the feasibility of using remote sensing to assist in project planning and development work.

2 METHOD

Thematic maps of generalized land cover types were produced for the LJR at OCRS basin using digital analysis and visual interpretation of LANDSAT satellite data. The maps were produced using a standard supervised multispectral classification technique (Kalensky et al 1981).

A key component in the OCRS process is its Applicon printing system which has the capability to print high quality black and white or colour maps at a wide variety of scales. The Applicon system and supplementary software provide a rapid and economical method of producing hard-copy land use maps. Colour separation masters for map publication can be produced within one hour. The inherent flexibility and relative economy of the printing system provided a chance to experiment with a large range of map formats and scales in the course of the study. The Applicon system allows the operator to interact with data to produce generalized or customized hard-copy maps. The ability of the system to isolate and map only those portions of an image within the river basin was a highly desirable feature. Opportunities to superimpose digitized information in both raster and vector format on a classified image (e.g., administrative and reservoir boundaries) also proved to be an effective tool.

2.1 Mapping results

The output of the LJR pilot project included eleven hard-copy map sets, at scales ranging from 1:50,000 to 1:500,000. The entire drainage basin was mapped according to generalized land cover types using a geometrically corrected ground resolution (picture element or 'pixel' size) of 50 m x 50 m (0.5 ha). All maps produced are geo-referenced and can therefore be used to complement, or in conjunction with, existing data and topographic maps.

Test applications were run on a smaller part of the basin (about 1000 km^2) which would be directly influenced by hydroelectric development, (i.e., the LJR between Mojikit Lake and Ombabika Bay - Figure 1) in order to test the capabilities of the technology, and to obtain more information relevant to assessing environmental effects. The following outputs were obtained:

1. 1:50,000 scale map of Ombabika Bay turbidity (suspended sediments);
2. 1:50,000 and 1:100,000 scale classified (and unclassified) theme maps of LJR, with and without elevation contours and flooded (proposed reservoir) area, including hard copies and one transparent overlay;
3. 1:50,000 scale black and white map of LJR only, with one theme (deciduous forest) highlighted in colour;
4. 1:50,000 scale map showing only forest areas, combined and segregated into themes; and
5. 1:50,000 scale map showing correlated land cover types and potential moose habitat over a black and white background.

2.2 Costs and accuracy

The project was evaluated in terms of its overall costs based only on what it cost to produce generalized land cover maps of the entire drainage basin at 1:250,000 and 1:100,000 scales (e.g., not to perform the test applications). Based on a total study area of about 15,000 km^2 (excluding Ombabika Bay), the approximate total cost to produce the maps was \$40,000 or \$2.67 per km^2 (CDN). Sears (1985) compared costs of other selected LANDSAT mapping projects and found that typical operational costs can be in the \$1.50 to \$2.50 km^2 range. Costs associated with conventional data collection methods (e.g., ground surveys and air photos) are significantly higher, and can range from \$8 to \$44/km^2 (Illinois EPA 1978) (Still and Shih 1985).

A detailed assessment of the accuracy of the resulting classified maps (e.g., ground-truthing) has not yet been undertaken due to budget restrictions, and due to the remoteness of the study area. Based on previous experience with this type of mapping, OCRS has estimated an accuracy of 80-95 percent for most cover types for the LJR project (Pala et al 1981). Detailed field investigations to be conducted to support current planning and development studies on the LJR should provide opportunities to better evaluate the accuracy of data derived in this pilot project.

3 APPLICATION TO PLANNING AND DEVELOPMENT STUDIES

Based on results of the LJR pilot project, the following sections discuss how remote sensing has or could potentially be utilized to support or complement planning and development studies for future hydroelectric projects (Figure 2). Potential problems or limitations related to certain applications are also identified.

3.1 River system planning

A major difficulty in carrying out broad-based assessments of large study areas, particularly in northern remote regions of Ontario, is obtaining complete and consistent data coverage for the area in question. The collection and management of baseline environmental and resource use information in Ontario is often highly fragmented and based primarily on administrative districts established by regulatory authorities. The scope, vintage and format of available data can vary substantially among districts. Data quality is often a function of administrative policy (e.g., district land use priorities) and/or budget-time constraints within a district. Some regional and provincial data bases do exist for certain land and resource use parameters in Ontario, but these tend to be relatively dated, one-time inventories (e.g., Forest Resource Inventory done in the 1940's).

The land cover mapping produced for the LJR is well-suited to broad river system level studies. The maps provide useful, up-to-date, generalized information on primary land uses, vegetative cover types, wetlands, forest cutovers and burns, and land-water ratios. These maps are useful in providing the planner with a feel for the "context" within which development will take place. Tests conducted during the pilot project suggest that these primary land cover data can be potentially extended to provide considerably more detail with regard to certain other resource uses within a river basin (e.g., wildlife habitat). A 1:250,000 scale was found to be the most appropriate format for mapping information for the LJR basin. The appropriateness of scale will be a function of river basin size.

Application of remote sensing represents an acceptable intermediate level of detail and cost between the extremes of "quick and dirty" evaluations, which try to piece together fragmented and diverse data sets, and detailed land use mapping which often involves expensive and time-consuming field data collection and air photo interpretation. A key benefit of developing information from remote sensing is that the data base derived becomes a dynamic entity, that can be manipulated to assess and highlight certain attributes; and that can be readily updated in the future. The ability to input digitized information (e.g., roads, park boundaries) to a classified image adds important context to the generalized maps produced.

An added bonus with the remote sensing based system is the availability of statistical summaries for the land cover attributes classified. Conventional data

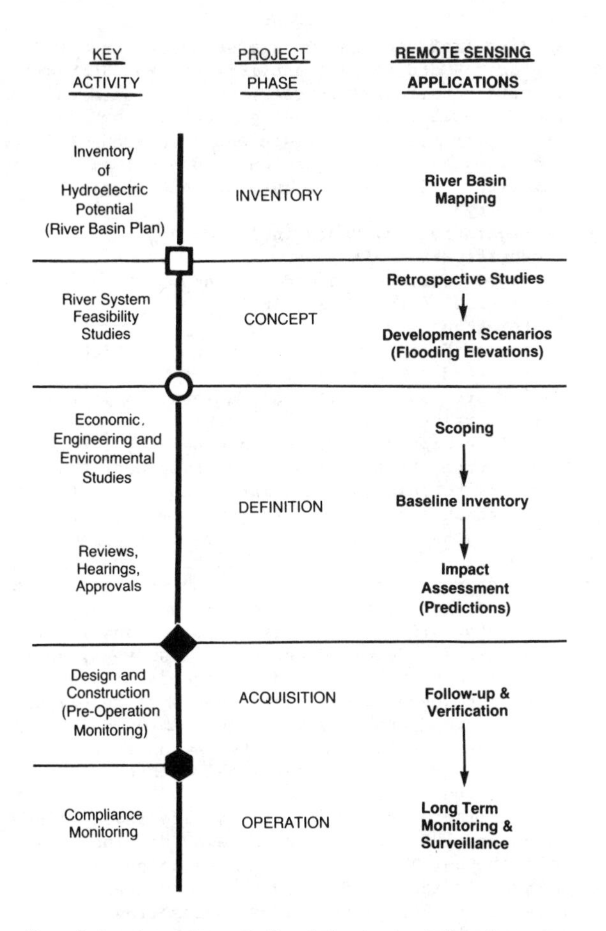

Figure 2 - Remote sensing applications in the planning and development process for hydroelectric generation in Ontario

collection techniques often tend to be "one-shot" efforts that remain static in time, and require extensive effort and expense to update. Recent successes in merging geographic information systems (GIS) and remote sensing based digital image analysis (DIA) systems are likely to vastly improve the utility of remotely sensed data for broad level planning applications in the future.

Based on the success achieved in the LJR mapping exercise, a second project has been undertaken with OCRS to provide land cover mapping for the Moose River basin (100,000 km^2) in northeastern Ontario. Mapping here will be used to improve broad level river basin planning, and as input to detailed hydroelectric project studies currently underway in a portion of the basin.

3.2 Conceptual assessments

Conceptual assessments are usually low budget, iterative planning exercises to determine if there are any major obstacles constraining future hydroelectric development in a given river basin. These studies typically look at a number of alternative concepts, and recommend preferred projects for more detailed engineering and environmental studies (in a Definition Phase). Environmental studies at this stage are largely qualitative in nature, and rely heavily on existing, readily accessible data, with a minimal provision for field reconnaissance. Broad categorization of cover type and land use is usually adequate for conceptual assessment purposes. LANDSAT imagery provides a convenient and readily updatable source of this type of information, which is well-suited to the conceptual level of detail. In remote northern areas particularly, it may be the only source of comprehensive data for an entire study area. LANDSAT imagery provides essentially a uniform data base for a river basin which is usually, or can be made to be, temporally consistent.

An important part of conceptual level assessments is a requirement to evaluate a large number of alternative development schemes. An analysis of alternative dam sites, as well as variations in flooding elevations at these sites, is particularly critical. While topographical contour information cannot be directly interpreted from LANDSAT imagery, the ability of the OCRS system to utilize digitized information makes it possible to manually input a series of flooding elevations, and derive statistical summaries of displaced vegetation and other resource uses for a large number of alternatives within a basin. Digitization can also be used to overlap, certain other spatial information (e.g., provincial park boundaries, sensitive areas) on a classified image to improve the data base available for feasibility assessment. It should be noted that manual digitization is time-consuming. Tests conducted during the pilot study suggest that time commitments may be a real controlling factor in determining the optimum number of alternatives that can be realistically examined at the conceptual phase. Evolving techniques that permit video camera input of topographic data (to a combined GIS and DIA system) should significantly simplify this process, and provide increased analytical capabilities at a relatively low cost.

In addition to potential environmental applications, generalized land cover mapping may have some value for preliminary techno-engineering studies at the conceptual level. For example, the availability of construction (e.g., aggregate) materials is a key consideration in judging the viability of future development. Maps produced for the LJR provide information on the location of active gravel pits within the basin, and suggest that the availability of readily accessible construction materials is somewhat limited. Some measure of aggregate potential may also be possible using remote sensing imagery (Ross and Singhroy 1983).

Another technical aspect that can be addressed is reservoir clearing. Vegetative and other resource use loss data for various reservoir areas in the basin can be used to broadly estimate clearing costs, major conflicting uses and potential compensation costs for single headponds or a series of reservoirs. The type and quality (e.g., merchantible vs non-merchantible) of forest vegetation in a reservoir can also be generally assessed; and this might influence clearing strategies for one reservoir vs another. While further detailed studies will be required to confirm a preferred reservoir clearing strategy, information provided by LANDSAT imagery appears quite suitable for determining the preliminary feasibility of alternative approaches.

3.3 Project scoping

An important part of early project planning for detailed EA studies (in the Definition Phase) is the definition of study area boundaries and the identification of critical or publicly sensitive environmental and other parameters within the study area. EA practitioners and regulators refer to this early planning phase as scoping. Scoping is viewed as a cost-effective method of focussing baseline field studies on critical components that are likely to be impacted by a proposed development.

Scoping workshops to consider natural environmental effects and to define purpose and scope of environmental field programs for the proposed LJR development were held coincident with the completion of this pilot project. Representatives from government and Ontario Hydro participated in the development and evaluation of a series of hypotheses of effect regarding potential impacts of project development on fisheries, water quality, water use and wildlife. The generalized satellite maps produced for the LJR were used during these workshops, and proved particularly useful in identifying areas with potentially valuable wildlife (moose) habitat. Mapping of sediment concentrations in Ombabika Bay also proved useful in assessing potential downstream sedimentation problems related to the project (Figure 3).

Government representatives at the scoping workshops were impressed with the format and quality of the satellite-derived maps. Similar baseline mapping has been recommended to support future scoping exercises with both the public and government.

3.4 Baseline inventories for EA studies

Ground access to the LJR basin is only available in the lower reaches of the river via a railway right of way, and to the lower Ogoki River using a forest access road. Therefore, float planes and helicopters must be used to carry out field investigations within the basin. This type of access restriction can complicate data collection and contribute significantly to the cost of baseline studies. Certain information available from LANDSAT imagery was assessed in terms of its suitability for documenting baseline conditions and reducing the need for ground-based surveys.

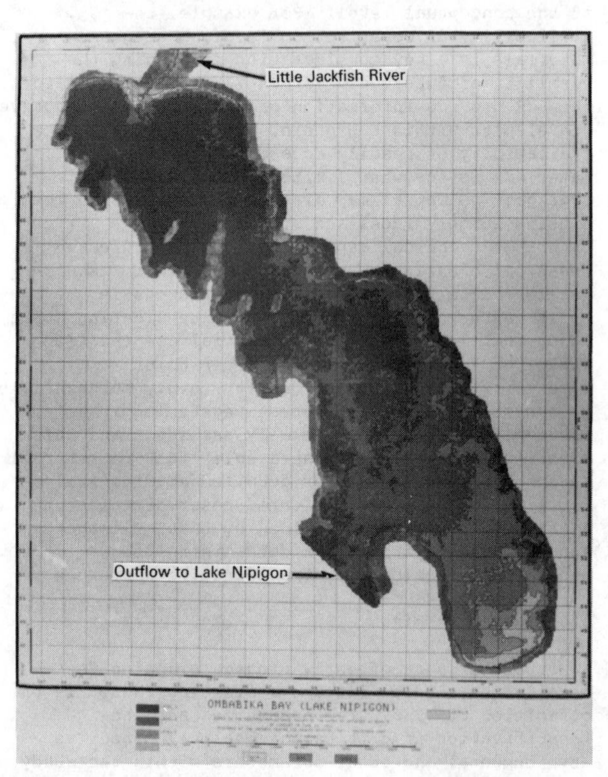

Figure 3 - Suspended sediments levels in Ombabika Bay (Lake Nipigon) based on water reflectance values recorded by band 5 of LANDSAT on June 29, 1984

Baseline field studies were previously conducted for Ontario Hydro in 1981 and 1982 in the LJR basin to support detailed EA studies, but were deferred due to reduced system need. The terrestrial component of these earlier field investigations cost about $50,000 and consisted of studies designed to provide:

- a history of forest disturbance and logging activity*
- a vegetation inventory including an identification of:
  - major tree species and wetlands*
  - distinct forest communities*
  - forest understory species
  - forest floor layer composition and thickness
  - rare and endangered flora
- a forest resource value assessment (habitat and timber value)*
- wildlife population census and density informatio

Results from the LJR pilot project suggest that much of the terrestrial information collected (and marked with an asterisk*) during the 1981 surveys could likely have been acquired more cost-effectively using remote sensing. Costs to perform the analysis and produce maps for this information in the LJR test area (downstream of the Ogoki division - Figure 1) would be about $6000 (Personal communication 1986). Clearly, certain detailed terrestrial surveys, such as a wildlife population census, require ground-based inventory; however, the effectiveness of these types of studies could probably have been improved using potential habitat assessment information that can be acquired quite readily from remotely-sensed vegetation and wetland data.

In establishing the scope of baseline inventories (in 1986 and 1987) to support renewed development studies on the LJR, the area to be studied was significantly increased (from the study area considered in 1981) such that some baseline information is now required throughout most of the LJR basin (Figure 1). Using remote sensing, the spatial coverage and update potential of acquired baseline information is increased significantly. Changes in study area size can be effectively accommodated with minimal increase in study time or cost. Also, a better "context" is provided for the data collected, since comparable baseline information is readily available for an entire river basin instead of just a small study area potentially influenced by a project.

Most baseline studies conducted for EA purposes extend over only a one-year period prior to EA review and approval. In such a short time frame, it is often difficult to distinguish any meaningful trends in many of the environmental parameters surveyed. In most cases, these trends will only be identifiable through more frequent sampling or by extending surveys over a series of years. Such routine monitoring is usually carried out by regulatory agencies responsible for environmental and resource management. In the absence of long-term, routine monitoring records, retrospective and repetitive surveys using LANDSAT imagery may provide an important means of improving the context and interpretative capabilities of some baseline inventory components.

3.5 Impact assessment and prediction of effects

A key activity in the planning and design phase is the development of an environmental assessment (EA) which predicts and evaluates the effects associated with a proposed development project. In Ontario, formal government review and approval is required at this stage before construction can be initiated.

In an EA for a hydroelectric facility, there are three general areas of impact that must be addressed when predicting and assessing potential construction and operation effects:

- reservoir effects (flooding and dam as a barrier)
- downstream effects (operating regime) and
- socio-economic effects (direct and indirect)

These effects are typically analyzed in terms of their spatial extent, duration, magnitude, significance and irreversibility. Given results of the LJR pilot project, it is felt that the accuracy and range of effect predictions in an EA can be enhanced using remotely-sensed data. Use of these data appears most useful in:

- estimating displacement of cover types/land or resource uses as a result of project development (e.g., flooded areas);
- putting this displacement in the context of a larger study area or region (e.g., percent of total habitat lost);
- performing retrospective studies (where past imagery exists) to establish historical trends of land use or cover type succession that may tend to complicate, accent or mask project-induced effects (e.g., vegetation damage, habitat changes).

The spatial resolution achievable using existing LANDSAT MSS imagery (0.5 ha) tends to be somewhat limiting in terms of its ability to examine and assess certain detailed project impacts. Improved resolution (about 80% better) provided by LANDSAT TM sensors and the new generation of satellites (e.g., SPOT) should progressively improve the utility of satellite imagery for detailed project assessment purposes.

The capability of the OCRS system to accept digitized data (e.g., reservoir limits) and to produce maps with rapid turnaround time, allows the EA analyst the freedom to investigate numerous alternative project flooding scenarios and derive comparative statistical summaries of displaced vegetation and other resource uses. Estimates of reservoir depths or volumes may also be possible (Hathout 1985). Maps produced at a 1:50,000 scale were found to be most suitable for analyzing LJR project-level impacts using LANDSAT MSS data.

In addition to providing quantitative information regarding the displacement of primary vegetative cover and land uses, LANDSAT data can be utilized to analyze impacts on certain resource uses. For example, some researchers have utilized satellite imagery to map potential wildlife habitat (Lunetta et al 1985). Habitat evaluation procedures are being advocated as a preferred method of assessing and quantifying development effects on wildlife. As part of the LJR pilot project, attempts were made to use generalized land cover maps for the basin to map potential moose habitat - moose being the most abundant and economically important large game species in the study area. Critical moose habitat requirements were determined from the literature (Cairns et al 1980) and used to establish correlations between vegetative cover and habitat potential. Although the satellite-derived habitat map provides a general picture of where potential habitat exists, it gives little information on the likelihood of moose being present in any particular area, or on the actual habitat suitability for moose. Detailed ground-based studies need to be integrated with these overview type assessments to provide a definitive indication of moose habitat quality. The mapped information developed in the pilot project can, however, be input to Habitat Suitability Index models which quantify the capacity of a given habitat to support moose (US Fish and Wildlife Service 1981).

Enhanced LANDSAT data have been used to examine near-surface conditions in the aquatic environment (Hathout 1985) (Hecky and McCullough 1984), including: suspended or floating vegetation (chlorophyl, algae), ice and snow cover, and suspended sediments. Due to the existence of highly erodable banks downstream of the proposed LJR dam site, potential changes in sedimentation patterns (particularly in Ombabika Bay in Lake Nipigon) are a concern for the LJR project. A previous study (OCRS 1977) established the viability of using LANDSAT imagery to study sedimentation in the LJR basin. Seasonal and spatial variations in sediment patterns within Ombabika Bay were qualitatively assessed. A test application conducted during this pilot project used LANDSAT imagery to provide a qualitative mapping of suspended sediment patterns in Ombabika Bay. Quantitative estimates of turbidity will require ground-based measurements of suspended sediment concentrations to be taken coincident with a LANDSAT overpass. The methods for establishing these field-satellite correlations are well-documented elsewhere (Hecky and McCullough 1984). Provisions have been made in the proposed 1986 LJR field study program to acquire water-based sampling of suspended sediment concentrations for selected future LANDSAT passes over the LJR basin.

3.6 Project follow-up and monitoring

In Canada, regulatory agencies are putting increasing emphasis on the need to follow-up project development with monitoring studies. Monitoring objectives include: demonstration that EA commitments are met, verification of models and predictions developed during the detailed impact assessment stage, demonstration of facility compliance with regulatory standards, and confirmation of the effectiveness of applied mitigative measures. Project follow-up can also be useful in identifying unforeseen effects resulting from project development, and in allowing suitable remedial actions to be implemented in a timely fashion.

The routine, repetitive nature of data acquisition provided by satellite imagery is ideally suited to temporal studies of land use change as well as changes in other environmental parameters. LANDSAT imagery can potentially be used to monitor a number of project activities and effects for hydroelectric stations as project implementation moves through the construction and operation phase.

In the short-term, LANDSAT imagery can be used to monitor construction progress and related activities. For example, downstream sediment patterns are likely to change significantly following reservoir inundation and these changes can be detected using satellite data. In addition to monitoring effects, it may also be possible to utilize LANDSAT imagery to judge the effectiveness of certain mitigative measures. For instance, project planning for the LJR development suggests that extensive rip rap protection may be required to stabilize banks in the lower reaches of the river to alleviate potential downstream erosion and sedimentation problems. Routine monitoring of sedimentation patterns in Ombabika Bay via satellite imagery can be carried out over several seasons or years, and used in conjunction with hydraulic models, to evaluate the effectiveness of bank protection and assess the need for operational controls. Retrospective studies which examine historical turbidity patterns can put project-induced effects in context.

Studies suggest that many of the effects associated with hydroelectric development take years to fully evolve, and that it may take decades for affected environments to stabilize (Hecky and McCullough 1984). This is particularly true in northern boreal environments like the LJR basin. Therefore, it may be necessary to carry out monitoring over an extended period of time to fully understand and evaluate the magnitude of project-induced changes. Field based studies are usually only carried out for a few years prior to and following project implementation. Long-term studies are extremely expensive. LANDSAT data have been used successfully to monitor the recovery and stabilization of sediment conditions in reservoirs in northern Manitoba over a seven-year period (Chagarlamudi et al 1979).

In addition to monitoring to assess system recovery, long-term satellite-aided studies may be useful in examining certain other aspects of reservoir dynamics. Better understanding of these dynamics may be important for reservoir management purposes. For example, the timing and sequence of ice cover break-up can be monitored (Hecky and McCullough 1984). Evaporation from reservoir surfaces has been successfully estimated using satellite thermal infrared data (Miller and Rango 1985). Interactions between reservoirs and adjacent groundwater regimes may also be assessed using remote sensing techniques (Rundquist et al 1985).

4 CONCLUSIONS

Results of the LJR pilot project confirm that remote sensing can be used effectively and economically to complement environmental studies conducted throughout the project life cycle for new hydroelectric generating stations. Applications show greatest potential benefits in remote northern areas where existing environmental data are often scarce, outdated, fragmented or difficult to obtain.

The dynamic, generalized land cover mapping produced for the LJR basin appears ideally suited to broad river system level planning exercises and conceptual assessments. Satellite data are also useful in scoping and conducting baseline inventories for detailed project assessments, particularly on large study areas where ground access is limited. Vegetative cover and derived wildlife habitat mapping proved extremely valuable for scoping terrestrial studies for the LJR hydroelectric project. While scoping and baseline inventory capabilities are strongest for terrestrial environmental studies, applications for studying near-surface aquatic environmental conditions are expanding.

Prediction and monitoring of project-induced effects can be aided using remote sensing techniques. Reservoir displacement effects can be quantitatively estimated for a range of cover types and resource uses. Certain changes in the aquatic environment (e.g., turbidity) can also be effectively assessed. Monitoring capabilities would appear to be most beneficial in carrying out extended surveillance, designed to track long-term project effects and study area stabilization times. The spatial resolution offered by LANDSAT MSS data makes it difficult to assess very specific project-related effects. It should be stressed that remote sensing-based studies do not displace the need for ground-based environmental studies, but should be used in a complementary way to improve the effectiveness of overall study programs at the detailed project assessment phase. The increased resolution provided by the new generation of satellite sensors (e.g, TM, SPOT), and the further integration of GIS and DIA systems, will only serve to improve the applicability of remote sensing for impact assessment purposes.

REFERENCES

Cairns, A.L. and E.S. Telfer 1980. Habitat use by four sympatric ungulates in boreal mixedwood forest, J. Wildlife Management 44(4):849-857.

Chagarlamudi, P., R.E. Hecky and J.S. Schubert 1980. Quantitative monitoring of sediment levels in freshwater lakes from LANDSAT, p 115-118. In V.V. Salomonson, and P.D. Bhavsar (ed.). The contribution of space observations to water resources management. Pergamon Press, New York.

Hathout, S. 1985. The use of enhanced LANDSAT imagery for mapping lake depth. Journal of Environmental Management. 20:253-261.

Hecky, R.E. and G.K. McCullough 1984. The LANDSAT imagery of Southern Indian Lake: A remote perspective on impoundment and diversion, Canadian Technical Report of Fisheries and Aquatic Sciences No. 1266. Winnipeg, Manitoba.

Illinois Environmental Protection Agency 1978. A land cover inventory from space. Staff report.

Kalensky, ZD., W.C. Moore, G.A. Campbell, D.A. Wilson and A.J. Scott 1981. Summary forest resource data from LANDSAT images Canadian Forestry Service, Information Report PI-X-5.

Lunetta, R.S., R.G. Congalton, A.M.B. Rekas, and J.K. Stoll 1985. Using remotely sensed data to map vegetative cover for habitat evaluation in the Saginaw River basin, Presentation to American Society of Photogrammetry.

Miller, W. and A. Rango 1985. Lake evaporation studies using satellite thermal infrared data. Water Resources Bulletin. 21:1029-1036.

Ontario Centre for Remote Sensing (OCRS) 1977. The impact of the Ogoki diversion or the erosion of the Little Jackfish River and on the turbidity of Ombabika Bay. Prepared for Ministry of Treasury, Economics & Intergovernmental Affairs.

Pala, S., T.J. Ellis and D.B. White 1981. Operational land cover type mapping in Ontario by LANDSAT based digital analysis and map production. Presented at 7th Canadian Symposium on Remote Sensing, Winnipeg, Manitoba.

Personal communication 1986. S.K. Sears (Ontario Hydro) with S. Pala (OCRS).

Ross, D.I. and V. Singhroy 1983. The application of remote sensing technology to the environmental assessment process, Prepared for Workshop on New Directions in Environmental Assessment: The Canadian Experience. Toronto, Ontario.

Rundquist, D., G. Murray and L. Queen 1985. Airborne thermal mapping of a "flow-through" lake in the Nebraska sandhills. Water Resources Bulletin. 21:989-994.

Sears, S.K. 1985. Remote sensing pilot project technical evaluation, Ontario Hydro Environmental Studies and Assessments Department. Report No. 852

Still, D.A. and S.F. Shih 1985. Using LANDSAT data to classify land use for assessing the basin - wide runoff index. Water Resources Bulletin. 21:931-94

US Fish and Wildlife Service (1981). Standards for the development of habitat suitability in index models, Ecological Services Manual 103-ESM.

Symposium on Remote Sensing for Resources Development and Environmental Management / Enschede / August 1986

Environmental assessment for large scale civil engineering projects with data of DTM and remote sensing

Taichi Oshima & Atsushi Rikimaru
Hosei University, Faculty of Engineering, Tokyo, Japan

Youichi Kato & Masaharu Nakamura
Pasco Corporation, Tokyo, Japan

ABSTRACT This study describes in Pre-surveying in what way the authors manege and analyze the numerical data of topography or DTM data and remote sensing data for project management of water resource development in the mountain district.
At first, the conditions such as topography, geology, vegetation and water resources of the projected area are analyzed together with DTM data and remote sensing data. Secondly, using the first step data, the estimation mapping for grade of weatherd granite and degree of landslide danger are analyzed by 'Maximum-likelihood method' and 'Similarity analysis'. Furthermore, the site proposed for re-development around the dam reservoir are selected and listed with the classified grade.

1.SPECIFICATION OF RESEARCH

(1)Using data
- Airborn MSS(M^2S)data
  - Observed date 1983.3.19
  - Flight altitude 2,000m
- LANDSAT MSS data
  - Observed date 1980.10.30
  - Path-121 Row-37
- Digital terrain data
  - Data intervals 25m X 25m
  - Data origin 1:10,000 scale map
  - Data input Digital dramscanner
- Aero photo Black & White
  - Scale 1:12,500 Taken in 1974
- Aero photo Color
  - Scale 1:10,000 Taken in 1981

(2)Methodology
- Heat condition index
- Cluster analysis
- Maximum likelihood method
- Similarity analysis
- Vegitation index
- DTM analysis
- Principal component analysis

(3)Test site
The basins of the Kaze and Jobaru rivers.Those areas are located in Kyushu Island in Japan.

(4)Structure of Research
This reserch consists of 4 units.
a. Basic data obtained
b. Basic analysis
c. Application analysis
d. Assessment
The block diagram of general flow is shown in fig.1 .

2.BASIC DATA OBTENTION AND ANALYSIS

2.1 Airborne MSS data

(1)Data acquisition Airborne MSS data was acquired on March 19,1983 and flight altitude was 2,000m , therefore ground resolution is about 5m at nadir point.This MSS is BENDIX Co.production ,and it has 11 bands,which are from blue-band to near infrared-band and also thermal infrared-band. MSS data was recorded by HDDT onboart with digital data.
(2)Radiometric correction Radiometric data correction is necessary to eliminate image distortion caused by the temporal variation of solar radiadion and the spatial variation of path radiance effect and the viewing angle effect of the scanner. This correction system refers to black body data,skylight data,standerd reference lamp data of the inside of the scanner and also ground-truth gray target spectral data.On the other hand, the thermal infrared ban,and temperature calibration is also necessary, because of atmospheric absorption of thermal radiation energy. This correction system refers to both high and low temperature black bodies,and also ground truth measurement data,which was mesured by a thermal infrared radiation thermometer at the same time of MSS operation.
(3)Geometric Correction and Digital Mosaic Geometric data correction is the basic process for spatial superposition of multi-stage image (Airborne MSS and LANDSAT MSS) and mapping of image onto the DTM data. The schematic description of the correction process, three courses of airborne MSS data were connected to one image with digital mosaic analysis.
(4)Thermal majority analysis image Using the majority process of the 3 X 3pixel filter, and normalization the small fluctuation.
(5)Heat Condition Index Specific heat, heat capacity,and heat conduction, these parameters show the heat condition of the meterials. And these factors are deeply related to thermal image distribution, especially characteristics of thermal continuity of plane space.Then heat condition process was done with the thermal stability model,which uses the standard deviation value of the 3 X 3 space filter.
(6)Vegetation Index Analysis Using the spectral characteristics of chlorophyl-a, the image was processed as near-infrared band/red band.

2.2 LANDSAT DATA

(1)Pre-processing Geometric correction was operated using several pairs of ground control data and resampling software system. The pixel size of result image is 25m X 25m . This size is just compatible to overlay on DTM data.

(2)Vegetation Index Analysis The process and idea are the same as airborne MSS data. The effective bands of LANDSAT are near-IR band 7 / red band 5.

(3)Lineament Analysis The processing method is 3 X 3 digital filter operation. And extracted lineament information is useful for application in topography,geomorphology and geology.

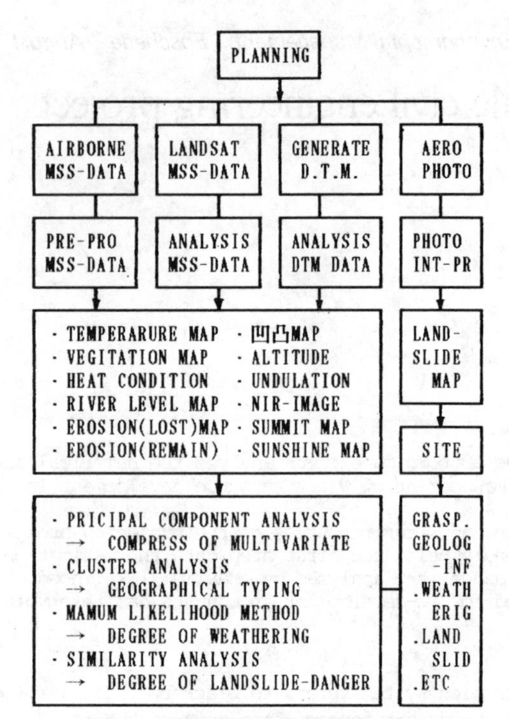

FIG-1 BLOCK DIAGRAM OF FLOW

2.3 Topographycal Analysis

(1)DTM data preparation Digital terrain model or DTM data is converted from a 1:10,000 topographic map . At first a color drum scanner system was used to read map-images and A/D conversion and input data to the computer. Second,image processing made X,Y,Z vector data from a luster image map, and the procedure is line-sharpness, data-normalize, luster/vector conversion, addition of attribute (example;altitude data).Third was arrangement of grid point data at intervals of 25m X 25m.

(2)Subject Map Preparation Using DTM data,several kinds of subject maps were prepared by topographical analysis,as follows.

Altitude contour map
Summit level map
River level map
Difference between summit and river level map
Erosion situation map (Remaining area)
Erosion situation map (Lost area)
Undulation map
Slope direction map
Slope gradient map
Unevenness map

2.4 Photo-Interpretation

The land-slide areas were extracted by photo-interpretation with 1:10,000 color and 1:12,500 black & white aero-photos. The result was plotted on 1:10,000 topographycal map, and classfies the geographical type of each extracted area. The extracted area amounts to 519 points. And the locations of these areas were input to the computer.

2.5 Groud Investigation

On the spot, the land-slide areas were checked, and the grade of weathered granit was observed along the road side slope. Then the weathering grade was classified into three types, A,B,C. A is a perfectly weathered area. B is weathered a little. C is not much wethered. These three types and their locations were input to the computer. The surveyed points amount to 444 areas.

Table-1 Training data for classification of wethering degree of granite

CH	Item	Unit	Class A Perfect-Wetherd mean	st.dv	Class B A little Wetherd mean	st.dv	Class C No-Whetherd mean	st.dv
1	Erosion (lost)	m	45.2	20.5	55.1	22.9	60.8	23.2
2	Erosion (remain)	m	31.4	20.5	34.9	21.2	42.2	30.8
3	Undu- lation	m	6.0	3.0	7.3	3.5	9.1	4.2
4	Slope gradient	tan	0.300	0.152	0.366	0.176	0.446	0.214
5	Uneven. +/convex	m	-0.1	2.7	-0.2	3.2	-0.3	3.7
6	Direct. (+E,-w)	tan	-0.004	0.242	-0.002	0.312	-0.076	0.318
7	Direct. (+N,-S)	tan	-0.006	0.252	0.074	0.274	0.114	0.382
8	LANDSAT PCA(1st)	PCA 1ST	153.8	23.7	156.3	24.5	158.7	24.8

Table-2 Training data for analysis of degree of land slide danger

CH	Item	Unit	Case 1 Convex Area mean	st.dv	Case 2 Concave Area mean	st.dv	Case 3 All Area mean	st.dv
1	Erosion (lost)	m	60.6	22.5	42.5	22.1	50.0	23.9
2	Erosion (remain)	m	46.6	22.5	65.8	23.9	57.0	24.8
3	Undu- lation	m	8.4	3.1	8.4	3.2	8.4	3.2
4	Slope gradient	tan	0.412	0.164	0.408	0.168	0.410	0.168
5	Uneven. +/convex	m	-3.2	2.1	+3.3	2.0	+0.5	3.6
6	Direct. (+E,-w)	tan	0.118	0.330	0.128	0.316	0.124	0.324
7	Direct. (+N,-S)	tan	0.172	0.246	0.164	0.260	0.168	0.254
8	LANDSAT PCA(1st)	PCA 1ST	142.6	22.8	140.3	21.5	141.7	25.8

3.APPLICATION ANALYSIS

3.1 Principal component analysis

Topographical analysis data originated as altitude data.So each topographical result data has some correlation to another result, and no result is independent. Principal component analysis (PCA) has some effect of extracting new variables and condensing the variables. Then the results of topographical analysis were condensed with PCA operation.

3.2 Cluster Analysis

This analysis is unsupervised classfication , and classifies the PCA data and remote sensed data. The results produce a geographical type map and some kind of vegitation condition map.

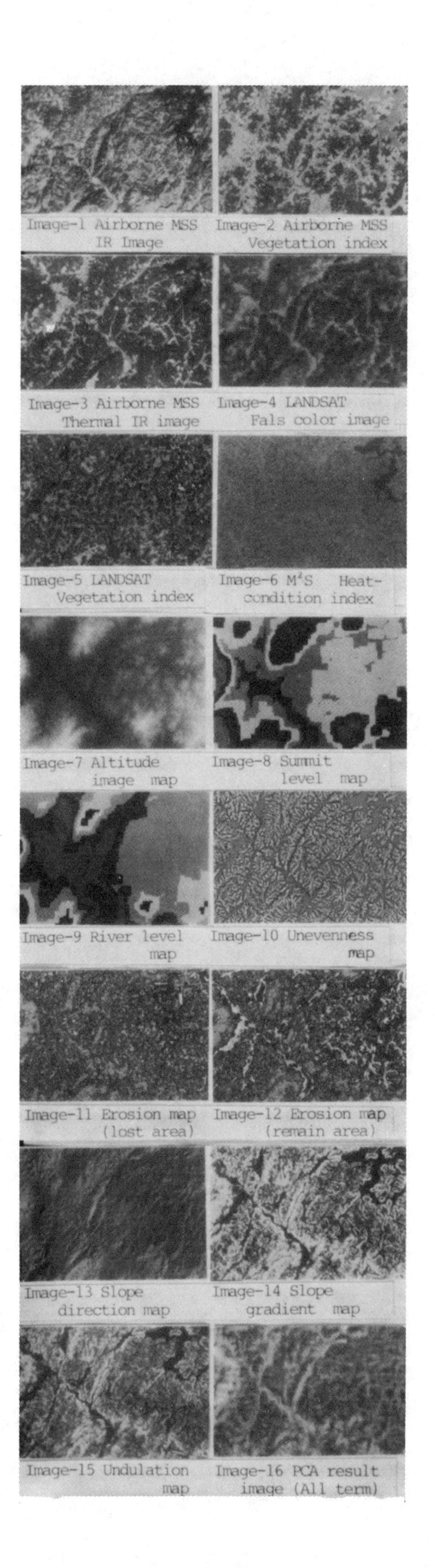

Image-1 Airborne MSS IR Image
Image-2 Airborne MSS Vegetation index
Image-3 Airborne MSS Thermal IR image
Image-4 LANDSAT Fals color image
Image-5 LANDSAT Vegetation index
Image-6 M^{4}S Heat-condition index
Image-7 Altitude image map
Image-8 Summit level map
Image-9 River level map
Image-10 Unevenness map
Image-11 Erosion map (lost area)
Image-12 Erosion map (remain area)
Image-13 Slope direction map
Image-14 Slope gradient map
Image-15 Undulation map
Image-16 PCA result image (All term)

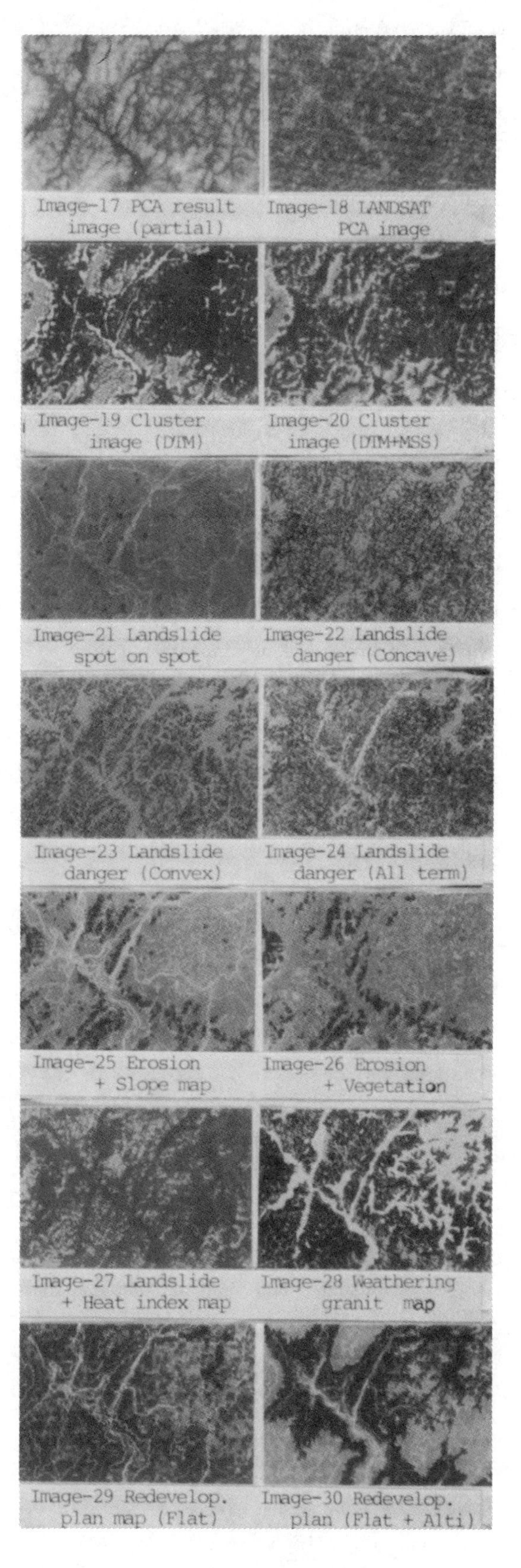

Image-17 PCA result image (partial)
Image-18 LANDSAT PCA image
Image-19 Cluster image (DTM)
Image-20 Cluster image (DTM+MSS)
Image-21 Landslide spot on spot
Image-22 Landslide danger (Concave)
Image-23 Landslide danger (Convex)
Image-24 Landslide danger (All term)
Image-25 Erosion + Slope map
Image-26 Erosion + Vegetation
Image-27 Landslide + Heat index map
Image-28 Weathering granit map
Image-29 Redevelop. plan map (Flat)
Image-30 Redevelop. plan (Flat + Alti)

3.3 Maximum Likelihood Method

The training area of supervised area was selected from ground investigation data and photo-interpretation results,and classified as to weathered granite type distribution on the image. It shows fine results compared with ground truth data.

3.4 Similarity Analysis

The training area or supervised area was selected from ground investigation data and photo-interpretation resuls. The similarity grade which is compared with condition of land slide area ,was analysed.

3.5 Redevelopment Planning

The people , who live in the dam reservoir planning area, must move to a redevelopment area, until the dam site construction is started. Then it must be simulated to find the optimum area for redevelopment and removal. The results of this study were used. And also the soil volume to construct a flat plane was estimated ,for some flat level area. For example how much for a 250m X 250m , or 150m X 150m flat plane ? After that,other kind of data were overlapped for example, altitude, weathered area map and so on.

4.CONCLUSION

It is possible to extract clealy the land slide danger area .
It was possible to identify the grade of weathered granite.
It was available to make use of simulation for extract optimum redevelopment area.
In the future , it is expected to be able to apply the large scale civil engineering field much more.

Symposium on Remote Sensing for Resources Development and Environmental Management / Enschede / August 1986

Sea surface temperature studies in Norwegian coastal areas using AVHRR- and TM thermal infrared data

J.P.Pedersen
University of Tromsø, Norway

ABSTRACT: This work presents an algorithm for deriving sea surface temperatures from infrared satellite data. The algorithm is based upon physical solution of the equation of radiative transfer. The theory for calculating the atmospheric transmittance and -radiance is briefly discussed. Calculated transmittances are compared to values reported by others. Results from applications of the algorithm on NOAA/AVHRR-data are presented. At last, sea surface temperature data derived from the Landsat/Thematic Mapper are presented. Due to lack of TM calibration data, the temperatures are derived from comparisons of digital values and in-situ measured temperatures at knownlocations in the data set.

1 INTRODUCTION

The thermal infrared data from the NOAA-series of satellites have been applied by different Norwegian institutes for studying sea surface temperatures and currents for many years. In recent years there has been a growing interest in the development of the natural coastal-zone resources in Norway. Today a lot of sea-farms are in operation all over the country, and a lot more are planned for the future. In selecting the most suitable location for a sea farm, it is often necessary to know about the annual variations of the currents and the temperatures in the actual area. One way of collecting this information is by using infrared data from satellites.

The NOAA-series of polar orbiter, sunsynchronous satellites, from which data are read out at Tromsø Telemetry Station, offers the opportunity to study surface phenomena in the Arctic regions with a high frequency of repetivity. The thermal infrared data from the NOAA-satellites are frequently used by different Norwegian institutes for studying currents and the sea surface temperatures (SST). The spatial resolution of the NOAA/AVHRR-dta of 1 km limits the applications to open ocean areas.

However, the new generation of satellites, represented by the Landsat/TM offers the opportunity to study surface phenomena at an increased spatial resolution. In studying currents and SST's, the 120 meter resolution of the thermal TM-channel is more adapable for coastal-zone applications, as compared to the NOAA/AVHRR.

2 SATELLITE INSTRUMENTATION

The primary surface observing sensor onboard the NOAA series of satellites is the AVHRR (Advanced Very High Resolution Radiometer). The AVHRR is a five channel radiometer observing at visible, near-infrared and thermal infrared wavelengths. The observing channels are 0.58-0.68 um, 0.7-1.1 um, 3.55-3.93 um, 10.3-11.3 um, and 11.5-12.5 um. The AVHRR scans the earth surface within +/- 55.4 degrees from nadir, representing a surface swath width of approximately 2500 km. The spatial resolution of the AVHRR channels is 1x1 km (at nadir). The satellite altitude is approximately 830 km, the orbital period approximately 102 minutes, which represents 14.1 orbits pr. day (Schwalb, 1978). Technical data for the NOAA-satellites are listed in table 1.

The TM onboard the Landsat satellites represents the next generation of earth observing sensors. TM is a seven channel radiometer observing at visible, near-infrared, and thermal infrared. The observing wavelengths are 0.45-0.52 um, 0.53-0.60 um, 0.63-0.69 um, 0.76-0.90 um, 1.55-1.75 um, 2.08-2.35 um, and 10.4-12.5 um. The TM scan angle is +/- 7.7 deg.,representing a swath width of 185 km at earth surface. The spatial resolution is 30x30 meter, except for the thermal channel which has a resolution of 120x120 meter. Satellite altitude is app. 705 km, the orbital period 99 minutes. The TM repeat cycle is 16 days. As for the NOAA satellite, the Landsat is a sunsynchronous, polar orbiter (NASA, 1984). Technical data for TM are listed in table 1.

For the purpose of this application, the thermal channel of theAVHRR and the TM have to be considered. The AVHRR channel 4 (eventually channel 5) and the TM channel 6 cover the same part of the electromagnetic spectrum, although the spectral width of the TM channel is twice the width of the AVHRR channel. Due to the spectral coincidence of the TM and the AVHRR channels, they are applicable for comparable studies of SST's. The predicted absolute accuracies are 0.5 K and < 0.12 K for TM and AVHRR respectively.

The most significant difference between the two channels is thespatial resolution, 120 meter for TM compared to 1 km for AVHRR. The spatial resolution limits the AVHRR applications primarily to medium scale phenomena studies in open oceans. For studies of small scale phenomena often observed in the coastal-zone areas and in the fjords, the TM thermal channel is the most suitable one.

Table 1. Technical data of NOAA/AVHRR and Landsat/TM.

	AVHRR	TM
Spectral bands (um)	ch1 0.58-0.68 ch2 0.7 - 1.1 ch3 3.55-3.93 ch4 10.3-11.3 ch5 11.5-12.5	ch1 0.45-0.52 ch2 0.53-0.60 ch3 0.63-0.69 ch4 0.76-0.90 ch5 1.55-1.75 ch7 2.08-2.35 ch6 10.4-12.5
Spatial resolution	app. 1 x 1 km (at nadir)	30 x 30 m ch6 120 x 120 m
Swath width	+/- 55.4 deg. app. 2500 km	+/- 7.7 deg. 185 km
Orbit	sunsynchronous, polar	
Data availab.	From Tromsø Telemetry Station	From Earthnet/ Kiruna

3 TEMPERATURE RETRIEVAL ALGORITHMS

Different algorithms have been proposed for retrieval of SST from infrared satellite data. These algorithms range from physical solutions of the equation of radiative transfer to algorithms based completely upon statistical regression analysis. The absolute accuracies vary from approximately 1 K for the single band direct solution algorithm, to a few tenths of a Kelvin for the split window algorithms (Prabhakara et.al., 1974 , McClain, 1980).

Weinreb and Hill (1980) describe an algorithm applying single band thermal infrared data. From the general equation of radiative transfer in an attenuating medium, the absolute surface temperature is derived after having corrected for the atmospheric influence. The atmospheric correction is calculated from atmospheric temperature- and humidity profiles from radiosondes. This algorithm is implemented for operational use at Tromsø Telemetry Station. In the next section, the thery for infrared SST techniques will be discussed in more deail.

3.1 The equation of radiative transfer

For a cloud-free atmosphere, infrared radiation emitted from the sea surface can propagate relatively unattenuated through the atmosphere at wavelengths for which gaseous absorption is small. Among these wavelengths, called 'atmospheric windows', the windows present at 3.5-4.0 um and 10-12 um are of interest for this application.

The major absorbing gas present in the atmosphere is water vapr. Depending upon the concentration, the transmission through the atmosphere can range from >90 % for a dry polar atmosphere (Pedersen, 1982), until <30 % for a humid atmosphere (5.5 cm water) (Njoku et. al., 1986).

The complete expression for the infrared radiation detected by an airborne sensor consists of the components from the sea surface, the atmosphere, and from downward radiation reflected at the surface into the direction of the sensor (Njoku et.al., 1986). However, in the thermal infrared the reflected component contributes <1 % to the total radiance (Maul, 1981). Therefore, neglecting this component will not introduce any significant errors in retrieving SST's.

Based upon the theory of Wark and Fleming (1966) and Weinreb and Hill (1980), a simplified form of the equation of radiative transfer can be derived.
At a given wavelength, λ, the equation of radiative transfer through a plane-parallell, non-scattering atmosphere in local thermodynamically equillibrium can be given from the following relation:

$$d\{R(\lambda,\theta)\} = \{-R(\lambda,\theta) + B(\lambda, T(z))\}\, k(\lambda)\, \rho(z) \sec(\theta)\, dz \qquad (3.1)$$

where $R(\lambda,\theta)$ = the spectral radiance at wavelength λ in wieving direction θ from the nadir
$k(\lambda)$ = absorption factor of the gas
$\rho(z)$ = absorbing gas density
$B(\lambda,T)$ = Planck radiation at wavelength λ and temperature T
z = altitude at local nadir
$T(z)$ = temperature versus altitude.

From different assumptions summarized in Pedersen (1982), equation 3.1 can be written on the following form:

$$R(\lambda,\theta) = e(\lambda,\theta) B(\lambda, T(p_0))\, t(\lambda, p_0, \theta) + \int_{t(\lambda, p_0, \theta)}^{1} B\{(\lambda, T(P))\}\, dt(\lambda, p, \theta) \qquad (3.2)$$

where $e(\lambda,\theta)$ = surface emissivity
p = atmospheric pressure
p = surface pressure
$t(\lambda,p,\theta)$ = atmospheric transmittance at wavelength λ, pressure p, in the direction θ.

Equation 3.2 expresses the spectral radiance at one wavelength λ. The sensor onboard the satellite detects the total radiance within a passband of finite width. The expression for the total, normalized satellite sensor passband radiance, R_i, is given from equation 3.2 weighted by the passband response function $H(\lambda)$:

$$R_i(\lambda) = \int_{\lambda_1}^{\lambda_2} R(\lambda,\theta) H(\lambda)\, d\lambda \Big/ \int_{\lambda_1}^{\lambda_2} H(\lambda)\, d\lambda \qquad (3.3)$$

where $H(\lambda)$ = the passband relative response at wavelength λ
$\lambda_{2,1}$ = the maximum and minimum passband wavelengths.

3.2 Atmospheric transmittance theory

To obtain the surface temperature, T_s, from equations 3.2 and 3.3 on an operational basis, the atmospheric corrections have to be made. From the absolute, atmospheric corrected radiance, inversion of Planck's radiation law will give the surface temperature if the emissivity is known.

For sea applications it is usual to assume the sea surface as a black body. This implies that the emissivity, $e(\lambda,\theta)$, in equation 3.2 is equal 1.0.
The next step is to calculate the atmospheric transmittance, $t(\lambda, p_0, \theta)$. This can be done from the system LOWTRAN 6 (Kneizys et.al., 1983). However, when the algorithm was implemented, the LOWTRAN system was not available at Tromsø. Instead a procedure based upon the theory of Weinreb and Hill (1980), Wark et.al. (1974), and Weinreb and Neuendorffer (1973) was developed.

The fundamental idea of this procedure is that the total atmospheric transmittance is given as the product of the transmittances for the individual absorbing atmospheric constituents water vapor, nitrogen and the uniformly mixed gases.

The real, inhomogenous atmosphere is treated as a succession of a number of homogenous layers, in each of which the pressure, p, temperature, T, and the water mixing ratio, w, are constants.

The total transmittance for the most important constituent, water vapor, can be treated as the product of spectral line- and two different continua transmittances. The spectral line transmittances are calculated from an approximation suggested by Weinreb and Neuendorffer (1973). This method assumes the transmittance as a known function of the amount of absorbing medium, U, temperature, T, and the total pressure, p, for each of the n-homogenous layers the real atmosphere is subdivided into.

The water continua transmittances are calculated according to theory discussed by Roberts et.al. (1976). The two different transmittances can be explained from collisions between water vapor molecules (sb), and collisions between water vapor and other atmospheric gas molecules (fb). From the theory of Roberts et.al. (1976) and Weinreb and Hill (1980), the continua transmittances can be given from the following expressions:

$$t_{sb}(\lambda,P) \propto C_0(\lambda) \sec(\theta) \int_0^P pr^2 \exp(T_0 (T-296)^{-1}))\, dp \qquad (3.4.a)$$

$$t_{fb}(\lambda,P) \propto C_0(\lambda) \sec(\theta) \int_0^P pr\, dp \qquad (3.4.b)$$

where $C(\lambda)$ = absorption coefficients at wavelength λ
P = total atmospheric pressure
θ = wieving angle from local nadir
r = mass mixing ratio of water vapor (g/kg)
$t_{sb,fb}$ = transmittance from top of atmosphere to a level of pressure P.

The Nitrogen transmittance can be derived from an

expression similiar to the water continuum transmittance (Weinreb and Hill, 1980):

$$t_N(\lambda,P) \propto C_N(\lambda,T=296\ K)\sec(\theta) \int_0^P p\ /\ T\ dp \qquad (3.5)$$

where the notation of equation 3.5 is similiar to the notation of 3.4. C_N is the Nitrogen absorption coefficients at a reference temperature T=296 K.

The transmittance of the uniformly mixed gases, t_U, is calculated from a simplified method of that used in the LOWTRAN system (Kneizys et.al., 1983, McClatchey et.al., 1972). The uniformly mixed gases are treated as one unit with fixed concentrations from McClatchey et.al. (1972) as input to the model.

Knowing the transmittances for all the atmospheric radiative absorbing factors, the total atmospheric transmittance at wavelength λ, pressure, P, and temperature T is given by:

$$t\ (\lambda,P,T) = t_w(\lambda,P,T)t_N(\lambda,P,T)t_U(\lambda,P,T) \qquad (3.6)$$

3.3 Emitted atmospheric radiance

From the equation of radiative transfer (3.2), an expression for the emitted atmospheric radiance can be derived (Dalu et.al., 1979). Equation 3.2 can be written on the following form:

$$R(\lambda_0) = B(\lambda_0,T_s)t(\lambda_0,p_0)+\{1-t(\lambda_0,p_0)\}\bar{B}(\lambda_0,\overline{T(p)}) \qquad (3.7)$$

The notation of equation 3.7 follows the notation of quation 3.2. $\bar{B}$ is the weighted, averaged atmospheric Planck radiance given from:

$$\bar{B}\{\lambda_0,\overline{T(p)}\} = \frac{\int_{t(\lambda_0,p_0)}^{1} B\{\lambda_0,T(p)\}\,dt}{\int_{t(\lambda_0,p_0)}^{1} dt} \qquad (3.8)$$

In equation 3.7 and 3.8, the radiance is calculated at the centre wavelengths, λ_0, for each of the passbands.

In numerical implementation of the algorithm, all the integrals are implemented as sums of finite width intervals across the actual passband. The weightfactor $H(\lambda)$ is given for each of the intervals: h=H(i), i=1, --.N, where N denotes the number of subintervals of passband.

3.4 Tests of method

In testing the algorithm, the transmittances and radiances were calculated for different wavelengths and atmospheric models. The calculated values were compared to corresponding values reported by others. The applied atmospheric models were taken from radiosondes data from Norwegian meteorological stations. In table 2, the comparisons for two different atmospheres are listed. The first atmosphere is dry atmosphere (0.5 cm water), and the second is an atmosphere containing approximately 2 cm water. These two atmospheres are fairly representative for atmospheres observed in the Arctic. All results in table 2 are for nadir viewing.

From the comparisons presented in table 2, there is an fairly good agreement between calculated and reported transmittances. Therefore, to conclude, the developed routines for calculating the total atmospheric transmittance seem to give reasonable results.

Knowing the transmittances and the atmospheric radiance given by equation 3.8, the absolute surface skin temperature can be derived from inversion of Planck's radiation law. The last section of this chapter will deal with calculations of transmittance versus scan angle for different atmospheres, and the effect upon the surface temperature.

In figure 3.1 are presented the total transmittance versus scan angle for three different atmospheres for the AVHRR channel 4. In figure 3.2 are illustrated the introduced errors on the surface temperature by ignoring the varying scan angle for the three atmospheres presented in figure 3.1.

Table 2. Comparison of calculated and reported transmittances.
$t_{w,c}$ - calculated water transmittance
$t_{w,WH}$ - water transmittance from Weinreb and Hill.
$t_{w,P}$ - water transmittance from Prabhakara et.al.
$t_{N,c}$ - calculated Nitrogen transmittance
$t_{N,WH}$ - Nitrogen transmittance from Weinreb and Hill

wave-length (um)	$t_{w,c}$	$t_{w,WH}$	$t_{w,P}$	$t_{N,c}$	$t_{N,WH}$
	0.5 cm water				
10.5	.979	.967	.946	1.0	1.0
11.5	.955	.933	.968	.945	.966
	approximately 2 cm water				
3.5	.880	.740	----	1.0	1.0
3.9	.970	.920	----	.958	.966
10.5	.841	----	.812	----	----
11.5	.748	----	.738	----	----

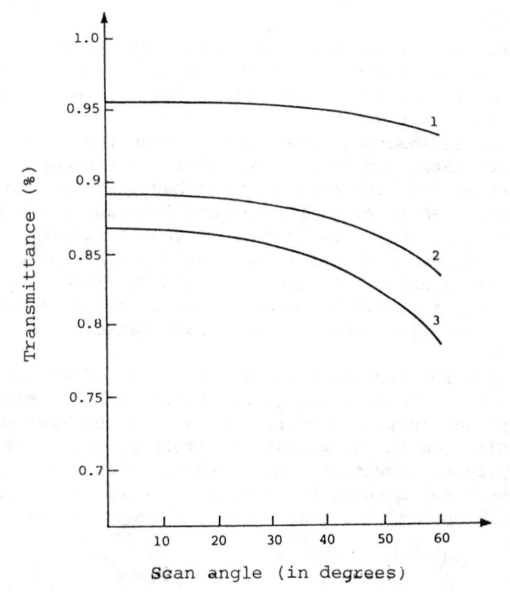

Figure 3.1. Transmittance versus scan angle for three atmospheres:
1: 0.5 cm vater content
2: 1.3 cm water content
3: 1.5 cm water content.

From the figure 3.1 it is seen that the transmittance variations are negligible for scan angles less than ~ 40 degrees from nadir, for the dry polar atmospheres. However, the effect is increasing for angles below 40 degrees as the water content increase from 0.5 to 1.5 cm water.

From the curves in figure 3.2, the observed error introduced by ignoring the across-scan transmittance variations is less than approximately 0.6 K for scan angles less than 60 degrees for a polar atmosphere with water content less than approximately 1.3 cm. The results agree well with results reported by Smith et.al. (1970). They report errors of order 0.7 K for a polar atmosphere and across-scan angles less than 60 degrees.

4 AVHRR-DATA SET APPLICATIONS

The major limitations in applying optical satellite

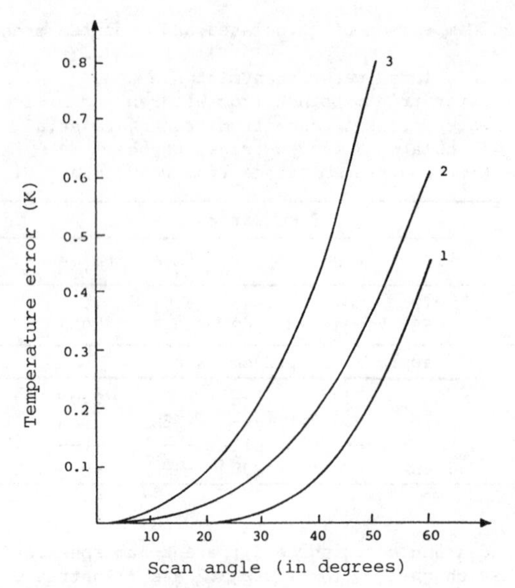

Figure 3.2. Error introduced on surface temperature by ignoring the effects from the varying scan angle. The numbers refer to the atmospheres applied in the calculations presented in figure 3.1.

data from the Arctic are introduced from the weather conditions. Very often the presence of clouds reduce the data availability. For testing the complete algorithm, a project which included collection of in-situ SST's and radiosondes data together with the NOAA-satellite data, was done summer 1981 off the coast of Northern Norway. Unfortunately the bad weather conditions resulted in no available satellite surface data.

Instead, for real application of the algorithm, a NOAA-6 data set from the island Jan Mayen, acquired at Tromsø Telemetry Station on August 20. 1984, was chosen. The approximate geographical coverage of the data set is illustrated by the square marked A in figure 4.1.

As input for atmospheric correction, a radiosonde profile from the meteorological station at the island was applied. In-situ measured SST's from the area were available from the Norwegian Institute of Marine Research (private communication). Although these measurements were taken about one week off the satellite data, they were applied for comparisons of temperatures.

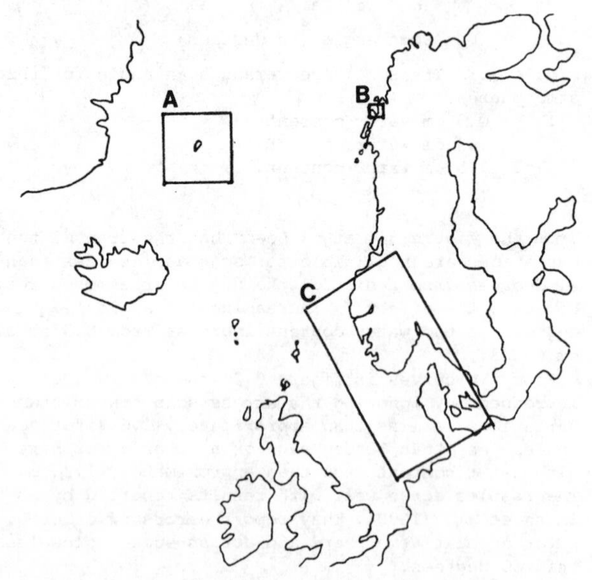

Figure 4.1. Approximate geographical coverage of the data sets applied in this report:
A: Jan Mayen
B: The Tromsø area
C: Southern Norway.

The water content of the applied atmosphere was app. 1.3 cm. From the atmospheric profile, an averaged transmittance, $t_a = 0.89$ was calculated for the NOAA-6/AVHRR channel 4. The error introduced if the atmospheric correction was not performed was of the order 0.5 K. In figure 4.2, the resulting atmospheric corrected SST image data for the Jan Mayen data set are presented.

At the top of the image in figure 4.2 there is a scale showing the correspondance between the grey levels and the temperatures. Comparisons of the satellite derived SST's and the in-situ measured temperatures indicate an average, absolute error of 1 K. Having in mind the one week difference between the compared data, this error agrees well with the commonly accepted error in deriving SST's from a single infrared band.

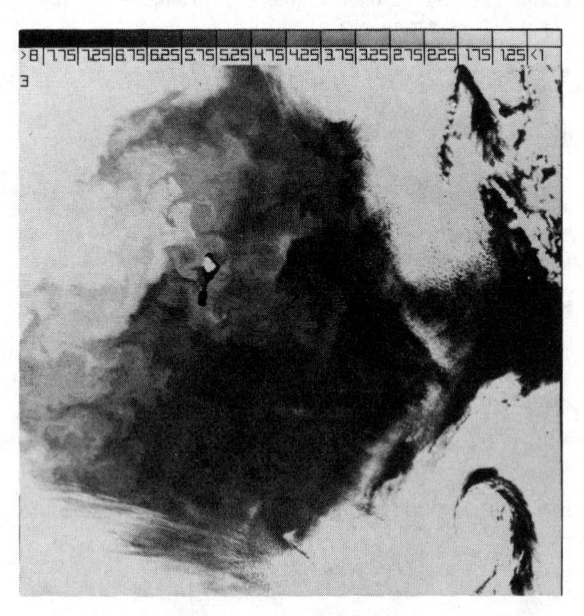

Figure 4.2. SST image from the Jan Mayen area derived from the NOAA-6/AVHRR data set. A temperature value of 1.25 or 1.75 means the actual temperature is within the range 1.-1.5 or 1.5-2.0 degrees Celsius respectively.

The data from the NOAA-satellites have also been applied for studying the strong currents often observed off the coast of southern Norway. A group at the Geophysical Institue at the University of Bergen has undertaken projects related to studies of the currents. In the worst case, the strong currents can be of danger to the activities at the important oil-fields located in the area of observed strong current gradients. In figure 4.3 an image showing eddies off the coast of Southern Norway is presented. The image is generated at Tromsø Telemetry Station from NOAA- 6 thermal infrared data acquired on April 13. 1981. Off the western coast of Norway a pattern of eddies is observed. The different grey levels in the pattern represent surface temperatures ranging from approximately 1.5 to approximately 6.0 degrees Celcius. For this application, there has been no atmospheric correction, and the derived temperatures have not been compared to in-situ measurements.

5. SST STUDIES FROM LANDSAT/TM

The increased spatial resolution of the TM thermal infrared channel, as compared to the AVHRR, offers the possibility of applying the operated SST-algorithm to coastal-zone studies from the TM.

During flight, the TM is calibrated against internal blackbodies. Although the calibration data is located in the spacecraft downlink datastream, at present

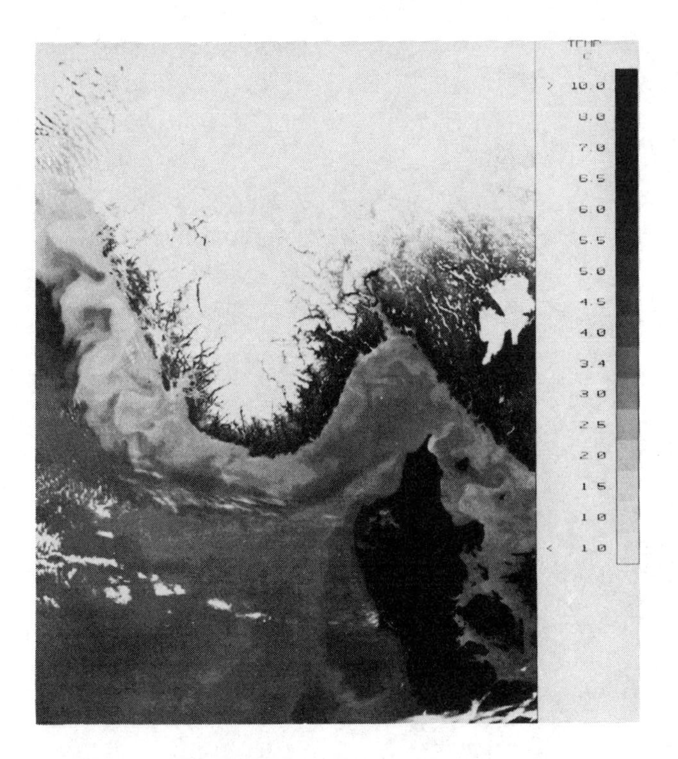

Figure 4.3. A NOAA-6 derived SST image covering the Sothern Norway and the North Sea. The upper part of the image is covered by clouds(Cfr. area C fig. 4.1).

they are not available to the users. Therefore, it is impossible to apply the SST-algorithm to the present TM data. The SST image presented here is generated from a look-up table based from comparisons of digital values and in-situ temperatures at known locations in the image.

The image data presented in figure 5.1 is a Landsat-5/TM channel 6 sub-scene from the Tromsø area (Path 197/Row 11) (Area B in fig. 4.1). Tromsø (69.6 N/18.9 E) is located in the upper right corner. The data were acquired at Kiruna on June 3. 1984, and have been processed at Tromsø Telemetry Station's image processing laboratory.

By combining the thermal and a near-infrared channel the land areas have been removed. Land is represented by the color black in the image. The different grey levels of the sea surface have been assigned a temperature as indicated by the scale at bottom left.

When the data were acquired, the current around the island of Tromsø was moving north (up the image), transporting relatively warm surface water (app. 10 deg. Celcius) northwards. The island is linked by two bridges, one to the east, and one to the west (not seen in the image). As the water flows through the bridges, the supports of the bridges cause turbulent mixing of the warm surface water with the colder subsurface water. This mixing is clearly identified in the image.

In the centre of the image there is an area where the surface temperature is approximately 2-3 degrees below that of the surrounding areas. This is caused by the upwelling of colder subsurface water due to the interaction of the small island and the local current pattern in the narrow sound. Also there is a cold snow-melt water outlet of a local river.

In the bottom right part of the image, a fjord which is an outlet for cold water can be seen. The cold water results from the river Målselv transporting cold snow-melt water from the local mountaineous ares towards the sea.

6 CONCLUSIONS

Although the Arctic weather conditions very often limits the applications of optical satellite remote sensed data in Norway, the results carried out by different Norwegian remote sensing institutes show the applicability of these services in research and to some extent also in (semi-)operational processing. With the development of the next generation of all-weather sensors, specially the operational processing of satellite data seems very promising.

The medium resolution AVHRR-data have been very useful in the research towards better understanding of oceanic processes. The eddies observed off the western coast of Norway (Cfr. figure 4.3) was fully discovered when satellite data became available. SST studies from airborne remote sensed data have also been very important in the understanding of the generation mechanisms for the eddies. In this case, the research has resulted in a system for forecasting the eddies from local SST studies.

The present algorithm operated at Tromsø Telemetry Station apply rediosondes data profiles for atmospheric corrections. Since there is a lack of spatial coverage from the radiosondes, the applications of these profiles assume a stationary atmosphere, which is a very idealized, never-occuring assumption. However, a combined application of AVHRR- and TOVS (Tiros Operational Vertical Sounder) data will offer the opportunities for taking the spatial atmospheric inhomogenitites into account when deriving atmospheric corrected sea surface temperatures. Therefore, the future activity in Tromsø regarding applications of NOAA-data will include registered AVHRR- and TOVS-data processing.

Compared to the 16 day repeat cycle of the Landsat, the high frequency of repetivity for the NOAA-satellites makes this system very useful for operational processing in Norway.

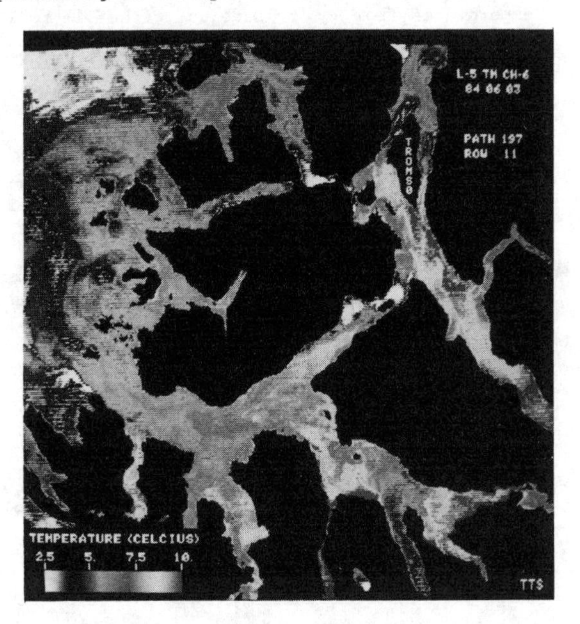

Figure 5.1. Landsat-5/TM derived SST image from the Tromsø area. The observed features are discussed in the text.

The high resolution thermal infrared TM data seem very suitable for studying surface temperatures and currents within the Norwegian coastal-zone. For a coastal-zone planner specially the applications in the fjords seem very interesting for the development of the natural resources.

At present there is a limitation in applying existing SST-algorithms to the TM, due to the lack of calibration data. However, this seems to be a temporary problem since the ESA/Earthnet plans to include the in-flight calibration data on the supplied data.

For studying other sub-surface parameters of interest to the coastal-zone planners, like the organic productivity and the salinity, there has been no effort in Norway until recently towards using satellite remote sensed data for this purpose. Results from abroad indicate that there is a correlation between these parameters and the surface temperature. Based upon these results, today there exists plans in the Tromsø remote sensing community for pilot projects related to studies concerning the suitability of TM data for such applications....

REFERENCES

Dalu, P. et.al. 1979. An improved Scheme for the Remote Sensing of Sea Surface Temperature, NASA Technical Memorandum 80322.

Kneizys, F.X. 1983. Atmospheric Transmittance/Radiance: Computer Code LOWTRAN 6, AFGL-TR -83-0187.

Maul, G.A. 1981. Application of GOES Visible-Infrared Data to Quantifying Mesoscale Ocean Surface Temperatures, J. Geophys. Res., Vol. 86, C9.

McClain, E.P. 1980. Multiple Atmospheric-Window Techniques for Satellite-Derived Sea Surface Temperatures, Proceedings of COSPAR Symposium, Oceanography from Space, Venice, Italy, May 26-30.

McClatchey, R.A. et.al. 1972. Optical properties of the atmosphere, AFCRL-72-0497.

NASA, 1984. A Prospectus for Thematic Mapper Research in the Earth Sciences, NASA Technical Memorandum 86149.

Njoku, E.G. et.al. 1985. Advances in Satellite Sea Surface Temperature Measurements and Oceanographic Applications, J. Geophys. Res., Vol. 90, C6.

Pedersen J.P. 1982. Atmospheric Effects on Infrared Satellite Measurements of Sea Surface Temperatures, Cand. thesis., University of Tromsø (in Norwegian).

Prabhakara, C. et.al. 1974. Estimation of Sea Surface Temperatures from Remote Sensing in the 11- to 13 um Window Region, J. Geophys. Res., Vol. 79, 33.

Roberts, R.E. 1976. Infrared Continuum Absorption by Atmospheric Water Vapor in the 8 - 12 um Window, Appl. Opt., Vol. 15, 9.

Schwalb, A. 1978. The TIROS-N/NOAA Satellite Series, NOAA Technical Memorandum NESS 95.

Smith, W.L. et.al. 1970. The Determination of Sea-Surface Temperature from Satellite High Resolution Infrared Window Radiation Measurements, Mon. Wea. Rev., Vol. 98, No. 8.

Wark, D.Q., Fleming, H.E. 1966. Indirect Measurements of Atmospheric Temperature Profiles from Satellites: 1. Introduction, MOn. Wea. Rev., Vol. 94.

Wark, D.Q. et.al. 1974. Satellite Observations of Atmospheric Water Vapor, Appl. Opt., Vol. 13, No. 3.

Weinreb, M.P., Hill, M.L. 1980. Calculation of Atmospheric Radiances and Brightness Temperatures in Infrared Window Channels of Satellite Radiometers, NOAA Technical Report NESS 80.

Weinreb, M.P., Neuendorffer, A.C. 1973. Method to Apply Homogenous Path Transmittance Models to Inhomogenous Atmospheres, J. Atmosph. Sci., Vol. 30., p. 662-666.

Symposium on Remote Sensing for Resources Development and Environmental Management / Enschede / August 1986

Satellite data in aquatic area research: Some ideas for future studies

Jouko T.Raitala
Jet Propulsion Laboratory, Caltech, Pasadena, Calif., USA (on leave from Dept. of Astronomy, University of Oulu, Finland)

ABSTRACT: Computer-aided digital remote sensing techniques were used to evaluate the usefulness of Landsat MSS data in aquatic area studies. These investigations unravelled some of the Landsat data potentials in monitoring factors critical to **limnology, aquatic botany, geomorphology** and **engineering**: 1) Besides depth (in relatively shallow Finnish lakes) the MSS data may also include useful information about Secchi disc values, humus content in water (colour, iron) and productivity (nutrients, chlorophyll). 2) Aquatic vegetation classification is possible only where vegetation units are big enough in respect to the 0.5 hectares ground resolution. Different life-forms (helophytes, nympheids, elodeids, bryophytes etc.) are mapped quite easily, but when using the supervised classification procedure even some minor nuances can be traced. 3) Multitemporal satellite imagery has been used to evaluate alterations within the littoral areas of some Finnish water reservoirs between successive periods of high water and also along the shallow coastal sea of the Gulf of Bothnia.

The limitations of the MSS data in exact small-scale interpretations are, however, so apparent that the most critical support of relevant water chemical and field data must not be underestimated. The use of MSS data with poor ground resolution and with too broad and too few spectral channels represents only a small range of the possibilities which will be gained by using the more advanced **TM** and **SPOT** data. The MSS data may nevertheless be important when trying to trace the changes in aquatic and other environments during the last decade.

1. INTRODUCTION

Different lakes and coastal sea areas are amongst the few most essential environments in Finland forming a natural surrounding for almost all activities of men. This means that aquatic areas are the object of various and often conflicting efforts of utilization. Research, investigation and planning of water areas for recreational and residential use and as a source of livelihood and raw water or as a dumping place for waste water all require accurate information, some of which can in principle be obtained and prepared by using Landsat or other satellite data and the computer-aided analysis of such data.

The Landsat MSS data includes information of aquatic areas down to the Secchi disc depth. The green and yellow wavelengths of channel 4 indicate variations within a few uppermost metres in brownish Finnish lakes and water features even down to tens of metres within certain open ocean areas. The red radiation of channel 5 usually comes from within the uppermost one metre water layer, and the near-infrared channels 6 and 7 include information from within the uppermost centimetres or millimetres, respectively. In addition the uppermost part of the water column is best represented within all recorded channels.

The present study describes some attempts to apply aquatic remote sensing to the preparation of parametric map-like presentations, quantitative evaluations and time-related investigations within various water areas in Finland.

2. WATER QUALITY

Laborious and expensive efforts are called for when attempts are made to inspect water quality and its changes over large Finnish lake areas by means of field work. Satellite images and their computer-aided analysis may help to locate the most critical areas where changes in water quality have taken place. Continuous satellite monitoring of water areas -- in terms of data availability -- can provide information for example of movements of different water bodies, changes in production rate, overgrowth, pollution and purification.

However, more basic studies and progress are still needed before remote sensing will be able to answer all questions unambiguously. The wavelength channels of the Landsat MSS sensor are too wide and too few to allow distinct parametric map-like presentations of only one of the aquatic variables to be performed. Indications of different phenomena are intermingled, resulting in difficulties and misinterpretations in monitoring. Naturally brownish Finnish lake waters are easily confused with waste water pollution and even a sparse nymphaeid vegetation may prevent recognition of water quality. We must, then, be extremely cautious when trying to generalize results obtained from one water area to another.

In applying the remote sensing technique to studies of aquatic areas the most critical support of the relevant water chemical data acquired by the Water District Offices of Finland makes it possible to find the main differences between various water areas in advance and to limit studies to within the same type lakes. Satellite data is thus not used to find major differences but rather to indicate minor variations within basically similar water units. Even certain small nuances and changes in water quality can be traced and followed once a critical choice has first been made of what factors can be monitored and when and where (Lindell 1980, Raitala et al. 1984c). Channel 7 for example displays very sensitively changes in the quality of the uppermost surface water and in vegetation above, on and just below the water surface. Near-infrared radiation is almost totally absorbed by even a thin water layer, while the existence of green vegetation, turbidity, algae production and surface-reaching shallows increase the quantity of reflected radiation.

Statistical correlation between certain water chemical values and Landsat record derivatives from within the Kuusamo and Kemijarvi areas in northeastern Finland indicated some significant-level correlations. Index values calculated from the MSS channel 4 and 5 values indicated information of lake depths, Secchi disc depths, water colour, iron and turbidity. Channels 6 and 7 added some information of such production factors as nitrogen, phosphorus and chlorophyll a. Because of the low level of nutrients only a qualitative approach was possible within these lakes (Raitala et al., 1984 a,b).

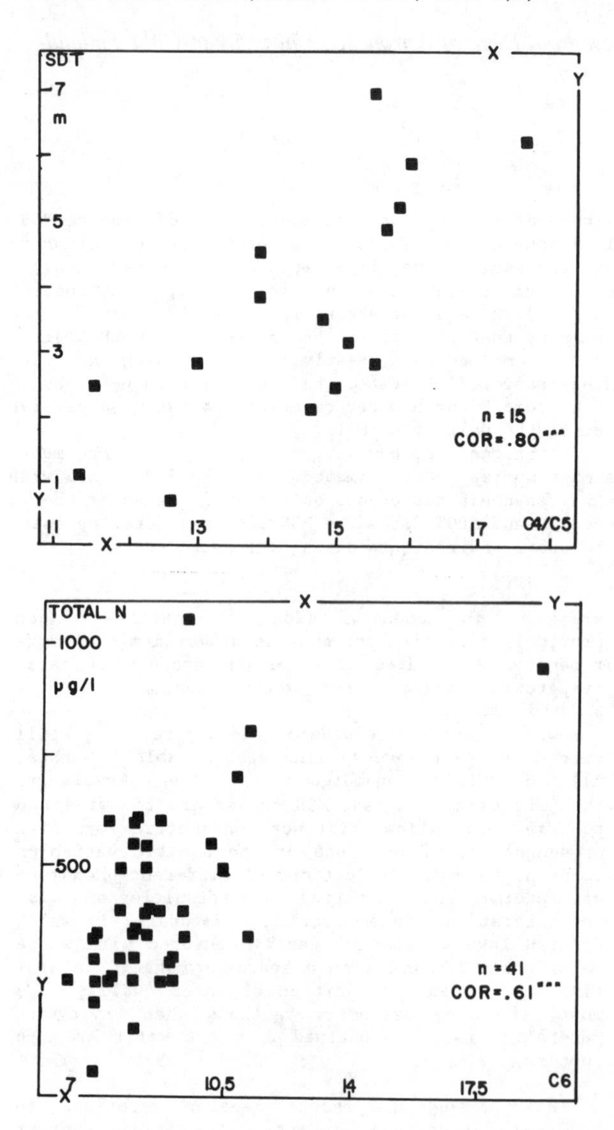

Figure 1. Landsat MSS derivative (channel 4/channel 5) vs Secchi disc transparency (SDT) values (upper) and channel 6 vs total nitrogen (lower) for lakes of the Kuusamo area (Raitala et al. 1984b).

3. VEGETATION MAPPING

According to differences in the reflected spectrum it is possible to distinguish different life forms and vegetation units from one another and likewise evaluate the share of turbidity and production areas within distinct water areas. Helophytic vegetation reflects near-infrared radiation more than nymphaeids and much more than bryophytes, elodeids and isoetids. The density of the vegetation unit is also one of the main factors which must be taken into account.

Several aquatic areas in different parts of Finland were classified using both supervised and unsupervised classification procedures. Once again attention must be paid to the critical value of field work when preparing and testing the classification. Because of the relatively rough pixel ground resolution (0.5 ha) the classification does not provide a neat correspondence with a uniconceptual parametric map but makes it possible to inspect different surface complexes and environmental gradients.

The spectral reflectance of aquatic areas depends on water depth (Hammack 1977, Arkimaa and Raitala 1984), bottom type (Raitala et al. 1984a), aquatic plants (Raitala et al. 1984 a,c, 1985), water quality and Secchi disc depth (Lindell 1980, Arkimaa and Raitala 1981, Raitala et al. 1984b). The effects of these variables may be substantially interwoven and intermixed. The most effective variables are most easily mapped by means of satellite classification, while the fainter ones can be identified only less distinctly. Computer-aided multispectral classification seems to provide an important tool for obtaining qualitatively different classes from within aquatic areas, but quantitative means of evaluation should also be developed further before this technique can be used in routine studies.

Figure 2. Classification of the coastal waters of the Gulf of Bothnia off Oulu. C = coastal open deep sea, D = open shallow sea, E = dilution and submersed vegetation, F = bottom and vegetation effects, G = discharge dilution and vegetation, H = discharge from a small river (Arkimaa and Raitala 1984).

In the most detailed studies it was found possible to obtain at least five distinct class categories from within quite simple water areas by means of supervised Landsat classification procedure. These categories consisted of 11 to 17 classes, some of which slightly overlapped in groups to form the categories. The five categories established may be quite reasonable, while the number of individual classes and their unambiguous validity somehow also depend on the complexity of the aquatic variables present and on fortunate circumstances in finding ideal representative reference fields for each of the classes. Usually there are several variables forming a surface complex and small number of reference fields must be chosen to cover all combinations of the surface variables. This may result in a small number (two to four) of related classes forming an association of intermingled classes or a category which includes a certain range of minor variations on the ground.

Landsat multispectral classification system over aquatic areas may be most advantageous when used in conjunction with ground-truth data. Scientists familiar with basic variables within particular water areas can use the method to yield rapid up-to-date information of the state of and changes in the aquatic environment in a valid and cost-effective manner.

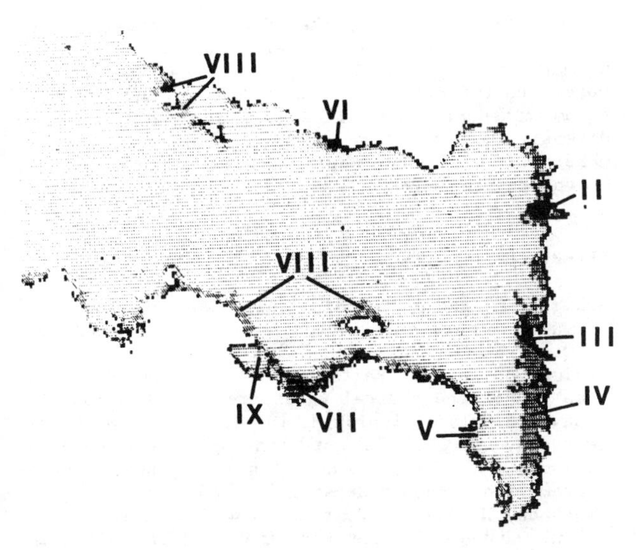

Figure 3. Detailed study of the shallow water areas of the Bay of Liminka (cf. lower part of Fig. 2) indicates different vegetation complexes as follows (Raitala et al. 1984c): II = Eleocharis acicularis, Eleocharis palustris, III = Eleocharis palustris, IV = Scirpus, Eleocharis acicularis, V = Eleocharis palustris, Potamogeton friesii, VI = Eleocharis palustris, Eleocharis acicularis on hard and clayey bottom, VII = Eleocharis palustris, Eleocharis acicularis on soft and stony bottom, VIII = Zannichellia, Limosella, IX = Eleocharis acicularis, Chara.

4. MULTITEMPORAL MONITORING

Because almost all of the variables affecting the reflected spectrum from within water areas are also time-related the most appropriate application of aquatic remote sensing is multitemporal monitoring of environmental changes. In Finland the recent littoral processes are most effective in two different places: around reservoirs with annually fluctuating water levels and along the seashore of the northern Gulf of Bothnia, with a maximum land upheaval of 1 cm a year.

The Porttipahta water reservoir displays an extremely intense vertical water level amplitude of 11 m, from very low in early spring to maximum flooding in autumn. The erosion of former terrestrial forest and bog vegetation and the underlying soil has been intense. Because only two images presenting lower and higher medium water levels were used two different presentations were prepared to display areas with different erosion effect between those two water levels and between the high medium-water level and the highest water level (Jantunen and Raitala 1984). The wind-induced fluctuations in the water level among the seashore off the city of Oulu are much smaller. Similar multitemporal monitoring evidently displays largely seasonal variations in vegetation within the shallow hydrolittoral areas, differences in water quality, but also some differences in depth relations.

Although only very simple arithmetic procedures were involved in comparing and manipulating the corresponding pixels, the results were encouraging and important enough to be evaluated in respect to further time-related monitoring of these and other related areas. The study areas were especially good for testing and evaluating the potential of multitemporal satellite remote sensing because environmental changes were and will continue to be so evident. Multitemporal satellite remote sensing will clearly offer a useful tool in monitoring environmental changes within other aquatic areas, too, although the exact outcome of this experimentation will nevertheless depend on future development of the Landsat and other satellite programs themselves (Beardsley 1986).

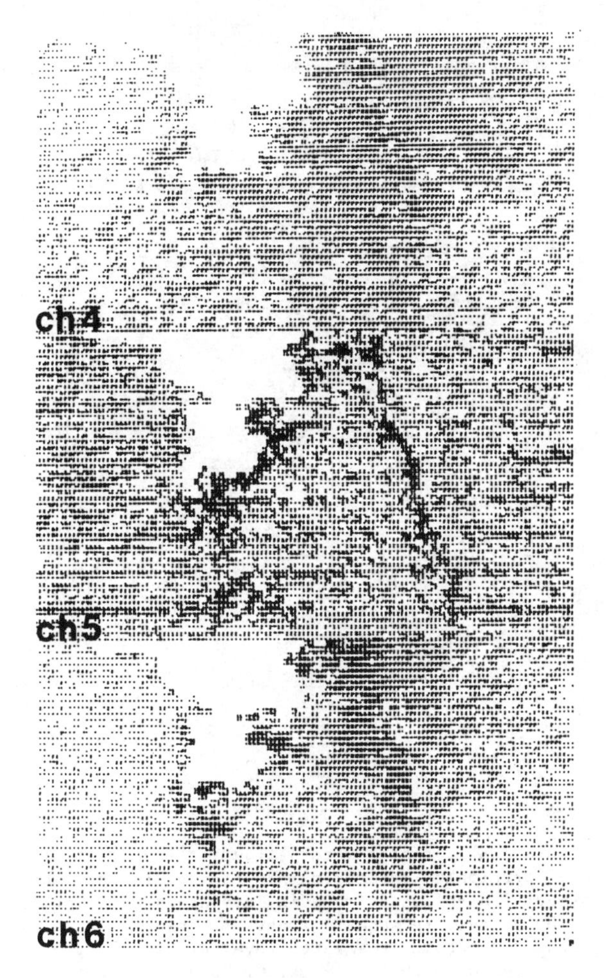

Figure 4. Multitemporal monitoring of the sea area off Oulu. The darker the symbols are the more distinct have seasonal and water level changes been.

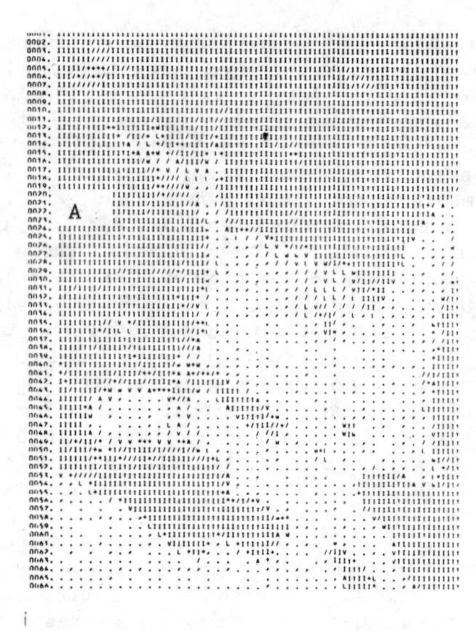

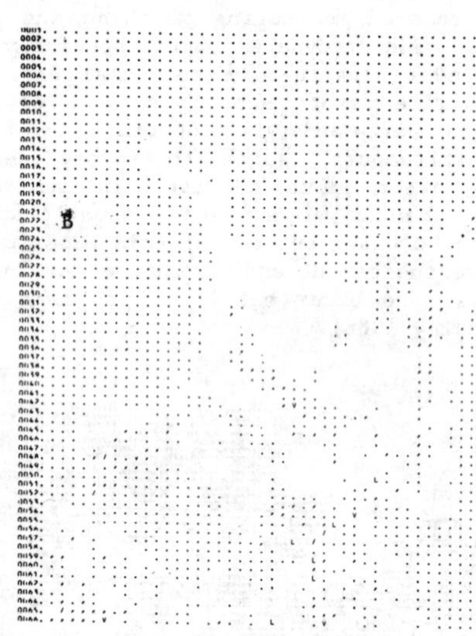

Figure 5 A. The high-water situation in a part of the Porttipahta reservoir in northern Finland. The symbol I displays surrounding dry areas while other symbols indicate aquatic areas. B. Lower littoral areas 2.3 m below the water level of Fig. 5A are displayed with white pixels. C. Upper littoral areas up to 5 m above the water level are indicated with various dark pixels (Jantunen and Raitala 1984).

5. DISCUSSION

The continuity of the Landsat program from 1972 until now has been very promising with respect to long-term studies. During the last 14 years the main focus of research has been to develop methods and approaches for practical applications of monitoring remote sensing. The latest satellite in this series, Landsat 5, is expected to stop transmitting data before replacement. This in itself increases expectations for the French SPOT satellite remote sensing, but there will still be abundant need of information especially on environmental changes over the last decade, and this need can be met only by using the old Landsat MSS data.

The use of previous Landsat data should not be diminished merely because of the availability of new satellite data with better spectral and ground resolution. Although remote sensing science itself will gain advantages with the increased accuracy of the radiation recordings many more practical demands of other natural and environmental sciences will derive advantage from the time perspective offered by the Landsat data. This is especially evident within the spheres of aquatic research and investigation because these repeated registrations may often constitute the only possibility of tracing the more original situation before the changes, appearing almost daily, have taken place.

There is no denying the charm of novelty with respect to new and ever-better satellite data. But we should also maximize the continued use of all satellite information and share the evident benefits of multitemporal satellite data with all possible diversified users, including scientists, decision-makers and ordinary people interested in the state of the environment.

REFERENCES

Arkimaa, H. and J. Raitala 1981. Landsat example of small lake classification. Aqua Fennica 11: 55-60.

Arkimaa, H. and J. Raitala 1984. Landsat classification of the coastal water areas of the Bothnian Bay off Oulu. Finnish Marine Research 250: 45-51.

Beardsley, T. 1986. Remote sensing. Nature 319: 4.

Hammack, J.C. 1977. Landsat goes to sea. Photogramm. Eng. Remote Sensing 43: 683-691.

Jantunen, H. and J. Raitala 1984. Locating shoreline changes in the Porttipahta (Finland) water reservoir by using multitemporal Landsat data. Photogrammetria 39: 1-12.

Lindell, T. 1980. Calibration of Landsat data for mapping of water quality in Mälaren, Sweden. Statens naturvårdsverk PM 1266, Uppsala, Sweden.

Raitala, J. and H. Jantunen and S. Hellsten 1984a. A Landsat-assisted study of the aquatic areas of the Lake Kemijärvi region, Northern Finland. Earth, Moon, and Planets 31: 183-216.

Raitala, J. and H. Jantunen and V. Myllymaa 1984b. Developments in the evaluation of small lake water quality from digital Landsat MSS data, Kuusamo, Northeast Finland. Earth, Moon, and Planets 31: 249-264.

Raitala, J. and J. Siira and H. Arkimaa 1984c. Landsat classification of the hydrolittoral areas of the Bay of Liminka (Gulf of Bothnia, Finland). Aquilo Ser. Bot. 20: 14-23.

Raitala, J. and J. Lampinen 1985. A Landsat study of the aquatic vegetation of the Lake Luodonjärvi reservoir, Western Finland. Aquatic Botany 21: 325-346.

Acknowledgement: The writing of this paper was performed while the author held a NRC - NASA Resident Research Associateship at the Jet Propulsion Laboratory, Calif. Institute of Technology.

Symposium on Remote Sensing for Resources Development and Environmental Management / Enschede / August 1986

Analysis of Landsat multispectral-multitemporal images for geologic-lithologic map of the Bangladesh Delta

A.Sesören
Geological Survey of the Netherlands (RGD), International Institute for Aerospace Survey and Earth Sciences (ITC), Enschede, Netherlands

ABSTRACT: Geological interpretation of Landsat images of delta areas is very difficult since the delta deposits are lithologically very similar, they do not produce identifiable morphologic features, and they are not even well exposed because of dense vegetation cover during most of the year. Nevertheless, it is demonstrated in this paper that lithologic-geologic interpretation of Landsat images of delta areas is possible by a proper planning of the interpretation.

1 GEOLOGIC-LITHOLOGIC INTERPRETATION OF THE BANGLADESH DELTA

The Bangladesh Delta (fig. 1) has been studied in order to verify whether a detailed geologic-lithologic map of a delta can be obtainted from Landsat multispectral-multitemporal images by a specially planned interpretation based on a step by step separation of the units present in the delta.

A hierarchical classification of the delta areas is applied based on genetic classes, geomorphological units, and on spectral, spatial and temporal characteristics of materials. A good background knowledge of the development mechanisms of deltas is required. This classification allows the distinction and mapping of lithologically similar but genetically different deposits, and provides detailed information about sedimentation processes and deposits (fig. 1).

1.1 Subdivision of the delta area into genetic classes

On the multispectral-multitemporal Landsat images of the Bangladesh Delta different drainage and landscape patterns can be recognized (fig. 2). They result from processes related to different flow regimes. Each type of flooding (e.g. river flooding, tidal flooding, rainwater flooding) can transport and accumulate other types of sediments, building up floodplains distinguished by their specific drainage and landscape patterns. By subdividing the delta into different types of floodplains, deposits that were developed under distinct conditions will also be separated, and genetic classes of sediments are obtained. Although a landscape pattern can not always be detected from Landsat images because of its small size or because it is masked by dense vegetation, however, a drainage pattern of delta areas - even through a dense vegetation - is always detectable on band-7 (near-infrared) images and therefore can be effectively used for the sub-division of deltas into floodplains areas.

By a combined analysis of the near-infrared images of three years (fig. 2), the drainage map of the Bangladesh Delta has been prepared (fig. 3).

On this map, area (A) it is identified as "tidal floodplain", as it is crossed by a close network of innumerable rivers and creeks, typically making "zigzag" band and interconnected with each other as well with the main rivers. Obviously, tides may advance far inland in dry seasons when the water level of main rivers is at a minimum, and retreat in wet seasons when main rivers are in flood.

Area (B), constituting a large belt parallel to the Ganges River and crossed by the Arialkhan and tributary rivers, is designated as "river floodplain". Typical features are a density of drainage lower then that of area (A), and "curved bends and meanders" of the rivers which are connected with the main rivers. Locally, abandoned channels and lops of cut meanders can be recognized.

Area (C) looks different from the areas (A) and (B) due to the absence of active important creeks. The drainage pattern, which is hardly visible, is produced by straight and geometrical shapes of man-made channels. Only one small river crosses the area in between its narrow floodplain. The area is located far from the main rivers and ative channels, because of its perennially wet nature and the occurrence of large peat areas, rainwater flooding is likely to be the dominant factor. Nevertheless, the development of area (C) may have undergone some influence by tidal flooding entering the area in dry seasons from one side, and by flooding of the main rivers in wet seasons from the other side. Therefore area (C) has been defined as a "composite floodplain".

1.2 Subdivision of river floodplain into geomorphologic classes

A major river in flood normally deposits different amount and types of sediments in different parts of its floodplain, in a well-known sequence ranging from sandy to silty to loam and clay, depending mainly on topography and velocity of flow.

Changes in landscape patterns and differences in spectral signatures allow a subdivision of the river floodplain into geomorphological sub-classes. An "upper river floodplain" or "meander floodplain" occurs along the Ganges River bank, where large meanders, lops of cut menaders, long and curved ridges and abandoned channels have produced a kind of meander landscape pattern (fig. 4a). Further downstream curved ridges and wide basins come closer together and smaller, abandoned channels, large meanders disappear, and the whole pattern looks like an "embroidery on canvas" (fig. 4b), hence this part of the floodplain is designated as "lower river floodplain" or "embroidery floodplain". Of course, the latter term is not an official morphological term, but it is used here because it very clearly describes the landscape pattern as seen on satellite images.

Some areas, which do not show either a meander or an "embroidery" pattern, are separated from the upper and lower floodplain areas and are interpret-

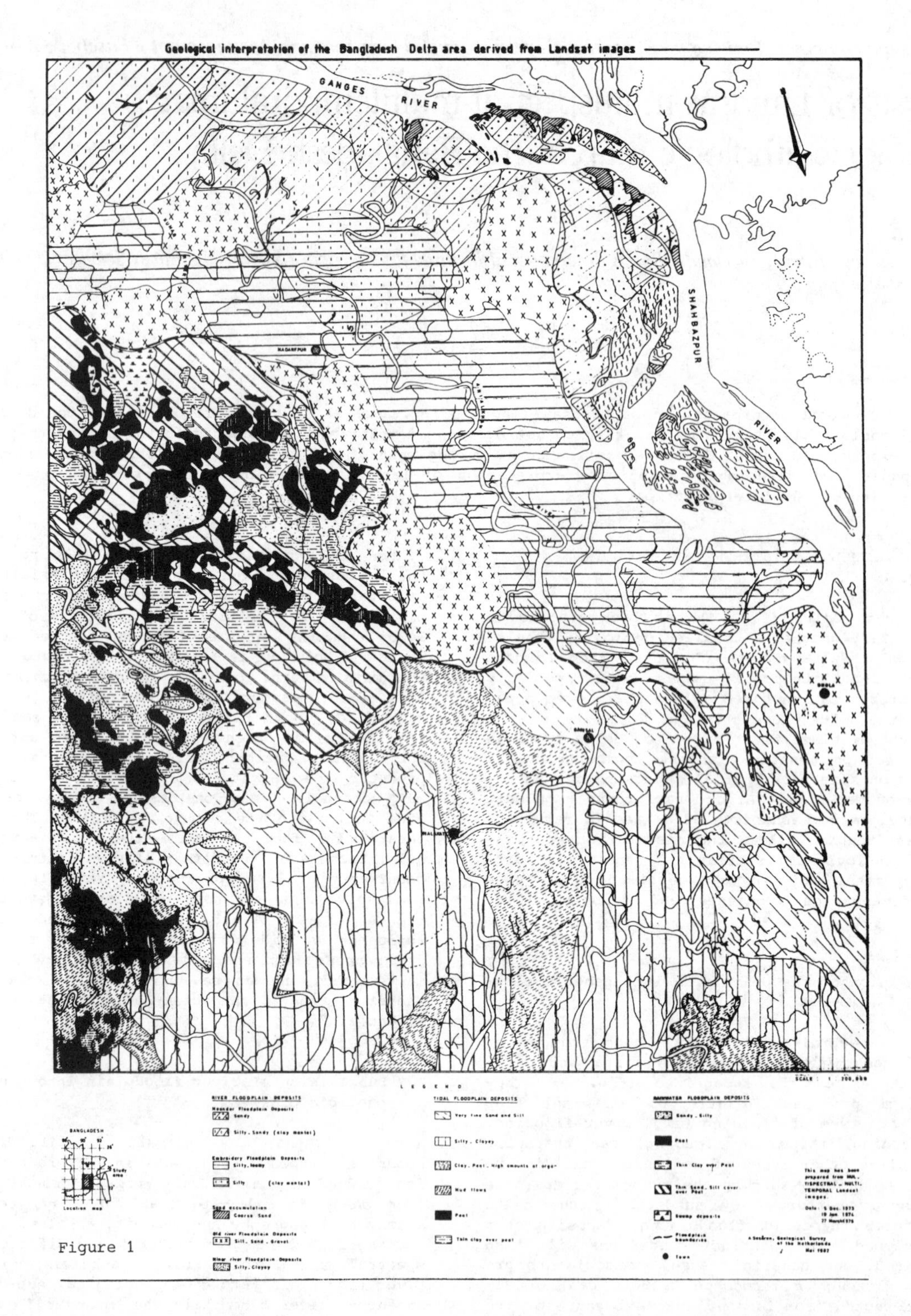

Figure 1

ed as "old river floodplain remains".

1.2.1 Subdivision of the upper river floodplain into lithologic units

Using differences in landscape patterns and spectral signatures the upper river floodplain is subdivided into two lothologic units: sandy deposits and sandy-silty deposits.

Sandy deposits: The meandering area along the Ganges River can be expected to consist of sandy materials deposited at flood times by fast flowing water (fig. 2). These areas appear white on band-5 (visible) and light-grey on band-7 (near-infrared) of the 1975 images, which corresponds to the fact that sand and silt reflect high amounts of energy in the electromagnetic spectrum.

Silty-sandy deposits under a clay mantle: In the areas further away from the Ganges riverbank, the basins appear uniform dark on the band-7 and light in colour on the band-5 images (fig. 2). Vegetation covered straight and curved ridges are clearly seen with their light colours occurring through dark coloured basins on band-7 and with their dark colours occurring through light coloured basins on band-5 images. A uniform dark appearance of basins on near infrared images suggests that they are covered by a clay mantle. This kind of clay mantle is often observed in floodplains where basins formed, long abandoned by the river and flooded again by very slow moving water. The basins under the clay mantle possibly consist of silty-sandy materials, deposited in a transition zone with decreasing water velocity.

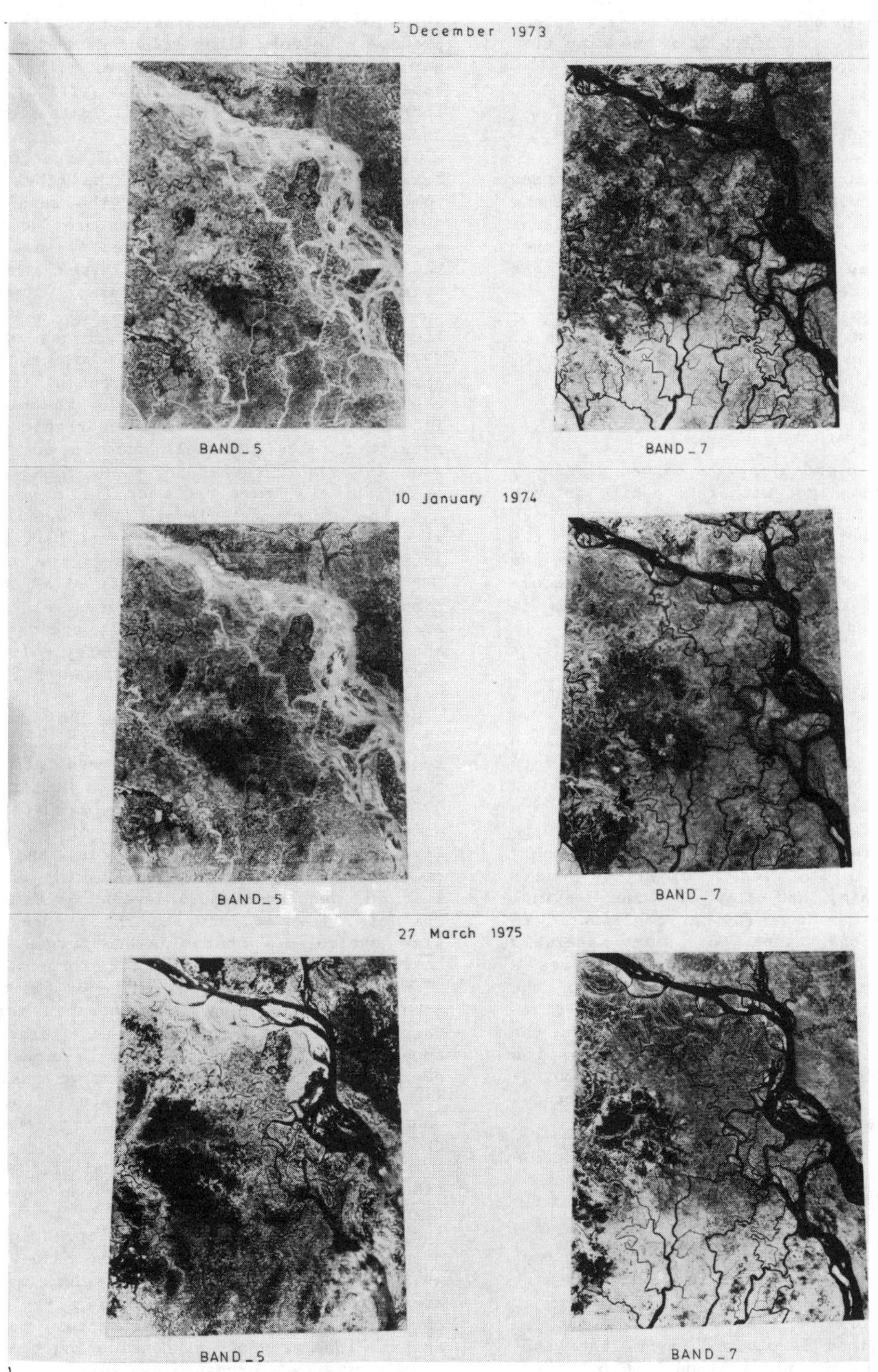

Figure 2 Multitemporal - multispectral Landsat images from the Delta - area of Bangladesh

1.2.2 Subdivision of the lower river floodplain into lithologic units

The "embroidery floodplain" pattern of the lower floodplain (fig. 4b), which can be seen on all band-5 images (fig.2) might have developed by the combination of very slow moving flood waters of the Ganges River in the North, the Kumar River in the Northwest and the Arialkhan River in the South. The basins and ridges of the lower river floodplain may contain finer materials such as silt, clay and loam. A sharp contrast produced by dark and grey colours of areas on the 1975 near-infrared images has allowed a subdivision into two units: silty deposits under a clay mantle and silty-loamy deposits.

Silty deposits under a clay mantle: As can be seen on the band-7 image, the clay mantle in the upper-river floodplain also extends somewhat into the lower river floodplain area (fig. 2). Here the crests of ridges coming out with light colours through the dark coloured clay mantle produce an embroidery pattern which is clearly recognizable. This pattern permits to trace the boundary between lower and upper floodplain under the clay mantle. The same boundary is seen on all band-5 images. The light coloured ridges visible on both band-5 and band-7 images are an indication that silty deposits might lay under the clay mantle.

Silty-loamy deposits: Apart from the dark coloured clay mantle area, the lower river floodplain appears light-grey on band-5 and slightly darker on band-7 (fig. 2). The same light-grey colour of the floodplain can also be seen on the 1973 band-5 image. This suggests the presence of sand or silt materials, which is in contrast with the dark colour in near-infrared images of silt, loam and clay with their high moisture content.

1.2.3 Old river floodplain deposits

On the band-5 images of all three years some areas in the river floodplain are separable with their sharply cut edges, smooth surfaces and different types of vegetation cover. They do not show either meander or embroidery landscape patterns. They might be older and, situated at a higher level above the floodplain, have not been effected by recent flooding of rivers. These areas are possible remains of old terraces build up by silt, sand and gravel.

1.2.4 Sand accumulations

A few small areas, in and along the Ganges riverbed, can be directly recognized with their white colours on band-5 and band-7 images of all three years. Both the very high reflection of these sediments in the visible and the near-infrared and their location are typical indications of thick and coarse sandy accumulations. It also indicates that new alluvium is still being deposited in the Ganges riverbed.

1.2.5 Minor river floodplain deposits

In the West of the study area downstream floodplain of a smaller river, the Madhumati river, extend into composite and partly tidal floodplains. Abandoned channels, lops of cut meanders, small ridges and narrow basins can be seen on the 1975 images. Slow running minor rivers such as this can deposit only finer materials (silt, loam, clay) on their downstream floodplain, and clay in lower basins. Rainwater flooding can have further contributed to clay deposition in the basins. More silty materials may have build up the ridges. The clay causes a reduction of the near-infrared reflectivity of the clay-silt combination of basins which appear medium-grey on the 1973 and 1975 band-7 images. In the visible part of the spectrum, however, silts dominating the reflectivity of a clay-silt combination cause basins to appear light-grey on band-5 1973 and 1975 images.

1.3 Subdivision of the tidal floodplain into lithologic units

In the tidal floodplain high tide water flowing back from the land into rivers during low tide cuts many deep drainage channels so that the whole area is crossed by a network of tidal creeks. These can provide only low flood levels and carry fine materials. Therefore, in the tidal floodplain only narrow levels of very fine sand and silt are formed, and silt or very fine silt and clay are deposited in the extensive basins. Due to a dark appearance of vegetated levees and a light appearance of bare basins, a very dense pattern of levees and basins is clearly seen on band-5 images. This pattern differs in detail in different locations mainly due to types of material in relation with topographical differences. Analysing the landscape pattern on band-5 images and the colours seen on both images of different dates, the tidal floodplain can be subdivided into a number of lithologic units.

A very fine sand and silt: On the 1973 band-5 image (fig. 2) a pattern produced by dark colours of vegetated densily developed small levees and light colours of bare, small and narrow basins, is visible in the areas in the North and Northwest of the tidal floodplain. The rest of the floodplain is more or less uniform dark in colour without a distinguishable pattern. Bare basins and vegetated levees produce a uniform light colour on the band-7 image, so that a pattern cannot appear either. The light appearance of basins on both visible and near-infrared images of different dates are typical indications of sand or silt.

Clay, peat and high amounts of organic matter: In between the light colour of the tidal floodplain some dark areas are seen on the band-5 image of 1974 (fig. 2). The dark colours are due to the dark appearance of both vegetated levees and bare basins. Although vegetated levees are white on near-infrared images of different dates, basins, however, appear partly dark-grey and partly as black patches on the band-7 images of 1974 and 1975. These dark-grey and black colours indicate an absorption of high amounts of energy in both visible and near-infrared regions of the electromagnetic spectrum which is characteristic for highly saturated delta materials such as peat, sediments rich in organic matter and clay. These areas might be relatively lower parts of the tidal floodplain and therefore peat layers and abundant organic material accumulated in the past, partly buried by later sedimentation, can be seen now only as dark patches. The later sedimentation might contain high amounts of clay which being always wet in the lower areas does not show remarkable changes with time and appears with the same dark-grey colours on both visible and near-infrared images of different years.

Silty-clayey deposits: After the identification of two small units in the tital floodplain the remaining area shows a nicely developed pattern on the band-5 image of 1974 and even better on that of 1975. This pattern is produced by white colours of wide basins and dark colours of vegetated long levees along the large channels, and light-grey colours of relatively narrow basins and dark colours of vegetated short levees far from the large channels. However, due to the gradual change of size and colours of basins and levees, a boundary between the two areas cannot be drawn. The basins along the large channels, white on visible images, appear also white on near-infrared images while dark-grey basins, far from the large channels, appear darker. It is therefore assumed that fine sediments accumulated in extensive basins of the tidal floodplain with silty basins along the large channels and clayey basins at a greater distance from the main channels.

Mud bands: Besides the tidal floodplain some tidal creeks can also be seen along the Meghna River. They are partly covered by green vegetation and partly bare. The bare areas appear uniform dark grey on band-5 and blackish on band-7 images of 1973 and 1974. The dark and very dark appearance of areas on images of different years typically indicate highly saturated organic matter, peat and clay as explained previously. Considering their location in the riverbeds, these areas might consist of "mud bands".

1.4 Subdivision of composite floodplain

The composite floodplain, which is found between the tidal and river floodplains, is situated far from the main rivers and active channels. Except the extension of the Madhumati River and a few tidal creeks, only a number of man-made channels can be identified because of their geometry. The dark colours on band-7 images of different years indicate its continuously saturated condition. As mentioned previously, rainwater must have played an important role in the development of the composite floodplain, besides the effects of tidal and river floodings. One of the consequences of flooding by rainwater is that organic matter rather than mineral sediments develop in perennially wet basins

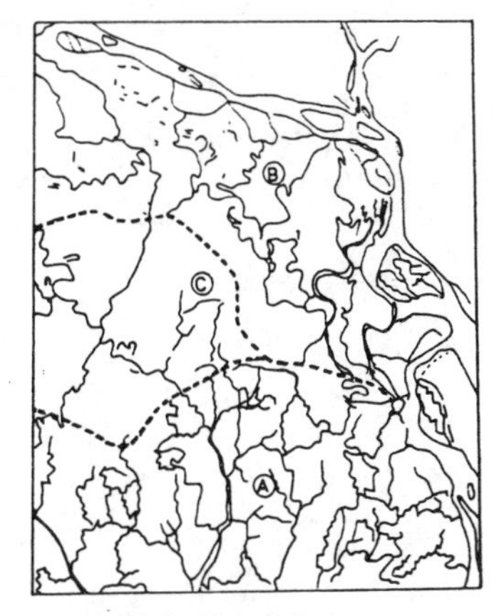

A - tidal floodplain
B - river floodplain
C - composite floodplain
Figure 3. Drainage pattern map of Bangladesh delta area

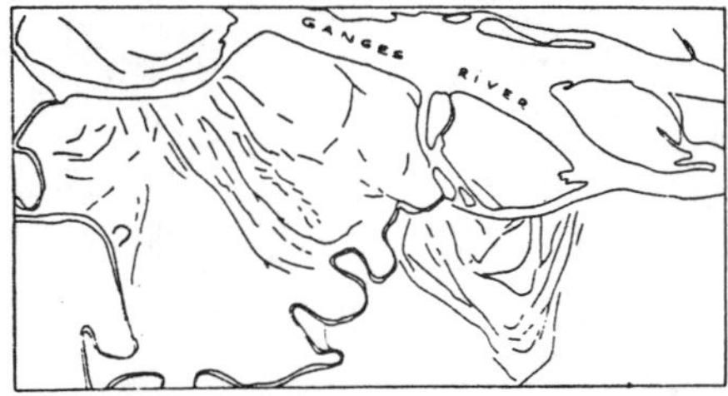

Figure 4a. Meander floodplain landscape Landsat image

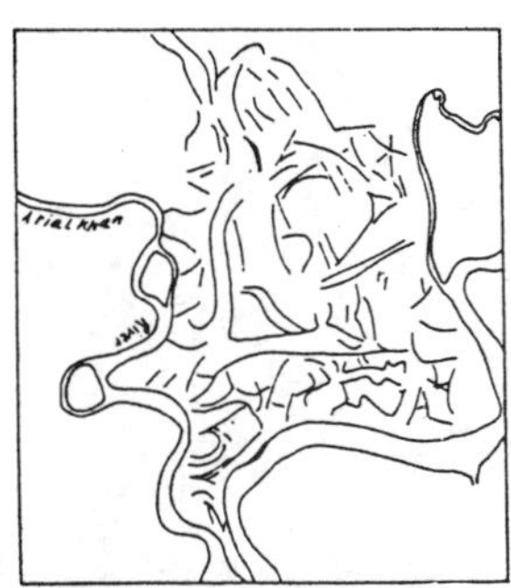

Figure 4b. Embroidery floodplain landscape Landsat image

far from active river channels. Moreover, rainwater flooding can only redistribute the finest sediments in the direction of low basins while leaving the relatively coarser ones such as sand and silt at the place where they were deposited by river or tidal flooding. According to differences in colour on both visible and near-infrared images of different years the composite floodplain is subdivided into smalle units as peat, thin clay cover over peat, thin sand-silt cover over peat, sandy-silty deposits and swamp deposits.

Peat: The largest part of the composite floodplain appears black and dark grey on the 1974 band-7 image. On the band-5 image, however, the black areas also appear black while dark grey areas are lighter, medium grey in colour. The black areas are also black on the band-7 image of 1975. The black colours on both images of different years are typical for very wet peat materials. These black areas must correspond to the lower parts of the composite floodplain where thick layers of peat could develop.

Due to differences in moisture content and vegetation cover at different times, peat areas do not appear with the same shape and size on images of two years. However, the real boundary of peat areas can be drawn by comparing the near-infrared images of these two years.

Thin clay cover over peat: The finest clay accumulates in the lowest parts of the floodplain. The areas around and between the peat layers appear medium grey on band-5 and dark grey on band-7 images of 1974. The clay materials normally show similar colours on visible and near-infrared images. When clay is highly saturated, as was the case during the 1975 recording, it appears black and blackish on near-infrared images just like the peat layers. By comparing the near-infrared images of 1974 and 1975 the boundary of clay deposits, which possibly cover parts of the peat layers, can be drawn.

Thin sand-silt cover over peat: The areas in the Southeast and in the North of the composite floodplain appear light grey on band-5 and medium grey on band-7 images of 1973 and 1974. These colours indicate that the materials are coarser and less saturated than clay accumulations. These areas, which are at the boundary with tidal and river floodplains, might consist of sandy-silty sediments partly transported and spread over the peat area by a few tidal and river creeks, and partly accumulated through rainwater flooding.

Thick sandy-silty deposits: A few small areas in between the black coloured peat layers can be easily distinguished by their light grey colour on band-5 and band-7 images of 1973 and 1974. The light colour indicates the presence of sandy-silty deposits. These areas, possibly having thick sandy-silty sediments, form relatively higher lying places in the lower parts of the composite floodplain.

Swamp deposits: Small areas in the Chitra and Madhumati River floodplains appear black in colour and similar in size and shape on band-7 images of 1973 and 1974. On the band-5 images of the same years, however, the same areas are dark grey. These colours are a typical indication of highly saturated peat, organic matter and clay materials. Considering the relations of these areas with the respective rivers, however, they are covered possibly mainly by water, organic matter and clay than by peat. They can be identified as "swamp deposits".

REFERENCES

Allum, J.A.F. 1966. Photology and regional mapping. Press Ltd., Oxford.

Bowers, S.A., Hanks, R.J. 1965. Reflection of radiant energy from soils. Soil Science, Vol. 100, no. 2, pp. 130-138.

Colwell, R.N. 1961. Some practical application of multiband spectral reconnaissance. Am. Scientist 49, 1, 9-36.

Higham, A.D. et al. 1975. Multisêctral scanning system and their potential application to earth resources survey. ESA/ASE Scientific and technical review 1. Editorial office: ESA. Scientific and technical information branch, ESTEC, Noordwijk, the Netherlands.

Holz, K.R. 1973. The surveillant science remote sensing of the environment. The university of Texas at Austin, 44 pp.

Kop, L.G. 1965. Moisture indication on grassland by vegetation and soils a comparison of maps. Neth. J. Agric. Sci., Vol. 13, no. 1.

LARS, Research Bulletin 1968. Remote sensing in agriculture. Vol. 3, Purdue University Agricultural Experiment Station, Research bulletin no. 844.

Meyers, V.I. 1970. Soil, water and plant relation, remote sensing with special ref. to agriculture and forestry, pp. 153-276.

Morgan, P.J. and McIntire, G.W. 1959. Quaternary geology of the Bengal Basin, East Pakistan and India. Bull. Geol. Soc. Amer., Vol. 70, pp. 319-342.

Pestrong, R. 1969. Multiband photographs for a tidal marsh. Photogram. Eng. 35, 453-70.

Reading, H.G. 1980. Sedimentary environments and facies. 1978 Blackwell Scientific Pub., 8 John Street, London.
Sesören, A. 1973. Application of remote sensing to a deltaic area (Zeeuws-Vlaanderen). NIWARS internal report, Kanaalweg 3, Delft, the Netherlands.
Sesören, A. 1984. Multispectral imagery and its potential application to geological studies in the Netherlands. Istanbul University, Institute of Marine science and Geography, Müsküle sok, Vefa, Istanbul, Turkey.
Sesören, A. 1984. Geological interpretation of Landsat imagery of the Bangladesh Ganges delta. ITC Journal 1984-3, pp. 229-232.
Sesören, A. 1985. Potential of remote sensing use in the Antalya region, Turkey. International symposium on karst water resources, 7-19 July, Ankara-Antalya, Turkey.
UNDP, FAO. 1971. Bangladesh soil resources, soil survey project. Techn. rep+ 3. AGL: SF/PAK 6, FAO, Rome, Italy.

Symposium on Remote Sensing for Resources Development and Environmental Management / Enschede / August 1986

Water quality monitoring of Lake Balaton using LANDSAT MSS data

H.Shimoda, M.Etaya & T.Sakata
Tokai University Research & Information Center (TRIC), Tokyo, Japan

L.Goda & K.Stelczer
Research Center for Water Resources Development (VITUKI), Budapest, Hungary

ABSTRACT: Water quality monitoring of Lake Balaton in Hungary was studied using LANDSAT MSS data. Ground truth measurements were done simultaneously with the data acquisition of MSS data and fourteen items of water qualities were measured on the lake. After certain pre-processing of MSS data, linear multi regression analyses were made between MSS data and ground truth data. Nine items among the water qualities showed correlations to the MSS data, especially transparency, chlorophyll-a, UV extinction and oxygen saturation showed sufficient and strong correlations. These four kinds of water quality patterns were clearly extracted.

1. INTRODUCTION

A water management has become one of the most important element for our life now. In this field, it is necessary to know the present status of water qualities periodically in order to control the water resources. A sattelite remote sensing is a powerful tool in this field.

In Hungary, studies of water managements are being done very actively. Lake Balaton is the largest water source in this country. The regulation of this lake is a continuous task of undiminishing importance. In this report, a joint research for water quality monitoring of Lake Balaton using LANDSAT MSS data by TRIC in Japan and VITUKI in Hungary is described.

2. STUDY AREA AND DATA ACQUISITION

In this study, the target area is the west part of Lake Balaton including Keszthely and Fonyod. System corrected LANDSAT MSS image covering the study area, acquired on the 2nd of July, 1981, was used for this study. Lake Balaton and the target area are shown in Figure 1 and Figure 2, respectively.

Ground truth measurements on the lake were done simultaneously with the data acquisition of MSS data. Fourteen items of water qualities were measured at thirty points shown in Figure 3. These items are as follows.

1) Transparency (TR)
2) Suspended solid concentrations (SS)
3) Chlorophyll-a (CH)
4) Water temperature (TM)
5) Oxygen saturation (OX)
6) Light energy on the water surface (LE)
7) Light energy reflected from the water surface (RF)
8) Light energy penetration (PE)
9) UV extinction (UV)
10) Acid soluble phosphorus concentrations (PS)
11) Acid soluble calsium concentrations (CA)
12) Acid soluble magnesium concentrations (MA)
13) The ratios of acid soluble calsium and phosphorus (CP)
14) The ratios of acid solbule magnesium and phosphorus (MP)

3. PREPROCESSING

The first step of data analyses was a preprocessing. In LANDSAT MSS data, there exists striping noises which were caused by the differences among responses of six detectors. This striping noise has big influences especially on a water quality monotoring because of low reflectances of waters. These noises were carefully eliminated using following three algorithms.

1) Mean and standard deviation matching
2) Histgram equalization
3) Random noises addition.

The method of 1) or 2) is generally used to eliminate striping noises from LANDSAT MSS image. However, these method can not fully eliminate scan line noises, because of quantization errors. In this study, these residual scan line noises were corrected by the method of 3) in order to improve the image quality. Random noises cancel the quantization errors and allow to make histograms continuously for each detectors. Figure 4 and Figure 5 show the original MSS image and destriped image.

4. IMAGE ANALYSES AND RESULTS

Image analyses were made by TIAS(Tokai Image Analysis System) 2000.

Multi regression analyses were made between fourteen items of ground truth data and four kinds of image signatures, i.e. original MSS values, mean values of 3 x3 pixels window, normalized values within 4 bands and ratios of band 4 and band 5.

Figure 1. Lake Balaton in Hungary.

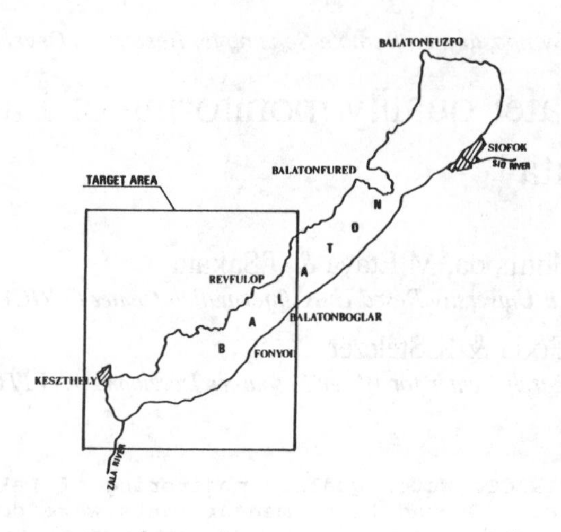

Figure 2. Target area.

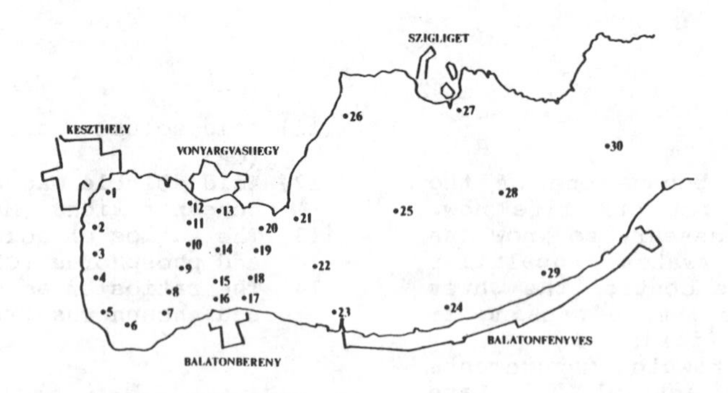

Figure 3. The position of ground truth measurements.

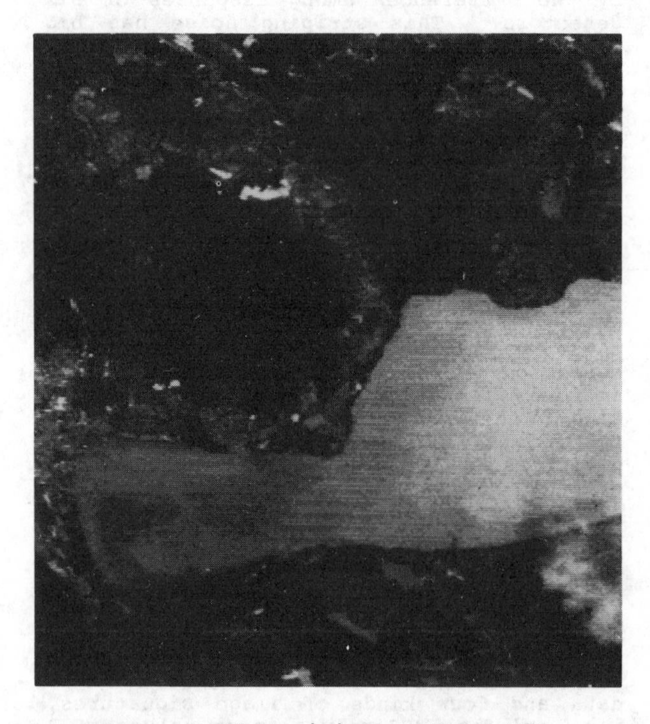

Figure 4. The original LANDSAT MSS image (band 4).

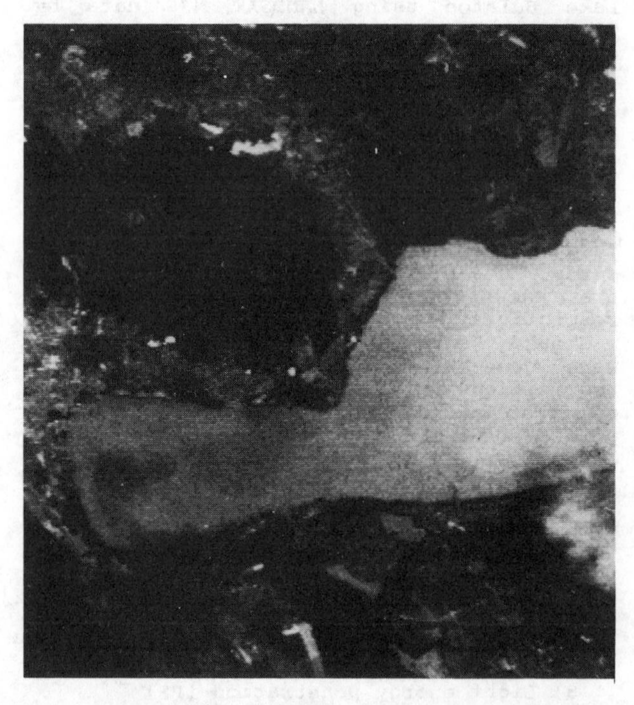

Figure 5. The destriped result of Figure 4.

Table I. Correlation coefficients between the ground truth data and four kinds of image signatures.

ITEM	ORIGINAL VALUE	MEAN VALUE (3x3 pix)	NORMALIZED VALUE	RATIO (B4/B5)
TR	**0.752**	**0.860**	**0.827**	0.301
SS	0.572	0.538	0.424	0.193
CH	**0.846**	**0.810**	0.667	0.681
TM	0.547	0.553	0.655	0.681
OX	0.502	**0.703**	**0.836**	0.524
LE	**0.773**	**0.786**	0.699	0.251
RF	**0.705**	**0.706**	0.681	0.277
PE	**0.845**	**0.773**	**0.736**	0.589
UV	**0.817**	**0.827**	**0.743**	**0.715**
PS	**0.803**	**0.757**	**0.759**	**0.747**
CA	0.448	**0.715**	0.667	0.129
MA	0.567	0.600	0.495	0.309
CP	0.684	0.681	0.595	0.579
MP	0.250	0.443	0.435	0.273

TR : Transparency
SS : Suspended solid concentrations
CH : Chlorophyll-a
TM : Water temperature
OX : Oxygen saturation
LE : Light energy on the water surface
RF : Light energy reflected from the water surface
PE : light energy penetration
UV : UV extinction
PS : Acid soluble phosphorus concentrations
CA : Acid soluble calsium concentrations
MA : Acid soluble magnesium concentrations
CP : The ratios of acid soluble calsium and phosphorus
MP : The ratios of acid soluble magnesium and phosphorus

____ : more than 0.8

Normalized values were calculated according to the following equation.

$$N_j = \frac{M_j}{\sum_i M_i} \times K \qquad (i=1,2,3,4) \quad (J=1,2,3,4)$$

Here, Mi is the mean value of 3x3 pixels window of band i, Nj is the normalized value, K is a constant, and suffix i correspond to band 4, band 5, band 6 and band 7. In order to avoid the influence of the cloud cover and its shadow in the image, ten points of ground truth data were eliminated and remaining twenty points named 1 to 20 were used for these analyses.

The resulting correlation coefficients and some examples of calculattion value which were obtained in this study are shown in Table I and Table II. As can be seen from Table I, nine items of water qualites,i.e. TR, CH, OX, LE, RF, PE, UV, PS and CA showed sufficient correlations to at least one of image signatures. Especially, TR, CH and UV had strong correlations more than 0.8 to two signatures. From this table, following results can be deribed.

1) Mean values of 3x3 pixels window showed meaningful correlation coefficients more than 0.7 to the whole of above nine items.
2) The general tendency of correlation coefficients between original values and ground truth data set was the same as mean values of 3x3 pixels

Table II. Some examples of multi regression analyses results.

(a) Transparency (MEAN)

FILE NAME EHRTR
COMMENTS : TRANSPARENCY 3 X 3 PIX. MEAN

- PARTIAL REGRESSION COEFFICIENTS -

CH00	107.591
CH01	.241650
CH02	-.634960
CH03	-1.41409
CH04	1.31702

- MULTIPLE CORRELATION COEFFICIENT -
.860

OBSERVATION NUMBER	OBSERVATION VALUE	ESTIMATE VALUE	DIFFERENCE
001	35.000	33.275	-1.7251
002	32.000	36.292	4.2916
003	29.000	29.615	.61539
004	33.000	32.891	-.10917
005	32.000	33.250	1.2503
006	50.000	49.786	-.21406
007	34.000	32.665	-1.3350
008	33.000	34.841	1.8406
009	43.000	42.020	-.98015
010	48.000	43.146	-4.8536
011	30.000	33.897	3.8972
012	41.000	34.054	-6.9465
013	52.000	48.867	-3.1329
014	37.000	40.379	3.3792
015	36.000	42.619	6.6192
016	40.000	36.333	-3.6673
017	35.000	37.389	2.3886
018	36.000	35.715	-.28546
019	36.000	37.523	1.5228
020	33.000	30.445	-2.5547

- VARIABLE -

OBSERVATION NUMBER	CH01	CH02	CH03	CH04
001	88.40	85.60	59.40	32.40
002	86.40	83.40	60.80	35.50
003	87.20	87.40	64.70	36.40
004	85.40	84.10	65.70	38.70
005	83.40	87.60	69.40	45.00
006	67.20	69.00	52.30	33.20
007	88.60	85.20	59.30	31.60
008	90.50	86.80	61.00	35.50
009	85.40	78.40	51.00	27.10
010	88.70	80.60	49.50	26.80
011	97.60	90.70	56.10	30.10
012	90.40	87.60	58.20	32.30
013	86.30	77.50	51.00	31.70
014	89.80	80.20	50.80	25.70
015	88.40	81.60	49.00	26.40
016	96.20	87.60	57.30	32.00
017	94.40	88.30	55.00	31.00
018	98.30	89.10	52.30	26.50
019	100.1	92.40	53.20	30.10
020	99.50	96.20	57.70	31.50

(b) Chlorophyll-a (MEAN)

FILE NAME : EHRCH
COMMENTS . CHLOROPHYLL-A 3 X 3 PIX. MEAN

- PARTIAL REGRESSION COEFFICIENTS -

CH00	4.45237
CH01	-.293197
CH02	-.264306
CH03	1.38374
CH04	-.453629

- MULTIPLE CORRELATION COEFFICIENT -
.810

OBSERVATION NUMBER	OBSERVATION VALUE	ESTIMATE VALUE	DIFFERENCE
001	21.400	23.406	2.0060
002	27.700	25.105	-2.5951
003	27.900	28.801	.90139
004	39.600	30.542	-9.0582
005	34.300	32.465	-1.8349
006	21.500	23.822	2.3218
007	19.700	23.678	3.9776
008	20.900	23.281	2.3808
009	19.400	16.969	-2.4306
010	23.800	13.481	-10.319
011	18.900	15.838	-3.0624
012	14.000	20.676	6.6758
013	9.6000	14.857	5.2567
014	12.500	15.562	3.0619
015	14.700	12.794	-1.9060
016	10.700	17.866	7.1660
017	14.000	15.480	1.4798
018	14.500	12.430	-2.0699
019	14.400	10.642	-3.7576
020	13.600	15.406	1.8058

- VARIABLE -

OBSERVATION NUMBER	CH01	CH02	CH03	CH04
001	88.40	85.60	59.40	32.40
002	86.40	83.40	60.80	35.50
003	87.20	87.40	64.70	36.40
004	85.40	84.10	65.70	38.70
005	83.40	87.60	69.40	45.00
006	67.20	69.00	52.30	33.20
007	88.60	85.20	59.30	31.60
008	90.50	86.80	61.00	35.50
009	85.40	78.40	51.00	27.10
010	88.70	80.60	49.50	26.80
011	97.60	90.70	56.10	30.10
012	90.40	87.60	58.20	32.30
013	86.30	77.50	51.00	31.70
014	89.80	80.20	50.80	25.70
015	88.40	81.60	49.00	26.40
016	96.20	87.60	57.30	32.00
017	94.40	88.30	55.00	31.00
018	98.30	89.10	52.30	26.50
019	100.1	92.40	53.20	30.10
020	99.50	96.20	57.70	31.50

(c) UV extinction (MEAN)

FILE NAME : EHRUV
COMMENTS : UV EXTINCTION 3 X 3 PIX. MEAN

- PARTIAL REGRESSION COEFFICIENTS -

CH00	.322390
CH01	-.756586E-02
CH02	.621260E-02
CH03	-.270545E-02
CH04	.109499E-02

- MULTIPLE CORRELATION COEFFICIENT -
.827

OBSERVATION NUMBER	OBSERVATION VALUE	ESTIMATE VALUE	DIFFERENCE
001	.10000	.86060E-01	-.13940E-01
002	.10500	.89611E-01	-.15389E-01
003	.10500	.99562E-01	-.54376E-02
004	.10000	.94333E-01	-.56674E-02
005	.14000	.13314	-.68633E-02
006	.17500	.16405	-.10948E-01
007	.80000E-01	.80816E-01	.81603E-03
008	.40000E-01	.79172E-01	.39172E-01
009	.50000E-01	.76709E-01	.26709E-01
010	.90000E-01	.68899E-01	-.21101E-01
011	.95000E-01	.52707E-01	-.42293E-01
012	.40000E-01	.86410E-01	.46410E-01
013	.70000E-01	.73025E-01	.30253E-02
014	.55000E-01	.52490E-01	-.25100E-02
015	.70000E-01	.77976E-01	.79762E-02
016	.40000E-01	.44395E-01	.43948E-02
017	.65000E-01	.66690E-01	.16897E-02
018	.50000E-01	.40930E-01	-.90699E-02
019	.50000E-01	.52200E-01	.22001E-02
020	.70000E-01	.70826E-01	.82599E-03

- VARIABLE -

OBSERVATION NUMBER	CH01	CH02	CH03	CH04
001	88.40	85.60	59.40	32.40
002	86.40	83.40	60.80	35.50
003	87.20	87.40	64.70	36.40
004	85.40	84.10	65.70	38.70
005	83.40	87.60	69.40	45.00
006	67.20	69.00	52.30	33.20
007	88.60	85.20	59.30	31.60
008	90.50	86.80	61.00	35.50
009	85.40	78.40	51.00	27.10
010	88.70	80.60	49.50	26.80
011	97.60	90.70	56.10	30.10
012	90.40	87.60	58.20	32.30
013	86.30	77.50	51.00	31.70
014	89.80	80.20	50.80	25.70
015	88.40	81.60	49.00	26.40
016	96.20	87.60	57.30	32.00
017	94.40	88.30	55.00	31.00
018	98.30	89.10	52.30	26.50
019	100.1	92.40	53.20	30.10
020	99.50	96.20	57.70	31.50

(d) Oxgen saturation (NORMALIZED)

FILE NAME : EHROXN
COMMENTS . OXYGEN SATURATION NORMALIZED VALUE

- PARTIAL REGRESSION COEFFICIENTS -

CH00	13824.7
CH01	-21.4930
CH02	-24.2083
CH03	-22.7489
CH04	-22.7185

- MULTIPLE CORRELATION COEFFICIENT -
.836

OBSERVATION NUMBER	OBSERVATION VALUE	ESTIMATE VALUE	DIFFERENCE
001	136.00	129.43	-6.5698
002	143.00	130.37	-12.633
003	125.00	118.92	-6.0820
004	120.00	125.92	5.9160
005	100.00	109.71	9.7129
006	82.000	84.787	2.7866
007	123.00	129.67	6.6719
008	124.00	126.36	2.3643
009	125.00	125.15	.14893
010	123.00	134.74	11.741
011	125.00	127.35	2.3457
012	130.00	123.85	-6.1450
013	140.00	138.49	-1.5142
014	124.00	133.81	9.8096
015	123.00	113.64	-9.3560
016	126.00	130.18	4.1753
017	128.00	116.28	-11.719
018	120.00	115.74	-4.2593
019	120.00	119.86	-.14063
020	122.00	124.83	2.8257

- VARIABLE -

OBSERVATION NUMBER	CH01	CH02	CH03	CH04
001	200.0	182.5	110.0	109.0
002	196.6	177.5	112.4	115.1
003	193.2	179.3	114.3	115.0
004	192.6	175.6	116.0	117.5
005	188.5	178.5	117.4	117.6
006	186.0	184.4	99.00	133.2
007	200.6	181.3	110.3	109.4
008	197.7	179.6	110.8	113.6
009	219.5	191.2	87.40	104.1
010	223.6	194.4	79.30	104.5
011	205.5	188.6	101.3	106.1
012	201.0	184.1	105.8	110.8
013	213.2	183.2	85.70	119.7
014	223.3	193.1	86.80	90.70
015	225.6	198.1	74.70	104.2
016	203.8	183.7	105.1	109.0
017	206.3	187.6	98.20	110.0
018	219.5	199.0	89.40	94.20
019	210.0	194.5	90.40	106.8
020	200.5	190.7	104.0	106.0

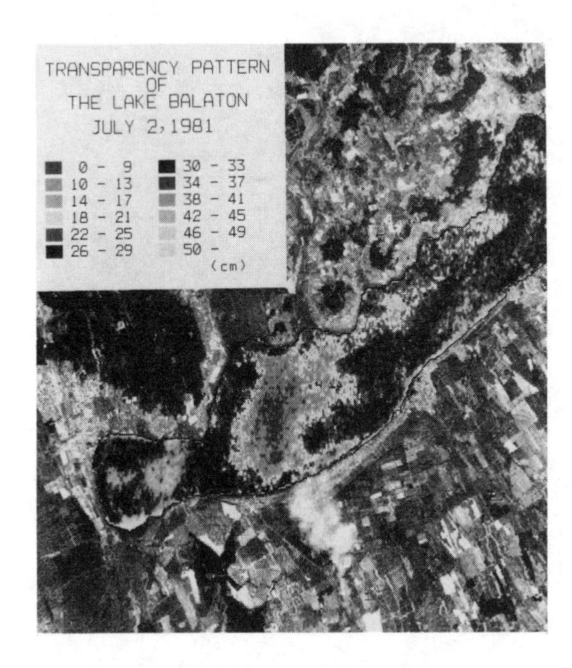

Figure 6. The extracted transparency pattern of Lake Balaton.

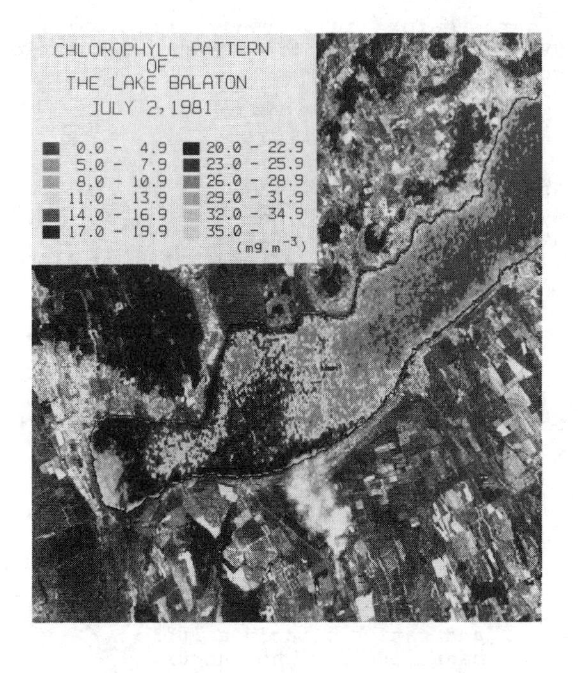

Figure 7. The extracted chlorophyll-a pattern of Lake Balaton.

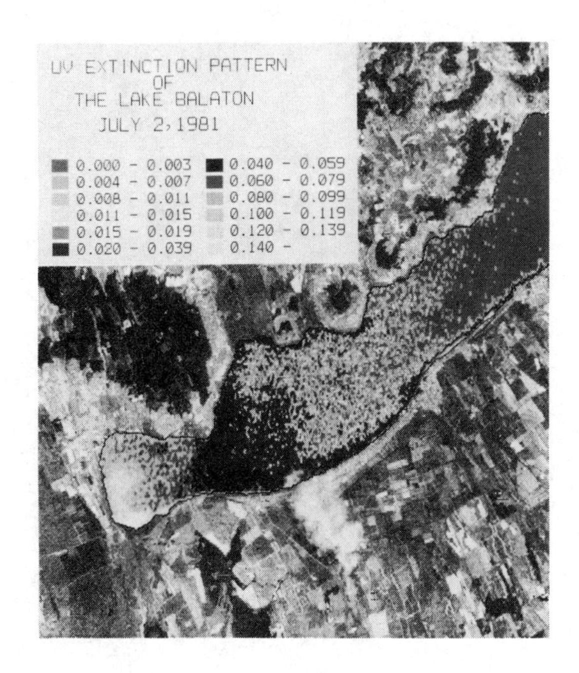

Figure 8. The extracted UV extinction pattern of Lake Balaton.

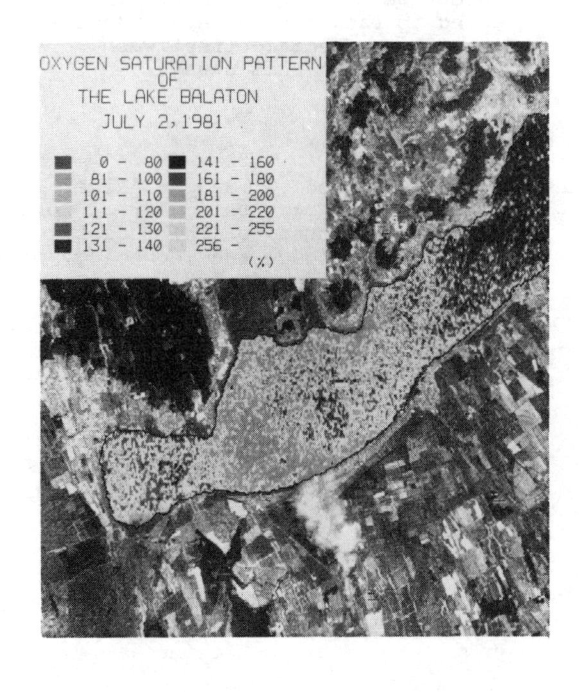

Figure 9. The extracted oxygen saturation pattern of Lake Balaton using normalized image.

window except OX and CA.

3) Only normalized values showed strong correlation to OX.
4) No strong correlations were obtained from ratios of band 4 and band 5.

There are four items, i.e. CH, PE, UV and PS, which have larger correlation coefficients than 0.8 using original values. However, these high correlations are considered not to be highly stable, because original data values are very sensitive to their positions on the image. Ground truth points addresses may not so accurate because of the errors by the geometric correction of the image and the errors in the position measurements of ground truth's.

Therefore, we have selected 4 items, i.e. TR, CH, UV and OX, which have larger correlation coefficients than 0.8 to mean values or normarized values for further data analyses. The best unbiased estimates for these four items were level sliced and color coded. The resulting water quality patterns of the lake are shown in Figure 6, Figure 7, Figure 8 and

Figure 9 for transparency, chlorophyll-a, UV extinction and oxygen saturation, respectively.

5. CONCLUSION

As a result of this study, the following conclusions were obtained for water qualities of Lake Balaton.

1) Transparency pattern extended along south shore of the lake.
2) Chlorophyll-a, UV extinction and oxygen saturation patterns generally extended from the river mouth of Zala to north-east direction.

Furthermore, for water quality monitoring using LANDSAT MSS data, the following conclusions are obtained.

1) The strongest correlation was obserbed between MSS data and transparency among those fourteen items.
2) Instead, no correlations were observed for suspended solid concentrations, water temperature and ratios of acid soluble magnesium and phosphorus.
3) Mean values of 3x3 pixels window can be considered to be the most desirable signature.
4) Although there are some reports about water quality monitoring using band to band ratios, e.g. band4 / band5, no special effect was recognized in this study.

6. REFERENCES

1) K.Fukue, H.Shimoda, T.Sakata "River Discharge Monitoring Using Aerial Color Photography", Jounal of The Society of Photographic Science and Technology of Japan, Vol. 44, No. 4, 1981.
2) K.Fukue, H.Shimoda, T.Hosomura, T.Sakata, "Improvement on Image Quality of LANDSAT MSS Images by Adding Random Noises", Jounal of The Japan Society of Photogrametry and Remote Sensing, Vol.21, No. 1, 1982.

Symposium on Remote Sensing for Resources Development and Environmental Management / Enschede / August 1986

Determination of spectral signatures of natural water by optical airborne and shipborne instruments

D.Spitzer & M.R.Wernand
Netherlands Institute for Sea Research, Den Burg, Texel

ABSTRACT: Interpretation of remote sensing imagery of watermasses requires knowledge on the spectral absorption and backscattering coefficients of the suspended and dissolved materials. Spectral (ir)radiance measurements are to be performed in order to determine the temporally and locally dependent signatures of diverse watertypes. Underwater, airborne and portable instruments ASIR (Advanced Spectral Irradiancemeter), CORSAIR (Coastal Optical Remote Sensing Airborne Radiometer) and PFC (Portable Four Channel radiometer) were developed, calibrated, tested and applied for numerous measurements in and above various coastal and oceanic watertypes. Large sets of data, both optical and seatruth, were obtained particularly during the IMERSE (Indonesian Marine Environment Remote Sensing Experiments) - Snellius II campaign.

RÉSUMÉ: L'interpretation des images, obtenues par la télédétection exige une connaissance des spectres d'absorption et de rétrodiffusion des matières suspendues et dissoutes. Des mesures de l'éclairement doivent être réalisées afin de déterminer les signatures, variables en temps et localité, des masses d'eau diverses. Des instruments aquatiques, aériens et portatifs ASIR (Advanced Spectral Irradiancemeter), CORSAIR (Coastal Optical Remote Sensing Airborne Radiometer) et PFC (Portable Four Channel radiometer) ont été développés, calibrés, testés et appliqués pour un grand nombre de mesures dans et au-dessus des diverses masses d'eau côtières et pélagiques. De grands ensembles de données, optiques ainsi que les concentrations des matières optico-actives, ont été obtenus, en particulier pendant l'IMERSE (Indonesian Marine Environment Remote Sensing Experiments) - Snellius II campagne.

1 INTRODUCTION

Investigation and experiments in the past decade have demonstrated applicability of the remote optical measurements for the study and monitoring of the distribution of the materials suspended and dissolved in the seawater. Diverse theoretical models, (semi)empirical approaches and statistical methods provide different algorithms for the retrieval of the concentrations of the phytoplankton and other suspended materials and of the dissolved organic matter. This is not surprising, considering the diversity of the aquatic conditions, algal populations and atmospheric and environmental influences. Since no general spectral behaviour of the substances suspended and dissolved in the seawater can be expected and predicted, the problem of the radiative transfer in and above water is not generally analytically solvable and thus no universal algorithms can exist.

Despite this, the optical remote sensing of the oceanic and coastal processes can supply striking results if supported by the bio-chemical and optical in situ measurements (sea truth) combined with the knowledge of the oceanography of the sensed region. Instruments must be employed specifically designed for the optical measurements in and above the waterbodies. Measuring conditions and specific objectives of the research determine spectral, spatial and time resolution of such instruments and data acquisition systems.

2 INSTRUMENTS

For the interpretation and correction purposes of satellite imagery, special requirements are put on the in situ measuring procedures. Design and application of the underwater and of the abovewater (low altitude) instruments substantially differ.

2.1 Underwater measurements

The investigations on the spectral properties (i.e. the absorption and scattering signatures) of natural waters, resulting into establishment of the relationships between the upwelling optical signals and the composition of the watercolumn, so called "colour algorithms", are preferably to be performed near surface underwater. Doing this, no influence of the surface reflection (glitter) and of the atmosphere is accounted, though experimental constrains are introduced by wave motion. Optimal depth of the measurements must be chosen, depending on the seastate, possible stratification of the watercolumn and on the solar conditions. Both, upwelling and downwelling spectral irradiance must be measured in real time, giving then the quasi-inherent reflectance. The spectral behaviour of the reflectance depends on the composition of the seawater (sea truth). Inversly, the absorption and scattering signatures and hence the concentrations of the suspended and dissolved materials can be derived from the reflectance. Short duration of a spectral scan is crucial, with respect to the horizontal and vertical instability of the watermass and to the variability of the incident solar radiation.

The developed Advanced Spectral Irradiancemeter (ASIR) can scan simultaneously 22 spectral channels between 400 nm and 720 nm within several seconds. Spectral bandwidth of each channel is within 10 nm. Radiation is collected by two cosine diffusers at each side (up and down) of the instrument mounted in gimbals. The upwelling and downwelling irradiance is simultaneously detected and recorded on board by means of an HP data acquisition and storage system controlled by microcomputer. Variations of the incident solar radiation are recorded by a separate instrument mounted at the top of the measuring platform (vessel). Irradiance depth profiles, characteristic for the structure of the watercolumn, can be recorded at a single chosen channel as well.

Fig. 1 shows an example of in situ spectral measurements by means of the ASIR. A depth profile is shown in Fig. 2.

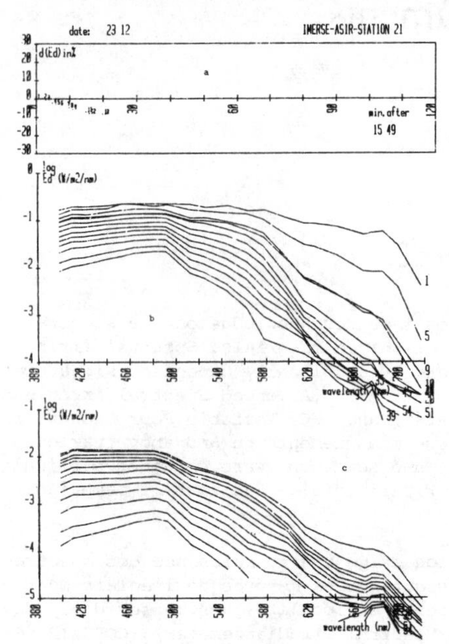

Figure 1. Record of in situ underwater irradiance measurements by means of ASIR. Variations of the incident solar irradiance during the whole measuring period (a), downwelling irradiance spectra (b) and upwelling irradiance spectra (c) are displayed at several depths, as measured at a station in Java Sea.

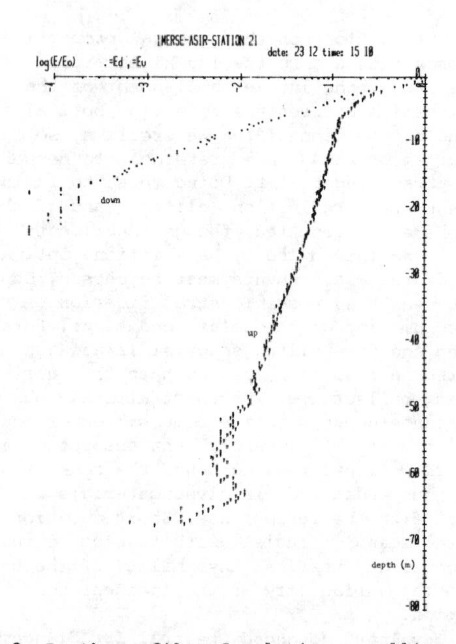

Figure 2. Depth profile of relative upwelling irradiance at 685 nm (fytoplankton pigment fluorescence maximum), as recorded by means of the ASIR.

2.2 Spectral measurements above water

When large spatial diversity of aquatic conditions and/or influence of the glitter and of the lower part of the atmosphere are investigated, measurements are to be performed above water. In order to allow the comparison between the underwater reflectance with the upwelling radiation recorded above the seasurface, rapid detection of downwelling irradiance and of downwelling radiance in the sun-detector plane is needed along with the measurement of the upwelling radiance. From a set of the collected radiance data, corrected for the glitter and atmospheric effects and related to the sea truth, algorithms and their variability can be studied. The instrument can be mounted on a stable platform (turret) above water, or preferably employed from a survey aircraft flying at low altitudes. Spatial resolution (angle of view) must be chosen accordingly to the expected horizontal inhomogeneity of the surface layer and to the wave amplitude and frequency. Fine spectral resolution of the measurements allows than final tuning of the algorithms.

Coastal Remote Sensing Airborne Radiometer (CORSAIR) was constructed employing a zoom objective lens for the radiance measurements, cosine collector for the downwelling irradiance measurements, liquid light guides, automatic optical switch and correction filters assembly, detector, polychromator, optical multichannel analyser, console and a microcomputer system providing rapid data acquisition and storage. Field of view and tilt of the objective can be adjusted. 125 channels between 400 and 720 nm can be scanned within 32 ms.

An example of the airborne measurements by means of the CORSAIR is presented in Fig. 3

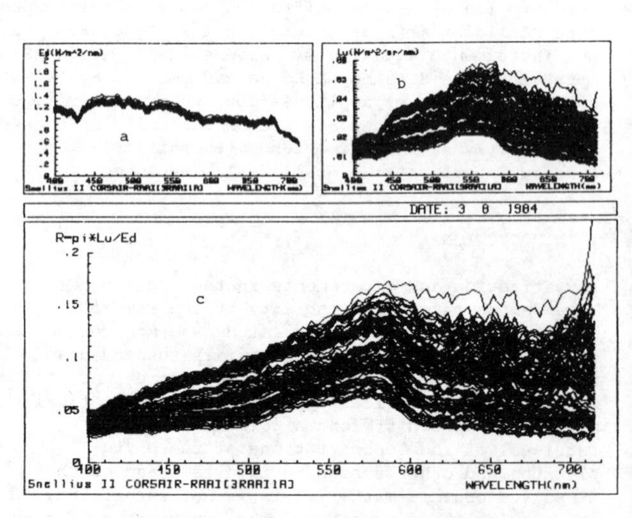

Figure 3. Airborne measurements by means of the CORSAIR as performed during a survey flight above the Strait Madura with highly variable suspended matter load. Downwelling irradiance (a), upwelling radiance (b) and their ratio (c) are displayed as function of the wavelength.

2.3 Broadband measurements above water

Some environmental processes and features can be remotely measured by using sensors with only a few broad spectral channels. For instance total suspended matter distribution or morphology in the coastal regions can be monitored by the Landsat MSS or TM instruments. In some cases even the AVHRR of the NOAA satellites can be applied. The relationships between the optical signals and the sea truth (algorithms) are generally dependent on the local and temporal (seasonal) conditions and should be determined prior to the interpretation of the relevant satellite imagery. The low altitude measurements, whether from a stable platform or from a low flying aircraft, should be preferably synchronized with the satellite overpasses. The algorithms are currently determined by statistical analysis of the low altitude (or under water) optical data in relation to the sea truth parameters sampled simultaneously.

Our Portable Four Channel radiometer (PFC), specially developed for the purposes of the establishment of the local algorithms, can be equiped with interchangeable sets of filters with spectral characteristic corresponding to the optical and near infrared channels of the Landsat MSS, TM and SPOT sensors. The upwelled radiation is collected by a zoom lens objective allowing choice of the spatial resolution dependently on the altitude of the measurements and on the spatial resolution of the satellite scanner of interest. The PFC is a lightweight battery operated instrument. Its four channels can be rapidly scanned and results stored by a microcomputer controlled data acquisition system or by a simple automatic data logger.

Figure 4 shows a relationship applicable for the determination of the pigment concentration as investigated in the western Wadden Sea for a possible interpretation of the Landsat TM data.

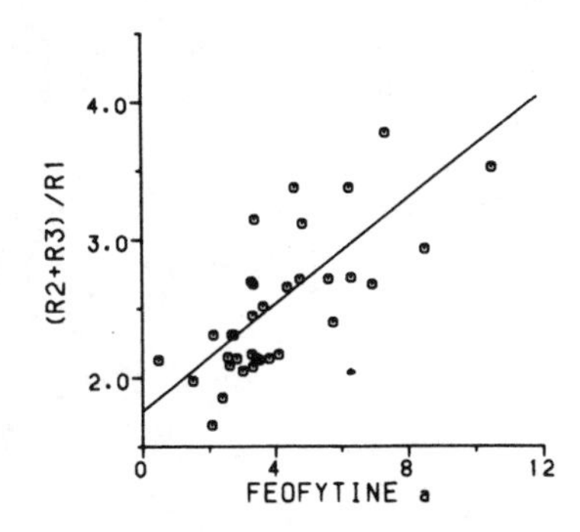

Figure 4. Combination of the upwelling radiance signals in the first three TM channels in relation to phytoplankton pigment concentration as determined from the measurements by means of the PFC above western Wadden Sea during September 1985.

3 INDONESIAN MARINE ENVIRONMENT REMOTE SENSING EXPERIMENTS (IMERSE)

Combination of the data obtained from the sets of measurements in water and above water at low altitudes is most advantageous for the final processing and interpretation of the satellite imagery. During the extensive campaign IMERSE performed in the coastal waters of north east Java, within the framework of the Snellius II expedition, numerous measurements were done of the underwater reflectance along with sampling of concentrations of the organic and inorganic materials suspended and dissolved in the water. Airborne measurements of the upwelling radiance and collection of multispectral imagery were provided simultaneously with the shipborne operations. Collected satellite imagery will be interpreted in cooperation between Indonesian, German and Netherlands scientists.

Preliminary results showing the particulate matter distribution as determined from the airborne measurements above the Strait Madura are presented in Figure 5.

The spectral signatures will be calculated from the underwater measurements in combination with the sea truth data using a radiative transfer model. From the absorption and scattering signatures, the algorithms will be determined. The airborne radiometer data will serve for the final tuning of the algorithms as well as for the evaluation of the atmospheric and glitter effects. The airborne imagery can be then processed and interpreted for the representative coastal and open sea regions. From these underwater and low altitude data the satellite imagery will be corrected for the atmospheric effects and interpreted in terms of the concentrations of suspended and dissolved materials representative for the whole area of interest.

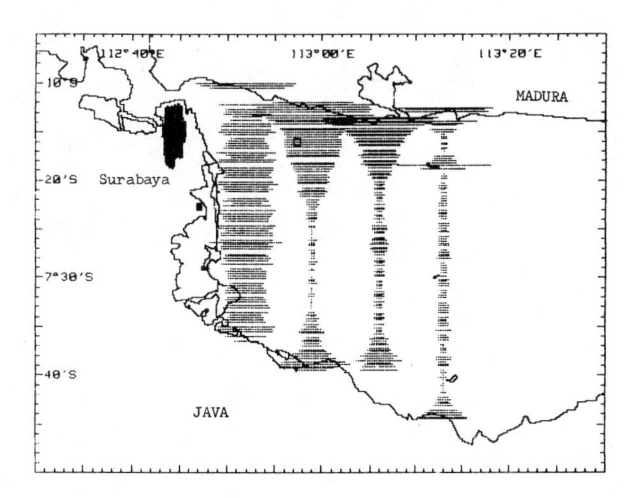

Figure 5. Distribution of suspended matter as assessed from the airborne measurements by means of the CORSAIR along four flight tracks above the Strait Madura. A ratio algorithm is applied. Length of the ticks corresponds with the concentration of the suspended matter varying between about 2 and 30 g/m^3.

REFERENCES

Participation in the Snellius II Expedition 1984. Joint Experiment Report, 1986. DFVLR, LAPAN, NIOZ.

Symposium on Remote Sensing for Resources Development and Environmental Management / Enschede / August 1986

Classification of bottom composition and bathymetry of shallow waters by passive remote sensing

D.Spitzer & R.W.J.Dirks
Netherlands Institute for Sea Research, Den Burg, Texel

ABSTRACT: Reflectance of solar radiation contains information about the composition of the watercolumn as well as about the bottom in areas where the visible light penetration depth exceeds the bottom depth. When the mapping of the bottom depth and composition is pursued, specific algorithms must be developed in order to remove the influence of the watercolumn on the upwelling optical signals. Calculations were performed relating the reflectance spectra to the parameters of the watercolumn and of the diverse bottom types. Measurements of the underwater reflection coefficient of sandy, mud and vegetation-type seabottom were performed. Two-flow radiative transfer model was employed, where the spectral signatures of suspended and dissolved materials and of the bottom were introduced as the input parameters. Several algorithms are proposed with respect to the application of Landsat MSS, TM and SPOT HRV scanners. Bottom depth and features appear to be observable down to 3-20 m, dependently on the water composition and bottom type.

RÉSUMÉ: La réflexion diffuse du rayonnement solaire détient des informations sur la composition de la colonne d'eau ainsi que sur le fond de la mer dans des zones où la profondeur de pénétration de la lumière visible excède la profondeur du fond. Pour dresser la carte de la profondeur du fond et de sa composition il nous faut des algorithmes capables d'éliminer l'influence de la colonne d'eau sur les signaux optiques ascendents.
Des calculs ont été faits pour estimer le rapport entre les spectres de réflexion diffuse avec les paramètres de la colonne d'eau et des catégories diverses du fond. Des mesures de la réflexion diffuse dans l'eau ont été effectuées sur des fonds de catégorie sableuse, boueuse et végétative. Le modèle de transfert radiatif des deux flux a été utilisé, dans lequel les signatures spectrales des substances dissoutes, des matières en suspension et du fond ont été introduit. Plusieurs algorithmes ont été proposés en rapport avec l'application de Landsat MSS, TM et SPOT HRV scanners.
La profondeur et composition du fond se trouve détectable jusqu'à 3-20 m, selon la composition de l'eau et de la catégorie du fond.

1 INTRODUCTION

In shallow waters and tidal areas solar radiation transmitted by the water layer and reflected by the bottom can substantially contribute to the upwelling optical signals. When appropriate absorption and backscattering signatures (coefficients) of both, the waterlayer and of the bottom, are introduced into a rigorous radiative transfer model, the contribution of the bottom reflectance can be evaluated and used for the remote determination of the bottom depth and type.
Spectral despendence of all the optical coefficients must be accounted with respect to the spectral characteristics of the remote (satellite or airborne) sensors. Recent investigations on the optical properties of the watermasses and on the radiative transfer in natural waters, generally initiated by the development of modern remote sensing techniques, allow calculation of the upwelling spectral radiance and hence establishment of useful algor ithms for bottom depth and bottom and watercolumn composition mapping.

Two-flow radiative transfer model in its reflectance form is used in this paper, with as boundary condition the (wavelength dependent) bottom reflectance. Variable concentrations of chlorophyllous and non-chlorophyllous particulate and organic dissolved materials are accounted as well as various types of the sea bottom. Reflectances within the spectral bands of the Landsat Multi-Spectral Scanner (MSS), Landsat Thematic Mapper (TM), SPOT High Resolution Visible (HRV) and the TIROS-N series Advanced Very High Resolution Radiometer (AVHRR) were computed in order to develop appropriate algorithms suitable for the bottom depth and type mapping. Ground resolution (20-80m) of the Landsat and SPOT instruments is more suitable for the coastal mapping purposes than the resolution of the Coastal Zone Color Scaner (CZCS, 800 m) and of the AVHRR (1100 m). However, the wide accessibility of the AVHRR data makes also this sensor attractive when large scale mapping is pursued. The orbital repeat cycles of Landsat and SPOT (16-26 days) are sufficient for observation of long term bottom depth and composition variability.

Linearity and high sensitivity of the algorithms for the quantities to be detected is desired along with low sensitivity of the other parameters.

2. TWO-FLOW RADIATIVE TRANSFER MODEL

Upwelling radiance measured by the remote optical sensors is proportional to the reflectance R. The reflectance dependence on the depth z and wavelength λ can be described by the radiative transfer equation

$$\frac{dR(z,\lambda)}{dz} = - b_d(\lambda) + (a_u(z) + b_u(z) + a_d(z) + b_d(z))\ R(z,\lambda) - b_u^2(z) R^2(z,\lambda)$$

where a_u, a_d are the absorption coefficients for the upwelling and downwelling irradiance respectively, and analogously b_u, b_d are the backscattering coefficients. For shallow waters, where the light penetration depth exceeds the bottom depth, the boundary condition $R(h,\lambda) = r(\lambda)$, where $r(\lambda)$ is the bottom reflectance, determines the solution of the radiative transfer equation. The supernatant waterlayer is assumed to be vertically homogeneous, i.e. the absorption and scattering coefficients are then depth independent.

For natural waters where the absorption dominates the scattering, solution for the near surface reflectance R_o can be approximated as

$$R_o(\lambda) = \frac{b_d(\lambda)}{a_u(\lambda) + a_d(\lambda)} + r(\lambda)\, e^{-(a_u(\lambda) + a_d(\lambda))h}$$

The first term of the above equation is depth independent and represents the deep-water reflectance. The radiance L detected by the remote sensing instruments can be calculated from an integral of the R_o multiplied by spectral sensitivity of each channel of the scanner.

Since the reflectance of the natural water is very low at the longer wavelengths, calculations can be limited to 700 nm and thus only the response in the first two (MSS, HRV) or three (TM) bands of the optical scanners can be further considered.

No substantial differences were found when using in the algorithms computations MSS band 4 (500-600 nm), TM band 2 (520-600 nm) or HRV band 1 (500-590 nm); and analogously MSS band 5(600-700 nm), TM band 3 (630-690 nm), HRV band 2 (610-680 nm) or AVHRR band 1 (580-680 nm).

Seawater absorption, scattering and bottom reflectance coefficients, either measured in our laboratory of tabulated previously (Prieur and Sathyendranath, 1981; Lyzenga, 1978), were used in the computations. Three bottom types: sandy, mud and vegetation were considered.

The applicability of the computed algorithms is limited by radiometric sensitivity (noise equivalent reflectance) of the instruments 0.5-0.8% (Robinson 1985).

3 RESULTS

Figures 1,2,3 show examples of the computed reflectance spectra for the sand, mud and vegetation bottom types respectively.

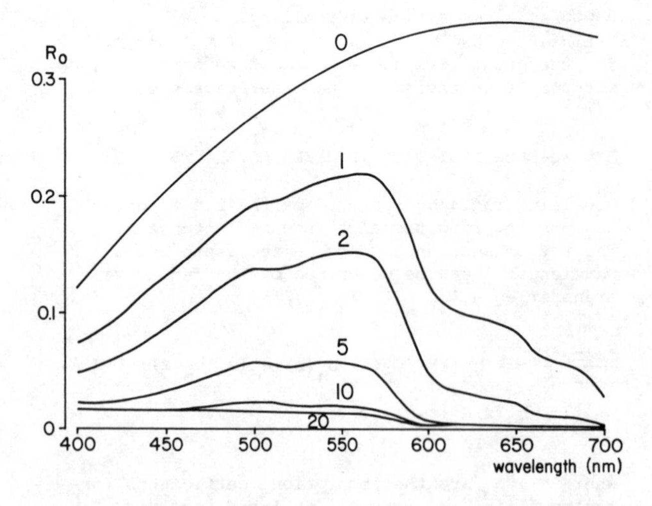

Figure 1. Reflectance spectra of natural water with sandy bottom. Depth in metres are indicated. The line at zero depth represents the bottom reflectance.

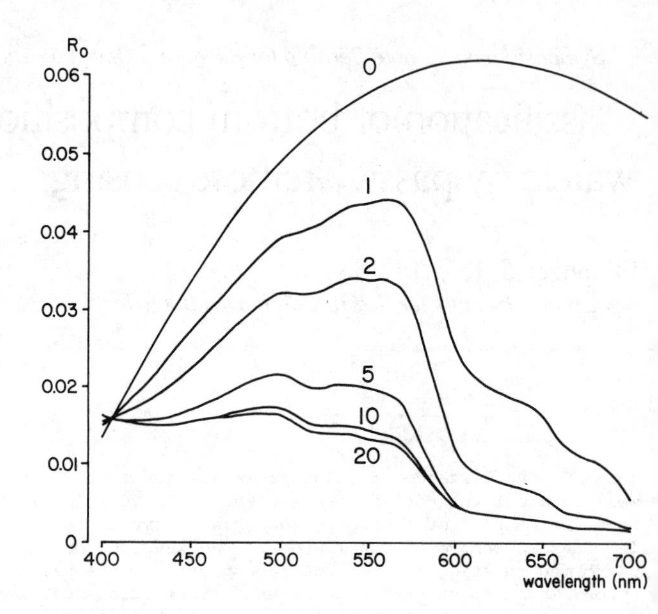

Figure 2. Reflectance spectra of natural water with muddy bottom. Depths are indicated.

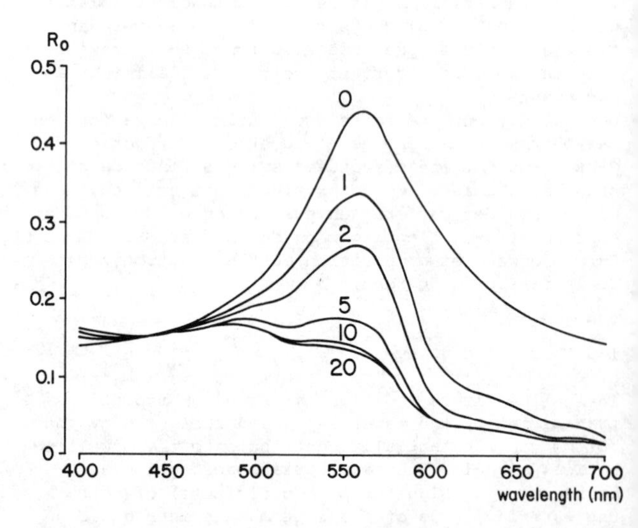

Figure 3. Reflectance spectra of natural water with vegetation-type bottom. Depths are indicated.

Algorithms for the bottom depth and for the bottom composition assessment were further investigated. Requirements for the applicability of each type of the algorithm differ. An optimal bottom depth algorithm should be insensitive to the variation of the bottom and watercolumn composition, while a bottom composition algorithm should be insensitive to the watercolumn structure including the depth. Logarithmic transformation of the differences between the shallow-water and the deep-water reflectances (or radiances) appears to be advantageous due to the exponential depth dependence of the irradiance (Lyzenga, 1978).

Several suitable algorithms can be proposed from the model calculations.

Generally the algorithms can be expressed as

$$A = \sum_i k_i \ln (L_i - L_{iw})$$

where L_i is the integrated radiance in the channel i,

L_{iw} is the deep water radiance, k_i are positive or negative constants determined from the spectral signatures of the water-bodies concerned. Obviously, knowledge on the local conditions, like turbidity of the water and bottom features will increase the accuracy of the mapping.

Examples of the dependence of the algorithms on the bottom depth and composition for both clear and turbid water regions are shown in Figures 4 and 5.

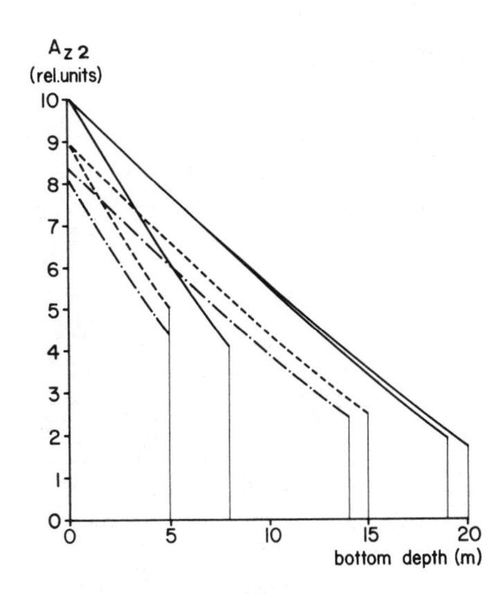

Figure 4. Bottom depth algorithm as function of the bottom depth for sandy (full lines), muddy (dashed lines ---) and vegetation (dashed lines -·-·-) bottom types. The constants $k_1 = 1$, $k_2 = 0.5$. Longer lines are representative for clear waters, shorter lines belong to turbid water. Limit values of the detectable depths are indicated.

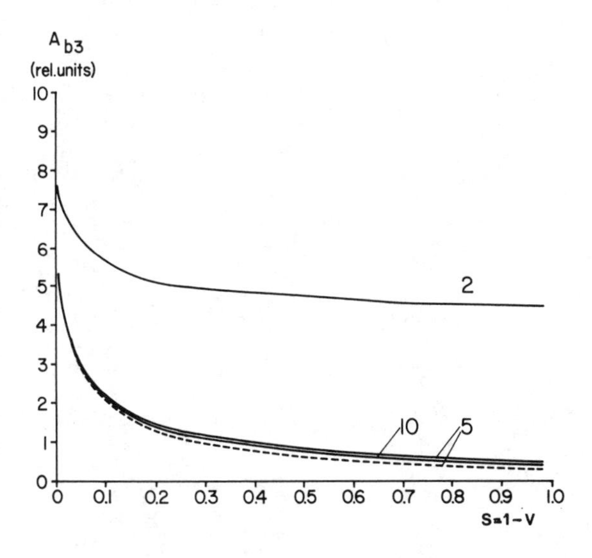

Figure 5. Bottom composition algorithm as function of the bottom reflectance for mixed sand-vegetation bottom types, S = 0 for fully overgrown bottom, S = 1 for pure sandy bottom. Lower lines are representative for clear waters with bottom depths as indicated in metres, upper line belongs to shallow turbid water region. $k_1 = -1.05$, $k_2 = 1$.

Final tuning of the algorithms, based on the local seatruth data should be done for each area to be mapped. A case study for North Sea coastal areas, using Landsat TM data, will be performed.

REFERENCES

Lyzenga, D.R. 1978. Passive remote sensing techniques for mapping water depth and bottom features. Appl. Optics. 17: 379-383.

Prieur, L. & S. Sathyendranath 1981. An optical classification of coastal and oceanic water based on the specific spectral absorption curves of phytoplankton pigments, dissolved organic matter, and other particulate materials. Limnol. Oceanogr. 26: 671-689.

Robinson, I.S. 1985. Satellite Oceanography. Chichester: Ellis Horwood.

Satellite remote sensing of the coastal environment of Bombay

V.Subramanyan
Indian Institute of Technology, Bombay

ABSTRACT: Remote sensing of the coastal environment of Bombay using the Landsat imagery (black and white and false-colour paper prints) makes it possible to trace the development and evolution of the coast. The coast is indented by many bays, promontories, creeks, tidal mudflats, stacks and beaches. It is thus one of retrogradation. All the creeks follow lineaments and there is a triple junction of such lineaments to the east of Bombay. These clearly indicate the erosional development of the region by fluvio-marine processes. A reconstruction of the initial configuration of the Bombay region is attempted. Marshes, raised beaches and littoral concrete point to the emergence of the region in the past. The bays and headlands will be eliminated towards the end of the current marine cycle and the coast, straightened. Three rows of cuestas of volcanic rocks trending NNE-SSW with westerly dipslopes traverse the island lengthwise; there are three aligned lakes.

1 INTRODUCTION

The Bombay coast, including that of the Salsette island to the north, forms a part of the Konkan Coast of the West Coast physiographic division (National Atlas 1964). The geomorphic environment of this coastal tract and the neighbouring Nhava-Sheva-Uran shoreline to the east was studied using the following data products of the Landsat: black and white paper print on MSS band 6 on the scale of 1:250,000 and false-colour composite paper print on MSS bands 4 (yellow), 5 (pink) and 7 (blue) on the scale of 1:500,000. The images were further magnified during the visual interpretation.

Geological and geomorphological field traverses were undertaken in selected localities for the collection of ground truth. The topographical maps on the scale of 1:63360 and the Town Guide Map of Bombay on the scale of 1:25,000 were also consulted for supplementary data.

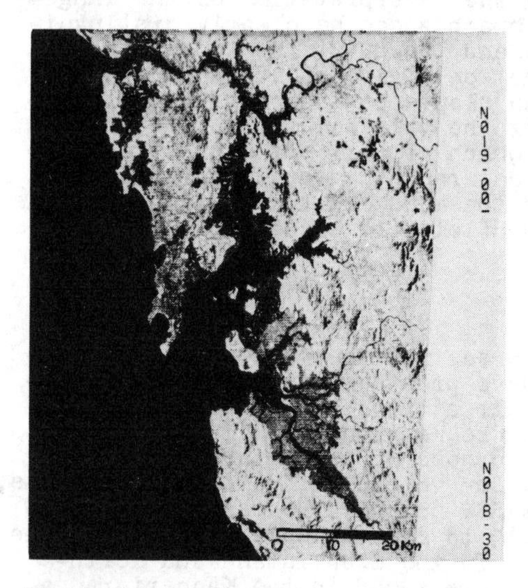

Figure 1. A Landsat view of the Bombay-Salsette island and the neighbouring country to the east and southeast on MSS band 6. The coastal and inland features show up very well.

2 SATELLITE DATA

The black and white image brings out clearly the configuration of the Bombay-Salsette coast and the Nhava-Sheva-Uran coast to the east. Three well-developed bays, namely, the Manori bay, the Mahim bay and the Back-bay and the broad embayment between Versova and Juhu show up along with their corresponding promontories (Fig.1). Sandy beaches can be readily recognised by the white tone around Manori, Versova, Juhu, Bandra and Girgaon. Rocky beaches with boulders, cobbles and gravels as well as wave-cut platforms can be identified at other places along the coast by the pale grey tone. The cuestas bordering the three lakes and continuing farther north display a light and shade effect on their escarpments and dipslopes. The morphological signature of the cuestas south of the Powai lake is a faint grey line whereas in the Kanheri Hills in the northern part of the island, the cuestas occupy a considerable area - about 25 sq. km. The big cuesta at the centre of the Trombay island in the east also shows up as a thick grey line. The mudflats that have formed in the tidal zone between the low and the high tides are readily recognised by their dark grey tone and their locations around the creeks. The lakes and the creeks appear black whereas the marshes in the Mahim and Chembur areas appear as medium grey patches.

The basalts are seen as pale grey areas in the Bombay island while the trachytes, rhyolites and tuffs in the Salsette island appear much paler. The salt fields have registered themselves as white streaks, expectedly. The urban, suburban and vegetated areas show a light grey tone.

The false-colour composite facilitates the

recognition of all the features described above. The sandy beaches show up as white strips, the tidal mudflats appear mixed areas of pink and blue and the lakes have recorded a deep blackksh blue shade. The marshes and the swamps are seen as blue areas. The basaltic cuestas show a paler shade of blue on the escarpments and a darker shade of blue on the dipslopes. The ridge crestline of the Vikhroli-Ghatkopar range shows up as a fine line whereas the Kanheri range is seen as a very wide dark pink patch. The urban and suburban areas appear as very light areas with patches of pink representing vegetated hills. Open green turfs like the race course are seen as deep pink areas. The rocky beaches in the Juhu-Bandra stretch show up as wavy blue lines along the shoreline. The salt fields are recognised by their white tone and the sandy spit at Versova is clearly picked up as a fine, white projection into the sea.

3 MAP DATA

The topographical maps on the scale of 1:63360 and the Town Guide Map on the scale of 1:25,000 were studied and these have added more local details while generally confirming the interpretation of the imagery. Rocky beaches can be clearly distinguished from sandy beaches. The topographical contours not only give an idea about the relief and slopes but also indicate the attitude of the volcanic rocks through the landforms which are all structural landforms. The tidal range is very clearly understood from the maps and the drainage details over the mudflats are exceedingly clear.

4 FIELD DATA

Field traverses for the collection of ground truth have provided data on the rock types and their attitudes. Three low ranges of volcanic rocks trending NNE-SSW traverse the island producing promontories bordered by cliffs. The volcanics, dominantly basalts, strike NNE-SSW and have westerly dips ranging from 11° to 23°. Trachytes and rhyolites are exposed in the Madh island and northward. Tuffs are found in the Kanheri range along with agglomerate. Intertrappean sedimentaries are exposed at a number of localities like Jogeshwari. Dykes of dolerites are met with in Trombay and in the Powai hills.

Geomorphological field investigations along the coast over the past ten years have revealed the recurrence of accentuated marine erosion in the area south of Versova. No other stretch along the coast has been or is undergoing erosion. Shetti and Subramanyan (1985) have studied the spit which is trending southwestward from the mouth of the Malad creek and have attempted to simulate its growth through a computer programme.

Field studies also indicate that the multistorey buildings have given rise to the densest concrete jungle in the Back Bay area in the southernmost part of the island and that urbanization is proceeding at a fast pace all along the coast in particular.

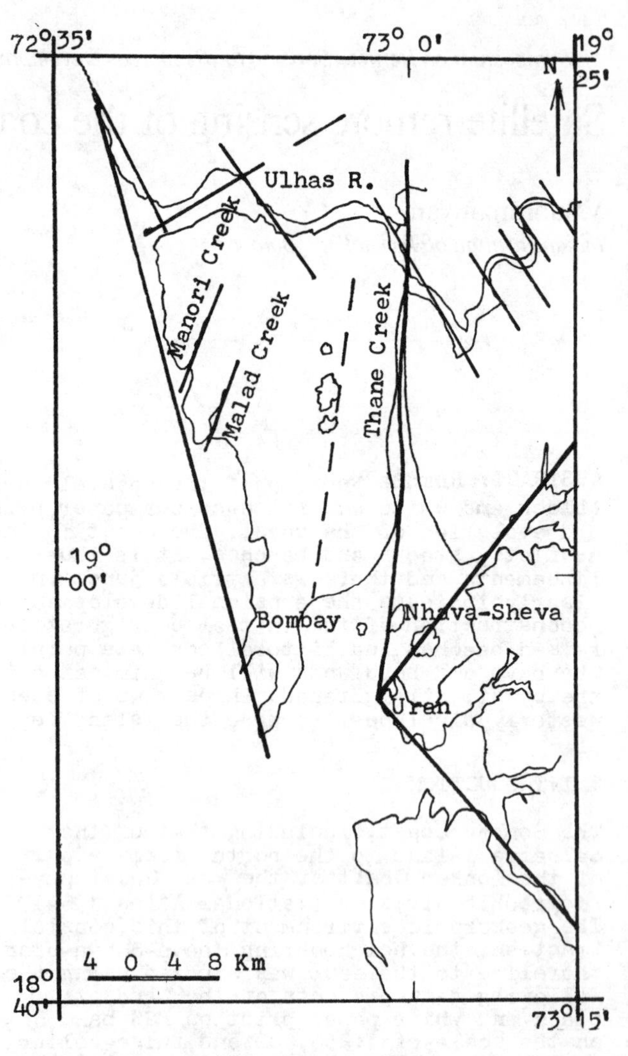

Figure 2. Map showing the lineaments in the Bombay region and neighbourhood which appear to have controlled the creeks and the river.

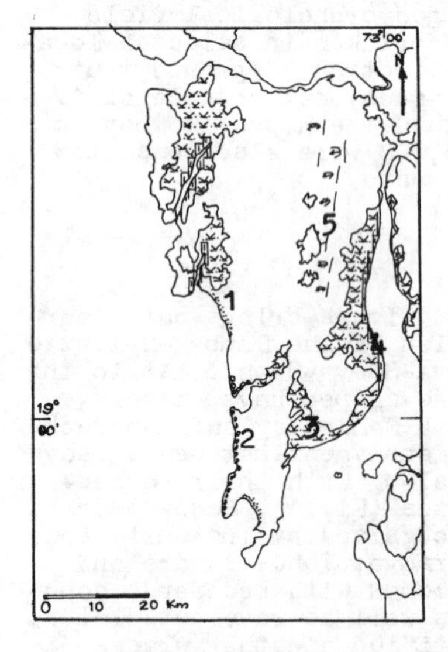

Figure 3. A geomorphic map of the Bombay region showing sandy beaches (1), rocky beaches (2), marshes (3), tidal mudflats (4) and cuestas (5).

5 DISCUSSION

The coastal scenery as depicted by the satellite imagery is one of a highly indented shoreline with creeks, promontories, cliffs, bays and beaches. There are also extensive marshy tracts; at a few places along the coast, raised beaches occurring 3 m. above the present sea-level and a band of littoral concrete are observed at Versova and Madh promontory respectively. The creeks keep to NE-SW, north-south and NW-SE trends.

The raised beaches, the littoral concrete and the marshes indicate a period of emergence of the Bombay region sometime in the past. This might have occurred consequent on the coastal faulting which is generally believed to have taken place all along the West Coast after the eruption of the Deccan Volcanics; Radhakrishna (1967) has given a Miocene age to this fault. The resultant coastline must have been straight.

This faulting, which represents a failure of the volcanics by rupture, has been preceded by plastic deformation of the volcanics which is reflected in the monoclinal flexure described by Auden (1949), the Panvel flexure; this is an important lineament of the Konkan region. Subramanyan (1981) has shown on the basis of field evidence that the Thane Creek has developed over the tension fracture trending north-south that has formed along the axis of this flexure at the southern end. Concurrently, shear fractures trending NE-SW and NW-SE appear to have opened out in the volcanics in the Bombay region. These shear fractures have controlled the courses of the creeks in the Bombay island and the adjoining Nhava-Sheva-Uran area in the east (Fig.2). In the latter area the lineaments intersect mutually to form a triple junction, cutting asunder the whole region and initiating the process of marine erosion.

A retrogradation cycle of marine processes had set in subsequently and gradually converted the initially straight coastline into a highly irregular, indented shoreline of the 'ria' type. Many bays have already been formed in the south - the compounded Back Bay and the perfect Mahim Bay. Due to intense urbanization in the Back Bay area, the erosional processes have been interfered with and they appear to have shifted their activity to the north. The intense erosion that is taking place south of Versova with a rhythmic regularity every year appears to be partly due to this human interference. The Madh promontory also appears to be refracting the waves on to this stretch of the coast which lies on the same latitude as the former. Significantly, the localities to the north and south of this stretch are free from erosional damage. The growth of a spit from the mouth of the Malad Creek in the Versova area is interesting; for, it may be able to prevent the severe erosion south of Versova if it gains height when it will take up the brunt of the wave attack, itself. From this point of view as well as from the point of view of the suitability of this stretch for developing into an harbour, if required later, this spit is being monitored by Shetty and Subramanyan (1985).

It is to be expected that, due to the operation of the retrogradation cycle, all the bays and promontories will be eliminated and that the coastline will restore its original straight configuration in a few million years.

6 CONCLUSION

Satellite remote sensing by itself has provided significant data adequate to trace the development and evolution of the Bombay coast. Supplementary data from the maps and field work have helped to complete the picture.

REFERENCES

Auden,J.B. 1949. Dykes in western India. Trans. nat. inst. sci. India. 3:123-157.

National Atlas Organization. 1964. India: Physiographic regions. plate 41.

Radhakrishna,B.P. 1967. The Western Ghats of the Indian peninsula. Proc. sem. geom. stud. India. univ. Sagar. 4-14.

Shetti,V. & Subramanyan,V. 1985. Spit simulation - an approach for Versova, Bombay. Sixth ann. meet. inst. Indian geogr. abs.

Subramanyan,V. 1981. Geomorphology of the Deccan volcanic province. Mem. geol. soc. India. 3:101-116.

Symposium on Remote Sensing for Resources Development and Environmental Management / Enschede / August 1986

A study with NOAA-7 AVHRR-imagery in monitoring ephemeral streams in the lower catchment area of the Tana River, Kenya

J.W.van den Brink
DHV Consulting Engineers, Amersfoort, Netherlands

ABSTRACT: This article presents a case study in which the possibilities of the NOAA-7 AVHRR system are evaluated for the extraction of hydrological information through the monitoring of the vegetation growth in the lower catchment area of the Tana River,Kenya. Emphasis is placed on the assessment of the hydrologic influence of the ephemeral streams in the lower catchment area on the Tana River. Due to the presence of severe cloud cover over the lower catchment no imagery was obtained by the NOAA-7 AVHRR system with sufficient quality to derive the desired information in a period of five months and a near daily coverage of the satellite system. However the imagery shows that the NOAA-7 AVHRR system forms an additional source of information for observation and monitoring of vegetation of large or even moderately sized areas, such as the secundary catchment areas in this study, with sufficient radiometric and spatial discrimination.

RESUME: Cet article présente unę étude dans laquelle les possibilités du système NOAA-7 AVHRR sont évaluées pour la déduction d'information hydrologique par l'enrégistrement de la croissance de la végétation dans la zone de captage inferieur du Tana fleuve au Kénia.
Beaucoup d'attention est donnée à l'obtention de l'influence causé par les cours d'eau secondaires (Laga's) qui, dans ce bassin ne produisent de l'eau que pendant une periode restreinte de l'année. Parce que Les études etaient exécutées pendant les saisons des pluies, dans deux périodes de 5 mois totales. Il existaint une couverture de nuages persistante dans ces periodes et le système NOAA-7 AVHRR ne procurait, malgré la répetition des observations quotidiennes par la satellite, pas d'images de qualité suffisante pour en déduire l'information désirée.
Cependant, les images montrent que le système NOAA-7 AVHRR constitue une source additionelle d'information pour l'observation et l'enregistrement de la végétation dans des zones par sa résolution radiométriques et spatiale même les zones de superficie surface moderée comme le cas des zones sécondaire de captage dans cette étude.

1 INTRODUCTION.

The monitoring study of ephemeral streams in the Lower Tana River catchment in Kenya with NOAA-7 AVHRR imagery formed part of the larger Tana River Remote Sensing Study. The Tana River Remote Sensing Study was financed by the Dutch Ministery of Education and Science and the Netherlands Remote Sensing Board (BCRS) and executed by DHV Consulting Engineers, in cooperation with the Delft Hydraulics Laboratry (DHL) and the Research Institute for Nature Management (RIN), all from the Netherlands, together with Landevco from Kenya. The study was carried out as part of the Tana River Morphology Studies in which the same organizations participate. The Tana River Morphology has the objective to analyse the effect of current and future interventions, such as reservoirs and irrigation schemes and the morphology of the Tana River and the riverine lands.

2 OBJECTIVES.

The lower catchment area of the Tana River is a very large and inaccessible area only inhabited by a small number of roaming pastoralists. The area contains some 90 smaller and larger ephemeral streams, called Laga's, which flow into the Tana River.
Only during the two rainy seasons, a period in may-june with long-rains when more than half of the annual precipitation falls and a period in october-december with short rains, it can be expected that water is carried by the Laga's.
It was the objective of the case study to evaluate the hydrologic influence of these Laga's on the Tana River. Given the constraints described above any research in the area require a lot of manpower and logistics. Therefore the use of satellite imagery was investigated for the extraction of the desired hydrological information.

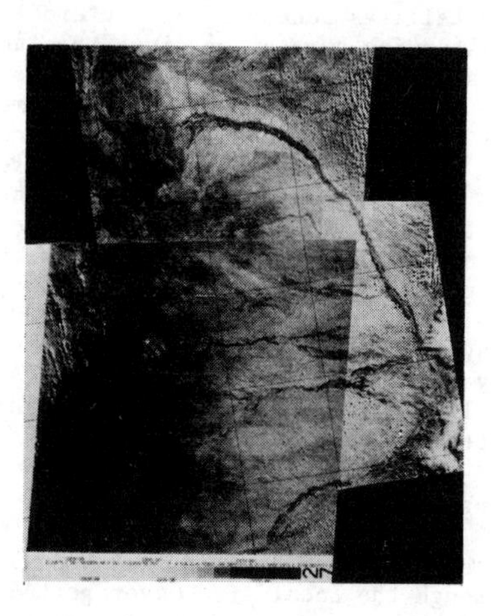

Figure 1. Mosaic of Landsat images of the Lower Tana River area(near-infrared channel, June 1975)

3 METHODS.

3.1 Outlines

The contribution in waterflow of 90 smaller and larger Laga's is an unknown factor, which is difficult to analyse given the vastness and inacessibility of the lower catchment area (55,000 km 2) and the ephemeral character of the outflow. To measure and monitor the flow in these Laga's in the field or even from the air in a systematic way is an operation which involves huge means in terms of logistics and manpower, and is therefore outside the scope of this study.
A study methodology principally based on the use of satellite remote sensing techniques offers in fact the only way to obtain data of the area in a systematic way.
Because of the dynamic nature of the Laga system a satellite monitoring program during the two rainy seasons in may-june and october-december in 1984 seems to meet best the objectives.

3.2 Selection of satellite imagery

The satellite systems available for civilian use and capable of imaging the area whith regular intervals are the Landsat system and the systems based on meteorological satellites. The current Landsat satellite passes over the area whith an interval of 16 days. Only on few occasions Landsat data of the area are being collected, while there also is a great chance that the area is hidden for the satellite behind the clouds during the time of observations. Since Landsat data do not offer a reasonable possibility for the collection of a satisfactory amount of data during the cloudy seasons, the study relies on imagery provided by the meteorological satellites.
Among the remote sensing instruments on board of the various meteorological satellites the Advanced Very High Radiometer (AVHRR) on board of the American satellite NOAA-7 offers the best possibilities given its characteristics in spatial and temporal resolution (described in detail in section 3.3.). Use is made of of the Local Area Coverage programme (LAC) for the acquisition of the NOAA satellite data. In this programme user request NOAA-7 AVHRR data is recorded onboard the satellite for subsequent playback as Local Areal Coverage (LAC) data.

3.3 The AVHRR system on board NOAA-7

The American NOAA-series of polar orbiting environmental satellites became a more interesting source for studying vegetation since the launch of the NOAA-6 satellite in june 1979. An improved sensor, the Advanced High Resolution Radiometer (AVHRR) was since then added to the basic configuration of the NOAA-satellites. The main features of interest for studying vegetation with the AVHRR system are the high temporal, near daily revisit of any given place on earth, associated with an increased spatial resolution of 1.1 km at nadir, a +/- 56 csan angle and a swath width of 2700 km. The AVHRR system on board of NOAA-7, from which the imagery for this study was obtained, records in 5 channels. The band width of these channels are presented in table 1.

3.4 Ordering of NOAA-AVHRR imagery

DHV requested NOAA/NESDIS to collect NOAA-7 AVHRR imagery through the Local Area Coverage(LAC) program during two periods of two months (May-June) and three months (October-December) in 1984. Quick-looks of the best two images of each week were produced by NESDIS. From these hard copies

Table 1. Band width of the NOAA-7 AVHRR system

Channel	Bandwidth (um)
1	.58 - .68
2	.725 - 1.10
3	3.55 - 3.93
4	10.5 - 11.3
5	11.5 - 12.5

DHV selected the images suitable for digital processing. Unfortunately clouds were always present over the study area during the first period of observation. Only two images of May 3 and June 20, with a relatively low cloud cover were worth digital processing. The digital tapes of these two dates were ordered for further investigation. During the second period of observation, data of only four heavily clouded days were archieved despite NOAA's adequate mission planning. So no suitable images for digital processing were made available during this second period of observation.

3.5 Image processing

The image processing facility at the Research Institute for Nature Management (RIN) was used for the processing of the NOAA tapes. The digital image processing routines that were used consist merely of image enhancements in order to obtain an optimal visual representation. Also the normalized difference images : (c2-c1)/(c2+c1) of the near-infrared and red channel resp. channel 2 and 1, (according to C.J.Tucker,1984) were calculated for both days. According to this author this so-called biomass index was presented as a succesful variable which correlate well with the green leaf biomass of vegetation.
Consequently no atmospheric corrections were deemed necessary.

4 RESULTS.

Due to the presence of severe cloud cover especially above the catchment area only parts of this area could be observed satisfactorily on the imagery of both days. Therefore no systematic analysis on the hydrologic influence of the Laga's on the Tana River could be made.
However, despite this severe obstruction the imagery shows that a considerable vegetation growth took place along the upper reaches, especially in the swamp areas, of the larger Laga's. No significant vegetation growth along the lower reaches of the larger Laga's and the other Laga's could be observed. Given the morphology of the Laga's as visible on Landsat-imagery, this indicates that no significant outflow from the Laga's took place during the period of observation.
The biomass image of May 3 shows consistently higher values of the biomass index in comparison with the biomass image of June 20. These observations are in accordance with the field observations and are caused by the fact that the only rains of this first rainy season of May-June fell in the period prior to May 3.
Although the NOAA-AVHRR system is especially suited for inventory of large areas, it can be concluded from the imagery that it offers sufficient detail in spatial and radiometric resolution to observe and monitor vegetation growth of small regions, like the area along the Laga's.

5 CONCLUSIONS AND DISCUSSION.

During a period of eight months it was not possible to obtain imagery of reasonable quality of the NOAA-7 AVHRR system, a system with a near daily revisit capability. This is due to the high cloud cover, which is often present in the Lower Tana River region. This demonstrate the weekness of optical satellite systems in operational aplication practices and the urge for non-optical systems like radar for areas with a high cloud cover such as the tropics and many deltaic regions in the world.
Despite the facts just mentioned, the AVHRR-images indicate their usefulness for investigation and monitoring of the vegetation growth in smaller regions. The radiometry and spatial resolution is sufficient to allow for such applications. Although no atmospheric corrections were performed on the imagery the enhanced images are in agreement with the field observations.

REFERENCES

Blyth, K. 1981. Remote sensing in hydrology. Report no. 74, Institute of Hydrology, Wallingford, U.K.
Colwell,R.N. 1983. Manual of Remote Sensing vol. 1 and 2, American Society of Photogrammetry,the Sheridan Press
Deutsch, M. & D.R. Wiesnet & A. Rango 1979. Satellite hydrology American Water Resources Association
DHV Consulting Engineers 1985. Experiences with microliight aircraft and small format aerial photography in the Tana River Area, Kenya
DHV Consulting Engineers 1985. Experiments to measure bank erosion along the Tana River with aerial photographs
DHV Consulting Engineers 1985. Land use along the Tana River in Kenya and its relation to river morphology and environmental conditions
DHV Consulting Engineers 1985. Monitoring experiments with NOAA-7 AVHRR satellite imagery of ephemeral tributaries of Tana River, Kenya
DHV Consulting Engineers 1983. Water Resources Study Tihama Coastal Plain Vol. 1 and 2, the Netherlands
Donker, N.H.W., R. Soeters 1979. Digital image processing subjects Preliminary lecture notes, ITC, the Netherlands
Lillesand, T.M., R.W. Kiefer 1979. Remote sensing and image interpretation, John Wiley and Sons
NOAA polar orbiting data users guide, 1984. U.S. Departement of Commerce. NOAA national enviromental satellite, data, and information service.
Tucker, C.J., J.A.Gatlin, 1984. Photogrammetric Engineering and Remote Sensing. Monitoring vegetation in the Nile delta with NOAA-6 and NOAA-7 imagery. pp 53-61

Symposium on Remote Sensing for Resources Development and Environmental Management / Enschede / August 1986

A simple atmospheric correction algorithm for Landsat Thematic Mapper satellite images

P.I.G.M.Vanouplines
Royal Museum of Central Africa, Tervuren, Belgium

ABSTRACT: An atmospheric correction algorithm for Landsat Thematic is developed on the basis of an algorithm proposed by Sturm. The algorithm needs only one meteorological variable, the horizontal visibility or meteorological range. The algorithm is tested with a sensitivity analysis, and shows to be well applicable in regions where the atmospheric conditions are stable, and well measured at observatories and airports.

RESUME: Un algorithme pour la correction atmosphérique des données Landsat Thematic Mapper est développé, basé sur un algorithme proposé par Sturm. Il ne fait usage que d'une seule variable météorologique, la visibilité horizontale. L'algorithme est testé avec une analyse de sensibilité, ce qui montre son applicabilité dans des régions où les conditions atmosphériques sont stables et bien mesurées dans les observatoires ou les aéroports.

INTRODUCTION

For many applications in water quality research on oceans and estuaria it is necessary to apply an atmospheric correction on the radiation received by the satellites. Such atmospheric corrections were developed and applied for the Nimbus-7 Coastal Zone Color Scanner (CZCS) and for some other satellites.

The disadvantage of these atmospheric corrections is that generally many meteorological variables should be known. If one obtains a satellite image, taken some months or years ago, it is often difficult to retrieve these variables. Therefore it should be interesting to have an atmospheric correction algorithm that needs only meteorological variables which are measured on a regular basis. In that case one has to retrieve the data from existing observation series in order to apply a more or less reliable atmospheric correction on a given satellite image.

Sturm's paper (1981) describes such a "simple" atmospheric correction model for oceanographic applications. The algorithm is called simple, since it needs only one meteorological variable : the meteorological range. This variable may easily be retrieved from meteorological stations and airports. Sturm developed the correction for the Nimbus-7 CZCS which has a resolution of 825 meter. Adapting this correction for satellites with a higher resolution, such as Landsat Thematic Mapper (TM) or the SPOT satellite, with resolutions of respectively 30 and 20 meter in the multispectral bands, will provide for applications in the field of surface water research. These higher resolutions allow for water quality studies in lakes, estuaria, larger rivers and canals.

In this paper the adaption of Sturm's atmospheric correction algorithm for TM will be developed.

1 GENERAL FORMULATION OF THE CORRECTION ALGORITHM

The signal L received from water surfaces by a remote sensor can be expressed as follows (see also figure 1)

$$L = L_{SG} + L_{HG} + L_{w} + L_{p} \quad (1)$$

where L_{SG} is the sunglitter, i.e. radiance due to direct solar radiance from the water surface, L_{HG} the skyglitter, i.e. radiance due to diffuse radiance reflected from the water surface, L_w the water leaving radiance and L_p the path radiance, i.e. radiance scattered from molecules and particles in the atmosphere.

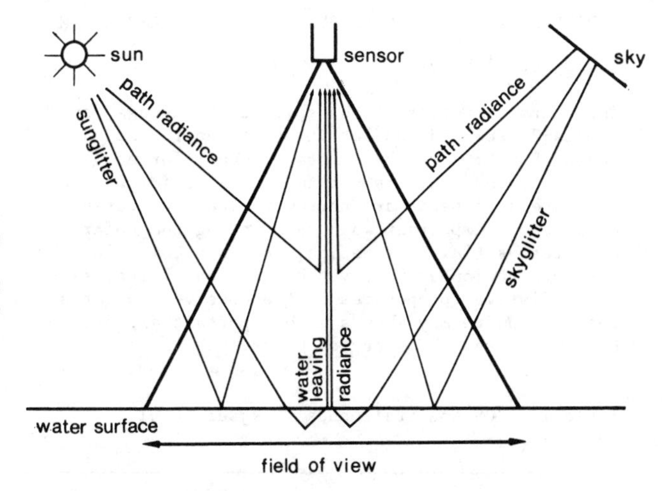

Figure 1. Processes in the atmosphere and at the water-atmosphere interface in remote sensing.

Equation (1) is valid for a remote sensor at height z, observing at a wavelength λ, with a view direction zenith angle μ, and a view direction azimuth angle ϕ.

A practical solution for NIMBUS-7 CZCS has been proposed by Gordon (1978). Following assumptions have to be made :

1 the sunglitter term is absent;

2 at a given reference wavelength the upwelling subsurface radiance L_p is zero;

3 the path radiance term L can be separated into a term due to air molecule (or Rayleigh) scattering (L_{PR}) and a term due to aerosol (or Mie) scattering (L_{PA}).

Equation (1) can then be written for a wavelength λ and a reference wavelength λ_0 as :

$$L^{\lambda} = L^{\lambda}_{HG} + L^{\lambda}_{w} + L^{\lambda}_{PR} + L^{\lambda}_{PA}$$

$$L^{\lambda_0} = L^{\lambda_0}_{HG} + L^{\lambda_0}_{PR} + L^{\lambda_0}_{PA}$$

Equation (2) supposes furthermore that the aerosol scattering path radiance at two wavelengths is proportional.

$$L_{PA}^{\lambda} = \alpha(\lambda,\lambda_0)\, L_{PA}^{\lambda_0} \quad (2)$$

where $\alpha(\lambda,\lambda_0)$ is the aerosol path radiance ratio at two wavelengths λ and λ_0.
Under these assumptions the water leaving radiance can be calculated from

$$L_w^{\lambda} = L^{\lambda} - (L_{PR}^{\lambda} + L_{HG}^{\lambda}) - \alpha(\lambda,\lambda_0)(L^{\lambda_0} - (L_{PR}^{\lambda_0} + L_{HG}^{\lambda_0})) \quad (3)$$

Equation (1) is to be solved for L_w. From this value the subsurface radiance just below the water surface L_u can be calculated with

$$L_u = \frac{n^2}{T}\, t\, L_w \quad (4)$$

where n is the refractive index of water, t the bi-directional transmission for a windroughened water surface and T the total tansmittance, see equation 9. Sørensen (1976:76) takes n=1.343; Austin (1974:320) takes n=1.341. t=1-ρ, where ρ is the Fresnel reflectivity of a plane water surface assumed to be constant (0.021) for zenith view angles smaller than 45° (Sørensen 1976:76), Austin (1974:323) gives for view angles between 0° and 10° ρ = 0,0211.

In what follows, the terms in equations (3) and (4) will be solved for Landsat TM.

2 DIFFERENCES BETWEEN NIMBUS-7 CZCS AND LANDSAT TM

2.1 Spectral bands

The characteristics of the spectral bands used by Nimbus-7 CZCS and Landsat TM are given in tables 1 and 2. For CZCS the bandwidth is equal for each spectral band. This is not the case for TM. Besides, CZCS spectral bands are narrower. These differences will become important when calculating the solar extraterrestrial irradiation, and when comparing the absolute radiances received by each satellite, since the radiative properties of the observed target and the atmosphere may show a more important difference over the larger TM spectral bands.

Table 1. CZCS spectral bands (Nykjaer et al., 1984:2).

spectral band number	frequency range (nm)	center frequency (nm)	band width (nm)	signal/ noise ratio
1	433-453	443	20	158/1
2	510-530	520	20	200/1
3	540-560	550	20	176/1
4	660-680	670	20	118/1

Table 2. Landsat TM spectral bands (Salomonson et al., 1983; NASA, 1982:4).

spectral band number	frequency range (nm)	center frequency (nm)	band width (nm)	signal/noise screen radiance minium	maximum
1	450-520	485	70	52/1	143/1
2	520-600	560	80	60/1	279/1
3	630-690	660	60	48/1	248/1
4	760-900	830	140	35/1	342/1

For CZCS, spectral band 4 (660 to 680 nm) was taken as a reference band where the upwelling subsurface radiance of "clear water" was supposed to be zero or negligable. For TM bands 1 to 4 are selected, where band 4 (760 to 900 nm) is choosen as reference band.

2.2 Radiometric performance

The signal to noise ratio is for TM worse than for CZCS (see tables 1 and 2). Tassan (1986) advises to cluster 4*4 pixels to obtain a comparable or even lower noise than the CZCS values. The resolution becomes 120 m, still a considerable advantage over the CZCS resolution.

2.3 Observation angles

Another important difference between CZCS and TM is that the CZCS sensor can be tilted. This possibility does not exist for TM. The CZCS was the first sensor developed for water surface observations. With the tilting sensor sunglitter may be avoided. Care should to be taken that no sunglitter occurs on a TM image. The advantage is that TM is always nadir-looking, which implies that equations taking in account oblique observation angles in Sturm's atmospheric correction will simplify for TM.

3 CALCULATION OF THE RAYLEIGH PATH RADIANCE AND SKY-GLITTER

3.1 Rayleigh path radiance

In a water-atmosphere system the Rayleigh path radiance can be seen in a first approximation as consisting of direct sunlight scattered by air molecules and by the water reflected sunlight, scattered by air molecules, both into the satellite's field of view.
The first contribution is given by equation (5)

$$L_{PR}^{(1)} = E_0 T^{ozone} \tau^{air} p^{M}(\psi_-) \quad (5)$$

in which E_0 is the extraterristrial solar spectral irradiance, T^{ozone} the ozone transmittance, τ^{air} the Rayleigh optical thickness (see section 3.3) and $p^M(\psi_-)$ the Rayleigh phase function.
The second contribution is given by equation (6)

$$L_{PR}^{(2)} = \rho E_0 T\, \tau^{air} p^{M}(\psi_-) \quad (6)$$

where ρ is the Fresnel reflectivity and T the total transmissivity.

The extraterrestrial solar spectral irradiance, which is function of the Julian day D and the wavelength λ, can be calculated from

$$E_0(D,\lambda) = \bar{E}_0(\lambda)\{1 + \varepsilon\cos\frac{2\pi}{365}(D-3)\}^2 \quad (7)$$

where $\bar{E}_0(\lambda)$ is the yearly average solar extraterrestrial irradiance at wavelength λ, ε the earth's orbit eccentricity (0.0167) and D the Julian day. The following values for the solar extraterrestrial irradiance are based on Neckel and Labs (1981:246-247) and made under the assumption that the TM sensor radiometric response curve follows a gaussian curve :

E (485) = 192 mW/(cm^2.sr.μm)
E (560) = 183 mW/(cm^2.sr.μm)
E (660) = 160 mW/(cm^2.sr.μm)
E (830) = 109 mW/(cm^2.sr.μm)

The Rayleigh optical thickness at wavelength λ can be approximated by equation (8) (Linke, 1956)

$$\tau^{air}(\lambda) = 0.00879\, \lambda^{-4.09} \quad (8)$$

All transmittances can be calculated from their corresponding optical thicknesses with

$$T^x = \exp(-\tau^x/\cos\theta_0) \quad (9)$$

where T^x is the transmittance, τ^x the optical thickness, θ_0 the solar zenith angle and x = air, aero ozone or total.
For the ozaone optical thickness (and hence for the ozone transmittance) a fixed value may be choosen for each wavelength. For TM in Belgian applications following values were choosen :

$$\tau^{ozone}_{485} = 0.015$$

$$\tau^{ozone}_{560} = 0.025$$

$$\tau^{ozone}_{660} = 0.014$$

$$\tau^{ozone}_{830} = 0.001$$

The Rayleigh phase functions can be calculated for nadir-looking satellite from

$$p^M(\psi_\pm) = \frac{3}{16\pi}(1+\cos^2\theta_0) \quad (10)$$

in which θ_0 is the solar zenith angle.

3.2 Skyglitter

The skyglitter is for a nadir-looking satellite given by equation (11)

$$L_{HG} = \rho E_0 T^{tot} \tau^{air} p^M(\psi) \quad (11)$$

The three equations (5), (6) and (11) are now combined in equation (12)

$$L_{PR} + L_{HG} = E_0 T^{ozone^2} \tau^{air} \{p^M(\psi) + p^M(\psi)(T^{MA}(\mu_0) + T^{MA})\} \quad (12)$$

where the total transmittance is written as

$$T^{tot} = T^{ozone}\, T^{MA}$$

The ozone transmittance is squared since the light has to pass the ozone layer twice.

3.3 Mie optical thickness

The visibility range or meteorological range V was introduced by Koschmieder (1938) and is related to the scattering or attenuation coefficient K (0), defined by Middleton (1957). This relationship is given in equation (13)

$$V = \frac{3.912}{K_\lambda(0)} \quad (13)$$

in which $K_\lambda(0)$ is the total scattering coefficient at height 0 m, and wavelength λ.
Since

$$K_{550}(0) = K^{air}_{550}(0) + K^{aero}_{550}(0) \quad (14)$$

where $K^{air}_{550}(0)$ is the air molecule scattering coefficient, at height 0m and wavelength 550 nm and $K^{aero}_{550}(0)$ the aerosol scattering coefficient, at height 0 m and wavelength 550 nm.
One obtains equation (15)

$$K^{aero}_{550}(0) = \frac{3.912}{V} - K^{air}_{550}(0) \quad (15)$$

Equation (15) is valid for standard conditions (temperature 15°C, atmospheric pressure 1013 mb), were $K^{air}_{550}(0) = 0.0116\ km^{-1}$

To obtain K^{aero}_λ, one takes in account the ratios of the effective cross sections of Mie particles (σ^{aero}_λ) as given in equation (16), and the particle density at height z ($N^{aero}(z)$).

$$\frac{\sigma^{aero}_{550}}{\sigma^{aero}_\lambda} \simeq \left(\frac{550}{\lambda}\right)^{3-\alpha} \quad (16)$$

One obtains now equation (17)

$$K^{aero}_\lambda(z) = \frac{N^{aero}(z)}{N^{aero}(0)} \left(\frac{3.912}{V} - 0.0116\right)\left(\frac{550}{\lambda}\right)^{\alpha-3} \quad (17)$$

The particle density $N^{aero}(z)$ can be approximated by the set of equations (18)

$$N^{aero}(z) = \begin{cases} 55\exp(-(z-5.5)/H_1) & z < 5.5\ km \\ 55 & 5.5 < z < 18\ km \\ 55\exp(-(z-18)/H_2) & z > 18\ km \end{cases} \quad (18)$$

where $H_1 = 0.886 + 0.0222\ V$ and $H_2 = 3.77$ km.
The set of equation (18) is developed by Mc Clatchey et al. (1972) which is based on 79 series of measurements by Elterman (1968, 1970).
The Mie optical thickness (by scattering of aerosol particles) is defined as

$$\tau^{aero}_\lambda(z) = \int_0^z K^{aero}_\lambda(z)dz \quad (19)$$

With equations (17), (18) and (19) equation (20) to calculate the aerosol optical thickness at a height (z >18 km) can be set up by integration

$$\tau^{aero}_\lambda(z) = \left(\frac{3.912}{V} - 0.116\right)\left(\frac{550}{\lambda}\right)^{\alpha-3} \{H_1(1-\exp(-5.5/H_1)) + 12.5\exp(-5.5/H_1) + H_2\exp(-5.5/H_1)(1-\exp(-(z-18)/H_2))\} \quad (20)$$

For satellite observations, z is equal to ∞, and one obtains

$$\tau^{aero}(\lambda) = \left(\frac{3.912}{V} - 0.0116\right)\left(\frac{550}{\lambda}\right)^{\alpha-3} \{H(1-\exp(-5.5/H)) + 12.5\exp(-5.5/H) + 3.77\exp(-5.5/H)\} \quad (21)$$

where V is the visibility range or meteorological range, λ the wavelength and H = 0.886 + 0.0222.V
α is normally equal to 4.
Data on the meteorological range can be obtained in Belgium from the Royal Meteorological Institute (KMI/IRM) for 20 stations (measured every three hours) in the monthly synoptical observations.

4 CALCULATION OF THE AEROSOL PATH RADIANCE RATIO

The aerosol path radiance ratio $\alpha(\lambda,\lambda_0)$ is first used in equation (2). It is possible to set up an equation for aerosol path radiance analogous to equation (5). The aerosol path radiance ratio is then expressed by equation (22), assuming that the phase function is independent of the wavelenght

$$\alpha(\lambda,\lambda_0) = \frac{\tau^{aero}(\lambda)}{\tau^{aero}(\lambda_0)} \frac{E_0(D,\lambda)\, T^{ozone}(\lambda,\mu,\mu_0)}{E_0(D,\lambda_0) T^{ozone}(\lambda_0,\mu,\mu_0)} \quad (22)$$

where

$$T^{ozone}(\lambda,\mu,\mu_0) = T^{ozone}(\lambda,\mu).T^{ozone}(\lambda,\mu_0)$$

and

$$T^{ozone}(\lambda,\mu) = \exp(-\tau^{ozone})$$

$$T^{ozone}(\lambda,\mu_0) = \exp(-\tau^{ozone}/\cos\theta_0)$$

(see equation 9).

5 COMPUTER IMPLEMENTATION NOTE

The values for the radiance received by TM have to be calculated from the digital number (DN) retrieved from CCT's. The absolute radiances can be calculated from equation (23) and tables 3 and 4.

$$L_b = \frac{(RMAX_b - RMIN_b)}{255} = DN_b + RMIN_b \qquad (23)$$

where L_b is the radiance received by TM in band b, $RMAX_b$ the minimum radiance required to saturate detector response (i.e. for DN = 255), $RMIN_b$ the spectral radiance corresponding to a DN_b = 0 and DN_b the digital number in TM spectral band b as obtained from CCT.

Table 3. Dynamic ranges of Landsat TM data processed prior to August 1983 (Scrounge System) (Barker,1984).

	spectral bands			
	TM1	TM2	TM3	TM4
RMIN ($mW.cm^{-2}\ sr^{-1}.\mu m^{-1}$)	-0.15	-0.28	-0.12	-0.15
RMAX ($mW.cm^{-2}\ sr^{-1}.\mu m^{-1}$)	15.84	30.82	23.46	22.43

Table 4. Dynamic ranges of Landsat TM data processed after January 15 1984 (Tips) (Barker, 1984)

	spectral bands			
	TM1	TM2	TM3	TM4
RMIN ($mW.cm^{-2}\ sr^{-1}.\mu m^{-1}$)	-0.15	-0.28	-0.12	-0.15
RMAX ($mW.cm^{-2}\ sr^{-1}.\mu m^{-1}$)	15.21	29.68	20.43	20.62

The computer implementation of the atmospheric correction algorithm is fast, since after calculation of the atmospheric variables, the correction consists of solving only equations (3) and (4) which have to be calculated pixel-per-pixel.
The atmospheric correction algorithm was implemented on the IBM 3081 computer of the Ministry of National Education of the French Community (Brussels) at the cluster of the Royal Museum of Central Africa (Tervuren, Belgium) in Fortran IV. The operating system is OS/VS. On a 512*512 image, less than 25 seconds CPU time was used.

6 SENSITIVITY ANALYSIS

6.1 Description of the variable ranges selected

The influence of changes in the following variables was examined :

1 visibility range;
2 solar zenith angle;
3 ozone optical thickness;
4 solar extraterrestrial irradiance.

While the influence of one variable was studied, the values of the other variables wad kept constant at a certain 'mean value'.

6.2 Analysis of the behaviour of the algorithm in relation to variable changes

Using equations (3) and (4) one obtains equation (24)

$$L_u^\lambda = \frac{n^2}{T_\lambda^{tot}} L^\lambda - \frac{C}{T_\lambda^{tot}} \{L_{PR}^\lambda + L_{HG}^\lambda - \alpha(\lambda,\lambda_0)(L^{\lambda^0} - (L_{PR}^{\lambda^0} + L_{HG}^{\lambda^0}))\} \qquad (24)$$

where $C = n^2 (1-\rho) = 1.7658$.
Consider now Fig. 2. Equation (25) is a linear function.

$$y = ax + b \qquad (25)$$

Coefficient a gives the slope of the line, b is the intersection with the y-axis, c is the intersection

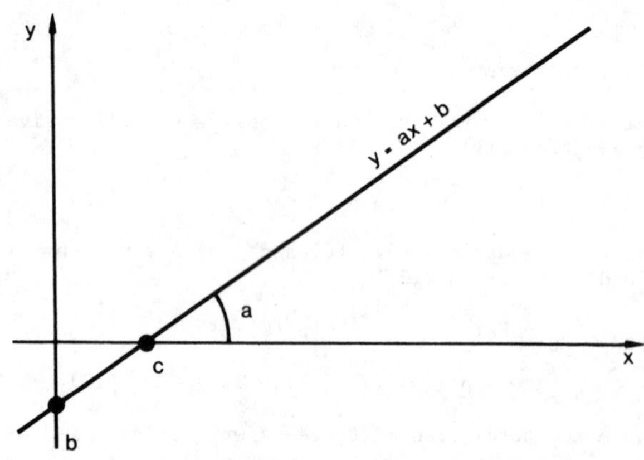

Figure 2. Characteristics of a linear function.

with the x-axis. Comparison of equation (24) with equation (25) gives

$$y = L_u^\lambda \qquad (26)$$

$$x = L^\lambda \qquad (27)$$

$$a = \frac{C}{T_\lambda^{tot}} = \frac{1.7658}{T_\lambda^{tot}} \qquad (28)$$

$$b = -\frac{C}{T_\lambda^{tot}} \{L_{PR}^\lambda + L_{HG}^\lambda - \alpha(\lambda,\lambda_0)(L^{\lambda_0} - (L_{PR}^{\lambda_0} + L_{HG}^\lambda))\} \qquad (29)$$

Coefficient a will only change if the total transmissivity $T(\lambda)$ changes, this means when the visibility range (V), ozone optical thickness ($\tau^{ozone}(\lambda)$) or wavelength (λ) change. In other words, if the total transmissivity does not change, one will obtain parallel lines if the results for different sets of input values are plotted on one graph, since the slope b doesn't change.

6.3 Results of the sensitivity analysis

The results of the sensitivity analysis are represented in table 5. Table 6 gives the digital numbers and their corresponding radiances for each spectral band, which were used to obtain table 5.
The digital number DN=11 for band 4 corresponds to the 'darkest pixel' or 'clear water' reflectance on

the Landsat TM image.
From table 6 it is clear that the selected values for the ozone optical thickness will not influence the results too much. However, the data for the meteorological visibility, the solar zenith angle and the solar extraterrestrial irradiance have to be choosen carefully. The first two variables can be retrieved easily and rather exactly from respectively the CCT and meteorological observations.

Table 5. Results of the sensitivity analysis.

Variable or parameter	low value		mean value		high value	
	input	L_w	input	L_w	input	L_w
V						
at 485 nm	5	5.147	23	2.675	100	2.308
at 560 nm	5	4.205	23	2.416	100	2.136
at 660 nm	5	3.017	23	1.899	100	1.715
θ_0						
at 485 nm	20	1.282	40	2.675	70	5.086
at 560 nm	20	1.824	40	2.416	70	3.553
at 660 nm	20	1.722	40	1.899	70	2.245
τ_{485}^{ozone}	0.005	2.325	0.015	2.675	0.020	2.850
τ_{560}^{ozone}	0.020	2.313	0.025	2.416	0.040	2.725
τ_{660}^{ozone}	0.005	1.792	0.014	1.899	0.020	1.970
τ_{830}^{ozone}						
at 485 nm	.0001	2.690	0.001	2.675	0.005	2.611
at 560 nm	.0001	2.427	0.001	2.416	0.005	2.370
at 660 nm	.0001	1.906	0.001	1.899	0.005	1.867
$\bar{E}_0(90,485)$	182	3.406	192	2.675	202	1.949
$\bar{E}_0(90,560)$	173	2.848	183	2.416	193	1.984
$\bar{E}_0(90,660)$	150	2.170	160	1.899	170	1.628
$\bar{E}_0(90,830)$						
at 485 nm	99	1.977	109	2.675	119	3.257
at 560 nm	99	1.909	109	2.416	119	2.939
at 660 nm	99	1.549	109	1.899	119	2.190

Table 6. Digital numbers (DN) and corresponding radiances used in table 5.

spectral band	frequency nm	DN_b	L^λ $\frac{mW}{cm^2.sr.\mu m}$
1	485	95	5.807
2	560	35	3.988
3	660	30	2.654
4	830	11	0.824

CONCLUSION

An atmospheric correction algorithm has been developed for Landsat Thematic Mapper. It will be rather easy to apply on images taken months or even years ago since the only meteorological variable, the horizontal visibility, can be easily retrieved from meteorological time series. For water quality applications it might be necessary to cluster 4*4 pixels to increase the signal to noise ratio, but even then the resolution is considerably better than for CZCS, namely 120 m. Future research will prove the suitability of this algorithm.

Acknowledgements

The author wishes to express his thanks to C. De Croly and P. Doyen from the Ministry of National Education of the French Community (Brussels) for their assistance in the computer work. The author also acknowledges S. Tassan from the Joint Research Center (Ispra - Italy) for the fruitfull discussion.

REFERENCES

Austin, R.W. 1974. The remote sensing of spectral radiance from below the ocean surface. In Jerrlov and Nielsen (eds.), Optical aspects of oceanography, p. 317-344. London, New-York : Academic Press.
Barker, J.L. 1984. Relative calibration of Landsat reflective bands. In NASA Conference Publication 2326, Landsat-4 Science Investigations Summary, Vol. I, p. 140-180.
Elterman, L. 1968. UV, visible and IR attenuation for altitudes to 50 km. Bedford Mass. : AFCRL.
Elterman, L. 1970. Vertical attenuation model with eight surface meteorological ranges 2 to 13 km. Bedford Mass. : AFCRL.
Gordon, H.R. 1978. Removal of atmospheric effects from satellite imagery of the oceans. Appl. Opt. 17 : 1631-1636.
Koschmieder, H. 1938. Naturwissenschaften 26 : 521.
Linke, F. 1956. Die Sonnestrahlung und ihre Swächung in der Atmosphäre. In F. Linke & F. Moeller, Handbuch der Geophysik, Kap. 6, Berlin: Gebr. Borntraeger.
McClatchey, R.A., R.W. Fenn, J.E.A. Selby, F.E. Volz & J.S. Gaving 1972. Optical properties of the atmosphere. AFCRL - 72-0497, Environmental Research Papers, No 411.
Middleton, W.E.K. 1958. Vision Through the atmosphere. Toronto: University of Toronto Press.
Nasa 1982. Landsat Data Users Notes, 23: 1-11.
Neckel, H. & D. Labs.1981. Improved data of solar spectral irradiance from 0.33 to 1.25 Å. Sol. Phys. 74: 231-249.
Nykjaer, L., P. Schlittenhardt & B. Sturm 1984. Qualitative and quantitative interpretation of ocean color - Nimbus-7 CZCS imagery of the Northern Adriatic Sea from May to September 1982. Ispra: Joint Research Centre.
Salomonson, V.V. et al..1983. Water resources assessment. In J.E. Estes & G.A. Thorley (eds. of volume II), Manual of remote sensing, p. 1497-1570. Falls Church: American Society of Photogrammetry.
Sørensen, B.M. 1979. The North Sea ocean color scanner experiment 1977. Ispra: Joint Research Centre.
Sturm, B. 1981. The atmospheric correction of remotely sensed data and the quantitative determination of suspended matter in marine surface layers. In A.P. Cracknell (ed.), Remote sensing in meteorology, oceanography and hydrology, p. 163-197. Chicester: Ellis Horwood Ltd.
Tassan, S. 1986. Private communication.

7 Human settlements: Urban surveys, human settlement analysis and archaeology

Chairman: W.G.Collins
Co-chairman: B.C.Forster
Liaison: P.Hofstee

Symposium on Remote Sensing for Resources Development and Environmental Management / Enschede / August 1986

The application of remote sensing to urban bird ecology

L.M.Baines & W.G.Collins
Remote Sensing Unit, University of Aston, Birmingham, UK

P.Robins
Operational Research, University of Aston, Birmingham, UK

ABSTRACT: The aim of this paper is to assess the feasibility of a remote sensing system to determine the carrying capacity of urban habitats for certain species, in this instance, woodland birds. Investigations were carried out into the relationship between variables which can be identified from the air photographs and bird populations. The major variables examined were habitat area and habitat dispersion. Evidence of an relationship between the number of bird species and habitat area was found, but as yet the results are too inconclusive to allow predictions.

1 INTRODUCTION

The value of green areas in the urban environment has been recognised by many offical bodies in recent years. "Everyday contact with the natural world is essential to a healthy physical, mental and spiritual existence". (Nature Conservancy Council, 1984). Increasingly statutory bodies and local authorities are aware of the role they play to ensure urban wildlife is considered in planning and land management decisions. Recently the now-extinct West Midlands County Council produced a valuable document "The Nature Conservation Strategy for the County of West Midlands" which stressed the need for the enhancement or creation of semi-natural open spaces to benefit urban wildlife and the residential population. This trend is not only evident in the UK, but throughout the 'developed world'. However semi-natural open space is only one of the competing land uses in the urban environment and therefore a compromise will always have to be made between the demand for green areas and the demand for development. As a consequence there is a need for a method to identify which areas are valuable ecologically and thus try, if possible, to steer development away from these to less valuable options. One method of allocating a value to urban open space is to identify the area's carrying capacity for a species of interest.

The aim of this paper is to assess the feasibility of aerial photography to determine the bird carrying capacity of an urban woodland. If it is possible to predict bird population numbers from air photographs the population rate of change due to development can be determined. Aerial photographs have been used in a number of studies concerned with bird population prediction, (Skaley 1981, Lancaster and Rees 1979), however either the technique has been too complex or the photographs have been used merely as an addition to field data rather than as a data source in themselves.

2 THE STUDY AREA

The Blackbrook Valley is situated in the heart of the industrial Black Country in the West Midlands. It is approximately 300 hectares of common land in a dense industrial and residential area. The heavy industry which once dominated this area has declined leaving high unemployment and dereliction. As a method to try and encourage industry back into the region, the northern section of the Blackbrook Valley was designated an Enterprise Zone in 1981. One of the attractions of an Enterprise Zone to industry is the lack of planning controls therefore industrial development has and is posing a threat to a valuable wildlife area. A local nature reserve has been established in the southern part of the valley - Saltswells Nature Reserve (41.2 hectares). This is comprised of a large area of mixed deciduous woodland planted at the end of the eighteenth century and an abandoned fireclay pit which has geological and ecological value. The anaylsis has concentrated on the nature reserve.

3 MATERIALS AND EQUIPMENT

The Nature Conservancy Council (N.C.C.) commissioned two sets of imagery of the study area. Large scale panchromatic, colour and colour infra-red photography (1:2500) was flown in the summers of 1981 and 1984 providing complete coverage of the area.

Interpretation was carried out using an old Delft stereoscope and the colour infra-red photography. This emulsion was found to provide the greatest amount of detail for this particular project. A monoscopic zoom transferscope was used to transfer the information extracted from the air photographs on to a 1:2500 scale base map.

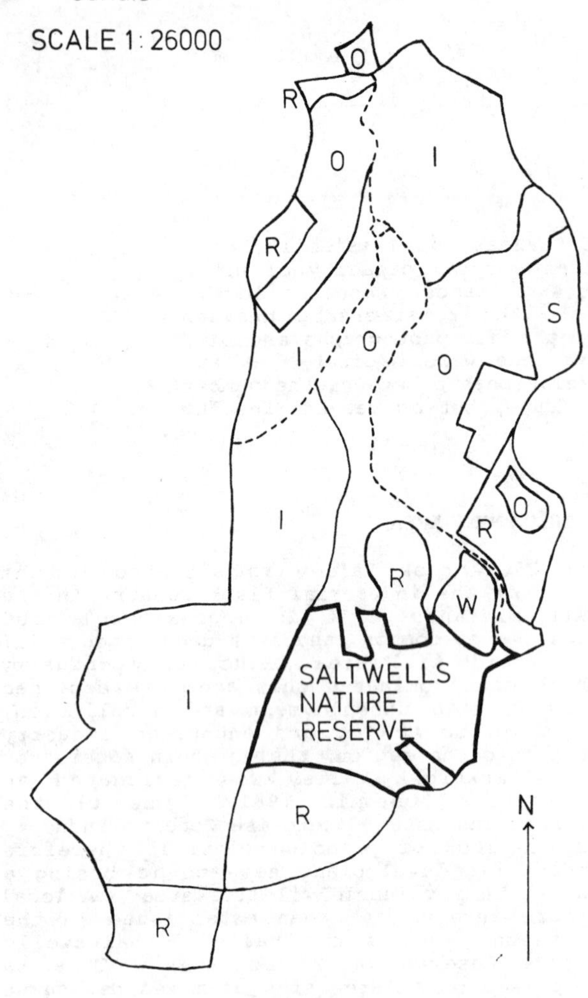

Fig 1. Map of the land-use in the Blackbrook Valley

4 DATA EXTRACTION - THE CLASSIFICATION

The photographic interpretation was based on a habitat classification devised for this project. Habitat mapping is well suited to remote sensing as an habitat unit has a larger surface area than individual species communities and therefore is more easily identified. The classification is divided into four groups:

1 land with no cover,
2 land with man-made cover,
3 land with water cover and
4 land with vegetation cover.

In this paper only land with vegetation cover will be discussed. The vegetation categories are based on the structure and density of the dominant growth forms. The distinction between open and closed woodland and shrubland categories were dictated by the distance between foliage canopies (see Fosberg and Peterken 1967).

28 Tree cover +3m height
29 Broadleaf communities
  30 Single tree
  31 Closed woodland
  32 Open woodland
  33 Linear woodland
  34 Woody shrub
35 Coniferous communities
  36 Single tree
  37 Closed woodland
  38 Open woodand
  39 Linear woodland
  40 Woody shrub
41 Shrub cover -3m
42 Broadleaf communities
  43 Closed shrub
  44 Open shrub
  45 Linear shrub
46 Coniferous, spikey communities
  47 Closed shrub
  48 Open shrub
  49 Linear shrub
50 Herbaceous cover
  51 Ruderal communities
  52 Tall herb and fern communities
  53 Rough tall grassland communities
  54 Wetland herbaceous communities
  55 Tall fragmentary marginal communities
  56 Smooth turf grassland (unmanaged)
  57 Smooth turf grassland (managed)
  58 Rough turf grassland
  59 Floating vegetation
  60 Submerged vegetation.

Figure 2. Classification categories

Although the idea of an universal legend is attractive no standard classification exists for ecological or urban surveys therefore new legends need to be devised for individual projects to ensure the detail necessary is obtained. The vegetation communities identified by the N.C.C. were the basis of this habitat classification, allowances being made for the difference in the survey's resolution.

The accuracy of the photographic interpretation and mapping was checked for 1981 and 1984. A pilot test had identified the categories which were frequently misclassified for example bracken. Previously this had been an individual legend category, but due to interpretation difficulties it was included in the tall herb and fern category. The accuracy level achieved for the 1984 analysis was better than the 1981 figure. This was a result of the time period between the dates of the 1981 photographs and the field work, 1983. A good deal of development had taken place in the area in this period. The 1981 interpretation was 89.5% accurate and the 1984 interpretation was 97.6% correct. The statistical tests (see Ginevan 1979 and Arnold 1985) suggest the number of errors did not occur by chance.

5 DATA ANALYSIS

In the literature, studies using air photographs to predict bird population numbers have not been very successful. Various researchers have concentrated on predicting bird species numbers, bird abundance and bird species diversity using variables measured in the field such as area of woodland, foliage height diversity, foliage canopy density and woodland isolation. It was decided to investigate if information about the variables used in these techniques can be obtained from air photographs, and therefore, used to predict bird carrying capacity. A bird census was conducted in Saltswells Wood in 1983 by Harrison and Normond. These local ornithologists identified the bird species which were breeding, their numbers and the approximate location of their territories. This field data was used as the control in this analysis.

One of the simplist and most common variables used for prediction is area of woodland. This relationship was first suggested by Arrhenius 1921. Woolhouse 1983 and Moore and Hooper 1975 have both conducted studies in British woodlands suggesting a linear relationship between the number of bird species present and the area of woodland.

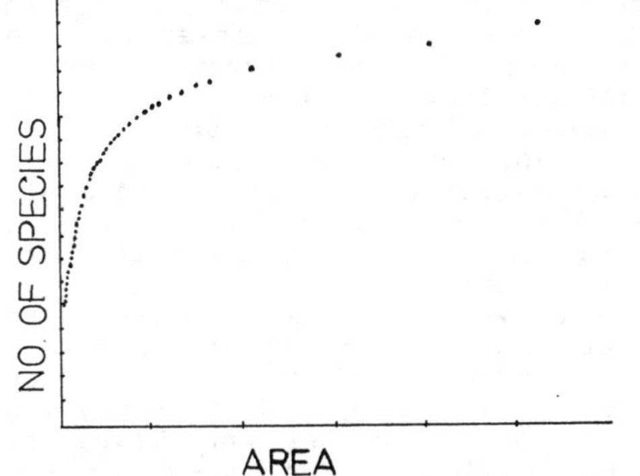

Hypothetical species-area curve
$S=cA^{z}$

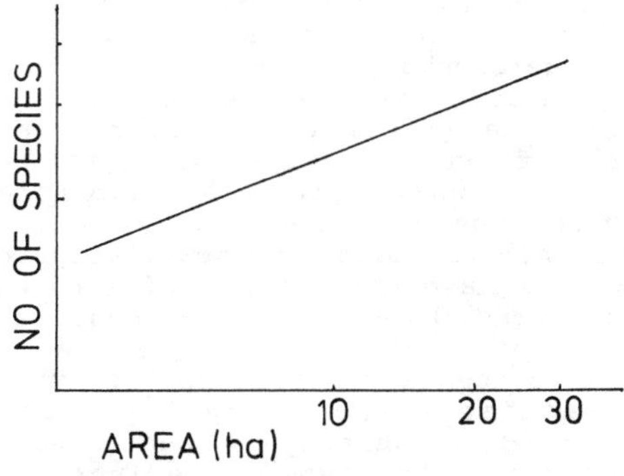

Species-area curve on a logarithmic scale $\ln S=c\ln A^{z}$
from Woolhouse 1981

FIG 3.

The area of the classifed habitat types was the major variable to be investigated. The area information was obtained by using a Hewlett Packard digitizer connected to a BBC micro computer using a digitising program.

The first stage was to identiy a list of bird species which were likely to be found breeding in Saltswells wood and relate their nesting and feeding requirements to the classification categories identified. Fifty-seven species commonly found in woodlands of a similar size and found breeding in the West Midland were earmarked. These birds were allocated one or more of the classification categories using the 'Handbook of British Birds' (Witherby et al 1965). For example a Blackbird is likely to be found breeding and feeding in categories 34/43/44/45, that is, in deciduous shrubby habitats. Some bird species are limited to a single category such as the Wood Warbler. This species was only allocated closed deciduous woodland - 31. Other species need areas of woodland and grassland to fulfil all their requirements. The system allowed for this, for example the Common Crow needs 32/33+51/52/53/58, that is, open deciduous woodland for nesting and grassland areas for feeding. As some species have similar requirements 32 habitat patterns were identified for the 57 species.

The equations relating bird species numbers to areas of woodland suggested by Woolhouse and Moore and Hooper were investigated. Different habitat combinations and their corresponding area totals were used in these equations to identify which, if any, habitat combination produced a result similar to 27 - the actual number of bird species found breeding in the area by the control study. For results see Table 1.

1 Woolhouse equation =
$$\ln S = 0.227 \ln A + 2.632$$

2 Moore and Hooper equation =
$$\ln S = 0.271 \ln A \pm 0.26$$

Where In S = Natural log of the number of species
In A = Natural log of area

Table 1.The relationship between woodland area and the number of bird species.

Area (ha)	Estimated number of species Eq.1	*1	Eq.2	*2
Reserve 40.2	32.2	+5.2	2.1	-24.9
Hab. 31 14.4	25.66	-1.34	1.6	-25.4
Hab. 34 7.0	21.59	-5.4	1.23	-25.7
Hab.31+34 21.4	**27.84**	**+0.84**	1.76	-25.2
Hab. 43,44, 45, 10.65	23.8	-3.2	1.46	-25.5
Hab.31,34, 43,44,45 32.1	30.56	+3.56	1.97	-25.0

* = the difference between the actual number of species,27 and the predicted number.

There is a large discrepancy between the number of bird species predicted by the two equations. The Moore and Hooper equation obviously under estimates the number of species expected in this area therefore only the figures produced by the Woolhouse equation were considered.

Of the different combinations of habitat units the total area of 31 and 34 provides the most accurate estimate of the number of bird species. Two apparent conclusions can be drawn.

The habitat categories 31 and 34 most accurately represent the conditions of the woodlands surveyed by Woolhouse. The total area of these habitats can thus be used to predict bird species numbers. However, these conclusions can only be accepted tentatively. This apparent relationship may have occurred purely by chance. Unfortunately field data only exists for one year therefore the relationship cannot be retested. A backup test was conducted. If the relationship between the total habitat area of categories 31 and 34 and the number of breeding species was real, one would expect all the bird species which breed to have either 31 or 34 as one of their habitat requirements. It was found that 20 of the 27 species known to breed in Saltswells wood had habitats 31 or 34 listed as one of their preferred habitat types. This leaves the breeding of 7 species unexplained. Two conflicting conclusions can be drawn. The apparent relationship between the actual number of breeding bird species and the number estimated using the Woolhouse equation has occurred by chance or the relationship is real but the habitat patterns allocated to each bird species are inaccurate. The results of this analysis remain therefore inconclusive and indicate that further work needs to be conducted.

Woolhouse, in the same paper also suggests that species abundance can be predicted from the area of woodland.

$$\ln I = 0.679 \ln A + 3.082$$

Where In I is the natural log of species abundance.

Using the total habitat area of categories 31 and 34 the expected number of individuals is 174.12. However, according to the field data 289 individual birds were recorded. This large discrepancy suggests that this equation may not be used with the area totals of 31 and 34 identified from the air photographs for urban woodlands. The field data collected by Woolhouse was obtained in rural woodlands. Various studies about bird populations in urban environments for example, Emlen 1974, Tomialojc and Profus, 1977 suggest that although species numbers are lower in urban environments than rural, actual bird densities are often higher in urban areas. As Saltswells wood is an urban wood this may be the cause of the discrepancy.

There is not enough evidence available to indicate whether the equation put forward in the literature concerning the relationship between area and bird species numbers can be used successfully with air photographs. The results of the analysis are inconclusive. However this does not necessarily suggest that it is impossible to predict the bird carrying capacity of a wood from habitat area data extracted from air photographs.

Further analysis was conducted by calculating the total amount of suitable habitat available in the reserve for each likely bird species.

The photographic area data suggested that the species which were identified as breeding in the field survey were those which had the largest area of suitable habitat available in the reserve. Two statistical tests, Chi^2 and Mann Whitney U test were conducted to investigate whether the available habitat totals for breeding and non-breeding species were from the same population or whether there was a significant difference between the two groups. The rejection level in both instances was 0.05.

Table 2. Species habitat area

Bird Species	Suitable Habitat categories	Total habitat area (ha)
Black bird	34/43/44/45	16.95
Woodwarbler	31	14.43
Willow Warbler	44/45	1.02

The Chi^2 test suggested there was no difference between the area totals but the Mann Whitney U test indicated that there was a significant difference, that is, the bird species which are found to breed in the reserve are those with the largest amount of habitat available to them. However this test does not indicate the magnitude of the difference, nor does it provide any indication of a critical value of available area above which bird species are likly to breed. The parameters needed for prediction have not been identified.

Further tests were thus felt necessary to discover whether any other differences could be identified from air photographs between the habitats of breeding and non-breeding species. It was believed the level of dispersion of suitable habitat types may have an influence on whether a bird species breeds. The previous tests have looked at the relationship between the numbers of breeding birds and the total amount of suitable habitat. However in reality, in the reserve, the habitat types are fragmented into a number of small units rather than individual large blocks.

This fragmentation may result in suitable habitat blocks being below the territorial requirements of an individual species so being of little use. It was therefore suggested that the bird species with the least divided habitat will be more likely to breed. They will have more chance of finding a suitable sized block of the required habitat type.

It is easy to identify the number of units of each habitat type from the air photographs. Mann Whitney U tests were carried out on data from breeding and non-breeding birds. The tests looked at:

a) whether any significant difference existed between the numbers of units of suitable habitat for breeding and non-breeding bird species,

b) whether any significant difference existed between the amount of area per unit and

c) whether any significant difference existed between the average number of units per habitat type for breeding and non-breeding species.

In each instance the rejection level was set at 0.05. In each case no significant difference was found. No conclusison could therefore be drawn about the affects of habitat unit dispersion and the likelihood of birds breeding in the urban woodland.

6 CONCLUSION

There is a great deal of discussion in the literature concerning the usefulness of different variables to predict bird species numbers, bird abundance and bird species diversity. Unfortunately there has yet to be any agreement. This makes any investigations using air photographs more difficult. Of the tests carried out a possible relationship was identified between area of woodland and number of breeding species. No estimates could be made concerning species abundance, the type of species actually breeding or their distribution in the woodland. However, the results are not totally negative, suggesting prediction may still be possible. The next stage of this research will be to look at all the variables which have been identified in the literature in woodland bird population studies and to investigate how many of these variables can be obtained from air photographs. It may then be possible to combine equations using suitable variables.

7 REFERENCES

Aronoff S 1985. "The minimum accuracy value as an index of classification accuracy". Photogrammetric Engineering and Remote Sensing, 51, 1:99-111.

Arrhenius O 1921. "Species and Area". Journal of Ecology, 9:95-99.

Emlen J T 1974. "An urban bird community in Tucson, Arizona: derivation, structure and regulation". Condor, 76:184-197.

Fosberg R, and Peterken 1967. "A classification of vegetation for general purposes. A guide to the check sheet for IBP areas". IBP Handboodk 4:73-120.

Ginevan M E 1979. "Testing land-use accuracy: Another look". Photogrammetric Engineering and Remote Sensing 45, 10:1371-1377.

Lancaster R K and Rees W C 1979. "Bird communities and the structure of urban habitats". Canadian Journal of Zoology, 57,12:2358-2368.

Moore N W and Hooper M D 1975. "On the number of bird species in British woods". Biological Conservation 8:239-249.

Nature Conservancy Council 1984 "Urban Nature Conservation" Unpublished position statement.

Skaley J E 1981. "Classifying avian habitats with aerial photographs." Ph D Cornell University U.S.A.

Tomialojc L and Profus P 1977. "Comparative analysis of breeding bird communities in two parts of Wroclaw and in adjacent Querco - Carpinetum forest" Acta Ornithologica Warszawa 29, XI:117-169.

West Midlands County Council 1985. "The Nature Conservation Strategy for the County of the West Midlands" Unpublished Report.

Witherby H F et al 1965. "Witherby's Handbook of British Birds" 5 Vols. H F and G Witherby Ltd. London.

Woolhouse M E J 1983. "The theory and practice of the species area affect, applied to breeding birds of British woods". Biological Conservation, 27,4:315-322.

Symposium on Remote Sensing for Resources Development and Environmental Management / Enschede / August 1986

Automatic digitizing of photo interpretation overlays with a digital photodiode camera: The ADIOS system

C.A.de Bruijn & A.J.van Dalfsen
ITC Department of Urban Survey and Human Settlement Analysis, Enschede, Netherlands

ABSTRACT: The rapid introduction of geoinformation system in various fields of planning and municipal administration creates an urgent need for efficient methods to input information obtained by visual photointerpretation.

The paper describes the ADIOS system for "automatic digitizing" of photointerpretation overlays, currently under development at the ITC Department of Urban Survey and Human Settlement Analysis.

A digital camera with 2048 x 2048 pixel resolution is used to scan landuse interpretation overlays. After some processing to improve the quality of the scanned lines, the landuse codes of each polygon are added, from a centroid file or during an interactive session at a color graphics terminal. Resulting data can then be sent to a geoinformation system like USEMAP for geocorrection and further processing.

1 INTRODUCTION

Aerial Photography can supply planners of fast-growing cities with thematic information on landuse, residential patterns, site suitability, traffic behaviour, etc.. Often such information is difficult to get by other means. Visual interpretation will, for the time being, remain the predominant technique for data extraction from airphotos, especially in urban areas with complex spatial structures. However, data obtained in this way have to fit in the geo-information systems that will increasingly be adopted by municipal authorities for their spatial datahandling, both thematic and topographic.

Since 1973 various methods to input photo intepretation data in geodatabases have been developed at ITC as part of the USEMAP software. Starting with manual gridencoding they now rely mostly on manual digitizing.
Valuable results have been obtained in various projects. The use is rapidly increasing and has created new possibilities in linking thematic survey data closer to their eventual applications.
Still, manual digitizing is far from optimal and as pointed out by TOMLINSON (1980) digitizing one of the main problems in the field of geodataprocessing. Better, faster and more accurate input methods are urgently required if airphoto interpretation is going to remain an efficient datasource.

2 VARIOUS METHODS OF DATA INPUT

Entries in a geodatabase consist usually of two elements: a spatial description (location of point or area) and attribute information (thematic information about the point or the area, e.g. landuse). The spatial description of an area can be in the form of lines (vectors) defining the limits of the area, or in the form of gridcells (pixels) that are marked as belonging to the area. Choice between lines and gridcells has been the source of much debate in the past, but the development of modern hardware has made it rather irrelevant, since it has become fairly easy to convert from one datatype to another. However, at present "large scale" topographic databases are usually vector databases, while "medium and small scale" thematic databases tend to be gridcell databases.

Interpretation data can be entered into a computer database by coding gridcells, (now rather obsolete), manual digitizing (the most common method) or scanning.
Manual digitizing during interpretation may distract the interpreters' attention and is certainly not feasible for all types of interpretation, while digitizing after interpretation means an additional step in the survey process, it may delay the final results and introduce additional errors.

Either way, digitizing remains a time consuming obstacle, when data are to be used in a digital environment and at some places a tendency has been observed to use lower quality satellite Remote Sensing data instead of data from airphoto interpretation "because the data are already in digital format". The introduction of more efficient data input methods is therefore of paramount importance to ensure the continuing usefulness of airphoto interpretation.

In 1981, based on earlier work by MEISNER (1981) the first author wrote a research proposal to develop at ITC a "digitizing machine" based on a high resolution CCD camera. The proposal called ADIOS (Automatic Digitizing of Interpretation Overlays System) was accepted in 1982 but acquisition of the required hardware was complex and time consuming due to the absence of representatives for this type of digital cameras in Europe. In October 1985 an Eikonix 78/99 digital camera arrived at ITC and development of application software and testing of various methods could eventually start in 1986.

3 EMPHASIS ON PHOTO INTERPRETATION OVERLAYS

The starting point of the ADIOS concept is that the normal visual stereoscopic photo interpretation should not be affected by subsequent digitizing procedures. The automatic digitizing should be based on the "interpretation overlay", the transparancy containing lines and codes which forms the end product of most regular interpretation jobs. (fig.1)

The size of an interpretation overlay is limited (23 x 23 cm) and the content is not very complex. Often it is only a single theme and a single line type. A typical landuse interpretation consists e.g. only of polygons that are not overlapping and which have a single attribute. Each piece of land can only belong to one polygon (one landuse). Automatic digitizing of such an overlay should be relatively easy.

4 COMPARABLE APPROACHES

MEISNER (1981) from the Remote Sensing Laboratory of the University of Minnesota has been using a medium cost image processing system to obtain 640 x 480 pixel video images of map sections. An operator positions the screen cursor inside a polygon and enters the (landuse) code for that polygon. The computer then scans from the indicated point to the left until a boundary is reached. An automatic line following routine is used to follow this boundary back to the starting point. The entire process takes 1 - 2 seconds for a typical polygon.

LEBERL (1982) describes the use of a rather expensive flatbed scanner, the Karto Scan, for the automatic digitizing of polygon data. He used the scanner only for scanning the polygon boundaries only. Centroids with polygon labels have been digitized separately on a manual digitizer. The scanned boundary data are vectorized, using Syscan software, (before scanning, labels had to be removed from the map in a predigitial preparation step). All further processing is in vector format.

At the 1986 ACSM conference N.R. CHRISMAN reported on the use of a scanning system, incorporating a Data Copy 1700 x 3800 pixel CCD camera. The system is used to digitize 9 x 15 inch soilmaps. After thresholding and thinning vector output is generated that is further processed with a "spaghetti and meatballs" approach and edited to obtain a topologically correct datafile. Scanning and vectorizing a soil map took 20-30 minutes. Chrisman believes that "devices like this scanner will alter the digitizing environment to create higher quality with lower expense".

Also recently published has been a description by FAIN (1985) of a commercial system, the Smart Scan System, developed by Energy Images Inc., that seems to use (according to the accompanying illustration) an Eikonix 4096 x 4096 CCD array scanning camera. The Smart Scan system outputs vector data and uses some artificial intelligence software to divide the vector data into various layers based on information how lines and symbols are represented on the document. Subsequent editing is carried out in an interactive vector graphics environment. The scanned raster data are displayed on the CRT screen with the derived vector data superimposed, so that errors can be detected easily.

5 THE ADIOS CONCEPT

5.1 How ADIOS will look to the user

It is assumed that the result of an interpretation will be in the form of a (transparent) 23 x 23 cm overlay, belonging to a particular airphoto, drawn with a rapidograph pen of0 .3 mm linewidth or larger. To a limited extent it may be possible to impose certain conventions to the drawing style of the interpreter (cf norms for technical drawings that have to be microfilmed). That means that the interpreter should draw the fiducial marks in a certain way, mark tiepoints where applicable, keep text and lines separate, close or overshoot polygons etc.

The geometry of the information as digitized is the same as the geometry of the airphoto that has been interpreted. In certain cases it is possible to digitize overlays from orthophotos where the geometry is already correct, but orthophotos should not be prerequisite for the system.

The overlay will be "push button" scanned in a raster mode, results are pre-processed and displayed on a high resolution color CRT. Next step is to indicate with a lightpen or cursor on the CRT lines or areas with identical codes and to enter those codes. That should be easy since all codes on the manuscript are also scanned and hence displayed on the screen.
An alternative approach is to digitize the centroïds of each polygon on a conventional digitizer together with its code and use this datafile for automatic coding.

When a code is assigned to an area the software will look for its boundaries and paints the area with the color allocated to that code. The colored map on the screen enables easy checking by the interpreter. Editing, (deleting, changing or entering areas and boundaries) can be done either by correcting the original overlay drawing and redigitizing it, or by interactive editing at the terminal.

When the colored map is correct, it may be sent to the appropriate database after applying the specified geometric correction(s).

5.2 Hardware

At present three technologies should be considered for a simple system
- videodigitizing
- digitizing with photodiodes or CCD's electronic camera
- low cost document scanners

Typical resolution for video digitizing is 512 x 512 or 640 x 480 while for photodiodes or CCD's 2048 x 2048 resp. 4096 x 4096 may be obtained.

The low cost document scanners introduced in 1986 have resolutions of approximately 2500 x 3500 , but little is known of their spatial accuracy as yet. They only handle up to A4 size originals and, like photocopying machines, they lack a viewfinder for orientation of the original. Some systems have the capability of handling color. Table I summarizes some characteristics.

TABLE I: A4 Image Scanners Hannover CEBIT fair 1986

Type	resolution dpi	grey levels	color	OCR
Ricoh IS 30	300,240,200,180	16	-	-
" CS 30	"	"	RGB	-
" FS 1	400,300,240,120	256	RGB	-
Compuscan PCS	200,150,100	?	-	+
Canon CS 220	300,200,150,75	16	-	-

dpi = dots per inch ocr = optical character recognition

MEISNER, already mentioned above, used a SPATIAL DATA EYECOM II Picture digitizer, a video digitizer with 640 x 480 resolution.
Applied to a 9 inch photo that gives resolution of 0.4 mm which is hardly sufficient if a 0.3 mm rapidograph is used and is considerably lower than the resolution of manual digitizers (0.1 to 0.025 mm).

For ADIOS, an Eikonix 78/99 digital camera has been selected with a linear 2048 element photodiode array mounted in a motor driven stage (fig.2) that moves in 2048 steps equal to the distance between the photodetectors (15 μm). The voltage of each photodiode is converted into a digital value having up to 12 bits and the information is passed to the computer, as parallel data (DWYER, 1985). Solid state array sensors have excellent geometric properties that are a direct consequence of the photolithographic method of fabrication. (NAGY, 1981). Eikonix claims a spatial precision of 1 pixel, corner to corner. This has not yet been verified, but ITC plans to do a full calibration of its digital camera in the near future. The system at ITC includes further a normalizer board to compensate for individual diode characteristics and a computer driven color filter wheel enabling color separation.

(Another Eikonix model uses a 4096 linear CCD array. The CCD however catches less greylevels and is less sensitive to color. A similar CCD camera is manufactured by Datacopy and was used by Chrisman in his earlier cited project.)

Using the 2048 pixel camera on a 9 inch airphoto overlay means scanning with a resolution of 9 pixel per mm or about 0.1 mm which is acceptable for a 0.3 mm line width and compares favourably with the effective accuracy of usual manual digitizing.

5.3 The current pilot system

Fig 3 gives an overview of the system as it is currently applied. The total system configuration used for ADIOS consists of the Eikonix camera, a PDP 11/24 and an AED 767 graphics terminal with 8 bitplanes.

For the experiments photointerpretation overlays are used (fig. 1) that were made by H.T.J. Lutchman in 1980 for a landuse change study (DE BRUIJN, 1981). They are typical for the output of such studies and were indeed made before there was any question of automatic digitizing. Hence it is expected that results obtained with those overlays will be representative for routine operational situations. The various steps of the procedure are as follows:

Step 1 scanning

In scanning the overlays a red filter has been used to eliminate some red wax pencil lines also present on the overlays but not relevant for the landuse interpretion. The filter reduces effectively the values of the red lines but they do not disappear completely and return to a certain extent after the edge enhancement in step 2.

Step 2 edge enhancement and thresholding

Although the scanned image looks quite good when displayed on the screen the numerical values of the pixels are affected considerably by
- unequal lighting
- varying density of overlay paper
- varying line intensity and width

In the present provisional setup a simple amateur photographers reprostand lighting system with 4 incandescent 60 watt bulbs is used and it has been found that lighting varies indeed considerably.

Some of the darker white areas have values below those of some of the black lines in the lighter areas on other scanlines. The values in the unprocessed scanfile vary approximately from 50 (dark lines) to 200 (light paper). When simple thresholding is applied the lines loose their connectivity or the white areas will get a lot of noise. (fig.4)

To "reconstruct" the linework a 3 x 3 px edge enhancement filter is used with the following weight factors

$$\begin{matrix} 1 & 1 & 1 \\ 1 & -8 & 1 \\ 1 & 1 & 1 \end{matrix}$$

On the resulting values a threshold is applied to separate the lines+noise from the paper; in this example:

black (lines+noise) $\geq$ 40
white (paper) $<$ 40

Step 3 separation of boundary lines and noise

Remaining noise and written landuse codes are now separated from the boundary lines using an AED polygon fill command. The cursor is moved on to point on a line and a polygon fill command to color the black line red is given. Eventually all lines connecting to that point will be colored red while disconnected linegroups, codes and noise remain black. In this stage gaps can be edited and the result is a bitmap of the boundary lines only.

Ideally all pixels of the boundary lines should form a connecting network (except for the island polygon boundaries). To apply the AED polygon fill it is however important that this connectivity is in the form of "edge-connectivity", i.e. connected pixels should have at least one edge in common.

To ensure this condition, a special 2 x 2 "edge connectivity" operator is applied prior to the actual separation. This operator recognizes cases of pixels that are only connected by one corner and adds an additional pixel to ensure that the pixels will be edge connected. The principle and some results of this operator are shown in fig. 6.

Step 4 digitizing fiducial marks

Fiducial marks are digitized with the AED cursor and their locations are stored to serve as reference point for later geocorrection procedures.

Step 5 entering land use codes

The operator can now enter the landuse codes by pointing with the cursor to an area, reading on the screen the (black) handwritten landuse code that was scanned together with the lines, and entering that code via the keyboard or a screen menu. The AED boundary fill is then used to give all pixels in the polygon the relevant landuse code. The operator can follow this as the polygon will be filled with the appropriate landuse code on the screen. In this interaction color is essential to minimize coding errors.
Results of the polygon encoding are shown in fig. 7.

An alternative is to digitize landuse centroids on a normal manual digitizer and use these points to fill the polygons. This could be an off line or an interactive procedure and further tests will have to show which method will be the most efficient.

Step 6 editing

The same applies to editing procedures. Gaps and other mistakes may be corrected in step 3 or in step 5 when a "leaking" polygon is detected. Depending on error types and user experience it may be preferred to correct the overlay and rescan it, rather than go through tedious editing procedures. On the other hand minor errors may be edited quicker and easier at the screen, while also a certain amount of automatic error correction may be carried out especially in cases where map contents and topology are well defined.

Step 7 transfer to GIS system

Results are read back from the AED into an output file with fiducial marks, codefilled polygons and boundary lines that can now be transferred to an appropriate pixel input programme of the selected GIS system.

At ITC, the first use of ADIOS will be to input data in the raster oriented,ITC developed, USEMAP geoinformation system.
In that case USEMAP software will
a) apply a geocorrection
b) eliminate boundary line codes by attributing line gridcells to the appropriate landuse polygon
c) resample to the regular gridcell size for the project area.

5.4 Geocorrection

In flat areas geocorrection will be based on digital rectification, defined by matching tiepoints on the photo with points on a map or from a list with known coordinates. Such data will be stored in a photo index file and after linking this file with the observed fiducial marks in the scanfile the geocorrection can be applied as part of the resampling procedure. If various overlays of the same photo have to be entered the same photoindex file can be used.

In gently rolling hilly areas digital facet plotting can be used in a similar way to digital rectification. In mountaineous areas also correction based on digital monoplotting can be used, provided that a digital elevation model can be made available.

It is also possible to vectorize the data first and then to apply the geocorrection on the vectorized data as has been done by LEBERL and CHRISMAN.

5.5 Improvements and extensions of the current pilot system

To enhance the quality of the "row" scanned data it is planned to improve the lighting system.
Automatic gap-closure is currently under study and some form of vectorizing will be implemented soon. Possible approaches in this field have been described by PEUQUET (1981) and HARRIS et al. (1982). It is expected that in the near future topologically encoded vector data will be derived from the scanfile after processing in the rasterdomain only.

6 DIGITIZING OF OVERLAYS VS INTERACTIVE DIGITIZING DURING INTERPRETATION

Several USEMAP programmes have been designed to digitize during the actual interpretation (DE BRUIJN, 1983).
Experience so far has shown that it is a good and efficient method for
a) "point" data e.g. houses
b) selective landuse data
For complex landuse interpretations that cover a full photograph there is a preference to interpret first on an overlay and then to digitize that overlay in a separate step.

One of the theoretical advantages of interactive digitizing during interpretation is that it should be possible to query the database e.g. to ask for a previous interpretation to assist in change detection. Due to lack of software development capacity this possibility has not been implemented so far at ITC.

When ADIOS becomes available it may be assumed that most general landuse type digitizing will be carried out automatically.

Apart from saving time and avoiding errors it has the big advantage that it does not interfere with traditional photointerpretation methods (the production of an interpretation overlay) and does not require digitizing training or the use of digitizer operators.

For certain corrections or certain types of change detection interactive digitizing will remain a valuable technique, but clearly complementary to automatic digitizing.

However, for "point" type date the production of a readable and reliable overlay is not without problems and there interactive digitizing seems for the time being the most efficient solution.

7 FURTHER APPLICATIONS OF THE DIGITAL CAMERA

The high number of greylevels that can be scanned with the Eikonix makes it possible to scan airphotos with a fairly good image quality. A typical 55 x 55 mm SFAP negative can be scanned with 37 lines/mm resolution, a 9 inch photo with 9 lines/mm.

This creates several new possibilities:

The photo can be scanned and then displayed (if necessary after geocorrection) as a background to the landuse data on an interactive workstation. This might be a good method for updating such data or for detection of errors.

Displaying a scanned photo on a screen enables also interactive interpretation where interpretation results can be displayed immediately. This can be useful in situations where airphotos are used to count items, like in housing studies and recreation surveys. The marks on the screen avoid double counting and allow an easy check on forgotten cases.

A step further is the production of digital rectifications or digital orthophotos. Using the geocorrection methods described earlier this is clearly feasible as has been demonstrated already by KONECNY in 1979, who used a photo scanner with a 0.1 mm pixel size. Low cost, medium resolution output, acceptable for a number of applications can be generated by the new generation of laser printers with 300 dpi resolution enabling 15-20 points/cm printrasters to render appr. 20 greylevels. An output quality that is better than that of the photocopies that are now increasingly used as field material and for provisional basemaps.

8 CONCLUSION

Although the results of the pilot system are still limited at the moment this publication had to be finalized, they show clearly that the approach is successful and that an operational system can be developed in a rather short time, using the available experience with USEMAP geodata processing, image processing and photogrammetry in ITC.

Developments in the computer industry where low-cost resolution graphic input and output devices are rapidly becoming available make it probable that such developments will have a considerable impact on the application of photo interpretation and its integration in modern geodata environments.

REFERENCES

De Bruijn, C.A. (1981). Analyzing Urban Development Issues using Sequential Land Use Data and a Geodatabase: A case study from Limburg, the Netherlands. Paper Harvard Computer Graphics Week 1981, Boston

De Bruijn C.A. (1983). Urban Airphoto Interpretation in a Geodataprocessing environment. Proc. 4th Asian Conf. on Remote Sensing. Nov. 10-15, 1983, Colombo, Sri Lanka

Chrisman, N.R. (1986). Effective digitizing: Advances in Software and Hardware. In: Proc 1986 ACSM-ASPRS Conv Vol. 1 pp 162-171

Dwyer, C.B. (1985). CAD/CAM & Digital Image Acquisition: Advance Textile Market. In: Computer Technology Rev Fall 1985 pp 159-162

Fain, M.A. (1985). Automatic Data Capture With AI. In: Computer Graphics World 85-12 pp 19-22

Harris et al. (1982).A modular system for interpreting binary pixel representations of line-structured data on maps. In: Cartographica 19/2 pp 145-175

Konecny, G. (1979). Methods and Possibilities for Digital Differential Rectification. In Photogr. Eng. and RS 45/6 pp 727-734

Leberl, F.W. (1982). Raster Scanning for Operational Digitizing of Graphical Data. In: Photogr. Eng. and RS 48/4 pp 615-627

Meisner, D.E. (1981). A low cost automatic line follower system for map encoding. Paper Harvard Comp. Graphics Wk 1981, poster session, 7 p

Nagy, G. (1981). Criteria for selecting automatic digitizers (optical scanners). Paper Harvard Comp Graphics Wk 1981, session #11, 27p

Peuquet, D.J. (1981). An examination of techniques for reformatting digital cartographic data/part 1: the raster-to-vector process. In: Cartographica 18/1 pp 34-48

Tomlinson, R.F. (1980). The handling of data for natural resources development. In: Proc. Wkshop Information Requirements for Development Planning in Developing Countries 1980, ITC Enschede.

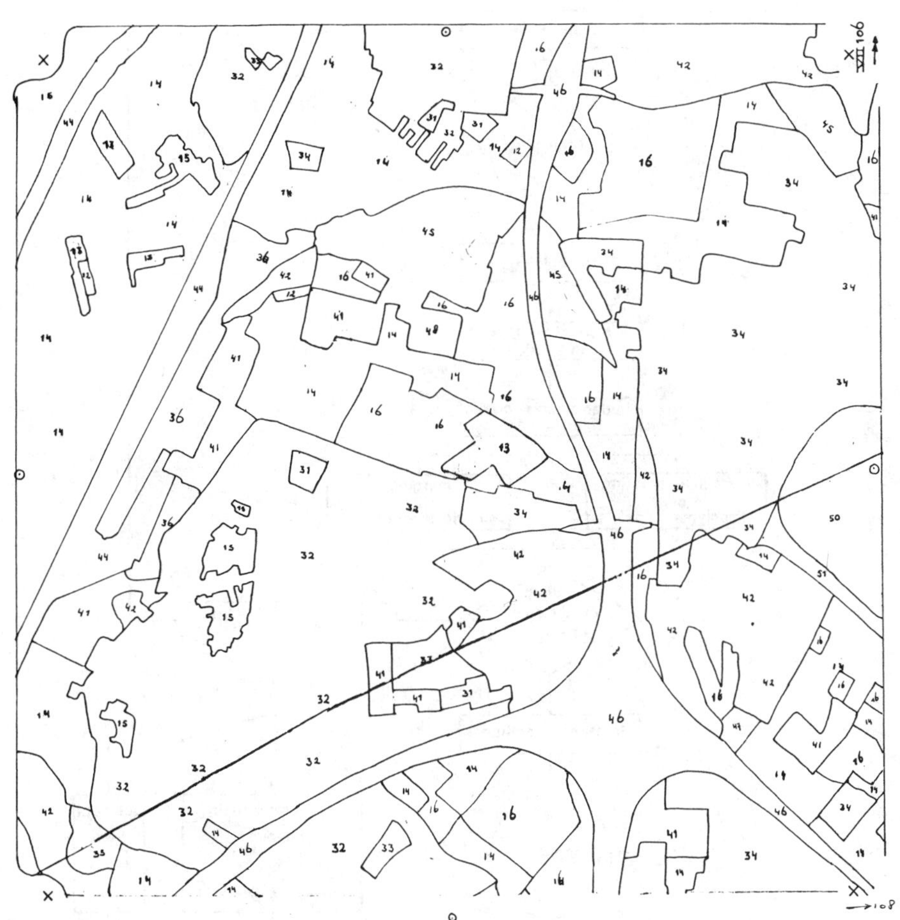

Fig. 1 Example of an interpretation overlay

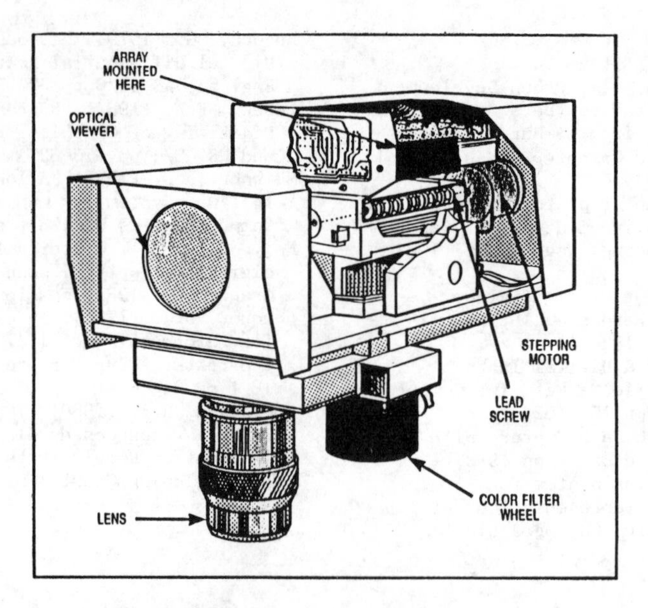

Fig. 2 Principle of the Eikonix photodiode camera

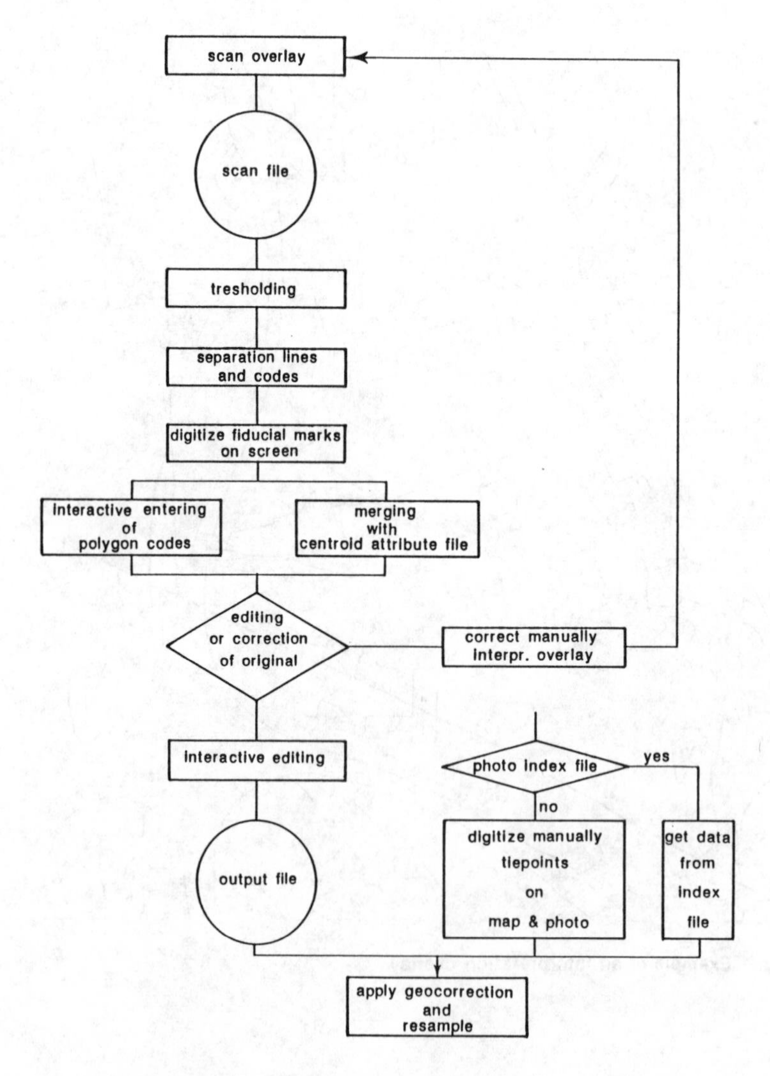

Fig. 3 Overview of the digitizing proces

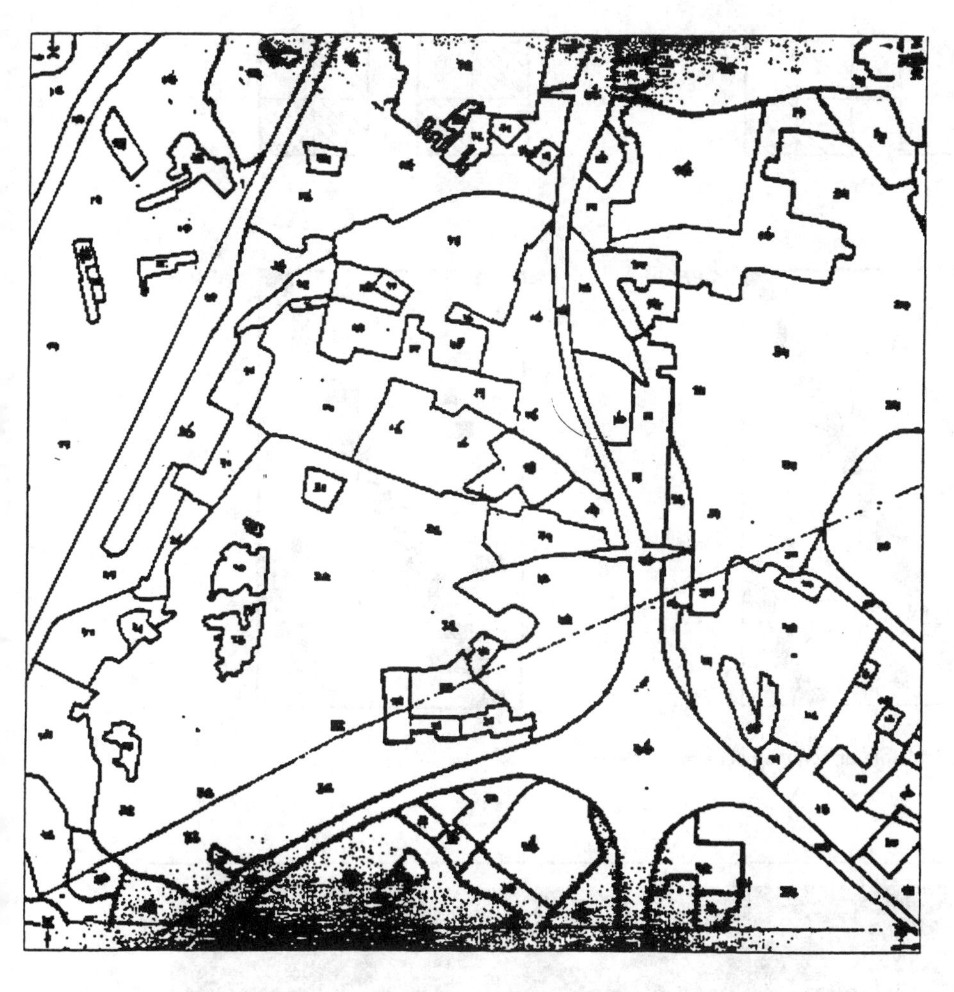

Fig. 4 The scanned file after simple tresholding

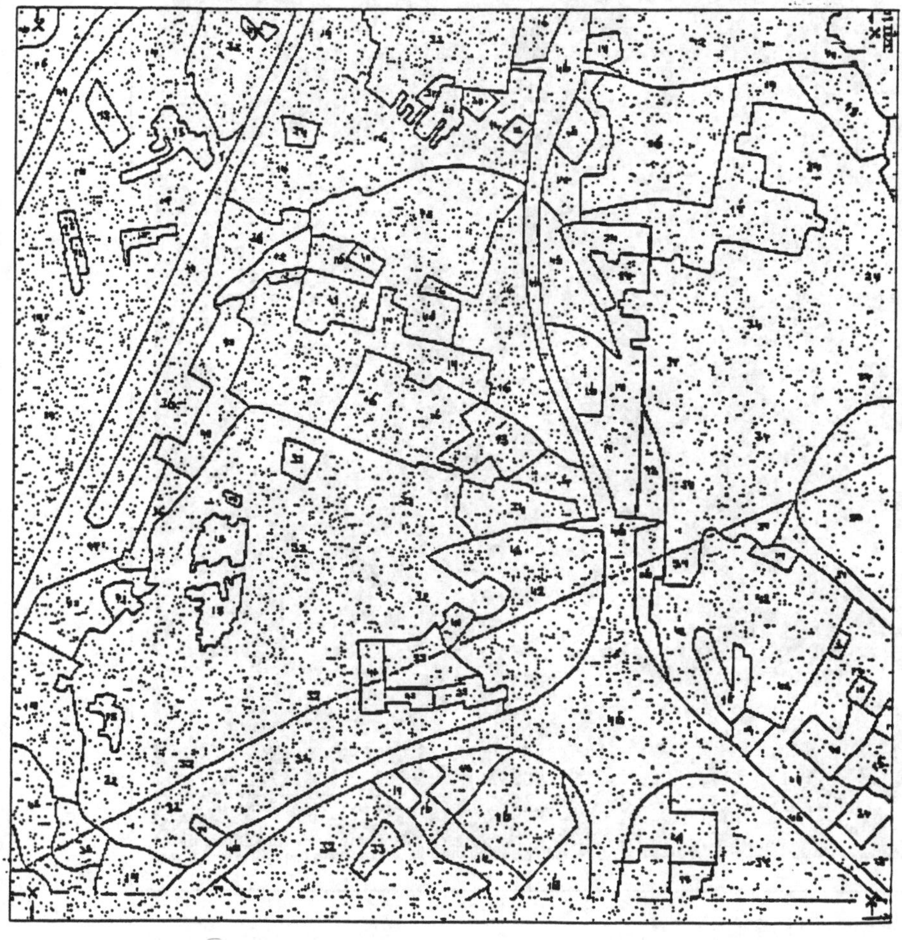

Fig. 5 Result of edge enhancement filter

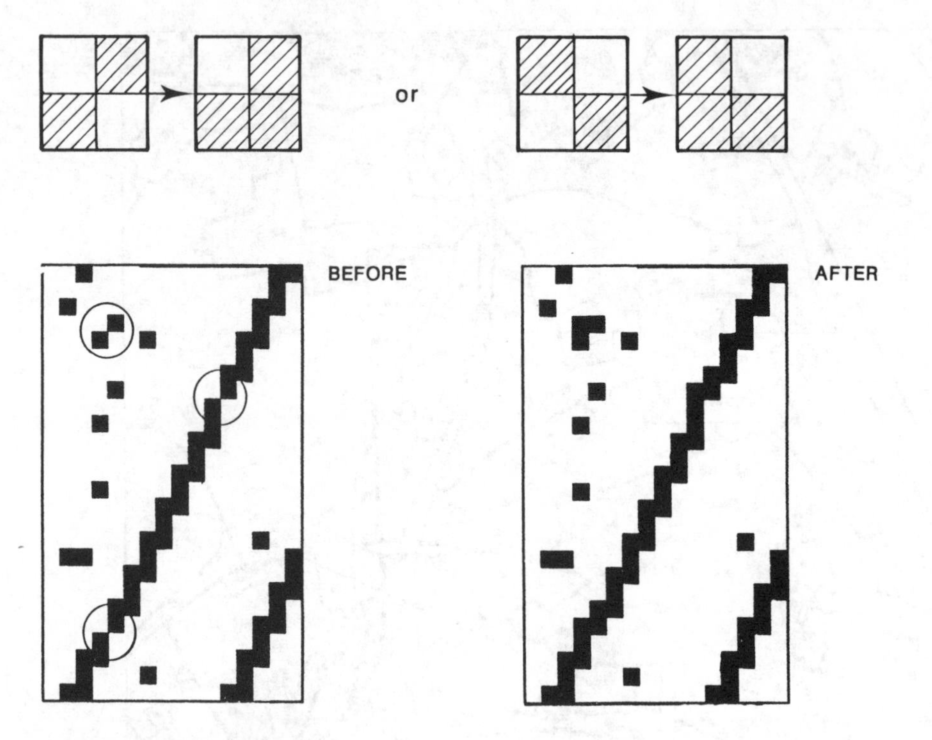

Fig. 6 Principle of connectivity operator

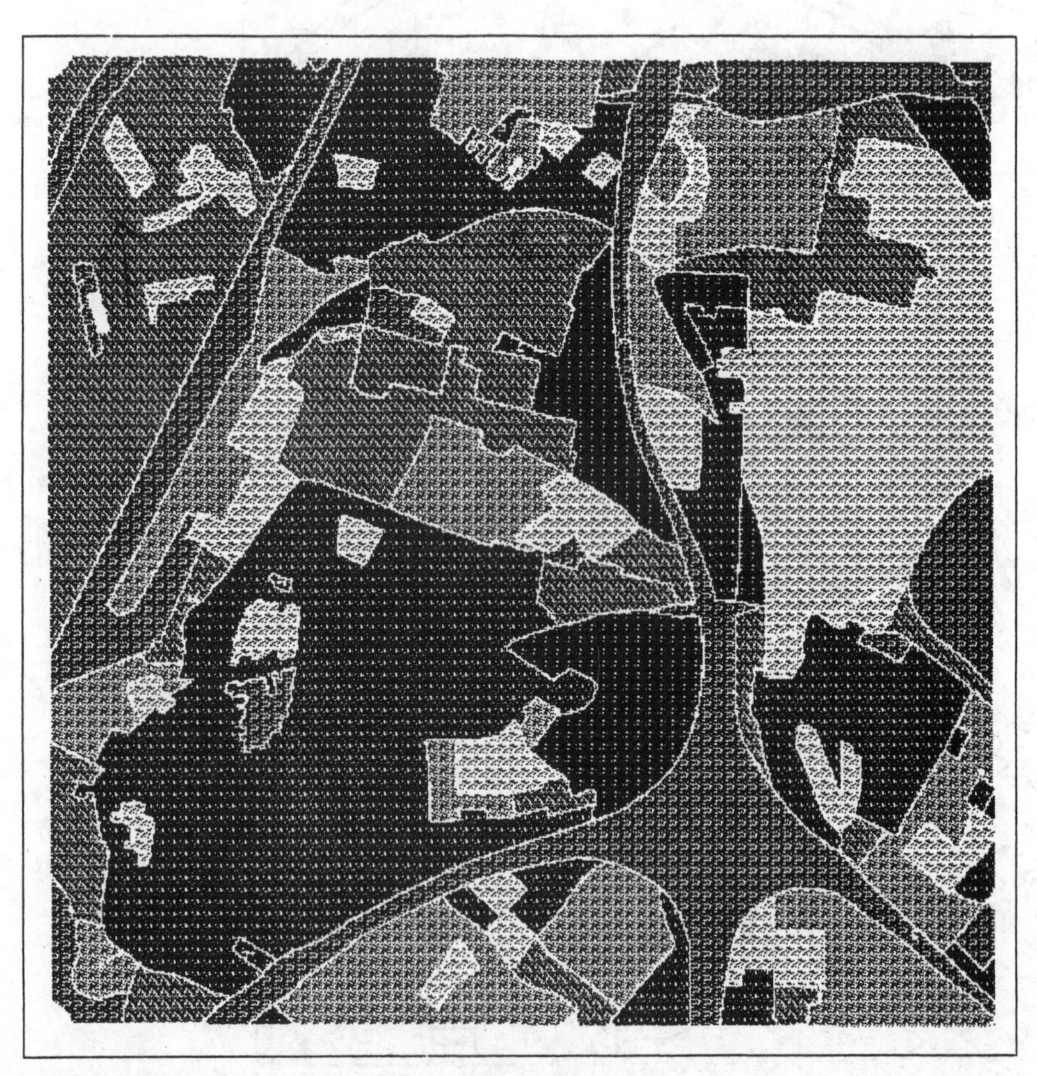

Fig. 7 Final results with encoded polygons

Symposium on Remote Sensing for Resources Development and Environmental Management / Enschede / August 1986

Visual aerial photograph texture discrimination for delineating homogeneous residential sectors: An instrument for urban planners

Maria de Lourdes Neves de Oliveira
Instituto de Pesquisas Espaciais/MCT, São José dos Campos-SP, Brazil

ABSTRACT: This paper presents a method for the definition of a geographical reference system to be used by planners in urban residential area analysis. The purpose of the method is to exploit, in the diagnosis, the spatial component of the town which is the concrete result of the interactions between its physical and social elements. The assumptions made are: a) the urban residential areas in Brazilian towns are extremely differentiated as a result of the social stratification existent in the country; b) there is a strong association between physical aspects of these areas and the socioeconomic characteristics of their resident population. The method preconizes the use of panchromatic aerial photographs at a scale of 1:10000 for the delimitation of homogeneous residential sectors. The procedures described are based on the human visual capability to descriminate different textures. In order to test its validity, survey data obtained through its application tq São José dos Campos, SP, Brazil, were analysed using cluster analysis technique. The results of the work showed that photointerpretation, besides being a quick and economical instrument for urban analysis, may be usefully applied for delineating homogeneous residential urban sectors, to be used by urban planners.

1 INTRODUCTION

Urban planning, as any decision-making process, depends on an efficient information system for support. Urban system related decisions, specifically, require the availability of a wide range of information, particularly those associated with an appropriate geographic reference, that allows the orientation of specially located actions.

Among these types of information we would mention that related to the residential differentiation on urban soil, the corresponding distribution of the different populational segments on this soil, as well as its socioeconomic features. Such information is essential for the planning of residential areas that involve, among others, decisions related to the placement of locally used urban equipment, as well as to the distribution of determined urban services.

The present paper suggests a method for defining the urban residential sectors aiming at assisting the planner in the analysis of the town residential areas that directs him towards the study of problems which require all of the above mentioned information at the same time.

This method was developed having in mind the Brazilian towns with their extremely diversified compositions of residential areas resulting from the pronounced stratification of their society

2 DESCRIPTION OF THE METHOD

This method is based upon visual photograph texture discrimination over panchromatic aerial photographs on the scale of 1:10.000. When town areas of residential use are investigated, the sectors with similar textures are delineated, thus forming urban units for data collection, analysis and storage.

Its result is a mosaic map, built up from small differentiated parts, in which neighboring sectors will show different textures.

This proposal is based upon the hypothesis that to these neighboring homogeneously textured sectors correpond different physical residential environments, to which correspond populational segments also differentiated according to their socioeconomic features.

The basic element of the aerial photographs for delineating homogeneous residential sectors is the texture, produced by the aggregation of small details, no longer analysed individually, but as a whole.

The capacity of the human visual system to perceive differences between determined textures is fundamental to the suggested sectorization process.

According to Haralick (1979) the texture is a phenomenon of area organization and has two basic dimensions:

a) one that refers to the primary elements that composes it; and

b) one that refers to the dependency among primary components, i.e., their spatial organization.

This being so, texture variation is defined by different primary components or by components of different sizes, and also by the density of these elements, their relative position and their spatial distribution. In residential areas this means that the textures of the sectors are type-depending: mansions are different from small houses; one-storey houses are different from apartment houses; high density areas are different from low density areas; arboreous areas are different from areas without vegetation; areas for exclusively residential use are different from mixed use areas. And so are their textures.

Basically, the process of delineating homogeneous sectors consists of the following:

a) Visual discrimination of the different texture areas through an overall perception of adjacent point sets.

b) Identification, within these areas, of the primary components of the textures as well as the spatial organization, investigating urbanistic and architetonic details of their elements, such as sizes of buildings and open areas, maps of the road system, existence of natural greenery, building density, nonresidential constructions.

c) Determination of whether there is differentiation between textures of visually discriminated areas based upon information collected from item (b).

d) Outlining of the limits that define the residential sectors of the same texture.

Figure 2.1 Homogeneous residential sectors. São José dos Campos, SP, Brazil, 1977.

0.5 Km

Figure 2.1 shows an example of the result of the delineating process for residential sectors through the differentiation of the photographic textures.

This aerial photograph clearly shows the differences between the textures 1, 2, 3 and 4 which are marked and allow the definition of four urban residential sectors.

Texture 1 is coarser than the other ones, i.e. its primary elements are larger. It is defined by a few large houses and large lots, many of which still unoccupied, spatially organized along the streets that constitute the road system of the sector that is regular, with parallel lines ending up in "dead ends".

Texture 2 introduces a more heterogeneous composition as far as its component elements are concerned. It is an urban reneval area in which can be seen large-size apartment houses, side-by-side with one-storey houses. Such elements are matched according to a spatial organization pattern that gives the area an overall homogeneous appearance.

Texture 3 is slimmer, i.e. made up from primary elements smaller than in the preceding texture. It is defined by one-family houses, smaller than the ones in the areas of the previous textures. The occupation is dense, not only because of the relationship, per area, between constructed and empty areas, inside the lot, but also because there are hardly any unoccupied lots left.

Texture 4 is defined by the presence of slightly larger houses than the ones described in the texture 3 area. Furthermore, there are many unoccupied areas, as well as a great deal of arborization, which gives the area an aspect that is well-differentiated from that of the previous areas.

The delineating of homogeneously-texture residential areas is a task that depends on the photointerpreter's capacity for discriminating between determined simpler textures and, by means of more developed cognitive processes, discriminating between more complex textures. The execution of this task requires not only the study of photointerpretation techniques, but also specific knowledge of architecture and urbanism.

3 VALIDATION OF THE METHOD

This validation consists of verifying if the residential sectors of different textures also includes groups of inhabitants different with respect to given socioeconomic variables.

Such activity was carried out within determined restrictions due to the impossibility of executing another flight over the test area, of redefining the homogeneous town-areas using new photographic products, as well as of executing a specifically planned field survey, in order tc obtain the validation.

In view of the impossibility, dictated by budgetary restrictions, aerial photographs, taken in 1977 of the urban area of São José dos Campos, were used, as well as the residential sectors of the same texture then defined (Oliveira et al., 1978), and the field survey effectuated for a research by DalBianco and Netto Jr (1979), based upon the division of the town into sectors.

In order to adjust the available material to the current interests, some residential sectors were eliminated from the study. This was done because, in some cases, the number of sample elements (residences) typifiying it was very small, or because there was no demonstrative difference between its texture and the texture of the neighboring sector, although the division had appeared to be coherent when it was proposed.

While the data collecting was done sector by sector, in the data analysis the sectors were compared by pairs.

Two sets of pairs of sectors of homogeneous texture were constructed: a) the first contained pairs of neighboring sectors, the comparison of which aimed at validating the process by which such geographic units were delineated. In this case 46 pairs of sectors were analysed; b) the second contained some pairs of sectors, defined by the photointerpretation, nonneighboring, and of a markedly differentiated texture. In this case the objective was the validation of the differentiation in texture as a standard for the differentiation of the residential population segments, as far as their position in the social structure of the city has concerned.

The processes used to analyse each of the sets of residences contained in each pair of sectors studied were the following:

Using the K MEANS algorithm implemented by Cappelletti (1982), in an adaptation of the algorithm introduced by Hartigan (1975), the residences of both sectors were reassembled according to field data referring to variables used as indicators of the social position: habitation standard, main householder's income and his schooling.

Hence for each set of two residential sectors, a data matrix was used with dimensions N x M, where N stood for the number of residences researched and M for the number of variables considered.

The algorithm aims at minimizing the sum of the squares of the Euclidean distances between units of a "cluster" and its center.

To the variables of habitation standard and schooling were associated the numbers 1, 2, 3 and 4, from the worst to the best habitation as well as from the lowest to the highest schooling.

The results of the utilization of the K MEANS algorithm determined the number of residences of each of the two sectors classified in each of the two clusters. The referring proportions being determined subsequently, according to the model presented in table 1, in which p. ij. stands for a proportion of elements of the sector i (defined through the photograph texture), assembled into cluster j (through the use of the algorithm).

Table 1. Data analysis model.

	1st Cluster	2nd Cluster
Sector 1	p. 11	p. 12
Sector 2	p. 21	p. 22

Based upon this table, a statistical test was carried out to verify if there was an expressive difference between the proportions of elements in each of the sectors classified in one of the clusters. This difference would mean that the populations of the sectors, where samples came from, were different with regard to their social position in the structure of the local urban society.

4 RESULTS

The differences between the proportions of elements in both clusters were examined in 46 pairs of neighboring sectors. The results showed an expressive difference among 29 of these neighboring sectors, at a significance level of $\alpha = 0.20$ and among 33 at a significance level of $\alpha = 0.30$.

During a new examination of the aerial photographs, it was discovered that those pairs of sectors, in which this difference did not prove to be expressive, were generally related to the pairs of sectors with less evident visual discrimination of the texture. There was only one exception that occurred in the case of one of those pairs.

In relation to the tests carried out with nonneighboring sectors having an outstandingly differentiated photograph texture, it was found that all of the 8 pairs compared showed a statistically expressive difference at a level of $\alpha = 0.01$ among the proportions of their elements classified in both clusters defined by the K MEANS algorithm. Six of them were expressively different at a level of $\alpha = 0.0007$.

Such results lead to the acceptance of the photograph texture differentiation of the residential areas as an appropriate standard for discriminating the different segments of the urban population according to their socioeconomic level.

Furthermore, it is to be emphasized that the sucess of this method will depend on the screening of only clearly differentiated textures.

5 CONCLUSIONS

The results in this study demonstrate that the visual photograph texture discrimination is an appropriate process for delineating residential town-sectors so as to become geographic references suitable to the purposes of urban planners.

Once, by means of the definition of these sectors, a set of geographic units sensitive not only to the physical differentiation of the residential environment, but also to the socioeconomic differentiation of the inhabitants, is obtained. These sectors may become a useful planning instrument. This may be possible specially if we take into account that the method involves a relatively simple process that can be carried out by a photointerpreter qualified for this task.

REFERENCES

Cappelletti, C.A. 1982. An application of cluster analysis for determining homogeneous subregions: the agroclimatological point of view. São José dos Campos: INPE (INPE-2490-PRE/173).

Haralick, R.M. 1979. Statistical and structutal approaches to texture. Proceeding of I.E.E.E. 67,5:786-804.
Hartigan, J.A. 1975. Clustering algorithms. New York. Wiley.
Oliveira, M.L.N.; Manso, A.P.; Barros, M.S.S. 1978. Setorização urbana através de sensoriamento remoto. São José dos Campos. Anais do 1º Simpósio Brasileiro de Sensoriamento Remoto. 2:436-451.
Dal Bianco, D.; Netto Jr, O.B. 1979. Um método para o planejamento de redes telefônicas urbanas de grande porte. São José dos Campos: INPE (INPE-1470-RPE/021).

Symposium on Remote Sensing for Resources Development and Environmental Management / Enschede / August 1986

Evaluation of combined multiple incident angle SIR-B digital data and Landsat MSS data over an urban complex

B.C.Forster
Centre for Remote Sensing, University of New South Wales, Sydney, Australia

ABSTRACT: As part of the NASA sponsored SIR-B experiment, digital data with incident angles of 17°, 36° and 43° were recorded over Sydney, Australia, and have been used in a study of radar imagery for urban purposes. The effect on radar backscatter of the multifaceted and oriented features found in urban regions has been examined as a means for improving urban discrimination when combined with Landsat multispectral data. Imagery at different incidence angles were registered to each other and to Landsat data and analysed using an image analysis computer system. While systematic interpretation of the radar imagery is complicated by the high response from urban features aligned at right angles to the incident radiation, the combined radar and Landsat images are shown to give good discrimination between sites cleared for development and those heavily urbanised. These areas show a similar Landsat response but are markedly different in radar. Moreover in older residential areas, with significant tree cover, the Landsat response is dominated by the vegetation signature, while radar is shown to provide an increased response from the underlying buildings.

1. INTRODUCTION

A number of previous studies have examined urban areas using radar imagery, these include Bryan (1979) and, Hardaway and Gustafson (1982). In general these studies have concentrated on the so called 'cardinal effect'. Here the intersection of roads and buildings tend to act as corner reflectors when they are aligned at right angles to the incident radiation. They showed that backscatter was very sensitive to street alignment, and reduced from a maximum to a constant value as the street orientation approached a threshold value of 20° to 25° away from the right angle relationship. Both these studies used single incident angle synthetic aperture radar data from earlier NASA missions, Seasat and SIR-A (Shuttle Imaging Radar-A), while the present study had the advantage of multiple incident angle data in a digital form compatible with digital Landsat MSS data.

Data over Sydney was acquired as part of the NASA SIR-B (Shuttle Imaging Radar-B) experiment launched on the 17th flight of space shuttle in October of 1984. This 8-day mission collected microwave data of many parts of the Earth's surface. Table 1 compares the characteristics of Seasat, SIR-A and SIR-B.

Table 1. Characteristics of NASA Spaceborne Synthetic Aperture Radars.

	Seasat	SIR-A	SIR-B
Spacecraft Altitude	800 km	260 km	225 km
Wavelength	23.5 cm	23.5 cm	23.5 cm
Polarisation	HH	HH	HH
Look Angle	20°	47°	15°-57°
Swath Width	100 km	50 km	14-44 km
Azimuth Resolution	25 m	40 m	25 m
Range Resolution	25 m	40 m	58-17 m

Data at three incident angles were acquired over the Sydney region (43°, 36° and 17°), as shown in figure 1, imaged from a path to the northeast of the city. Part of the swath of the 36° data take is shown in image form in figure 2.

2. IMAGE REGISTRATION AND PREPROCESSING

The SIR-B data was available in a digital form at 12.5 m pixel centres. Because this was substantially less than the inherent resolution of the data, and to reduce speckle effects, the data was resampled to a pixel size of 25 m using a cubic convolution resampling procedure. Following this a total of 50 control points were selected in an overlapping area of the 43° and 36° incident angle images (with overlap dimensions of approximately 10 by 40 km). These were naturally occurring points (road and stream intersections, centres of small parks, water/land features, etc) arrayed in banks of five across track. Their location and number were designed to give maximum registration accuracy, particularly across track where the maximum errors due to geometry would occur. Using a 2nd order polynomial transformation and holding the 36° data fixed, registration errors of 2.2 pixels across track and 0.9 pixels along track (standard error of the estimate) were determined. A nearest neighbourhood resampling procedure was used to register the 43° incident angle data to the 36° data.

Data from Landsat bands 7 (near infrared) and 5 (visible red) were viewed simultaneously with the radar data and a total of 12 common points were determined. Selection of common points was difficult

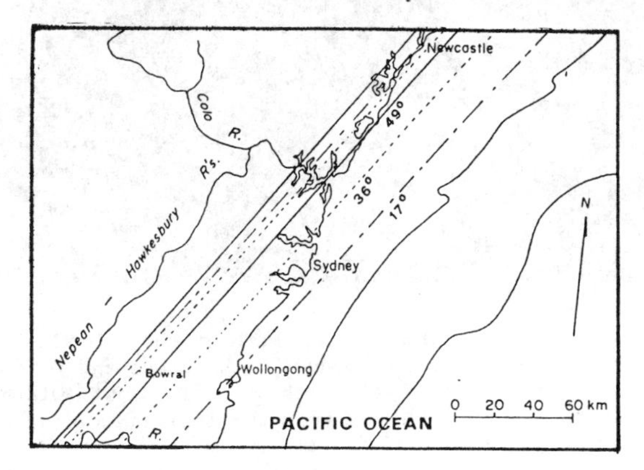

Figure 1. Coverage of SIR-B radar swaths at 17°, 36° and 43° incident angles.

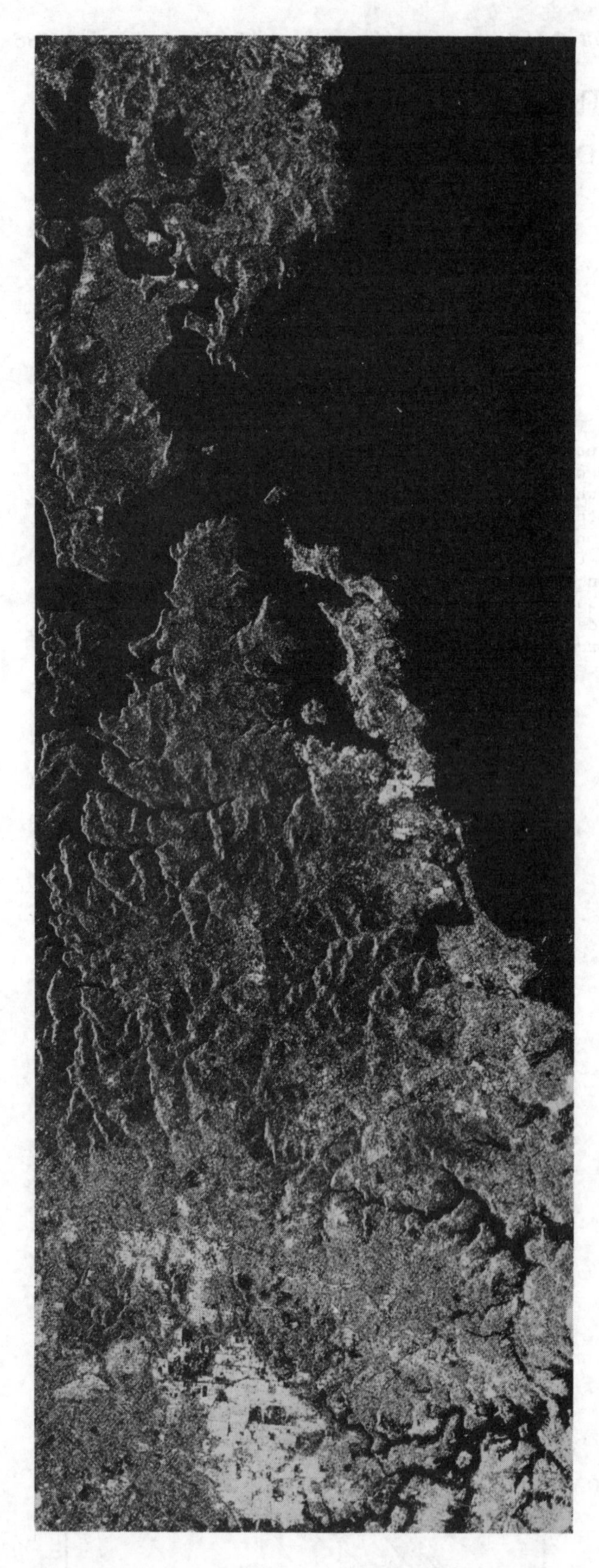

Figure 2. One section of the swath of the 36° SIR-B data take, over part of Sydney with forest and natural waterways to the north west. The swath width is approximately 20 km with the incident radiation being from the left of the image.

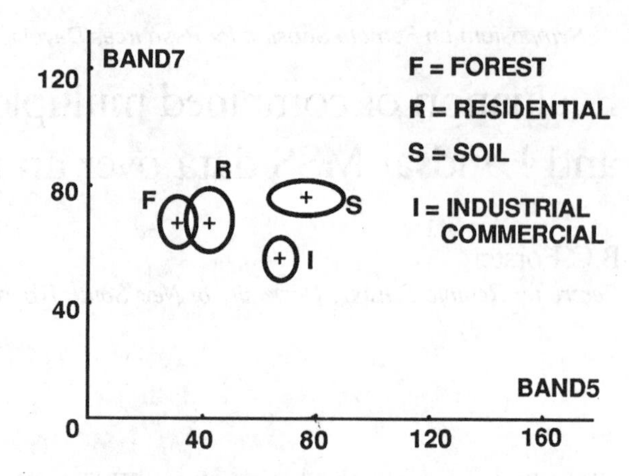

Figure 3. Scattergram of Landsat Bands 7 versus 5 for various urban surfaces. Mean response values of forest (F), mature residential (R), soil (S), and intense urban (I) (commercial and industrial), are shown with an ellipse representing two standard deviations (derived from 20 values).

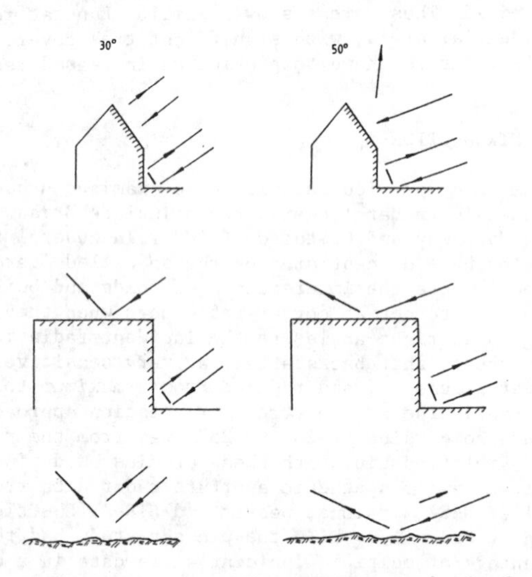

Figure 4. Illustration of the potential changes in backscatter and backscatter geometry with incident angle, for various idealised urban surfaces; residential, industrial and cleared development site.

due to the inherent difference in the imaged response. Neverthless the Landsat data was resampled to 25 m pixels using a cubic convolution procedure and registered to the 36° incident angle data with a standard error of the estimate of approximately 60 metres.

3. COMBINED MULTISPECTRAL RESPONSE

Landsat data acquired over urban areas typically give a similar multispectral response for areas recently cleared for development (particularly when soil colours are light) and areas that are heavily urbanised (typically industrial and commercial land use). Classification of these areas as similar cover types is a gross error, that leads to logical inconsistencies when monitoring change over a period of time. A scattergram of these two cover types is shown in figure 3 (Landsat band 5 versus 7) to illustrate the problem. It was considered that

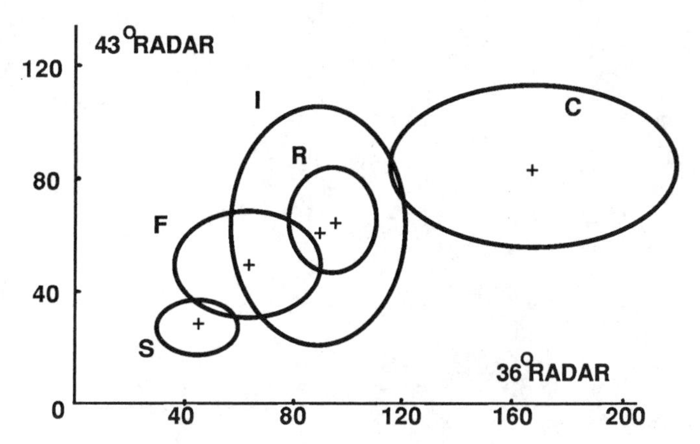

Figure 5. Scattergram of radar (SIR-B) incident angle 43° versus 36°. Mean response values of Forest (F), mature residential (R), soil (S), intense urban (I) (commercial and industrial), and cardinally aligned residential (C) are shown with an ellipse representing two standard deviations.

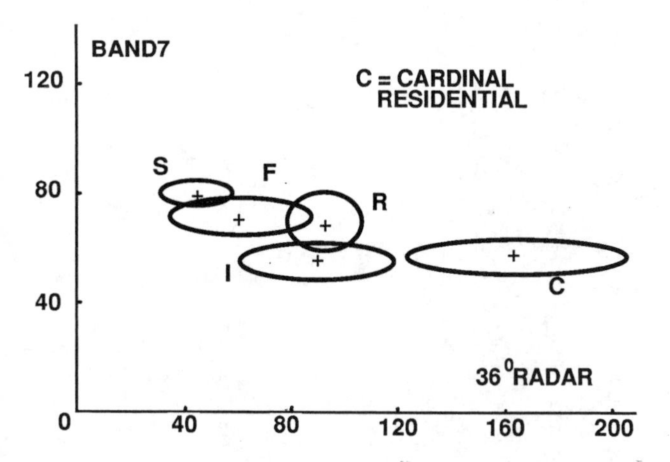

Figure 6. Scattergram of Landsat Band 7 versus radar (SIR-B) incident angle 36°. Mean response values of Forest (F), mature residential (R), soil (S), intense urban (I) (commercial and industrial), and cardinally aligned residential (C) are shown with an ellipse representing two standard deviations.

the response from different incident angle radar could discriminate between urban classes due to the differing combinations of specular, diffuse and corner reflector back scattering. Possible variations in backscatter with incident angle, are illustrated in figure 4 for various idealised urban surfaces. The scattergram of incident angle 36° versus 43° is shown in figure 5, and Landsat 7 versus 36° in figure 6, for cleared and heavily urbanised surfaces.

Similar difficulties with Landsat data have occurred when attempting to discriminate older residential areas, that typically are surrounded and overhung by mature vegetation, from low density forest which has a predominantly vegetation signature, modified by soil and understorey, in the visible and the near infrared. Because the SIR-B wavelength of 23.5 cm (L-Band) has a reasonable penetrating capacity it was expected that an increased response from the buildings underlying the trees would result, allowing the separation of these confused cover classes. Examples of their response from Landsat and SIR-B are also shown in the scattergrams, figures 3, 5 and 6. Note that units used in the scattergrams are count values on a 0 - 255 scale as recorded.

4. DISCUSSION OF RESULTS

From the scattergrams (figures 3, 5 and 6) it can be seen that the radar response, for both incident angles, has a much greater range of values than the equivalent Landsat response, and the spread of values for individual cover types is again much greater for radar. The greater overall range of radar values is due to the increased number of surfaces that act in a specular manner. Whilst the majority of surfaces in a Landsat urban scene respond in a near diffuse manner, for high sun angles, surfaces of roughness variation less than approximately 3 cm, for L-Band radar, will cause specular reflection. Such surfaces can include grass, concrete, bitumen and buildings. Thus, depending upon the relative angle between the incident radiation and the surface, either no response or a saturated response can result. This extreme dependence on the alignment of the surfaces also holds true for individual surface classes, where varying roof and building facets, and orientation of leaves and branches can cause considerable variation in radar backscatter from a surface that appears essentially homogeneous to Landsat.

Examining the plot of Landsat Bands 5 and 7 as shown in figure 3, it can be seen that forest and mature residential surface classes overlap and their mean response values are only marginally separated by approximately 10 count values in Band 5, making it difficult to define a decision surface between them for classification purposes. However in figures 5 and 6, while there is still some overlap, the mean values have a greater separation. The introduction of radar response has thus aided class discrimination. The situation with soils and the intense urban class of commercial and industrial land use, is not so clear. Whilst they are relatively well separated in the two dimensional space of Landsat Bands 5 and 7 (figure 3) this may not be the case when darker soils are involved because the Band 7 on Band 5 ratios of these surfaces response are very similar, and thus a darker soil response would move closer to, and possibly overlap, the intense urban response cluster. This is not the case in figure 6 where the lowered radar response of soils, due to its specular reflection away from the receiving antenna, and the high radar response of the intense urban surfaces allows for easier class separation with little possibility of overlap due to their significantly different ratios.

A further surface 'cardinal residential' is also shown in figures 5 and 6. This represents tree covered residential areas where major street patterns are aligned at right angles to the incident radar radiation, resulting in the so called 'cardinal effect' discussed early. Because of the significantly higher backscatter from these areas, the separation between forest and residential is even more pronounced, even though the spread of values is considerably greater than for 'non-cardinal' residential. A similar effect would result from 'cardinally' aligned intense urban (industrial and commercial) surfaces, resulting in greater separation from soils, but examples of these were not available in the study area.

It is clear from the scattergrans that the greatest difficulty in using radar backscatter in the classification of urban surfaces is the wide range of values displayed by each class. This internal class 'noise' could be reduced by the use of the mean value determined over say a three by three neighbourhood. The resultant reduction in variance would significantly improve classification accuracy and have the added advantage of being more spatially compatible with the Landsat data, but with the resultant disadvantage of loss of spatial detail for interpretation.

5. CONCLUDING REMARKS

While the results from this study are limited at

this stage and require further research in the application of various classification routines, it has shown that the introduction of radar backscatter either with Landsat data or from different incident angles does tend to separate classes that were previously difficult using Landsat alone. Concern at the range of backscattered response values from individual urban surfaces has been brought out, with the suggestion that mean values over a cell with lower spatial resolution may be more appropriate. This would also overcome the relatively high registration errors between the radar scenes, which were of the order of one and two pixels along and across track. Smoothing after registration would reduce these to 0.3 and 0.7 of a pixel if a three by three mean filter were used, which would be more comparable with the errors of 0.5 pixel obtained when different Landsat scenes are registered.

Not considered in this paper, but an area of further research, is the use of the variability of the data as a positive factor in classification. As can be seen from the scattergrams the synthetic urban surfaces of road and building, when compared to the natural surfaces of forest and soils, have a greater variability. This textural information, as measured by standard deviation or derived using a high pass filter, may contain additional means of separating urban classes. Further work also needs to be undertaken on the relationship between building spacing and backscattered response, and the effect of change in incident angle on the radar view, as projected at right angles to the incident radiation, of built urban surfaces.

REFERENCES

Bryan, M.L. 1979. The effect of radar azimuth angle on cultural data. Photogrammetric Engineering and Remote Sensing. 45:1097-1107.

Hardaway, G. & G.C. Gustafson 1982. Cardinal effect on Seasat images of urban areas. Photogrammetric Engineering and Remote Sensing. 48:399-404.

An analysis of remote sensing for monitoring urban derelict land

E.C.Hyatt, J.L.Gray & W.G.Collins
Remote Sensing Unit, Aston University, UK

ABSTRACT: The use of aerial photography and the potential of Thematic Mapper satellite imagery for monitoring urban derelict land areas are considered. Aerial photography is revealed to be an accurate assessor of derelict and waste land sites at a mapping scale of 1: 10 000, comparing favourably, and being compatible with trends determined by local authorities.
The role of satellite data and their problems and prospects are reviewed in the context of urban area studies.

Introduction

The aims of this paper are to review the use of aerial photography in derelict land studies of the West Midlands Metropolitan County area, and to provide an assessment of the potential in satellite monitoring of this problem.

The evaluation of a site in the Dudley County Borough of the Dudley Metropolitan District Council is presented as an example of the utility of aerial photography in change detection and accurate monitoring of derelict and waste land use patterns in a dense urban conurbation.

A remote sensing system for derelict land monitoring using Landsat 5 Thematic Mapper (TM) imagery is proposed and its application critically evaluated with regard to the findings of previous urban studies employing satellite data.

Dereliction in historical perspective

Birmingham was revealed to be the third worst area in the UK for dereliction, as a percentage of total County area, by the Department of the Environment's (DOE) 1974 Survey. This amounts to 1.7% under the DOE (1975) classification, but nearer 4.4% using the West Midlands County Council (WMCC) classification, the difference is explained in a later section. The WMCC's 1980/82 survey revealed that by the end of 1980 nearly 5%, 4800 hectares (ha) out of 86,664 ha, could be classed as derelict. Representing an increase of nearly 1000 ha, or 26%, on the 1974 figures, 75% of that increase was within the Black Country District of Dudley, Walsall, Sandwell and Wolverhampton.

According to Cave (1983), the amount of land becoming derelict in the County exceeds that being reclaimed by 29%, while the DOE (1977) state that nationally land becoming derelict exceeds that being reclaimed by 1500 ha/year. It seems ironic that this should be the case, with Birminghams potential considered by virtue of its location at the centre of the national road and rail networks.

The Black Country has a history of extractive and heavy industries, making it inevitably an area of acute dereliction. However the legacy of despoiled land left from over 500 ninteeth century collieries was substantially reduced from 5,760 ha in 1903 to 3933 ha in 1946 and to 1554 ha in 1965, a net decrease of over 60%. Beaver (1945) suggests that this reclamation was for housing and industrial development during the inter and post war periods, however little attention was paid to planning.

The aim of The Town and Country Planning Act of 1947 was to produce development plans that discouraged the mixing of industrial and other land uses. The West Midlands Advisory Plan of 1948 attempted to prevent excessive industrial growth in the districts. Since this was never published, there was no regional framework for local authorities to produce their development plans. The high levels of industrial decentralization never materialized, despite the regulated development encouraged by the Town Development Act of 1952. Instead as Johnson (1958) noted, enclaves of industry, surrounded by housing, had spread across the entire area, with the emphasis on industrial strip development along the dense rail and canal network. Derelict land was consumed by industry to the extent that between 1948 and 1964, and 1964 to 1975, 43% and 36% respectively of all land used for development was formerly classed as derelict.

The economic slump has taken its toll of the traditional manufacturing industries. Coupled with the formation of the West Midlands Metropolitan County Council (WMMCC) in 1972 which ensured the differentral decentralization of people and firms through comprehensive redevelopment and planning policies, the relative locational advantages of the Black Country were lost, assisting the closure of many firms and increasing the number of redundant industrial properties. Cave's (1984) survey for the WMCC reveals that 765 vacant properties in the Black Country accounted for an area 1689 ha large, and that 28 of the largest properties comprised over 50% of the total site area vacant, which will probably encourage aggregation.

In an attempt to remedy the situation, the Land Reclamation Programme, which previously concentrated on open space schemes by virtue of the Derelict Land Grant from central government is now being increasingly geared towards the requisition of sites for housing and industry in priority areas.

Any upsurge in the economy would now favour a move towards Hi-tech industries that would benefit both aesthetically and in terms of communications from a green belt locality. Necessarily this encourages migration from city centre areas, which are being reclaimed for housing, in itself a slow process.

Dudley seems to have the least flexibility with regard to reallocation of its industrial land and premises. Cave's (1984) Planning Survey reveals that only 4% of existing industrial premises are less than 5 minutes 'drive time' away from the motorway network, whilst 26% are more than 15 minutes away. Nearly 75% of Dudleys industrial sites are subject to this marketability of vacant property proportionally affected by accessibility. In as much as current market conditions dictate the need for an alternative use for vacant industrial land, over 50% of the properties surveyed in the Black Country appear to be unsuitable for residential development.

A need is revealed for future trends in dereliction to be monitored carefully, especially since the WMCC which initiated policies to arrest the problems has now been abolished and therefore decentralized. In terms of monetary and temporal economy, opposed to traditional ground survey methods, remote sensing lends itself to such information gathering, presenting derelict land data in a site and situation context.

The definition and classification of derelict land

Derelict land is problematic in that it suffers from many different definitions. A precise definition has proved elusive for some time and many variations exist, (Oxenham 1966, Thomas 1966). These are reviewed by Collins and Bush (1969), to arrive at 'land which has beens so damaged by extractive or other industrial processes that it gives offence to the eye, and is likely to remain so until subjected to special forms of restoration'. This definition has been refined by the DOE (1975) to 'land which has been so damaged by industrial and other development that it is incapbable of beneficial use without treatment'.

Changing, and more specific, definitions affect classifiction formats; thus the revision and extension of the DOE's 1974 classification of derelict land required further data collection. Many local authorities began to use aerial photography to cope with the extra information requirements.

When reviewing local authority aerial surveys from 1968-1973, Denton (1973) found that 45% used aerial photography for derelict land surveys, and that this application was eigth out of nineteen listed uses.

The Remote Sensing Unit at Aston has been awarded county contracts for aerial surveys of the West Midlands, Merseyside, Glamorgan and Nottinghamshire. The operational system that has evolved clearly demonstrates the effectivness of aerial photography in such surveys.

It should be noted that the classification scheme used throughout this paper is that of the WMCC. They have departed from the DOE definition and consider 'any land without a beneficial use' as derelict. Waste land in rural areas is exempt, but potential future areas of dereliction, such as active extractions and tipping sites are included. Table 1 gives an example of how this definition affects statistical returns.

Table 1 Comparison of how differing definitions affect classification figures.

Derelict Land Classification (hectares)	1974	1980/82
WMCC Derelict/Waste Land		
DOE classification	3793	4779
WMCC classification	12029	10547
Dudley Derelict/Waste Land		
DOE classification	951	1428
WMCC classification	2841	2445

The amount of derelict and waste land is underestimated with reference to the WMCC definition and classification by 31% and 45% in 1974 and 1980/82 repectively. Such diversity suggests that both definition and classification are in need of revision or standardization, and in consideration of the WMCC abolition, both the new district councils and the DOE should perhaps pay heed to Dueker (1971) who states that 'if however decentralization is necessary due to budget considerations, compatible classification schemes must by utilized'.

Method

The methodology consists of several stages and has been developed to fulfil a two-fold application, with a preliminary classification by aerial photography being used to gauge the classification accuracy of satellite data:

1. A review of data quality
2. The definition of the study area
3. Classification set-up for the aerial survey.
4. Area classification from aerial photography for 1971 and 1980
5. Change detection between 1971 and 1980
6. Training site selection from aerial photography for satellite data
7. Classify area with the satellite imagery
8. Compare classifications and assess their relative accuracies.

Accurate dyeline copy ground survey maps are available for the whole of the WMMC at a scale of 1:10000. These are the results of 2 surveys;

1. June 1980. Derelict Land in the WMCC; County Planning Department; AA Cave (1983).
2. December 1984. Vacant Industrial Property in the Black Country; County Planning Department; AA Cave (1984).

The sites, identified on field survey maps at a scale of 1: 2500, have been transferred to 1:10000 scale clear film maps, together with the site reference number. Site data is maintained on file on the WMCC mainframe computer, analysis is carried out using the FILETAB, GIMMS and SPSS packages.

The aerial survey utilized 2 sets of low and medium altitude panchromatic aerial photography, totalling 78 prints, flown in

October 1971 and July 1980 at scales of 1:12000 and 1:6000 respectively. Both sets are of good quality, being flown in late morning/midday to minimise shadow effects of buildings. However the 1980 set suffers from 16 missing prints which covered the north-west quadrant of the area, therefore figures for this area come from the WMCC ground survey.

Cloud free, 30m resolution satellite imagery from the Landsat 5 TM is used for the second classification stage of this study. The image was obtained on 26 April 1984, Path 205/Row 25. Only 6 channel data was requested. Band 6, the thermal channel, was not required for this study. The image was obtained form the National Remote Sensing Centre archive at Farnborough, UK. The characteristics of the TM sensor will not be discussed here and are well documented in IEEE Transactions on Geoscience and Remote Sensing, Volume GE-22, Number 3, May 1984 and Photogrammetric Engineering and Remote Sensing, Volume LI, September 1985, Number 9.

As yet little analysis of data quality has been possible, however from preliminary investigations it seems that the combinations of channels 1 or 2, with 4 and 5 may produce the best results.

The study area, part of the Dudley Metropolitan District Council in the south-west of the West Midlands Black Country Conurbation was chosen for a variety of reasons. The availability of ground survey, aerial photography and satellite data presented no problem. The area has changed dramatically in a short time and features extractive and heavy industries, extensive rail and canal networks and there have been attempts at reclamation. It is relatively close to Aston University which facilitates field-study and communication with the local council. An eight point classification modified from the DOE 1974 Derelict Land Return System by the WMCC is used for the aerial photographic survey. Land without any beneficial use is divided into:

1. Spoil heaps
2. Excavations and pits
3. Military and other Service Dereliction
4. Disused Rail Land
5. Disused Sewage Works and Installations
6. Disused Waterways Land
7. Neglected Waste Land
8. Other

The 8 basic characteristics were further divided by the WMCC for their ground survey to give 23 sub-divisions. As many of these are not apparent from aerial imagery they will not be exhaustively listed here.

A photo mosaic was compiled using the panchromatic aerial photography. These were examined stereoscopically and areas of dereliction were plotted on acetate overlays placed on alternate photographs within the flight strips. The vertical strips overlapped by approximately 60% and adjacent photographs had a 25% overlap.

A minimum site mapping unit of .25 hectares was specified, this corresponds to an area of 2.7 TM pixels (TM resolution being approximately 0.1 ha). It was considered that this would give sufficient accuracy for the TM classification, as according to Wilson and Thompson (1982), MSS data, with 0.5 ha resolution, can be used for map scales down to 1:25000. Sites were identified, classified according to type and transferred to and recorded on 1:10000 clear film maps of the area.

This procedure was followed for both the 1971 and 1980 photography, leading to a third overlay noting change between the 2 dates.

The resultant polygons were digitised and the perimeter length and area of all the sites was recorded.

Vegetation maps are currently being produced by the same method, using a different system, to aid in the satellite classification.

The aerial classification map for 1980 will be used in conjunction with the 1980 and 1984 WMCC ground survey maps of derelict and waste land and vacant industrial property. These maps will be amalgamated and used to choose training areas for the computer classifier. This will also provide a means of visual comparison with hard copy of the satellite classification.

Computer classification of the TM data is the next planned stage in the research. Preliminary work has involved demarcation of the study area, requiring the definition of a 167 x 167 pixel extract from a 512 x 512 sub-scene. It is the intention to classify only the derelict and waste sites via a thresholding exercise. Training sets will be identified from the aerial photography reference maps, whilst in order to maximise classification accuracy only large and relatively homogenous areas will be used. Areas of urban open space other than spoiled land, such as golf courses and parks, will be masked so they do not interfere with classification. The reference maps will also be used as a measure of classification accuracy.

It is intended to investigate the further incorporation of the classification, land use and any ancillary data into a Geographic Information System (GIS), this together with the potential of satellite imagery for urban studies is discussed later.

Analysis of aerial photograph classification

The results of the aerial photographic survey of derelict and waste land are presented in Tables 2, 3, and 4. The most meaningful results are those in Tables 2, dealing with the number and size of derelict sites in the mapped area.

In 1971, 571 out of 2500 ha, 23% of the map area were derelict. This figure had been reduced in 1980 by 119 ha, or 5%, to 18% of the map area, 453 ha. The reduction in size of the total derelict area is very encouraging, reflecting the WMCC success with reclamation. However, a more worrying trend is the number of sites that are derelict in the Dudley area. The increase in site number by 59% from 117 in 1971 to 185 in 1980 suggests a more disseminate distribution. The number of sites in the 0-0.5 ha and 0.5-1 ha range has increased by 69 between the two dates and 48% of all the sites in 1980 fit this category.

The reason behind the rise in site number is an acceleration of the decline in traditional manufacturing industries, which reached epidemic proportions in the early 1980s. Cave (1984) reveals that the amount of vacant industrial floorspace in the Black Country increased by over 400% from 60 ha in April 1979 to 258 ha in April 1982. 16 ha of this is in Dudley. For the same period unemployment rose 290% from 77,000 to 225,000. It seems reasonable in the light of this evidence to conclude that the closure of factories and other premises accounts for the disproportionate rise in small derelict sites.

Table 2. Number and size of sites in the Dudley area.

1971 Dudley Derelict

hectares.	no. sites.
0-0.5	2
0.51-1	18
1.01-1.5	19
1.51-2	14
2.01-3	17
3.01-4	9
4.01-5	6
5+	32
total	117

1980 Dudley Derelict

hectares.	no. sites.
0-0.5	42
0.51-1	47
1.01-1.5	27
1.51-2	11
2.01-3	21
3.01-4	12
4.01-5	5
5+	20
total	185

The number of sites in the 1-5 ha range seems to have remained reasonably stable, with a reduction of 11 sites over the period being evident. However, with reference to sites over 5 ha a significant reduction in number has occurred from 32 to 20 sites. Larger sites were generally of the neglected waste land category, the type of land that is relatively easy to reclaim. It appears that much of the land has been returned to private development and recreational use. The move towards these uses seems reasonable considering the Countryside Review Committee (1976) statement concerning reclaimed land that 'the original agricultural quality is hardly ever recovered'.

Tables 3 and 4 reveal the classification results for 1971 and 1980, these are complete for the 1971 survey. The absence of 16 air photographs means that 64 sites on the 1980 map have not been identified with regard to type of dereliction, these are indicated as 'ground truth' in the classification and the aerial data has been abstracted from the WMCC 1980 ground survey. The author is currently performing categorization of these sites, therefore, at the present time, it would be fatuous to attempt any comparison between the data sets and they are presented merely as a guide.

Table 3. 1971 classification scheme.

1971 Dudley Derelict Land

	sites	hectare	%	mean
spoil heaps	11	29.4	5.1	2.7
excavations/pits	14	137.4	24	9.8
military	0	0	0	0
disused rail land	6	14.7	2.6	2.5
disused sewage works	0	0	0	0
disused waterways	4	8	1.4	2
neglected land	82	382.1	66.9	4.7
other	0	0	0	0
total	117	571.6	100	

Table 4. 1980 classification scheme.

1980 Dudley Derelict Land

	sites	hectare	%	mean
spoil heaps	11	13.2	2.9	1.2
excavations/pits	15	39.3	8.7	2.6
military	0	0	0	0
disused rail land	4	6.6	1.4	1.6
disused sewage works	1	0.8	0.2	0.8
disused waterways	3	5.6	1.2	3.1
neglected land	85	260.6	57.6	3.1
other	2	3.9	0.9	2
ground truth	64	122.5	27.1	1.9
total	185	425.2	100	

It is apparent that the air survey methods outlined in this paper are efficient and accurate, with the study results reflecting county and district changes. It has been demonstrated by Collins and Gibson (1980) that aerial surveys have substantial advantages over traditional ground based surveys, which take up to 12 times as long, cost 4.5-8.5 times as much and do not locate as many sites.

Satellite monitoring in urban areas

The previous section has demonstrated the accuracy and benefits of using aerial photography in the assessment of specific urban land covers. One of the main problems with such surveys is that councils obtain coverage on an aperiodic basis only. In considering reclamation, Bullard (1983) explains the philosophy behind this, 'The cost of monitoring may be a small percentage of the total cost of reclamation but like any 'service' they are only fully appreciated when they can show significant changes taking place in one or more of the conditions'. It is unlikely that the mentality of local authorities will be changed and therefore another source of monitoring data needs to be utilized. Satellite imagery is readily available and being multitemporal is suited to the needs of local authorities. The use of such data is well documented for agricultural and other land uses, however interest is increasing in using high spatial and spectral resolution systems, such as SPOT 1 and Landsat 5 TM in urban studies. Necessarily there are problems with the new data types and what follows is a brief evaluation of their suitability for qualitative and quantitative urban monitoring.

Landsat MSS imagery has been used in the urban environment, but only with limited success. Wang (1984) notes that studies in the urban environment concentrate on those areas requiring less resolution, such as boundary changes, and that few have realised the harder task of identifying urban land use categories.

The TM appears to have advantages over MSS data in terms of data quality that may figure significantly in urban area analysis. Anuta (1984) reveals that using the same clustering and merging sequences, TM data exhibits 42 classes as opposed to 21 from MSS. Additional evidence of the increased dimensionality in TM opposed to MSS is recognised when Principal Component (PC) images are examined. Quattrochi (1982 and 1985) has recognised that using PC images in ratio and photographic forms facilitates detailed examination of urban structures. Additional dimensionality over MSS and the ability of the third PC of 4 band data for detection/discrimination of built up areas/bare soil has been noted by Sadowski (1983).

Increased spatial and spectral resolution do not necessarily mean that this data provides an uncomplicated answer to urban monitoring problems. The nature of urban cover must be considered, Owe et al (1984) state that heterogenous urban features often cause classification error. The reflectance from mature trees, which are often higher than residential units, has been found by Baumann (1979) to influence classification accuracy. Other investigators, Bryant (1971) and Forster (1981 and 1982) have highlighted the problems presented by abrupt changes of urban land use over short distances. These take the form of considerable inter and intra pixel differences, which seriously alter the reflection from one cover class that is surrounded by classes giving dissimilar readings. The difficulties associated with such conditions are summarised by Clark (1979) who states 'Physical or spectral conditions of a land use do not always divide as sharply as the cultural definitions of the land use'.

Boundary definition is very important in the urban area, however sharp contrasts are infrequently seen. Instead pixels lie along boundary lines and introduce cover mixing effects. Merickel et al (1984) consider that up to 60% of the pixels in some Landsat scenes are mixed, whilst Owe et al (1984) say that with the TM, more pixels per unit area do not lessen the chance of boundary features being crossed. A programme, 'CASCADE', introduced by Merickel et al (1984), specifically to combat the mixed pixel effect, assigns pixels to homogenous regions in a neighbourhood after judging that region responsible for the mixing effect.

A further influence on category discrimination, particularly in heterogenous regions, is the sensor Point Spread Function (PSF). Acting over a 3x3 pixel area, this seriously affects the signature from cover classes with dissimilar neighbours. According to Forshaw (1983) a better representation of resolution would be a deconvolved PSF rather than a pixel only estimate. Forshaw considers that as a term, 'spatial resolution' is 'poorly defined and improperly used' and is artificially high in order to compensate for the rapid data sampling rates of modern satellite systems. The views on the increased spatial resolution of sensors are as mixed as those on the improved spectral range. Irons (1984) suggests that a 'stalemate' is evident, where increasing resolution does not affect accuracy because; a) category spectral variability increase hinders classification; and b) a decrease in mixed pixels (by up to 24%) enhances classification. A 'Point of diminishing return is reached and 30m IFOV should be the best for multispectral classification' is the stated view of Clark (1979) in an analysis of multi resolution TM Simulation data in an urban environment. Clark also recognised that as resolution decreased, classification accuracy actually increased, probably due to the heterogenous nature of urban sites being 'smoothed' out.

Finally, to summarize Forster (1982), higher resolutions will, in urban areas;

1. Reduce mixed pixels
2. Aid contextual identification
3. Aid in registration
4. Reduce the PSF effect
5. Higher data redundancy will allow more accurate judgement of surface percentages
6. Higher pixel homogenity will aid in clustering procedures
7. Texture studies may be implemented

The increase in information content of TM data has proved problematic, not least in terms of data handling, but also regarding classification techniques. PC and canonical analysis have been used as data reduction and feature extraction techniques by a number of workers, Brumfield (1981), Jackson (1984) and Sadowski (1983), non traditional methods such as canonical analysis can result in upwards of a 20% improvement in classification accuracy. The need for new classification algorithms for TM imagery has been recognised by Irons (1984), and classification schemes currently being developed at the Natural Environment Research Council by Jackson et al (1984) exploit per-pixel, textural and contextual algorithms. Wang (1984) considers that such techniques need careful development because the 'Averaging process smooths out a certain amount of the data's unique qualities', while Forshaw (1983) implies that resampling may limit resolution to 2 times the pixel size. It is evident that although significant classification accuracy can be obtained with TM data, better results can be expected and a hiatus exists with current methods unable to realise the full potential of TM data.

Much of the work being done on TM imagery and urban environments originated from the USA, which features a different urban make-up to the UK. Jackson et al (1984) define the problem with a statement on the quality of the first TM images, 'in rural areas, there is a very significant improvement over the MSS, whereas in urban areas the improvement is much less marked, probably as a result of the high density of English urban development'. For this reason the possibility of incorporating TM data into a GIS is currently being investigated with regard to derelict land. Although cartographic fidelity does not appear to be a problem with TM data (Welch et al 1984), data transformation does alter pixel values, damaging their essential qualities.

The difficulties in combining digital data with different sorts of ancillary information are well known, and to quote Brooner (1982), 'A new generation of information systems whose design bridges both cell and polygon inputs, characteristics of 'conventional' systems is needed'. The problem is compounded by the quality of present and proposed satellite data, which should not be excluded from a comprehensive GIS.

Clark (1979) suggests 30m resolution as the optimum for urban studies, Forshaw (1983) considers that 'resolutions rather better than 10m will be necessary for consistently high recognition accuracies', and Jackson et al (1984), in recognition of the dense nature of English development states 20m as a minimum resolution requirement. Forster's (1982) theory is that TM should be used to determine surface types, while SPOT panchromatic data, 10m resolution, could provide high resolution cartographic and contextual information.

It is the authors opinion that while these suggestions are valid, it is essential to include data from aerial photography in a GIS to facilitate the provision of qualitative and quantitative information, the accuracy of which is demonstrated in the early part of this paper. It is acknowledged that if aerial photography is incorporated into such a system, operator expertise will still be required, necessarily cutting down on the automated stages of analysis, but not necessarily data retrieval.

References

Anuta,P.E.,L.A.Bartolucci.,D.F.Lozano-Garcia, J.A.Valdes & C.R.Valenzuala 1984. Comparison of Classification schemes for MSS and TM data.1984 Mach. Proc. of Remotely Sensed data Symp.

Baumann, P.R. 1979. Evaluation of 3 techniques for Classifying Urban Land Cover Patterns using Landsat MSS Data. NASA Report 178 790100 E81 10127, TM 82323, NASA Earth Resources Labs. Bay. St.Louis. Miss.

Beaver, S.H. 1945. Report on Derelict land in the Black Country. Ministry of Town and Country Planning, London, 46pp.

Brooner, W.G. 1982. An overview of remote sensing input to Geographic Information Systems. Proc.7th Pecora Symp. p.318-329.

Brumfield,J.O., H.H.L.Bloemer & W.J.Campbell 1981. An Unsupervised Classification Approach for Analysis of Landsat Data to Monitor Land Reclamation in Belmont County, Ohio. 1981 Mach. Proc. of Remotely Sensed Data Symp. p.428-433.

Bryant, N.A. & A.J.Zorrist 1976. Integration of Socioeconomic Data and Remote Sensing Imagery for Land Use Applications. Proc. 2nd Ann. WT Pecora Mem. Symp. Oct 1976. p.120-130.

Bullard, R.K. 1983. Monitoring Reclaimed Land by Remote Sensing. Remote Sensing for Rangeland Monitoring and Management. Proc. 9th Remote Sensing Society Conf. Sept 1983. Silsoe. p.161-165.

Cave, A.A. 1983. Derelict Land in the West Midlands County. WMCC, County Planning Office. June 1983.

Cave, A.A. 1984. Vacant Industrial Property in the Black Country. WMCC, County Planning Office. Dec 1984.

Clark, J. 1979. Landsat D TM Simulator in an Urban Environment using Aircraft MSS Data. Remote Sensing Quarterly. 1:17-32.

Collins, W.G. & P.W.Bush. 1969. The Definition and Classification of Derelict Land. J.Town.Pln.Inst.March 1969.

Collins, W.G. & L.Gibson 1980. Derelict and Degraded Land Surveys: An Evaluation of the Cost Effectivness of Air Survey Methods. Proc. 14th Int.Cong.of the ISPRS,Hamburg 1980. p.204-213.

Countryside Review Committee. 1976. The Countryside-Problems and Policies. London, HMSO.

Denton,J.P.W. 1973. Air Survey and Environmental Planning. Unpublished Dissertation, Oxford Polytechnic.

D.O.E. 1975. Resulte of the 1974 Survey of Derelict and Despoiled Land in England. Mineral Working Section, London, D.O.E. 93pp.

D.O.E., M.A.F.F., S.A.G.A., 1977. Joint Agricultural Land Restoration Experiments. Department of the Environment.

Dueker, K.J. & F.E.Horton. 1978. Towards Geographic Urban Change Detection Systems with Remote Sensing Inputs. Proc. 37th Ann. Meeting of the American Society of Photogrammetry, March 7-12th, Washington DC, ASP. p. 204-218.

Forshaw,M.R.B.,A.Haskell,P.F.Miller,D.J.Stanley,J.R.G.Townshend 1983. Spatial resolution of remotely sensed imagery A review paper. Int.J.Remote Sensing. 4:497-520.

Forster, B.C. 1981. Prediction of urban surface reflection from Landsat data using mixed surface models. 1st Coll. on Spectral Signatures of objects in remote sensing, Avignon, France,Sept 8-11. p. 569-578.

Forster, B.C. 1982. Overcoming urban monitoring problems with the new generation satellite sensors. Photo.Eng.Rem.Sens.Int.Symp. 1:889-896.

Irons, J.R. 1984. The utility of TM sensor characteristics for surface mine monitoring. Mach.Proc.of Remotely Sensed Data Symp.pp. 74-83.

Jackson, J.R. Baker, J.R.G. Townshend, J.E. Gayler, J.R. Hardy 1984. The use of TM data for landcover discrimination, preliminary results from the UK SATMaP programme. Landsat 4 Sci. Characterisation. Early Results, NASA. 4:369-385

Johnson,B.L.C. 1958. The distribution of the factory population in the West Midlands conurbation. Trans.Inst.Brit.Cart. 25:209-223.

Merickel, M.B., J.C.Lundgren,S.S.Shen 1984. A spatial processing algorithm to reduce the effects of mixed pixels and increase the separability between classes. Pattern Recognition, 17,5:523-533.

Owe, M., J.P.Ormsby 1984. Improved classification of small scale urban watersheds using TM simulator data. Int.J.Remote Sensing. 5:761-770.

Oxenham, J.R. 1966. Reclaiming Derelict Land, Faber,204pp.

Quattrochi, D.A., et al 1982. An initial analysis of Landsat 4 TM data for the classification of Agricultural, Forested wetland and Urban land covers. NASA Report No. NAS 1 15 85499 E84 10073. 50pp.

Quattrochi, D.A. 1985. An initial analysis of Landsat 4 TM data for the classification of Agricultural, Forested wetland and Urban land covers. Landsat 4 Sci.Invest.Summ. 2:111-112.

Sadowski, F.G., et al (Barker J. ed) 1983. Study of TM and MSS. Data Applications. NASA Report No.NAS 1 15 85499 TM 85499. Landsat 4 Sci.Invest.Summ. 2:129-132.

Thomas, T.M. 1966.Derelict land in South Wales. Town Planning Review, July, 1966.

Wang, S.C. 1984. Analysis methods for TM data of urban regions. Mach.Proc.of Remotely Sensed Data.Symp. 1984:134-143.

Welch, R., E.L.Usery 1984. Cartographic accuracy of Landsat 4 and TM Image data. IEEE.Trans.GE-22. 3:281-288.

Wilson, C.L., F.J.Thompson 1982. Integration and manipulation of remotely sensed and other data in Geographic Information Systems. Proc.7th. Pecora Symp. 1982:303-317.

Symposium on Remote Sensing for Resources Development and Environmental Management / Enschede / August 1986

The Nigerian urban environment: Aerial photographic inventory and mapping of land use characteristics

Isi A.Ikhuoria
Department of Geography & Regional Planning, University of Benin, Benin City, Nigeria

ABSTRACT: The Nigerian urban environment has experienced three great periods of development: pre-colonial, and colonial, and post-colonial. Each stage gave rise to a special type of urban land-use (or city) that reflects its experience. These developments which are the results of traditional and modern development mechanisms have remained largely uncontrolled, unmonitored and unmapped. Consequently, the management and mapping of urban land resources is, seemingly, not in tune with the needs of current and future generations on the one hand, while the haphazard developments have posed intractable problems for planning on the other. In this paper, a practical application of aerial photographic remote sensing in inventorying and mapping land-use characteristics in three Nigerian urban centres, Lagos, Benin City and Warri is made. The result wholly provides the necessary land-use information and maps for planning and indepth analysis. It shows more glaringly the unequal impact of our development efforts from pre-colonial period to date. Especially relevant is that the research illustrates the effectiveness of aerial photographic remote sensing in providing the data needed to meet urban land resources and management constraints in developing countries.

1 INTRODUCTION

Within the past half a century, most Nigerian urban environments have experienced three great periods of tremendous developments: pre-colonial, colonial and post-colonial. Each stage gave rise to a special type of urban land use which reflects the cities' experiences. These developments manifested by intensive land use activities, population concentrations and environmental changes, which are due to both traditional and modern development mechanisms have largely remained uncontrolled, unmonitored and unmapped. Consequently, they have posed intractable problems for urban environmental planning and management.

For example, Sada (1980) pointed out the existence and the need to arrest and improve decarying Nigerian rural and urban environments and: Mabogunje (1986), noted that "it is clear that our cities are the places where environmental stress... has achieved the greatest salience in recent times."

The situation is compounded by the lack of up-to-date thematic maps, land use and population data for effective control. Perhaps, the cause is the inadequacy of current traditional methods of urban survey (that is, field enumeration and questionnaires) which do not incorporate practical application of better procedures of determining urban growth, land use, population concentration and urban environmental quality changes. Thus a development of better ways to merge information acquired by non-conventional methods, for example, airborne and satellite remote sensing with existing data need to be explored.

Indeed, having been afflicted with such environmental stress as urban blight (Ikhuoria, 1986), urban rejuvenation which process poses intractable problems for planning (Sada, 1975), rapid population concentration and high rate of migration (Sada, 1984) and, land use and environmental dereliction (Omuta, 1985), Nigeria is now in a position to, and should, appreciate the need for adopting all necessary measures to monitor and register through mapping these urban spatial structures and environmental changes. Such measures must be accurate, reliable, timely, cost-effective and available as and when required. Remote sensing, particularly aerial photography is, in the third world context, ideal for mapping urban environments. Thus the specific objective of this paper is to identify and map the land use characteristics in Lagos Island, Benin City, and Warri; (Fig. 1) as well as seek theoretical explanations of their internal patterns.

2 CONCEPTUAL FRAMEWORK

Firstly, in acquiring data for urban land use analysis, the level of aggregation or resolution and descrimination strongly influences the quality of the conclusions which may be drawn. These concepts, so familiar to users of land use and census statistic are equally important in photo interpretation. Thus the four dimensions of resolution which include space, categories, intensity and time which were hitherto acquired by conventional field records of situational, personal and socio-economic indicators of urban structures have their parallels in remote sensing. These are spatial, spectral, radiometric and temporal resolutions (Lintz and Simonett, 1976). Also, the concept of discrimination embodies three levels of assessment in remote sensing: detection, identification and analysis. Contained within these notions of discrimination are the probabilities of correct detection, identification and analysis. These probabilities vary from area to area and application to application. Urban planners and geographers, as a general rule, require information at the first two levels of detection and identification (Lintz and Simonett, 1976). And, for urban land use application, the acceptable probability of correct identification is 85% (Anderson et al, 1976).

Secondly, the explanation of the urban organization of any society is a function of the socio-economic activities of man (Sule, 1982) which consequently create some spatial patterns. A land mark in urban land use theory was Burges (1923) concentric zonal theory which says that urban land use is patterned in a number of concentric zones. A second theory (Hoyt, 1939), asserted that urban growth occurs in sectors along major traffic arteries. Thirdly, (Ulman, 1945) developed the multiple nucli model in which a city is shaped from a number of focal points where land uses of a similar type are concentrated.

In the African context, Akin Mabogunje (1968) provided the Twin-centre concept in which he asserted that Nigerian cities grow from an amalgamation of two different urban processes (the traditional centre and the colonial centre) each of which has its centre of intense activity.

3 METHODOLOGY

3.1 Data sources

The land use patterns were interpreted from mosaics of black and white aerial photographs. Mosaics of 1981 photographs at 1:5000 was employed for Lagos, 1979 photographs at 1.6000 for Benin City and 1977 photographs at 1:10,000 for Warri. Ancillary data and base maps were obtained from the Federal and State Departments of Surveys. In addition the interpretation and field checks were aided with terrestrial photographic reconnaisance using a cannon AE1 camera.

3.2 Classification

As in any body of organized knowledge, classification is vital to the study of land use (Northam, 1979) but most developing countries including Nigeria, lack pertinent urban land use classification schemes. Consequently, Adeniji's (1980) classification scheme was adopted with minor modifications (See table 1).

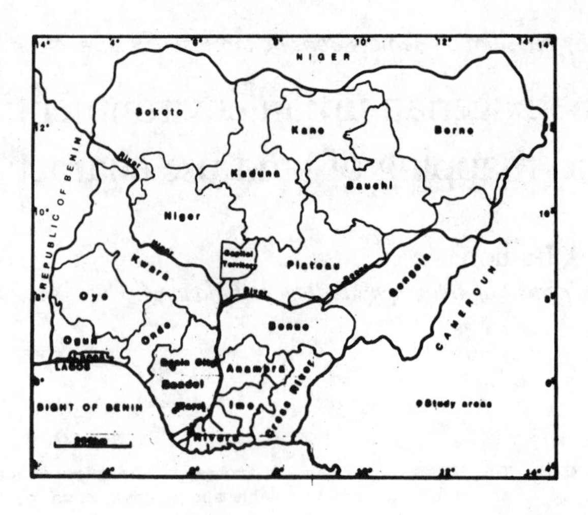

Figure 1. Location map

Table 1. Urban land use classification scheme

Category		Map Designation	Description
1 RESIDENTIAL			
	Low density	10	Large plot, 1 and 2 storey (flat) buildings with or without vegetated open spaces
		11	Apartment buildings (3 storey and above)
	Medium density	12	Medium plot, mixed 1 and 2 storey flat buildings with small individual or common open spaces
		13	Single storey row houses with moderate single or common open spaces
	High density	14	Mixed, traditional court-type buildings and traditional rooming buildings interspersed with modern 1 to 3 storey buildings
		15	Mixed traditional rooming buildings and modern 1 to 3 storey (flat) buildings
	Developing	16	New developing residential areas with completed and uncompleted residential structures in close juxtaposition and interspersed with undeveloped plots
2 COMMERCIAL		20	Main commercial centres
		21	Scattered roadside developments
		22	Shopping centres
		23	Traditional markets
3 INDUSTRIAL		30	Industrial complex areas
		31	Sawmills
4 INSTITUTIONAL		40	Educational (schools and colleges)
		41	Hospitals
		42	Government offices and public establishments
		43	Police establishments
		44	Military establishments
		45	Other institutional premises
5 COMMUNICATIONS AND UTILITIES		50	Airport premises
		51	Highway rights-of-way
		52	Motor parks
		53	Utilities
6 RECREATIONAL AND OPEN SPACES		60	Indoor recreation areas
		61	Sport grounds
		62	Parks
		63	Cemeteries
7 VACANT LAND		70	Undeveloped (dry) vacant land (usually cleared)
		71	Undeveloped bare ground
		72	Undeveloped (dry) vegetated land
		73	Undeveloped (wet) non-forested land
		74	Undeveloped (wet) forested land
8 NON-URBAN LAND		80	Undifferentiated rural villages
		81	Agricultural plantations
		82	Farmland
		83	Forested wetland
		84	Non-forested spottily vegetated wetland
		85	Shrub and secondary forest areas

Table 1. Urban land use classification scheme (continued)

Category	Map Designa-tion	Description
	86	Sandy areas other than beach
	87	Sand and gravel pits
9 WATER	90	Open water body

A standard manual or visual photographic interpretation of land use types was made using an 0.25 ha minimum mapping unit. Such photographic interpretation elements as size shape, tone texture as well as cultural and traditional cover type attributes were employed. Equipment and materials include a hand lens, magnifying glass pocket stereoscope F.71 mirror stereoscope and acetate transparent film. The stereoscopes were employed in a preliminary examination of the 1979 and 1977 prints of Benin City and Warri before they were mosaiced. In the case of Lagos Island, the mosaic had been compiled by the Lagos State Survey Department from where they were procured.

Interpreted land use patterns were delineated on acetate overlays. The overlays information were field checked and photo mechanically transferred to the base maps with the aid of an optical pantograph. Finally a Klimsch commodore cartographic camera was used to reduce the base maps to the publishable sizes.

4 RESULTS

The results show that not only can the land use categories be delineated from aerial photographs but that the diversity of land use characteristics manifested on the aerial imagery are positive explanations of urban spatial patterns. The distinct analysis of these characteristics are made under the following headings:

4.1 Land use in Lagos Island

Lagos Island dating from the 15th century, is part of metropolitan Lagos which comprises Ikoyi Island, Victoria Island and, Apapa, Yaba, Ikeja, Mushin, Oshodi, Agege and other settlements on the mainland which have been subsumed by urban sprawl. Situated on 6°, 20'N and 3° 20'E it is the nucleus of Federal Government administrative activities in Nigeria as well as its main commercial centre. It is one of the spectacular Nigerian cities which owe their growth and development largely to European influence (Mabogunje, 1968).

The spatial organization of land use activities in Lagos Island is shown in Fig. 2. Three distinct zones (districts) are recognizable: the vacant lands in the north central parts; the residential district which extends from the north-west through the central parts to the north-east; and the central Business District (CBD) in the south. Although this paper is mainly concerned with demonstrating and advocating the acquisition of urban land use data from aerial photographs, it is interesting to note the general residential and central Business District characteristics.

Three residential subdistricts are spatially distinguisable in Lagos Island. These are the low-grade high residential density subdistrict in the north-west: the medium grade, medium residential density district in the north-east and; the high grade, low residential density district in the south-east.

The north-western residential zone comprises the old traditional centre of Lagos city and the Oba's (King) palace. It constitutes the oldest and densest (47 houses per ha) part of the city with narrow, confused lanes, poor housing conditions and a predominantly Yoruba ethnic composition (Mabogunje, 1968, Adeniyi, 1976). The medium grade residential district in the north-eastern part initially had Portuguese style houses which constituted "an oasis of planned layout in a wilderness of confused housing" (Mabogunje, 1968). Today, however, the preponderance of petty traders, and make-shift extensions to the houses have created a situation where the housing conditions are indescribably squalid. The high grade, low residential density zone in the south-east is characterized by well planned (class 10) layouts where the houses (2 to 4 houses per ha) stand in the midst of well-kept lawns.

The central Business District is a well developed and highly specialized zone. It consists of two main parts: the commercial zone to the west and the institutional zone to the east. The commercial part which developed around the twin arterials of the Marine and Broad Streets has a multiplicity of functions which include warehouse/whole saling, retailing and, financial activities. Many of these commercial activities are housed in multi-storeyed veritable sky scrapers of over 25 floors which, with the 'departmental store' now comprise the most characteristic form of land use in Lagos Island. The institutional zone consists of the educational and religious institutions adjourning the commercial district and the municipal and Federal Government administrative offices as well as educational institutions. Among these are the cathedrals of Anglican, Methodist and Baptist faiths, Convent School, Colleges, the Defence, Communication and Justice Ministries.

In addition to the CBD, there are outlying business centres and business thoroughfares that are manifestations of lower order polynucleations of business formations which occur as a complementary adjunct to local markets. They remain the real centre of neighbourhood economic and cultural life (Mabogunje, 1968). Table 2 shows the proportional uses of land. The estimated total urban area is 367.4ha excluding vacant lands. Of this 36.1, 22.6, and 18.4 percents are used for residential, commercial and institutional purposes respectively. Communicational and recreational/open spaces occupy 6.6 and 16.6 percents respectively.

4.2 Land use in Benin City

Benin City is an ancient and "rejuvenated" modern centre of trade, commerce, industry, government and education (Sada, 1975). It is the capital of Bendel State in the midwestern portion of southern Nigeria (see Fig. 1). The geographic coordinates of the city limits are within latitudes 6° 17' and 6° 26' N and longitudes 5° 35' and 5° 41' E. The history of Benin dates back to the 12th century when it was the headquarters of the Benin Kingdom. Before the British conquest in 1897, Benin Kingdom was ruled by a dynasty of kings called "Obas". A mid-15th century Oba is credited with designing and developing the city and the moats and patterns of roads radiating from the Oba's palace plaza (now a ring road) are still observable today.

Benin city experienced remarkable growth after it became the capital of Bendel State in 1963. The present high annual growth rate has created planning problems especially when up-to-date maps are not available (Ikhuoria, 1983). Fig. 3 shows the general

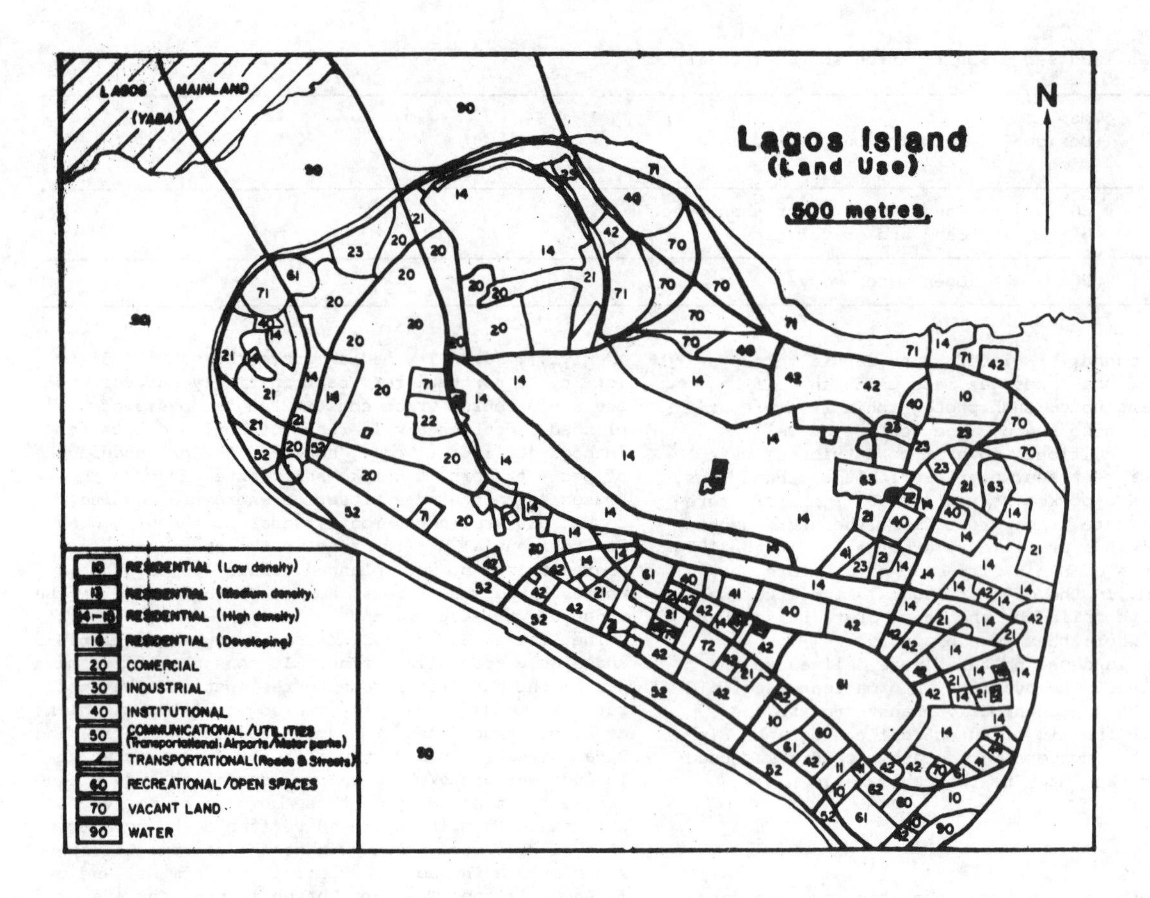

Figure 2. General patterns of land use in Lagos Island

patterns of land use in Benin City. Conspicuously evident are the radial-concentric form of development, the zonal/outlying nucleation of commercial areas, and the polynucleation of industrial estates.

Table 2. Lagos Island land use statistics

Classification	Area (ha)	Area as percentage of Urban and non-rban use
Residential (low density)	9.2	0.9
Residential (high ")	123.1	12.2
Commercial	83.2	8.2
Institutional	67.6	6.7
Communicational and tilities	24.1	2.4
Recreational and Open Spaces	59.8	5.9
Vacant Land	26.7	2.6
Water	615.0	61.0
T O T A L	1008.7	100.00
Total residential area	132.3	13.1
Urban: built up area excluding vacant and non-urban lands	367.0	36.4
Urban: built up area including vacant lands	393.7	39.0

Benin City, as the map shows, consists of a central open space core from which radial and concentric thoroughfares connect the other parts. The core is surrounded by concentric nucleations of traditional and modern businesses, and extensive high density residential neighbourhoods and, sectoral medium, and low density residential neighbourhoods. It also includes outlying nucleation of commercial centres, principally at the intersections of the radial and concentric arterials, and peripheral ribon nucleation of industries.

The high density residential areas comprise two major residential classes in both inhabitants and dwelling structures. The first is residential class 14, which is made up of the indigenous population. The area is characterized by dense (20 to 24 houses per ha) regularly arranged but monotonously rectangular housing units. The housing conditions, housing density, street and flooding conditions indicate the indigenous core is a blighted neighbourhood (Ikhuoria, 1986).

The other group of inhabitants includes migrants living in houses of approximately 460 m^2 and a density of 20 to 22 houses per ha. These are generally either rooming or flat type residential structures (class 15) partitioned into separate self-contained dwelling units. The polynucleation of residential areas based on social status, is manifested in the medium density (12 to 20 houses per ha) residential areas. Also the impact of colonialism on Benin residential patterns, is manifested in the low density residential class 10. This area known as the "Government Reservation Area" (GRA) was developed into high grade, low density (4 tc 10 houses per ha) residential quarters. The GRA low density residential area consists of large plot residential buildings which occupy an average area of 2 000 m^2. The structures

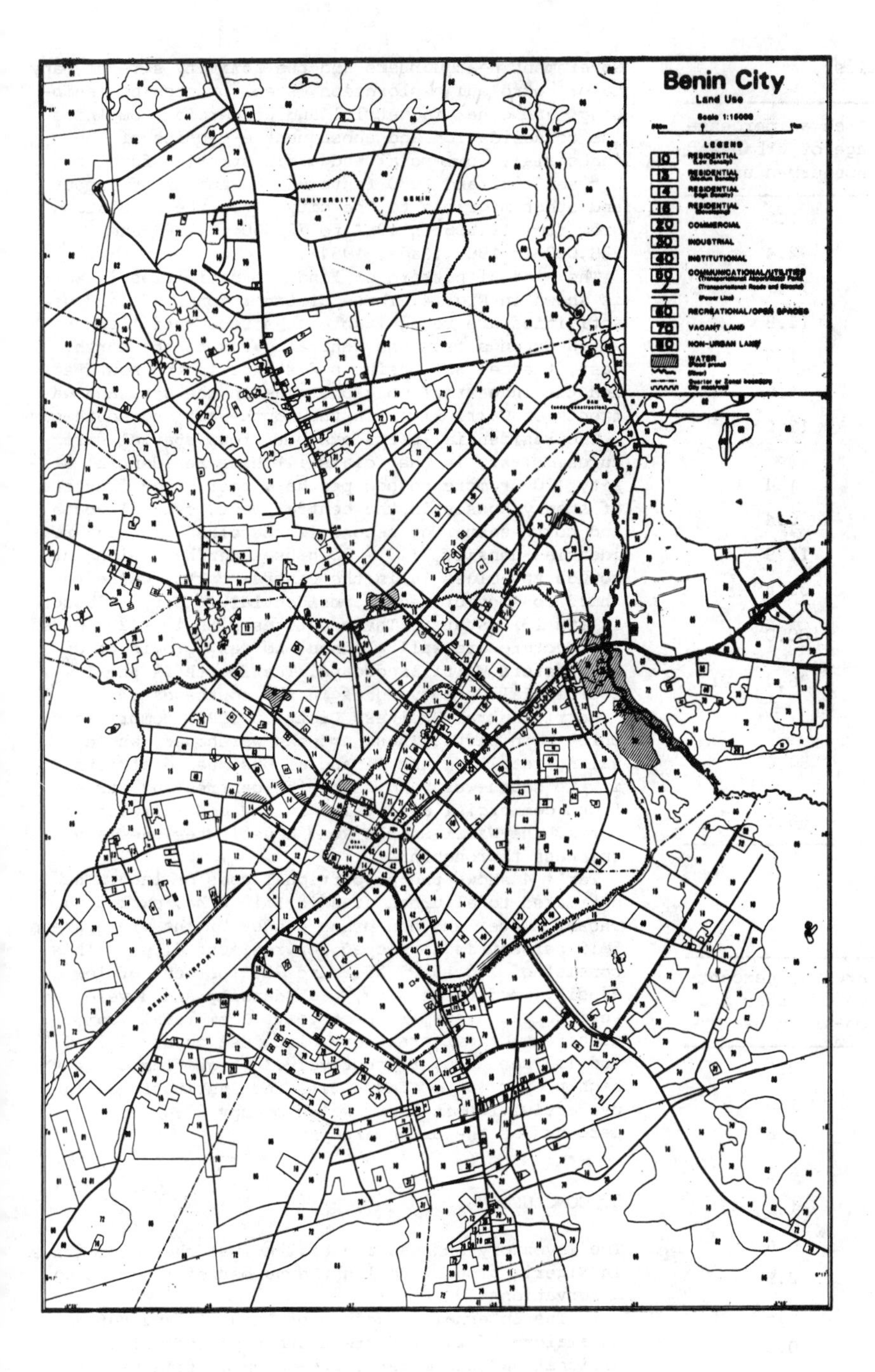

Figure 3. General patterns of land use in Benin City

reflect colonial and, in recent times, modern architectual designs.

Table 3 shows the proportional uses of land. The estimated total urban area is 6091.2 ha, excluding vacant and non-urban lands.

Of the urban areas, 66.7 percent is developed for residential use. This consists of 2.8, 4.4, 32.3 and 27.1 percent of low density; medium density, high density and developing residential areas, respectively. Commercial and industrial uses occupy 2.4 and 3.4 percent, respectively, while institutional and recreational/open space use is 24.1 and 0.8 percent, respectively. Communication/utilities occupy 2.6 percent. This amount does not include the street, road and powerline spaces, however, which are linear features on the map.

The total map area is 11 152.2 ha. Of this, 54.6 percent is under urban use, 11.9 percent is vacant and 33.5 percent is non-urban. Benin City exhibits a highly developed western model central Business District (CBD) "Plaza" with a central open space ringed by a five lane circular arterial "ring Road". In the circle is a green lawn, flower garden, civic centre, cenotaph and a museum. On the northern part of the CBD is a conglomeration of financial, wholesale/retailing enterprises and public service institutions. Cultural and traditional institutions are concentrated on the west and south-west while the east and south-east consist of another conglomeration of shopping centres, financial, health, religious, public and political institutions. In addition to the CBD, there are Business thoroughfares and outlying Business centres along the major radial routes. Also there are polynucleation of lower order street-front commercial centres at major street intersections. Some of these commonly develop adjunct to traditional markets in order to take advantage of the convergence population in the area.

Table 3. Benin City land use statistics

Classification	Area (ha)	Area as percentage of urban and non-urban use
Residential (low density)	168.2	1.5
Residential (medium ")	267.0	2.4
Residential (high ")	1975.9	17.7
Residential (developing)	1648.1	14.8
Commercial	145.6	1.3
Industrial	208.6	1.9
Institutional	1470.9	13.2
Communications/ utilities	158.8	1.4
Recreation/open spaces	48.1	0.4
Vacant land	1322.5	11.9
Non-urban	3738.8	33.5
T O T A L	11152.5	100.0
Total residential area	4059.2	36.4
Urban: built up area excluding vacant and non-urban lands	6091.2	54.6
Urban: built up area including vacant lands	7413.7	66.5

Table 4. Warri land use statistics

Classification	Area (ha)	Area as percentge of urban and on-urban use
Residential (low density)	20.1	0.5
Residential (medium ")	65.1	1.6
Residential (high ")	281.1	6.9
Residential (developing)	266.0	6.5
Commercial	76.6	1.9
Industrial	259.5	6.4
Institutional	118.1	2.9
Communicational/Utilities	10.4	0.2
Recreational/Open Spaces	13.9	0.3
Vacant Land	317.7	7.8
Non-Urban	2292.0	56.2
Water	356.0	8.7
T O T A L	4076.5	100.0
Total residential area	632.3	15.5
Urban: built up area excluding vacant, water and non-urban lands	1110.8	27.2
Urban: built up area including vacant lands	1428.5	35.0

4.3 Land use in Warri

Warri is situated 6.1 metres above sea level. It is a sea port in the Niger delta on 5°, 31' N and 5°, 44' E (Fig.1). The city has experienced remarkable growth since it became a provincial and later Local Government Headquarters and the activity site of many major petroleum exploration, steel processing, shipping, and other industrial and commercial companies. The consolidation and subsequent expansion of the industrial, commercial and administrative functions between 1952 and 1980 fostered the influx of labour and other migrants to the city. Thus the population grew from 19,526 in 1952 to 72,000 in 1963 and 228,000 in 1982 (Sada, 1984).

The spatial pattern of land use activities in Warri is shown in Fig. 4. Three distinct zones of land use activities are recognizable: the residential zone which extends from the central parts to the north-east, the central Business district in the south-east and the industrial conglomerates in the west and water fronts of Warri River in the south-west and south-east.

The residential parts exhibit four spatial structures. First, is the (class 14) high residential density (30 or more houses per ha), poor grade clusters of shanty houses in the central west, central south and north-east. The second is the class 15, high residential density, (20 - 30 houses per ha) which surrounds the class 14 in the central parts. While the third consist of medium residential density (classes 12 and 13) estates. These estates with 10 - 20 houses per hectare are well planned and very prominent on the eastern Warri landscape. The fourth is the low residential density zones in the south-west.

Warri central Business District (CBD) comprises of three main parts. These are the warehouse, whole saling/retailing and financial business districts in the south-west and, shopping centres and street front commercial activities in the central/eastern parts. The Warri-Sapele Road constitutes the main business thoroughfare.

Table 4 shows the proportional uses of land. The estimated total urban area is 1110.8 ha, excluding vacant, water and non-urban lands. Of the urban areas 56.9 percent is developed for residential use. This consist of 1.8, 5.9, 25.3 and 23.9 percent of low density, medium density, high density and developing residential areas, respectively. Commercial, communicational and recreational land use occupy 6.9, 0.9 and 1.3 percents, respectively. Of significance is the high proportion of industrial and institutional land use activities in Warri which occupy 23.4 and 10.6 percents of the urban area.

5. CONCLUSIONS

The inventory and mapping of land use characteristics in Nigeria brought to light a number of interesting observations:

1. The internal structure of Lagos Island reflects an amalgam of two different urban processes; the traditional and the modern. Pre-colonial traditional land use specialization was predominantly cultural and residential. In the core of the traditional centre is the King's palace which is flanked by royal chiefs' palaces and surrounding indigenous and migrant residential neighbourhoods. Commercial land use specialization centred around traditional markets. Residential development is very dense with a squalid environment and communicational arterials are too poor. In contrast the southern parts of Lagos Island reflect modern urban planning and land use specialization. This is manifested in the layout of streets and the spatial organization of commerce, institutional and residential activities.

2. Benin City's internal structure reflects a juxtaposition of traditional, colonial and modern urbanization processes. It consists of a central open space core from which radial and concentric thoroughfares connect the other parts. The core is surrounded by concentric nucleations of traditional and modern businesses (CBD), indigenous and migrant residential neighbourhoods, and developing modern residential areas. Furthermore it consists of sectorial high social class residential neighbourhoods; outlying nucleation of commercial centres at the intersection of

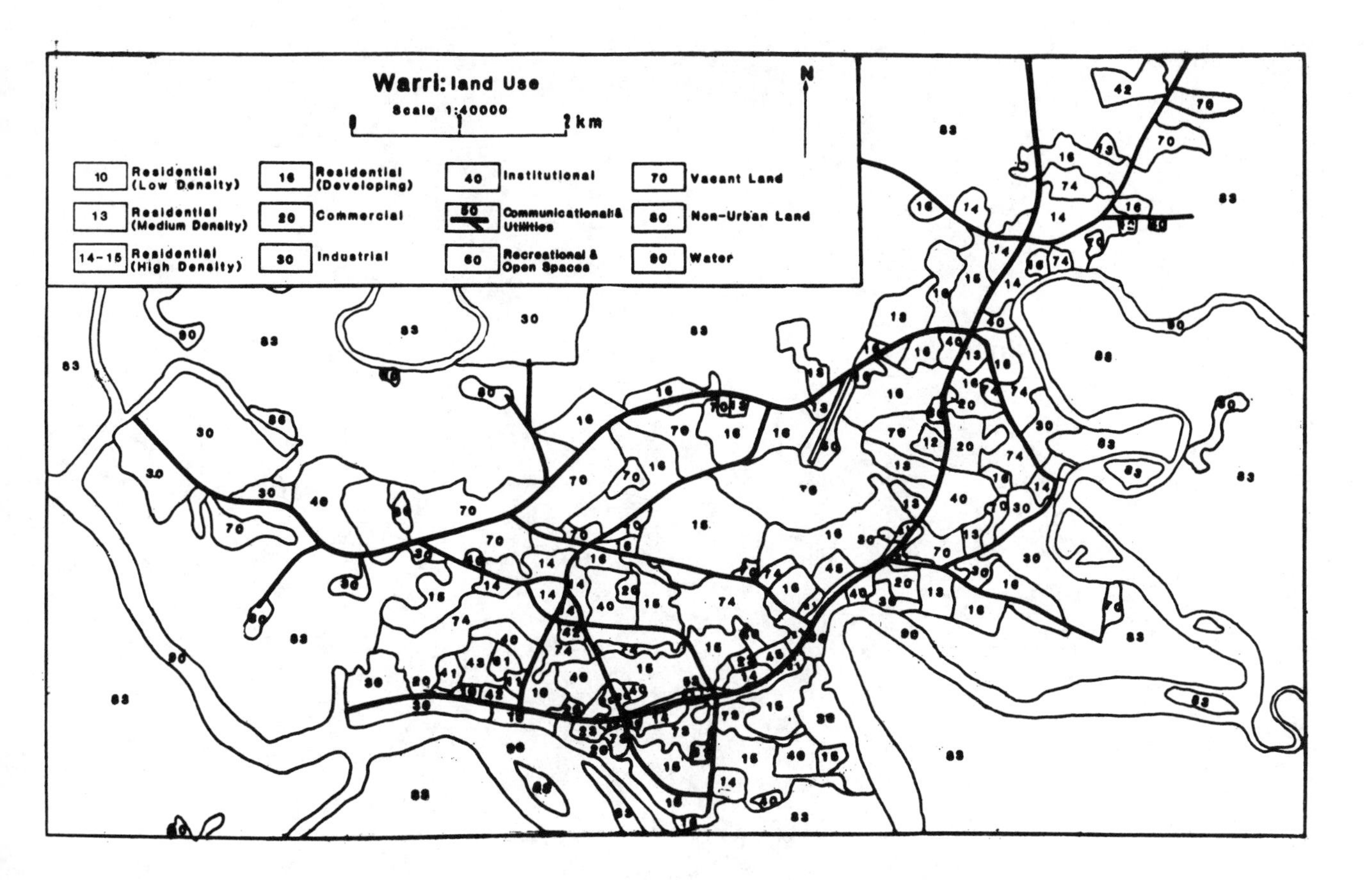

Figure 4. General patterns of land use in Warri

the radial and concentric arterials and peripherial nucleation of industries. By and large, the present land use pattern in Benin City is "polymeric": partly responding to the sectorial process and partly to the forces of concentric formations. The sectorial forces include the social identification of people of the same status, while the concentric pattern is mainly a product of geographic inertial. And, the polynucleation of outlying and peripheral commerce and industry is partly due to economic and accessibility forces (Ikhuoria, forthcoming, January 1987).

3. The internal structure of Warri reflects admixture nucleations of residential neighbourhoods based on ethnic, social and economic considerations in different parts of the city, agglomeration of commerce and industry due to accessibility forces, well planned and highly developed middle class and high class residential estates.

4. The indigenous cores of Lagos Island, Benin City and Warri exhibit serious indications of urban blight.

5. The diversity of land use patterns manifested on the landscape can be inventorized and mapped from remotely sensed imagery and the data serve as positive explanations of urban spatial structure.

6. In creating an effective environmental resource management and monitoring systems in third world countries, the utilization of remote sensing (particularly aerial photography) to acquire the necessary data should be encouraged.

REFERENCES

Adeniyi, P.O. 1976. Application of aerial photography to the estimation of the characteristics of residential building. Nigerian Geogr.

Adeniyi, P.O. 1980. Land use change analysis using sequential aerial photography and computer techniques. Photogrammetric Eng. and Remote Sens. Nov. 1447 - 1464.

Anderson, J.A. et al. 1976. A land use and land cover classification system for use with remote sensor data. Geological survey Professional Paper No. 964.

Ikhuoria, Isi, 1983. Urban land use inventory and mapping from semi-controlled photomosaics. ITC Journal, 4.

Ikhuoria, Isi. 1986. Use of aerial remote sensing to determine urban blight. In P.O. Sada & F. Odemerho (eds.), Environmental issues in management in Nigerian development. Ibadan, Evans.

Ikhuoria, Isi. A. (forthcoming). Urban land use patterns in a traditional Nigerian city: a case study of Benin City. Land Use Policy.

Lintz, J. & D.S. Simonett. 1976. Remote sensing of environment. Reading, Addison - Wesley.

Mabogunje, A.L. 1968. Urbanization in Nigeria. London, Univ. London.

Mabogunje, A.L. (in Press). The debt to posterity: reflections on a national policy on environmental management. In P.O. Sada & F. Odemerho (eds.), Environmental issues in Management in Nigerian development. Ibadan, Evans.

Northam, R.M. 1979. Urban geography. New York, John Wiley.

Omuta, G.E.D. 1985. Land use and environmental dereliction in the urban fringe: the case of Benin City, Bendel State of Nigeria. Socio-Economic Science 19:5:303-311.

Sada, P.O. 1975. Urban housing and the Spatial pattern of modernization in Benin City. Nigerian Geogr. Journal 18: 39-55.

Symposium on Remote Sensing for Resources Development and Environmental Management / Enschede / August 1986

Urban change detection and analysis using multidate remote sensed images

Chen Jun, Guan Zequn, Zhan Qinming, Sun Jiabing & Lu Hueiwen
Wuhan Technical University of Surveying and Mapping, China

Zheng Zhixiao
The Research Institute of Urban Planning and Design of Hubei Province, China

ABSTRACT:: In this paper, the methods and experiments of urban change study using the remote sensing techniques are introduced. A series of Hanyan's urban change maps were compiled after the interpretation and analysis of the sequential airphotos (1955, 1965, 1977, 1981). On the basis of the remote sensed data and auxiliary data, the authors have made a quantitative and qualitative analysis of the change of Hanyan's urban morphology, landscape and touristic resources in the last thirty years. The implementation of the urban master planning was also examined by comparing the sequential airphotos and urban master plans.

I. Urban Change Mapping Using Multidate Airphotos

The airphotos are real images of land surface at the moment of photographing. A great deal of information of urban area, such as landuse, environment, population density etc., could be derived from airphotos. The relevant information of urban growth and change could also be obtained using the sequential airphotos. Four year's airphotos (table 1) are used in the detection and analysis of Hanyan's urban change and a series of urban change maps was compiled on the basis of photointerpretation and analysis. It aims mainly at representing the historical process of urban expansion and the dynamic changes of each urban element. The main contents of each map are listed on table 4.

Table 1. The airphotos used in Hanyan's urban change detection and analysis.

No	year	Type	scale
1	1955	Black white pancramatic 18x18	1:60000
2	1965	Black white pancramatic 18x18	1:25000
3	1977	Black white pancramatic 18x18	1:20000
4	1981	false color infra-red 23x23	1:8000

It is worthing mentioning that an expert knowledge and supplementary data are necessary to delineate the different urban landuse. Since the airphotos of 1955, 1965 and 1977 offer a different picture of what presently existing on the ground, a special method of interpretation was considered. Considering the fact that some urban landuse types remain unchanged during certain years, the photos taken several years ago were interpreted by making reference to the current urban landuse and the topographic maps and other auxillary data. By comparing the sequential airphotos and the sequential urban landuse maps (compiled after photointerpretation), the change of different urban landuse, urban morphology and some urban elements are easily derived and are compiled into the corresponding urban change maps. The present public facilities, touristic resources, urban pollution sources were also interpreted with some ground surveys. Furthermore, by comparing the urban master plans (1965, 1977) and the corresponding airphotos (1977, 1981), two maps were made to reflect the differences between the urban master planing and the real urban expansion (1965-1977, 1977-1981).

II. Hanyan's Urban Change Analysis

The series of urban change maps as well as the sequential airphotos are used to the analysis of urban change of Hanyan, which is one of the three cities of WUHAN. Figture 1. gives the basic philosophy of the study.

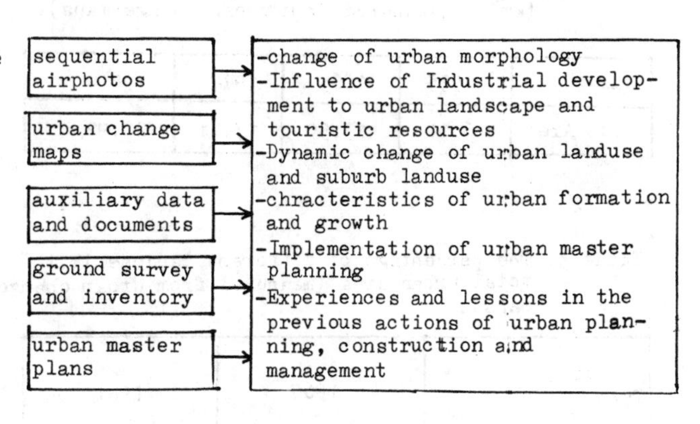

Fig 1. Philosophy of urban change analysis of Hanyan using sequential airphotos.

1. Change of urban morphology and landuse.

It is seen from the urban change maps that Hanyan city grew very fast. The urban area of Hanyan increased 14.81 kilometer square from 1955 to 1981. The main motive force of the urban growth is the passthrough of railway from Beijing to Guangzhou and the construction of a number of large entreprises and key projects in Hanyan. The layout of the industrial landuse and storehouses played an important role to the scale, formation and function of the city.

As indicated in table 3, the percentage of industrial landuse and storehouses to the total urban area in Hanyan was 26.93% in 1965 and 34.63% in 1981.

The city not only expanded to the outside, out also developped to the space. According to the results of photointerpretation, the height of buildings in Hanyan increase continuonsly since 1950s. In 1950s, the residencial buildings were almost the low brick-wood slope buildings which could be distinguished easily in the airphotos. High residential buildings were built between 1965 and 1977. And the number of floors of new residential building increase between 1977 and 1981 (Fig 2).

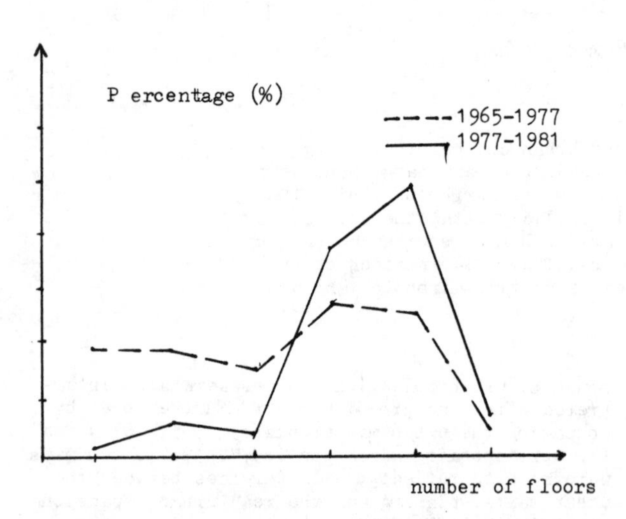

Fig 2. The percentage of the new buildings with different floors (height) from 1965 to 1981 (measured from airphotos).

Table 2. The area of built area in sequential years (km^2) (measured from urban change maps).

Year	1955	1965	1977	1981
Built Area	6.11	13.50	17.21	20.92

Table 3. The percentage of different landuse to the total urban area (measured from urban change maps).

Year / Type	1965	1981
Industry	22.5	28.19
Store	4.43	6.62
Traffic	9.69	8.62
Public facility	4.41	9.56
Residential	15.58	18.47
Water	9.51	9.73
Green land	10.45	4.65
Vegetable plot	14.61	4.65
others	8.82	6.84

Table 4. Series of Hanyan's Urban Change Maps Compiled with Sequential Airphotos (scale 1:10000).

No	Map	Main contents or classification
1	Urban landuse (1955, 1965, 1977, 1981)	Industry, store, residential landuse, public facilities, infrastructure, roads and greenland, water, vegetable, agricultural landuse etc.
2	Residential change	Structure, material, floor number & year of set up of every residential building; Increase or decrease of residencial buildings
3	Industical landuse change	Spatial distribution of the present industral branches; Limits of industrial landuse in the sequential years etc.
4	Traffic landuse change	Traffic network of the sequential years; Type of pavement, order of roads, etc. Increase or decrease of roads
5	Garden and green land change	Spatial disitributions, species of green land of the sequential years; Transformational relationship with other urban landuse (For instance, green land changes into industries etc.)
6	Suburb agricultural landuse change	Spatial distribution of the nonirrigated farmland, vegetable plot, paddy field in the sequential years; Transformation with other landuse (For instance, the nonirrigated farmlands changes into vegetable lands, the vegetable polts change into residential use etc.)
7	Water change	Boundary of water (rivers, lakes) in the sequential years); Transformation with other landuse (water body change into industrial landuse, the paddy fields change into dishponds etc.)
8	Population density	Spatial population density distribution in the daytime and in the night (represented in different classes)
9	Urban pollution	Types and Spatial location of urban pollution sources (such as chemenies, escaped wasted water, solid waste etc. Spreding way and influence of the waste disposals
10	Public facilities	Spatial location of public facilities and the relationship with other urban landuses
11	Landscape & Touristic reresources	Spatial location of the main landscape and touristic spots; Relevant facilities to tourism, Influence of urbanization to landscape and touristic in the past years
12	Execution of urban master planning (1965-1977, 1977-1981)	Non confirmed area (non planned urban use); Differences between planned landuse and real landuse in the planned area

2. Influence of urban industrial expansion on the landscape and touristic resources.

The influence of industrial expansion in Hanyan on its famous landscape and touristic resources had been studied using the airphotos and urban change maps (Fig 3). The main conclusions are summed up as follows:

1) The local landscape around the main landscape and touristic spots have descended.

2) The quality of the touristic enoiroment was getting worse.

3) The area of some lakes and collines, greenland, which are among the main touristic spots, have been occupied partly by factories or other buildings (table 5).

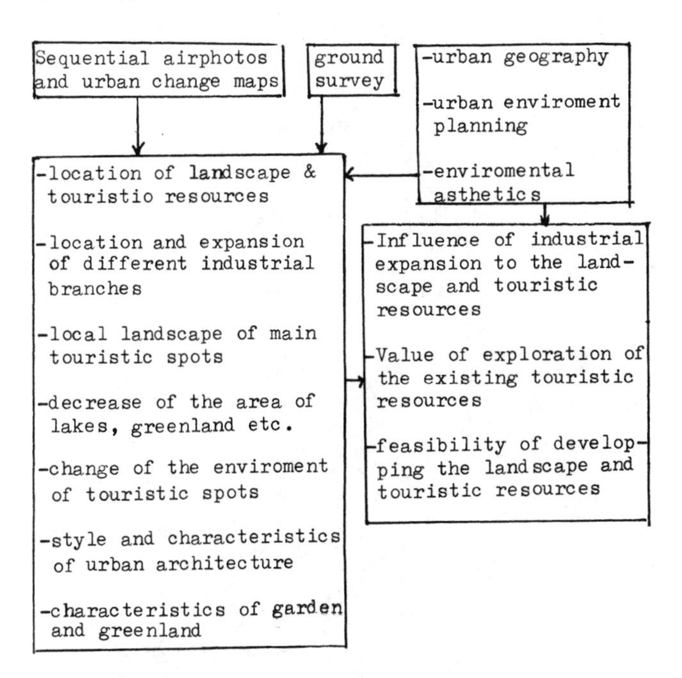

Fig 3. The study of the influence of industrial expansion to the landscape and touristic resources.

Table 5. Change of the Area of some lakes (hectare).

No	name	1950s	1980s
1	West Yue Lake	72.56	11.11
2	East Yue Lake	96.80	73.40
3	Ma Cang Lake	43.43	41.41
4	Lian Hua Lake	11.62	9.09

III. The examination of the implementation of Hanyan's master planning.

The urban master plan of Hanyan in 1965 was compared with the urban landuse map and the airphotos of 1977 And the same method was applied to the urban master plan of 1977 with the airphotos in 1981. It is found that differences exist between the urban planning and its implementations. Not only the non confirmed urban area exist but also the actual urban landuse types in some planned area are not the same as the planned landuse types. This fact indicates the problems in the urban management and the implementation of urban master planning.

IV. CONCLUSION

1. The results prove that a great deal of static and dynamic urban information could be derived easily and quickly from airphotos.

2. The sequential airphotos analysis is very useful for urban monitoring and urban change study.

3. The combination of the remote sensing technics and the disciplines of urban sciences (urban planning and management, economic geography etc.), will provide with the chinese urban planners and managers an efficient and powerful tool in urban planning and urban management.

Symposium on Remote Sensing for Resources Development and Environmental Management / Enschede / August 1986

Abandoned settlements and cultural resources remote sensing

Aulis Lind
University of Vermont, Burlington, USA
Noel Ring
Big Ben Community College Overseas Program, New York, USA

ABSTRACT: It is becoming increasingly clear that resource planning and management in many countries of the World includes a range of settlement features having origins from prehistoric times. Such features, which are collectively termed cultural resources, may be endangered by the continued development of other resources and thus need to be included in monitoring aspects of land use studies. In addition, there is the need to survey developable and remote areas for such resources relating to former human settlements whether agglomerated or dispersed. Remote sensing applications involve the discovery, survey, mapping, and analysis of the abandoned landscape elements as well as the current functional elements. The existing variety of remote sensing tools provides the promise of identifying new cultural resource landscapes or districts. At the same time, archival sources of data, such as old aerial photographs, are particularly valuable. Thus, thematic mapping with emphasis on cultural resources, becomes a multisensor task with numerous other remote sensing dimensions including data from satellites, aircraft and baloons. Examples of applications from the U.S., Europe and S.E. Asia are presented to illustrate the complexity and scope of the remote sensing challenge.

1 INTRODUCTION

News of an exciting discovery has recently emerged (Begley S. and S. Katz, 1986) which describes aNew World Pompeii on the slopes of the Costa Rican volcano Arenal, where airborne radar, lidar and color-infrared photographic sensing were employed to locate abandoned settlements and appropriate sites for archaeological excavations. The advantages of the aerial perspective and aerial photography to document abandoned settlements are well known to students of past landscapes. As the search for abandoned settlements and the study of relict landscapes progresses, the roles played by remote sensing continue to expand due to ongoing research, development and application of new remote sensing tools and methods. Satellite remote sensing programs such as LANDSAT, the Spaceborne Imaging Radars-SIR A, B, and the newly launched SPOT system offer some new and challenging directions for past landscape analysis (Olsen, 1985).

Ancient settlement patterns are often submerged in the landscape of recent history, and these overlap an environmental matrix of resource significance. The inventory and analysis of these spatially overlapping patterns forms a significant focus for several fields of study including, for example, environmental archaeology, historical-cultural geography and human ecology. Emerging from the application of remote sensing techniques within these fields, and within site-oriented archeological surveys, is an expanding and increasingly significant area of investigation which has appeared as "cultural resources remote sensing". The application of satellite remote sensing, radar, multispectral imaging and the tools of the computer and space age offer to greatly increase information about bygone populations and their works. Perhaps more importantly, abandoned settlements as well as such features as ancient agricultural tracts, early irrigation and transportation canal networks, fortifications, and religious centers are considered to be resources which specifies that they are a matter of inventory, conservation, and management as meaningful parts of the economy of nations, states, regions, or locales. Conserving and managing the current mass of cultural resources (Reichstein, 1985) and providing for new organizational and planning schemes to handle such features as historic districts and relict landscapes (Melnick, 1984) are of major concern due to the destructive threats of such processes as urban expansion, reservoir construction, mining and the landscape manipulations associated with modernization of agriculture and forestry.

Remote sensing's role within cultural resources management will continue to receive increasing attention since its approach is archaeologically and environmentally non-destructive and this is an increasingly important attribute. Moreover, Federal laws in the U.S. at least require surface surveys to be made. But, the amount of effort expended and the rate of progress achieved ultimately depends on the amount of funding available, the interest level of investigators and the relative importance given to cultural resources as a whole by society. Cultural resource managers sometimes view remote sensing studies with skepticism, but that is mainly the result of improper or misapplied use and interpretation of remote sensing data, or simply a lack of knowledge regarding remote sensing methods and techniques. Both of these problems become a matter of education and training within the appropriate sub-disciplines.

What are the remote sensing elements and challenges in the development and management of cultural resources that may offer global societies some understanding of their evolution? Posing this question from time to time seems necessary to assess the nature of the challenges lying ahead, especially as rapid advances in remote sensing technology ultimately impact on methodological approaches. This paper attempts to examine that question by: a) surveying major cultural resource tasks within the context of a scaled approach and b) exploring the nature of remote sensing application potentials within selected regional contexts.

2 SCALING OF CULTURAL RESOURCES TASKS

Application of remote sensing techniques to cultural resources problems involves a potentially broad spectrum of scales and sensing tools. It might be difficult to find another field of study which requires a range of remotely sensed data stretching from scales of about 1/1000 to 1/1,000,000. An examination of scale in a bivariate context is a convenient way to identify application dimensions, along with other characteristics of cultural resources remote sensing. This is attempted in Table 1 to follow. While scale speaks for itself, definition of the major tasks and an indication of the tools generally used is required.

Table 1. Bivariate Matrix of Scale Factors and Cultural Resources Remote Sensing Tasks.

	Tasks A	B	C	D	E	F
Scale Ultra-large 1:1000	1	2	3	3	3	2
Large 1:5000	2	1	1	2	3	1
Medium 1:25,000	3	3	2	2	1	1
Small 1:50,000	3	3	2	2	1	2
Ultra-samll 1:500,000	3	3	2	2	1	3

Task Key: A=Site element survey/analysis
B=Site plan survey
C=Landscape survey, site detection
D=Spatial analysis, sub-regional survey
E=Regional geographic information
F=Management of cultural resources
Applicability Key: 1=High
2=Moderate
3=Low or None

Six major tasks in cultural resources remote sensing may be identified as follows, and virtually all (except f.) result in plans or maps as direct output.

a) Site element survey and analysis: Use of ground-based platforms, low altitude photography from baloons, kites, miniature drones, and helicopters almost exclusively photographic and frequently involving photogrammetric assessments. Interest is focused specific feature, i.e. structures, monuments and habitation areas.

b) Site plan survey and internal organization: Low altitude aerial photography from aircraft, helicopters and baloons often involving photogrammetric interest. Micro-topography, dimensions, plan and arrangement of whole settlements, internal characteristics (i.e. style) and functional organization are interest foci. Detection of small cultural features such as grave sites and markings, house foundations and pathways.

c) Landscape surveys, site detection and survey: Aerial photography and multispectral scanning from aircraft. Multispectral video systems are emerging to greatly reduce turn-around times for data analysis with both digital and analog data. Terrain characteristics involving geomorphology, soils, vegetation, land use and cultural elements (i.e. roads, canals, moats, agricultural field patterning, etc.) in relation to abandoned settlements and their immediate surroundings are major topics. Detection, identification and mapping of new or potential cultural resources of major importance. Delineation of historic districts.

d) Spatial analysis and sub-regional survey: High altitude aerial photography and multispectral scanning as well as radar, including high resolution satellite data. Investigation of environmental distribution (including terrain characteristics from above) pertinent to surveying cultural resource concentrations, and as input to geographic information system base. Spatial analysis of cultural resource interrelationships, networks and survey of sub-regional patterns of diverse cultural resources.

e) Regional geographic information: Satellite multispectral scanners, radar, and photography. Investigation of terrain characteristics (as above) and large-area environmental distributions showing temporal variations that impact on cultural resources (i.e. flooding, land use changes).

f) Management: Multiple remote sensor applications. Planning processes, property delineation, natural hazards assessment, monitoring forestry and agricultural practices, land use changes (i.e. urbanization), riverbank/shoreline erosion, intervention of other resource developments as for example, reservoir and highway construction, archeaological plundering and malicious destruction, and pollution effects.

Table 1 does not address the details of methodological problems characteristic of this area of remote sensing applications. These have been variously covered by other authors, such as Ebert (1984), Normann (1985), and Vogt (1974) and more and are beyond the scope of this paper. The table presents a sample of the kinds of tasks that characterize cultural remote sensing and a subjective, albeit necessary scalar classification.

Also, an "applicability rating" is shown which is based on both field and literature survey. It becomes readily apparent that the most applicable scales for cultural resources tasks are the large and meso-scales, which encompasses the range where aerial photography reigns while the extreme scales are more specialized in their application although this does not diminish their value in providing meaningful data for use in the overall scheme of things.

A voluminous inventory of imagery already exists at the most applicable scales with virtually every area of potential cultural resource significance covered, and in some cases multiple coverages embracing a time scale of some five or six decades. The archives at satellite scales are also burgeoning with data, but all in all, there is still disparity in both quantity and quality in the availability of the most applicable scales between the developed and the developing world, while it is encouraging to note that satellite coverage is available the world over, although there are severe local problems in quality. These disparities result in numerous difficulties in the methodological realm requiring investigations to shift to more costly and less efficient means of survey.

Few investigators have the opportunity to acquire dedicated large scale coverages and must resort to available imagery acquired for some other purpose. However, the vast archives of historic, semi-recent and recent large and meso-scale coverages in the storage boxes of various agencies around the world are potentially of considerable value and deserve attention. Also, the ingenuity of investigators should not be underestimated as cheaper alternatives to contract aerial photo acquisition have recently been shown in the popular and professional literature, such as tethered baloon (Myers and Myers, 1985), drone-model airplane and kite. (Normann, 1985, Holm and Stridsberg, 1985).

3 MANAGEMENT AND REMOTE SENSING

Lipe (1984:2) defines the philosophical basis for cultural resource management as follows: 'all cultural materials, including cultural landscapes, that have survived from the past, are potentially cultural resources-that is, have some potential value or use in the present or future'. Management implies that a rational policy exists based on detailed survey and inventory of cultural resources. Cleere (1984), summarizing a range of World cultural resource management problems from a survey conducted in twelve countries points to a need for aerial and ground survey which should obviate hasty rescue operations. Another problem area increasingly recognized by scholars is the matter of context. Due to the long-standing and still prevalent focus of archaeology on sites, archaeologists fail to appreciate the more diffuse cultural elements imprinted on the landscape at large (Butzer, 1982), thus, the resource content of the rural landscape can be overlooked. This site-oriented focus is showing signs of changing largely through the application of aerial survey methods and the emergence of a more problem-oriented approach to studying the past. On a world-wide basis, economic development and social progress invariably impacts on the cultural resource base as already indicated, and in the developing nations these pressures are extreme to the point where many potentially rich resources are destroyed. Looting and wanton destruction are also common in these countries, and they are problems that still require some attention in the developed nations.

Against this background, a range of examples has been selected pointing to various dimensions of cultural resource remote sensing applications representing a historic to prehistoric time span (Figs. 1-3).

4 THE CHALLENGES AHEAD

As with any developing sub-discipline in its youthful stage, there are few established rules, and insufficient time has elapsed for the "trickling down" of advanced methods and techniques. Emerging from this attempt to investigate some of the significant elements in cultural resources remote sensing, we identify the following challenges for the remote sensing community:

1. The first and foremost challenge involves the training and education of personnel, not only in the area strictly concerned with cultural resources, but in cognate areas in the geo-sciences and bio-sciences. Just recently, the American Anthropological Association provided a special session on remote sensing applications at its annual meeting, and at the University of Vermont, a course on "Remote Sensing of Past Landscapes", was offered for the first time with considerable interest on the part of graduate students from geography, archaeology, history and historical preservation. Workshops, seminars, and courses at the University level need to be developed. A mechanism to promote education in this field may be possible through international means, as for example through the United Nations University. If any progress is to be made in this area, it must start with increasing the number of enlightened scientists, resource managers and educators. Developing and incorporating teaching units that could fit into existing courses in geography, history, archaeology, ecology, civil engineering and geo and bio-sciences would go a lon way toward increasing awareness and could serve to feed in-depth courses. Cleere (1984) also stresses the importance of the educational process ranging down to the elementary school level and the incorporation of such topics within the context of the traditional school subjects of geography and history.

2. A second challenge focuses on facilitating survey work at all appropriate scales through providing easier access to the imagery inventory. The major detective work required to locate imagery often discourages application attempts. Many archaeological and historic preservation agencies do not use imagery of any kind, and where used, it may only be applied as a field tool after which it is forgotten. Large scale orthophoto maps, which are increasingly being produced make excellent bases for inventorizing cultural resources and for comparative work with other remote sensing imagery, including findings from high altitude or satellite. In all countries there is a need to find a way of declassifying the storehouses of obsolete, now unused imagery acquired by various military establishments. A system of reporting new image acquisition in the civil realm, such as the international effort used to report radiocarbon dating, would be extremely useful and might be possible under the auspices of an organization such as ISPRS.

3. A third challenge focuses on the area of increasing interdisciplinary and interagency cooperation and the removal of barriers between the several disciplines that overlap in this field to circumvent duplication and conflict. Remote sensing can serve as the tie that binds. It would seem appropriate that remote sensing experts should take every opportunity to communicate findings that have potential value in the cultural resources field whatever their discipline. Multi-disciplinary resource inventory and management projects offer the benefits of a more comprehensive view and should be encouraged.

4. Increasing public awareness and support has been identified by Cleere (1984) and Riechstein (1984) as a critical element in the management process. This is no less true regarding remote sensing applications. The treatment in the American magazine Newsweek cited at the outset is indeed an extremely rare event. Media attention is critical and is much more prevalent in Europe than in the United States. Items worthy of public notice found during remote sensing data analysis can be called to the attention of the mass media with the cooperation of appropriate management authorities. Making contacts with local authorities and historical and archaeological clubs is another fruitful approach.

5. The involvement of remote sensing in the management of cultural resources offers a fifth challenge. The numerous aspects of managing such resources has been outlined by Riechstein (1984) and at least two-thirds of the 46 listed threats to cultural resources can be monitored by remote sensing means. Staff and funding limitations are a problem so the solution is not simple. Again education plays a role, perhaps involving government officials. In the land-use planning, process inclusion of cultural resources cannot be too strongly emphasized (Cleere, 1984). Countries such as Denmark and Sweden and several other countries of Northern Europe have advanced significantly in this area, while in the developing countries, the landscape is likely to be sacrificed for the immediate economic result with considerable loss of cultural resources.

6. The sixth challenge rests with the rapidly developing geographic information system (GIS) approach to resource management and planning in which remote sensing inputs play an increasingly significant role. Its application offers a valuable aid in delimiting cultural landscapes. GIS, with the inclusion of remote sensing data provides a means of assessing land-use change which is one of the major factors affecting cultural resources. The application of the GIS approach, although gaining in use, lacks widespread implementation in which cultural resources are an element. Denmark, along with other Northern European countries have made important strides in this direction by digitizing various levels of landscape information including their register of monuments and sites. Aerial imagery is employed in many aspects of their work. An interesting example involves route planning for a major pipeline system in Denmark involving some 2000 km. of right-of-way (Kristiansen, 1984). Using a GIS approach it was possible to plan a route of least destruction.

7. The research challenge cannot be stressed too strongly. Although multi-spectral studies regarding the detectability of certain cultural resources are developing in the United States, Canada, a few European countries and in scattered locales elsewhere, as for example in the Costa Rican case under American auspices mentioned at the outset, there is a critical need for research that would point to some predictable results. Links between university research

Figure 1. The concept of an historical cluster (U.S.), consisting of several buildings, essentially organized around a churchyard provides one type of focus. The cluster may also reside in an historical landscape, and both are readily definable on aerial imagery. Large scale imagery is used to delimit the area of interest, and where othophoto mapping has been implemented, as in some states, the precise location is readily assessed and can readily be incorporated within a GIS. Smaller scale imagery may be used in subsequent management to assess land-use change threats, and ultra-large scale imagery may be obtained as part of a larger program to assess internal conditions of the site.

Feature Key: Black line outlines historic cluster including church buildings and cemetery.
Photo scale = 0 50 m. (USDA)

Figure 2. The Visbecker Braut area (F.R. Germany), a famous prehistoric, megalithic site, illustrates a number of problems facing cultural resource managers. The autobahn rest area is but 125 meters from the site, and the autobahn itself traverses an agricultural landscape in which soil/crop marks relating to potential cultural resources close to the right-of-way exist. While it is apparent that the megalithic site is protected, the bulk of the remaining area is not and is subject to the threats of continued agricultural disturbance. Aside from the rescue operations needed to prepare for construction, the main site is now within reach of a potentially large audience, and therefore might be additionally enhanced and developed for public enjoyment. Large scale aerial survey in the adjacent landscape geared for site detection and future planning seems appropriate. While modern transport routes may not be denied, every effort should be made to provide for the least amount of destruction along the route.

Feature Key: A. Prehistoric megalithic site. Photo scale = 0 100 m. (Landesvervaltungsamt, Hannover)

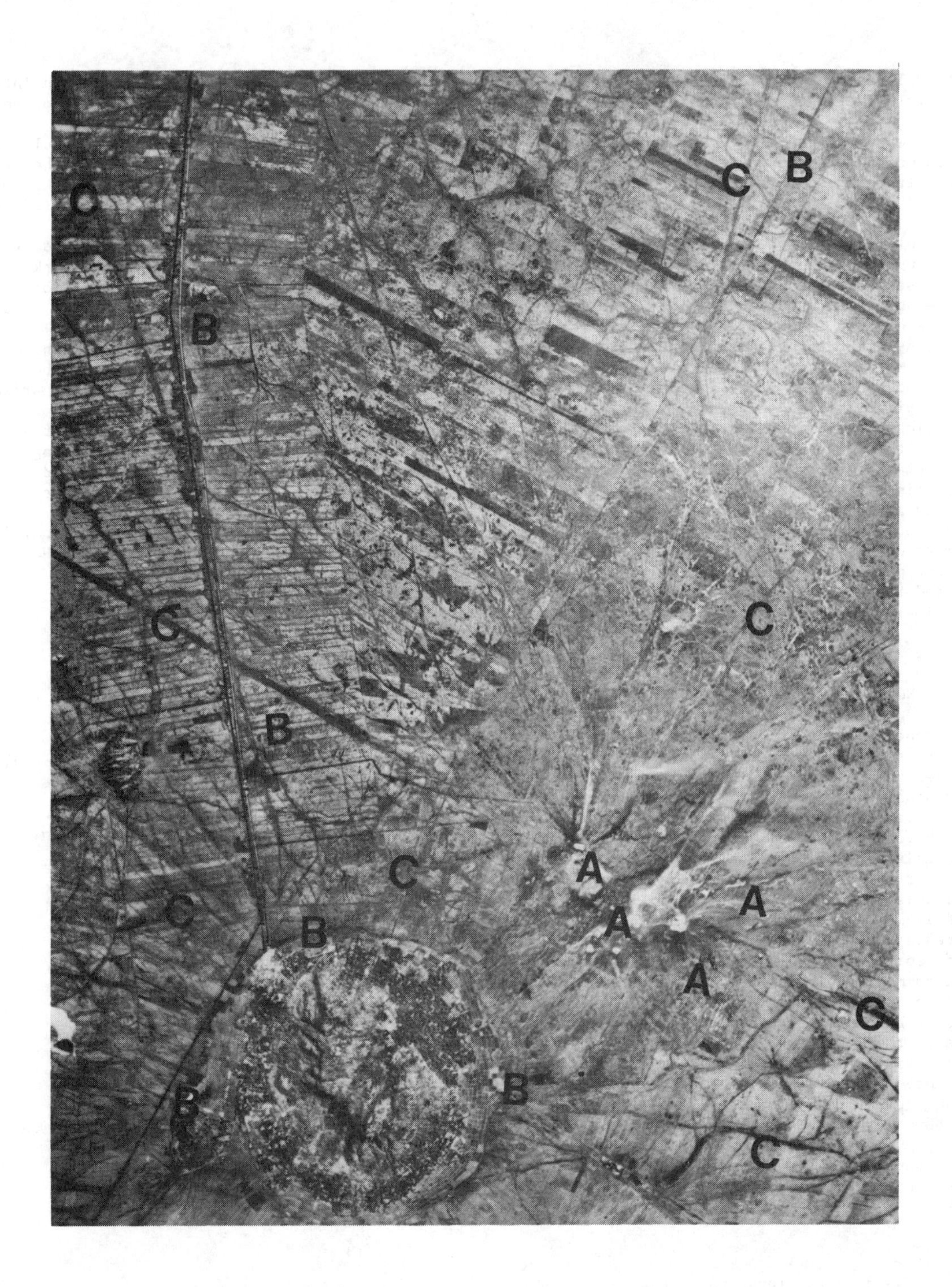

Figure 3. A multitude of sites and a landscape of considerable archaeologic and historic significance in Southeast Asia (Vietnam) is shown. The main site consists of the ruins of the capital city of Funan, Oc-eo (1900-1300 BP). As the site was excavated in part by French archaeologists in the 1940's during exploration activities, its contents and those of adjacent sites is still not fully known. Considerable quantities of precious metals and jewels were discovered, including coinage of Roman, Greek and Indian origin. The impact of looters is visible over much of the site. In the process of studying the ancient canal system covering the general landscape and clearly shown in the imagery, early agricultural fields associated with Funanese occupation were also found. The site and landscape are not developed and these features in general are not within the current planning activities of the Mekong Secretariat (U.N.). As Vietnam is among the poorest countries of the World, the situation regarding these resources is desparate. The canals are also discernible on LANDSAT imagery (Lind, 1981). Since the area is affected by major flooding, the satellite data provides a major aid for potential planning and management. The application of remote sensing is especially relevant in this area due to its remoteness and difficulty of access.

Feature Key: A. Oc-eo ruins, B. Other sites, C. Ancient canals. Photo scale = 0 400 m.

and cultural resource management are especially important. In the developing countries, such research still appears to be a luxury since there is a general paucity of manpower and all human resources are geared to coping with inventorying and cataloging the mass of cultural resources.

8. For developing nations, the challenge is multi-faceted, involving many of the items above. Vitually all of these countries are mainly in the survey stage of assessing their resources and have neither the mechanisms nor the funds to employ the more sophisticated management techniques practiced in European countries. Moreover, much of the survey work is accomplished by foreign investigators and is sporadically applied in time and space. Remote sensing applications planned for these areas might well include cultural resources considerations and investigators planning work there would do a great service by attempting to incorporate cultural resources approaches along with the traditional ones relating to land-use, water resources, mineral exploitation, and forest or grassland management. Training programs geared for developing countries have generally not approached these resources in any meaningful way, so the time is ripe for planning the introduction of the topic.

5 CONCLUDING REMARKS

The non-renewable nature of cultural resources and the numerous threats converging on their well-being in the decades to come makes the matter of resource management all the more critical. Although their value may vary from place to place, all nations and peoples have some recognition that the rights of the past to exist is philosophically continuous with a respect for the rights of the future (Lipe, 1984). All scientists, resource managers and indeed all human beings, have a stake in the business of cultural resources management. We have attempted to identify some of the challenges facing the remote sensing community within the cultural resources realm in a broad way with the hope that they will ultimately become matters of routine concern in all corners of the World. We have also suggested some of the complexities involved with a few examples. The more remote areas of the planet are no longer hidden from the view of remote sensing "eyes". As the earth is the home of mankind, it is appropriate that we should apply relevant remote sensing technologies to discover and manage its ancient or historic cultural features and landscapes.

REFERENCES

Begley S. and S. Katz. 1986. Unearthing a culture: a New World Pompeii. Newsweek. 107:48.

Butzer, K.W. 1982. Archaeology as human ecology: method and theory for a contextual approach. London: Cambridge University Press.

Cleere H. (ed.) 1984. Approaches to the archaeological heritage: a comparative study of world cultural resource management systems. London: Cambridge University Press.

Ebert, J.I. 1984. Remote sensing applications in archaeology. Chap. 5. in Advances in archaeological method and theory V. 7. New York: Academic Press.

Holm, J. and S. Stridsberg 1984. Flygarkeologi från drake och från modellplan. Popular arkeologi 2: 30-32.

Kristiansen, K. 1984. Denmark. Chap. 3 in Approaches to the archaeological heritage: a comparative study of world cultural resource systems, H. Cleere (ed.). London: Cambridge University Press.

Lind A. 1981. Applications of aircraft and satellite data for the study of environment and archaeology: Mekong delta, Vietnam. Proc. fifteenth international symposium on remote sensing of envirnment. 1529-1537.

Lipe, W.D. 1984. Value and meaning in cultural resources. Chap. 1 in Approaches to the archaeological heritage: a comparative study of world cultural resource systems, H. Cleere (ed.). London: Cambridge University Press.

Melnick, R.Z. 1984. Cultural landscapes: rural historic districts in the national park system. Washington: U.S. Department of Interior.

Myers, J. and E. Myers. 1985. An aerial atlas of Crete. Archaeology 38: 18-25.

Normann, J. 1985. Flyg arkeologi. Stockholm: Widlunds.

Olsen, J.W. 1985. Application of space-borne remote sensing in archaeology. University of Arizona remote sensing newsletter. 85:1.

Reichstein, J. 1984. Federal Republic of Germany. Chap. 4 in Approaches to the archaeological heritage: a comparative study of world cultural resource management systems, H. Cleere (ed.). London: Cambridge University Press.

Vogt, E.Z. (ed.). 1974. Aerial photography in anthropological field research. Cambridge: Harvard University Press.

Symposium on Remote Sensing for Resources Development and Environmental Management / Enschede / August 1986

Human settlement analysis using Shuttle Imaging Radar-A data: An evaluation

C.P.Lo
University of Georgia, Athens, USA

ABSTRACT: The detectability of human settlements from Shuttle Imaging Radar-A images was determined with reference to the radar system geometry and physical and cultural characteristics of the environment in four specific geographic regions of the United States represented in five strips of images. The usefulness of the settlement area data directly measured from the images for population estimation was also evaluated. It was concluded that Shuttle Imaging Radar-A data could produce accurate population estimates of individual settlements and complement other forms of high-resolution space data in human settlement analysis.

1 INTRODUCTION

In recent years, high-resolution imagery obtained from space platforms which can be usefully employed in human settlement study becomes available. The most notable examples are Thematic Mapper data (ground resolution 30 m/pixel), SPOT data (ground resolution 20 m/pixel in the multi-spectral mode and 10 m/pixel in the panchromatic mode), NASA Large Format Camera photography (resolution 80 lp/mm), the Metric Camera Photography of the European Space Agency (resolution 40 lp/mm), and the Shuttle Imaging Radar-A data (ground resolution 40 m/pixel) (Doyle, 1984; Cimino and Elachi, 1982). Despite the relatively poorer spatial resolution of the Shuttle Imaging Radar-A (SIR-A) data, its employment in human settlement analysis is invaluable because of its all weather imaging capability which makes it suitable for use to monitor changes of the environment at any time and in different parts of the world. However, the interpretation of these radar image data is more complicated than that of photography or imagery obtained within the visible portion of the electromagnetic spectrum (0.4-0.7 μm). Research is required to examine problems associated with this type of space data and to evaluate objectively their utility in human settlement analysis. Human settlement analysis is taken here to refer to the identification of the shapes and sizes of individual settlements and their pattern of spatial distribution. This paper reports on some preliminary findings of such a research.

2 NATURE OF THE SIR-A DATA

The SIR-A data were acquired by the Space Shuttle Columbia on 12 November, 1981 with a side-looking synthetic aperture radar using horizontally polarized microwave radiation transmitted at L-band (1.278 GHz) from an altitude of 259 km (Cimino and Elachi, 1982; Ford et al., 1983). The depression angle of the antenna varied from 46° for the near range (southward edge of the film) to 40° for the far range (northward edge), which produced a swath width of 50 km on the earth surface. The SIR-A data employed for human settlement analysis were optically processed and tilt corrected two-dimensional image film which was amenable to visual interpretation. The scale of the image was 1:500,000. Preliminary investigations carried out by the author in connection with settlement pattern analysis in the North China Plain has confirmed good planimetric accuracy of the SIR-A data as compared with 1:250,000 scale topographic map (Lo, 1984). The quality of these radar data was rated excellent.

3 OBJECTIVES AND METHODOLOGY

The application of remotely sensed data to human settlement analysis involves a study of the following points: (a) the detectability of individual settlements, (b) the accuracy with which the shape of each settlement can be determined, (c) the accuracy with which the areal extent of each settlement can be delineated, and (d) the accuracy with which the population size of each settlement can be estimated.

In the case of synthetic-aperture radar (SAR) data, the answers to these questions are clearly related to the nature of backscatter from the terrain, which is affected by the following factors: (a) the depression angle of the antenna, (b) the incidence angle of illumination, (c) the terrain slope, (d) the properties of the structures, (e) the orientation of the structures in relation to the illumination, and (f) the spatial resolution of the imaging system.

In order to evaluate the significance of these various factors, five strips of the SIR-A imagery covering four distinct regions of the United States of America were selected. These were: (1) St. Joseph, Missouri, (2) Mobile, Alabama -- Hattiesburg, Mississippi, (3) Tallulah, Louisiana, (4) Louisville, Kentucky, and (5) Sterling, Colorado, which represented Interior Plains, Gulf Atlantic Coastal Plain, the Appalachian Plateau, and Great Plains in morphological regions (Fig. 1). The settlements were visually detected and their areas measured using 1-mm square grids directly from the images. These results were then compared with those obtained from the U.S. Geological Survey 1:250,000 scale topographic maps. Visual comparison was also made on the shapes of these settlements. In addition, the areas of these settlements as measured from the imagery were correlated with the population data obtained from the 1980 census.

Apart from these observations and measurements, the effects of depression angle, land-surface forms, soil types, vegetation cover, and land use on the detectability and size determination accuracy of these settlements from the SIR-A data were also investigated.

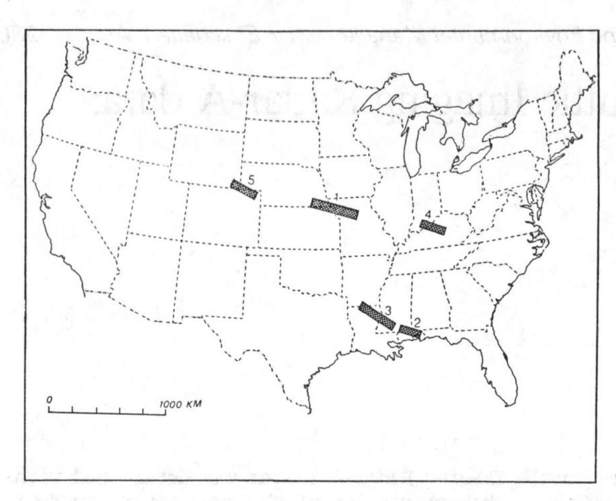

Figure 1. Location of the five SIR-A data strips: (1) St. Joseph, Mo., (2) Mobile, Al., (3) Tallulah, La., (4) Louisville, Ky. and (5) Stirling, Co.

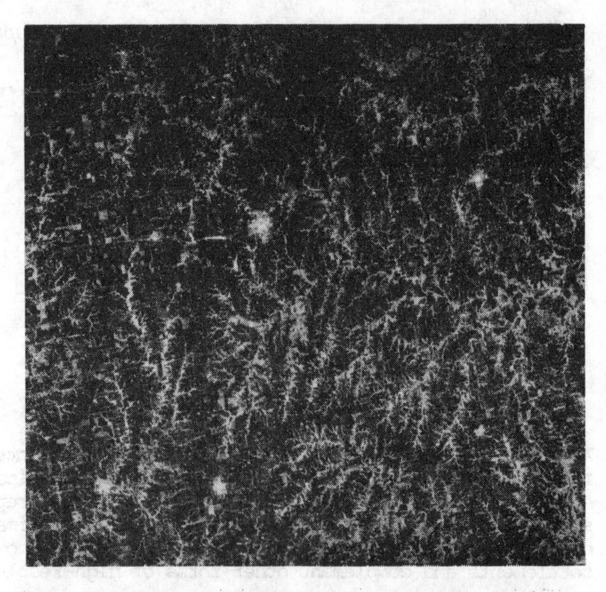

Figure 3. An extract from St. Joseph, Mo. strip. Note the variations in building density detectable in some settlements.

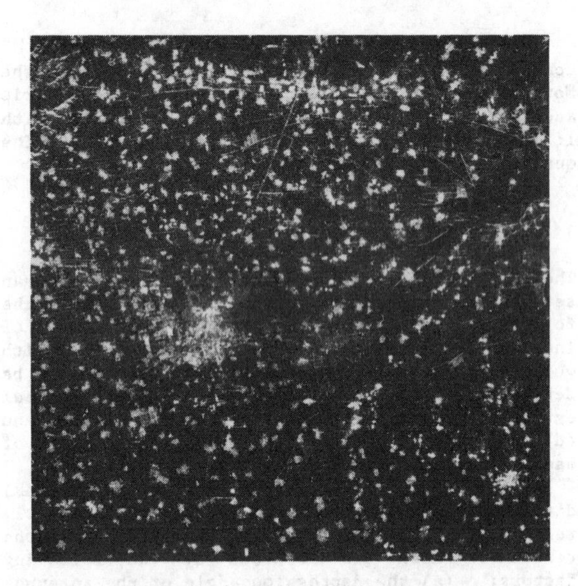

Figure 2. An extract showing settlements of the Baoding area in North China as recorded on the SIR-A image.

Figure 4. An extract from the Mobile, Al. strip. Note the city of Mobile is difficult to delineate. Notice also the bright return from the industrial area and the weak return from the air fields.

4 RESULTS

4.1 Detectability of human settlements

It is interesting to note that the detectability of human settlements from the SIR-A data varied considerably from region to region. A striking difference was observed if human settlements in the North China Plain were compared with those recorded on the five different United States strips. The human settlements in the North China Plain exhibited such strong backscatter that a complete range of them could be detected (Fig. 2). Even the very small villages could be mapped without much trouble. A comparison with the 1:250,000 scale topographical map for this part of China suggested near 100 percent success rate of detectability. It is also noteworthy that these Chinese settlements are highly compacted and mostly walled in the past. The plain is so very flat that it is almost at sea level -- the danger of flooding was revealed by the levees built along the rivers which were readily visible from the images (Fig. 2).

In contrast, the United States environment appeared to show a greater diversity and variability in the detection of human settlements from the SIR-A data (Figs 3-7 and Table 1).

The settlements normally distinguished themselves by the very strong backscatter which may have been caused by the corner reflector effect of some structures. It has long been recognized that the orientation of the radar antenna in relation to the cultural features being imaged can affect the strength of the microwave backscatter, hence variations in the grey tone of the objects in the images (Bryan, 1979; Hardaway et al., 1982). It was noted that when linear features were oriented perpendicular to the radar beam very strong backscatter called cardinal effect occurred. Henderson and Anuta (1980) also observed that settlement detectability was significantly influenced by radar azimuth angle on

Table 1. Settlement detectability in five SIR-A strips of the U.S. environment

Strip	Description	Region	No. detected	Detectability (%)	Look direction azimuth
1	St. Joseph, Mo.	Interior Plains	14	87	6°30'
2	Mobile, Al.	Gulf Atlantic Coastal Plain	10	63	32°
3	Tallulah, La.	-ditto-	9	60	27°
4	Louisville, Ky.	Appalachian Plateau	14	82	19°
5	Sterling, Co.	Great Plains	7	100	1°

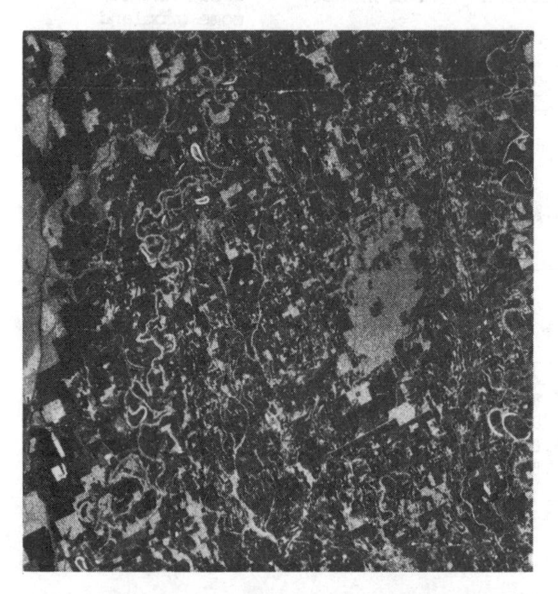

Figure 5. An extract from the Tallulah, La. strip. Note the poor contrast between settlements and the alluvial plain of the Mississippi River.

Figure 7. An extract from the Stirling, Co. strip. Note the large, regular fields and the center pivot irrigated circular plots.

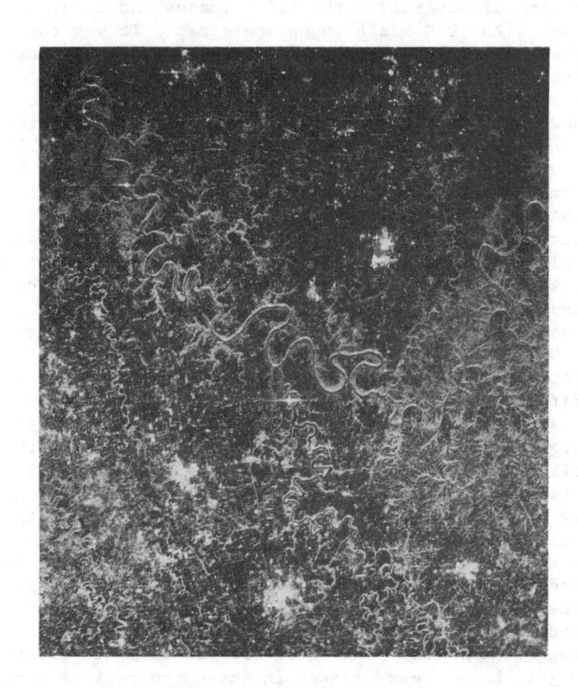

Figure 6. An extract from the Louisville, Ky. strip. Note the cardinal effect.

aircraft imagery in which the settlement was oriented parallel to the flight line. In the case of satellite imagery such as the Seasat SAR, the effect was also noticeable. As for the SIR-A imagery which was flown at much lower altitude than that of Seasat (259 km versus 800 km), one should expect the applicability of the same principle. Indeed, from Table 1, it is obvious that when the look direction of the SIR-A antenna is oriented more towards the North direction, a higher degree of detectability of settlements occurs. In other words, as the settlements are oriented more orthogonally to the radar antenna look direction, strong radar echoes result, thus confirming the previous observations of Henderson and Anuta (1980). Some cardinal effects can actually be observed (Fig. 6).

An important observation of the present research is that the detection of large settlements (population over 100,000) such as Mobile, Al. (Fig. 4), Louisville, Ky., and Monroe, La. was much more difficult than the small ones. This is because the large settlement spreads itself out over a much larger area and is less compact. The orientation of buildings showed a greater degree of irregularity, thus the strength of radar backscatter was not uniform, as exemplified by the images of the three large cities mentioned above (Fig. 4). This observation suggests that detectability of settlements in radar images is probably not as much affected by the size factor as in other types of remotely sensed imagery. It

Table 2. Terrain and land cover characteristics of the five study areas (Source: U.S. Department of the Interior, 1970, The National Atlas of the United States of America).

Area	Mean height (m)	Land-surface form	Soil type	Vegetation cover	Land use
1. St. Joseph, Mo.	346	irregular plains	Mollisols (Udolls)	Oak-hickory	mostly cropland
2. Mobile, Al.	67	irregular plains	Ultisols	pine	forests and woodland grazed
3. Tallulah, La.	30	flat plains	Inceptisols	oak-gum-cypress	cropland with pasture
4. Louisville, Ky.	247	open hills	Alfisols	Oak-hickory	woodland with some cropland and pasture
5. Sterling, Co.	1,369	irregular plains	Mollisols (Ustolls)	pine	grassland and grazing land

Table 3. Population estimation using area as input to a linear regression model and an allometric growth model

Area	Linear: $P = a + bA$ r*	a	b	Allometric: $\log P = \log a + b \log A$ r*	loga	b
(1) St. Joseph, Mo.	0.648	-3921.41	27.78	0.896	-0.5854	1.5940
(2) Mobile, Al.	0.973	-2516.13	116.87	0.828	1.1595	1.2620
(3) Tallulah, La.	0.706	291.65	14.82	0.742	1.3861	0.9004
(4) Louisville, Ky.	0.728	3431.46	8.05	0.867	2.3190	0.5781
(5) Sterling, Co.	0.985	-254.99	14.87	0.814	0.2057	1.2724

*r = correlation coefficient, all significant at a level of 5 percent or below

helps to explain why in the case of the North China Plain even very small villages can still be detected (Fig. 2). The high degree of compactness of these formerly walled Chinese settlements made them good corner reflectors to radar signals.

Another observation is that the detectability of the settlements appeared to be affected also by the nature of the geographic region (Table 1). The Gulf Atlantic Coastal Plain region came out to be the worst of all four regions while the Great Plains region was the best. To assist further in understanding the effect of terrain characteristics and land cover types on the detectability of settlements. Table 2 was compiled. It appeared that mean terrain heights, soil types, and land use were important factors. High terrain, irregular plains, Mollisols (soils with nearly black, organic-rich surface horizon) with grassland and grazing land of the Sterling, Colorado strip in the Great Plains (Fig. 7) seemed to provide favorable conditions for the detection of settlements. On the other hand, the low, forest covered terrain and the lowlying alluvial plain of the Mississippi river covered with cropland on Inceptisols (wet soils with weakly differentiated horizons) were unfavorable (Fig. 5). These environmental conditions have probably affected the settlement-background contrast, thus making the detection difficult.

4.2 Accuracy of settlement area measurement and population estimation

An important application of the SIR-A data is to determine the area and population size of the human settlements detected. A commonly employed method is to measure the areas of these settlements and then input them into a mathematical model linking area (A) with population (P). A popular model is the allometric growth model in the form of $\log P = \log a + b \log A$ (Lo and Welch, 1977). In the present research, the area of each settlement was first measured with 1-mm square grids directly from the SIR-A images and then from the 1:250,000 scale topographic map. It was found that the image area and map area of the individual settlements exhibited a very strong correlation of 0.92 at 0.01 per cent level of significance. However, it was observed that all the measured image areas were exaggerated by a factor of 1.5X from the actual map areas. This may be caused by some human errors in measurement, but careful inspection revealed that more significantly the strong radar backscatter had produced a glare which tended to exaggerate the size of the settlement. It was fortunate that this exaggeration appeared to be constant and could be easily corrected.

Despite some discrepancy in time, the 1980 population figures of these settlements in different regions were correlated with measured image areas first in the form of a linear regression model and then in the form of the allometric growth model mentioned above. The results (Table 3) indicated overall strong relationship between population and area in the allometric growth model for all regions. It is noteworthy, however, that in Mobile, Al. (Fig. 4) and Sterling, Co. (Fig. 7) regions much stronger relationship existed with the linear regression model than the allometric growth model, a suggestion that the rate of settlement growth might have been faster in these two regions than in the others. These results indicated that settlement population estimation using settlement area as an independent variable could produce reasonably accurate results.

4.3 Shapes and internal structures

It is obvious from the above that the shape of the settlement registered on the SIR-A data may not be correct because of the strong backscatter. This is particularly so for large and less compact settlements. However, it was possible to identify the density of buildings inside a settlement under the favorable imaging conditions. St. Joseph, Mo. is the best among the five strips in displaying distinctly the shapes and internal structures of some settlements (Fig. 3). Apparently, the radar azimuth (N 6°30'E) has much to do with this result. In large settlements, it was possible to identify industrial areas and airfields easily, the former usually exhibiting bright return and the latter giving very low return from their smooth surfaces, as exemplified in Mobile, Al. (Fig. 4).

5 CONCLUSIONS

This research has indicated that great potential exists in the use of SIR-A images for human settlement analysis because the settlements give strong backscatter, which renders them easily detectable. It was observed that the radar viewing geometry has a great impact on the settlements' detectability, and a preliminary observation seems to suggest the importance of orienting the radar look direction orthogonally to the settlement to give a better detection capability.

Cultural factors have also affected the detectability. The highly compact Chinese settlements which were formerly restricted by walls in the North China Plain can be more easily distinguished, irrespective of size, than their American counterparts.

Within the American environment, the detectability of settlements varies, not only because of differences in radar azimuths but also because of differences in terrain height, soil type, vegetation and land cover type. It appears that high terrain, cropland or grassland cover, and Mollisols soil with an irregular plain landform are favorable factors that enhance image-background contrast in settlement detectability. The Gulf Atlantic Coastal Plain emerges as the worst of all four regions in settlement detectability by a combination of poor environmental contrast and an unfavorable radar azimuth angle. It is also noteworthy that large settlements with a population over 100,000 and having a large spatial spread are more difficult to detect than small settlements.

An important finding from this research is the very strong relationship that exists between map areas and image areas of settlements, although the image areas consistently exaggerate the actual areas of the settlements. The image areas can be usefully employed in the allometric growth model for population estimation in all four geographic regions. Although the shapes of the settlements are not always correctly recorded on the SIR-A images, one can still detect building density and other functional units of the settlement in cases when the optimum conditions of imaging have been achieved.

To conclude, the SIR-A data in the present optically processed and tilt corrected two-dimensional image form possess adequate resolution and imaging quality for use in human settlement analysis. They also complement nicely other forms of high-resolution space data in yielding timely data of the settlements under adverse weather conditions. Already SIR-A data have been successfully combined with Landsat MSS data to provide improved spatial data for human settlement analysis (Welch, 1984).

6 ACKNOWLEDGEMENTS

I wish to thank Dr. Charles Elachi of Jet Propulsion Laboratory, California Institute of Technology, Pasadena, California and the National Space Science Data Center for providing me with the SIR-A data, which makes this research possible.

REFERENCES

Bryan, M.L. 1979. The effect of radar azimuth angle on cultural data. Photogrammetric Engineering and Remote Sensing 45:1097-1107.

Cimino, J.B. & C. Elachi 1982. Shuttle Imaging Radar-A (SIR-A) experiment. Pasadena, California: Jet Propulsion Laboratory.

Doyle, F.J. 1984. Surveying and mapping with space data. ITC Journal, No. 4, 314-321.

Ford, J.P., J.B. Cimino & C. Elachi 1983. Space shuttle Columbia views the world with imaging radar: the SIR-A experiment. Pasadena, California: Jet Propulsion Laboratory.

Hardaway, G. & G.C. Gustafson 1982. Cardinal effect on Seasat images of urban areas. Photogrammetric Engineering and Remote Sensing 48:399-404.

Henderson, F.M. & Anuta, M.A. 1980. Effects of radar system parameters, population, and environmental modulation on settlement visibility. International Journal of Remote Sensing 1:137-151.

Lo, C.P. 1984. Chinese settlement pattern analysis using Shuttle Imaging Radar-A data. International Journal of Remote Sensing 5:959-967.

Lo, C.P. & R. Welch 1977. Chinese urban population estimates. Annals of the Association of American Geographers 67:246-253.

Welch, R. 1984. Merging Landsat and SIR-A image data in digital formats. Imaging Technology in Research & Development, July, 11-12.

Symposium on Remote Sensing for Resources Development and Environmental Management / Enschede / August 1986

Urban-land-cover-type adequate generalization of thermal scanner images

Peter Mandl
Institute of Geography, University Klagenfurt, Austria

ABSTRACT: A thermal infrared image showing the urban area of Klagenfurt, Austria on a cloudless day in autumn 1979 is generalized by forming radiometric and spatial classes and by clustering temperature-value-histograms of these generalized data. The results are plotted in map resembling form and interpreted according to thermal urban-land-cover-types. The errors appearing in such a generalization are analysed in detail. Some of these complex land cover classes are investigated in detail trying to explain their temperature distributions by the percentages of overbuilt and vegetated areas within these different urban housing quarters. The results of this study can be used for urban climate investigations and town planning purposes.

1 INTRODUCTION

There are three points which characterize modern aerial remote sensing and which distinguish it from classical aerial photo interpretation. The first point is the possibility to acquire images in the non-visible parts of the electromagnetic spectrum like the microwave or the thermal infrared region. The second point is the sensing of discrete and calibrated measurements (e.g. by scanner) which allows the drawing of measurement maps and is also the base for the third point, the utilization of digital computers to use methods of pattern recognition, statistics, etc. for handling and analysing the huge amounts of data. These features of modern remote sensing have extended the applications of this technology in the last two decades.

One field of research which got many impulses by utilizing these new features of remote sensing (particularly by the introduction of computerenhanced and analysed thermal infrared images) is micro climatology, especially urban climatology and the application of the findings there in town and regional planning. A very good review on urban climate research is D.O.Lee (1984). One main problem of these sciences is to get measurements of climatic parameters covering the terrain more or less continuously. The classical method to solve this problem is to drive measurement traverses over the test area or to have a fix sensor network and to interpolate the data between the measured traverses or points (e.g. Oke & Hannell 1970 or Nübler 1979).

Now remote sensing can provide urban climatology with continuous terrain covering measurements on radiation temperature. Unfortunately this parameter is of no direct use for climatology and a conversion of radiation temperatures to surface temperatures is complicated because of the differnt error sources that influence the remotely sensed data. Nevertheless special processed, enhanced and printed images of the thermal infrared emission of urban areas have been used to find out "heat islands" in various towns, fresh air areas in the surrounding of towns or fresh air canals into the central towns and to give basic informations for planning purposes.

The aims of this study now are to show special aspects of the heat emission field of the urban area of Klagenfurt, Austria on a cloudless day at 10 o'clock in the morning in autumn. The primary data for this were thermal infrared images (8 - 13 μm wavelength) acquired by a Bendix M^2S 11-channel scanner on the 13th of September 1979. A correction of the radiometric errors in the image which are intern relative errors (0,1-0,2° C), errors due to the atmosphere between the objects and the sensor (minus 1 - 3,5° C for the correction from radiation to surface temperature), errors due to the emission coefficients of different surface cover types and influences of the different viewing angles in the image on the thermal emission distribution of various areas seemed not necessary for the purposes of this study.

In a first part of this study images of the primary data, geometrically rectified from the panoramic distortion, sliced into five temperature classes were produced (Fig. 1). The pixels then were stepwise combined to "macropixels" consisting of 10x10, 25x25, 50x50 and 100x100 "micropixels" by calculating the arithmetic mean value of every macropixel. An example of the resulting images is given in Fig. 2. All the results of this first part of the study are described in detail in Seger & Mandl 1985.

In the second part of this study which will be described in this paper three main problems are dealt with:

1. The determination of the temperature distributions of typical urban-land-cover-types and their relations to the proportions of vegetation, asphalt and houses in test areas, to find out if thermal data can be a good indicator for such "pure" land cover types.
2. An error investigation of different generalizations of the radiometric and the spatial domain of the thermal image data.
3. The attempt of a regionalisation of the thermal image using temperature-value-histograms of the macropixels as variables and cluster analysis as classification method.

2 SOME ASPECTS OF THE GENERALIZATION PROBLEM OF REMOTE SENSING DATA

Digital remote sensing data can be characterized by four different types of resolution. In this study the temporal and the spectral resolution of the data is held constant, a thermal infrared image (9 - 13 μm) at one acquisition date is used. The spatial resolution of the primary data is 2,3m x 2,3m at the nadir point and the radiometric resolution is 0,12° C temperature intervall from one data value to the other.

The thermal image of Klagenfurt consists of 803 x 3.200 pixels. Each pixel has a precision of 256 possible values which means that the amount of information in the image is greater than 20,5 million bits. In order to seperate the user important information (signal) from the unimportant one (noise) a very well-considered and well-balanced correspondence between the objectiv,the degree or level of generalization and the methods used for this have to be found. There are

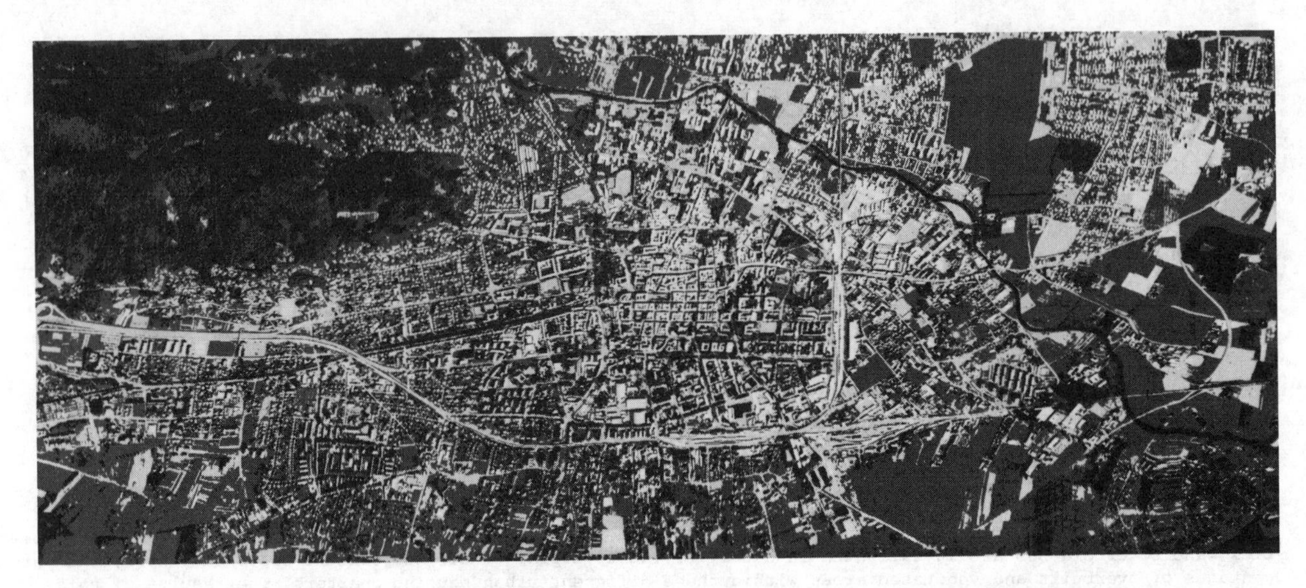

Figure 1. Radiation temperature image of Klagenfurt, micropixels, 2,3m x 2,3m ground resolution, five temperature classes: white = over 30° C, light grey = 26,6 - 30° C, middle grey = 23,3 - 26,6° C, dark grey = 20 - 23,3° C, black = under 20° C.

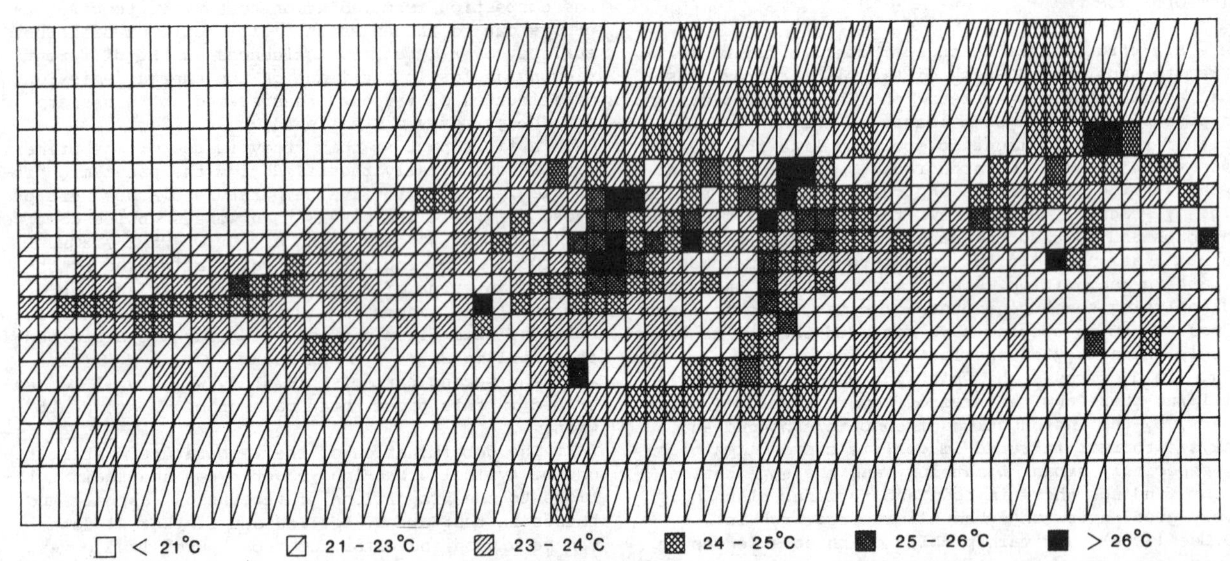

Figure 2. Generalized radiation temperature image of Klagenfurt, macropixelsize 50 x 50 pixels, 116m x 116m ground resolution in the middle of the image, six temperature classes.

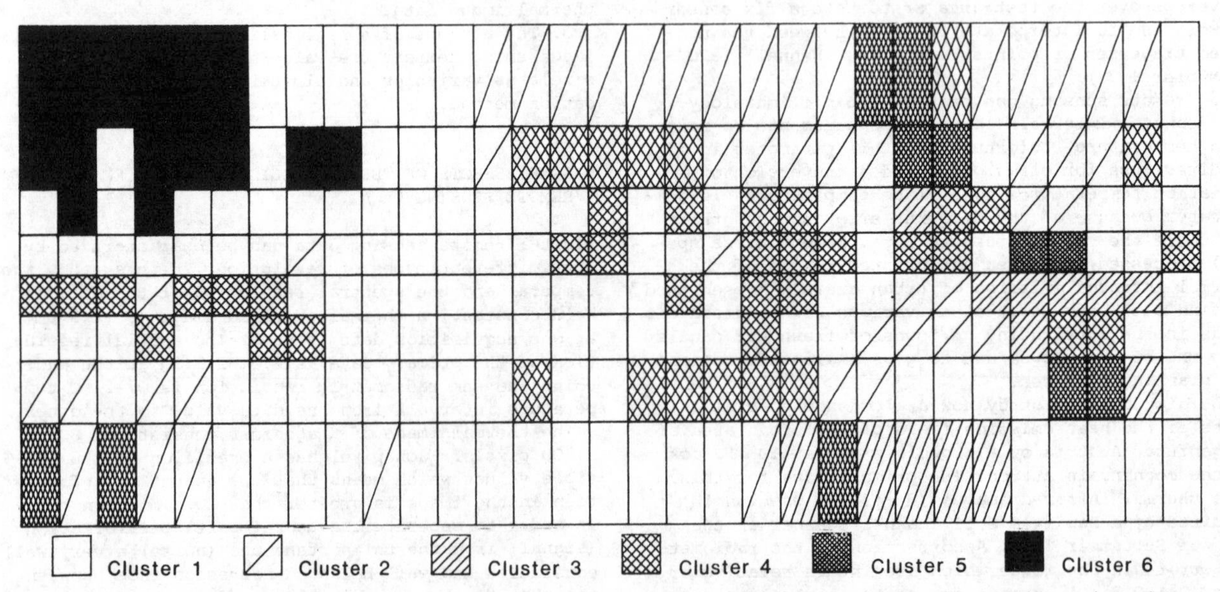

Figure 3. Clustered radiation temperature macropixel image of Klagenfurt, macropixelsize 100 x 100 pixels, 232m x 232m ground resolution, six clusters of temperature-value-histograms of the macropixels, interpretation of the clusters in the text.

Table 1. Portions of overbuilt and vegetated areas and mean values of the radiation temperature of different urban-land-cover-types in % of the total test area; area delineation and calculation by Karin Fera.

Test area	area in ha	over-built area	vege-tated area	buil-dings	traf-fic areas	other overb. areas	trees	bushes	mea-dows	mean radiat. temper.
1 Medieval inner city	8,8	88	12	56	21	11	5	1	6	26,03°C
2 Renaissance inner city	8,3	95	5	52	33	10	3	0	2	25,29°C
3 Old suburb buildings	1,6	73	27	42	11	20	11	3	13	25,57°C
4 House-blocks end 19th century	3,6	69	31	43	4	22	15	3	13	24,82°C
5 Better single-family-house area	7,9	33	67	14	10	9	21	11	35	22,90°C
6 House-blocks 1939/40	12,0	46	54	22	11	13	14	5	35	21,42°C
7 Simple single-family-house area	5,1	35	65	17	10	8	24	8	33	21,35°C
8 New suburban house-blocks area	17,4	34	66	12	14	8	5	2	59	22,66°C
9 Old village kernel with new single-family-houses	7,9	30	70	14	6	10	18	6	46	23,16°C

only very few rules of thumb, no procedures and it is mostly left to the scientist's knowledge and luck to find the right mixture of these three things to get the desired results.

3 RADIATION TEMPERATURE DISTRIBUTIONS OF DIFFERENT URBAN-LAND-COVER-TYPES

Pure land use classes covering large areas like forest or fields are easy to delineate in thermal images because of their homogeneneous temperature distributions. Not so are the manyfold mixed structures in urban environment. To find out something about these complex classes nine test areas representing nine types of urban-land-cover were selected. With the help of largescale infrared aerial photographs the areas of the pure cover classes overbuilt areas (subdivided into buildings, traffic areas and other overbuilt areas) and vegetated areas (subdivided into trees, bushes and meadows or grass) were delineated and measured. The results are listed in table 1. On the other hand the same test areas were delineated in the thermal image and radiation-temperature-value-histograms aggregated into one degree temperature classes (30 classes because the sensor was calibrated between 10 and 40° C) and the arithmetic mean values were calculated (table 1).

Applying cluster analysis (SPSS-X procedure Cluster and Quick Cluster) to group the nine test areas into urban structure types using three variable groups (temperature-value-histograms, percentage overbuilt and vegetated area, percentages of the six pure cover classes) no corresponding classification could be found (see table 2). This fact is due for the most part to the effects of shadows, different roof materials and differing viewing angles. To the variable groups "percentage overbuilt - vegetated area" and "percentage of pure cover classes" the three error reasons are not applicable. The differences in cluster membership between these two variable groups are due to the varying percentages of the pure classes in the vegetated areas. In this example the level of generalization of the 30 temperature classes in the histogramm data, which were strongly influenced by the three error sources mentioned above, was not adequate to answer the question if it was possible to explain the overbuilt - vegetation proportion of urban-land-cover-types by the radiation temperature distribution of the same test area. Also the method (the cluster procedure) was not optimal.

A better method and a better level of generalization to show such a relation is to use the arithmetic mean value of every test area and to correlate it with the proportions of the land-cover-classes. The results of this procedure are listed in table 3. We see that the proportions of overbuilt and vegetated areas in the test regions are highly correlated with the mean radiation temperature values. The six pure cover classes are one generalization step too fine to explain the mean temperature values in an optimal way. So we see that the optimal description parameters of the selected test areas and the affiliated level of spatial generalization are the mean radiation temperature values and the percentages of the rough urban-land-cover-types "overbuilt" and "vegetated".

Table 2. Clustermembership of the urban-land-cover-types named in table 1, using different variable sets for clustering into 4 groups.

No. of ur-la-co-type	temperature value-class histogramms	% overbuilt and vegetated area	% of six pure land-cover types*
1	1	1	1
2	1	1	1
3	1	2	2
4	1	2	2
5	2	3	3
6	3	4	3
7	3	3	3
8	4	3	4
9	2	3	4

* buildings, traffic areas, other overbuilt areas, trees, bushes, meadows.

Table 3. Correlation between mean radiation temperature values of the 9 test areas and the 8 land-cover-class percentages (see table 1).

Land-cover-class	corr. coeff.	R2	significance
Overbuilt area	0,86	0,74	0,00145
Vegetated area	-0,86	0,74	0,00145
Buildings	0,86	0,75	0,00132
Traffic areas	0,41	0,17	0,13853
Other overbuilt areas	0,49	0,24	0,09243
Trees	-0,53	0,28	0,07036
Bushes	-0,60	0,36	0,04349
Meadows	-0,72	0,52	0,01412

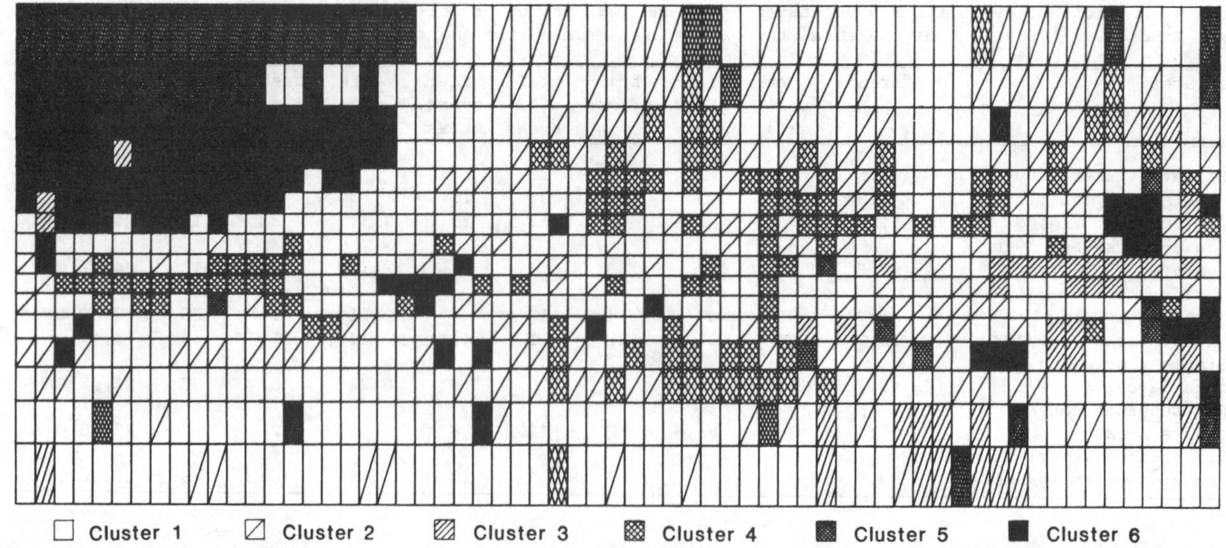

Figure 4. Clustered radiation temperature macropixel image of Klagenfurt, macropixelsize 50 x 50 pixles, 116m x 116m ground resolution, six clusters of temperature-value-histograms of the macropixels, interpretation of the clusters in the text.

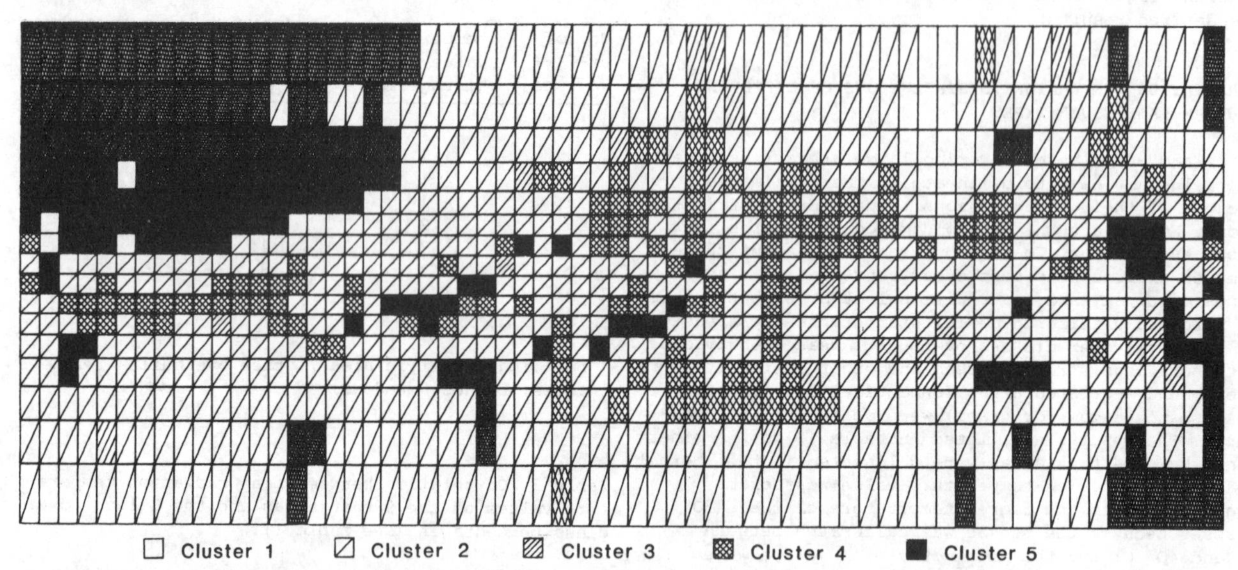

Figure 5. Clustered radiation temperature macropixel image of Klagenfurt, macropixelsize 50 x 50 pixels, 116m x 116m ground resolution, five clusters of temperature-value-histograms of the macropixels, interpretation of the clusters in the text.

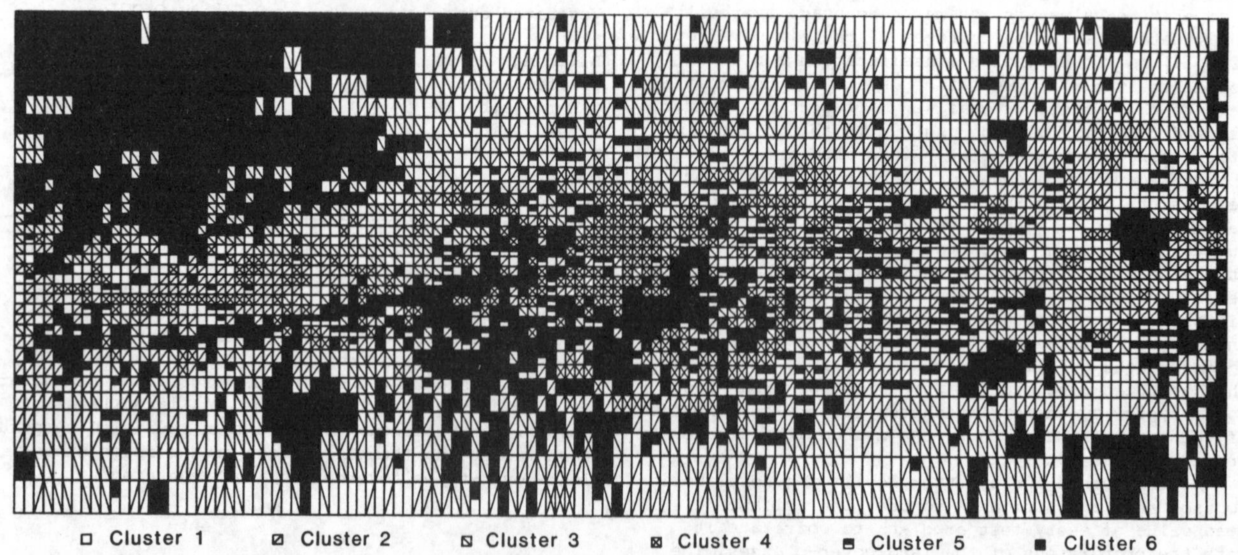

Figure 6. Clustered radiation temperature macropixel image of Klagenfurt, macropixelsize 25 x 25 pixels, 58m x 58m ground resolution, six clusters of temperature-value-histograms of the macropixels, interpretation of the clusters in the text.

4 ERROR INVESTIGATION OF DIFFERENT GENERALIZATIONS OF THE THERMAL IMAGE

In this chapter we want to extent the concept of the adequate generalization level from non-spatial feature spaces of temperatures and land cover classes to the spatial domain of a remote sensing image. We do not yet want to classify the image but only do an enhancement by generalization to show the main thermal structures like heat islands and cool air valleys in the temperature field of Klagenfurt.

For this the original pixels (micropixels) were spatially generalized to larger areas by calculating the arithmetic mean values of these macropixels. Then the new temperature values were classified into six temperature classes and the resulting macropixelfields were plotted to visualize the result. An example of this procedure, which is fully described in Seger & Mandl 1985 is shown in Fig. 2. To illustrate the differences between the various micro- and macropixel radiation temperature fields some statistical parameters shown in table 4 were calculated. There we see that the standard deviation decreases with increasing macropixelsize, which is due to the calculation of mean values. These mean values nearly are the same. The small differences can be explained by the increase of the standard error with increasing macropixelsize. The increasing median and modus values show, that with increasing macropixelsize the centre of the distribution is going towards the higher temperature values which is due to the positive skewnesses of all the distributions. We can also see that the range of the temperature values in the micro- and the macropixelfields varies so that the temperature classes in the micro- and macropixelfields have to be changed in order not to get empty classes plotted (compare Fig.1 and Fig.2).

A variation in the range and the locations of these temperature classes can also be used to stress special aspects of the image like the heat islands in the inner city, when the classes are moved to the upper region of the temperature scale and the inner differentiation of the forests and green areas, when the classes are shifted to the lower parts of the scale.

Finally it was analysed which effect has a shifting of the macropixelgrid relativ to the original macropixelgrid when all other parameters (size of the macropixels and temperature classes) are held constant. The 50 x 50 macropixelgrid was moved 10, 20 or 30 micropixels in one direction parellel to the shorter side of the image. The structure of the temperature field is not changed very much and in table 5 the percentages of the macropixels which are classified into other temperature classes are listed. We can see that in a 30 pixel shifted grid more than 60% of the macropixels fall into the same temperature class they had fallen in the original image which is due to the twofold generalization.

All these investigations were done to analyse the effects of spatial generalization on the special urban temperature field in Klagenfurt. The generalization in Fig.2 (50 x 50 pixel macropixelfields, six temperature classes of this special location in the temperature scale) describes the desired picture of the town best. We see the heat islands (inner city, industrial regions, cut fields), the fresh air valleys and areas (woods around the built up areas, fields with maize crop, green belts around the town) and the mixed areas mainly the single-family-houses region which are characteristic for the town structure of Klagenfurt.

5 REGIONALISATION OF THE RADIATION TEMPERATURE IMAGE USING CLUSTER ANALYSIS

In chapter 4 spatial and radiometric generalization was used as image enhancement method to emphasize the thermal structures of the urban area of Klagenfurt. The next step of information reduction and generalization is classification. Using an unsupervised clustering algorithm (SPSS-X procedure Quick Cluster) a regionalisation of the temperature field trying to

Table 4. Statistical parameters of the micro- and macropixelfields

Pixelfield size	number of rasterfields	Min.	Max.	Range	Median	Modus	arithm. mean value	stand. error	stand. devia.	skewness
1 x 1 = 1 5,4 m^2	2,568.797	1	256	255	100,3	93	107,3	0,017	26,8	1,4
10x10 = 100 538 m^2	25.839	17	200	183	103,7	95	105,8	0,113	18,1	0,7
25x25 = 625 3364 m^2	4.191	66	161	95	105,2	100	105,6	0,227	14,7	0,3
50x50 = 2.500 1,3456 ha	1.071	66	147	81	105,7	109	105,4	0,397	13,0	0,1
100x100 = 10.000 5,3824 ha	279	71	137	66	105,8	118	105,1	0,694	11,6	- 0,2

Table 5. Matrix of the change of class membership of the macropixels after shifting the macropixelgrid in one direction, in % of all macropixels in the image, macropixelsize 50 x 50 pixels, 1071 rasterfields, constant temperature classes.

Class membership change from - to	classified 3 temp.	2 classes	1 lower	no change	classified 1 temp.	2 classes	3 higher
original data - 10 pixel shift			8,6	82,8	8,4	0,1	
10 pixel shift - 20 pixel shift			8,4	83,1	8,3	0,1	
20 pixel shift - 30 pixel shift		0,2	10,0	81,2	8,5	0,1	
original data - 20 pixel shift		0,8	13,2	71,8	13,5	0,7	
10 pixel shift - 30 pixel shift		1,0	14,5	70,2	13,6	0,7	
original data - 30 pixel shift	0,2	2,2	17,3	61,6	16,9	1,8	0,1

find out thermal-land-cover-types was attempted. The data used for clustering were the temperature-value-frequency-histograms (30 one-degree temperature variables) of the micropixels being united in the macropixels of 25x25, 50x50 and 100x100 pixels size. As input for the clustering algorithm only the number n of the desired clusters has to be specified and in an iterative process all the macropixels were classified to the nearest of the n cluster centers. As distance measure the Euclidian distance was used. The best interpretable results were obtained by choosing six cluster centers. The cluster-macropixels were plotted in the same manner as the temperature-macropixels and Fig.3 to Fig.6 show the results.

In the 100 x 100 macropixelfield only the dense woodland (cluster 6), large fields with maize (cluster 5) and harvested fields or asphalted areas (cluster 4) can be identified quite well. The rest are more or less mixed areas and not clearly interpretable.

In the 50 x 50 macropixelfield (Fig. 4) the clusters can be interpreted in the following way:
Cluster 1: fields cultivated with maize, single-family-houses with large vegetated areas around.
Cluster 2: single-family-houses , small regular structure, quarters with modern house-blocks and large vegetated areas in between.
Cluster 3: green fields or meadows.
Cluster 4: bare soil areas, harvested fields, industrial areas, railway and other traffic areas, motorway.
Cluster 5: mixed areas with high rate of vegetation.
Cluster 6: wood, large parks, water.

In Fig. 6 the 25 x 25 macropixelfield possesses the following clusters:
Cluster 1: centers of large green fields, meadows, mixed fields.
Cluster 2: single-family-houses with regular structure.
Cluster 3: green fields, edges of fields, meadows, clearings.
Cluster 4: inner city without vegetation, harvested fields, motorway, house-blocks with warm roofs, railway areas.
Cluster 5: harvested fields, industrial areas, railway areas, sports-grounds.
Cluster 6: wood, house-blocks with large shadows, partly gardens and tree groups.

When we compare Fig.4 and 5 where only five clusters were determined we can see the mixture in cluster ordination (see table 6).

Table 6. Crosstable representing the number of macropixels classified in the clusters of Fig. 4 and 5.

Clusters Fig. 4	Clusters Fig. 5					
	1	2	3	4	5	sum
1	45	388	2	16	29	480
2	0	185	8	7	0	200
3	44	0	0	0	0	44
4	0	0	1	116	1	118
5	0	0	13	0	0	13
6	0	0	0	0	153	153
sum	89	573	24	139	183	1008

When we do the same comparision using figures 3, 4 and 6 no trend is discernible. We can see this fact also in the different descriptions of the clusters from Fig. 3, 4 and 6. That means that the different levels of spatial generalization represented by the different macropixelsize are more important for an urban-land-cover-type adequate classification of the thermal infrared image than the actual number of clusters in a certain range. For this reason the verification of the cluster-images (figures 3 to 6) were only done qualitatively and not quantitatively by correlating the maps with a handdrawn ground truth land use map.

6 CONCLUSION

It is not possible to classify thermal images into very special land use classes like we can do with multispectral data because most of the land use classes which can be separated very well in visible and near infrared light have overlapping sinature intervalls in one dimensional thermal infrared data. This can be seen very clearly from a table where different land cover classes and their thermal signature are listed (see Seger & Mandl 1985, p.73).

So only thermal more or less homogenious classes or urban cover types can be separated. This separation can be done for instance by image enhancement methods namely generalization in the radiometric and/or spatial domain or unsupervised classification using clustering of simple texture parameters like the temperature-value-histogram parameters in our study.

The extracted classes should only be characterized in a qualitative way and a correlation with a ground truth map showing conventinal land use classes will only bring good results when the classes have very rough definition.

We have to choose a certain level of generalization in all domains (spatial, radiometric and temporal) and choose corresponding methods to answer the question of the special work.

REFERENCES

Lee, D.O. 1984. Urban climates. Progress in Physical Geography 8:1-31.
Nübler, W. 1979. Konfiguration und Genese der Wärmeinsel der Stadt Freiburg. Freiburger Geographische Hefte, Heft 16. Freiburg i. Br.
Oke, T.R. & F.G.Hannell 1970. The form of the urban heat island in Hamilton, Canada. In WMO, Urban Climates. Technical Note No.108. Geneva.
Seger, M. & P.Mandl 1985. Strahlungstemperaturbilder als Beitrag zur Stadtklimatologie. In M.Seger (ed.), Forschungen zur Umweltsituation in Klagenfurt, p.59-93. Klagenfurt.

Symposium on Remote Sensing for Resources Development and Environmental Management / Enschede / August 1986

Small format aerial photography – A new planning and administrative tool for town planners in India

P.Misra
Human Settlement Analysis Group, Indian Institute of Remote Sensing, Dehradun

ABSTRACT: Small format (35mm and 70mm) aerial photography (SFAP) was flown in Delhi in July, 1985 for the first time in India. In carrying out the development of the technology, the requirements of major physical inputs for the town planners were kept in view, although other disciplines like vegetation, soil, land use, and agriculture would also benefit from SFAP.

The SFAP has been received very well by the profession of town planners. At present, it has been used for small areas eg. slums, clusters of small hutments and unauthorized encroachments on govt. land.

Case study of Rohini city development project under Delhi Development Authority has been described. The technique is highly suitable for updating an existing city base map even when non-metric cameras are used. Experiment is continuing with metric 70mm camera for producing quick base map for small towns and for providing information for environment impact-assessment which is so very essential for starting any physical project in India.

A simple technology like SFAP which is cost and time effective will always be more acceptable in developing countries.

1. INTRODUCTION

Conventional aerial photography is taken by sophisticated cameras (WILD, ZEISS etc.) costing more than 150-200 thousand rupees. The manufacturers have tried to make their lenses almost distortion free, resulting in accuracies of much better than 1/10,000 of flying height. Such a technology is also utilised in the field of photo-interpretation (aerial remote sensing) where high geometric (metric) quality is not required. For interpretation, the stress is on high resolution which will enable a town planner to pick up 'dwelling units' or 'encroachments' with good discrimination.

Further, the process of getting the conventional large format aerial photography is not economical for small areas. If the information is required for a small area like 'slums' and unauthorized colonies, the conventional flight mission becomes very costly and time consuming. A way out is now available in the use of small format aerial photography.

The use of 35mm and 70mm camera has found many applications in 'Urban Planning and Monitoring'. Some of the uses are :

a) monitoring changes over a period of time
b) updating information expeditiously
c) location of exact size, composition, distribution and physical characteristics of marginal settlement eg. slums, squatter area and other low-income settlements
d) supplementing regular census surveys
e) traffic and parking studies
f) vegetation/tree inventory and other thematic information gathering.

It may be mentioned that aerial photography is governed by the security measures/procedures as laid down by the Ministry of Defence and under the Aircraft Act in India. However, bonafide aerial photography continues to grow in India. Over the last 3-4 years, the possibility of utilization of small format aerial photography has been explored by the Human Settlement Analysis Group of Indian Institute of Remote Sensing. It may be mentioned that HUSA Group has been established under a joint ITC-IIRS project which started in India in 1982. Case study of Rohini Project is the first case in which SFAP has been used in India for urban planners

2. CASE STUDY OF ROHINI PROJECT OF NEW DELHI

The location of our study area falls within the overall project area of Rohini, a special colony development and housing project of Delhi Development Authority. The project is designed to cater to the needs of about 2.5 million persons on the NW extent of existing capital New Delhi. It was reported by the Project Manager that unauthorized construction of structures/buildings is going on especially near the villages which have been acquired for the project.

2.1 Objectives of the study

a) to get experience of actually carrying out aerial flying including navigation with small format (35mm and 70 mm) cameras; performances of the pilot-airc raft and the photo-elements of camera-film combinations
b) since the flying was to be done over real problem areas, to get feed back from user on problems of encroachment on public land, slums etc.
c) to judge the capability of small format aerial photography (SFAP) for updating the base map which was available on 1 :10,000 and 1:5000 scales
d) to evaluate SFAP for providing information on themes like tree cover, soil characteristics, drainage, land use etc. in the urban environment.

2.2 Platform (aircraft) and camera mount

It was originally planned to make use of 'SHADOW' (U.K.) Microlight* (a bare-minimum sports aircraft weighing less than 250 lbs). The procurement of this aircraft was, however, delayed and we had to resort to another aircraft called 'PUSHPAK' belonging to the Delhi Flying Club, a

* Note : Microlight aircrafts are a new development in aviation sports. These are very low in cost (about $6000). Most of them can be dismantled and can be carried over top of a trailer/big car and can be assembled in less than one hour. Another important characteristic is that they do not require an area bigger than a football field for take-off and landing. These can be operated without the consideration of a big aerodrome in the vicinity of area to be surveyed. It is our intention to test this aircraft as an additional survey facility.

PLANIMETRIC MAP FROM SMALL FORMAT AERIAL PHOTOGRAPHS OF MK– 70

SCALE 1:2,000

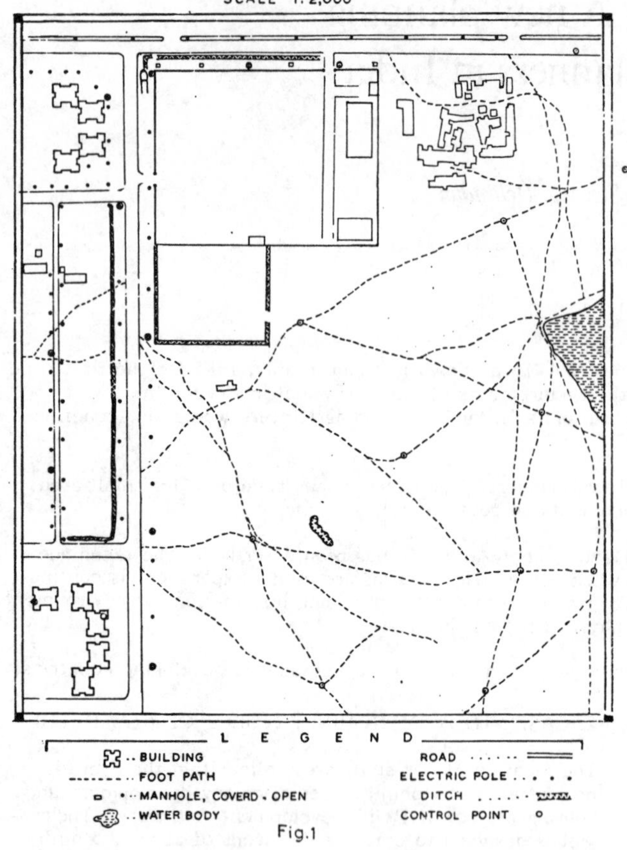

Fig.1

Semi-Govt. Organisation. Pushpak is 2-seater high wing, light weight aircraft which is used for civil pilot training. This aircraft is suitable for hand-held camera as the doors could be easily taken out for ease in aerial photography. Another aircraft named REVATHI, again belonging to the Delhi Flying Club was also commissioned. A hole of about 10 inches was opened in the belly of the aircraft just behind the co-pilots seat. (See Figure-1). A camera mount was fabricated which could hold Hasselblad camera 500 EL/M or Hasselblad MK 70 (metric) camera. There is s provision to give upto $\pm 30^{\circ}$ of correction for crab.

2.3 Cameras

Following table gives the information about the types of cameras used for the study. 35mm cameras were used in hand-held mode while Hasselblad was fixed in the mount (as described). 70mm camera with its film spool is quite heavy for hand-held and it is patently risky to venture it this way inspite of best precautions. Vertically of the h hand-held camera was judged visually.

2.4 Films and filter combinations

Several types of film-filter combinations were tried. All the films were commercially available as such, no storage problems were present. Table-2 provides information about the films etc.

Exposed 70mm black and white films were processed in DK-50 developer using rewind equipment. 35mm colour films were processed in commercially available automatic machines. Colour enlargements of 35mm were obtained on post-card size in the first instance for interpretation.

2.5 Constraints - flying and season

It was the month of July, 1985 when monsoon rains had just set in Delhi. There were, however, some spells of good weather in which aerial flying was possible. Rohini area comes within the funnel of approaching international airlines to Delhi airport. This gave us difficulties of getting flying clearance from control tower. In many cases, the base of clouds became our upper limit to which the aircraft could go although we desired to fly higher upto 8000 feet to generate small scale pictures which will naturally incorporate more area.

A pleasant experience, however, was that SFAP could be done even in rainy season or cloudy days, if the flying height is kept anywhere from 500 feet to 3000 feet. During July, the cloud base, as given by Met. Office, generally stayed at 3000 feet.

Table 1. Types of small format cameras used

Camera make	Format Category	Focal length	F number	Exposure meter	Film advancing	Remarks
ASAHI PENTAX MX	35mm	50mm	1.4	TTL	Motor drive	non-metric SLR
ASAHI PENTAX SPOTMATIC	35mm	50mm	1.4	TTL	Manual	non-metric SLR
NIKON	35mm	50mm	1.4	TTL	Manual	non-metric SLR
NIKON	35mm	28mm	2.8	TTL	Manual	non-metric SLR
HASSELBLAD 500 EL/M	70mm	80mm	2.8	Kodak aerial exposure meter	Motorised	non-metric SLR
HASSELBLAD MK-70	70mm	60mm	5.6	Kodak aerial exposure meter	Motorised	metric

Table 2. Film/Filter/Exposure data

Film make	Film type	Film speed ISO	Negative format	No. of exposures	Filter used	Exposure settings
Kodak colour	Colour negatives	100/21°	24x36mm	36	sky light	f/8 1/500 sec.
Kodak colour	"	200/24	"	"	"	f/8 1/500 sec.
"	"	400/27	"	"	"	f/11 1/500 sec.
"	"	1000/30	"	"	"	f/16 1/500 sec.
Sakura colour	"	100/21	"	"	"	f/11 1/500 sec.
"	"	200/24	"	"	"	f/8 1/500 sec.
Fuji colour	"	200/24	"	"	"	f/5,6,1/500 sec.
Agfa Superpan	B&W Negatives	200/24	"	"	" UV	f/16 1/200 sec.
Agfa-chrome	Colour diapositives	100/21	"	"	"	f/8 1/500 sec.
ILFORD HP5	B&W negatives	400/27	56x56mm	100	Yellow	f/5,6,1/500 sec.
ILFORD FP3	"	200/24	55x55mm	80	"	"

3. INTERPRETATION AND UTILISATION OF DATA

Readable colour enlargements, upto 17 times, were made from 35mm negatives. These enlargements were studied for

- encroachments on government land - several areas were marked
- cluster of dwellings/temporary huts/slums

Slide-1 illustrates the presence of a large number of huts. The area is fast becoming a slum even in a new colony
- unauthorized construction of boundary walls on many plots of land near Mongolpur village. The area south east of village has a large number of boundary walls. These are all unauthorizedly constructed with a view to build temporary hut inside the premises. The modus-operandii follows the pattern: boundary wall, temporary structures, perhaps selling of plot at this stage, permanent structure. Slide-2 refers.
- condition of storm water drains in Rohini area and soil erosion near the drain
- progress of construction in the area, quantity of construction material stocked on the road sides - Slide-3.
- location of borrow pits for construction of roads in Rohini colony
- general land use status in the area.

The Project Manager was so impressed with the extense and intensity of information on a synoptic basis that he desired photography every 3 months or so for proper monitoring of land-based activities in his area. He also realized that in case of unauthorized structures where he foresaw legal action, the aerial photographs provide irrefutable evidence in any court of law.

3.1 Difficulty in navigation was experienced

- Mosaic of post card size enlargements done quickly on a soft board showed the quality of flying to the pilot. This aspect itself kept the pilot on best possible state of his flying performance. It is desirable to have basic navigation aids even if the aircraft is very small of 'microlight'. Navigation remains most important 'bug' in small format aerial photography.

4. EXPERIENCE WITH SFAP FOR PHOTOGRAMMETRIC SURVEY OF A SMALL AREA

Several requests are made by urban authorities, industries, slum upgradation officers, wherein they wish to quickly get surveyed plans of a small area. Generally, it is planimetry in which they are most interested.

We used MK-70 Hasselblad metric camera black and white film to generate a plan on 1:2 ,000. MK-70 negatives were enlarged 4 times and by affine plotting we got 1:2,000 plan of the area. The plan was checked on the ground for planimetry by using EDM instruments. Although, statistically, the sample was not large, we got an accuracy of R.M.S. of 0.2 metres. We propose to carry out these experiments further and ascertain the errors more thoroughly.

5. CONCLUSION

The experiment of SFAP over Rohini has shown promise and the town planning community is very keen to adopt this technology for regular monitoring of selected urban phenomenon, specially technolegal situations on the ground. It is also our opinion that for small areas, we should be able to generate survey plans. However, at present, we advocate the use only for updating the base maps.

SFAP is a technology of generating information which can be produced quickly and very economically.

The author acknowledges the assistance of Shri R.P.Kala in photographic operations.

The author also acknowledges the base provided by the earlier experience of SFAP namely, small format flying in Bangkok by Royal Thai Air Force under the guidance of Prof.M. Juppenlatz of ITC (6) and SFAP in Kenya (Tana River study) by D.H.V. Consulting Engineers of Netherlands (12), and experiments of Mr. Paul Hofstee of ITC at Bandung, Indonesia (5).

REFERENCES

1. Aziz, Lukman. The use of small format aerial photographs in the production of large scale maps (e.g. 1:2,000) of urban areas in regions of moderate relief, with particular reference to the requirements of developing countries. M.Sc. Thesis. Enschede, The Netherlands. May,1982.
2. Clegg,R.H. 1975. A comparison of 9 inch, 70mm and 35mm cameras. Photogrammetric Engineering and Remote Sensing, Vol.41, No.12,pp 1487-1500.
3. Everitt,J.H. and Nixon,P.R. 1985. Using colour aerial photography to detect camphorweed infestations on south Texas range lands. Photogrammetric Engineering and Remote Sensing,Vol.11,No.11, pp.
4. Graham,R.W., Read,R.E. and Kure J. 1985. Small format microlight surveys. ITC Journal 1985-1 pp 14-20.
5. Hofstee, P. 1984. Small format aerial photography for human settlement surveys, the technique - paper yet to be published.
6. Juppenlatz,M. et al. An innovation in aerial photography for development planning. 1985. Case study - Metropolitan Bangkok. Joint exposition with Thai officials.
7. Killmayer,A. and Epp,H. 1983. Use of small format aerial photography for landuse mapping and resource monitoring. ITC Journal 1983-4. pp 285-290.
8. Lord,J. 1980. Oblique and vertical aerial photography with the Hasselblad, aerial photography. Hasselblad publication.
9. Misra,P. 1985. A plea for small format aerial photography in India - a note. Technical paper - National seminar on remote sensing for planning and environmental aspects of urban and rural settlement, Vishakhapatnam, India.
10. Nancy,L.M. and Merle,P.M. 1981. Application of 35mm colour aerial photography to forest land change detection. Eighth biennial workshop on colour photography in plant sciences and related fields. The American Society of Photogrammetry.
11. Small format aerial photography for urban planners. May, 1983. The use of aerial photographs for town planners. Annual conference. Institute of Town Planners, India.
12. Tana River Remote Sensing Study, Kenya by D.H.V. Consulting Engineers, May, 1985.

Symposium on Remote Sensing for Resources Development and Environmental Management / Enschede / August 1986

Notes on the geomorphology of the Borobudur plain (Central Java, Indonesia) in an archaeological and historical context

Jan J.Nossin & Caesar Voute
ITC, Enschede, Netherlands

ABSTRACT

In the second half of the Quaternary, a lake originated at the place of the present Borobudur plain, as a result of blocking of the Progo channel by Merapi fluviovolcanics piling up after slipfaulting against the Menoreh Hills. The lake has existed long enough to leave deposits which are locally 10 metres thick. Faulting in a northeasterly trend, along the present course of the lower K. Tangsi, has initiated uplift of the southeastern block. The existence of the fault is derived from the behaviour of the drainage pattern , notably a ninety-degree turn in the K.Tangsi, which then follows a clear lineament implied to be the faultline. This lineament exercises strong influence on the drainage pattern of the area.

The uplift of the southeastern block has initiated the draining of the lake and led to steep incision of the Progo and Sileng rivers. This incision is marked by at least two, but most probably three main terrace phases.

Considerable time therefore has elapsed since the onset of the final draining of the lake. It is therefore concluded that the lake had disappeared long before the Borobudur temple was constructed. The slipfaulting in the Merapi deposits which may have initiated the whole train of events, must be much older of age than the 1006 AD catastrophic eruption of the Merapi.

1. THE BOROBUDUR PLAIN

The Borobudur temple stands on a faultblock of volcanic rocks including the Gunung Gandul-Gunung Sipadang hills (Benschop Koolhoven, 1929). This outlier of intrusive 'younger andesites' (Rahardjo c.s., 1977) overlooks a plain toward the S, SE and E, at an average height of 240 - 250 metres a.s.l. This plain, into which the K.Sileng sinks its incised meanders, is made up of alluvial deposits. It is bordered to the S by the andesites of the Menoreh Hills reaching over 900 metres a.s.l. To the E, it is bordered by the young fluviovolcanics of the G. Merapi.

The plain continues north-and northwestward, where it is not made of alluvial deposits, but of young deposits derived from the G.Sumbing. The plain here lies somewhat higher, around 265 - 270 metres a.s.l.

Nine shallow water borings near the Borobudur temple, in the plain, have shown a thickness of over 10 metres of sandy/clayey lake deposits (Purbohadiwidjojo and Sukardi , 1966; Voûte, 1969).

2. OBJECTIVE, METHODOLOGY

The objective of this paper is, to analyze the geomorphic setting and trace the evolution of the (lake)plain, in order to verify certain archaeologically significant hypotheses on the presence of an ancient lake and on the impact of major volcanic events.

Air photo interpretation was carried out on three sets of photos:

- colour-IR photos of 1972, at an approximate scale of 1:10.000;
- colour-IR photos of 1982, at an approximate scale of 1:30.000;
- black/white panchromatic photos of 1982 ,at approximately 1:50.000.

Additional data were obtained during field observations in 1985, and also from the geological map of the Yogyakarta Quadrangle (Rahardjo c.s., 1977), and from published papers.
The interpretations of the air photos are presented in the form of sketchmaps.

3. THE ARCHAEOLOGICAL AND HISTORICAL CONTEXT

Borobudur has engaged the attention of scholars for over a century, resulting in more than 500 learned studies in several languages. Moreover, preservation and restoration work has been carried out on the monument in successive phases, again extending over a period of more than 100 years, culminating in the world-wide known Unesco-supported campaign of the years 1968-1983 (Soekmono, 1972; Voûte, 1983). Especially during this last phase much additional research covering a wide range of subjects was carried out.
But still a number of mysteries exist, surrounding this ancient temple, as reflected even in the title of one of the books on Borobudur (Bernet Kempers, 1976).

One of these concerns the often debated existence of a former lake around the temple, involving two utterly different and conflicting assumptions, and another the probable reason for the apparently sudden eclipse, in the 10th-11th century, of the old Mataram Kingdom of Central Java.

3.1. The Borobudur Lake

The discussion about the existence of an ancient lake near and around Borobudur was started in 1931 by the famous painter W.O.J.Nieuwenkamp, who had spent several years of intensive studies on Hindu-Buddhist architecture, in particular on the Island of Bali, and who also visited Borobudur several times. As an artist he visualized Borobudur, sitting on top of its hill, as a "lotus flower drifting on a lake, on which the new-born Buddha was seated". Support for this idea is found in the conventional lay-out of Buddhist sanctuaries, very often surrounded by a moat symbolizing the cosmic ocean from which the earth was created. Typical examples of such moats exist around the various temples of the Angkor Vat complex in Cambodia, whereby it should be remembered that the Buddhist Khmer civilization of Cambodia was much inspired by the older Javanese Hindu-Buddhist traditions. Modern examples are the ponds with their lotus flowers, so often found in front of the Balinese temples.

Several geologists took part in the debate which was carried around 1933 in the monthly journal "Nederlandsch-Indie Oud en Nieuw ", and the daily paper "Algemeen Handelsblad", Amsterdam, in particular the two friends Prof. Dr. L.M.R.Rutten and Dr W.Nieuwenkamp, the latter son of the painter. Furthermore, persuaded by W.O.J.Nieuwenkamp , Cr.Ch.E.A.Harloff and Dr. A.J.Pannekoek carried out a serious geological and geomorphological study, first reported in 1937 and finally published in 1940 (Harloff and Pannekoek, 1940). The subject was studied again in 1966 by two Indonesian hydrogeologists (Purbohadiwidjojo and Sukardi, 1966). These studies confirmed that a lake could have existed here in pre-historic times, filled up by at least 10 metres of sandy-clayey alluvium.

It is true, the hill on which Borobudur stands, consists largely of man-made fill requiring a borrow-area of fair dimensions. This could have created a new, although artificial small lake at the time of construction of Borobudur (Soekmono, 1969, 1976). However, the study of large-scale air photos of the Borobudur site, undertaken in 1968, failed to reveal traces of such a major borrow-area, which once could have formed a large pond or small lake corresponding to the moats found in Angkor Vat or the ponds in front of the Balinese temples (Voûte, 1969, 1981).
Moreover, soil samples taken from archaeological trial trenches on Borobudur hill and on the plain immediately south of it, were analyzed for their content of pollen and spores, in order to obtain information on the vegetation of the area surrounding Borobudur at the time of its construction and immediately thereafter. No indications were found of a vegetation characteristic of an aquatic environment (lake, pond or marsh); on the contrary, Borobudur at the time of its construction appears to have been surrounded by agricultural land and palm trees as is still the case today (Thanikaimoni, 1977, Dumarçay, 1977).

3.2 The eclipse of the Mataram Kingdom

The second theory refers to an entirely different phenomenon. Formulated in its final form by Dr. Ir. R.W. van Bemmelen in 1949 after a first publication in 1941, it is based originally on a translation and interpretation of ancient Hindu-Javanese sanskrit texts by the archaeologist van Hinloopen Labberton (Van Bemmelen, 1941,1949). In a study published in 1921 the apparent destruction of the old Mataram State in the Sjaaka year 928 (equivalent to AD 1006) was assigned to a "Maha Pralaya" (natural calamity of terrible proportions), interpreted as a major volcanic eruption accompanied by intense rainfall and very extensive flooding, during which King Dharmawangça and many people of high standing and noble birth died. The King's son-in-law, 16 years old, escaped, and under the name of King Erlangga ("He, able to escape the floods") , founded a new kingdom in east Java, which under the name Daha (Kediri) continued the Hindu-Javanese traditions in a modified form. This found its apogee in the 14th century in the empire of Modjopahit (also in East Java). Meanwhile, during two centuries the ancient sources kept an absolute silence on once flourishing and densely populated Central Java (Van Hinloopen Labberton, 1921).

Van Bemmelen, who had formulated new hypotheses about geology, including the role of gravity-induced large-scale sliding movements in relatively recent times, interpreted this calamity as a cataclysmic outburst in 1006 AD, of Merapi Volcano, depopulating, desorganizing and destroying the prosperous Hindu state of Central Java , and converting its fertile fields into grim deserts of ashes and mudflows (lahars). This outburst would have been accompanied by the collapse of much of the western slopes of Merapi, which slided down along slipfaults, pushing up the Gunung Gendol hills south of Muntilan village, blocking the Kali Progo valley, and through this damming up, flooding the Borobudur area by a large lake (van Bemmelen, 1941, 1949).

The volcanic cataclysmic theory has remained very popular with the Indonesian archaeologists as a rational explanation for the shift of political power and architectural/cultural activities from Central to East Java (Boechari,1976, 1976/82, 1977/82; Moendardjito, 1978/82; Soekmono, 1976). They were convinced to have found further confirmation when the various archaeological trial trenches made in the surroundings of Borobudur from 1973 onward showed the presence in several places of an ash layer from a few millimetres to a few centimetres, and locally also a 30 cm. thick layer of sand, below which there occurred to a depth of 1 metre, occupational soil with many potsherds of Indonesian and Chinese manufacture (Boechari, loc.cit; Mandardjito, loc.cit ; Voûte, 1981).

However, this sand layer could very well be the product of local downwash and sheetfloods from the Menoreh Hills somewhat farther South. In any case the accumulation of soil and debris of a magnitude as found here, need not put an end to human occupation, as is quite evident from the fact that the layers with potsherds and other artefacts, sign of regular religious festivals over a period of about two centuries, are rather thick.

Some other evidence also pleads against the deposition of large masses of materials (pyroclastics and lahar deposits) during the ill-fated year of 1006 AD of the old chronicles and descriptions. In the districts immediately surrounding Borobudur no less than 207 archaeological sites have been inventorised, including the remains of 43 temples (Voute, 1975/82) , none of which are deeply buried. At Mendut temple, east of Borobudur, several feet of ashes and lahar deposits had to be cleared away to reach the original surface of the temple square,

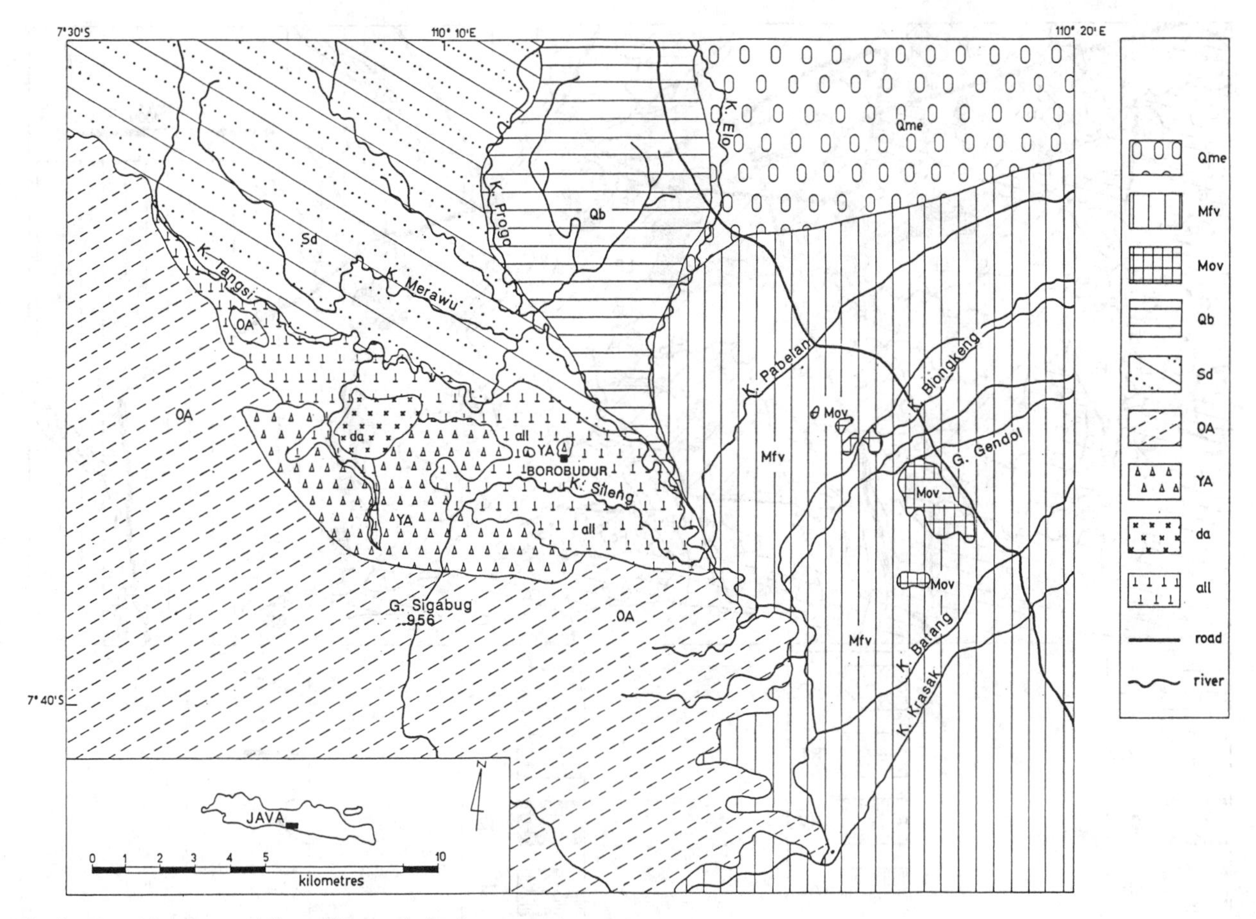

Fig.1. Location map with outline of drainage system and geology

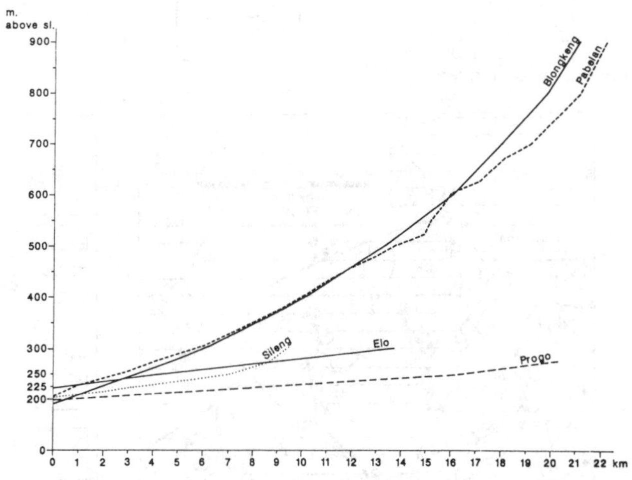

Fig.2. Long profiles of rivers from the area

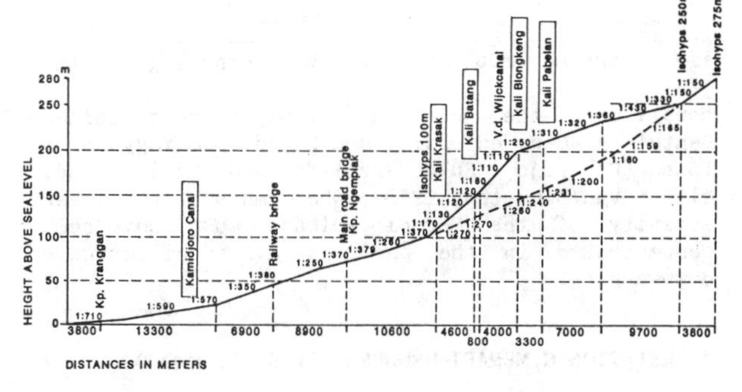

Fig.3. Long profile of the Progo river, after Schmidt(1934)

while in some places the topsoil here even reaches up to 3 m. (Brandes, 1902, 1903 ; van Hinloopen Labberton, 1921).
Moreover, Mendut already lies on the lower Merapi slope, where easily some thickness of sediment accumulation, in the absence of adequate cleaning-up and maintainance, can occur through " normal" volcanic activity and lahar outwash. An accumulation of considerable amounts of dust, rubbish and organic matter, and the falling down of parts of the mortar-free masonry at the Borobudur site during eight to nine centuries of neglect, accellerated by the gradual invasion of shrubs and trees, need not surprise either and does not require exceptional disasters for an explanation.

Many authors have mentioned this partial cover of Borobudur, starting with Sir Thomas Stamford Raffles in his "The History of Java "(London, 1817), the Baroness U.S. Baud-van Braam in her diary describing her travelling through Java in 1834 (published in the original French version in "De Indische Gids, Maart 1939, pp. 198-224) and S. van Kinsbergen in 1873 (Notulen van de Algemeene en Bestuursvergaderingen van het Bataviaansch Genootschap voor Kunsten en Wetenschappen, XI, 1873, pp.71-74).

This is quite in contrast with the situation east of the town of Yogyakarta, where during the last 25 years several important archaeological finds, inclusing the beautiful Sambisari Temple, were found underneath 2-8 metres of lahar deposits, in an area where there are almost no surface finds of antiquities of the Hindu-Javanese period.This area

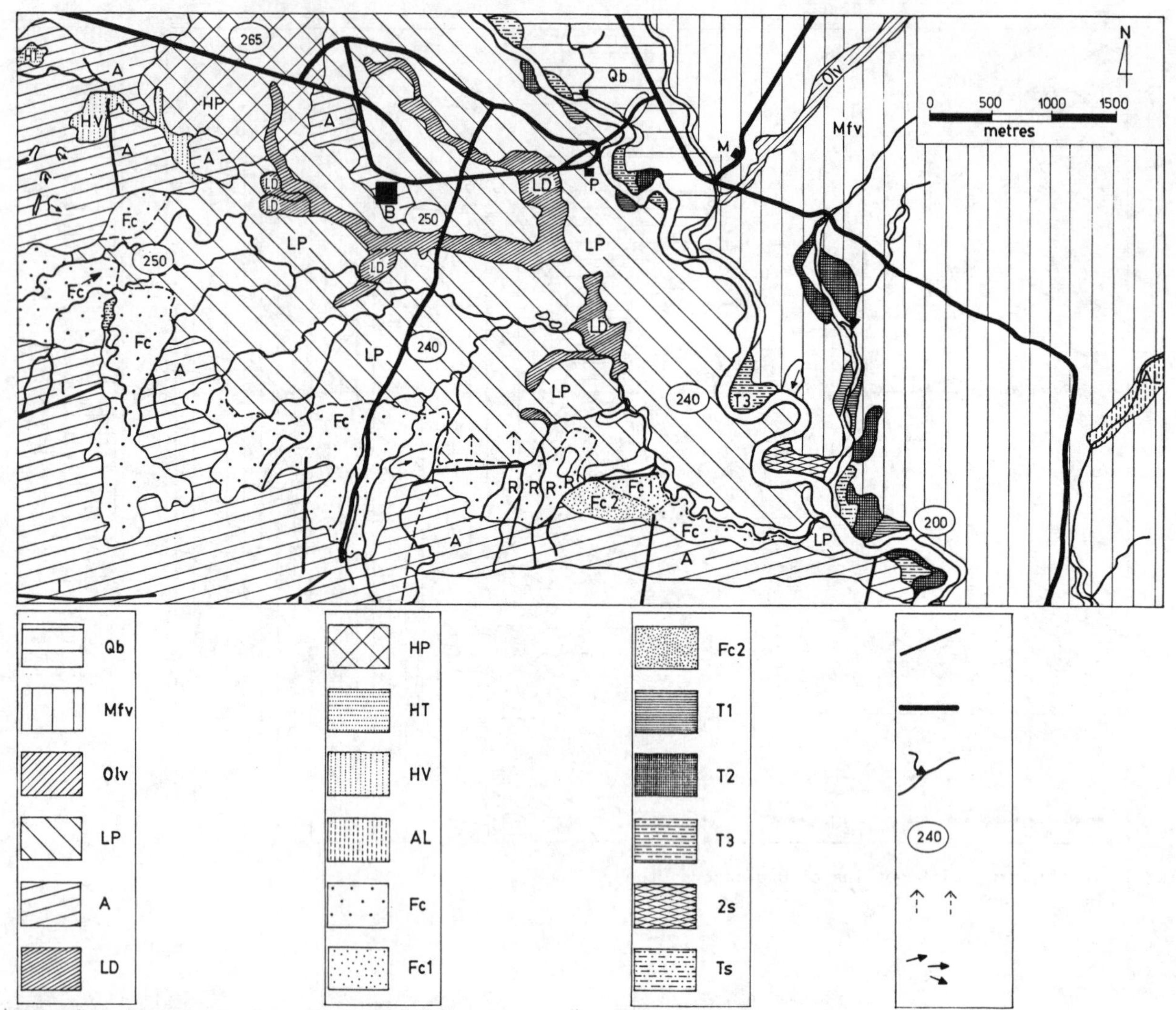

Fig.4. Sketch map showing the morphology of the Borobudur plain area

constitutes the northern portion of the so-called Bantul graben, extending southward from Yogyakarta town (Rahardjo, Sukandarrumidi and Rosidi, loc. cit.; Van Bemmelen, 1949), where much of the lahar activity of the Merapi volcano must have been concentrated in the past in a zone of tectonic downwarping.

4. RELATION G.MERAPI-MENOREH HILLS- K. PROGO.

The fluviovolcanics of the Merapi meet with the andesite rocks of the Menoreh hills (also referred to as the West Progo Mountains, van Bemmelen 1949) just south of the confluence of the K. Sileng and the K. Progo , the main stream of the area (fig. 1).

The Merapi volcano produces material at a high rate; van Bemmelen (1949) estimates it at about 6.4 million m3/year averaged over the last 120 years, and its notorious lahars have an excessive carrying capacity.
In fig. 2, some long profiles are shown of river sections in the area of study; the steepness of the K.Pabelan and the K.Blongkeng, both deriving from the G. Merapi, with respect to the other rivers ,is illustrative.

The growth of the Merapi mountain body has gone accompanied by slipfaulting towards (a.o.) the southwest (van Bemmelen, 1949 ; Bahar, 1985), causing crumbling and folding of the mountainfront

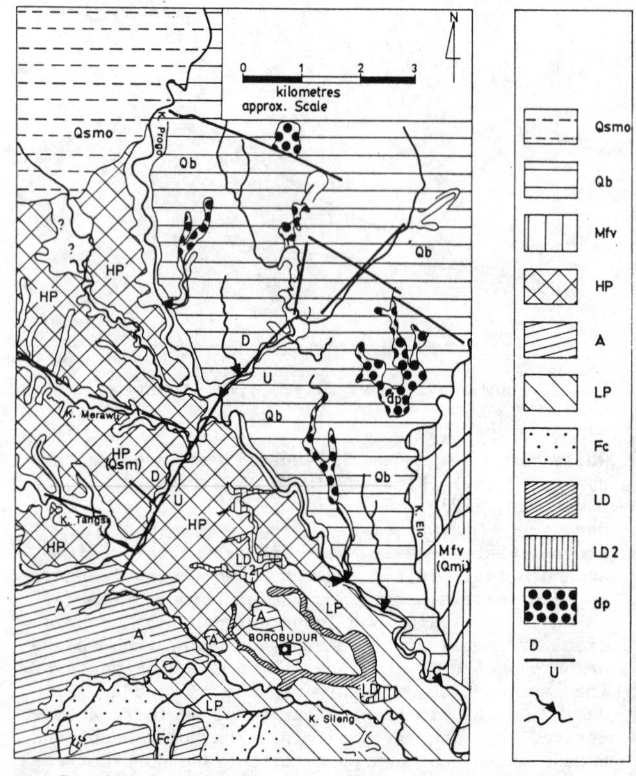

Fig.5. Sketch map showing the setting north and northwest of the Borobudur plain

of the andesite hills possibly accompanied by local upwarping. A blocking of the main drainageway, the K.Progo, has been the result.

This is borne out by a survey of the long profile of the K.Progo by Schmidt (1934), reproduced in fig. 3 , showing a distinct bulge upstream of the debouchure of K.Krasak, by as much as 50 metres. Van Bemmelen had little doubt that this phenomenon has been accountable for the origin of a 'Borobudur Lake' , ponding the water upstream of the blockade.

The lake may have had various expansions and retractions as a result of posterior changes in the ponded runoff. At times, its extent may have been considerable. From the present morphology it can be judged that these expansions and retractions have mainly caused shifts in the north bank of the lake. The area south of the present Borobudur temple has been covered for the longest period- as is also borne out by the thickness of the lake deposits.

5. PRESENT DRAINAGE CHARACTERISTICS

5.1 K.Sileng

The plain stretching south of the Borobudur is drained by the K.Sileng, which flows out of the Menoreh Hills on a northerly course. It soon bends eastward, receiving tributaries only from the south, from the Hills. This plain is marked 'LP' (for Lake Plain) in fig. 4 and 5; its level is 240 metres a.s.l in its eastern and central parts.

East of the centre, the K.Sileng sinks into the lake plain in a course of incised meanders, while assuming a southeasterly direction that brings it back to the foot of the andesite hills.
Finally it breaks through the expanse of the plain in a gorge of some 30 metres depth and debouches into the Progo river.

5.2 The LD drainageways

Affluents from the north into the K.Sileng are absent (with one exception still deriving from the northern andesite hills). Instead, some wide, flatfloored and very shallow channel patterns are discernable, barely below the LP level. These are designated LD (for Lake Drainage) channels in fig. 4, and are considered to represent a system of drainageways from a late phase of the dwindling lake.

If any well-defined main channels have ever existed in these vales at all, they are not distinguisable anymore because all the water is presently diverted into the irrigation systems of this intensive rice cultivation area.

5.3 The K.Progo

The Kali Progo displays incised meanders to a depth of about 30 metres below the LP level; along its course through the area, at least two terrace levels can be distinguished, marking corresponding phases of rest in the incision.

The incision of the meanders of the Sileng and Progo rivers points to an uplift of the area after the emptying of the lake, or perhaps starting in a late phase of its existence.

Corresponding terrace levels are observed in the lower reaches of the K.Pabelan, one of the lahar rivers from the Merapi. However, its very nature of lahar river renders a correlation of its terraces with those of the Progo river, somewhat hazardous. The Pabelan debouches with a steepened gradient into the Progo (figs. 1 and 4). Other Merapi-derived rivers show the same phenomenon of steep, even hanging, debouchure. The K. Batang shows a downward jump of over 10 metres at its debouchure into the Progo.
This should not be ascribed to excessive apport of material from the Merapi, which would cause blocking and ponding rather than a hanging debouchure, but rather to uplift on a local scale in this sector of the Progo river.

5.4 Kali Tangsi

The Tangsi river, entering the area of study from the northwest (figs 1 and 6), displays a curious phenomenon. **About 3 km northwest of the Borobudur, it takes a 90-degree bend to the northeast,** and continues on this course until it debouches into the Progo river. Its channel in this tract is some 15-20 metres below the level of the plain (here at 270 m. a.s.l and designated 'HP' (for Higher Plain) in figs 4 and 5).
On the way, it receives the K. Merawu, which also comes from the northwest and debouches at a 90-degree angle into the K. Tangsi.

This trend taken by the K. Tangsi is also visible on the opposite bank of the Progo river, where an affluent from the northeast takes roughly the same trend, only in the opposite sense.
As can be seen in figs. 1 and 5 , this phenomenon leads to the concentration of runoff from the northwest and north, at one single point into the Progo river.

5.5. Terraces of the Sileng and Progo rivers

Clearly discernable terraces are present along both the Sileng and the Progo rivers. They are separable into **two main levels,** possibly a third one can be distinguished.

These two or three main levels point to as many phases of rest in the incision of the rivers. This incision is, as will be discussed later, ascribed to uplift on a regional or local scale. The Kali Sileng breaks through the LP level in an antecedent gorge which shows that the river had its present course with respect to the lake plain, prior to the uplift-cum-incision.
A notable feature is also, that the LD drainageways drain into the Kali Sileng and not into the Kali Progo.

6. STRUCTURAL IMPLICATIONS

The lineaments along the K.Tangsi point to a structural (=fault(line)) control of the tract of the river that runs to the northeast. Its counter part from the northeast, on the other bank of the Progo, is likewise structurally controlled.

The relative configuration of the HP and LP plains points to a relative uplift on the southeast side of the Tangsi lineament. This is corroborated by the straight and gorge-like incision of the K. Progo downstream of the Tangsi debouchure, in contrast to its meandering course in a wider (though also incised) valley upstream of this point.

It is worthy of note that there is no elevation between the lake plain and the Progo gorge. Therefore, this incision can only have started after the beginning of the emptying of the lake. The Progo river must have been able to (re-) establish its course at its present position, before the formation of this (antecedent) gorge.

Further downstream, towards the debouchure of the K. Elo, the Progo river shows incised meanders, but still in a gorge-like setting. The uplift has affected at least the sector of the K.Progo from the K.Tangsi to the K.Sileng debouchures. As pointed out earlier, the 'bulge' noted by Schmidt (1934) starts even further downstream, at the Krasak debouchure.

From this discussion it will be clear that we consider the uplift leading to the incision into the lake plain, different from and younger than the uplift ascribed to slipfaulting of the Merapi deposits against the Menoreh Hills.

As stated in para 5, also the Sileng had established its course prior to the uplift. Its incised nature is clearest in the eastern part of the plain and decreases in the upstream direction.

This uplift may also be responsible for the southward bends in the lower reaches of the Blongkeng, Pabelan and Elo rivers (fig 1).
These bends actually increase the length of waterway to the local base level (K.Progo), to which the channels' debouchures are not adjusted either. This anomaly is likewise explained by this uplift.

7. PHASING

Based on the aforegoing, the following morphogenetic phasing in relation to the 'Borobudur Lake' is arrived at:

1. **Merapi fluviovolcanics reach the andesites of the Menoreh Hills.** As the deposit grows in extent and thickness, this leads at one stage to local folding-crumpling of the front of the mountainfoot of the Menoreh Hills, which were already buried under the Merapi fluviovolcanics. This may be considered the result of slipfaulting in the Merapi body (van Bemmelen, 1949).
2. **Blocking of the drainage creates a lake at the place of the present Borobudur plain,** limited to the north, ultimately, by the Sumbing deposits of that time.
3. The lake has had **various phases of expansion and retraction,** and may at times have been quite extensive.
4. The lake must have existed- at least south of the Borobudur- for a **considerable period**: borings near the Borobudur show more than 10 metres of lake deposits (Purbohadiwidjojo and Sukardi, 1966).
5. **Local faulting trending northeast** followed at the place of the p esent sector of the K.Tangsi running northeastward. **Relative uplift in the south-western block** caused the straight incision of the Kali Progo. This was preceded by a reduction of the lake whereby K.Progo - thus far flowing into and out of the lake- could reestablish its course.
6. The late-lake phase, really a stage in its draining, gives rise to the **wide drainageways** shown as LD in figs 4. In places, two different levels can be discerned, slightly differing in height. The lowest is then indexed -1, the other, higher one, -2. These drainageways are very characteristic for the present Borobudur plain, and their pattern resembles that of flatfloored gullies in a lowland coastal plain as e.g. found in northern West-Java.
7. **The uplift noted in 5) took place in three major phases,** causing incised meanders of K.Sileng and K.Progo upstream of their confluence: at least two major terrace levels can be distinguished, marking places of relative rest in the uplift. The uplift also caused the southward bends in the courses of the K. Pabelan and K. Blongkeng, from southwest to due south.
8. The events as outlined above have led to the present morphological appearance of the area.

8. DISCUSSION

Van Bemmelen (1949) postulated that the Gendol Hills (fig.1) must have slid down from the Merapi crater area in a catastrophic eruption which he places in 1006 AD. He considered that the slipfaulting in the Merapi body might be of the same age.

From what was said in para 7, it will be evident that we consider the whole train of events leading to the origin and later dwindling of the 'Borobudur Lake' as much older. The whole sequence must, however, be of Quaternary, and probably late-Quaternary age, as it started only when the Merapi deposits had extended as far as the foot of the Menoreh Hills.

Assuming a linear relationship between the height of the Merapi and the areal spread of its deposits in time (which may be debatable), **the Merapi would have had a height of about 2100 metres a.s.l at the time when its deposits reached the foot of the Menoreh Hills** (i.e. the position of the Gendol Hills), as against its present height of 2911 metres.
If a further assumption is made for parallel growth, in time, of all sections of cone, midslope and footslope of the volcano, then the total volume of the Merapi body, at that time, would have been **roughly half its present volume.**

Accepting van Bemmelen's postulation that the further growth in mass and weight has ultimately led to slipfaulting and to folding of the front of the Menoreh Hills (now buried) with associated uplift or upwarping, then this event has to be placed well into the second half of the lifespan of the Merapi, i.e. the second half of the Quaternary.

The question of whether the Gendol hills have slid from the summit area in the catastrophe of 1006 AD, or much before that time, does not take a central position in the context of the present paper. However, events as reconstructed here point to a much older date for this event also.

The draining of the lake, in its turn, was initiated and/or followed by a phase of uplift with incision and terrace formation. **In other words: a considerable length of time has elapsed after the draining of the lake, before the present morphology came into shape.**

These conclusions do not preclude the possibility that the political and socio-economic fabric of the ancient Central Javanese Mataram kingdom was weakened sufficiently by volcanic activity, com-

Fig.6. The Borobudur

bined with violent stormfloods, to permit a competing alternative centre of power to develop successfully in Eastern Java, as suggested by the Old-Javanese Erlangga inscription (Van Hinloopen Labberton, loc. cit).

Summarizing: Borobudur temple has never stood by a lakeside; the lake had disappared long before the construction of the temple was started.

REFERENCES

Bahar, Iwan, 1984 , Contribution à la Connaissance du Volcanisme Indonésien: Le Merapi (Centre-Java); cadre structural, petrologie-géochimie et implications volcanologiques; Thèse Doctorat Université de Montpellier, 2 Fevr. 1984.

van Bemmelen, R.W., 1941 , Bull. East Indian Volcanological Soc. for the year 1941 (Nrs 95-98) II, Special Part , No 41 pp. 70-71; Bandung (published 1949 due to World War II).

van Bemmelen, R.W., 1949 , The Geology of Indonesia. 3 Vols; in particular Vol I a, p. 560-562 ;The Hague, Martinus Nijhoff.

Benschop Koolhoven, W.C., 1929 , Geology of Gandoel Hill near Borobudur, Central Java; Fourth Pac. Science Congr., Exc. D 1.

Bernet Kempers, A.J., 1976 , Ageless Borobudur-Buddhist Mystery in Stone-Decay and Restauration - Mendut and Pawon - Folklife in Ancient Java ; Servire, Wassenaar,

Boechari, 1976, Some consideration of the problem of the shift of Mataram's center of government from Central to East Java in the 10th Century A.D.; Bull. Research Centre Archaeol., No 10, Jakarta.

Boechari, 1976/82, Preliminary Report on some archaeological finds around the Borobudur Temple; reported 1976, published 1982 in Pelita Borobudur, Seri CC No 5, Departemen Pendidikan dan Kebudyaan, Jakarta, p. 90-95.

Boechari, 1977/82, Further remarks on preliminary report on some archeological finds around the Borobudur Temple; reported 1977, published 1982 in Pelita Borobudur, Seri CC No 6 , Departemen Pendidikan dan Kebudayaan, Jakarta, pp. 43-47.

Brandes, J., 1902, in : Notulen Directievergadering Bataviaansch Genootsch. van Kunsten en Wetenschappen 17 nov. 1902 , Bijlage CLIII.

Brandes, J., 1903, ibid., 4 Aug., 1903, IV a, p. 76.

Dumerçay, J., 1977, Histoire Architecturale du Borobudur; Publ Ecole française d'Extrême Orient, Mem. Archaeologiques, XII, Paris.

Harloff, C.E.A. and Pannekoek, A.J., 1940, De Omgeving van den Boroboedoer, Tijdsch. Kon.Ned. Aardr. Gen. Dl, LVII, No 1 p. 13-23.

van Hinloopen Labberton, D., 1921 Oud-Javaansche gegevens over de vulkanologie van Java ; Natuurk. Tijdschr. Ned. Indie, Vol 81, p. 124-158.

Moendardjito, 1978/82, Preliminary Report on pottery found in the Borobudur site; reported 1978, published 1982 in Pelita Borobudur, Departemen Pendidikan dan Kebudyaan, Jakarta, pp 46-47.

Purbohadiwidjojo, M.M., and Sukardi , Tentang ada atau tidak adanja suatu danau lama didekat Borobudur (Laporan geologi sedimentary); Direktorat Geologi- Bagian Geoteknik Hydrologi, No 1514, unpubl. rep. dat 12 Jan 1966.

Rahardjo, Wartono, Sukandarrumidi and H.M.D.Rosidi, 1977, Geologic Map of the Yogyakarta Quadrangle (1408-2 and 1407-7), 1: 100.000, with explanatory text; Geological Survey of Indonesia, Bandung.

Schmidt, K.G, 1934, Die Schuttstrome am Merapi auf Java nach dem Ausbruch von 1930; De Ingenieur in Ned. Indie, 1933 Nos 7, 8, 9, 69 pags.

Soekmono, 1969, New Light on Some Borobudur Problems; Bull. Arch. Inst. Indonesia, No 5, Jakarta.

Soekmono, 1972, Riwajat Usaha Penjelamatan Tjandi Borobudur (sampai achir 1971); Pelita Borobudr, Laporan Kegiatan Proyek Pemugaran Candi Borobudur, Seri A No 1, Jakarta.

Soekmono, 1976, Candi Borobudur, A Monument of Mankind; Van Gorcum, Assen/ Amsterdam, Unesco Press Paris.

Thanikaimoni, G., l'Analyse pollinique des debris archaeologiques du Borobudur; Annexe I in Dumercay (1977).

Voûte C, 1969, Geological and Hydrological Problems involved in the Preservation of the Monument of Borobudur; Paris, UNESCO, Doc.FR/TA/Cons, Serial No 1241, MSS.RD/CLT.

Voûte,C., 1975/82, Landscaping of the surrounding area of Borobudur; Reported 1975, publ 1982, Pelita Borobudur, Seri CC No 4, Departemen Pendidikan dan Kebudyaan, Jakarta, p. 125-128.

Voûte, C., 1981, Borobudur, the Restauration Works; Proc. Int.Symp. Candi Borobudur Sept 25-27 1980, Kyoto Japan, Engl. ed., Kyoto News Enterprise, Tokyo, p. 8 -21.

Voûte, C., 1983 , The Preservation of Borobudur; Explorers Journal, Vol. 61 No 4 p.174-185.

LEGEND for all figures

symbol	explanation
Fc	Colluvial footslope
Fcl, Fc2	Locally steep upper zone [Fcl] and real footslope [Fc2] [arrow shows direction of colluviation]
LP	Lake Plain principal surface; elevatio 245 (E) to 260(W) m.a.s.l
HP	Higher Plain surface north of Borobudur, elevation > 265 m.a.s.l
LD (1,2)	Lake Drainageways representing late lake stage; sometimes two levels
T 1,2,3,..,s	Terraces, indexed from riverbed upward; inedexed s if steps not discern-able
===>	After incision, LD hanging over principal valleys
--------- f 1	Faultline, or lineament
U --------- D	Lineament implied as fault(line) D: down U: up
R	Ravining
d	depression associated with ravining
OLV	NW of Progo : old lahar valleys
AL	active lahar valleys
dp	depressed areas NE of Progo in Merapi footslope
HT	High terraces in andesite hills
HV	High valleyfloors in andesite hills
Mfv	Merapi fluviovolcanics; footslope
Qb	Volcanic breccia ; undulating plain
Sd	Sumbing deposits ; footslope
all	alluvial deposits ; plain
YA	Younger Andesites ,intrusive; steep, densely dissected hills (Menoreh mts)
OA	Older Andesite formation; rugged hillsl, steep slopes, characteris-tic sheetjointing (Menoreh mts)
A	Andesite (hills) , non-differentiated
da	Dacites, intrusive
Mov	Merapi older volcanincs (Gendol hills, folded cone material
Qsmo	Old deposits of Guning Sumbing
Qme	Volcanic deposits of Gunung Merbabu

symbol	explanation
	lineament
	road
	hanging debouchure
240	spot height
	directions of mass movements

Symposium on Remote Sensing for Resources Development and Environmental Management / Enschede / August 1986

Photointerpretation and orthophotograph at the study of monuments in urban areas

E.Patmios
Aristotle University of Thessaloniki, Greece

ABSTRACT: Photogrammetry and Remote Sensing essentially contribute to all the faces of urban planning, because of the wealth of the metric and qualitative information which offer. Specially sensitive aspects many times come in the general considerations for civil areas. From this point of view, the preservation of the cultural inheritance in general, and especially in civil areas, they are of particular interest. Aspects as the above mentioned, are examined in this paper for the civil area of Thessaloniki, which interests a lot from the point of monuments and historic centers. The study is mainly based on photointerpretation and orthophotograph for relatively small scales and it is the preliminar stage for further efforts of photogrammetric study of monuments of Thessaloniki.

1 INTRODUCTION

The study of monuments and historic centers is satisfactorily faced, among others, by photointerpretative and photogrammetric methods.

The study is done from various aspects and it may lead to considerations of a systematic facing at which, beyond others, local particularities (amount and kind of monuments, existing substructure etc.) are seriously taken in mind. (18).

The photointerpretative study is usually a preliminar stage useful for the conception of the general physiognomy of the monument and historic center (5, 11, 13, 17). In cases of wide areas, which include an amount of monuments, the photointerpretative work may demand the creation of a suitable photomosaic (7). Photointerpretation during the time attendance (4), exploitation of relatively small scale airphotographs for graphic restitution of limited accuracy (8) and simultaneous multiple photogrammetric exploitation (graphic, numeric, photographic documents) (12) are aspects of particular interest according to the case. The preparation of a kind of an atlas of monuments and historic centers, including stereoscopic pairs of airphotographs of different times, simple photogrammetric restitution preferably graphic and relevant explanations, presents interest (6). Studies on special forms as castles, towers etc. (3), use of specific techniques as DTMs (16, 17) as well various terrestrial takings and restitutions (phototheodolites, stereocameras, graphic-analytical-photographic restitution, takings of totality, external faces, inside takings etc.) essentially offer to create archives of monuments and historic centers. (9, 10, 11, 14).

The study of monuments and historic centers in urban areas in connection with their wide urban surrounding (1, 2) is a characteristic case. In this way the monument is not considered isolated from its surrounding. Several urban considerations, problems of excavation are simultaneously served and generally efforts of preservation and surrounding protection can be effectively served.

The study of uses (Land Uses - Space Uses) is very important at such works. A combination of photointerpretative study of orthophotograph and of field control was examined (1, 2) and found effective.

To continue the above experiences, we attempt (Patmios E., Lasaridou M., Halkias Th., Farakou A.) a corresponding study on the city of Thessaloniki, which includes a great amount of monuments and historic centers of different kind. In this paper some subjects from the first stage as photointerpretative study, use of orthophotograph, location of monuments, are presented.

2 PHOTOINTERPRETATIVE STUDY, ORTHOPHOTOGRAPH, LOCATION OF MONUMENTS

Airphotographs in scale 1:20000, taken in 1978, on which stereoscopic study was done, were used for the study. Orthophotograph in scale 1:10000 was produced from the same airphotographs. Bibliographic study about the monuments of the city (19) and relevant local information were done. Based on the above, the following monuments were located at the orthophotograph (Fig. 1). 1 Byzantine Fortified Walls, 2 Trigonion or Alysos Tower, 3 White Tower, 4 Heptapyrgion, 5 Roman Agora (Forum), 6 Rotonda, 7 Galerius Arch of triumph (Kamara), 8 Galerius Palace, 9 Octagon Building, 10 Roman Hippodromos, 11 Saint Demetrius church, 12 Acheiropoietos church, 13 Hagia Sophia church, 14 St. John's Baptistry (Nymphaeum), 15 Panagia Chalkeon church, 16 Holy Apostles church, 17 St. Catherine church, 18 Prophet Elias church, 19 St. Nicolaos Orphanos church, 20 Vlattades Monastery, 21 Church of the Transfiguration of the Saviour, 22 St. Panteleimon church, 23 Taxiarchae church, 24 Byzantine Bath House, 25 Ypapanti church, 26 Panagouda church, 27 St. Haralambos church, 28 St. Athanasios church, 29 St. Minas church, 30 St. Theodora church (Monastery), 31 St. Ypatius church (Panagia Dexia), 32 St. Constantine church, 33 Nea Panagia church, 34 St. Laodigitria church, 35 St. Georgios church, 36 St. Gregorios Palamas church (Metropolis), 37 Greek Consulate during the Turkish Occupation,

Fig. 1 Orthophotography in scale 1:10000 of area of Thessaloniki, Greece and location of monuments

38 Dikitirio (Konàki), 39 "St. Demetrius" Hospital (Dimotiko), 40 Central (Kentriko) Hospital, 41 Frankish church, 42 Bezesteni, 43 Sydrivani.

3 CONCLUSIONS

The first stage of the study presented in paper was done based on airphotographs and orthophotographs of relatively small scale. In this way aspects of more general supervision of monuments in their wide urban area are served, significant for general urban considerations. The acquisition of more detailed information about the partly elements of the monuments is not of course anticipated and pursued at this stage. Besides for this reason we did not attempt any drawing but simple location was enough.

The stereoscopic recognition of monuments, as it is possible from the scale of airphotographs, was satisfactory while their appearance at orthophotograph in some cases includes obscurities because of the limitations of the scale, of tonal alternations etc.

The city of Thessaloniki includes a great amount of most remarkable monuments of different times, different form, size, grade of ruinning. The presentation of all was studied by stereoscope and for a great number of them the location is being done at the orthophotograph. The general study and the preservation of monuments of Thessaloniki are faced with great interest, especially after the earthquake in 1978. In this direction we make our effort which is continued with directions of more detailed study based on the exploitation of photographic material with possibilities of greater detail and on other photogrammetric methods.

REFERENCES

1 Patmios E., Preparation of a space use map with orthophoto. Photointerpretation and field check. Aus Iena Review 1980.

2 Patmios E., Sur la préparation d'un plan à buts multiples. International Society of Photogrammetry Commission IV Vol. XXII-4, pp 586-593, International Symposium New Technology for mapping, 2-6 October 1978, Ottawa-Canada.

3 Patmios E. Examples sur la contribution des méthodes Photogrammétriques et photointerprétatives à l'étude des châteaux et Forteresses. Congrès International de l'I.B.I. (Internationale Burgen Institut), 10-14 Mai 1978, Ouranoupolis-Mont Athos.

4 Patmios E. Quelques aspects pour une étude en global de l'Asclepeion sur l'île Cos. V International Symposium for photogrammetry in Architecture and conservation of monuments, 9-12 October 1978, Sibenik - Yugoslavia.

5 Patmios E. Les photos aériennes dans l'étude des sites Archéologiques, contenant de Basiliques. Xe Congrès International d'Archéologie Chrétienne, Thessaloniki, 28 Septembre - 4 Octobre 1980.

6 Patmios E., Tsakiri-Strati M. Photogrammetric methods on environmental studies. Basic photogrammetric elements and Photointerpretation. International Geographical Union. Brazil National Commission. Rio-Janeiro 9-21/8/1982.

7 Patmios E., Tsakiri-Strati M., Georgoula O., Lasaridou M. Photogrammetric Methodolody to littoral studies. Photointerpretation in Athos Peninsule, Greece. Fifth International Symposium on Computer - Assisted Cartography and International Society for Photogrammetry and Remote Sensing Commission IV. Virginia, 22-28/8/82.

8 Patmios E. Small scale air photographs in researching monuments and historic centers. (Mystras, Peloponnese, Greece and Palace of Faistos, Crete, Greece). Symposium and Exhibition: Photogrammetry applied in architecture, monument Preservation, archaeology, science of arts, Wien 16 to 18 September 1981.

9 Patmios E., Tsakiri-Strati M., Chalkias Th. Methodology of Terrestrial Photogrammetric takings on churches faces and parts of fortress at historical centre Mystras (Greece). XVI International Byzantine congress. Vienna 4-10 October 1981, Austria.

10 Patmios E. Terrestrial photogrammetric takings and rectification in researching Neoclassical monuments (Villa Alatini, Thessaloniki, Greece). Symposium and Exhibition: Photogrammetry applied in architecture, monument preservation, archaeology, science of arts, Wien 16 to 18 September 1981.

11 Patmios E., Tsakiri-Strati M., Georgoula O. Etude Photogrammétrique sur les centres historiques de Dylos et de Mystras (Greece). International Symposium on Photogrammetric contribution to the documentation of historical centres and monuments. Siena, 18-20 October 1982, Italy.

12 Patmios E. Methodology of taking multiple data for monuments. Photogrammetric Society of Greece. Athens, 3 June 1981.

13 Patmios E., Tsakiri-Strati M., Chalkias Th., Georgoula O. Photogrammetric studies on islands of Aegean Sea, Thira-Mikonos. VII International Symposium of Aegean, Kos 27-31/8/1981.

14 Patmios E. Photogrammetric study of statues. XVth Congress of International Society for Photogrammetry and Remote Sensing, Vol. XXV, tome A5, pp 605-611, Rio de Janeiro, 1984.

15 Patmios E. Photogrammetric study of curved surfaces of monuments. Examples of the historical center of Mystras, 1985.

16 Patmios E., DTM on drainage studies, 1985.

17 Patmios E., Tsakiri-Strati M. Photointerpretation on the study of historic centers. Photogrammetric Society of Greece. Athens 3 June, 1981.

18 Patmios E., Halkias Th., Lasaridou M. Photogrammetry and photointerpretation on the study and preservation of the cultural Inheritance.

19 Papagianopoulos A. Monuments of Thessaloniki. 1983.

Symposium on Remote Sensing for Resources Development and Environmental Management / Enschede / August 1986

Remote sensing in archaeological application in Thailand

T.Supajanya
Geology Department, Science Faculty, Chulalongkorn University, Bangkok, Thailand

ABSTRACT : Remote sensing particulary aerial photograph is a powerful tool in archaeological application. Thailand has taken this benefit to survey and map man-made features which are evidences of ancient settlement such as moat and barai. Other man-made structures in the past such as; canal, road, rampart, water reser voir and etc are also mapped and recorded. The total 1,300 sites of ancient settlements which have evidences manifested on aerial photograph are discovered. Locality maps and aerial photographs of each ancient settlement are prepared for an inventory which is now benefits for national management of cultural resources, and is a paramount uses for historical and archaeological researches and conservation planning. Informations achieved from photo-interpretation of each ancient site are made possible to be recorded basing on GIS with an aid of computer system. Selected examples of remote sensing approaches in archaeological in Thailand are described and demonstrated.

1 INTRODUCTION

Archaeological approaches of remote sensing practice currently, can be described as a technique using remote sensing in an area of the landscape that shows evidence of past human activity; a portion of the environment used by people (Knudson 1978). Appropriate aerial photograph and other remote sensing are used as a tool in surveying, measuring, and recording of man-made features, which are moat, rampart, canal, road, water reservoir and etc. This has been proved in many parts of the world and leads to the success in management of cultural resources for both conservation and research.

Thailand is situated in Southeast Asian region. The northern part is on the Asian mainland while the southern part stretches, as a narrow strip, down the Maley Peninsular facing to the Pacific Ocean on the east,and the Indian Ocean on the west. The territory is approximated about 513,000 square kilometer, and has been proved by several archaeological expeditions to be occupied by human in long time back to early prehistoric period, and being rich in archaeological evidences on ground (Charoenwongsa, 1983).

First moated archaeological site to be recognized on aerial photograph was honoured to The Royal Thai Air Force back to 1922 by Seidenfaden (1950). Uses of aerial photograph in archaeology in Thailand became known to public after the publishing of William-Hunt's articles "An introduction to the study of archaeology from the air", where ancient sites in Thailand were demonstrated (William-Hunt, 1949), and "Irregular earthworks in Eastern Siam: An air survey" (William-Hunt 1950). Unfortunately that, the set of aerial photographs used have never been available in Thailand. By personal communication with E.H.Moore, who is now working on Ph.D. Thesis for the Institute of Archaeology London, basing on these aerial photograph indicated that, these aerial photographs have had been left untouched in a store room of Pitt Rivers Museum, since the passing away of William Hunt in 1953.

Since 1953, whole Thailand have been covered by aerial photographs of the scale 1:50,000 and later by 1:15,000 and are available to govermental agencies and researchers. They are used for discovering of ancient sites and suggested for working in Town plan ning (Hinchiranan, 1964). In 1972, the author's artical on "The need for an inventory of ancient sites for anthropological research in northern Thailand" (Supajanya, and Vallibhotama, 1972) was published by basing on the author's collection of aerial photographs manifesting moated ancient settlements, they are discovered during his works on teaching and research since 1964 after his post-graduated from I.T.C., the Netherlands. The total of 500 moated sites were discovered in Thailand and 300 sites are situated in northeastern Thailand at that time. However, there were no response to that articles untill the Toyota Foundation, Japan, has supported his projects on "The inventory of ancient settlements in Thailand on aerial photographs" (1982-1984) and "Data base on ancient settlements in Thailand: preparation and establishing of the data centre" (1984-1986). Recently, archaeological information achieved under these projects are being used through nation wide, and benefits in academic, conservation and development purposes.

2 AN INVENTORY OF ANCIENT SETTLEMENTS IN THAILAND ON AERIAL PHOTOGRAPH

2.1 Ancient settlements on aerial photograph

Ancient settlements are recognized on aerial photograph through evidences such as; moat and water reservoir which are the most simplest form to be seen on imagery (figure 1). And they are probably the last evidences to be seen on the ground, particulary when other constructions are made of wood. A medium scale of aerial photograph approximately 1:50,000 is found sufficient for site selecting. Selected sites then are studied under stereoscope, and larger scale aerial photograph of 1:15,000 is used to confirmed the foregoing interpretation. Through this method, most of ancient settlements having moat and barai (big pond) can be discovered. The total of approximately 2,000 were selected from aerial photographs of the scale 1:50,000 and about 1,300 sites are approved by using aerial photograph of the scale 1:15,000. Accordingly, several small ancients sites cannot be detected, and need to be surveyed by using larger scale of aerial photograph; Co-ordinate of the centre of each ancient site is recorded and refered to map sheet. UTM coordinate is use as a code for that particular sites.

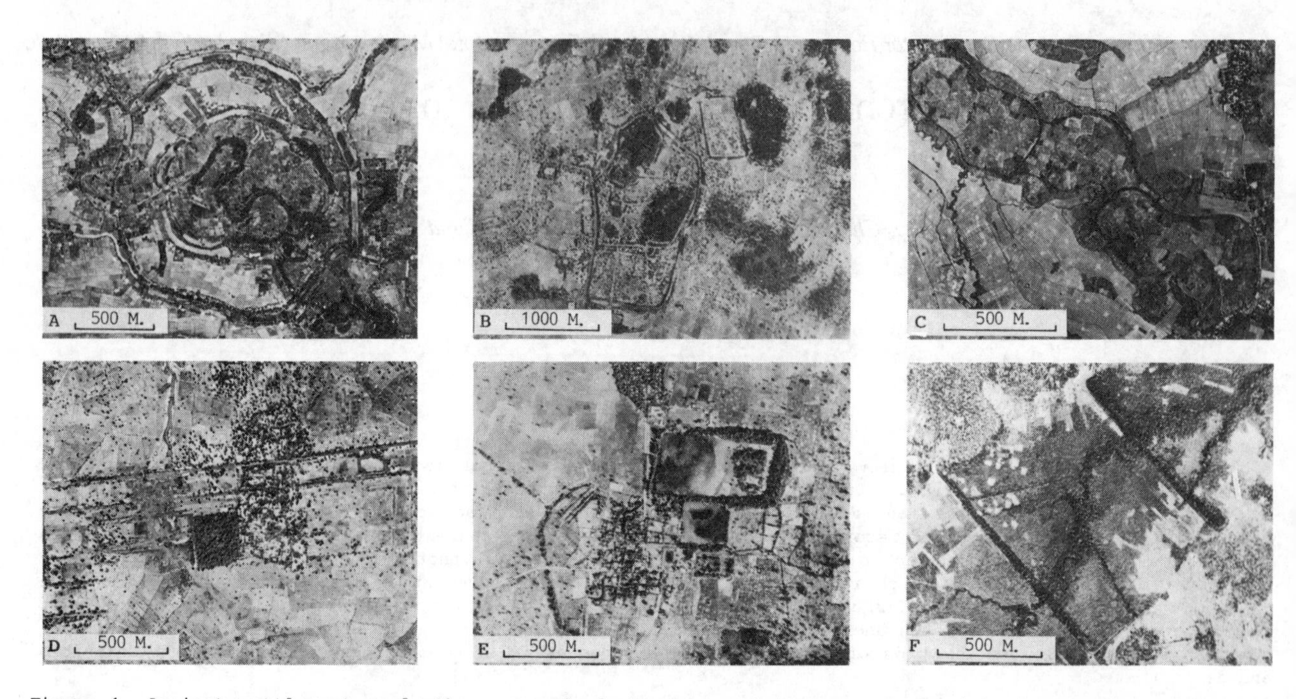

Figure 1. Ancient settlements and other man-made features, as being seen on aerial photographs: a) double moats having circular shape; b) moats surrounding two small hills and bounded together by larger moat classified as curved corner moat pattern; c) moat pattern having topographic controlled, being construced at the edge of hills; d) moat pattern of angular corner; e) ancient settlement indicated by moat and barais; f) ancient dam structures built across river flood plain.

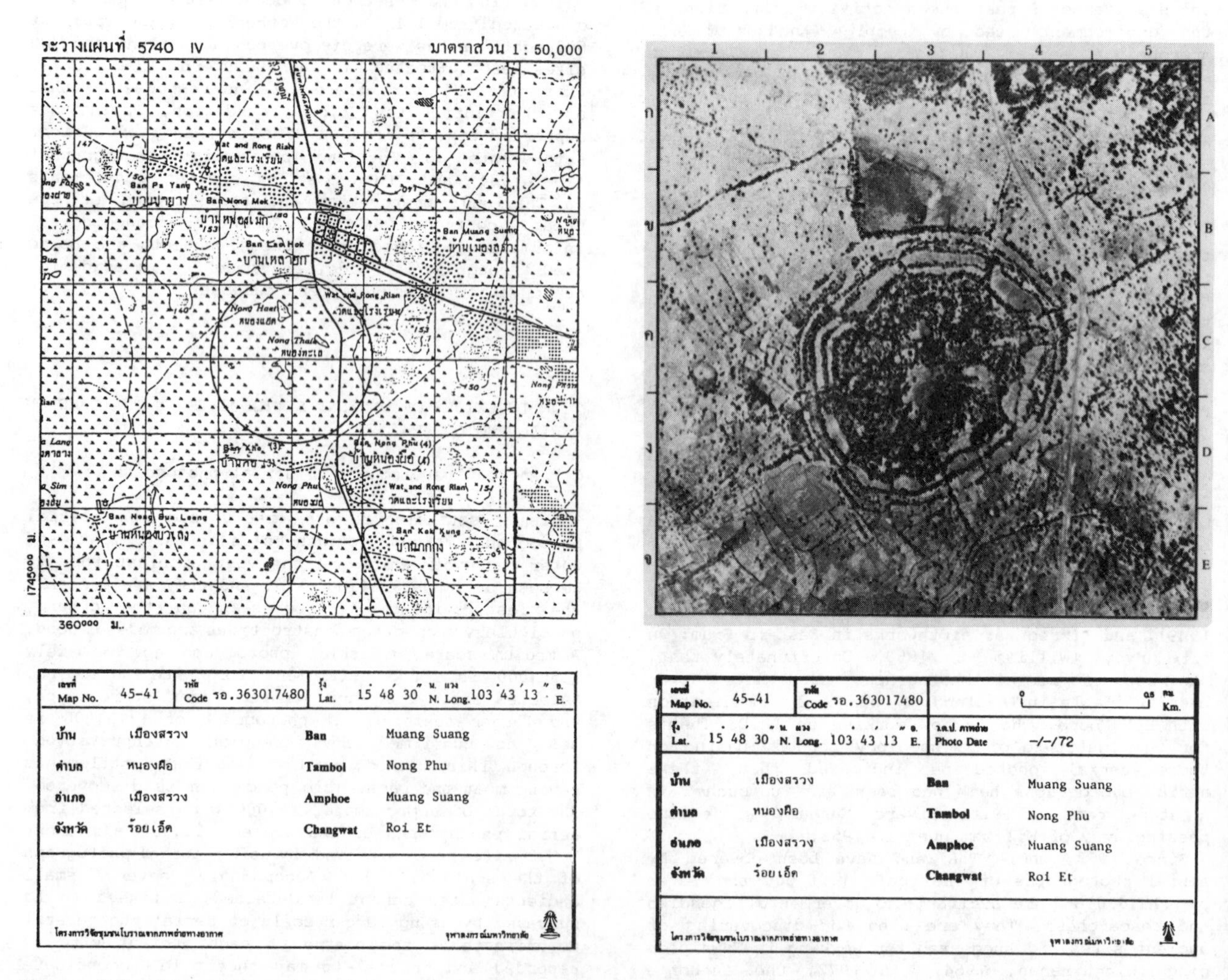

Figure 2,3. Map and aerial photograph of ancient settlements as prepared in "The Inventory of Ancient settlements in Thailand on Aerial Photograph".

2.2 Inventory preparation

Inventory is made in a form of aerial photograph and map of each ancient site (figure 2,3), with information of refered map sheet and aerial photo graph. Geogrphical information of each ancient site such as; geographical co-ordinate, UTM co-ordinate, and administrative names (Tumbol, Amphoe, Changwat) are also indicated. These informations are stored with an aid of computer (Compatable PC-16) using software available in market, basing on the GIS and Thai-English languages.

2.3 Uses of inventory

Inventory is made available especially for research and conservation. Uses of an inventory benefits particulary in managing of cultural resources, some of examples are mentioned here-in:

2.3.1 Site investigation: There are numbers of site discovered,so that ground survey is limited to some ancient sites. Site investigation is planned to cover all ancient settlements. Inventory is arranged and handed to teacher colleges, according to The total of 26 local teacher colleges distributing all over Thailand. Teacher college again organize local school teacher, working in an area situated within the vicinity of ancient site, to visit and record general information such as: surface finding artifacts, present condition of earthwork, landuse and etc. Accordingly, not only all ancient sites can be investigated at the same time, but they are made known to the local which leads ultimately to conservation of ancient site.

2.3.2 Locating a zone for conservation and development: Recent development, such as; construction of transportation routes, irrigation and water reservoir,and expansion of agricultural land and settlemental area, has caused ruin of ancient settlements. The National Environmental Board has used the inventory in locating zone for conservation and managing of cultural environment.

2.3.3 Management of historical and archaeological data: Thailand have limited sources of historical data, and archaeological informations have been loosely recorded and correlated. Not only aerial photographs that serve a new data, but also the inventory prepared in GIS serves as a mean in managing and correlating of all data. Lists of historical place-names,temporal grouping of ancient settlements, and the others that can be analysed basing on historical geographic method, are being organized by using the inventory (Li-Sheng & Supajanya, 1985)

3 MOAT PATTERNS

The total of 1,300 sites of ancient settlements are exposed by aerial photographs and 900 sites are surrounded by moats. Moat pattern elements such as: shape, size and orientation, are studied basing on aerial photograph of the scale 1:15,000. Figure 4 demonstrates moat patterns which are classified according to shape and topographical relationship (Supajanya, 1985). First group, moat patterns have their shape not being controlled by topography, and are subdivided according to geometrical characteristics. Attempt on an evolutional study of moat pattern was suggested (Supajanya & Vanasin, 1979). Second group, moat patterns have their form being controlled by topography. Detailed study basing on photo-interpretation, has made possible to sub-divide this group, and each of them indicate its own

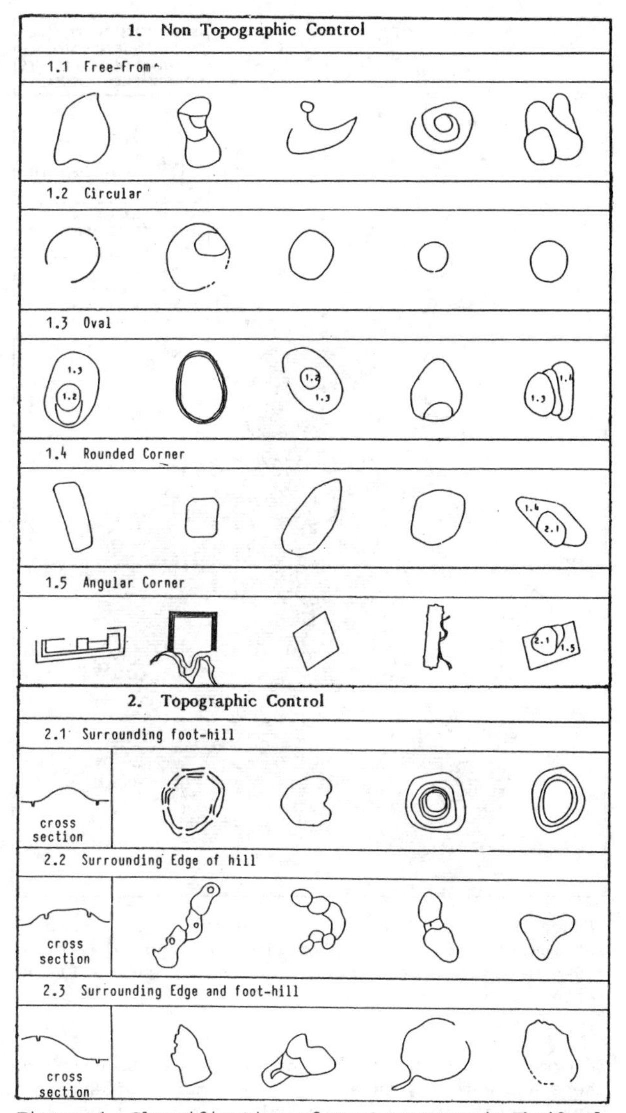

Figure 4. Classification of moat patterns in Thailand exposed by aerial photograph (after Supajanya, T. 1984).

functional uses of moat. Spatial distributions of each type are found also limited to a paticular area (Supajanya, 1986). A settlement having moat pattern surrounding at the foot of hill is found only in the northeastern region, namely E-san, while those surrounding at the edge of hill are limited only to the northern region, namely Lanna Thai.

4 ENVIRONMENTAL STUDIES OF ANCIENT SETTLEMENT

Information interpreted from photo-interpretation of ancient site is a prime basic data for environtal study. Aerial photograph and LANDSAT imageries were used to correlate ancient settlements exposed in coastal area along shoreline of approximately 2,200 kilometer with former shorelines of different elevation. All ancient settlements dated more than thousand years ago are situated above former shoreline at the elevation of approximately 3.5-4 meter (Supajanya, 1983). Figure 5 demonstrates the correlation of ancient site and former shoreline in the area of the Lower Chao Phraya Plain. (after Vanasin and Supajanya, 1980). An aerial photograph were used to study the remainded ramparts aligning along a distance of 123 kilometers, and they have been believed to be constructed as a road.

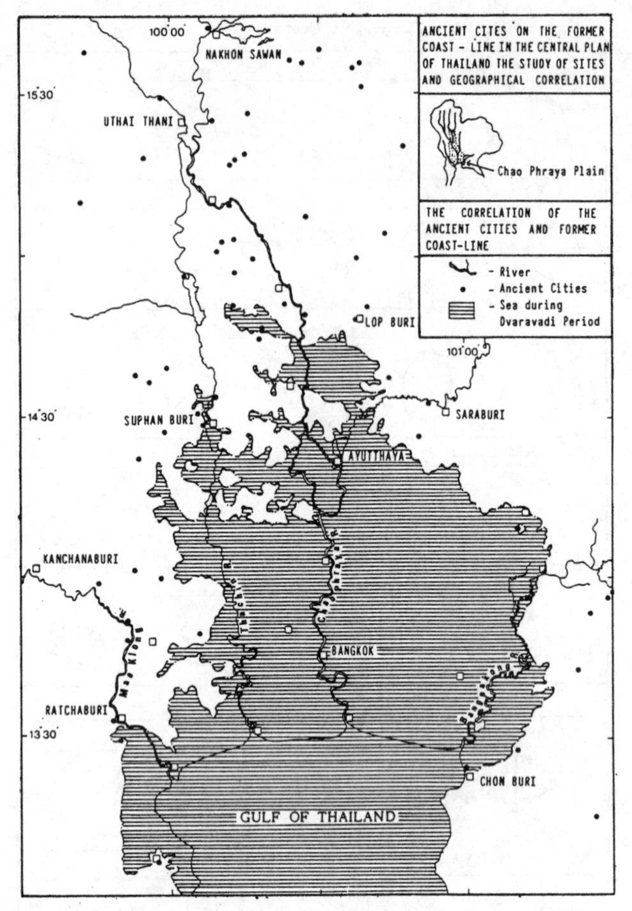

Figure 5. Map showing a correlation of ancient settlements and former shoreline at the elevation 3.5-4 meter in the area of the Lower Chao Phraya Plain (after Vanasin, & Supajanya, 1980).

Informations gained from photo-interpretation prove these ancient structures to be a remains of ancient canal (Supajanya, 1984). Several other old structures in Thailand which are remained on the surface and they have not been recognized before, were now recorded basing on the interpretation of aerial photograph.

5 CONCLUSION

Remote sensing in archaeological application in Thailand has been proved successfully. The inventory of ancient settlements and other man-made features are prepared. Basing on which cultural resources can be managed for academic and conservational purposes. However, detailed working on each ancient site needs to be done basing on aerial photographs, and training personal to be able to scope with instruments and techniques becomes necessary. The current, work was done basing on manual interpretation and basic instruments, hopefully that, the modern techniques in remote sensing may be applied in the future.

ACKNOWLEDGEMENT

Completion of the inventory and presentation of this paper are supported by The Toyota Foundation, Japan, through the Ancient Settlement Research Project, Chulalongkorn University, Bangkok.

REFERENCES

Charoenwongsa P., 1983. The identily of Thailand: Tracing back the past. SPAFA Digest, Vol. IV, No. 1, p. 4-10.

Hinchiranan, N., 1964. Public park along moate and wall. Town Planning Magazine, Vol. 3, p. 12-38 (in Thai).

Knudson S.J., 1978. Cultural in restrospect: An introduction to archaeology. Chicago, Rand-McNally.

Li-Sheng, D., and T. Supajanya, 1985, Place names in Chinese documents related to Thai history. Non publised report, Research Project on Ancient Settlements, Chulalongkorn University, Bangkok.

Seidenfaden, E., 1950. Note on William-Hunt's article. Antiquity, Vol. 24, No. 93, p.35-36.

Supajanya, T. and S. Vallibhotama, 1972. The need for an inventory of ancient sites for anthropological research in northeastern Thailand. The Southeast Asian Studies, Kyoto Vol. 10, No. 2, p. 284-297.

Supajanya, T, and P. Vanasin, 1979. The patterns of moats in the lower Chao Phraya Plain, Thailand. Geographical Journal, Geograph. Assoc. of Thailand, Vol. 4, No.3, p.53-56 (in Thai).

Supajanya, T., 1983. Tentative correlation of old shorelines around the Gulf of Thailand. p. 96-105 in Proceedings of the First Symposium on Geomorphology and Quaternary Geology of Thailand Bangkok, compiled by N. Thiramongkol

Supajanya, T., 1984. Moat patterns in Thailand: Classification for data base system. Non published report. Research project on ancient Settlements, Chulalongkorn University Bangkok.

Supajanya, T. and Others 1984. Tho Poo Phraya Ruang: Ancient canal in Sukhothai period. Presented in the First Historical Symposium of Kamphaeng Phet, Kamphaeng Phet, Thailand, 33 pages (in Thai).

Vanasin, P. and T. Supajanya, 1980. Ancient cities in the lower Chao Phraya Plain, Thailand: The relationship of sites and the former shorelines. p. 93-99 in Proceeding of Historical Geography 24th Internat. Geographic. Cong. Section 9, Compiled by T. Tanioka & T. Ukita, Japan.

Williams-Hunt, P.D.R., 1949. An introduction to the study of archaeology from the air. Journal of the Siam Society Vol. XXXVII part 2, p.86-110.

Williams-Hunt, P.D.R., 1950. Irregular earthworks in Eastern Siam: An air survey. Antiquity, Vol. 24, No. 93, p.30-36.

Symposium on Remote Sensing for Resources Development and Environmental Management / Enschede / August 1986

Application of physiographic photo interpretation technique to analyse the enigmatic drainage problem of the Hyderabad Metropolitan Region, Pakistan

M.N.Syal
Soil Survey of Pakistan, Lahore
I.E.Schneider
EMPLASA, São Paulo, Brazil

ABSTRACT: The Hyderabad Metropolitan Region is represented by flat Rocky Plateaus, Flood Plains and Piedmont Plans situated along the Indus River. Under natural conditions, the region has quite suitable physical environments for urban expansion and for maintenance of existing civic infrastructure. But as a result of population implosion and lack of proper urban planning, the region's natural drainage has been subjected to progressive deterioration in the past. The present situation has become quite alarming, but the development strategies are not yet attuned according to the natural lay of land. Apparently, the drainage conditions have become enigmatic for the planners. In order to give a scientific solution for the drainage problem, the physiographic technique of aerial photograhic interpretation was employed. Causes of drainage deterioration were correlated with physiographic processes. The rate of deterioration was qualitatively determined by comparing the interpretation results of two sets of aerial photographs taken with a time interval of 23 years. It was observed that the spatial growth and the degree of deterioration could precisely be attributed to the technically fragile physiographic positions. This experience was used to predict the behaviour of similar site conditions observed elsewhere in the region.

1 INTRODUCTION

This study represents a practical application of physiographic aerial photo interpretation technique to determine clues for soil drainage deterioration in the Hyderabad Metropolitan Region, for which a comprehensive Master Plan is under preparation (PEPAC,1986). It was carried out to assist the Hyderabad Development Authority, the planning and execution agency for the region. The objective is to highlight the principal physical factors affecting drainage conditions.

The study predicts the behaviour of available areas for urban expansion based on the performance of similar site and material conditions existing in the areas already under use. Further it is aimed at providing proposals for recovery and optimal utilization of affected areas. The contribution of Physical factors towards drainage deterioration needs quantitative verification based on spatial distribution of the problem areas.

The description of the region enumerates its physical characteristics employed during the process of physiographic photo interpretation. Most of these were studied during Reconnaissance Soil Surveys of the adjoining regions (Beg etal., 1970 and Mushtaq Ahmad etal., 1971). The magnitude of the present drainage conditions was largely drawn from the later studies (WAPDA, 1979).

2 THE REGION

The region has certain peculiar physical environments which, one way or the other, characterize the sequential deterioration of drainage conditions. The following are the relevant components.

2.1 Spatial setting

An area of about 100 square kilometers occurring along the flanks of the Indus River below Kotri Barrage and falling between 68° 15' - 68° 30' E and 25° 15' - 25° 30' N, represents the Hyderabad Metropolitan Region with a population of nearly one million people (Government of Pakistan, 1980).

The region is bounded in the north and east by a system of three parallel canals i.e. Lined Channel, Pinyari and Fulleli, taking off from the river, upstream of the barrage located on the north boundary of the region. The southern boundary runs across the Ganjo Takkar rock outcrops and the confluence of Baren steam and the Indus River. In the west the region's boundary roughly demarcates the watershed of western rock plains.

The region includes three well-connected sister towns: old Hyderabad city situated at the lest bank and Kotri and Jamshoro towns at the right bank of the river. It is situated at a distance of about 160 K.M. from Karachi, the largest city of Pakistan and has good communication linkages with it. The junction of two national arterial roads running along either sides of the Indus is also located within the region.

The level parts of the Flood Plan are used for intensive agriculture except for the parts which are salt-affected. The salt-free areas have a very high agricultural value. The saline flats, basins and channel remnants are unused and serve as recipient sites for runoff and uncontrolled effluent from the adjoining areas. The land price of these area is low and hence returns high income to the land speculators. The Rock outcrops, Rock Plains and Piedmont Aprons are generally used for poor grazing. The land price of these landscapes is very low because of their location being away from the city infrastructure.

The river attains deltaic character in the region and its course has been properly trained by construction of earthen embankments on either side. As a result of continuous sedimentation by the river on areas falling between the embankments, the river course lies at a somewhat higher elevation than the adjacent Flood Plains. Consequently it appears that the river course is located on a sort of extensive ridge.

2.2 Climate

The climate of the region is subtropical semi-desert type (Ahmad, 1951). It is characterized by low but highly erratic rainfall, very dry season for larger part of the year and larger diurnal and seasonal fluctuation of temperatures.

The region lies in the rain shadow area. The mean annual rainfall is about 130 mm which is mainly received during the short rainy season of about 10 days during July and August. Though the average rainfall is low, the region receives high rainfall occasionally when 120-250 mm of precipitation may occur within 24 hours.

Evaporation generally far exceeds the rainfall. Relative humidity varies considerably. It is highest in August (62 per cent) and lowest in April (36 per cent). Wind occurs throughout the year. Mean monthly maximum temperature fluctuates between 39.4C° and 42.3C° for about seven months from April onwards.

2.3 Geology

The eastern half of the region represents a part of the Lower Indus Alluvial Plains formed over Tertiary consolidated sediments, which crop out abruptly as two monoclinal hills with fairly flat tops. Of these the smaller one occurring in the north, represents the seat of the old city of Hyderabad, while the larger one situated in the south is called as Ganja Hills locally known as Ganjo Takkar. The western half of the region lying west of the Indus River is largely composed of consolidated beds of moderate to low dip (10-2°). The river bed lies in a vast syncline, the corresponding anticline of which lies near the western boundary of the region (LIP, 1965-66).

The exposed rocks are of Middle Eocene age Laki Series and consist of alternating beds of limestones and shales. The Limestones are lenticular in structure with intercalations of Marl. The exposed beds are weathered upto a depth of 5 meters from the ground surface. The weathered part, though porous, cannot hold or transmit sufficient quantity of water due to its Marly nature. The lower beds of limestones have intercalated chalk and cavities filled with calcite crystals (Wadia, 1966). The shales are extremely hard posing difficulty in drilling. A test hole drilled in the saddle between the monoclinal hills showed that the alluvium is resting on a bed of shale about 8 meters thick and occurs at a depth of 3 to 6 meters from the land surface (WAPDA, 1979). In the north-western corner of the region Lower Eocene deposits (Ranikot Series) are exposed. The sediments consists of alternating beds of shales, sandstones and limestones. These beds are extremely dissected. A section of the regional geological map (Abu Bakr and Jackson, 1964), is presented in Fig.1.

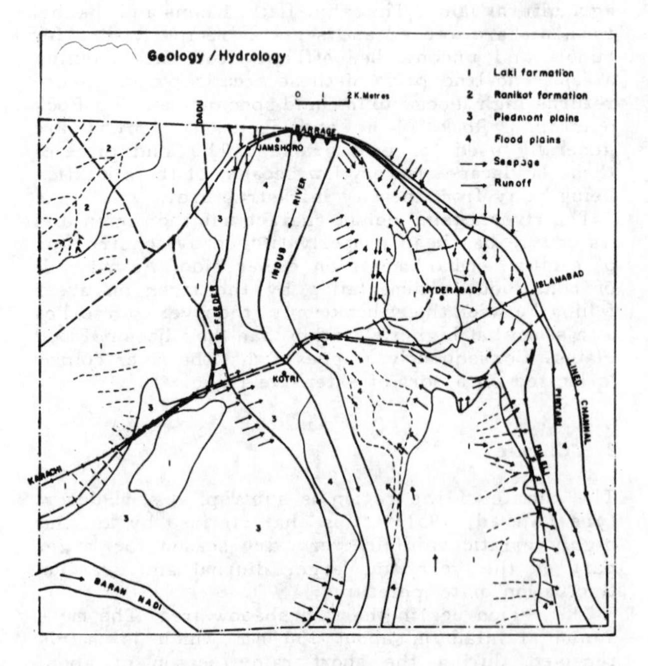

Fig.1 Adapted Regional Geological Map with hydrological conditions as studied through aerial photo interpretation.

3 REVIEW OF PREVIOUS STUDIES

Before 1979 no concrete affort had been made to study the urban drainage conditions of the region. Some earlier reports illustrated the drainage conditions keeping in view largely the agricultural aspects (Beg, etal., 1970). Soon after independence in 1947, a satellite town was planned on the Flood Plains adjoining rock outcrops to accomodate the great surge of migrating population from India. Apparently no site specific studies were undertaken for urban planning.

3.1 WAPDA Southern Zone Planning Studies

Keeping in view rapid deterioration of drainage, the Southern Zone Planning (WAPDA 1979), conducted a comprehensive study. According to the main findings of the study "three serious difficiencies, which both directly and indirectly, have aggravated the drainage problem of the city. Firstly, the city has no surface drainage system for disposal of storm water creating critical conditions during unusual downpour of high intensity in most monsoon years. Secondly, due to lack of surface and subsurface drainage the watertable in Latifabad (a satellite town of the region) has come up close to the surface causing heavy damage to the residential buildings and other infrastructures. Thirdly, the sewage system for Latifabad and the city area is under-designed and that too has not been properly maintained, resulting in chocking of main sewerage lines and blockage in the connecting sewerage pipes".

The study of soils was carried out during monsoon season when surface ponding and subsoil wetness were at peak. The observations were recorded on rigid grid pattern, at an intensity of 16 observations per square kilometer in newly developed area and 8 observation per square kilometer in remaining areas. Using Earnest's auger hole method only 21 hydraulic conductivity tests were carried out at the rate of one test for every 250 hectares. The soil profiles were though examined in respect of texture, structures, porosity and ground watertable, but overall emphasis had been on texture with which all the findings were ultimately correlated. The textures in turn, were grouped in four broader classes: coarse, coarse medium, coarse fine and fine. The soil profile information was recorded according to prefixed depth intervals of 25 cms. Texturally 82 percent of the newly developed area and 67 percent in the remaining parts of the region, other than rock outcrops, was declared to have good drainage conditions.

For estimation of surface drainage, infiltration rates were presented in relation to textural distribution. Accordingly, about two-third of land was described as having good surface drainage. Most of the remaining area was reported to be with moderate drainage conditions. The extent of soils with poor drainage was found to be not more than 15 percent.

The internal drainage of the region was determined from an array of hydraulic conductivity tests, despite great variation in results. The K-value of 60 percent of the profiles was less than 0.1 meter per day, indicating poor internal drainage. But unfortunately, the results are presented as average of all the sites studies, which came to 1.53 meter per day. Through the average values the diverse hydraulic conductivity results were described to be supporting the textural representations. The causes of variation in K-value within the generalized textural groups were not explained.

3.2 Soil Survey Inventories

By the year 1979, adequate technology was available in Pakistan in respect of modern physiographic soil surveys based on aerial photo interpretation. Through this technique, reconnaissance soil survey of the areas around the region had already been conducted and results of studies published. Reference knowledge gathered from these surveys had revealed certain physiographic and morphological aspects which have an important bearing on the study of drainage conditions of the region.

Soils were identified and mapped as natural bodies having well defined landscapes and unique sets of profile characteritics. Soil drainage as a land quality was inferred both from the landscape characteristics (landform, relief, landuse, and surface salinity), and profile characteristics (texture, structure, porosity, watertable/saturation). According to these studies, the region is largely a young floodplain, profusely marked by riverain features. The surface has a complex relief changing at short distances. Soil salinity is found associated with well-defined landscape positions.

The sediments are estuarine in nature and are deposited by river floods of abruptly decreasing velocity. They are extremely sorted and are dominant in very find sand and silt fractions (Jalal-ud-Din, etal, 1970). As a result, the sediments are characterized by inherent capillary porosity due to which water is held in the soil profile against the gravity. The ground water or perched water pockets have created saturation fronts, far above their actual levels. From the wet spots the moisture is attracted even laterally towards the unsaturated parts. Further, under stress conditions the saturated very find sandy and silty soils have very low bearing capacity (Syal, 1973).

It is also understood that presence of heavy clay lenses within the profile (upto a depth of 250 cm) may create perched watertable conditions due to percolating rain, irrigation or sewerage water. Thickness of such layers usually does not matter much, even a thin layer may create situation equal to a thick layer. Similarly, the topography of the underlying rocks may be too uneven to ensure prompt movements of the underground flow of water. All of these factors locally affect the internal drainage conditions to varying degrees.

Dense saline-sodic soils are unsuitable for economic agricultural development, bur they are readily available for non-agricultural uses. Some special measures are needed to use these soils for housing or industry (Mian and Brinkman, 1970).

3.3 Soil drainage: Conditions and requirements

Soil drainage represents the rapidity and extent to which a soil is capable of disposing off its surplus water externally (by run off) and internally (by leaching) (Soil Survey Staff, USDA, 1954). The concept holds good regardless of land use of the region—agricultural or non-agricultural. How easily and effectively a soil is drained depends on depth to watertable or of ponding, soil permeability, depth to bedrock, flooding and slope (SCS., USDA, 1983). Drainage conditions are best studied under stereoscopes by landforms, relief graytone and site conditions. Relation of these photo elements with soil/site condition is high (Buringh, 1960). A coarse loamy soil on a concave position and the same texture on convex slope surely has different surface drainage conditions, which could be easily readable on the aerial photographs and delineated on a map.

A very detailed physiographic soil map representing soils at phase level, coupled with deep auger observations (5-10 meters) at representative sites could be the most useful document to group the soil phases according to site conditions to be evaluated for urban expansion (Tahir and Mushtaq, 1977). The natural drainage conditions of a site dictate that either it should not be brought under a township or adequate drainage should be provided and adapted architectural design made in the light this knowledge (Mian and Brinkman, 1970).

3.4 Main short falls of the studies

The studies of Southern Zone Planning (WAPDA, 1979) did not employ the systematic terrain analysis technique of photo interpretation. Conversely, it appears that photographs were used only as base maps for field guidance. Soil mapping was done on the basis of rigid grid sampling procedure which is not adequate in Flood Plain areas wherein vertical and horizontal veriations in soil materials and landscape conditions are rather too frequent. Similarly, instead of using available soil maps or the aerial photographs, topographic maps of 1:50,000 scale were used to delineate specific drainage development areas. In doing so, relief inaccuracies were encountered. Consequently, very precise network of open and tile drains could not be proposed. Further, the effect of effluent from the high-lying old city and runoff from the denuded rock plains could not be estimated properly. As physiographic processes responsible for terrain evolution were not studied, drainage deterioration could not be attributed to real processes on site-specific basis.

Effect on subsoil drainage has not been duly considered, of the impervious shale and its internal topography observed at shallow depth under the alluvium. Instead, the problem has been solely attributed to unfavourable nature of the overburden. Further, it is assumed that the canals system has no appreciable seepage zone to act as a barrier for natural internal drainage. This does not appear very convincing in case of inherent high capillarity of the soils of this region. Similarly, the role of seepage water brought in the region by old abandoned river creeks entering the area in the north has not been duly assessed. Water column in the barrage may also contribute seepage water to the region internally.

Soil Survey reports of the country have not provided interpretation of land resource data for urban and engineering uses. Stress on interpretations for agricultural development has lead the urban planners to assume that the available soils information is inadequate. The scale of soil mapping (1:250,000) is also not enough for making detailed predictions about soil behaviour for urban development.

4 PHYSIOGRAPHIC PHOTOGRAPHIC INTERPRETATION

4.1 Material and method

Semi-controlled photo mosaics of 1:40,000 scale were studied visually to identify broad photo-patterns representing main landtypes of the region. Each landtype was analysed in detail under the Ziess-Aerotopo stereoscope using panchromatic, black-and-white, semi-matt aerial photographs taken in 1953. The resulting terrain units were described in detail. Using another set of aerial photographs of scale 1:30,000 taken in 1976 (during the same season), similar photo interpretation study was carried out objectively. The resulting terrain units were compared in respect of changes in photo image characteristics.

The systematic physiographic procedure of aerial photographic interpretation (Vink 1963) was employed leading to understanding of the type of sediments, their mode of occurrance in the terrain and of the factors affecting the drainage conditions in the region. "The theme of physiographic analysis is to find and describe features of the stereo photo image which are characteristics of certain physiographic processes; can be used to identify these processes and so on, in turn, will provide important clues for delineating the soil pattern" (Goosen, 1967). Deductions about photo image characteristics based on local reference knowledge were the most significant aspects of the physiographic photo interpretation procedure.

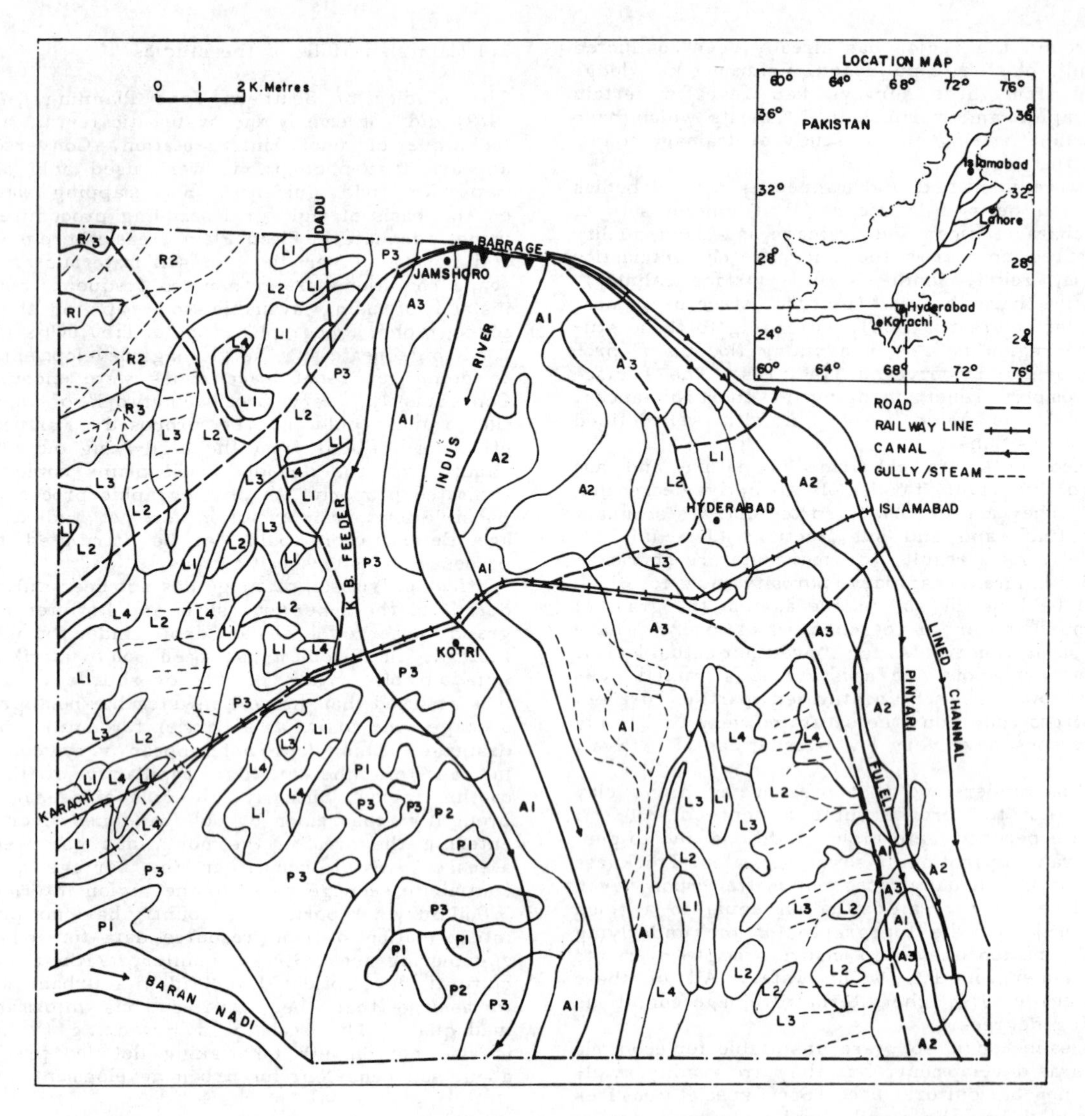

FLOOD PLAINS	PIEDMONT PLAINS	ROCK PLAINS (LAKI FORMATION)	DISSECTED ROCK PLAINS (RANIKOT FORMATION)
A1 Levees	P1 Piedmont Aprons/ Valley Floors	L1 Mesas	R1 Mesas
A2 Level Plains	P2 Gently Undulating Plains	L2 Higher Terraces	R2 Higher Terraces
A3 Basins/ Channel Infills	P3 Undulating Plains	L3 Lower Terrace Remnants	R3 Lower Terrace Remnants
		L4 Terrace Footslopes	

Figure 2. Physiographic aerial photo interpretation map, Hyderabad Metropolitan Region, Pakistan.

From the nature of sediments and other environmental characteristics the behaviour of various terrain elements was predicted for urban expansion and improvement. The results of the study were matched with the reference knowledge of the physiographic processes studied outside the region.

4.2 Terrain analysis

The region represents several physiographic units significantly varying in landform, relief and elevation. Indus Flood Plains, Rock outcrops, Piedmont Aprons, Dissected Rock Plain and the main river channel, including annually flooded land, are the main components. The following subdivisions of the main physiographic units were identified. They are shown in Figure 2.

4.2.1 Flood Plains

The Flood Plains make up the matrix of the Indus left bank part of the region. The area is generally flat but small-scale unevenness associated with the riverain features is very common. Based on photo image characteristics and physiographic processes involved in evolution of the landscape, the following three main units were identified.

A1 Levees. This unit is characterized by high relief, lighter graytone, profuse river activity in the form of numerous small channels perpendicular to the river streams and by seasonal vegetation. The unit is formed by bank-topping floods of the Indus river. Presence of rock outcrop at shorter distance from the river modified the drainage pattern parallel to the river, leaving sandy deposits besides the rock outcrops. The resulting convergent stream pattern south of the railway line indicates structural influence of rocks, as it is contradictory to the normal sedimentation pattern. A faint stream pattern originating from the southern rock outcrops also runs across the river pattern. This modified drainage pattern shows runoff activity from the rocks.

A2 Level plains. With medium to dark graytone, nearly level relief and low river activity, this unit represents parcels of typical agricultural areas on 1953 photographs. Adjacent to A1 unit the graytone is somewhat lighter indicating sand at shallow depth. On the east and south-east of the old city, this unit displays somewhat darker graytone indicating intensive land use and/or clayey nature of soils or both. On 1976 photogrphy the graytone is found brighter with sharp parcels on the river side, revealing intensive agriculture, whereas the tone of the eastern part appeares darker but diffused. Some ponded sites are clearly observed. Cultivated area has reduce in extent.

A3 Basins/Channel Infills. The lowest parts of the Flood Plains are included in this unit. It is medium gray with whitish patterns but no parcels. The area is saline. Conspicuous meander scars are present with open water at lowest sites and untraceable courses towards tips. They receive runoff from the rock outcrops. The basin along Hyderabad-Jamshoro road is a back-swamp, receiving seepage water from the canals, barrage and runoff from the old city of Hyderabad. On 1976 photographs, the overall graytone is darker, meander scars got enlarged enought to be extrapolated almost to their entire course.

4.2.2 Piedmont Plains

Mapped at the foot of the rocky plateaus on the right bank of the river, the Piedmont Aprons appear as continuous belt. They are easterly sloping with drainage pattern perpendicular to the flow of the Indus river. The south-western corner of the region has been reshaped by wind action into low and high dune patterns. The following are the subdivision of the landscape.

P1 Piedmont Aprons/Valley Floors. They are level to very gently sloping with light to medium graytone. Locally they are cultivated with torrent water or lift irrigation from the K.B. Feeder canal.

P2 Gently Undulating Plains. They represent parts of piedmont plain having low dunes, scattered low vegetation and some interdunal valleys. Dunes appear stable, in general. No moisture accumulation is apparent.

P3 Undulating Plains. Areas with high dunes and shrub vegetation are included in this unit. Inter-dunal valleys are common. Some dune are unstable and occur along the Baran Nadi.

4.2.3 Laki Rock Formation

The rock of this series are monoclinal in nature. They have light graytone with a sub-parallel drainage pattern. Gully density is medium. Extremely eroded parts display slightly darker graytone, dendritic drainage pattern and higher gully density. No appreciable differences in drainage pattern are noticed on the two sets of photographs. The landscape is divided into the following four components.

L1 Mesas. They occupy the highest position in the area with extremely level central parts. The peripheries are very gently sloping with covex slopes. No gullies are apparent.

L2 Higher Terraces. Generally this unit occurs at a lower elevation than the Mesas. It is linked with Mesas through a sharp bluff. On more dissected parts the terraces attain the form of Lower Mesas but their surface is not so flat. Gully density is low.

L3 Lowest Terraces Remnants. They are severely eroded with high gully density. Flat surfaces occur on top of steep pinnacles. Graytone is generally a complex pattern of light and very gray shades indicating exposed rock strata.

L4 Terrace Footslopes and Aprons. This unit generally occurs below the dissected terraces. It is characterized by dendritic drainage pattern. Gullies are shallow with gully bottoms having some vegetation. Landscape elements like terrace escarpments, pediments and colluvial footslopes are included in this unit.

4.2.4 Ranikot Rock Formation

Rocks of this formation are not extensive in the region. They are moderate to strongly dipping. Shales are erosive but sandstones are sticking out as prominent features. Graytone is darker than Laki Formation. Gully density is much higher and as such the formation has very limited flat surfaces in the form of Mesas or Terraces. The subdivisions are, however, similar to Laki Formation i.e. R1, R2, and R3.

4.3 Discussion of results

Physiographic analysis reveals that the drainage problem is complicated in the region because of interaction of several forces contributing surplus moisture (See Fig.1). Of these, erratic rainfall and runoff collection are significant. River activity in the flood plains, continued erosion on the rocky terrain and wind action on the piedmont aprons have complicated the surface configuration of the region and hence the surface drainage. Presence of contrasting stratified soil material and rocks at shallow depth and their internal topography are factors worth a mention in connection with internal drainage. Under-designed civic infrastructure and irregular urban growth are still other factors making the drainage problem as apparently enigmatic. These factors are discussed below as illustrated by photo image characteristics.

4.3.1 River Activity

The saddle between the two rock outcrops roughly divides the broad photo-patterns. The darker eastern part consists of mostly clayey sediments possibly deposited in a relatively vast basin by the river creeks taking off from the river about 10 km. north of the region. The whitish tone in the western half of the saddle indicates mainly sandy sediments deposited along the main course of the river. The photo-pattern is a characteristics channel-levee complex with diverse sediments as displayed by striated moisture distribution pattern on 1953 photographs (See Fig.3, Stereogram). The saddle itself displays a marked meander scar which does not appear to be linked with main channel remnants or the main river trunk. Absence of drainage outlet in the saddle reveals irregular topography of the basement rocks. A shallow basin is apparent north west of the old

A shallow basin is apparent north west of the old city, extending along the Hyderabad-Jamshoro road. This basin appears to receive moisture from several sources. The cut-off river creecks may contribute recharge internally from the river while in spate. Some infilled river channels in the north extend across the canal system. They appear to add seepage water from the canals even if closed or lined. Pockets of wet areas around the rock outcrops are indicators of shallow depth to rocks.

4.3.2 Runoff Collection

The rocky areas are either built-up or devoid of any soil cover and vegetation. The drainage pattern is sub paralled on rock outcrops and western rock plains and is dendritic in the north western corner of the region. The surface being generally sloping, most of the rain water contributes to runoff. The direction of runoff flow is radial on the outcrops and easterly on the western rock plains. In either case the runoff flows over the flood plain areas against the riverain features.

4.3.3 Wind action

South-western corner of the region is a vast piedmont plain over which extensive sand dunes have formed as a result of wind action. The sources of sand are apparently the beds of Indus river and the Baran Nadi. Interdunal valleys are small. Sand dunes appear to be stable. Presence of K.B. Feeder canal along outskirts of these dunes restricts the flow of runoff (1976 situation). Along the left bank of the Feeder canal, south of Jamshoro, the areas are ponded by seepage. But due to presence of sand cover in the remaining left bank area, the situation is not clear on the photographs.

4.3.4 Under-designed infrastructure

Several instances of progressive spreading of wet bodies are studied on the two set of aerial photographs. On 1953 scene, no appreciable ponding or seepage is observed along the railway line, flood protection embankments and canal system. However in 1976 scene, surplus water stagnates along these features and occasionally permeates across them. This situation indicates that the urban infrastructure is largely under-designed. In addition, around the peripheries of the rock outcrops sizable pockets of open water are observed on 1976 photography. The earlier situation displays wetness only on limited patches.

Extensive wet areas were observed west of K.B. Feeder. From the adjacent rock plains, these areas receive runoff which occasionally accumulates along the canal bank creating temporary ponding. This ponding is not apparent on 1953 photographs.

It was understood (WAPDA, 1979) that the canals system in the region does not create seepage zones along their courses due to entrenchment of three canals below the ground surface and lining of the fourth one. The study of 1976 photographs, however, reveals that seepage zone does restrict subsoil flow of water and contributes to the basinal areas through old partially covered channel remnants. The entire wet zone occuring along the western bank of the canals is mainly due to restrictive action of the seepage zone and inadequacy of surface water disposal. On 1953 photographs, this area appears well drained.

Fig.3 Stereogram showing complex altered drainage pattern as a result of flooding, runoff and structure of the basement rocks.

5 PROPOSALS FOR URBAN DEVELOPMENT

Most of the open areas available for urban expansion adjacent to the existing civic infrastructure are associated with drainage problems. The experience gained from development of Latifabad has not been used as guidelines for future developments. The development schemes along Hyderabad-Jamshoro road are being implemented on a back-swamp area which is being fed by seepage from canals and the river and by runoff from the old city. The area has high watertable and highly sloughing type of soil material. It has been difficult to lay sewerage pipe in the subsoil (Fig 4). Without provision of adequate surface and subsurface drainage, this area would behave exactly like parts of Latifabad Project where most of the sewerage system has collapsed as a result of sloughing action.

Urban expansion schemes along Hyderabad Tando Mohammad Khan road are located on areas which are flat. But may have rocks at shallow depth.

Fig.4 Photograph indicating difficulty in excavation of trenches for laying sewerage pipes in very fine sandy and silty sloughing material along Hyderabad-Jamshoro highway.

These areas are saline and receive runoff from the adjoining rocks. Subsoil water flow would be restricted by the seepage zones of the canals in the east. In addition, presence of restrictive clayey lenses in the subsoil would locally create perched watertable conditions. Open drains with proper arrangements for disposal of drainage water are required.

On the right bank of the river, the Piedmont Plains are quite suitable for urban expansion provided adequate arrangements are made to dispose off the storm water and effluent. Hinderance posed by K.B. Feeder canal embankments and its seepage zone, calls for proper measures prior to expansion of the urban land use on the piedmont areas.

Based on the behaviour of the site conditions, relatively flat areas (L1, L2 and L4 units, see physiographic map), available on the Ganjo Takkar rock outcrops and on the western Rock Plains along the Jamshoro-Karachi Super Highway, are best sites for urban expansion. According to rough estimates, the land price of these areas is far less than the areas with even lowest agricultural economic potential, situated in the flood plains.

The areas with a high agricultural economic potential (physiographic unit A2) represent nation's non-renewable resource which should not be misused for urban expansion.

In order to develop the open areas around the saddle between the rock outcrops, natural drainage channel remnants have to be developed throughout their courses, which are traceable on the aerial photographs. The collector drains should be located along the natural surface water flow lines. Large-scale aerial photography would help in locating these lines. Detailed soil survey and topographic survey would also help in locating the small shallow basins which need to be linked with the main drain through appropriate interceptor drains. The deepest parts of the meander scars could be developed as lakes for recreational activities. These sites have rocks at shallow depths and hence may not pose seepage problems.

As soil drainage is associated with particular kinds of soils, variations in moisture contents are best studied on large-scale aerial photographs (1:10 000). Photographs of atleast two seasons should be studied for comparison.

For quantification of the drainage parameters, tests should be located on representative sites of the selected physiographic units. Soil samples required for particle size analysis should be taken from the natural soil horizons, especially the soil horizons which affect soil drainage. Sampling according to pre-established depths would always lead to erroneous results.

REFERENCES

Abu Bakr, M. and Jackson, R.O. 1964. Geological Map of Pakistan, Scale 1:2 million Geological Survey of Pakistan, Quetta.

Ahmad, K.S., 1951. Climatic Regions of West Pakistan. Pakistan Geographical Review. Vol VI.

Beg, M.S., et.al., 1970, Reconnaissance Soil Survey Hyderabad. Soil Survey of Pakistan, Lahore.

Buringh, P. 1960. The applications of Aerial Photographs in Soil Surveys, Manual of Photographic Interpretations, Washington D.C.

Goosen, D., 1967. Aerial Photo-interpretation in Soil Survey. Food and Agricultural Organization of the United Nations, Rome.

Government of Pakistan, 1980. Population Census of Pakistan. District Census Report of Hyderabad and Karachi. Government Printers, Karachi.

Jalal-ud-Din, Ch. Brinkman, R. and Rafiq, M. 1970. Soils of the Indus Delta. Soil Survey of Pakistan, Lahore.

L.I.P., 1965-1966 (different volumes). Lower Indus Report, West Pakistan WAPDA. Hunting Technical Services Ltd. and Sir Mcdonald and Partners.

Mian, M.A. and Brinkman R., 1970. Soil Surveys and their Uses in Town Planning and Siting and Design of Buildings. Engineering News, Lahore.

Mushtaq Ahmad, et.al. 1971. Reconnaissance Soil Survey, Nawabshah. Soil Survey of Pakistan, Lahore.

PEPAC, 1986. Physical Environments studies, Greater Hyderabad Master Plan.

Soil Conservation Service (USDA), 1983. National Soils Handbook, Application of Soil Information (SMSS, USAID).

Soil Survey Staff (USDA), 1951. Soil Survey manual, Hand Book No.18. Government Printing Office, Washington, USA.

Syal, M.N., 1973. Study of soils for designing highways in Larkana District, with a special reference to causes of subsidence of Larkana-Ratodero road. Soil Survey of Pakistan, Lahore, Pakistan.

Tahir, M.A and Mushtaq Ahmad, 1977. Soil Map, an Aid to Civil Engineering, Pakistan Scientific Society.

Vink, A.P.A., 1963. Aerial Photographs and the Soil Science, UNESCO. Paris.

Wadia, D.N., 1966. Geology of India. The English Language Book Society of Macmillan and Co., London.

WAPDA, 1979. Planning Report, Hyderabad City Drainage Project. Project Planning Organization, South Zone, Hyderabad.

Symposium on Remote Sensing for Resources Development and Environmental Management / Enschede / August 1986

Spatial resolution requirements for urban land cover mapping from space

William J.Todd
Lockheed Missiles & Space Company, Sunnyvale, Calif., USA

Robert C.Wrigley
Ames Research Center, National Aeronautics & Space Administration (NASA), Moffett Field, Calif., USA

ABSTRACT: Very low resolution (VLR) satellite data (Advanced Very High Resolution Radiometer, DMSP Operational Linescan System), low resolution (LR) data (Landsat MSS), medium resolution (MR) data (Landsat TM), and high resolution (HR) satellite data (Spot HRV, Large Format Camera) were evaluated and compared for interpretability at differing spatial resolutions. VLR data (500 m - 1.0 km) is useful for Level 1 (urban/rural distinction) mapping at 1:1,000,000 scale. Feature tone/color is utilized to distingish generalized urban land cover using LR data (80 m) for 1:250,000 scale mapping. Advancing to MR data (30 m) and 1:100,000 scale mapping, confidence in land cover mapping is greatly increased, owing to the element of texture/pattern which is now evident in the imagery. Shape and shadow contribute to detailed Level II/III urban land use mapping possible if the interpreter can use HR (10-15 m) satellite data; mapping scales can be 1:25,000 - 1:50,000.

1 INTRODUCTION

The payloads of earth resources satellites offer great potential for urban mapping, but data is now available in a wide range of spatial resolutions -- ranging from 10 m to 1.1 km -- and may or may not be suitable for a particular application. Our objective was to examine data from a wide range of resolutions to evaluate interpretability of urban land cover and derive a set of spatial resolution requirements for urban land cover mapping.

2 ANALYSIS OF DATA COLLECTED FROM SPACE OVER URBAN AREAS

The example data which we compared is shown graphically in Figure 1. Very low resolution data was collected by the Advanced Very High Resolution Radiometer (AVHRR) with 1.1 km resolution carried on NOAA's TIROS satellite, Coastal Zone Color Scanner (CZCS) with 800 m resolution on NASA's Nimbus-7, Operational Linescan System (OLS) with 600 m resolution on the Defense Meteorological Satellite Program's (DMSP) Block 5D, and the Heat Capacity Mapping Radiometer (HCMR) with 500 m resolution, part of NASA's Heat Capacity Mapping Mission. While Nimbus-7's Scanning Multichannel Microwave Radiometer (SMMR) has spatial resolution measured in tens of kilometers and did not deserve discussion in this paper, it is included on the chart as an example of a passive microwave sensor.

Low resolution data is represented by the 80 m resolution Multispectral Scanner (MSS), which has been carried on all five of NASA's Landsat satellites. Medium resolution data includes the Landsat-3 Return Beam Vidicon (RBV) 40 m resolution imagery, and Landsat-4 and -5 Thematic Mapper 30 m resolution multispectral data.

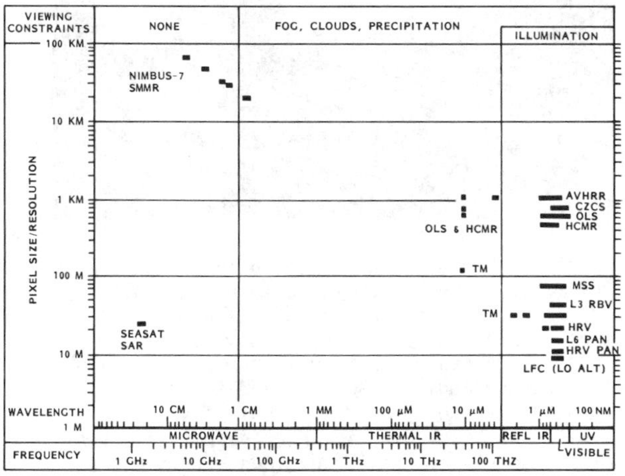

Figure 1. Spatial resolution and electromagnetic spectrum characteristics of selected satellite sensors useful for mapping urban areas.

Figure 2. Digitally enhanced AVHRR thermal infrared image (Band 4, 10,300-11,300 nm) of central and southern California collected on August 17, 1984. The San Francisco-Oakland-San Jose metropolitan area is in the upper part of the 1.1 km resolution image, and Los Angeles the lower section.

Figure 3. Digitally processed Landsat-1 MSS subscene of NE part of Kansas City collected 13 August 1972. Left, visible red (Band 5, 500-600 nm); right, reflective infrared (Band 7, 800-1100 nm). The City Center is in the left central portion of the 80 m resolution image; moving eastward, one passes through older housing and then newer housing and into the suburb of Independence. Large agricultural fields are located in the floodplain of the Missouri River.

Finally, examples of high resolution data are 25 m resolution color photography taken with the S-190B Earth Terrain Camera carried on NASA's Skylab, 9.5 m resolution B&W photography taken by the Large Format Camera (LFC) from NASA's Shuttle Mission 41-G, High Resolution Visible (HRV) 10 and 20 m resolution multispectral data carried on the Systeme Probatoire d' Observation de la Terre (SPOT) satellite, proposed 15 m TM panchromatic data to be collected by Landsat-6, and Seasat Synthetic Aperture Radar (SAR) with 25 m resolution.

2.1 Very Low Resolution Data

A large urban region with an area of 100 square kilometers would be covered by over 80 AVHRR pixels, while over 400 pixels would be collected by the HCMR. The resolution of AVHRR (Figure 2) is adequate for generalized, Level 1 land use mapping as reported by Gervin et. al. (1983). Anderson et. al. (1976) suggest that "Urban and Built-Up Land" be a Level I land use class, in a hierarchical scheme; Gervin calls the analogous urban category "Developed".

Urban settlement patterns may be detected using HCMR data. Bonn et. al. (1981) reports that even small towns with a population of 15,000 are "clearly visible as hot spots".

Another application of very low resolution data collected over urban areas is in the area of energy utilization analysis. Welch and Zupko (1980) found that 3.6 km resolution, visual OLS data collected during the evening hours correlated highly with both reported urban energy consumption (kwh) and population.

2.2 Low Resolution Data

Five Landsat satellites have carried the 80 m resolution MSS, and many researchers have investigated its utility for urban applications. The 0.46 ha pixels collected over important land use types have different proportions of land cover -- concrete, asphalt, grass, shrubs, bushes, bare soil,

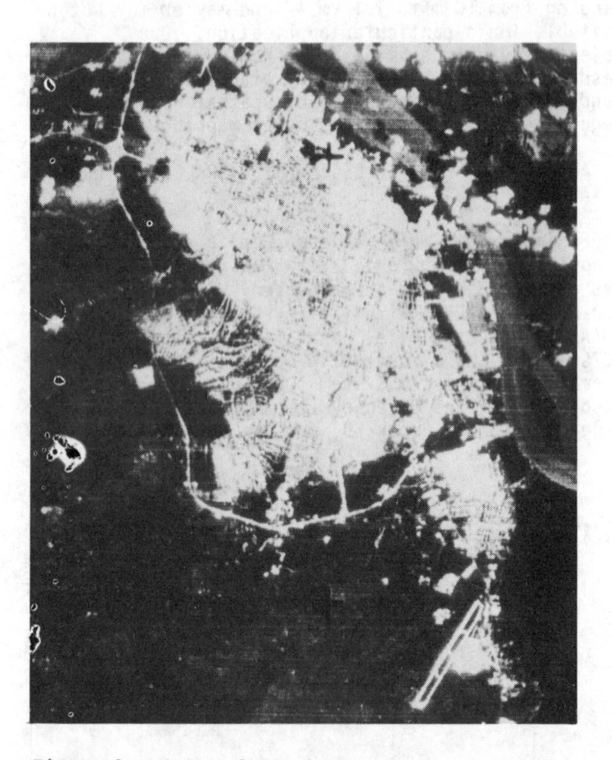

Figure 4. Landsat-3 RBV image of Barranquilla, Columbia. Ernesto Cortissoz International Airport is in the southern portion of the 40 m resolution image; immediately to the north is the suburb, Soledad. A newly-constructed bypass highway skirts the western edge of the city and the Magdalena River forms the eastern boundary. Extensive port facilities, including an inner harbor, are visible along the river. Small cumulous clouds obscure some detail in the northern part of the city, but a great many streets can be detected throughout the urban area.

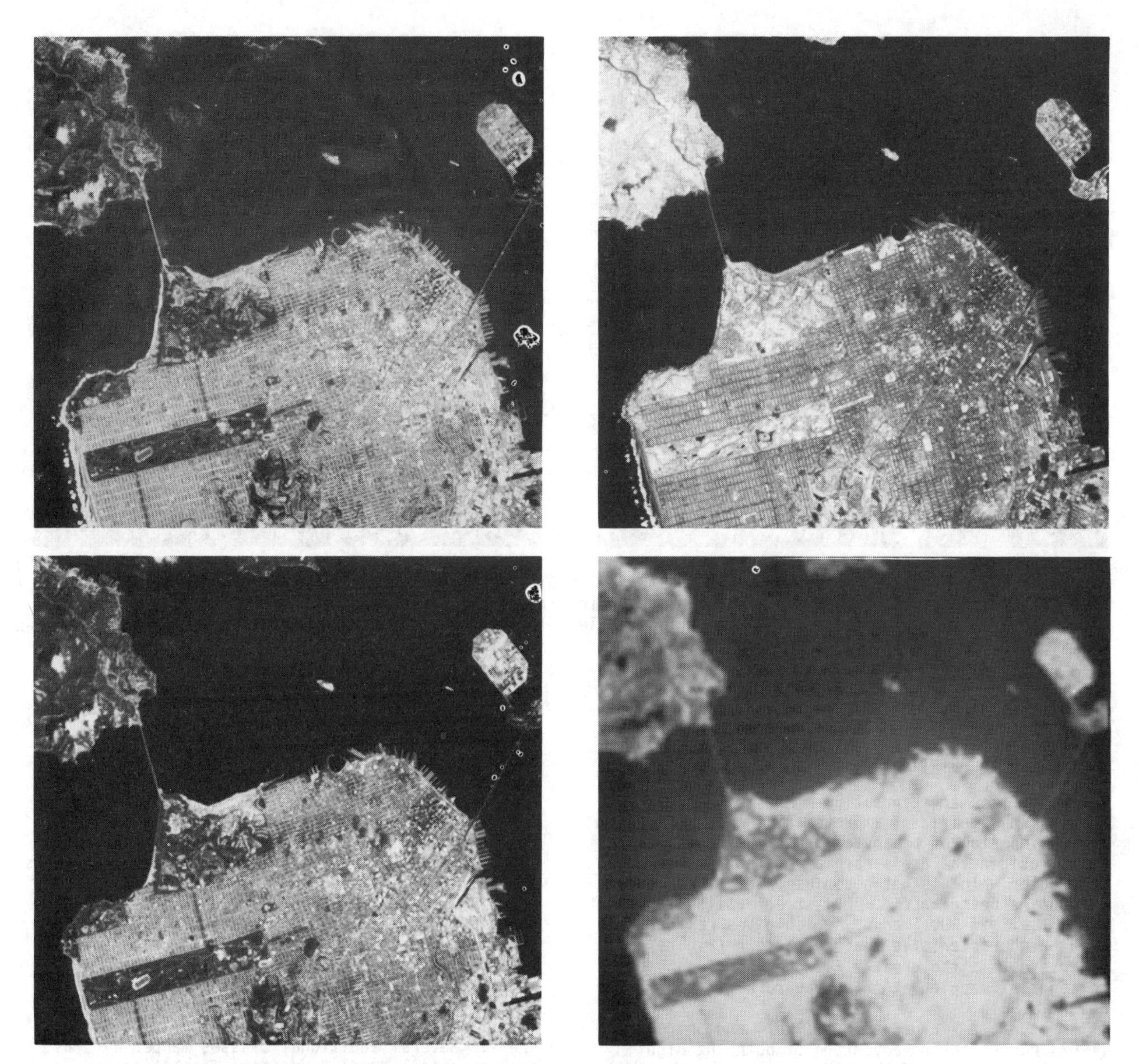

Figure 5. Digitally processed Landsat TM subscene collected over San Francisco on May 2, 1984. upper left Band 3, visible red, 630 - 690 nm; upper right Band 4, near reflective infrared, 760-900 nm; lower left, Band 7, short-wave infrared, 2080-2350 nm; lower right; Band 6, thermal infrared; 10,400-12,500 nm. Streets and land cover patterns are apparent in the 30 m resolution imagery, as are golf course fairways, piers, and large industrial buildings.

water, etc.-- which results in different reflectance values (Figure 3). In a computer classification of Landsat MSS data collected over the Seattle - Tacoma area (Gaydos & Newland, 1978), land cover classes included Commercial-Industrial, Residential, Pasture-Grass, Cropland, several types of forested land, Wetland, Barren Land, and Quarries-Transitional. Analysis of Washington, D.C. Landsat data yielded similar urban classes, including Commercial-Industrial-Services, Paved Surfaces, Older Residental, Newer Residential, Disturbed Land, Improved Open Space, Agriculture, and Forested Land/Brushland (Gaydos & Wray, 1978).

The urban-rural fringe presents a problem to the 80 m MSS data, for two reasons. First, residential and other land use developments are in a transitional state. From land clearing through construction through landscaping through maturing of vegetation and graying of concrete, a new development has a changing appearance both spectrally and spatially. Secondly, the rural land use adjacent to the urban-rural fringe is in either of three states: natural vegetation (forested land, shrub, grassland), agriculture/truck farming, or vacant land. Natural vegetation is the most predictable, spectrally, of the three states, and usually provides the best contrast with urban land uses. The agricultural land, unfortunately, undergoes significant signature changes as the land changes from plowed to emerging crop to mature crop to harvested crop to fallow land. In the Great Lakes region of North America, agricultural land contrasts well with urban areas when the crops are at the height of the growing season. The third state, vacant land, may not contrast well with an adjacent urban development. Forster (1980) provides detailed insight on residential land cover, and Jensen & Toll (1982) suggest that texture measures may be helpful in mapping the urban-rural fringe zone.

While transition at the urban-rural fringe causes difficulty for land cover computer-assisted classification and mapping, scraping the earth for new development is detectable under most data collection circumstances. An application of Landsat to urban land cover change detection (using Landsat

Figure 6. B&W reproduction of September 18, 1973 Skylab-3 S-190B image of Chicago's O'Hare Airport (left) and Loop (right). Many details of the airport, surrounding industrial parks, and residential areas are readily discernable in the 25 m resolution O'Hare image, while details of Chicago Harbor, Grant Park, the Navy pier, and Meigs Field can be detected in the Loop image.

scenes collected two years apart) is described by Todd (1977).

2.3 Medium Resolution Data

Advancing from the Landsat 80 m MSS to 40 m RBV results in significant improvement in identification and delineation of urban features. Using the MSS data, residential areas have a mottled texture and tone/color which is often confused with land cover categories being mapped. But street patterns are evident within RBV imagery, which gives residential areas a characteristic texture/pattern (Figure 4). Of interest is the greatly improved confidence of residential mapping at the urban-rural fringe (Lauer & Todd, 1981).

Two studies -- Lauer & Todd (1981) and Snyder (1982) report taking advantage of both the high spatial resolution of the RBV and the multispectral attribute of the MSS data. In the first study, the MSS and RBV data were spatially registered, the MSS data were resampled to the smaller pixel size of the RBV data, and then color composite images were created for interpretation. The authors report that the RBV data can be used alone, without the MSS, for urban land cover mapping, but that the "combined RBV/MSS color composite image is easier and quicker to work with than the MSS or RBV image alone." In his analysis of MSS and RBV imagery collected over Soviet cities, Snyder (1982) found that the RBV provides "better delineation of boundaries," but MSS gives "better categorical accuracy."

The 40 m RBV data was only a forerunner of the 30 m Thematic Mapper (TM) data, which became available with the launch of Landsat-4 in July, 1982. Both Toll (1985) and Quattrochi (1983) discuss the significant advantages of TM over the MSS. Bernstein et. al. (1984) notes that high-contrast, linear features as narrow as 7.6 m (about 0.25 pixel) can "be easily discerned."

Similar to the RBV data, residential areas exhibit a characteristic texture/pattern which greatly assists in detection and delineation (Figure 5). The urban-rural fringe can be outlined, and confidence in identification of other features is increased. For example, a hint of within-feature texture/ pattern aids in mapping industrial/ commercial areas, transportation/port facilities, and densely vegetated urban land uses such as parks, golf courses, vacant land, and low density residential.

2 High Resolution Data

Relatively high resolution photography was collected from space over urban areas in the early 1970's by Skylab (Figure 6). Evaluation of Skylab's S-190B Earth Terrain Camera photography was done by Welch (1974) and an urban mapping experiment is described by Lins (1976). The 25 m S-190B photography was interpreted to yield urban land use categories, including single-family residential, multifamily residential, industrial and commercial complexes, highways and other transportation facilities, improved open space, and transitional areas. Several additional classes mapped by Lins such as retail trade, education facilities, religious facilities, and government/administration/ services were probably identified with either ancillary data or field work.

More recently, the Large Format Camera (LFC) has been flown on the Shuttle, and has taken high resolution photographs of urban areas (Doyle, 1984 & 1985). The 9.5 m spatial resolution of the LFC photograph shown in Figure 7 illustrates that shape and shadow are additional image characteristics which can be used in image interpretation, permitting detailed (Level II/III) land use and land cover interpretation. Residential areas (several types may be identified) have detailed and unique texture and pattern, which allows accurate detection and delineation. Shapes of structures and transportation features assists in identification of industrial, commercial, services, and transportation land use patterns.

Similar resolution is available with SPOT data, although SPOT also has the 20 m multispectral data (Figure 8). SPOT analysts commonly utilize digitally enhanced SPOT color composite imagery -- combinations of the 10 m panchromatic and 20 m multispectral data. Welch (1985) reports that "classification accuracies in excess of 80 percent can be realized for selected level II/III urban classes." Following is the Urban portion of the classification system used (after Anderson et. al, 1976) to analyze SPOT simulation data:

1 Urban or Built-Up
- 11 Residential
  - 111 Single Family Housing
  - 112 Multiple Family Housing
- 12 Commercial and Services
  - 121 Commercial
  - 122 Institutional
- 13 Industrial
- 14 Transportation and Utilities
  - 141 Road, Highway, Railroad
  - 142 Utilities
- 17 Other Urban or Built-Up Land
  - 171 Urban Grassland
  - 172 Golf Course

Colwell and Poulton (1985) achieved equally good success in using SPOT simulation data for identification of urban features. The authors note that detailed image interpretation "depends very largely on resolution detail that preserves the integrity of building shape, ... and allows detection of major and minor streets and secondary roads."

Welch tried computer-assisted multispectral classification of SPOT data, but with unsuccessful results. Interestingly, successful 80 m Landsat MSS classification results cannot be duplicated with the SPOT data because of the greatly increased texture and complex image information level of the latter (Ballut and Nguyen, 1984).

Microwave data collected at lower frequencies has the highly desirable characteristic of pentrating clouds. Bryan (1982) describes applications of 25 m resolution Seasat SAR for urban mapping. He was able to distingish urban land cover types which are similar to categories researchers have been able to detect using multispectral data collected in the visible and infrared portions of the spectrum. A future source of 30 m resolution radar data is the planned European Space Agency's Remote Sensing Satellite (ERS-1) C-Band SAR (Duchossois, 1984). ESA plans to launch ERS-1 in April 1989.

Figure 7. Large Format Camera photograph of central Boston collected 7 October 1984 by Shuttle Mission 41-G. With an altitude of 231 km, ground resolution is about 9.5 m. Large aircraft, ships, piers, buildings (including their shadows) and details of Logan International Airport are all visible within the image.

Figure 8. Digitally enhanced SPOT simulator visible red image (Band 2, 610-680 nm) of Santa Cruz collected in June, 1983. Street patterns are clearly visible in the 20 m resolution image, as well as the piers in the elongated small boat harbor which occupies a dredged slough.

Table 1. Empirical Evaluation of Satellite Imagery Collected over Urban Areas.

CATEGORY	IMAGE GROUND RESO-LUTION	PRIMARY IMAGE CHARACTERISTICS FOR IMAGE INTERPRETATION	CHARACTERISTICS OF URBAN FEATURES
VERY LOW RESOLUTION	500M-1 KM	TONE/COLOR	• URBAN VS NON-URBAN
LOW RESOLUTION	80M	TONE/COLOR + RELATIVE LOCATION	• URBAN LAND COVER TYPES • MAJOR TRANSPORTATION/ COMMERCIAL ARTERIES • LAND CLEARING DETECTABLE
MEDIUM RESOLUTION	30M	TONE/COLOR + RELATIVE LOCATION + SOME TEXTURE/PATTERN	• RESIDENTIAL HAS SOME TEXTURE/PATTERN OF STREETS • URBAN/RURAL FRINGE DISTINCT • VERY LARGE BUILDINGS DETECTABLE • URBAN LAND USE/LAND COVER • FEATURES DETECTED AND DELINEATED WITH HIGHER CONFIDENCE • NEW CONSTRUCTION (LAND SCRAPING) EVIDENT
HIGH RESOLUTION	10-15M	TONE/COLOR + RELATIVE LOCATION + DETAILED TEXTURE/ PATTERN + SHAPE/SHADOW	• DETAILED LEVEL II/III URBAN LAND USE MAPPING POSSIBLE • LARGE AND MEDIUM SIZED STRUCTURES (AND SHADOWS) DETECTABLE • ALL TRANSPORTATION FEA-TURES EVIDENT • VEGETATION HAS DISTINCTIVE TEXTURE

3 SPATIAL RESOLUTION ANALYSIS SUMMARY

A summary of the utility of the sensors for urban mapping is listed in Table 1. As the spatial resolution of the data increases, more image characteristics are available for image interpretation. With very low resolution data, tone/color is available to the interpreter, but only urban/rural differentiation may be made. Moving to low resolution, the analyst uses tone/color, along with relative location of features, to distingish major urban land use and land cover types, major transportation arteries and commercial strip development, and land clearing for new construction. A degree of image texture/pattern is available when using medium resolution data, permitting urban land cover types--particularly at the urban-rural fringe--to be detected and delineated with higher confidence than using low resolution data. Finally, the high resolution data from space reveal shape and shadow of urban features; when that image interpretation aid is added to tone/color, relative location, and detailed texture/pattern the interpreter can produce detailed Level II/III maps.

Table 2. Summary of Spatial Resolution Requirements for Urban Mapping from Space.

URBAN MAPPING REQUIREMENT	MAP SCALE	EXAMPLE USER	CATEGORY/ IMAGE GROUND RESOLUTION
• URBAN/RURAL DIFFERENTIATION (LEVEL I) • ENERGY UTILIZATION	1:1,000,000	• UTILITY PLANNER • DEMOGRAPHIC ANALYST	VERY LOW RESOLUTION/ 500M-1 KM
• GENERALIZED LAND USE/LAND COVER • LARGE LAND CONVERSION (NEW CONSTRUCTION) MONITORING	1:250,000	• STATE AND REGIONAL LAND USE PLANNERS • TRANSPORTATION PLANNER • ENVIRONMENTAL PROTECTION ANALYST	LOW RESOLUTION/ 80M
• GENERALIZED LAND USE/LAND COVER • URBAN/RURAL FRINGE DEMARCATION • NEW CONSTRUCTION MONITORING	1:100,000	• REGIONAL AND CITY LAND USE PLANNERS • URBAN TRANSPORTATION PLANNER • ENVIRONMENTAL PROTECTION ANALYST • WATER RESOURCE PLANNER • DISASTER WARNING PLANNER	MEDIUM RESOLUTION/ 30M
• LEVEL II/III LAND USE (DETAILED) • NEW CONSTRUCTION MONITORING	1:25,000-1:50,000	• REGIONAL AND CITY LAND USE PLANNERS • CARTOGRAPHERS • URBAN TRANSPORTATION PLANNER/ENGINEER • ECONOMIC AND COMMUNITY DEVELOPMENT PLANNER • ENVIRONMENTAL PROTECTION ANALYST • WATER RESOURCE PLANNER • DISASTER WARNING PLANNER • INDUSTRIAL/COMMERCIAL LOCATION ANALYST	HIGH RESOLUTION/ 10-15M

4 CONCLUSIONS: Spatial Mapping Requirements

Agency mapping requirements vary greatly in level of detail needed, timeliness of data required, desired map data output products, area of responsibility (acreage to be mapped), and agencies vary in ability to purchase and process data.

Notwithstanding, we attempted to use the empirical data from Table 1 to derive the spatial resolution requirements listed in Table 2. For generalized, large-area, Level I urban-rural differentiation, the very low resolution data might be applicable. Example map scale is 1:1,000,000, and potential users are utility planners, demographers, and federal land use planners. For generalized land use/land cover maps of urban areas at the scale of 1:250,000, low resolution (80 m) data could be used. State and regional planners, transportation planners, and environmental protection analysts could all be placed in the potential user category. Land use/land cover maps at the scale of 1:100,000 with clear demarcation of the urban-rural fringe could be used by city and regional planners, as well as a wide range of urban and resource planners/analysts. The requirement for Level II/III land use mapping can be met by interpretation of high resolution (10-15 m) data collected from space. Maps with scales ranging from 1:25,000 to 1:50,000 can be made, and many types urban analysts, planners, and engineers are potential users, including regional and city planners, cartographers, transportation planner, city engineer, urban demographer, and industrial/commercial location analyst.

REFERENCES:

Anderson, J.R., Hardy, E.E., Roach, J.T. & Whitmer, R. E. 1976. A Land Use and Land Cover Classification System for Use with Remote Sensor Data. U.S. Geol. Survey Prof. Paper 964, 28 p.

Ballut, A. & Nguyen, P.T. 1984. Potential Applications for SPOT Data in the Paris Region from the Results of 1981 & 1983 Simulation Studies. Proc. 18th Int. Symp. on Remote Sensing of Environment. 2:693-703.

Bernstein, R., Lotspiech, J.B., Myers, H.J., Kolsky, H.G. & Lees, R.D. 1984. Analysis and Processing Landsat-4 Sensor Data Using Advanced Image Processing Techniques and Technologies. IEEE Trans. on Geoscience & Remote Sensing. GE-22:192-221.

Bonn, F., Bernier, M. & Brochu, R. 1981. Visual and Digital Analysis of H.C.M.M. Data over Eastern Canada. Proc. 15th Int. Symp. on Remote Sensing of Environment 3:1449-1463.

Bryan, M.L. 1982. Analysis of Two Seasat Synthetic Aperture Radar Images of an Urban Scene. Photogrammetric Eng. & Remote Sensing. 48:393-398.

Colwell, R.N. & Poulton, C.E. 1985. SPOT Simulation Imagery for Urban Monitoring: A Comparison with Landsat TM and MSS Imagery and with High Altitude Color Infrared Photography. Photogrammetric Eng. & Remote Sensing. 51:1093-1101.

Doyle, F.J. 1984. The Economics of Mapping with Space Data. ITC Journal, p. 1-9.

Doyle, F.J. 1985. The Large Format Camera on Shuttle Mission 41-G. Photogrammetric Eng. & Remote Sensing. 51:200.

Duchossois, G. 1984. ERS-1: Mission Objectives and System Description. Proc. 18th Int. Symp. on Remote Sensing of Environment. 1:145-157.

Forster, B.C. 1980. Urban Residential Ground Cover Using Landsat Digital Data. Photogrammetric Eng. and Remote Sensing. 46:547-558.

Gaydos, L. & Newland, W.L. 1978. Inventory of Land Use and Land Cover of the Puget Sound Region Using Landsat Digital Data. Journ. Research U.S. Geol. Survey. 6:807-814.

Gaydos, L. & Wray, J.R. 1978. Land Cover Map From Landsat, 1973, with Census Tracts, Washington Urban Area, D.C., Maryland, and Virginia. Folio of Land Use in the Washington, D.C. Urban Area, Map I-858-F. U.S. Geol. Survey. Scale 1:100,000.

Gervin, J.C., Kerber, A.G., Witt, R.G., Lu, Y.C. & Sekhon, R. 1983. Comparison of Level I Land Cover Classification Accuracy for MSS and AVHRR Data. Proc. 17th Int. Symp. on Remote Sensing of Environment. 3:1067-1076.

Jensen, J.R. & Toll, D.L. 1982. Detecting Residential Land-Use Development at the Urban Fringe. Photogrammetric Eng. & Remote Sensing. 48:629-643.

Lauer, D.T. & Todd, W.J. 1981. Land Cover Mapping with Merged Landsat RBV & MSS Stereoscopic Images. Proc., ASP-ACSM Fall Tech. Meeting, p. 68-89.

Lins, H.F., Jr. 1976. Land-Use Mapping from Skylab S-190B Photography. Photogrammetric Eng. and Remote Sensing. 42:301-307.

Quattrochi, D.A. 1983. Analysis of Landsat-4 Thematic Mapper Data for Classification of the Mobile, Alabama Metropolitan Area. Proc. 17th Int. Symp. on Remote Sensing of Environment. 3:1393-1402.

Snyder, D.R. 1982. Integration of Landsat RBV and MSS Imagery to Produce Land Use Maps of Soviet Cities. Proc., Pecora VII Symp. Remote Sensing: An Input to Geog. Info. Systems in the 1980's, p.94-103.

Todd, W.J. 1977. Urban and Regional Land Use Change Detected by Using Landsat Data. Journ. Research U.S. Geol. Survey. 5:529-534.

Toll, D.L. 1985. Landsat-4 Thematic Mapper Scene Characteristics of a Suburban and Rural Area. Photogrammetric Eng. & Remote Sensing. 51:1471-1482.

Welch, R., Jordan, T.R. & Ehlers, M. 1985. Comparative Evaluations of the Geodetic Accuracy & Cartographic Potential of Landsat-4 and -5 Thematic Mapper Image Data. Photo. Eng. & Remote Sensing. 51:1249-1262.

Welch, R. & Zupko, S. 1980. Urbanized Area Energy Utilization Patterns from DMSP Data. Photogrammetric Eng. & Remote Sensing. 46:201-207.

Welch, R. 1974. Skylab-2 Photo Evaluation. Photogrammetric Eng. 40:1221-1224.

Welch, R. 1985. Cartographic Potential of SPOT Image Data. Photogrammetric Eng. & Remote Sensing. 51:1085-1091.

Symposium on Remote Sensing for Resources Development and Environmental Management / Enschede / August 1986

Analysis and evaluation of recreational resources with the aid of remote sensing

D.van der Zee
International Institute for Aerospace Survey and Earth Sciences (ITC), Enschede, Netherlands

ABSTRACT: Everywhere in the world the pressure on the recreational resources is increasing, proper management of these resources is dearly needed, and for this good information is indispensable. Airphoto interpretation can give an important contribution to such information. Analysis of the present situation may reveal what recreational activities are attracted by which recreational resources. By a study of sequences of airphotos the process of developments can be analysed,the impact of the recreation on the landscape assessed, and an indication of the spatial behaviour of recreationists obtained. Analysis of this impact and of the spatial behaviour can give a more detailed understanding of what use is made of which specific (parts of) recreational resources. Next to physical suitability and accessibility of the terrain, the general attractivity of the area in which the recreational activities take place can be evaluated by special methods, in many of which airphoto interpretation is an important tool.

RESUME: De partout dans le monde la pression sur les ressources de loisirs augmente, une aménagement adéquate de ces ressources est indispensable et nécessite de bonnes sources d'information. L'interprétation des photographies aériennes peut contribuer fortement a l'obtention de cette information. L'analyse de la situation présente peut révéler par quelles ressources sont attirées certaines activités de loisirs.
Par une étude à l'aide de photographies aériennes séquentielles, le processus de développement peut être analysé, l'empreinte des loisirs sur le paysage peut être étudiée, ainsi que la conduite spaciale des utilisateurs des activités de loisirs.
L'analyse de cet empreinte et cette conduite spaciale permet une compréhension plus détaillée de l'utilisation de quelles (intégrales ou partielles) ressources spécifiques de loisirs. En plus des propriétés physiques et l'accessibilité du terrain, l'attrait général de la région dans laquelle les activités de loisirs ont lieu peut être évalue par des méthodes spéciales. Dans plusieurs de ces méthodes l'interprétation des photographies aériennes joue un role important.

INTRODUCTION

Next to landevaluation for all kinds of agricultural, grazing, forestry or other landuses, landevaluation for recreational landuses is becoming more and more relevant also in developing countries. Increasing numbers of people take part in recreational activities and the income from tourism is a welcome support to the economy of many a country or region.
Everywhere in the world pressure on the recreational resources is increasing. Therefore proper management of these resources is dearly needed. For such proper management good information is indispensable and can be best presented in a land(scape)evaluation procedure. In the inventory and analysis phase of such a procedure remote sensing techniques - so far mainly airphoto interpretation has been used - are very important.

SOME BASIC CONCEPTS

Landevaluation is a method or procedure in which specific land uses or "land utilisation types" (lut's) with their requirements are confronted with "land (mapping)units" (lu's) with their characteristics and qualities in order to establish which land units are in what degree suitable for which land utilisation types (FAO, 1977).

Before starting the discussion on the application of this procedure to the topic recreation, this term, and some related terms, may need some definition in order to avoid misunderstandings. Out of the many definitions of recreation the following definition is preferred: "recreation": refreshment of body or mind by activities or a planned inactivity undertaken because one wants to do it, without any moral, economical, social or other pressure (Zee 1971: 6; 1985: 1).
It is a rather wide definition of recreation and comprises a large variety of activities from watching television to mountain climbing. But since the majority of these activities concentrates in and around the home (Cosgrove 1972: 22), only a few are relevant to consider as land utilisation type.

Recreation in almost all cases is associated with "leisure" or "leisure time". Leisure is the time left after the economic, social and other obligations are fulfilled, it is discretionary time, to be used as one chooses (Clawson 1966: 12).
Although leisure and recreation are highly correlated, they are not the same. Leisure is time of a special kind; recreation is activity (or inactivity) of special kinds (Clawson 1966: 12).

Another term recreation is frequently related to is "tourism". Tourism can be considered as recreation, but not all recreation is tourism. Tourism can be defined as those types of recreation for which one leaves the home environment for shorter or longer duration. Many definitions of tourism emphasize the staying one or more nights away from home and the using of facilities and services (Büchli 1962: 23-28; Robinson 1953: 91; Defert 1952: 127), and clearly are formulated from the point of view of the "industry" that provides travelers (= tourists) with food, lodging and entertainment and indeed sometimes this industry itself is defined as "tourism" (Pearson 1961: 448). The difference between recreation and tourism therefore seems not to be a difference in activities between recreationists and tourists, but rather the distance away from home at which these activities are persued. For the purpose of landevaluation it is the type of activities that determine the character of the lut, and the differrence between recreation and tourism seems to be less relevant.

In general, if one is planning to satisfy the recreational needs and wants of the own population one will use the term recreation; if one is planning to attract recreationists from elsewhere tourism will most likely be the term used.

The providing of goods, services and entertainment to visitors who come to an area for recreational activities can be of considerable economic importance for that area (Krapf 1962: 93; MacConnel 1969: 686; Pearson 1961: 448; Peppelenbosch 1973: 53; Zee 1983: 270-271). If these visitors come from abroad, tourism may become an important source of foreign exchange (Krapf 1962: 91) and can be seen as "invisible" export (Krapf 1952: 3).

The relevance of landevaluation for recreation, or in this case rather tourism, for developing countries is mainly seen in this context (Robinson 1972: 561). But the recreational needs and wants of the local population should not be overlooked. Their rates of participation in recreation are rapidly increasing as are the resulting impacts on the landscape (Robinson 1972: 561).

IDENTIFICATION OF RECREATIONAL RESOURCES

In landevaluation land utilisation types are confronted with land units. In landevaluation for recreation the land utilisation types are different "types of recreation" (boating, swimming, hiking, riding, etc.), and the land units can be interpreted for their "recreational resources". The first thing to do will be to identify what are relevant recreation types (lut's) and what are their requirements with respect to the characteristics of their resources (lu's).

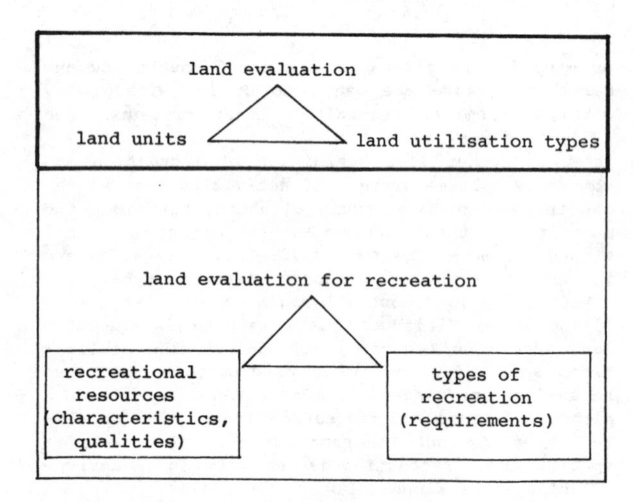

Which types of recreation are preferred will depend not only on the socio-economic and cultural characteristics of a population, but in addition to that also on fashion, taste and changing lifestyles (Cosgrove 1972: 16). With respect to the developing countries in this context Robinson's analysis of recreation patterns in South Asia is interesting (Robinson 1972). Which types of recreation out of this preference then are actually carried out depends in part also on the availability and accessibility of resources and facilities.

Recreational resources often are associated with the natural environment, with the rural landscape. But not all natural areas or rural landscapes are in the same degree suitable or attractive for recreation. It is not only the natural qualities but also the ability and desire of man to use them, that makes a resource out of what otherwise may be a more or less meaningless combination of rocks, soil and trees (Clawson 1966: 7). Still there is an apparent preference for certain characteristics when choice is possible (Clawson 1966: 141). It is these characteristics that need to be identified in order to be able to carry out a landevaluation for recreation.

It is not only natural features of the landscape that can be considered as resources for recreation. There may be also certain man-made attributes (reservoirs, dams, monuments, historical sites, picturesque settlements, etc.) that attract recreationists (Christaller 1955: 5). In general neither natural features nor these man-made attributes are created nor maintained in the first place for recreation, even though at present they may attract many recreationists. Therefore they can be considered as "original recreational resources" (Zee 1971). Man-made attributes specially created and maintained for the sake of recreation are the "recreational facilities" or "recreational infrastructure", that are often closely related to, and impossible to exist without, the original resources (Zee 1971).

For identifying relevant recreational land utilisation types one may use all kinds of (sample) enquiry surveys on recreational preferences and participation rates. But most of them require a lot of time and effort. A more direct approach is to make an inventory of the number, type and capacity of all kinds of touristical, or recreational, infrastructure, in the way of the "touristical possibilities" approach (Defert 1954: 111-112). From such an inventory, which can be easily carried out by airphoto interpretation (see e.g. MacConnel 1969; Zee 1982), a general impression can be obtained of what types of recreational activities occur in an area and to what specific elements in the landscape they apparently are related, that is, what are their original resources.

Most recreation areas are complex in the sense that different types of recreation do occur side by side (Defert 1966: 100), and make use of the same or of different elements of the landscape.

Once the original resources have been identified out of the analysis of actual recreational use, these resources can be inventoried also for areas where no actual recreational use is observed. This can give an indication of the available potential that may be developed. Comparison of these resources in both types of area may reveal which factors at the moment determine the use or non-use by recreation, that is, the present suitability rate.

It may be simply a question of distance with respect to the nearest concentration of potential users (Clawson 1966: 146), or of accessibility, but recreational use may also be frustrated by other uses of the same resources. Water pollution may impede or restrict the recreational use of waterbodies or streams (Mac Connel 1969: 687).

Or, resources that at first sight seem to have the same characteristics, in closer observation show some differences that causes a difference in suitability too. Also, when the basic physical requirements of a recreation lut are met, additional characteristics may determine a difference in the overall attractivity of an area and thereby influence the actual use pattern.

It may be selfevident that not all characteristics that are important in determining the suitability for a certain type of recreation can be interpreted from airphotos, and that types of recreation, that do not need specific physical facilities in the area, will also escape the attention. Nevertheless, airphotos can be a powerful tool for landevaluation for recreation.

ANALYSIS OF RECREATIONAL DEVELOPMENTS

When it is possible to analyse not only the most recent pattern of recreational use of an area, but also the use patterns in one or more situations in the past by interpreting older sets of airphotos, the factors determining the suitability and/or attractivity of recreational resources may become more apparent than when only one situation can be studied. Examples of such sequential airphoto interpretation studies are that on the Skanör peninsula in south Sweden (Rasmussen 1962), the Connecticut River Valley (MacConnel 1969) and the Proserpina Lake area in south-western Spain (Zee 1982).

From such studies it can be deduced where recreational development started, which parts of the area were first occupied by recreational uses and which parts were incorporated in later phases. This may give a clear indication on which type of resources or resource elements have (or had) the highest preference and which parts are more of a second or third choice. Also the character of the area at the time before the recreational development took place can be analysed. In this way the requirements of the recreational lut's may thus be more narrowly defined.

But from such a comparative study also information can be abstracted about which types of recreation were apparently more favourite in the earlier period and which types of recreation are more recently in the focus of interest, thus about changes in preference or fashion.

It may also reveal in which period the bulk of recreational development has occurred and thus whether the area is a rather established, traditional recreation area, or whether it is a new emerging recreation area. Both types of area may require their own approach in landevaluation for recreation.

IMPACT OF RECREATION ON ITS RESOURCES

In an already established recreation area the impact of recreation on its environment may have become evident,and analysis of this impact may give indications of what impact can be expected in newly developing recreation areas and how possibly to avoid or restrict negative effects.

The impact of recreation on its resources, which are embedded in the landscape, can be manifold. What type of impact and how strong it will be,will depend on the type of recreational activity on the one hand and the type of landscape on the other hand.

In many developing countries the impact of recreation on its environment as yet may be very slight only, but will certainly become more obvious as standards of living rise, and with that the rates of participation in recreation (Robinson 1972: 561).

The impact of recreation needs not always be negative. On many of the European Wadden Sea islands,for example, where recreation became an important source of income, as a consequence the agricultural pressure - in particular on the most vulnerable areas which were less suitable for agriculture anyway - could be reduced considerably. This positive influence, however, often is balanced or even outweighed by the negative impact (Zee 1983: 270).

A very clear effect on the landscape is caused by the physical facilities for recreation, whether permanent, semi-permanent or temporary. This effect makes it possible to inventory the actual recreational use by airphoto interpretation. The aspect of the landscape can change drastically if it is occupied by summer cottages, a caravan or camping site, a large parking place, a marina, etc. Access to that part of the landscape may be restricted to the owners or users of those facilities, excluding other recreationists and other types of recreation.

The presence of the recreational facilities will affect the characeristics of the very resources that attracted them.

The same sequential studies as mentioned earlier are also good examples of the analysis of the impact of recreation on its environment.

But it is not only by constructing physical facilities that the landscape is affected by recreation. Since for many types of recreation the natural landscape is the major original resource, it can be expected that recreationists visit this landscape, walk in it, sit in it, play in it, throw around litter in it, in short, display a behaviour that normally is not destructive in intention but is damaging in effect (Zee 1983: 273).

The effects of recreation are most pronounced when the feet of repeatedly passing recreationists create a network of tracks and paths and even areas of bare soil. This is the ultimate stage of a process that starts with a change in the vegetation composition and a general degeneration of the vegetation (Ittersum 1977: 67-68; Zee 1983: 273). Because the process is very gradual it is often not easily recognized. Changes in vegetation composition and early stages of degeneration are hard to detect in the field, leave alone on airphotos, but the effects of "recreational erosion" can often be assessed, and especially by comparing sequences of aerial photographs the results of the gradual process can be clearly demonstrated.

Measuring the increase in length of path-network or the increase in area of bare soil is a relatively easy way to quantify the impact of recreation on the environment (Ittersum 1977: 75-76; Zee 1983: 273).

Which parts of the natural landscape are influenced mostly,depends on their vulnerability, their location with respect to villages and concentrations of recreational facilities, as well as on their attractivity for certain types of recreation (Ittersum 1977: 71; Zee 1983: 273).

SPATIAL BEHAVIOUR OF RECREATIONISTS

An analysis of the impact of recreation on the landscape, as mentioned in the previous section, may already give an indication of the spatial behaviour characteristics of recreationists. The recreational use is not evenly spread over an area and not all parts of the area are used with the same objectives. In the path-network created by recreational erosion several distinct patterns can be detected reflecting different types of spatial behaviour (Ittersum 1977: 68-70; Zee 1983: 273-274).

The study of spatial behaviour of recreationists can be done by means of direct observations and questionnaires. But if impacts of recreation can be detected on airphotos, a good, and above all an overall and comprehensive, impression of spatial behaviour also can be obtained.

Knowledge about this spatial behaviour can help not only in further specifying the physical requirements and preference of a recreation type, but also in determining which factors apart from physical suitability and accessibility make up the attractivity of a landscape for recreation. It also can be very useful in planning for proper management of recreational resources.

Therefore it would be interesting to also be able to analyse this spatial behaviour if no impacts can be detected on airphotos. If the recreationists or their vehicles can be easily detected, analysis of a series of airphotos taken during one day can be a solution.

In the Netherlands this has been done especially for watersports and for beach and shoreline recreation. Both oblique and vertical airphotos have been applied in various case studies (CD&PW 1970; RWS 1977; PWF 1977; RWS 1979; see also Dodt 1984).

For the watersports it revealed which parts of a lake or lake-system were more frequented than other parts, which shores were used more than other sho-

res. Also it showed that there are certain peaks in time with respect to the movements of boats, and that only a relatively small part of all boats actually was sailing at any given time.
In this way this specific recreational lut could be further specified and even subdivided and the requi rements defined in more detail. It helped in design ing a plan for a proper management of the recreation area.

Also in the analysis of recreation along beaches and shorelines specific patterns of behaviour in space and time could be observed (RWS 1979; and also Dodt 1984).

The "border effect", that is the result of people's preference for selecting a place that permits a clear view over the open space but at the same time gives some backcover, appears rather clear (Jonge 1965: 1873; 1968: 13).
Distance from parking place or entrance gate is another important factor.

This type of information makes it possible to design a recreation area in such a way that a maximum number of recreationists can be satisfactorily accomodated on a minimum of area. It also makes it possible to anticipate an expected distrubution over an area with measures to protect some sensitive parts of the area.

MAIN APPROACHES TO LANDEVALUATION FOR RECREATION

In the previous sections it has been discussed how to identify and define - in more or in less detail- relevant recreational lut's and the qualities of land units (resources), that are able to satisfy the requirements of these lut's, by various methods in which airphoto interpretation can play a major role. Now it is time to try and discuss how these methods can be applied in real landevaluation procedures.

Different approaches are possible on different levels of detail. The first approach could be called the "recreation approach". In this approach the starting point is an apparent (often rapidly increasing) demand for recreation that is exerting an increasing pressure on the available resources. After identification of the major demands and properly defining them as recreational lut's with their requirements an inventory can be made of the land units, landscape elements or resources, that are in varying degrees suitable for these lut's. Actual use can then be compared with the potential of the resources and can give an indication of the possibilities for further development. These possibilities can be the new development of yet unexploited resources or measures to achieve an optimal use of the presently used resources.

The second approach could be called the "tourism approach". In this approach the starting point is the notion that a certain recreational (or touristical) resource is available and that development of that resource might have a positive influence on the economy of the region. After a first exploratory definition of the resource the potential demand should be identified. That means:for what type of recreational lut's would this resource be suitable, where are concentrations of demand for this lut and what alternative competing supplies of resources for this demand are available? In other words, what is the chance that development of the resource will attract sufficient numbers of tourists to make the investment worthwile?

A third approach could be called the "conservation approach". In this approach the central issue is, that recreationists are attracted by resources that are also considered to have high value from a point of view of nature and/or landscape conservation and that may be damaged by (too high) a recreational pressure. After a first inventory of the resources involved the recreational lut's attracted by these resources can be inventoried and analysed, especially with respect to their impact on the resources and also to the spatial behaviour patterns. Actual use and potential use can be compared and possible future developments identified. These future developments can be anticipated and guided or deflected making use of the knowledge of factors influencing the spatial behaviour. Without banning recreation completely, with proper management the main conservation aim may be achieved as well.

A variation on this approach is to only identify which recreational lut's, with respect to their impacts, could be tolerated in which parts of the area without objection to the major aim of conservation. This could be called the "permissive approach". Examples of the use of this approach are the Meyendel case in the Netherlands (Meulen 1985) and the Štiavnické Vrchy landscape area in Czechoslovakia (Krajčovič 1985).

Of course there may be more approaches or variations on these approaches possible. One major further subdivision could be made with respect to the level of detail. Just as with landevaluation for all other kinds of lut's also here a distinction can be made between reconnaissance, semi-detailed and detailed level (FAO 1977: 6).

Thus,in the recreation approach in a reconnaissance survey a first inventory of resources can be made with a rough suitability grading. Then a further semi-detailed suitability analysis can be made for only the most promising resources, after which a detailed analysis would be necessary to establish the proper management of the finally selected resources. An example of a multistage approach of landevaluation for recreation is given by Dill (1962) for the north-eastern USA. Land evaluation for recreation in more or less single stages with the aid of airphoto interpretation are described by Olson et al. (1969) for Michigan, USA, and by MacConnel (1969) for the Connecticut River Valley, USA.
British examples of landevaluation for recreation are that for Snowdonia National Park (Gittins, in: Rodgers 1973: 483-484) and that for the North York Moors (Statham 1972). In these examples airphoto interpretation was not actually used, but could have been used very well.
A example of the German approach to landevaluation for recreation is the study carried out for Sauerland (Kiemstedt 1975).

In the tourism approach the reconnaissance phase would have to give the answer on the question whether the available resource will be able to attract enough tourists or not. If yes, then further analysis can be carried out to determine where best to concentrate the development of what type of facilities.
So if, for example, a country like Botswana wants to exploit its major touristic resource, which is wildlife, it has to realize that with respect to the main sources of tourists, Western Europe and North America, it has heavy competition from East Africa where wildlife viewing can be combined with a beach holiday, and that nearby concentrations of demand, mainly from South Africa, are relatively small (Zee 1985).

Only in the conservation approach it may be expected that analysis at semi-detailed or even detailed level is directly required. This approach may be, or better, ought to be called in at this level in the two other approaches.

With respect to the methodology involved, the reconnaissance level could be based on interpretation of a single (small scale) airphoto coverage only, or may be even of a satellite image. For semi-detailed analysis larger scales and sequential coverages might be necessary to give more information about trends and about the impact on the environment. The detailed level may require special purpose photography to assess spatial behaviour characteristics.

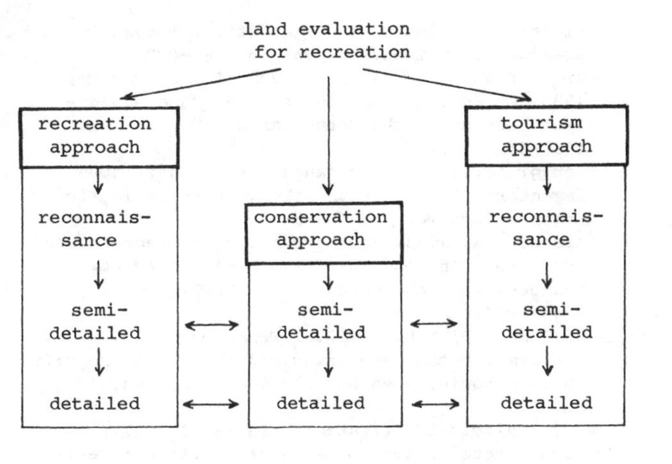

ACCESSIBILITY

Physical suitability of a piece of land for one or another type of recreation alone is not sufficient. Accessibility is another important factor, that wil determine in what degree recreational use can actually be made (Defert 1954: 113).

With respect to accessibility distinction should be made between the "external accessibility", that is the long distance accessibility to the outside world, linking a recreation or tourist area with the main population centres, and the "internal accessibility", that makes it possible for the recreationists to move into and around in the recreation area (Zee 1983: 275; see also Defert 1966: 106).

Modern transportation, of course, makes the whole world accessible if one has enough time and, above all, money. For most recreationists however the potential zones for day, weekend, and vacation recreation are restricted (Zee 1982: 362). These zones do not have absolute limits but are more like ranges within which major parts of the recreationists or tourists will stay, and that only will be passed by small percentages.

Thus, in the recreation approach of landevaluation one may start defining the range to which one might restrict the procedure, in the tourism approach one first has to establish whether the resource concerned falls within the ranges of certain major sources of tourists and whether there are no competing "intervening opportunities".

Location and distance, measured in time and effort of travel rather than in mere kilometers, are important factors determining the external accessibility, but have always to be considered in relation to the available means of transport.

The type and lay-out of the internal accessibility of a recreation area may strongly determine the actual spatial behaviour pattern of the recreationists and the resulting impact on the environment. Manipulating accessibility can be an important instrument in the management of a recreation area.

GENERAL ATTRACTIVITY OF THE LANDSCAPE

Not only the physical suitability and accessibility are important to determine whether a potential recreational resource will be used or not. Almost as important is the overall attractivity of the area, that is, the aesthetic qualities of the environment, the landscape, in which the recreational activities take place.

The area to be evaluated for this general attractivity cannot be restricted to the site of actual recreational use. Large parts of the landscape that are the scene of little activity are nevertheless highly valuable and sometimes indispensable to the heavily used areas by serving as a bufferzone or merely as the background against which an activity takes place. A campground on a city block would never provide the same experience as a campground in the forest even though the sites were physically identical and the user never left the small site in the forest (Clawson 1966: 155).

The analysis of this general attractivity is the subject of "landscape evaluation".

Preferences for specific types of scenery or landscapes are largely subjective, even though most people would agree that some areas are inherently more attractive and outstanding than others (Clawson 1966: 38). Therefore it is very difficult to define the features or characteristics that determine such preferences and with which the quality of the landscape can be assessed. It is even more difficult to express this in quantitative terms to make comparison between areas possible.

Still, analysis of the spatial behaviour of recreationists may reveal which elements in the landscape apparently are more attractive than others, for example, the borderzones. Especially in Germany methods for the evaluation of the landscape for recreation have been developed, in which the overall attractivity is related to the visual structure of the landscape, expressed, amongst others, in terms of the length of borderzones per unit of area (Kiemstedt 1967, 1972, 1975). Although the values are commonly expressed as values per map gridsquare, for the inventory of borderzones and other landscape structuring elements airphoto interpretation can be a very useful tool.

With respect to the analysis of the visual structure of the landscape several Dutch approaches can be mentioned (Ham 1971; Werkgroep Helmond 1974; Grontmij 1975), in which, especially in the last one, the interpretation of airphotos plays an important role. For more hilly or mountainous areas the Dutch method of landscape analysis is not the most appropriate one, but in other countries other methods have been developed, for example the "viewshed" method (Aguilo 1981), and the method applied by Baumgartner(1981).

For each region a method has to be developed that is tailored to the specific type of landscape. But where some of the methods originally are based on the analysis of topographic maps and fieldobservations, an increase in efficiency can be obtained if airphoto interpretation is used.

CONCLUSIONS

Land(scape)evaluation procedures can be very useful in providing the necessary information for proper planning and management of recreational resources. These procedures can, in general, be made much more efficient when airphoto interpretation can be used in the inventory and analysis phase. Since each recreation area has its own specific characteristics the landevaluation procedure has to be tailormade for that area within the framework of generally accepted principles of landevaluation.

REFERENCES

Aguilo, M. & A. Ramos 1981. Viewshed and Landscape Morphology. In: Proceedings of the International Morphology. In: Proceedings of the International Congress of the Netherlands Society for Landscape Ecology, Veldhoven, the Netherlands, p.310-311. Pudoc, Wageningen.

Baumgartner, R. 1981. Inventory and Evaluation from the Visual/Aesthetic Perspective. In: Proceedings of the International Congress of the Netherlands Society for Landscape Ecology, Veldhoven, the Netherlands. p.318-319. Pudoc, Wageningen.

Büchli, H. 1962. Zur Terminologie des Fremdenverkehrs und der Fremdenverkehrswerbung. Zeitschrift für Fremdenverkehr, no 1: 23-28.

CD&PW (Cultuurtechnische Dienst & Provinciale Waterstaat Friesland) 1970. Luchtfototellingen op een aantal Friese meren. Leeuwarden.

Christaller, W. 1955. Beiträge zu einer Geographie des Fremdenverkehrs. Erdkunde, Band IX, Heft 1: 1-19.
Clawson, M. & J.L. Knetsch 1966. Economics of outdoor recreation. Resources for the future, Inc., Washington D.C. (2nd edition 1969).
Cosgrove, I. & R. Jackson 1972. The geography of recreation and leisure. Hutchinson University Library, London.
Defert, P. 1952. Les fondements géographiques du tourisme. Zeitschrift für Fremdenverkehr, no 4: 126-132.
Defert, P. 1954. Essai de localisation touristique. Zeitschrift für Fremdenverkehr, no 3: 110-118.
Defert, P. 1966. Der touristische Standort - Theoretische und praktische Probleme. Zeitschrift für Fremdenverkehr, no 3: 99-108.
Dill, H.W. 1963. Airphoto Analysis in Outdoor Recreation: Site Inventory and Planning. Photogrammetric Engineering, vol XXIX: 67-70.
Dodt, J. & D. van der Zee 1984. Möglichkeiten der Anwendung von Luftbildinterpretation in der räumlichen Freizeit- und Erholungsplanung. In: Angewandte Fernerkundung. Methoden und Beispiele, p.65-70. Akademie für Raumforschung und Landesplanung. Vincentz, Hannover.
FAO 1977. A framework for land evaluation. ILRI Publication no 22, Wageningen.
Grontmij NV,afdeling recreatie en landschapsarchitectuur. 1975. Midden Holland - het landschap en zijn bebouwing. I. De ruimtelijke opbouw van het landschap. Intergemeentelijk overlegorgaan Midden Holland, Gouda.
Ham, R.J.I.M. van der & J.A.M.E. Iding 1971. De landschapstypologie naar visuele kenmerken. Methodiek en gebruik. Afdeling landschapsarchitectuur, Landbouwhogeschool, Wageningen.
Ittersum, G. van & C. Kwakernaak 1977. Gevolgen van de recreatie voor het natuurlijk milieu. In: Eilanden onder de voet, p.59-103. Werkgroep Recreatie van de Landelijke Vereniging tot Behoud van de Waddenzee. Harlingen.
Jonge, D. de 1965. Structurering van de ruimte in recreatiegebieden. Bouw, no 49: 1872-1875.
Jonge, D. de 1968. Plaatskeuze in recreatiegebieden. Bouw, no 1: 13-15.
Kiemstedt, H. 1967. Zur Bewertung der Landschaft für die Erholung. Beiträge zur Landespflege, Sonderheft 1. Institut für Landesplanung und Raumforschung der TH Hannover.
Kiemstedt, H. 1972. Erfahrungen und Tendenzen in der Landschaftsbewertung. In: Zur Landschaftsbewertung für die Erholung. Forschungsberichte des Forschungsausschusses "Raum und Fremdenverkehr" der Akademie für Raumforschung und Landesplanung. Forschungs- und Sitzungsberichte, Band 76. Raum und Fremdenverkehr 3, Hannover.
Kiemstedt, H. et al. 1975. Landschaftsbewertung für Erholung im Sauerland. Schriftenreihe Landes- und Stadtentwicklungsforschung des Landes Nordrhein-Westfalen, Landesentwicklung, Band 1-008/I.
Krajčovič, R., J. Šteffek, H. Hilbert & P. Mudry 1985. Möglichkeiten der Anwendung des Landep-Verfahrens für die Landschaftsschutz und Naturschutzplanung. (Die Nützung des Landep bei der Lösung von Problemen der Natur- und Landschaftsschutzes am Beispiel des Modellterritoriums CHKO Stiavincke Vrchy.) In: VIIth International Symposium on problems of Landscape Ecological Research. Panel 1, volume 2, part 1.6; 21-26 October 1985, Pezinok, Czechoslovakia.
Krapf, K. 1952. Fondements de la récherche scientifique du tourisme. Zeitschrift für Fremdenverkehr, no 1: 1-5.
Krapf, K. 1962. Le tourisme, facteur de l'économie moderne. Zeitschrift für Fremdenverkehr, no 3: 90-98.
MacConnel, W.P. & P. Stoll 1969. Evaluating Recreational Resources of the Connecticut River. Photogrammetric Engineering, vol 35: 686-692.
Meulen, F.van der, E.A.J. Wanders & J.C. van Huis 1985. A landscape map for coastal dune management. Meyendel, the Netherlands. ITC Journal 1985-2: 85-92.
Olson jr, C.E., L.W. Tombaugh & H.C. Davis 1969. Inventory of Recreation Sites. Photogrammetric Engineering, vol 35: 561-568.
Pearson, R.N. 1961. The terminology of recreational geography. In: Papers of the Michigan Academy of Science, Arts and Letters, Vol XLVII, 1962, p.447-451.
Peppelenbosch, P.N.G. & G.J. Tempelman 1973. Tourism and the developing countries. Tijdschrift voor Economische en Sociale Geografie, 64, no 1: 52-58.
PWF (Provinciale Waterstaat Friesland) 1977. Luchtfototellingen op een aantal Friese meren. Planologie, verkeer en recreatie onderzoek. Leeuwarden.
Rasmussen, G. 1962. From poor heath to flourishing seaside resort. A comparative airphotostudy of some land use changes on the Skanör peninsula, Sweden. In: Symposium Photointerpretation, Working Group 5, p.285-294. Delft.
RWS (Rijkswaterstaat, Dienst Zuiderzeewerken 1977. IJsselmeer in beeld. Een onderzoek naar patronen en intensiteiten van de recreatievaart op het IJsselmeer, Markermeer en IJmeer aan de hand van luchtfoto kartering. ZZW-RFO, nota nr 291, Lelystad.
RWS (Rijkswaterstaat, Dienst Zuiderzeewerken 1979. Randmeren in beeld. De resultaten van een luchtfoto onderzoek naar de recreatie op de randmeren van Flevoland. Lelystad.
Robinson, G.W.S. 1972. The recreation geography of South Asia. The Geographical Review, vol LXII, no 4: 561-572.
Robinson, R.A. 1953. Methods of tourist market research. Zeitschrift für Fremdenverkehr, no 3: 89-96.
Rodgers, H.B. et al. 1973. Recreation and Resources The Geographical Journal, vol 139, part 3: 467-497.
Statham, D. 1972. Natural resources in the uplands. Capability analysis in the North York Moors. Journal of the Royal Town Planning Institute, vol. 58, no 10: 468- 478.
Werkgroep Helmond 1974. Landschapsonderzoek Helmond. Hoofdstuk 7. Visuele aspecten. Afdeling landschapsarchitectuur, Landbouwhogeschool, Wageningen.
Zee, D. van der 1971. Rekreatie in en vanuit vaste buitenverblijven in het Drie Provincien Gebied. Planologisch Studiecentrum Rijksuniversiteit Groningen, 1974.
Zee, D.van der 1982. An analysis of recreational development using sequential aerial photographs. ITC Journal 1982-3: 362-366.
Zee, D. van der 1983. Man's activities and their impact on the natural landscape of the islands. Part 8.1 of chapter 8: Man's interference. In: K.J. Dijkema & W.J. Wolff (eds): Flora and vegetation of the Waddensea islands and coastal areas, p.270-279. Balkema, Rotterdam.
Zee, D.van der 1985. Tourism and Environmental Conservation. Aid or Threat? Public lecture given for the Botswana Society on June 20th, 1985 at the National Museum, Gaborone.

Symposium on Remote Sensing for Resources Development and Environmental Management / Enschede / August 1986

Spectral characterization of urban land covers from Thematic Mapper data

Douglas J.Wheeler
Utah State University, Logan, USA

ABSTRACT: Using Salt Lake City, Utah, as a test case, this study evaluates the capabilities of Landsat-5 Thematic Mapper (TM) digital data for distinguishing urban land cover materials. This was accomplished by using a newly developed hierarchical clustering algorithm which statistically derived spectral classes from TM channels 2, 3, 4, and 5 (visible, near infrared and middle infrared). The relationships between spectral groups were further analyzed using three statistical evaluations: principal components analysis, cluster analysis, and discriminant analysis. Through the use of component scores, cluster linkage diagrams, and canonical discriminant function scatter plots; as well as TM spectral curves, aerial photography, and ground investigation; the spectral classes were grouped into twelve predetermined land cover categories. The accuracy of classification was assessed at approximately 80 percent (0.05 significance level). A significant improvement in classification accuracy (91.5 percent) was achieved by stratifying the multispectral classification with thresholds from the TM thermal channel introduced as ancillary data.

INTRODUCTION--DERIVING SPECTRAL CLASSES FROM TM

With a high proportion of the world's population living in cities it is increasingly important to understand the very complex ecological interactions that are taking place within the urban environment. One key to better understanding these relationships is to characterize the land cover materials that influence radiational and micro-climatological balances. By monitoring successional changes in land cover materials one may observe its effect on urban ecosystem processes.

The value of Landsat's multispectral scanner (MSS) in detecting land cover has been established (Todd 1978). Landsat 5's advanced multispectral scanner called Thematic Mapper (TM), with improved spatial and spectral resolution over MSS, could be an even better tool for detecting land cover characteristics in complex urban environments. Although very little is published at the present time on the use to TM data in urban analysis, preliminary studies using Thematic Mapper Simulator (TMS) data flown from aircraft indicated that the 30 meter pixel size of TM might be ideal for mapping urban land cover elements (Clark 1980, Welch 1982). Two studies using TM data for Mobile, Alabama, show promising results (Quattrochi 1983, Wang 1985).

The urban region of Salt Lake County, Utah, was chosen as the study area for this evaluation of TM digital data due to the diversity of land activity found within a relatively small area. The majority of Salt Lake County's 1984 population of 650,000 is found within a 15 kilometer wide strip stretching north-south along the base of the Wasatch Mountains. Salt Lake City contains the diversity of land activities usually associated with metropolitan areas of a much larger size. There are many heavy and light industrial and commercial activities; diverse multifamily, single-family, and rural residential areas; extensive irrigated and nonirrigated agricultural lands; as well as natural vegetation and water features distributed throughout the county. The central valley portion of the county consists of several urban areas that are coalescing due to ongoing urban development and suburbanization. Even within the city limits of the communities in the Salt Lake valley there are many agricultural and natural areas that are being filled in with urban land use activities.

A strip 15 kilometers wide and 25 kilometers long was selected for the study area, extending through the urban corridor of the valley. From within this urban study area, 10 test windows were selected to be used for generating spectral signatures from the raw TM data. The principal data source for analysis was the digital tape of a Thematic Mapper scene dated July 27, 1984. Using test windows to break up the study area into manageable size units made the field investigation and computer processing more cost effective. The 10 windows were carefully chosen from aerial photography to represent the broadest spectrum of cover materials found within the Salt Lake urban environment. The 10 test windows accounted for approximately 16 percent of the study area.

After the test windows were selected, several trips to the field were made to identify land cover materials. At that time, ocular estimates were made of the percentages of surficial materials found in association with one another, and which combinations comprised various cover types. It was decided that 12 particular land cover classes would be desirable to detect from the TM data of Salt Lake City. The 12 preliminary land cover classes decided upon for this study included: (1) open water, (2) light inert materials (e.g., bare soil, concrete, and reflective metals and glass), (3) coal and slag, (4) dark inert materials (e.g., railroads and blacktop surfaces), (5) light asphalt-gravel surfaces, (6) mixed pixels with mostly inert cover and little vegetation cover, (7) mixed pixels with high vegetation cover, (8) senesced weeds and natural grass, (9) healthy moist vegetation, (10) drier or sparse vegetation with soil showing, (11) short cropped grasses (lawns), and (12) trees and shrubs.

A large portion of the computer processing of TM data was accomplished using ELAS digital image processing software obtained from NASA's Earth Resources Laboratory. The TM data tape was reformatted into an ELAS data file format to be processed on a Prime 400 computer. Of the seven original TM channels it was decided to generate statistic files from data in spectral bands 2, 3, 4, and 5, giving representation from the visible, near infrared, and middle infrared wavelengths. The thermal channel (band 6) was not used in determining spectral classes because of its lower spatial resolution (120 meters compared to 30 meters) and the lack of differentiation in spectral values. While there appear to be many opinions on which TM bands are optimum for processing, there is considerable support for using bands 2, 3, 4, and 5 for generating spectral statistics on urban areas (Sheffield 1985; Chavez, Guptill, & Bowell 1984; Wang 1985).

Spectral classes were derived using an ELAS module named "CLUS" (cluster), a hierarchical clustering process based on Ward and Hook's clustering algorithm (Ward 1963). This process works on a pixel-by-pixel basis to build a similarity matrix. This matrix is computed by summing the squared differences in spectral value between each possible pair of pixels or groups of pixels. The two groups with the most similarity are merged into one group at each stage of the clustering process and a new "Stat" (mean spectral value of class) is computed. In all, there were 67 "Stats" generated from the hierarchical clustering algorithm. A classification map (named CLUS67) was made for the study area using a minimum distance classifier which assigned each pixel to the class in the "Stat" file that had the nearest mean value to that pixel in feature space. The algorithm used Euclidean distance measurements to assign pixels to their particular classes.

2 CHARACTERIZATION OF LAND COVER BY SPECTRAL CLASSES

In many instances, the mean spectral values of satellite data are not sufficient for characterizing urban land cover conditions alone. It is valuable to examine the relationship between spectral classes through statistical analysis. In order to better understand the relationship between spectral classes and the information contained in those classes, several statistical analyses were performed on the file of 67 "Stat's" generated for this study.

The first analysis derived principal component factor scores from the statistical means. The first component accounted for approximately 67 percent of the variance found within the data, and the second component accounted for 28 percent. The first component was very highly correlated to the visible TM channels (bands 2 and 3) and the second component was very highly correlated to the near infrared channel (band 4). Channel 5 was spaced between the two factors.

The principal component factor scores for each class were then entered into a clustering analysis which grouped spectral classes according to a similarity index. This clustering algorithm printed a tree-linkage pattern showing which spectral classes had means that were similar and calculated their amalgamated distances. From studying this tree-linkage diagram the groups of spectral classes that were most similar were assigned group numbers. Discriminant analysis used these spectral group numbers and factor scores from the principal components to determine canonical discriminant function scores for each class.

The most useful product of the discriminant analysis was a scatter diagram which plotted a symbol for each spectral class onto a graph according to their discriminant scores. The feature space within this two-dimensional graph may be divided into regions or groups of signatures that correspond to particular ground cover types. This procedure allows the analyst to concentrate on particular classes of interest while signatures of lesser interest may be grouped or discarded. The distribution of classes on the discriminant function scatter diagram show two distinct axes (Figure 1). The first axis displays a range of varying brightness from dark signatures to light signatures, spreading from class 54 to class 41. This is commonly referred to as the "brightness axis." The second axis, which is highly correlated with the near infrared band, stretches from class 61 to class 40, and is called the "greenness axis." The "Greenness axis" is related to percent vegetation cover and plant vigor.

The next step in the land cover classification process involved the description of specific land cover that was characterized by individual spectral

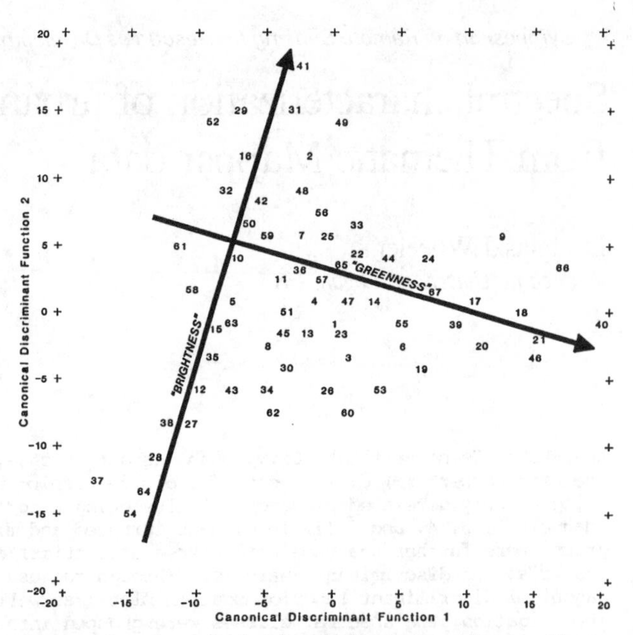

Figure 1. Canonical discriminant function scatter diagram for CLUS67 spectral classes.

classes. This was accomplished by sequentially highlighting individual classes on an image display manifesting variations in cover density or brightness. It should be noted that there is a gradation between surface cover materials, so different land cover classes may display very similar spectral curves. Several of the classes contained pixels that could be placed into more than one cover category, especially when there was confusion caused from different surficial materials yielding similar signatures. By looking at the shapes of the signature curves, the tree-linkage cluster diagrams, the discriminant scores, and the aerial photography, it was determined which cover category was best suited for each spectral class.

In an attempt to demonstrate characteristics of urban land cover categories, the spectral classes the for CLUS67 map are represented in Figure 2 by "families" of similar spectral curves. Ground investigation gave descriptions of surficial materials that were manifest in the spectral signatures. Figure 2(a), for example, represents the curves associated with open water (bottom), and light inert materials, such as bare soil, concrete, metal, and glass (top). Differences in cover conditions contributed to the reflective variations in both of these categories. Spectral classes 49, 2, and 48 displayed a more pronounced horizontal component between channels 3 and 4 than the rest of the light inert classes. These classes mostly represented fields that had been plowed, yet had a small amount of crop stubble or vegetation remaining. Classes 52, 32, and 61 demonstrated a very steep drop between channels 3 and 4 and a flatter curve between channels 4 and 5. These classes represented many of the flat gravel roofs in commercial areas as well as saline or mineral soils which were devoid of vegetation. Classes 41, 31, and 29 represented bare soil with the cover vegetation scraped or plowed off, and were generally found in construction areas of dry farms (non-irrigated fields in fallow condition, usually used for wheat). Classes 16 and 42 were mixed between bare soil and some newly completed concrete sections of the I-215 freeway. Classes 50 and 42 contained combinations of roof tops, construction sites, and transportation corridors. Many of these classes seemed to confuse roof tops in trailer parks and commercial areas with plowed or scraped land. This is understandable, since the light colored soils are similar components to the bright sand covered asphalt shingles on the roofs.

Water was expected to have a low spectral response in all four channels, especially in band 5, the water-absorption band. There was difficulty, however, in distinguishing between water and other dark cover categories by four channel mutlispectral classification alone. Class 37 displayed a typical pattern for open water and was found in lakes, sewage treatment ponds, and in other deep ponds. Class 54, though, was comprised of equal proportions of open water and coal piles. By training on individual pixels in both water and coal cover categories it was noted that the spectral responses from the visible and near infrared bands were essentially identical, while the middle infrared channel (band 5) showed a slightly lower response for the water. This, however, did not present enough variability to keep the two classes from being merged in the hierarchical clustering process. Classes 64, 28, and 27 also showed a high propensity for water within their spectral classes. The spectral curve for class 43 displayed a rather odd shape for water, yet it characterized the Jordan surplus canal with it silty water and its mix with embankment materials.

The very dark materials in Figure 2(b) were primarily considered to be open coal or slag piles, with occasional sites of tar or very black asphalt. As mentioned earlier, there was considerable difficulty in clearly distinguishing these cover types from water surfaces by spectral characteristics alone. Other methods were later tested to better differentiate these cover types. Class 54 represented the darkest coal and slag piles in a smelter location. Lighter colored materials were represented by classes 64 and 28 and were found in scattered coal piles throughout the industrial areas of Salt Lake.

Asphalt surfaces cover much of the urban Salt Lake area and are good indicators of commercial and transportation land uses. A distinction was made between light and dark asphalt surfaces since they often represent different land use activities or have different effects on the urban environment. The dark inert materials cover type mostly included blacktop areas and dark material mixed with soil (such as railroad yards). Classes 12, 38, and 27 represented this cover category although class 27 also characterized some water bodies. The other classes were also occasionally confused with water in the surplus canal. The light asphalt surfaces were generally asphalt mixed with gravel to form roads and parking lots. Several of the pixels in these classes also represented roofing materials in the commercial areas composed of tar and gravel. Classes 58 and 10 especially characterized these areas devoid of vegetation. Classes 59 and 5, on the other hand, had a higher proportion of vegetation cover mixed in, as evidenced by the flatter slopes between channels 4 and 5 in their spectral curves shown in Figure 2(d). Classes 15 and 35 primarily represented parking lots and road networks. Although class 63 was also mostly transportation, there was a little mix with natural grass areas.

There were large portions of the study area with mixed pixel responses. This generally occurred in residential areas where a large variety of heterogeneous surface materials were spaced very closely together, usually into an area smaller than the spatial resolution of a pixel. The mixed responses from surfaces such as lawns, concrete, asphalt shingles, trees, metals, etc. made the hybridized signature curves displayed in Figure 2 (e-f). While these classes occupied virtually the same region on the discriminant function scatter diagram as other non-mixed land cover categories, they reflected distinct differences in the shapes of their spectral curves. This was observed by comparing Figure 2 (e-f) with Figure 2 (g-i). Class 56 had a very bright response and was found in many trailer courts. The major contributor to this response was the shiny roofs of trailers with sparse lawns and asphalt mixed in. This class also expressed some confusion with agricultural fields having bare ground and stubble remaining. The shape of the spectral curve for class 36 was almost identical to class 56 yet it was slightly darker in reflectance. Class 36 was also primarily shingle roofs with small mixes of lawns and trees. This class was found among light roofed condominium complexes and was also occasionally confused with stubble fields. Classes 8 and 34 were found in condominium complexes and other high density residential areas fringing the central business district (CBD). There were three distinct cropped fields, however, that were also identified as class 8. Class 45 was not as commonly associated with residential areas. This class more often represented the mixed pixels where asphalt borders on natural grass areas (e.g., along freeways, railroad tracks, and in commercial or industrial areas).

Other mixed pixel locations showed a higher percentage of vegetation contributing to the spectral response. Classes 65 and 4 represented surface materials that were approximately 50 percent covered by vegetation (usually lawns and trees). Class 1 contained about 60-70 percent vegetation cover, while classes 14 and were generally over 75 percent healthy vegetation. These mixed pixels with high vegetation components were again found primarily in residential locations where surface cover was very heterogeneous within a small area. These mixed classes also showed a very large within-class spectral variance (over 6.5 times the mean variance for all remaining land cover classes).

A large portion of the study area was not developed and was covered by senesced annual grasses and weeds. A wide variety of species and cover densities were grouped within a few signatures for this category. Figure 2(g-h) illustrates two distinct patterns in these natural grasslands. Classes 44, 47, 13, and 30 displayed a pattern of healthier vegetation, as indicated by the steepness in curves between channels 3 and 4. These were usually weedy fields or vacant lots where the plants were not completely senesced. In classes 13 and 30 the soil was often wet, contributing to the darkness of the response. There was some confusion between this cover category and fields in areas where crops were newly planted and the soil was still contributing a major portion of the response. Classes 33 and 22 still had a considerable amount of vegetative response and were quite reflective, containing a high proportion of light soil. Classes 51 and 57 represented darker colored soils in natural, undeveloped areas, while classes 25, 7, and 11 were often found as idle fields in dry farm or irrigated areas.

Cropped or sparsely watered agricultural fields represented in Figure 2(i), were typified by mixed soil and vegetation responses. The surface materials contributing to these spectral curves mostly included cropped fields with some stubble and some new growth showing through; newly planted or young crops; pasture areas, irrigated or subirrigated, with weeds or bare patches; and occasionally short cropped grasses with some soil showing through. Since this category was a blend of the vegetation and soil responses there was some confusion with other similar cover categories. Pixels representing these spectral classes were often found in areas classed as lawns, natural grasses, or in residential areas. Classes 23, 55, and 67 were usually correctly identified as pastures or cropped fields, while classes 24 and 9 were frequently confused with sparse lawns.

Distinguishing lawns from other vegetative surface materials was also rather difficult. Various lawns were quite different from each other in terms of moisture content, amount of thatch, shortness of the grass, and grass vigor. Fairways of golf courses and school playfields were most characteristic of the lawn cover type. Classes 66 and 17 from Figure 2(j) showed the short cropped, but healthy grass found at parks, playgrounds, and many golf courses. Class 39 identified lawns that often had trees or shrubs nearby. Several other spectral classes were

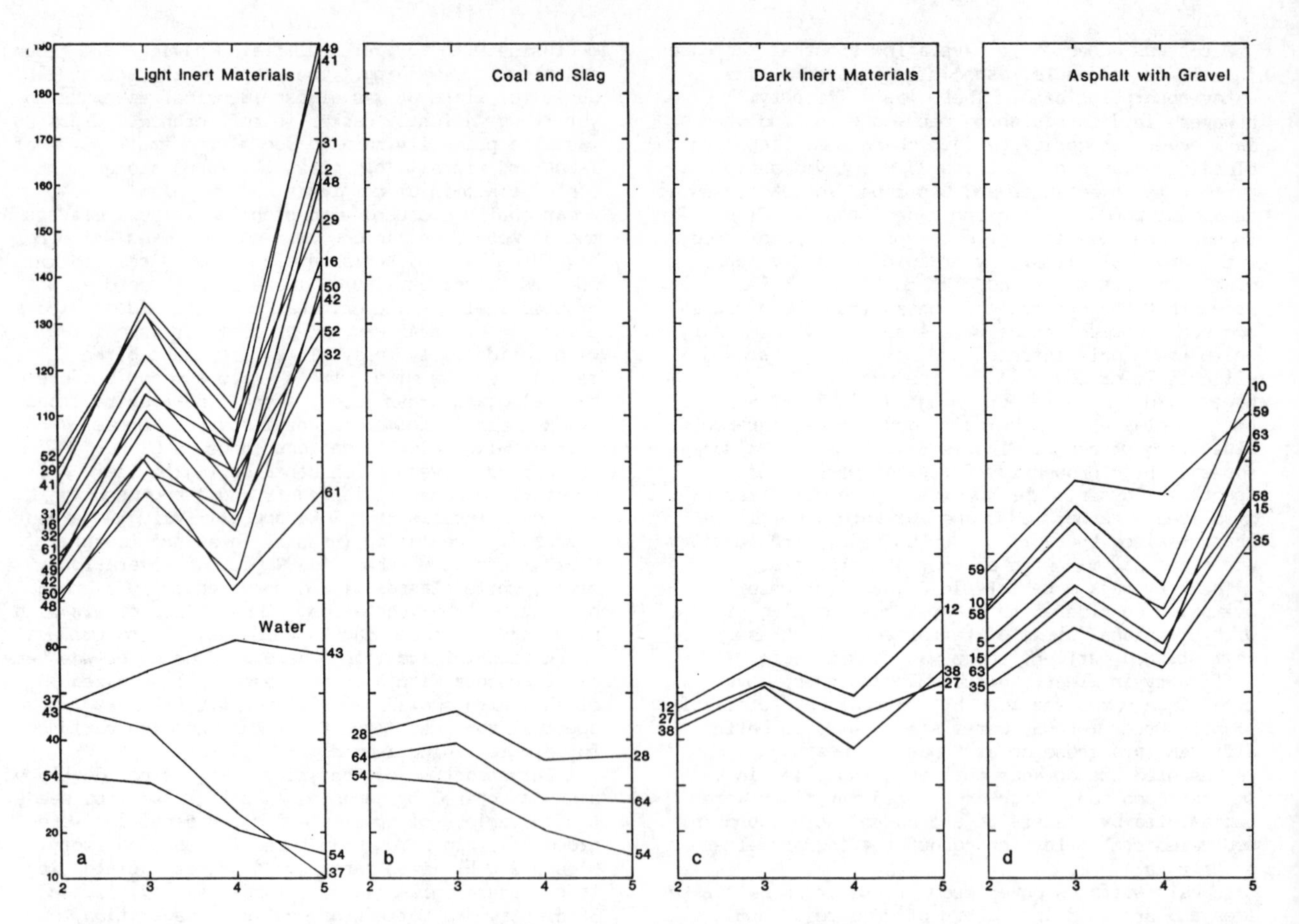

Figure 2. Families of CLUS67 spectral curves for TM bands 2, 3, 4, and 5, representing land cover categories.

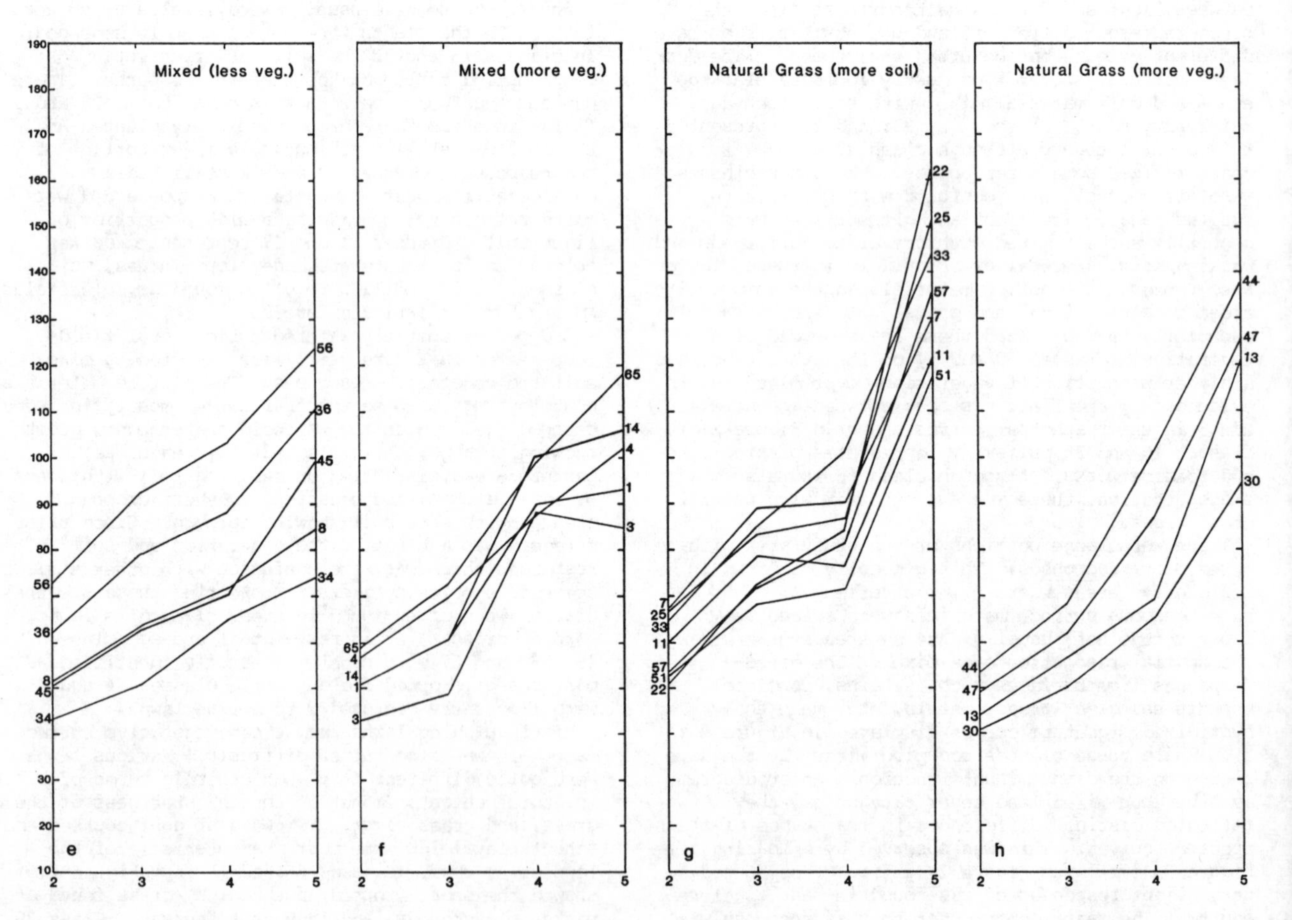

Figure 2. Continued.

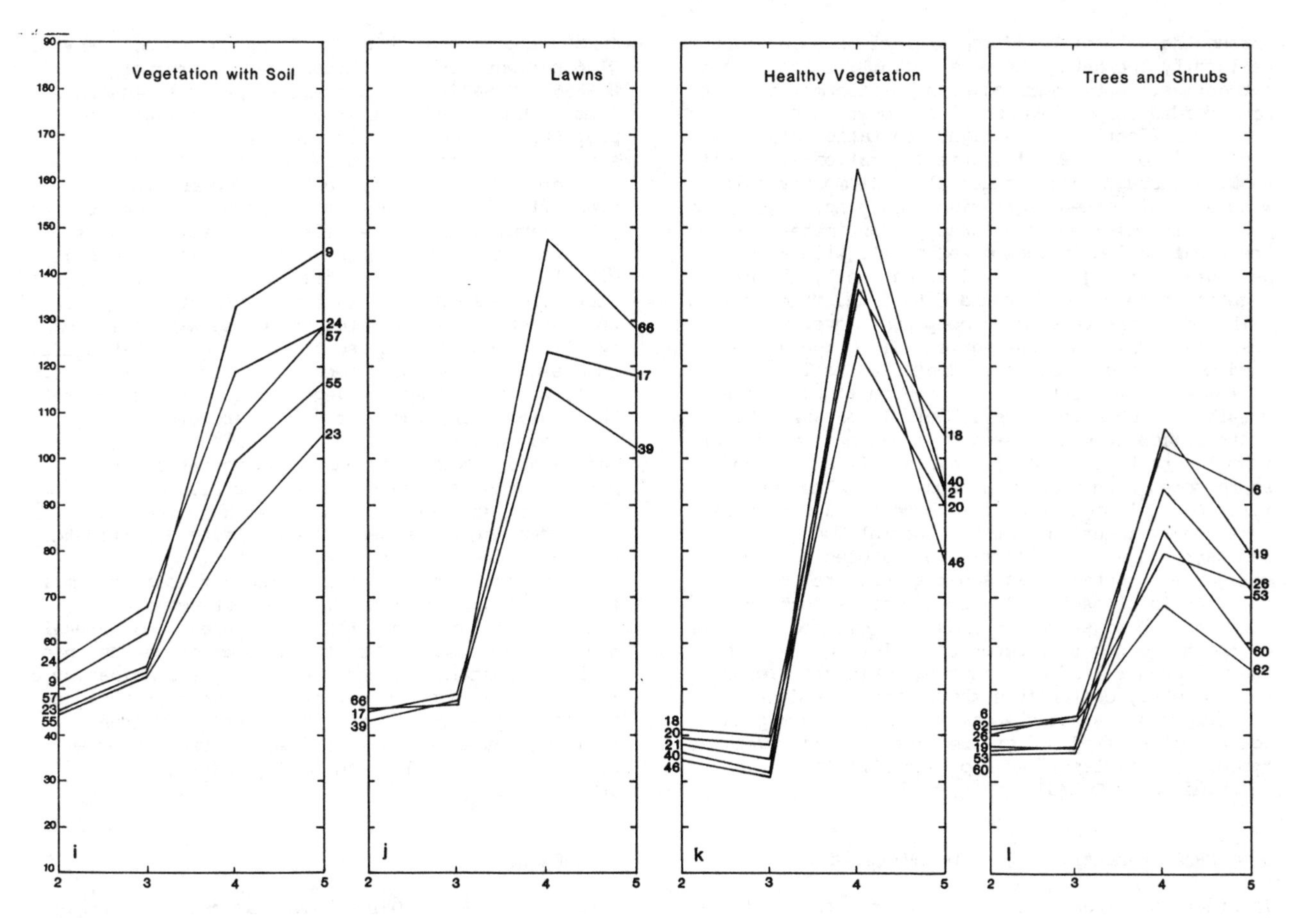

Figure 2. Continued

classified as lawn areas but were primarily drier grasses. There was some class confusion between lawns and vegetation/soil mixes (sparse cover) and healthy vegetation (long, thick grasses). Lawns were also often associated with mixed pixels, either with inert materials in residential areas or with trees and shrubs in parks, cemeteries, or landscaped areas.

Healthy, frequently watered vegetation was characterized on the signature plots by a steep rise between channels 3 and 4 and a substantial drop in channel 5, as shown in Figure 2(k). On the discriminant function scatter plot (Figure 1) this category was very high on the "greenness axis." Class 40 showed the most vigorous vegetative response of the spectral classes in CLUS67. Most of class 40 involves healthy alfalfa fields in the agricultural areas, although some healthy lawn areas were combined into this class. Classes 20 and 21 also represented alfalfa fields of varying plant densities, while class 46 usually indicated corn fields. Class 18 was on the borderline between very thick and healthy grasses (often found in lawns or golf course roughs) and the slightly drier crops found in some agricultural fields.

Trees and shrubs had a very similar spectral pattern to both lawns and moist vegetation, except that the response in the near infrared (channel 4) was usually not as high. Since trees are not as large as the TM sensor's IFOV (30 meters), there was usually some mixing between tree canopies and the understory materials, with both contributing to the pixel's response. Often, the density of tree cover was difficult to observe. Aerial photography that is slightly off nadir will show oblique views of trees, which are in turn hiding other surface cover materials, making tree canopies appear as the predominant land cover. With a completely vertical view it became apparent that tree cover density in the urban setting is actually quite a small percentage, with other cover materials contributing a major proportion of an individual pixel's response. For this reason, the land cover in virtually all of the tree and shrub category was mixed. However, trees with shrubs or dense weedy materials were major contributors to the spectral response. Classes 53 and 60 were the most representative signatures in this category for densely wooded tree cover. Class 19 was a borderline class between trees and other healthy vegetation. It often represented areas where grass was showing through the trees, as in city cemeteries or parks. Class 19 also represented clumps of shrubby trees and marshy weeds. Classes 26 and 62 were primarily located in residential areas and most often represented treelined streets or back yards with large trees. Class 62 contained a slightly higher proportion of inert material than the other classes in this category.

3 USE OF THERMAL CHANNEL SIX DATA IN RECLASSIFICATION

In the past, very little use has been made of the Thematic Mappers thermal band (channel 6) in land cover analysis due to its coarser resolution (120 meters) and low range of spectral variation. This is unfortunate, since the two parameters most responsible for variability of surface temperatures are surface moistness (moisture availability) and diurnal heat capacity (Carlson & Boland 1978). These two factors are highly related to the nature of surficial materials in the urban setting.

It was observed in this study that many of the land cover categories that were being confused in multispectral classification were actually very different in terms of thermal properties. For example, coal and asphalt were classified interchangeably as water, and cropped agricultural fields were often confused with residential areas or natural grass. For this reason, TM channel 6 was used as an ancillary data layer to set thermal thresholds within the 67 spectral classes in order to reclassify the data into more definitive land cover categories. After working with the thermal data it was decided to add two more land cover categories to the classification. The light vegetation class was divided into a class of sparse vegetation with soil showing through, and another class of mostly soil with a little green vegetation remaining. The mixed pixels were divided into three groups instead of two. These categories include mixed pixels with a low percentage of vegetation (10-40 percent), medium vegetation response (around 0.50), and mixed pixels with a dominant vegetation component (over 0.75). These two additional categories raised the number of land cover types that were identified to 14.

From a thermal channel 6 histogram it was often possible to observe "breaks," or places where bimodal distribution curves crossed. It was mentioned, for example, that class 54 of the classification map represented both open water and coal. The mean multispectral signatures for these two cover types were nearly identical yet the thermal band demonstrated a clear distinction between the two categories. Other class separations were not as clear as in class 54. A more typical situation is class 56, where stubble fields were classed the same as the mixed cover in several trailer courts. The thermal channel indicted that agricultural fields were slightly cooler than their urban residential counterparts. After changes were made the map was again reclassified and grouped into the 14 cover types. The reclassified map with 122 classes is referred to as THRM67.

4 ACCURACY EVALUATION OF LAND COVER MAPS

In order to compare the utility of the two land cover maps (THRM67 and CLUS67) an accuracy assessment was made on each. The first step was to geometrically correct the maps so that each classified pixel corresponded to a particular Universal Transverse Mercator (UTM) coordinate on the ground.

A random sample stratified by map categories was selected as the procedure for assessing map accuracy. An ELAS module named "RANS" (uniform random sample) created a new data file from the geometrically corrected THRM67 classification map. This file contained the original classification data minus 350 sample locations (25 samples each from the 14 land cover categories). A cursor was trained on each sample pixel to determine the precise UTM coordinates for each sample. These coordinates were then plotted onto USGS 1:24,000 topographic maps. When all 350 samples were plotted, an investigation commenced to determine the "ground truth" category for each of the sample sites. In an attempt to eliminate the bias, the map categories were not included on the maps taken to the field for ground investigation.

When the "ground truth" determination was completed, the observed cover categories were compared to the map categories and a confusion matrix was constructed. Percentages were calcuated for the number of pixels correctly classified, pixels that were classified in the wrong category (errors of commission), and pixels that were not placed in the correct categories (errors of omission). The map category marginal proportions were calculated in order to remove the over-represenation of small categories from the stratified random sample (Card 1982). These marginal proportions were multiplied by the percentages of correctly classified pixels and then summed to provide the overall map classification accuracy (estimated probability correct).

The same 350 samples derived from the THRM67 classification were also used to build a confusion matrix for the CLUS67 map. The CLUS67 map demonstrated an overall map accuracy of 80.2 percent for the fourteen land cover categories, with a 0.05 confidence interval between 0.7581 - 0.8459. The THRM67 map, on the other hand, showed an accuracy of 91.6 percent with a 0.05 confidence interval of 0.8866 - 0.9434. The overall accuracy assessments from each map were compared using a "difference of proportions" test to determine if there was a significant difference between them. There definately was a significant improvement in classification accuracy using thermal channel 6 data

The mapping accuracies for each class were also compared and tested for significant differences at 0.05 level. In these particular cases a "t" statistic was computed rather than a "z" value since the number of sample sites per class was lower than 50. The results of these tests indicate that there is a significant improvement in detecting asphault, coal, water, and sparse vegetation by using thermal thresholds. The most dramatic improvement was observed in the water and inert classes. CLUS67 maintained a high degree of classification error between water and coal. Water was also commonly classified as dark inert material and asphault. These descrepancies were almost entirely eliminated by use of the thermal data.

Another significant advantage of using thermal thresholds was in discriminating between cool agricultural areas and the warmer urban and natural grass cover types. Even though a minor confusion of classes is still evident after using a combination of multispectral and thermal classification techniques, the land cover map of the Salt Lake study area derived from this study is a substantial improvement over previous TM, TMS, and MSS classifications of urban areas.

REFERENCES

Card, D.H. 1982. Using known map category marginal frequencies to improve estimates of thematic map accuracy. Photogrammetric Engineering and Remote Sensing 48:431-439.

Carlson, T.N., & F.E. Boland 1978. Analysis of rural-urban canopy using a surface heat flux/temperature model. Journal of Applied Meteorology 17:998-1013.

Chavez, P., Jr., S.C. Guptill, & J.A. Bowell 1984. Image processing techniques for Thematic Mapper data. ASP-ACSM Technical Papers Annual Meeting, p. 728-743.

Clark, J. 1980. The Effect of Resolution in Simulated Satellite Imagery of Spectral Characteristics and Computer-Assisted Land Use Classification. NASA, Jet Propulsion Laboratory Report 715-22, Pasadena, CA.

Quattrochi, D.A. 1983. Analysis of Landsat-4 Thematic Mapper data for classification of the Mobile, Alabama metropolitan area. Proceedings 17th International Symposium on Remote Sensing of Environment. Ann Arbor, MI, p. 1393-1402.

Sheffield, C. 1985. Selecting band combinations from multispectral data. Photogrammetric Engineering and Remote Sensing 51:681-687.

Todd, W.J. 1978. A Selective Bibliography: Remote Sensing Applications in Land Cover Inventory Tasks. Technicolor Graphics Services, Sioux Falls, SD.

Wang. S.C. 1985. The potential of Landsat Thematic Mapper data for applied research. University of New Orleans, New Orleans, LA. (Manuscript for article).

Ward, J.H., Jr. 1963. Hierarchical grouping to optimize an objective function. Journal of the American Statistical Association 58:236-244.

Welch, R. 1982. Spatial resolution requirements for urban studies. International Journal of Remote Sensing. 3:139-146.

8 Geo-information systems

Chairman: J.J.Nossin

Symposium on Remote Sensing for Resources Development and Environmental Management / Enschede / August 1986

How few data do we need: Some radical thoughts on renewable natural resources surveys

J.A.Allan
School of Oriental and African Studies, University of London, UK

ABSTRACT: The experience of applying remote sensing techniques to renewable natural resources surveys and evaluation has been to show that the necessary is too expensive and the affordable inadequate. The promise of remote sensing has been that it could provide overviews sufficiently frequently to monitor and predict agricultural change and production. In the event the remotely sensed data sets assembled for predictive purposes, such as that which provided a proportion of the data used in the LACIE experiment proved to be unwieldly even though the samples handled were only a tiny fraction of the data sensed.

It is argued here that the time has come for remote sensing specialists to identify economic activities which attract returns to institutions which will benefit by advance warning of agricultural performance at regional level, and then to determine the minimum level of sample coverage and frequency necessary to provide predictive information at a price which the economic activities will bear. The experience of users of the general purpose remotely sensed satellite data available to date for renewable natural resources surveys (viz Landsat data) has led many users, and especially potential users, to acquire extravagant expectations; they have accepted the marvel of the comprehensive high resolution overview enabled by the satellite perspective without recognising the consequences of needing in addition high radiometric and temporal resolution data to detect information of concern. It will be concluded that any economically viable operational system will require economies in spatial cover through sampling, and suggests principles according to which such economies might be decided. The implications for sensor design and data preocessing are also discussed briefly.

Keywords: sampling, remotely sensed data, land cover, crops

1 INTRODUCTION

The fifteen years of handling satellite remote sensing data for renewable resources studies have been more remarkable for the research papers generated than for the economic benefit of the activity. Such initial disappointment in a new area of technology is not unusual, however, and there are recent precedents for example in a similar sphere, that of satellite space communication. The 1950s and the 1960s were decades when the optimism held out by the proponents of satellite communication was not confirmed by effective technological innovation. In applying remotely sensed data to renewable natural resources studies we are currently in a similar phase where the up-take of remote sensing data capture and data processing systems by professionals in relevant fields has been very partial. Impediments to the adoption of remote sensing procedures have been partly:

- technological,
- to do with the inertia of potential adopting institutions
- the unreasonable expectations of the users of general purpose satellite systems
- the lack of sampling and data reducing procedures for the vast data sets generated in studies of the dynamic environments of concern to renewable natural resource surveyors.

This last aspect will be the main concern here; the other issues will be referred to in the discussion but will not be treated systematically.

It has been shown elsewhere (Allan 1984) that with the launch of the Landsat satellites in the 1970s it became possible for the first time to address the demanding temporal features which have to be taken into account if land-cover, agriculture and cropping are to be surveyed comprehensively . The same the study emphasised that there was a clear data-volume- threshold at which current computing technology, and even future technology, would be inadequate to handle the huge volumes of digital information generated by sensing systems resolving the multi-dimensional remotely sensed data in the spatial, spectral, radiometric and temporal domains. The study also showed that most of the useful applications in renewable natural resources studies, namely crop area and production monitoring, were on the wrong side of the data volume threshold because relatively high spatial, temporal and radiometric resolutions were necessary to enable the reliable detection of such features as crop extent and condition.

The only rational approach to the type of study which will inevitably generate huge data sets is some method of sampling or data reduction. The most obvious method is to relax the spatial resolution reducing the data volume in this domain by acquiring the data at a much larger pixel size and the 1980s have seen some important developments in the use of AVHRR data (one to four kilometres spatial resolution depending on whether captured at the nadir or the swath edge) as described by Justice et al. (1985) At the one kilometre spatial resolution the use of such data represents a reduction of 226 times in the volume of data in the spatial domain compared with Landsat MSS; at the four kilometre level its use represents a 3616 times reduction. Meanwhile the specification of the AVHRR sensor brings other economies in that it only has three spectral bands. On the other hand it has very much higher temporal resolution in that daily coverage (16-18 times the Landsat frequency) is acquired at radiometric resolutions four times those familiar on the Landsat sensors (1024 levels instead of the 256 levels of Landsat MSS). It is clear in these comparisons that it is in the spatial domain that the largest economies are being made.

One very important area of research in the past twenty years has been into the spectral properties of objects and their detectability by remote sensing instruments. The research to date into the spectral properties of the complex soil, vegetation and crop surfaces which are the concern of those who survey and manage renewable natural resources shows that spectral information alone will not be sufficient to discriminate many types of cover of great significance to managers. The basis of this very broad statement is the findings of the series of meetings convened to exchange information on the spectral signature of objects (INRA/CNES 1982 and 1984, ESA 1986). The results of the ten years of detailed research into data derived from ground radiometer, as well as from airborne and satellite instruments, show that only major categories of cover at what have come to be known as Anderson's Level 1 land cover classification for remotely sensed data can be reliably detected. Otherwise only tracts which have been managed to a high degree of homogeneity and which are large in extent, that is many times the pixel size of the sensor system, can be distinguished from eachother and even some of these homogeneous tracts may be difficult to discriminate. This is particularly the case when such crops as wheat, barley and grass are the crops to be detected. It has been conclusively demonstrated that reliable discrimination of cover will require multi-date information and probably improved radiometric resolution. Happily the management of large tracts (that is tracts bigger than one hectare) to a high degree of homogeneity of cover is normal in a large proportion of the rural areas of the world and the major thrust of the argument of this paper is that advantage should be taken of this normal circumstance when designing a data capture system for the detection of agricultural land cover. The discussion will be restricted to the utilisation of visible and near infrared data of the Landsat MSS/TM types as these data have proved to be most useful in rural surveys to date. At the same time the rich information environment which the MSS and TM types of data represent enable the comparison of various sampling strategies in a search for an economical approach to the potentially prohibitively expensive and data engorged problem of acquiring and handling comprehensive information on the land surface.

2 RELEVANT APPLICATIONS AND INSTITUTIONAL ENVIRONMENTS

The majority of the land surface which it is likely to be worth surveying for information on soil, water, vegetation including forest resources and crops falls into two categories. First those tracts which are managed by man intensively and secondly those tracts which are managed less intensively which grade into the regions enduring low rainfall and which are as a consequence of no interest from a resource management viewpoint unless they can be commanded by irrigation water. The first type of land, the intensively managed tracts, to which we shall refer to as HISTAB Land (high intensity with stable parcels) is characterised by:

- high or at least adequate rainfall for one season or perennially.
- a limited range of crops grown in one or more growing seasons.
- a relatively static arrangement of parcels (agricultural fields) in which crops are managed. The parcels will generally be more than one hectare, but there are many areas, especially in the tropical world where parcellation is much more fragmented than the one hectare level.
- a high level of inputs and outputs suggesting the usefulness of monitoring activities and an economic justification for expenditure on agricultural censuses, including remote sensing.

The second type of land, the less intensively managed land, to which we shall refer as LOSTAB Land (low intensity wih unstable parcels), is characterised by:

- low and unreliable rainfall with the intensity of use determined by the level of rainfall. There are usually very clear gradients in the intensity of use which change slowly over long distances.
- a limited range of crops and livestock usually restricted to one season.
- irregular and ill-defined parcels for crop and livestock management except in those regions where fencing has been installed.
- low levels of inputs and outputs and no economic justification for the deployment of expensive monitoring including high resolution remote sensing systems.

The exact global extents of the above two types of land are impossible to estimate but it is suggested that these two types of environment include ninety percent or more of the agriculturally (including rangeland) managed environment. The other ten per cent or so of land falls into an intermediate type which for reasons of fragementation and low intensity of use falls outside the areas susceptible to monitoring by remote sensing. The discussion here does not refer to this type of cover which albeit comprises a minority of global cover.

The second type of land, the lostab land, is associated with low levels of output and any type of monitoring, including remote sensing, can only be justified if it is inexpensive. The gradual nature of the changes in land cover characteristic of these areas indicate that only low resolution systems of the AVHRR type will be appropriate and economically viable in these areas probably in association with low level aerial survey tecniques. (Watson 1981)

It is in surveys of the the first type of cover that remote sensing is most likely to be afforded and where at the same time remote sensing is most likely to have the greatest impact. This type of cover produces at least eighty per cent by value of the world's crop and livestock production although by area it probably comprises only twenty per cent of total agricultural and rangeland cover. One matter that has become clear as a result of the research of the past twenty years is that because of the fragmentation which is characteristic of a substantial proportion of these intensively used rural areas high spatial resolution data are needed. By high resolution we mean at least 30 metres and to be certain of acquiring pure pixels and not mixels a resolution of 10 metres would be appropriate. Since such high spatial resolutions put a great strain on data transmission, archiving and processing systems there is an urgent need to determine the extent to which economies can be made in data capture and handling procedures.

3 APPROPRIATE ECONOMIES IN DATA PROVISION

Spatial data can be classified according to a number of criteria and for the purposes of the argument here we shall adopt the folowing:

Vector data (point, line and area information)
Raster data (information by individual grid squares)

Static or relatively static spatial data - changes after 5 years
Dynamic spatial data - changes weekly

The place of data on renewable natural resources falls in the position shown below in the following matrix:

	Vector	Raster
Static	Topographic mapping	--
Dynamic	--	Renewable natural resource data

All of the above types of data have an important role in renewable resource surveys but they can only be utilised if there is some means of merging them. The following section deals with a method by which this could be achieved.

4 SAMPLING PRINCIPLES FOR RENEWABLE RESOURCES STUDIES

It is unfortunate that the types of data relevant to the detection and mapping of renewable natural resources surveys fall into different spatial data domains, that is the vector data concerning the posiitons of the relatively static parcel boundaries and the raster data relating to the land cover. Yet it should be possible to merge the two types of data to maximise the effectiveness of the raster data which contain the dynamic land cover information but which are not effective in locating the precise position of linear features. It is the static information on the extent of parcels which is essential as control on to which some or all of the dynamic data can be registered.

In the introduction it was emphasised that there was a need to economise in the volume of data handled especially as increased temporal resolution would be needed in future for reliable discrimination. The question should therefore be asked how few spatial data can suffice in discriminating parcels of particular crops or cover. In order to answer this question remotely sensed TM data of an area in southern England were used of a tract with a limited range of crops and substrate, namely two types of grassland and bare soil. The date of the imagery was 4 February 1983.

The data were analysed to determine the effect of the level of sampling on the spectral information as such a measure would seem to indicate the extent to which spectral discrimination would be possible as well as to reveal the effectiveness of sampled data. The 30 metre spatial resolution of Landsat TM provided data which showed a large number of pixels in this particular study area where the fields were all a number of hectares in extent.

Field	S1	S2	G1	G2	L1	L2
Area (ha)	7.2	11.16	3.69	8.82	10.26	10.98

S1 and S2 were bare soil
G1 and G2 were under grass
L1 and L2 were under another grassland type

The samples from the indivdual parcels were built up from five randomly selected pixels within the parcel. Boundary pixels (mixels) were not used in the samples. A total sample for each parcel of thirty pixels was built up by taking five further sets of five random pixels. The results were as follows:

Grassland type 1 - mean & standard deviations of dns for TM bands 3 and 4

Sample size	Spec. band(TM)		Parcel G1	Parcel G2
5	4	mean	38.47	35.07
	3		18.83	16.03
	4	s.dev	2.46	1.87
	3		0.25	0.31
10	4	m	38.43	35.87
	3		18.91	16.26
	4	sd	2.79	2.10
	3		0.26	0.39
15	4	m	39.54	36.92
	3		19.8	19.15
	4	sd	3.0	2.38
	3		0.31	0.44
20	4	m	38.23	36.08
	3		18.71	16.76
	4	sd	3.37	2.66
	3		0.35	0.51
25	4	m	38.57	37.11
	3		19.21	17.24
	4	sd	3.71	3.20
	3		0.40	0.57
30	4	m	39.12	37.24
	3		18.65	17.30
	4	sd	3.86	3.72
	3		0.44	0.65

A number of features are evident in the above tables. First the three types of land cover are easily distinguishable in the spectral data, even the two very similar types of grass cover. Secondly the cover types can be as easily distinguished on the basis of five pixels as on the basis of 30 pixels or total cover.

The total areas of the parcels ranged from 3.69 hectares to 11.16 hectares. In the former the number of TM pixels which could fall in the parcel would be 41 and in the latter 124. The results indicate that in areas where the parcels size is as large as 10 hectares it is possible to

Grassland type 2 - mean & standard deviations of dns for TM bands 3 and 4

Sample size	Spec. band(TM)		Parcel L1	Parcel L2
5	4	mean	30.18	31.12
	3		17.54	17.55
	4	s.dev	1.30	2.04
	3		0.22	0.30
10	4	m	29.98	31.17
	3		17.43	17.58
	4	sd	2.17	2.59
	3		0.28	0.34
15	4	m	30.97	30.35
	3		18.00	17.11
	4	sd	2.28	3.04
	3		0.31	0.43
20	4	m	31.52	30.86
	3		18.32	17.40
	4	sd	2.68	3.31
	3		0.38	0.49
25	4	m	32.17	31.53
	3		18.70	17.78
	4	sd	2.92	3.72
	3		0.43	0.59
30	4	m	31.23	31.76
	3		18.15	17.91
	4	sd	3.24	4.23
	3		0.48	0.66

Bare soil - mean & standard deviations of dns for TM bands 3 and 4

Sample size	Spec. band(TM)		Parcel B1
5	4	mean	24.87
	3		21.42
	4	s.dev	1.66
	3		0.29
10	4	m	24.14
	3		20.98
	4	sd	1.75
	3		0.48
15	4	m	25.03
	3		21.22
	4	sd	1.94
	3		0.54
20	4	m	23.57
	3		20.65
	4	sd	2.11
	3		0.65
25	4	m	24.21
	3		20.76
	4	sd	2.56
	3		0.72
30	4	m	22.92
	3		19.14
	4	sd	2.83
	3		0.89

Source: data assembled by Mr Jin-King Liu, April 1985

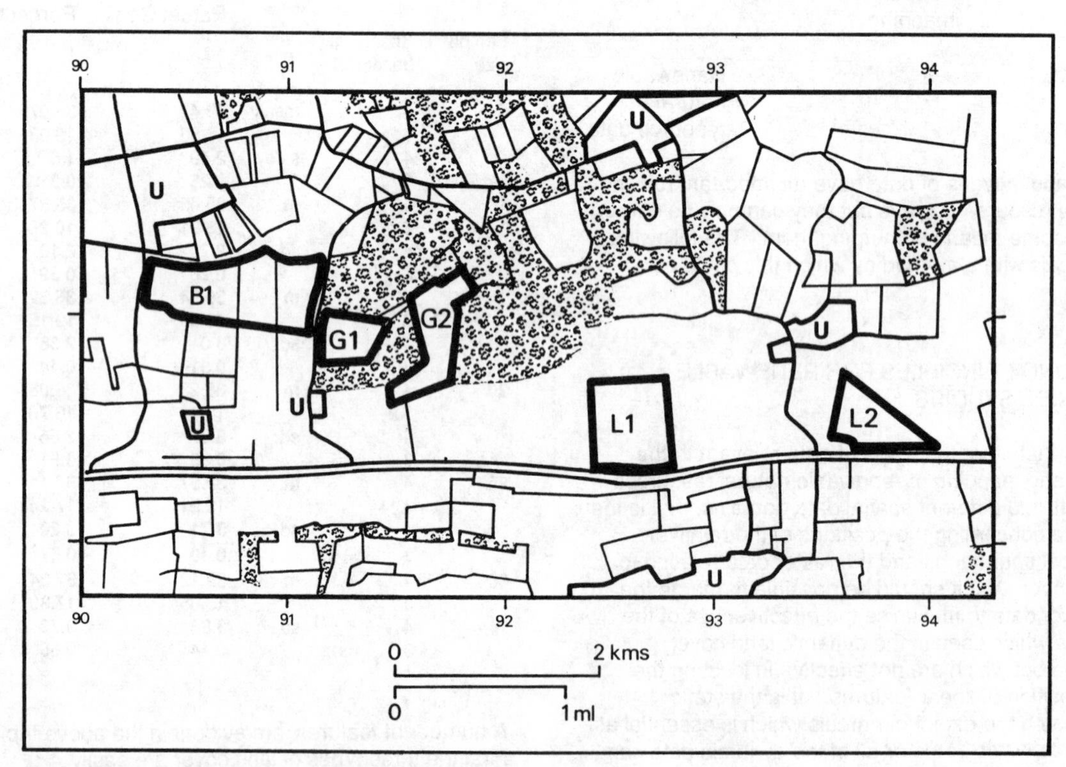

Guildford area : position of parcels

detect cover types with less than five per cent of the total cover. It would seem that only one pixel per ten hectare plot would be sufficient to provide adequate discriminative power representing a reduction in data of one hundred times. Such economies are very attractive both to those designing data transmission systems as well as to those handling and archiving data.

It should be emphasised that the precision with which the estimate can be made on the basis of the sample is not because the discriminative power of the sensor is special in spectral terms. The confidence with which different cover can be detected is because each parcel has been managed to a relatively homogeneous crop cover. It is an awareness of this behavioural feature which should

Standard deviations of dn values of three cover types showing the usefulness of sampling: Guildford area UK

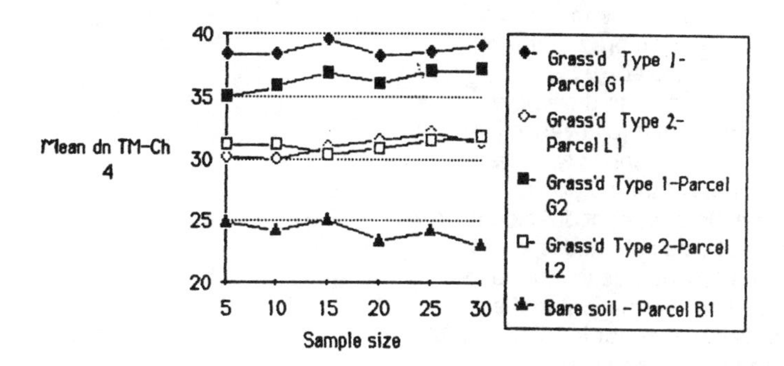

be incorporated into the strategy of remote sensing detection. Those aspects of the rural environment which are predictable should be the starting point of such strategies. They are:

- parcel boundaries are relatively static
- farmers tend to manage parcels to a uniform cover
- farmers tend to raise a limited number of crops in a particular region

The above circumstances are relevant to detection strategies with respect to data acquisition and data processing.

5 CONSEQUENCES FOR SENSOR DESIGN FOR RURAL LAND COVER DETECTION

If spatially sampled data, only one tenth or even one hundredth as voluminous as total cover, can be shown to be as predictive of land cover as total cover it is time to pose the question can sensors be designed to acquire such economical sampled data? And alternatively can on-board processing be arranged to communicate such spatially sampled data at an appropriate level? Data reduction procedures have been discussed for a decade and a half on the assumption that total cover was desirable and that data compression would reorganise the spectral information or at least reduce the number of spectral dimensions. Principal components analysis is a good example of such a procedure where it is assumed that a satisfactory approach is to reduce the dimensions of the spectral data without compromising the spatial record. The main purpose of this paper is to question whether most major applications in the field of renewable natural resources need total cover. Some of the major applications referred to are crop monitoring and the survey of the extent of irrigated land. These are the applications which are likely to generate the revenue which could support the costs of monitoring and they are the applications which governments would regard as essential for national and regional agricultural management. These are the types of data which government agencies already spend heavily to acquire.

On the basis of the evidence in section 3 above in which it has been shown that spatially sampled data is as effective as total cover for the detection of rural land cover in managed environments. It is appropriate, therefore, to ask whether it would be possible to construct an instrument which would capture data according to an appropriate sampling frame. Ideally the pixels should be able to be placed at random within each parcel but it would be difficult to construct such a sensor. Much easier to construct would be a sensor which acquires a systematic spatial sample with a regular spatial frequency. The disadvantage of this type of sample is that a proportion of the pixels would fall on parcel boundaries. For example in a five per cent systematic sample of the study area ten per cent of the sampled 30 metre pixels fell on parcel boundaries. However, it was still possible to detect differences in crop cover even when these points were discounted. It is suggested that the probability of encountering boundary pixels should be reduced by decreasing the pixel size to ten metres.

Here it is important to emphasise that the position of the field boundaries is known from the topographic record. What is required is the ability to overlay the sampled spectral data on the static topographic record to determine which pixels should be discounted. It is presumed that the identification of boundary mixels should be achieved by spatial registration but another approach which could be useful would be to discount those pixels which differ from the mean spectral reflectance for the parcel by more than two standard deviations. This last is an approach which has yet to be tested.

An ideal solution to the problem of maximising the useful data and avoiding the acquisition of information on confusing boundary mixels would be the development of a sensor which could point accurately to within one pixel. If in addition it was possible to programme the sensor so that particular pixels were recorded along specified scan lines so that boundaries could be avoided then this would be the optimum record with respect to the volume of data initially acquired and subsequently transmitted. The author has no idea whether such a data filtering strategy would be feasible in terms of on-board programming but is confident that a procedure such as this would be an ideal one for the user in most circumstances where the agricultural environment is heavily managed.

6 CONSEQUENCES FOR DATA PROCESSING IN RURAL LAND COVER DETECTION

The discussion in the previous section raises some important data processing problems. Spatial registration is a notoriously demanding process in terms of computing capacity. It is for this reason that the notion of pointing the sensor accurately has been raised.

The presentation of results would also involve unfamiliar procedures in that if the results were to be mapped rather than providing information in a list then programmes would have to be written which would apply the classification deduced from the sample to the whole parcel. Such a procedure would be a relatively minor programming task and well within the capacity of many existing GIS systems. The whole of the proposed innovation lies clearly in the main-stream of development involving the merging of spatial data sets and is a good example of the need to create procedures to enable the merging of static vector data with information on the dynamic features of land cover on a raster format. In this case the raster data are sampled but spatially identifiable.

REFERENCES

Allan, J. A. 1984. The role and future of remote sensing in Remote Sensing Society, Satellite remote sensing: retrospect and prospect, Remote Sensing Society, Proceedings of the Society's Conference, Reading. pp 23-30.

ESA. 1986. The spectral signatures of objects in remote sensing, Proceedings of the Les Arcs (1985) symposium convened by ISPRS Commission VII, Working Group 3, ESA, INRA, CNES, Paris

INRA/CNES. 1982 & 1984. The spectral signatures of objects in remote sensing, Proceedings of the Avignon (1982) and Bordeaux (1983) symposia convened by ISPRS Commission VII, Working Group 3, INRA/CNES, Paris,

Justice, C.O., Townshend J. R. G., Holben B.N. and Tucker C.J. 1985. Analysis of the phenology of global vegetation using meteorological satellite data, International Journal of Remote Sensing, 6: 1271-1318.

Watson, M., 1981. Down-market remote sensing, Remote Sensing Society, Matching remote sensing techniques and their applications. Proceedings of the London Conference, Remote Sensing Society, Reading, p 5-36

Symposium on Remote Sensing for Resources Development and Environmental Management / Enschede / August 1986

The potential of numerical agronomic simulation models in remote sensing

J.A.A.Berkhout
Centre for World Food Studies, Wageningen-Amsterdam, Netherlands

ABSTRACT: Long-term experience with numerical agronomic simulation techniques has resulted in a methodology that is specifically application-oriented. Incorporated in a geographic information system, which consists of static data on land, the simulation model has proven to be a valuable tool in a quantitative land evaluation (i.e. the identification and quantification of possible land use developments). Introduction of Remote Sensing Images will introduce the possibility to monitor actual crop growth situations (land use, crop development). The ability to monitor actual crop growth using RS images enhances the detection of 'problem areas' and thus the system can be used for early warning purposes. The paper will describe a GIS system including crop growth models. The possible uses of RS images in such a system and the possibilities of such a system as a partial substitute of ground truth in RS analysis will be developed and demonstrated in the paper.

1 Introduction

Satellite Remote Sensing (RS) provides a scientific tool to analyze the spatial variability in the momentary state of the surface of a defined area on the earth, and applying the multi-temporal possibilities of Remote Sensing, the differential change of the state in space and time. The usefulness of the information obtained by RS depends on the one hand on the type of RS techniques applied, including their spatial resolution and on the other hand on the availability of real information about the observed region, i.e. the ground truth, regarding the actual land use and vegetation, the landscape and soils, the hydrological and meteorological conditions, etc.

This paper discusses the possibilities for reducing the required ground truth data for RS image interpretation purposes by the introduction of numerical simulation models on plant- and crop growth within the framework of a Geographic Information System (GIS). The disciplines involved are treated and it is argued that all applications will benefit from a close cooperation.

2 A systems approach to plant- and crop growth

To study the complex, continuous reality of the world, a meaningful (as related to the goal of the study) section of the reality has to be identified and separated from its environment. To obtain relevant and useful results, such a section, i.e. a so called system - in the systems approach terminology - must be sufficiently complex to exhibit a high degree of internal coherence. But on the other hand, it must be simple enough for comprehension and investigation (Chorley, Kennedy, 1971). For a regional, agricultural land use analysis the conceptual model of the system must at least cover the two main objects involved, i.e.;

i. land, which comprises according to the F.A.O. description (Brinkman, Smyth, 1973) all the earth-related features such as landscape, soil, hydrology, weather, vegetation and man made structures; and

ii. the farming system, or all the human activities that are directly and indirectly, related to agriculture in the region.

The selected features or object attributes, including their observed or assumed relations, determine the type of model applied. A schematic distinction can be made between:

i. stochastic models, that contain statistical relations between some relevant and perceptible attributes of the system and lead indirectly to the required results. The functioning of the system in terms of the flow of energy, mass and information within the system is considered a black box.

Specimens of this approach are the commonly used (multiple) regression models, such as that by Wiegand and Richardson (1984) and Ambroziak (1985). The former describing the relation between incident photosynthetic active radiation, leaf area index and yield of defined crops. Ambroziak estimates yields using actual monthly total precipitation and the condition of the crop as deduced from its reflectance, in combination with historical records on the performance of a crop in the selected regions.

A disadvantage of this type of models is their inherent specific character with respect to the site and crop type; they do not separate causes and cannot be applied in evaluation modules, assuming possible changes within the current agricultural system.

ii. deterministic models, where at a predefined level of generalization the variables and their relations are formulated and quantified, based on insight in and knowledge of the underlying basic processes. Such models are applicable under a wide range of conditions after a sound validation and calibration procedure (van Keulen, 1976).

The Centre for World Food Studies has developed such a 'cause-and-effect' model for agricultural production, following a hierarchical approach (van Keulen, de Wit, 1982). A top-down approach is applied to generate production estimates (expressed in kg/ha of dry matter of various components of the crops as roots, stems, leaves and storage organ) for specific crops, cultivated at specific locations, with specific growth periods (van Keulen, Wolf, 1986; Rappoldt, 1986). The latter are characterized by their specific soil and weather conditions. For any crop, characterized by its genetic and physiological properties, the model (fig. 1) starts with the calculation of the potential production as a function of the incident photo-synthetic active radiation (PAR, roughly 50 percent of the global irradiance) and the temperature only.

This potential can subsequently be reduced by the negative influence on crop production of the lack or excess of water (using a water balance model with time steps of one day), the lack of plant nutrients and the occurrence of weeds, pests and diseases. Fig. 2 shows some model results: for example production of Pearl Millet calculated for Dori, Burkina Faso, using long-term mean monthly meteorological

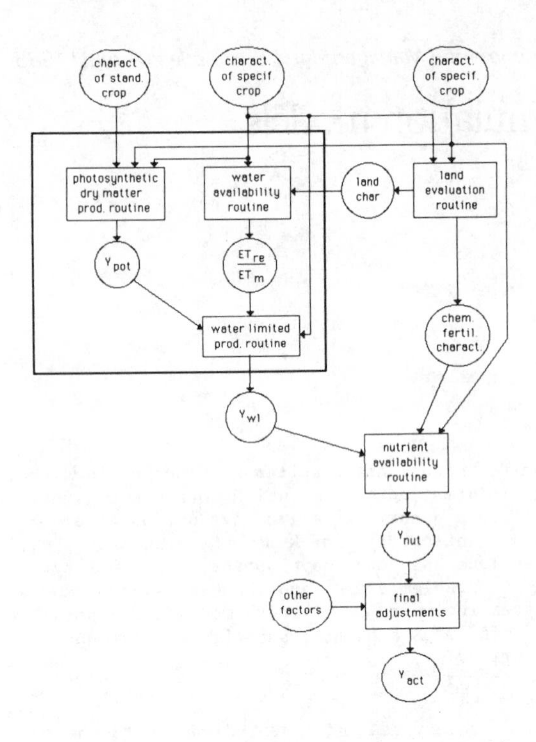

Figure 1. Hierarchical analysis of the main factors determining the physical production of a specific crop at a specific site.

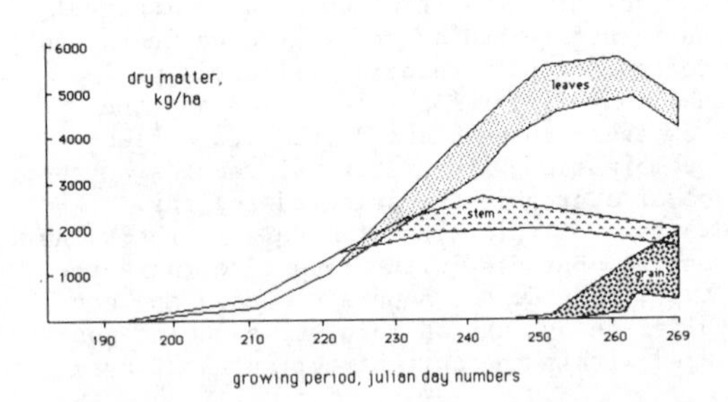

Figure 2. Model result: production of Pearl Millet, calculated for Dori, Burkina Faso (shallow, sandy/clayey sand, long-term mean monthly meteorological data).

data. The variation in the partial results is due to differences in rainfall distribution created by the use of a random number generator to distribute precipitation over each month. This procedure suggests a partially stochastic approach.

The calculations, based on physical and chemical principles, are in the following analysis linked with the current agricultural management practices (farming systems), which are strongly influenced by the knowledge and skills of the farmers and their socio-economic and institutional environment.

An approach useful for management and agricultural planning, must take explicitly into account such physical and socio-economic factors (van Keulen et al., 1986).

3 Satellite Remote Sensing

The operational RS-satellites (table 1) detect and process the electro-magnetic radiation, reflected and/or emitted by an observed surface of the earth. The wavelength of the scanners in these satellites is developed in accordance with the diverse spectral properties of the constituents of the atmosphere (i.e. water, carbon dioxide, ozone and dust) and of the surface, as waterbodies, soil types and the state of the vegetation.

Interpretation of the acquired data sets are assessed by the wavelength and by the spatial resolution (surface area of an individual pixel). Currently, due to the spatial resolution, data obtained from operational satellites prohibit detailed agronomic analyses, based on detailed vegetation or crop reflectance models as applied by means of ground and/or near-ground scanner-systems (as described e.g. by Badhwar et al., 1985). The strength of satellite RS lies in its synoptic possibilities, enhanced by multi-temporal applications, i.e. the possibility to combine various data sets covering the same area (see e.g. Tucker et al., 1985).

In the domain of the agricultural sciences, crop and crop area identification may be regarded an operational RS interpretation technique, that generates results within acceptable statistical limits. The quality of these results based on multi-spectral and the multi-temporal capabilities of RS, are highly dependent on:

- the availability and quality of data obtained in the field (i.e. ground truth) such as the current landuse pattern, crop characteristics, including crop development in the course of the growing season;

Table 1. Comparison of the important operational satellites - sensor characteristics.

	METEOSAT	NOAA/AVHRR	LANDSAT - MSS	LANDSAT - TM	SPOT HRV-CCD
Attitude	36000 km	833 km	(1,2,3) 920 km (4,5) 725 km	725 km	832 km
Orbits/day	-	14.2	14	14	142
Temporal resolution	4 hours	eff. 2-3 days	(1,2,3) 18 days (4,5) 16 days		variable
Spatial resolution	2.5 km (vis) 5 km	1.1 km	56x79 m	30x 30 m 120x120 m (band 6)	20x20 m 10x10 m (p)
Spectral resolution	0.4- 1.1 μm (1) 5.7- 7.1 (2) 10.5-12.5 (3)	0.58-0.68 μm (1) 0.72-1.10 (2) 3.55-3.93 (3) 10.5 -11.5 (4) 11.5 -12.5 (5)	0.5-0.6 μm (1) 0.6-0.7 (2) 0.7-0.8 (3) 0.8-0.11 (4)	0.45-0.57 μm (1) 0.52-0.60 (2) 0.63-0.69 (3) 0.76-0.90 (4) 1.55-1.75 (5) 10.4 -12.5 (6) 20.8 -2.35 (7)	0.51-0.73 μm (p) 0.50-0.59 (1) 0.61-0.68 (2) 0.79-0.89 (3)
Radiomatic resolution	8 bit (256 levels)	10 bit (1034 levels)	6/7 bit (64, 128 levels)	8 bit (256 levels)	8 bit (256 levels)

- the spatial resolution of the RS satellites compared with the pattern of the land use(e.g. layout of the arable fields); and
- the occurrence of (nearly) cloudfree periods during the growing season. These pre-conditions limit the use of these interpretation techniques to the more developed areas of the world and/or to areas having a semi-arid to semi-humid climate.

Related to crop identification is crop monitoring and yield forecasting. These techniques include the acquisition and interpretation of synoptic meteorological data in relation to the development of the 'green' surface. Assuming a quantitative relation between yield and the occurrence of meteorological anomalies (compared with an average year), particularly water stress of the crop or vegetation due to lack of rainfall, statistical models may be applied to estimate the actual yields on the basis of rainfall predictions (see e.g. Ambroziak, 1985). If such models do not also include other variables as the meteorological variables, such as the natural fertility of the soil, the possible application of fertilizer (see section 4), the occurrence of weeds, pests and diseases, the results may become to be regarded as doubtful. It is therefore necessary that such models incorporate a thorough knowledge of the existing farming systems in the area under study.

4 The combination of remote sensing and simulation models

To link numerical simulation techniques with RS, it is necessary to determine those attributes that could be observed and quantified (or approximated) with RS and could act simultaneously as forcing variables in the models, or at least could be used as calibration variables. Although the operational satellites have not been designed specifically for the quantitative assessment of most of the relevant variables, such as the rainfall and soil-moisture, the following variables could be derived from RS information (based on the current state of the art):

i. actual precipitation
Rainfall mapping requires sequences of satellite images, ground measurements of precipitation and operational definitions of cloud types (Griffiths, et al., 1978). The classification of cloud types is based on the reflected radiance (albedo) in the visible spectral channel(s) and the emitted radiance (cloud top temperature) in the infra-red spectral channel(s). Applying proper threshold values, depending on for example the actual temperature profile of the atmosphere, rain clouds could be identified (Rosema, et al., 1985). To obtain a more accurate discrimination between the various cloud types, an iterative cluster analysis procedure could be introduced (Seldon, Hunt, 1985). Problems may always occur, due to e.g. the layered structures of certain cloud systems (as frontal systems).

The relation between the presence of (rain) cloud types and observed spatial extent and amount of precipitation remains obscure. Intensive ground observations are required to establish statistical relations between the precipitation patterns and the RS images (Milford, Dugdale, 1983). This emphasizes the region specific character of such rain mapping procedures.

ii. actual irradiance
The amount of solar radiation reaching the ground depends on the cloud cover and the degree of the absorbtion and scattering in the atmosphere. Major absorbers are water vapour and aerosols. The radiation is scattered by atmospheric molecules and aerosols, producing diffuse light; the presence of clouds increases largely the scattering.

Irradiance estimation procedures (daily, hourly values) using RS data are based on quantitative relations between actually observed irradiance and standardized cloud-free irradiance, the cloudiness, possibly augmented by the cloud type (RS data) and an atmospheric depletion characteristic, such as the atmospheric surface pressure (Tarpley, 1979). Again, such procedures are region-specific; slight modifications to be included are site-specific data on slope and exposition (Cappellini, et al., 1982).

iii. air temperature
Analogous procedures could be applied to estimate from satellite images the actual daily range of air temperatures and air temperature profiles. Region-specific relations have to be derived from observed shelter temperatures (i.e. the minimum and maximum air temperatures in a shelter at a given height above the surface) and the surface temperature, as derived from satellite measurements in the infrared spectrum, after correction for atmospheric attenuation (Davis, Tarley, 1983).

iv. actual evapotranspiration (LE)
The amount of water transpired by the plants and - in case of an incomplete cover - evaporated from the soil surface is a function of the moisture content of the top soil layer, the energy balance at the surface and the aerodynamic drying power of the atmosphere. The calculation procedure could be simplified (e.g. Soer, 1980, Menenti, 1984) for defined regions and crops by the introduction of calibrated relations between several variables, especially those specifying the aerodynamic component in the evapotranspiration equation and RS measurable variables, such as the surface temperature (Ts, see iii.) and the compound reflection coefficient α of the surface and the crop/vegetation

$$\alpha = \alpha(veg) \times Sc + \alpha(soil) \times (1-Sc),$$

where Sc is the coverage of the soil by the crop, $\alpha(veg)$ and $\alpha(soil)$ the reflection coefficient of the vegetation and the soil.

The crop cover could be deduced from the leaf area index (LAI), which in turn could be approximated with the RS vegetation index (VI) (Tucker, 1977). Such a procedure would require an a-priori availability of a crop identification procedure to determine cropping patterns.

5 Geographic Information Systems

Data Base Management Systems (DBMS) are developed to facilitate handling of large quantities of data in a computer system: it consists of facilities to store and change data, retrieve information by means of formal query rules and to present information in any type of format. If at least one entity in a DBMS is defined and identified by its spatial attributes (e.g. sets of geographical coordinates) and if facilities are available to map such a spatial entity as points, lines or bounded regions, the DBMS is regarded a GIS, i.e. Geographic Information System (Abel, 1983).

In accordance with the main objective of a DBMS, the organization of the data within a GIS must allow the user to extract and analyse data sets in order to derive new information in respect to the selected geographical location or areas. In a standard GIS such procedures are relatively simply structured (Tomlin, 1983); for the purpose of in-depth agronomic analyses a possible combination with simulation models could be fruitfull.

To exemplify such an approach, reference is made to a recent study of the Centre for World Food Studies, in which the effects of fertilizer application on food production were analyzed in a number of African countries. One of these countries was Burkina Faso. First a selection was made of the crops and cultivars relevant to the study, i.e. several varie-

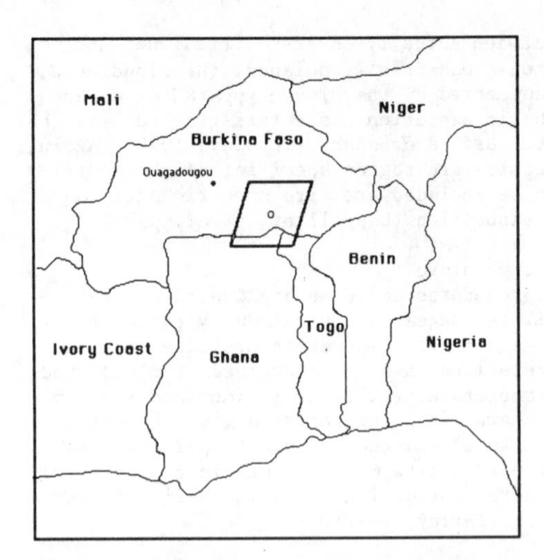

Figure 3. Example of the application of a Geographic Information System and agronomic simulation model (see colored figures). Explanation see text.

ties of maize, millet and sorghum. Next, the data set, defining the selected crop cultivars was calibrated against experimental data. Subsequently, the crop growth simulation model was used to calculate the yield level, constrained by irradiance, temperature and water availability. These calculations were performed for each unique combination of the various attributes, explicity specified in maps derived from the digitized soil - climatic region map of Burkina Faso, based on the soil map of Boulet (1976).

a. The climatic regions (map A), defined on the basis of the natural vegetation zones. For each zone a meteorological station, considered to be representative for the zone was selected.

b. Soil moisture characteristic or pF-curve (map B), based on the dominant top-soil texture.

c. The soil depth (map C), including the stoniness and the existence of sub-soil layers preventing root penetration. The results of the calculations are presented in map D, showing the water-limited production potential of millet over Burkina Faso.

The production potential constrained by the natural availability of plant nutrients from the soil was calculated using the QUEFTS system, developed for Kenya (Janssen et al., 1986). A natural fertility map (map E) showing classes of the expected average maximum yield in kg/ha of dry matter in the marketable product of the selected crops, was also derived from the original soil map, using additional information from fertilization trials and soil chemical analyses.

By subtracting the nutrient-limited production potential from the water-limited production potential the effect of fertilizer application can be shown (map F). The map indicates that application of fertilizer will considerably increase yields over large areas in Burkina Faso. This conclusion confirms the results of other studies in the Sahelian region (e.g. Penning de Vries, Djiteye, 1982).

The study illustrates the potential of incorporating of numerical simulation models in a GIS, e.g. for land evaluation purposes. A serious disadvantage of this analysis is that long-term average weather data have been used, so that the temporal variability likely to govern farmer's behaviour to a large extent is lost in the process (fig. 4). Besides, the generalized character of the input data and consequently of the results, prohibit a comparison with the actual land utilization (e.g. see the detailed vegetation pattern on the Landsat frame, compared with the other maps of fig. 3.). This is regarded mainly resulting from the lack of data on the actual land use pattern.

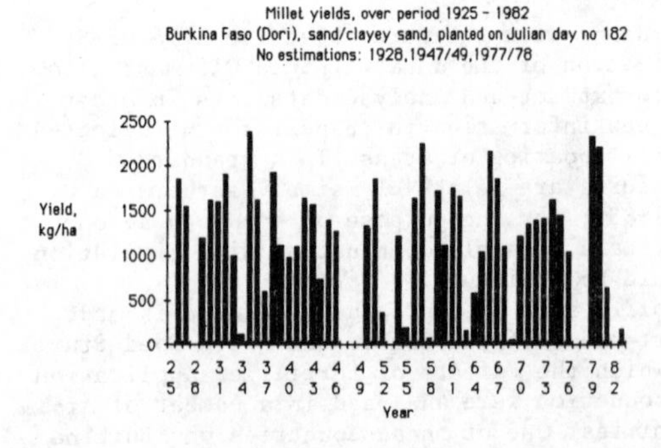

Figure 4. Variability of the model results, using historic rainfall data.

6 Conclusions

The transformation of the framework described in the previous section into any RS-monitoring system as e.g. a tool in Early Warning Systems requires at least (i) additional knowledge on the actual land use pattern, (ii) synoptic time series of the actual weather (e.g. occurrences of rainfall), and (iii) more detailed and specific knowledge on the physiography, including soil characteristics and hydrology (as runon-runoff, groundwater).

This information can partly be obtained from RS imagery. Particularly crop identification procedures and satellite weather elements monitoring procedures appear to be ready for implementation. However, in contrast to some systems, regarded to be operational (e.g. Merrit, et al., 1985) we feel that extensive field observations are and will remain to be required to quantify the combined effect of all land entities and farm practices on the cultivation of specific crops. The number of observations could be reduced with the introduction of numerical simulation models after thorough calibration over a number of years.

Acknowledgements

The author wishes to thank Dr. M. Menenti and Ir. H. van Kasteren for the discussions on the subject, Dr. D. Faber and Dr. H. van Keulen for their comments on the original version of the manuscript, and Drs. H. Huizing for providing the RS imagery.

REFERENCES

Abel, D.J. 1983. Towards a relational database for geographic information systems. In proceedings of the workshop on databases in the natural sciences. Brisbane, Queensland: CSIRO Cunningham Lab.

Ambroziak, R.A. 1985. Global crop monitoring: a integrated approach. Recent advances in civil space remote sensing, 1984. SPIE. 481:238-244.

Badhwar, G.D., W. Verhoef & N.J.J. Bunnik 1985. Comparative study of suits and SAIL canopy reflectance models. In remote sensing of environment, pp. 179-195.

Boulet 1976. Notice des Cartes de Resources en Sols de la Haute-Volta. Paris: OSTROM.

Brinkman, R. & A.J. Smyth (eds) 1973. Land evaluation for rural purposes. Wageningen: ILRI Publications No. 17.

Cappellini, V., C. Conese, G.P. Marachinni, F.P. Miglietta & P. Pampaloni 1982. Agro-ecological classification by remote sensing. In proceedings

Map A. Climatic regions in Burkina Faso

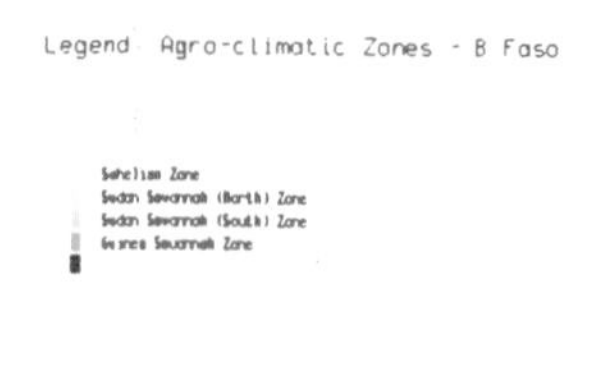

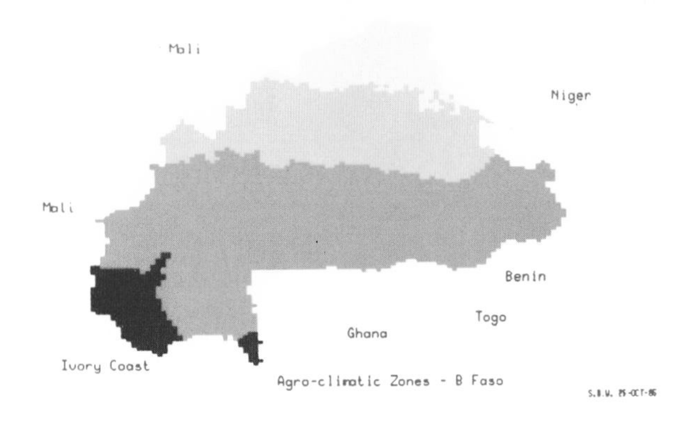

Map B. Soil moisture characteristics in Burkina Faso

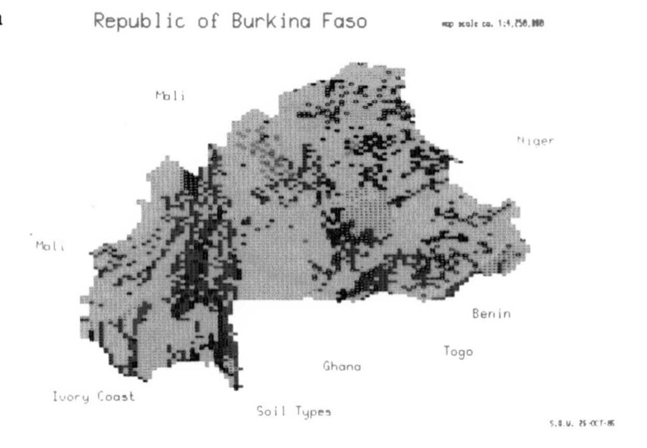

Map C. Soil depths in Burkina Faso

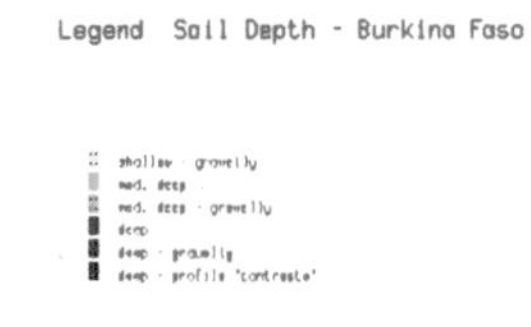

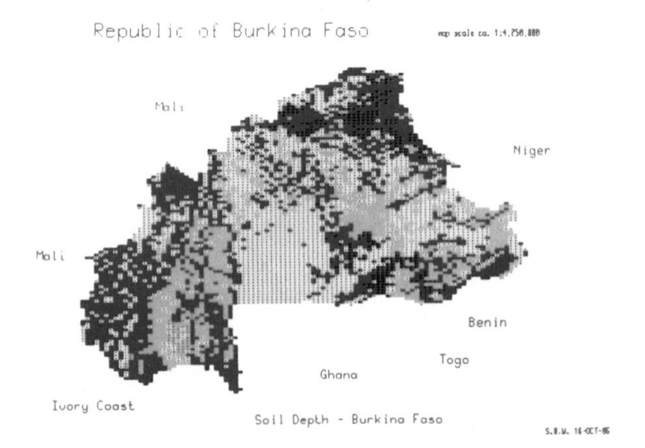

Map D. Water-limited yields of millet in Burkina Faso

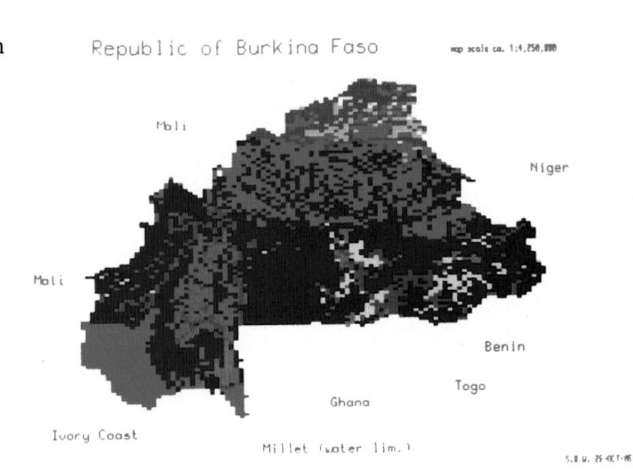

Map E. Soil fertility classes and related nutrient-limited yields in Burkina Faso

Legend Soil fertility - Burkina Faso

- class D
- class E1
- class E2
- class E3
- class E1 / 50%
- class E2 / 50%
- class E3 / 50%
- class F / no yield

Soil fertility class	Nutrient-limited yields, kg ha^{-1}		
	Maize[1)]	Millet[1)]	Sorghum[1)]
D	1600	1000	1400
E1	1200	750	1050
E2	800	500	700
E3	400	250	350

1) Grain yields, 12% moisture.

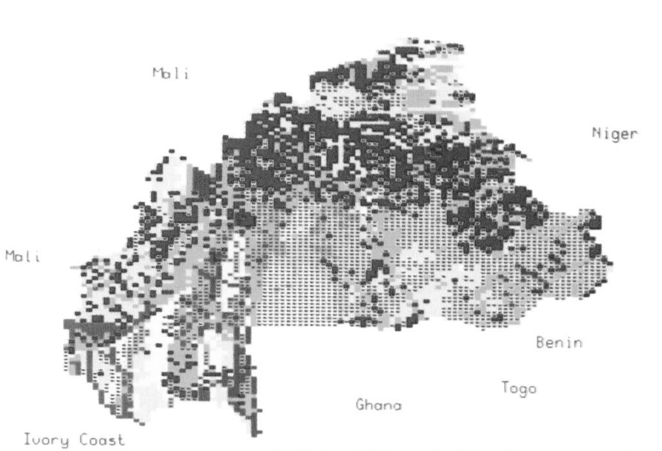

Map F. Difference between water-limited and nutrient-limited yields of millet in Burkina Faso

Legend Difference Millet: water - nf

- no increase
- 1-250 kg/ha
- 250-500 kg/ha
- 500-1000 kg/ha
- 1000-1750 kg/ha
- 1750-2500 kg/ha
- 2500-5000 kg/ha

Yields: dry matter in grains

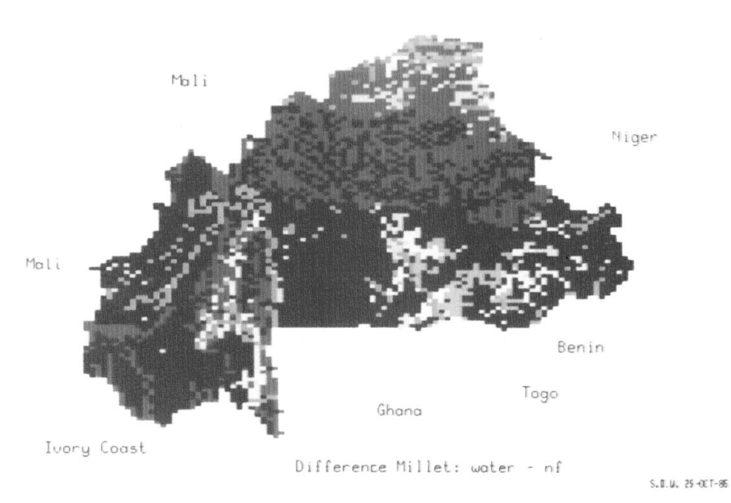

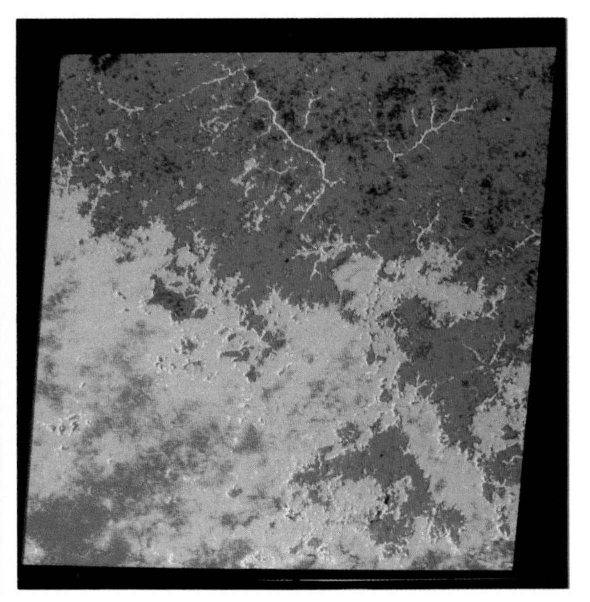

Map G. Vegetation pattern at the end of the growing season (6 Nov. 1975), S.E. of Ouagadougou, Burkina Faso (Landsat, frame right: false colour, frame left: vegetation index)

EARSeL-ESA symposium, 20-21 April, pp. 111-129. Igls, Austria.
Chorley, R.J.& B.A. Kennedy 1971. Physical geography, a systems approach. London: Prentice Hall International Inc.
Davis, P.A. & J.D. Tarpley 1983. Estimation of shelter temperatures from operational satellite sounder data, 1983. Journal of Climate and Applied Meteorology. 22: 369-376.
Griffith, C.G., W.L. Woodley, P.G. Grube, D.W. Martin, J. Stout & D.N. Sikdar 1978. Rain estimation from geosynchronous satellite imagery-visible and infrared studies. Monthly Weather Review. 106:1153-1171.
Janssen, B.H., F.C.T. Guiking D. van der Eijk, E.M.A. Smaling & H van Reulen 1986. Quantitative evaluation of fertility in tropical soils (QUEFTS). In prep.
Keulen, H. van & C.T. de Wit 1982. A hierarchical approach to agricultural production modeling. In G. Golubev & I. Shvytov (eds.), Modeling agricultural-environmental processes in crop production, pp. 139-153. Vienna: IIASA.
Keulen, H. van 1976. Evaluation of models. In Arnold, G.W. & C.T. de Wit (eds.), 1976, Critical evaluation of systems analysis in ecosystems research and management, pp 22-29. Simulation Monographs. Wageningen: Pudoc.
Keulen, H. van & J. Wolf (eds.) 1986. Modelling agricultural production: weather, soils and crops. Simulation Monographs. Wageningen: Pudoc.
Keulen, H. van, J.A.A. Berkhout, C.A. van Diepen, H.D.J. van Heemst, B.H. Janssen, C. Rappoldt & J. Wolf 1986. Quantitative land evaluation for agro-ecological characterization, 1986. Paper presented at the intre-centre workshop on agro-ecological characterization, classification and mapping, Rome, 14-18 April.
Menenti, M. 1984. Physical aspects and determination of evaporation in deserts applying remote sensing techniques. Report to ICW. Wageningen.
Merrit, E.S., L. Heitkemper & K. Marcus 1985. CROPCAST - a review of an existing remote sensor-based agricultural information system with a view toward future remote sensor applications. Recent Advances in Civil Space Remote Sensing (1984). SPIE. 481:231-237.
Milford, J.R. & G. Dugdale, G. 1983. Notes on operational processing of Meteosat data to give moisture budgets in the semi-arid tropics. Reading.
Penning de Vries, F.W.T. & M.A. Djiteye, M.A. (eds.) 1982. La productivite des paturage saheliens. Une etude des sols, des vegetations et de l'explotation de cette ressource naturelle. Agric. Res. Rep. 918. Wageningen: Pudoc.
Rappoldt, C. 1986. Crop growth simulatation model WOFOST documentation, version 3.0. Wageningen: Centre for World Food Studies.
Rosema, A. 1985. Group agromet monitoring project (GAMP). Executive summary report.
Seddon, A.M. & G.E. Hunt 1985. Segmentation of clouds using cluster analysis, 1985. Int. J. Remote Sensing. 6:717-731.
Soer, G.J.R. 1980. Estimation of regional evapotranspiration and soil moisture conditions using remotely sensed crop surface temperatures. Remote Sensing of Environment. 9:27-45.
Tarpley, J.D. 1979. Estimation incident solar radiation at the surface from geostationary satellite data, 1979. Journal of Applied Meteorology. 18:1172-1181.
Tomlin, C.D. 1983. Digital cartographic modelling techniques in environmental planning. Ph. D. thesis. New Haven, Connecticut: Yale University.
Tucker, C.J. 1977. Use of near infrared/red radiance ratios for estimating vegetation biomass and physiological status. XX, pp.493-494.
Tucker, C.J., J.R.G. Townshend & T.E. Goff 1985. african land-cover classification using satellite data. Science. 227:369-375.
Wiegand, C.L. & A.J. Richardson 1984. Leaf area light interception, and yield estimates from spectral components analysis. Agronomy Journal. 76:543-548.

Symposium on Remote Sensing for Resources Development and Environmental Management / Enschede / August 1986

Recording resources in rural areas

Richard K.Bullard
The National Remote Sensing Centre, Chelmsford, Essex, UK

ABSTRACT: In the developed and the developing countries of the world, land in rural areas is under continuing pressure. Many of the European countries are attempting to replace the traditional and existing cadastre with a Multi-Purpose Cadastre (MPC) in an attempt to more adequately record resources. With the advent of the improved resolution of satellite imagery (2nd and 3rd generation) the possibility of recording many of the resources required for a MPC in rural areas will be considered as well as with more conventional imagery.
Many of the parcels of land in Europe are uneconomic and the need for reapportionment has become a major activity of the land surveyor.
This paper will consider the applications of satellite imagery and aerial photography in conjunction with the cadastre to determine the resources of individual parcels of land in rural areas with a view to their improvement and future land use.

1 INTRODUCTION

Rural land has traditionally been under pressure, ever since man has changed his life style from a nomad to a pasturalist, and with his subsequent settlement into communities. The village became the nucleus for the future urban area.

In the context of this paper rural refers to that part of a country, particularly in Europe, which has a 'low' population density and is largely given over to agriculture, forestry, national parks, wilderness and mountainous areas. Urban refers to that part of the country which has a 'high' population density and is given over to housing, industry, transportation, service centres, energy plants, termini, etc. The 'low' and 'high' densities must be considered as variables in each country, for example, in countries with large population densities the respective density figures will be more than in countries with less population density

Resources in the context of this paper refers to the items that will be recorded and listed in a multi-purpose cadastre, that is those factors which have a direct bearing on land.

To be able to record resources in rural areas it will be necessary to relate them to a spatial dimension. In those parts of Europe that have a numerical cadastre the spatial dimension can be related to the parcel boundaries and this provides a density of control which fits well into the multi-purpose cadastre. In countries that do not have defined cadastral boundaries the need for densification of national control will have to be undertaken.

Many of the land parcels in Europe are uneconomic and in an attempt to consolidate and re-apportion them into economic units there will be a need to establish the resources that they contain.

The advent of the Digital Terrain Model (DTM) enables the resources of the land parcel to be considered in three-dimensions and such variable factors as aspect, and sun's elevation may be taken into account when changes are proposed into new property boundaries.

2 PRESSURE ON RURAL LAND

As described in the introduction, rural land is that land which is less densly populated and is presently used for agriculture etc. In most countries the rural region forms the largest area which in developing countries often contains the largest percentage of the population. In developed countries the rural region usually contains the smallest percentage of the total population.

2.1 Population movement and increase

Even though there is a movement of population in rural areas into the major urban areas, more so in developing countries, there is a reverse move from urban to rural areas now occuring in developed countries. It could follow that there is more outward pressure in developed countries. However, the developing countries have the additional problem of population growth, sometimes excessive, and although the movement here is presently towards urban areas the total numbers are still on the increase in rural areas.

2.2 Food production and agriculture

The need to increase the production of food in rural areas for an increasing population, largely occuring in developing countries, puts additional pressure on rural land. With some of the agricultural techniques adopted in devloping countries there is a trend for diminishing return per unit area and in certain circumstances no return at all. Examples are the short term benefits of clearing tropical forests for agriculture, often after only 2 years the land is no longer useable. This cleared land will not revert back to forest because the thin top soil has been eroded and only poor secondary growth will occur.

2.3 Forestry

The increasing demand for timber, both for energy and for manufacturing paper and wooden objects has involved more land being allocated to forestry. In Europe forests where not already established are planted on poorer soils, in developing countries forests are being indiscriminately cleared putting more pressure on remaining forests in rural areas.

2.4 Housing

With expanding populations in developing countries increasing amounts of land are allocated to housing. Some of the housing will be built by squatters and even when these people are moved away the sites remain derelict until clearance takes place.

2.5 Allocation of land

In many developing countries, small parcels of land have been allocated to the previously landless people. While this allocation has occured because it meets certain political and other promises, the impact is to reduce the land available for commercial agriculture and to leave these people at or below subsistence level. It has also meant that large areas of land are put under pressure with erosion and other environmental problems occurring.

3 RECORDING THE SPATIAL DIMENSION

As stated previously control can be provided from existing data or it may have to be specially surveyed as a framework to control the multiple data.

3.1 Position of the cadastral boundary

There is a need to use numerical control to enable the position of the resources to be established. The cadastral boundary can provide adequate control especially where the land holdings are small.

3.2 Extension of national trigonometrical control

Where there is no numerical control provided by the cadastre then there will be a need to densify the trigonometrical network. The positioning of control at a spacing of better than 800 metres (National Research Council 1980) will be required.

3.3 Determining control

Where the control is not adaequate and has to be provided a Global Positioning System (GPS) can be used for determining horizontal control to better than one metre accuracy. This control will be adequate for many purposes and may well be able to control the aerial photographs and satellite images especially if it is in the form of pre-marking or can be recognised on the image and used for post-marking.

Vertical control will be required for controlling the aerial photographs and satellite images (especially SPOT) and will also be needed for resources that are recorded in height, water heights, water table, flood limits etc.

4 THE MULTI-PURPOSE CADASTRE

The multi-purpose cadastre will contain a variety of information in addition to that required for boundary demarcation, and will assist in the efficient administration of the land (Bullard 1981). The land parcel is the basic unit in a MPC.

4.1 The land parcel

The land parcel is the smallest piece of land that is recorded in a land register. The size of parcels vary from the very large, the 'latifundi' (Jacoby 1971) to the very small, the 'minifundi'.

Where possible, resources should be related to each parcel of land. With the 'minifundi' this may not be possible because the resolution of satellite imagery is not adequate; this will be achieved with aerial photography where the scale is suitable.

4.2 The multi-purpose cadastre

The MPC can be thought of as a series of planes each containing land related data. Figure 1 (Archer 1980) shows an example of the MPC depicted as 6 planes that

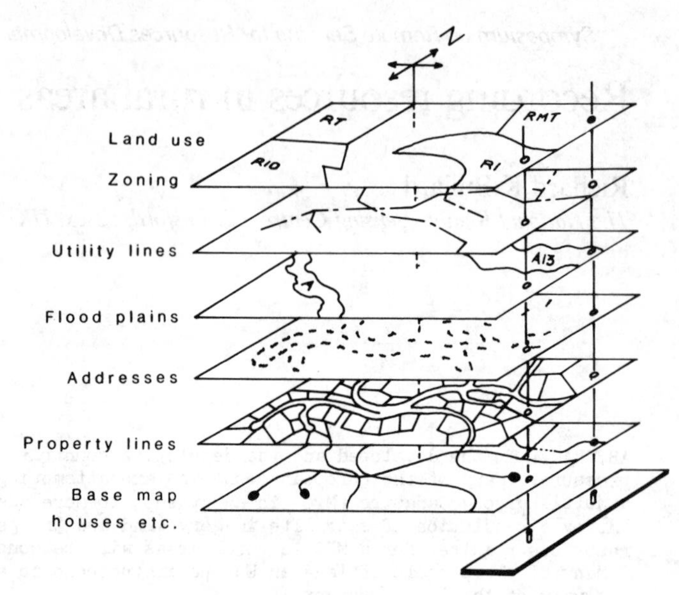

A REGISTERED OVERLAY SYSTEM.

are registered by control.

The property lines are the cadastral boundaries that depict the individual land parcels. The addresses or the property descriptions must be such that they provide a unique description such that there is no duplication within a property register and, ideally, within a country.

The base map shows the improvements, houses, and the surface construction, streets etc. The utility lines show the position, three dimensionally recorded, of the network of services.

Flood plains will be determined with the aid of vertical control.

Land use is recorded as well as zoning which will relate to the legislation that is enforced to control activity that may be carried out on a parcel or parcels of land. The boundary (artificial) between the rural and urban zones will be of particular interest.

Dependent on the imagery some of the above data may be established except for adresses. The recording of property boundaries will be limited to large scale aerial photographs or where they are depicted by tonal changes on airborne imagery.

5 AERIAL PHOTOGRAPHY AND SATELLITE IMAGERY

5.1 Aerial photography

The limitation of the aerial photograph for determining rural resources is that it is in analogue format while much of the other data in an MPC is in digital format. It is possible to convert the aerial photograph into digital format by scanning but with a resultant loss in resolution. The photograph is also usually only available in single band format, mainly panchromatic. The aerial photograph is, like the other imagery, weather dependent.

5.2 Airborne thematic mapper

It is suggested that the Airborne Thematic Mapper (ATM) will provide data sufficient for many of the rural resources. An example is the Daedelus scanner which can provide digital data with a resolution of better than one metre and in eleven bands including thermal infra-red. Unlike the scanner carried in a satellite, the airborne model is flexible in that it can be transported to an airfield nearest to a site and await suitable flying conditions. Cost may, however, be a limiting factor in using the ATM.

5.3 Satellite imagery

With the increasing resolution of satellite imagery, for example with SPOT, this has increased to 10 metres in the single band. This data will be increasingly suitable for obtaining details of resources of rural areas.

The SPOT imagery has the added advantage that it has stereoscopic potential and therefore the vertical dimension can be established.

6 RECORDING RESOURCES FOR THE MULTI-PURPOSE CADSTRE

The imagery referred to in the previous chapter will be suitable for recording those resources that exist on the surface of the earth. The limitation will be the resolution of the imagery, the size, shape, and the spectral signature of the object being recorded. All the descriptive data, names, property description, and ownership will have to be obtained from other sources.

6.1 Using aerial photography

The aerial photograph would primarily be used for producing the topographic or cadastral map where the boundaries are visible from the air. The position of houses and other improvements on the land can be established from the aerial photograph. With photographs in the thermal band it will be possible to detect water boundaries as well as the position of flood plains provided the data is collected close to flooding or the 'marks' are retained.

6.2 Using airborne thematic mapper

With the ATM having multiple bands a wide range of resources can be recorded. The thermal infra-red band can be used for detecting services and the image shown was used for this purpose.

Thermal infra-red image showing surface and indications of underground features

6.3 Using satellite imagery

Many of the natural resources in rural areas can be recorded with satellite imagery. The second generation satellites have improved resolution and have stereoscopic capabilities. However, because of the resolution of the satellite imagery, compared with aerial photography, resources may only be recorded over larger areas than the average sized parcel.

7 DATA FOR IMPROVEMENTS TO LAND PARCELS

The data that is collected together for an MPC, which contains the resources related to land parcels, can be used for improvements.

7.1 Digital terrain models

By combining the three dimensional model, or Digital Terrain Model (DTM), with other details of the land parcel it will be possible to consider improvements. Such details as aspect, sun elevation and shadow, relative elevation, can be incorporated with the DTM. From the model it will be possible to establish the agricultural crops as related to the above details. The anticipated runoff and hence the storage of water in a catchment area can be determined. Information on the soils, vegetation has to be combined with data on slopes to determine runoff, these are some of resources that will need to be obtained.

7.2 Consolidating land parcels

When two or more land parcels are combined to make a viable unit the DTM can assist in establishing the new boundaries. The MPC can supply all of the data for each of the constituant parcels and the 'planes' of data can be added to the DTM to see if they are compatible.

8 CONCLUSIONS

* There will be a need for a dense network of boundary or network control within which the resources can be recorded.

* The use of the airborne thematic mapper will enable more resources to be recorded and at a greater resolution.

* That imagery in digital format, compatible with much of the other data is the most acceptable and manageable.

* That data of resources in analogue format will have to be digitised to make it compatible.

* Resources and services are contained below as well as above the surface of the earth and imagery to detect them is required.

* The multi-purpose cadastre is an ideal data base in which information on rural resources can be stored.

* By incorporating selected data from the multi-purpose cadastre with a digital terrain model improvements can be achieved for the land parcel.

REFERENCES

Archer, A.J.A. 1980. Unified approach for mapping in Prince William County, Virginia. American Congress on Surveying and Mapping, Bulletin 71. p.17-19.

Bullard, R.K. 1981. Multi-purpose cadastre and remote sensing. Matching remote sensing technologies and their applications. Proceedings of an International Conference of the Remote Sensing Society, London, December 1981. p.339-346.

Jacoby, E.H. & C.F.Jacoby 1971. Man and land. London: Andre Deutsch.

National Research Council, 1980. Need for a multi-purpose cadastre. Washington, D.C., National Academy Press.

Symposium on Remote Sensing for Resources Development and Environmental Management / Enschede / August 1986

Evaluation of regional land resources using geographic information systems based on linear quadtrees

James Hogg, Mark Gahegan & Neil Stuart
School of Geography, University of Leeds, UK

ABSTRACT: Evaluation of regional land resources involves the integration and analysis of geographic data which comes from a variety of different sources and in many different forms. This paper describes results of a pilot study using a computerised geographic informations system (GIS) based on linear quadtrees to integrate and analyse geographic data for evaluation of regional land resources near Matlock in the Peak District of Derbyshire, England. Results are presented which show the response to queries involving set logic operations on binary raster images and are discussed in relation to methods of regional land resources evaluation. The paper concludes that GIS based on linear quadtrees provide a flexible, powerful analytical tool for geographical research involving integration of geographic data from various sources, including remote sensing.

RESUME: L´ evaluation des ressources régionales d´i terrain comprend l´integration et l´analyse de données géographiques provenant de différentes sources et se présentant sous des formes differentes. Cette recherche decrit les résultats d´une étude-pilote utilisant un système d´information géographique sur ordinateur, de facon a établir une évaluation des resources régionales d´un terrain situé près de Matlock, dans le Peak District du Derbyshire, en Angleterre. La recherche conclut que le Système d´Informations Geographiques est bases sur des "quadtrees" lineaires, et fournit un outil d´analyse flexible et efficace pour la recherche géographique concernant l´intégration de données géographiques provenant de sources diverses, y compris les images satellites.

INTRODUCTION

Evaluation of land resources involves the study of geographic data from many different sources and in many different forms. Geographic data can be efficiently integrated and analysed using computerised geographic information systems (GIS). These are data base management systems which allow users to store, retrieve, manipulate, analyse and display geographic data at their request. The concept of GIS has evolved over the past two decades (Tomlinson 1984). Its origin lies in the computerised data banks which were created to store locational data such as the coordinates of points for specific applications in surveying and mapping. It has now broadened and expanded rapidly to embrace sophisticated computerised systems for modelling and decision-making in land management (Dangermond, 1984; Estes et. al. 1985).

The purpose of this paper is to describe the characteristics of a pilot GIS that we are developing and to demonstrate its use for land resources evaluation in an area near Matlock in the Peak District National Park, Derbyshire. Results are presented and discussed in relation to traditional methods of land resources evaluation and the need to integrate geographic data from different sources, with various levels of resolution and accuracies.

In order to place the current work in context, we begin by outlining recent changes in the approach to land resources evaluation and trends in the development of integrated GIS. Then we describe a quadtree data model for encoding images and present and discuss results.

LAND RESOURCES EVALUATION

Land resources evaluation is concerned with making assessments about man´s potential use of land for purposes such as agriculture, forestry, recreation, urban planning or engineering (Christian and Stewart, 1968). It involves analysis of the capabilities and constraints imposed by the physical characteristics of a region and is usually conducted in support of some decision-making process in land management. It is carried out by scientists from many different fields of study but, in many cases, they adopt a similar approach, though of course the level of detail and specific requirements and methods usually differ (Mitchell, 1973). Moreover they draw typically upon a common core of information about the land. The extent of this common core is usually substantial - all require basic information about topography, geology, soils, climate and land use. While there are usually minor differences in specific requirements, the major difference often lies in the level of detail required.

In conducting land resources evaluations, land is usually characterised by a distinctive assemblage of attributes and interlinking processes in space and time (Townshend, 1983). The attributes include topography, soils, water, climate, vegetation, and fauna as well as the results of human activity. Mitchell (1973) stresses that land evaluation is a broad term which encompasses analysis, classification and appraisal of information from a variety of sources for a potential land use. Analysis involves selecting characteristics which have importance for a particular application and compiling land characteristics. Classification relates to the organisation of characteristics which distinguish one area from another and which characterise each. Appraisal uses these characteristics, along with other properties, to assign a value to a piece of land, expressed either by a numerical value or by a judgement of its worth in qualitative terms.

A land resources evaluation system has several basic requirements. Mitchell (1973) identifies three:

1. a means of answering queries from users;

2. a means of acquiring, storing, analysing and displaying information about the land and its potential uses;

3. a means of retrieving and manipulating information;

The traditional approach to fulfilling requirements for land resources evaluation has been by preparing manually various maps and transparent overlays showing features, such as slope, aspect, soils, drainage and other characterisitcs and by preparing statistical and textual reports. Visual comparison and interpretation of maps and reports leads to an evaluation of regional land resources for a particular application. The basic source of information for all these maps has usually been aerial photographs, though other forms of remote sensing are increasingly being used to aid sub-division of the land. Computers are used increasingly to store, process and retrieve at least some of the data and GIS have been developed using a fixed-cell size grid or polygon respresentation, but much geographic data is still stored in analogue maps because these have provided access much more quickly than existing GIS when large volumes of geographic data are involved.

GEOGRAPHIC INFORMATION SYSTEMS

Marble and Peuquet (1983) describe the development of GIS and observe that a GIS is designed to accept large volumes of spatial data, derived from a variety of sources including remote sensing, and to store, retrieve, manipulate, analyse and display these data. The development of intelligent GIS in which the concepts and techniques of artificial intelligence and database systems are integrated represents a major new field of research (Smith and Pazner, 1984a; 1984b; Smith and Peuquet, 1985; McKeown et al. 1984).

In designing a GIS, a critical decision is the choice of data model. This is the abstraction that is used to represent properties which are considerd to be relevant to the application in the computer. Peuquet(1974) reviews the different types of spatial data models that have been used in GIS and compares their performance. Geographic data have been represented using many different types of data models, but a basic difference is between vector and raster types.

1 Vector type

In this type of data model, the basic logical unit in a geographical context corresponds to a line on a map. It is recorded as a series of x-y coordinates with a heading describing the feature. Vector data is widely used in cartographic GIS and many other types which have been developed for specific projects.

2 Raster type

This type of data model uses a fixed-sized square cell or raster to represent geographic data in a binary array or grey-scale image. The development of data models based on raster has been largely driven by advances in the technology of remote sensing and computing over the past decade (Marble and Peuquet 1983). The use of MSS scanning systems in satellite remote sensing has been a major influence. At the same time, there have been significant advances in the technology of raster scan and video digitising systems. These have accelerated digitising maps and related documents. Because all these systems use a square cell or raster, it is generally agreed that this is the only practical tiling or tessellation. A number of other possibilities exist which may be theoretically better than the regular tessellation (Bell et. al., 1983).

Peuquet (1984b) discussed the main advantages of raster type of data models. Apart from the practical benefits of being able to get massive sets of raster data from satellite remote sensing, and raster scanning of maps, it is compatible with array data structures and various hardware devices for input and output. Peuquet (1984b) and McKeown (1984) argue that existing vector and raster data models are limited however by two basic factors:

1. the rigidity and narrowness in the range of applications and types of geographic data which can be accommodated;

2. the unacceptibly low levels of efficiency for storage and response to queries for the current and anticipated volumes of geographic data.

These factors restrict the potential of automated GIS based on the use of vector or raster data models to cope with the variety of different forms of geographic data and the massive volumes. For these reasons, attention has recently focused on another data model known as the quadtree.

QUADTREE DATA MODEL

A data model which has become increasingly important in recent years is the quadtree, which is based on the concept of recursive decomposition of a grid. The idea of the quadtree was formulated by Klinger (1971) but has been developed by many others, including Klinger and Dyer (1976), Hunter and Steiglitz(1979), and Samet (1980,1981,1984). Research into the theory and applications of quadtrees has broadened and expanded during the 1980's. They have become a major focus of interest in computer science for applications in image processing, graphics, robotics and GIS. Samet (1984) provides a comprehensive review of the quadtree and related hierarchical structures. He points out that there are now many different types of quadtree, such as the point, line, and regional quadtrees, that they are all based on the principle of recursive decomposition of an image but that they do not all share the same properties or ease of implementation. The efficiency of quadtrees for representation of regions and for interchangeability with more common representations such as vectors, chain codes, arrays and rasters has been studied intensively by computer scientists.

Quadtree encoding

A quadtree is constructed from a square binary array of pixels which represent an image. We refer to the set of black pixels in the image as the region. If we assume that an image comprises a $2^n \times 2^n$ binary array of pixels, then a quadtree encoding represents this image by recursively sub-dividing it into four quadrants until no further sub-division is necessary. This occurs when we obtain square blocks (possibly single pixels) which are homogeneous in value (i.e. either all black or all white) or when we reach the level of resolution that we require. This process is represented by a tree with four branches or sons in which the root node corresponds to the entire binary array of pixels or image, the four sons of the root node correspond to the four quadrants, which in our case are labelled North West, North East, South West and South East. The terminal or leaf nodes of the quadtree correspond to the homogeneous blocks for which no further sub-division is necessary. The nodes at level n (if any) represent square blocks of size $2^n \times 2^n$. Thus, a node at level 0 corresponds to a single pixel in the image, whereas a node at level n is the root node of the quadtree. An example is given to illustrate these concepts in Fig 1. The region in Fig 1a is represented by the binary array in Fig 1b. The resulting square blocks for Fig 1b are shown in Fig 1c and the tree in Fig 1d.

Initial work on regional quadtrees was carried out using pointers to represent the tree structure. Each node was represented by a record which consisted of five pointers: four to sons, one to an ancestor and a field for the colour of the node. This technique was used by Rosenfeld et.al.(1982,1983,1984). Although quick search times may be achieved using pointer based structures, a large amount of storage is taken up by the pointers. According to Stewart (1986), assuming that each pointer uses 16 bits of memory and the node descriptor 8 bits, then the pointers take up nearly 90% of the memory space used.

Regional representation using linear quadtrees

Gargantini(1982) proposed a data structure to represent quadtrees which was more economical in its requirements for memory space. Known as a linear quadtree, it is in the form of a linear list consisting of the quadtree nodes in some order of traversal of the tree. A number of different forms of keys is available to map a set of ancestors to a numeric key (Gargantini 1982; Abel and Smith 1983). In our case, we follow the form proposed by Gargantini(1982) but number the quadtrants 1, 2, 3 and 4. Thus nodes are arranged so that the quadrants are in order 1,2,3,4 corresponding to the NW, NE, SW, SE quadrants of an image. Only black nodes are recorded. Thus a region in an image corresponds to a node which has a unique key derived from its ordered list of ancestors (Fig 2).

Linear quadtrees offer several advantages over quadtrees based on use of pointers. Gargantini(1982) demonstrates how arithmetic operations on the key of the node can be evaluated to determine various properties such as determining the relative x, y coordinates of a node, adjacency of nodes, ancestor or descendent relationships and translation and rotation of images. Where disk resident quadtrees must be considered, the linear quadtree can readily be indexed to a B-tree memory management system to provide efficient access to nodes for such elemental operations as examinations of the neighbours of a given node (Abel 1984).

IMPLEMENTATION OF GIS USING LINEAR QUADTREES

As a part of a one year pilot study, we are developing a GIS based on linear quadtrees to determine its potential for geographical research in a range of different applications. Programs have been written in the C programming language on a VAX 11/750 computer running Berkeley UNIX 4.2.

Although we are continuing to develop and expand the range of functions that are available to the user, we have functions for input, manipulation, analysis and display of images. Input can be from binary array images, vector polygons or vector segments, the last two being converted to binary arrays. We are also implementing an algorithm for direct vector to quadtree encoding which was devised by Mark and Abel (1985). Analysis functions include traversing an image, finding the colour of a node, finding a neighbour to a node, finding the perimeter of regions, measuring distances, labelling the separate regions or components, determining geometric properties of regions such as size, shape or orientation, forming windows into an image at larger scales, generation of statistics about number of nodes, areas involved and correlations between geographic features and set logic operations on quadtrees (union, intersection, complement) (Mark and Abel, 1985). These functions allow users to perform the two basic operations

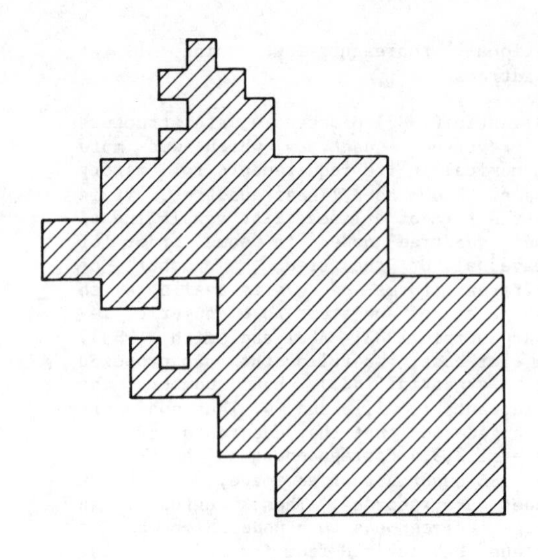

(a) Region.

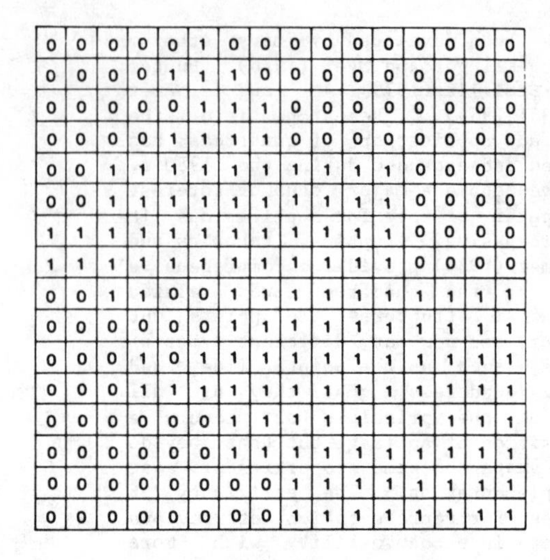

0	0	0	0	0	1	0	0	0	0	0	0	0	0	0	0
0	0	0	0	1	1	1	0	0	0	0	0	0	0	0	0
0	0	0	0	0	1	1	1	0	0	0	0	0	0	0	0
0	0	0	0	1	1	1	1	0	0	0	0	0	0	0	0
0	0	1	1	1	1	1	1	1	1	1	1	0	0	0	0
0	0	1	1	1	1	1	1	1	1	1	1	0	0	0	0
1	1	1	1	1	1	1	1	1	1	1	1	0	0	0	0
1	1	1	1	1	1	1	1	1	1	1	1	0	0	0	0
0	0	1	1	0	0	1	1	1	1	1	1	1	1	1	1
0	0	0	0	0	0	1	1	1	1	1	1	1	1	1	1
0	0	0	1	0	1	1	1	1	1	1	1	1	1	1	1
0	0	0	1	1	1	1	1	1	1	1	1	1	1	1	1
0	0	0	0	0	0	1	1	1	1	1	1	1	1	1	1
0	0	0	0	0	0	1	1	1	1	1	1	1	1	1	1
0	0	0	0	0	0	0	0	1	1	1	1	1	1	1	1
0	0	0	0	0	0	0	0	1	1	1	1	1	1	1	1

(b) Binary array.

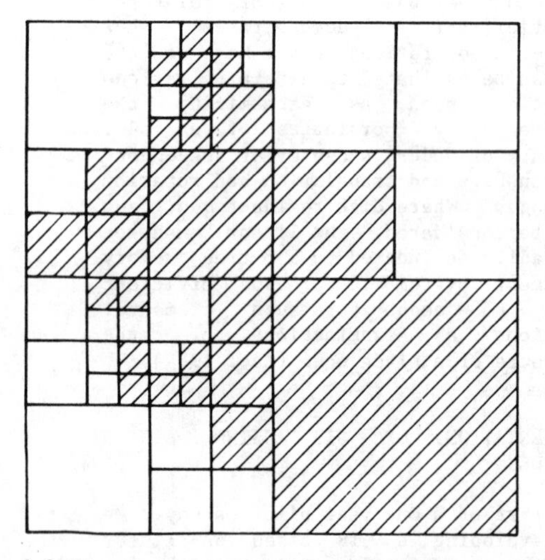

(c) Block decomposition of region in (a).

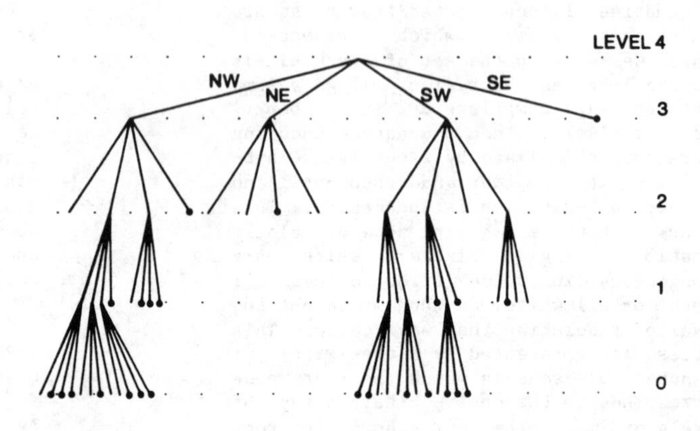

(d) Quadtree representation of the blocks in (c)

Fig (1) Quadtree region representation.

x-axis

y-axis	NW 0	1	2	3	4	5	6	7 NE
0	111	112	121	122	211	212	221	222
1	113	114	123	124	213	214	223	224
2	131	132	141	142	231	232	241	242
3	133	134	143	144	233	234	243	244
4	313	314	321	322	411	412	421	422
5	315	316	323	324	413	414	423	424
6	331	332	341	342	431	432	441	442
7	333	334	343	344	433	434	443	444

SW SE

Fig (2) Addressing scheme for locational keys

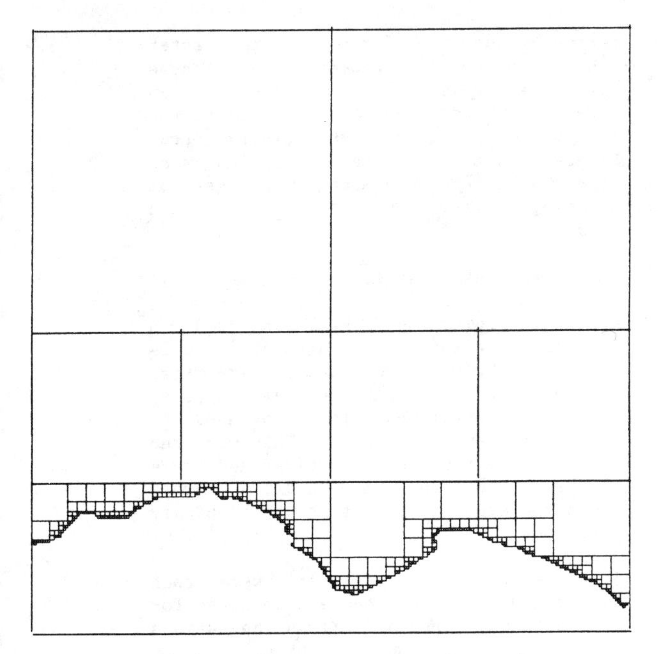

Fig (3) Areas underlain by grit.

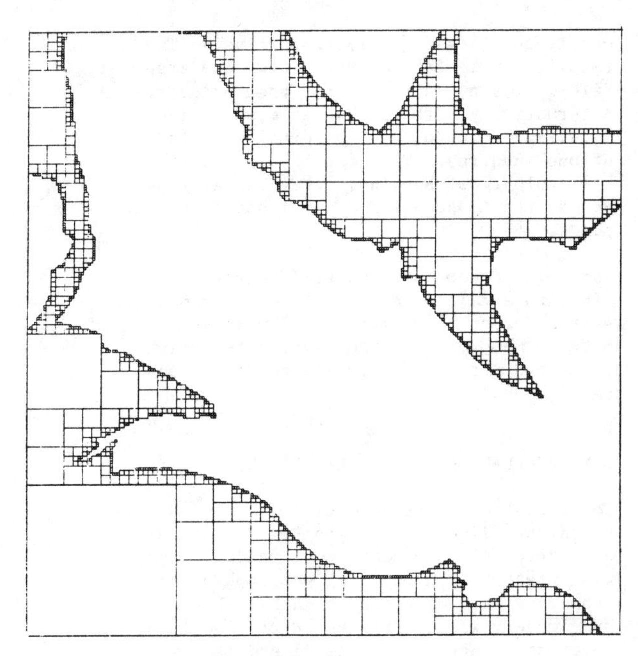

Fig (5) Brown Earth soils.

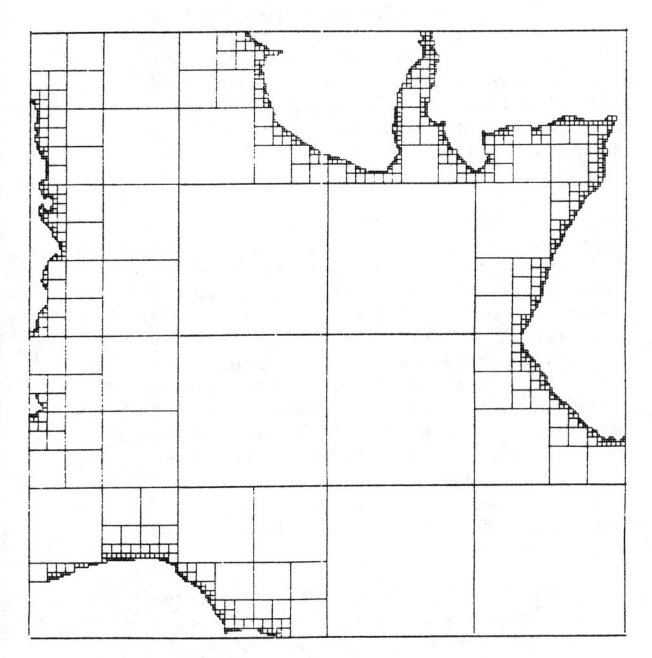

Fig (4) Land below 800 ft.

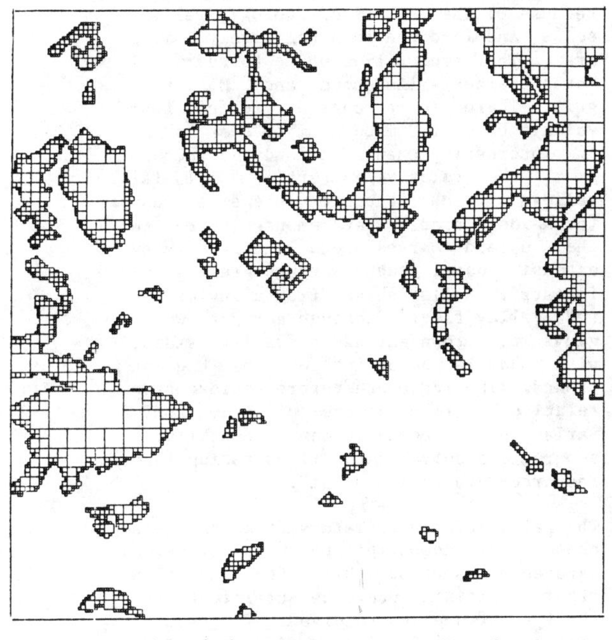

Fig (6) Areas of woodland

on GIS identified by Smith and Peuquet(1985):

1. find what is at a specified location;

2. find all occurrences of a specified entity or subset of entities.

Functions for output allow users to file results, plot maps and generate printed output of statistics and related diagnostic information about the number of nodes involved in operations.

The interface to the user is by a query language which allows users to issue a range of commands in abbreviated form. For example, if we wish to find the intersection of grit and land above 500 feet, we enter the query:

grit I N(height300_400 U height400_500)

where grit, height300_400 and height400_500 are the names of quadtree encoded files representing specific geographic features, and the operators I, U and N stand for intersection, union and complement (logical NOT) respectively. The results can be formed into a new quadtree file or sent directly to the printer. Alternatively, there is a menu driven interface which allows the detailed study of one quadtree. This can be used to find the colour of a point, traverse a line across the image or find a neighbour to a point.

Work is now being done to integrate the GIS into a relational data base management system for storing regions and objects. A B-tree memory management system is being developed to store very large quadtrees (Abel, 1984).

LAND EVALUATION IN THE PEAK DISTRICT

To illustrate the potential of the GIS based on linear quadtrees for regional evaluation of land resources, a data base was created by digitising geographic data from maps and aerial photographs. Landsat MSS imagery was classified to provide land cover at a much coarser level and is being included in the GIS.

Description of study area

A study area was selected close to Matlock, Derbyshire. This was chosen because of the variety of geology, relief, soils and land use in a relatively small area. The region lies on a boundary of Carboniferous limestone and Millstone grit. There is a range of relief from valleys below 100 m to summits over 350 m. An interesting change in land use occurs from the relatively fertile alluvial valleys which support intensive use (orchards, arable, settlements., etc), to the upland areas characterised by extensive uses, such as rough grazing and forestry. Soils change from alluvium on the valley floor, through surface water gleys to brown earths on valley sides, with podsols occurring on the highest ground. The region therefore represents a relatively rich source of geographic variation and contains many identifiable sub-regions suitable for illustrating land resources evaluation by GIS.

The area of approximately 25 sq km was chosen for compatibility with existing mapsheets, and so that the resolution after digitising would be sufficient for the type of queries received. The square region selected corresponded to O.S. mapsheet SK 26 SE at 1:10560 scale. This was one of the older O.S. maps produced in 1971 so heights were recorded in feet. The resulting raster grid of 256 x 256 pixels lead to a ground resolution of about 20 x 20 m, whilst the LANDSAT MSS data was resampled to give a ground resolution of 80 x 80 m pixels.

Data Capture

Two different methods were used to build arrays of spatial data for quadtree encoding:

1. Manual Digitising

A GTCO digitiser linked to an IBM PC/XT was used to digitise selected features from the O.S. map at a scale of 1:10560. A series of functions for polygon to raster conversion form part of the GIS. These allow vector data to be mapped on to an array of a specified size. Functions for filling regions then produce binary array images. A set of 10 binary array images of size 256 x 256 were made for images of geology, soils, land use and climatic data.

2. Video digitising.

A video digitising system was used to produce 256 x 256 8 bit images of contours from the O.S. map. The contours were first traced onto transparent overlays and each contour interval was coloured by hand to produce a grey scale image. This was then captured by the video digitiser and frame grabber then transformed to fit a square 256 x 256 array. The result was 10 binary array images of relief.

All the binary array images were then quadtree encoded to produce a database for the area for testing various operations on quadtrees and land evaluation.

RESULTS

Figures 3-8 show selected results from a land resources evaluation for the Matlock area. Parameters considerd were geology, elevation, soils and land use. Although several more parameters are to be included into the database, this example indicates how a GIS approach could be used to identify areas most suitable for agriculture.

Fig (3) shows those areas where there is grit parent material. This large area highlights the way in which the quadtree representation saves space by storing maximal blocks at different levels within the tree. The northern three quarters of Fig (3) is stored as only 6 quadtree nodes for the 6 maximal blocks. The complex boundary of the grit area requires considerably more nodes to be accurately represented. A characteristic feature of quadtree encoded images is the tendency for smaller blocks at the edges of regions, and larger ones towards the centre.

Fig (4) shows land below 800 ft above sea level. This map was produced by the union of all height maps for intervals above 800 ft, to produce the map 'height_above800'; this was then complemented to produce areas not above 800 ft. Following each union operation, maximal blocks are formed, since four small blocks with a common ancestor may be produced. Because the combined image often contains more space-saving larger blocks, the union image of Fig (4) need not necessarily contain more nodes than any single image from which it is derived. Fig (4) contains many larger blocks, showing how the quadtree stores large regions compactly, while being accurate enough to closely approximate the contours which bound the region.

Fig (5) indicates areas of acid brown earth soils, which occur in two regions on the valley sides. Note that four smaller blocks are not always merged into a single

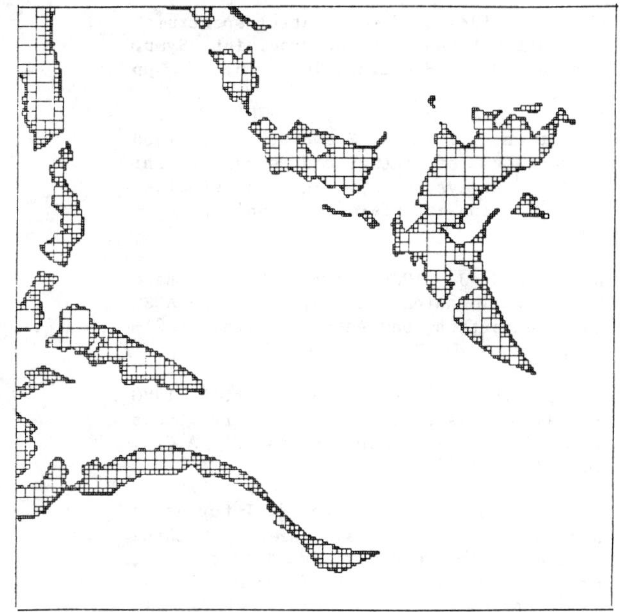

Fig (7) Brown earth soils below 800 ft., over grit, not covered by woodland.

Fig (8) 300-400 ft. elevation, generalised to 128x128 resolution

large block, if they do not have a common ancestor in the quadtree structure.

Fig (6) indicates woodland areas. This is a fairly complex image since it contains a large number of small polygons. A relatively large number of nodes are required to store this image, many of which are nodes at the lowest level. This results in the quadtree structure being less efficient at storing many small regions, rather than a few extensive areas.

Fig (7) shows only those areas which possess all the above characteristics. It displays only regions with grit geology and brown earth soils which are within the height limit, and which are not covered by woodland. This map was produced by the GIS query:

P brown_earth I grit I height_below800 I N woods

The P operator stands for Plot map. The N operator produces the complement of the woods map, since we are interested in those areas not covered by woods; this is then intersected with the height map, the result of this is intersected with grit geology areas, and finally an intersection is made with the soils map. At each stage those areas which do not share all the attributes are excluded. One of the main advantages of using linear quadtrees is the speed at which set operations such as these are performed; the above query taking only a few seconds.

Fig (8) shows areas between 300 and 400 ft generalised by omitting some details. In this case the smallest size blocks, present in the other images, are omitted. This illustrates the variable resolution of the quadtree structure, since it is equivalent to a descent of the tree which stops one level above the terminal or ´leaf´ nodes. The smallest blocks in Fig (8) have a ground size 80m square, and the coarseness of the picture is readily apparent. This is the level of resolution of Landsat imagery, which is being integrated into the database at this level. This shows how the quadtree structure accommodates data of different resolutions.

The above example showed how a user can query the GIS to identify areas most suitable for agriculture in the study region. Brown earth soils have the best drainage, water retention and textural characteristics of the soils present in the study region. The podsolic soils are too acid and poorly drained, whilst lower level gleys are very badly drained and have a massive structure making them difficult for agriculture. The valley floor alluvial soils are shallow and have poor water retention characteristics.

The height limit of 800 ft was selected to exclude areas considered to be too high for agriculture, because of exposure of crops and increasing rainfall likely to produce leaching or erosion.

Areas of grit geology were selected because this parent material permits relatively free drainage while allowing reasonable soil development. Shales occurring in the region tend to impede drainage, while limestone areas in the southern part of the image usually promote shallow soils and stony surfaces, which are suitable only for grazing purposes.

The result of this example land evaluation shows (Fig 7) that areas with the above characteristics are found in several locations on both valley sides. They occur mostly not as small residual areas, but as identifiable regions which could be productively used as field units. The total and individual areas of these land units can be calculated by block summation. The fact that after several intersections, the resulting areas still have reasonable size, shows there is quite a common area of overlap between all four images. This is being studied further, together with climatic and remote sensing data, as it suggests a correlation between the parameters.

CONCLUSIONS

The approach to land resources evaluation using GIS based on linear quadtrees has been discussed in relation to a study of the Matlock area, Derbyshire. Results show that linear quadtrees have an important practical and analytical role to play in regional resources evaluation. The practical role relates to (1) the integration of spatial data from a variety of different sources in one database which can be interrogated by users and (2) the production of land information at scales which suit the needs of users. The analytical role relates to spatial analysis of selected parameters and the generation of new information and reports about regional resources from information stored in the GIS. The ease, speed and efficiency with which this can be carried out on large volumes of geographic data is a major advantage of the use of the linear quadtree.

Results from this study have not only helped to clarify concepts and demonstrate the potential of the GIS based on linear quadtrees but have shown that they are a flexible analytical tool for geographic research which deserves further study.

ACKNOWLEDGEMENTS

Maps used in this study were based upon the 1971 Ordnance Survey 1/10560 map with the permission of the HMSO, Crown Copyright Reserved. The University of Leeds provided a research grant for M. Gahegan and a post-graduate studentship for N. Stuart to facilitate this research.

REFERENCES

Abel, D.J.& Smith, J.G. 1983 A data structure and algorithm-based on a linear key for a rectangular retrieval problem. Computer Vision,Graphics and Image Processing,Vol.24,pp1-13

Abel, D.J. 1984 A B-tree structure for large quadtrees. Computer Vision, Graphics & Image Processing, Vol 27, pp 19-31.

Abel, D.J. 1985 Some Elemental Operations on Linear Quadtrees for Geographic Info Systems. Computer Journal, vol 28 ,No1 pp73-77.

Bell,S.B.M. Diaz,B.M. Holroyd F.C.Jackson, M.J. 1983 Spatially referenced methods of processing raster and vector data. Image and Vision Computing. 1 (4): 211-220.

Chen, Z.T.1984 Quadtree spatial spectrum: generation & application. Proc. Int. Symp. Spatial Data Handling,1984, Zurich, pp 218-236.

Christian, C.S. & Stewart, G.A. 1968 Methodology of integrated surveys. In: Aerial surveys and integrated studies. Proc. Toulouse Conference, 1964, pp 233-80.

Dangermond J. 1984 Geographic Data Base Systems. Technical Papers, 1984 ASP-ACSM Fall Convention, San Antonio, Texas, p201-211.

Dyer, C.R. Rosenfeld, A Samet, H 1980 Region representation : Boundary Codes from Quadtrees.Communications of A.C.M. vol 23 No3 pp171-179.

Estes, J.E. 1984 Improved Information Systems - A Critical Need. Machine Processing of Remotely Sensed Data,Symp, Proc,10th, W.Lafayette,In,June 12-14, 1984, pp2-8.

Estes, J. E. Star, J.L Cressey, P.J. Devirian, M. 1985 Pilot Land Data System Photogrammetric Eng & Remote Sensing, Vol 51(6):703-709

Gargantini, I. 1982 An Effective Way to represent Quadtrees. Communications of A.C.M. vol 25 No. 12 pp905-910.

Gargantini, I. 1983 Translation, rotation and superposition of Linear Quadtrees. Int. Journal of Man-Machine Studies, (18), pp 253-263.

Hunter, G.H.& Steiglitz, K. 1979 Operations on images using Quadtrees. IEEE Trans on Pattern Analysis and Machine Intelligence,Vol.PAMI-1,No.2,145-153

Klinger, A 1971 Patterns and search statistics, In: Optimising Methods in Statistics, J.S. Rustagi, Ed, Academic Press, New York, p303-337.

Klinger, A Rhodes, M.L. 1979 Organisation and access of image data by areas.IEEE Trans on Pattern Analysis and Machine Intelligence Vol.PAMI-1,No.1,50-60

Klinger, A. & Dyer, C. 1976 Experiments on picture representation using regular decomposition. Computer Graphics and Image Processing.Vol.5,Part 1,pp68-105

Marble D.F. & Puequet D.J. 1983 Geographic information systems and remote sensing. Chapt 22 In: Manual of Remote Sensing, Vol 1, American Society of Photogrammetry,Falls Church, p923-957.

Mark, D.M. & Abel, D.J. 1985 Linear Quadtrees from Vector representation of polygons. IEEE Trans. on Pattern Analysis & Machine Intelligence, vol PAMI-7 No 3, pp343-349.

Mitchell, C. 1973 Terrain Evaluation. Longman Group, London pp 220

Peuquet, D.J. 1984a Data structures for a Knowledge - Based GIS. Proc. Int. Symp. Spatial Data Handling, 1984, Zurich, Switz.

Peuquet, D.J. 1984b Data Structures for Spatial Data Handling. Background Materials to Workshop W2, International Symposium on Spatial Data Handling, 1984, Zurich.

Rosenfeld, A. Samet, H Shaffer, C. Webber, R.E. 1982 Application of hierarchical data structures to geographical information systems TR-1197 Computer Vision Laboratory,University of Maryland.pp 160

Rosenfeld, A. Samet,H. Shaffer, C.A Nelson, R.C. Huang, Y.G. 1984 Application of hierarchical data structures to geographical information systems: Phase III
CAR-TR-99 CS-TR-1457 Center For Automation Research,University of Maryland.

Rosenfield, A Samet, H Shaffer, C.A
Samet, H. 1980 Region representation : quadtrees from boundary codes. Communications of the ACM Vol 23,No 3, pp163-170

Samet, H. 1981 An algorithm for converting rasters to quadtrees. IEEE Trans on Pattern Analysis & Machine Intelligence Vol.PAMI-3,No 1,93-95

Samet, H. 1984 A Geographic Information System using Quadtrees. Pattern Recognition, vol 17, No 6, pp647-656

Samet, H. 1984 The quadtree and related hierarchical data structures. ACM Computing Surveys 16, 2:187-260.

Samet, H. Rosenfeld, A. Shaffer, C.A. 1984 Use of hierarchical Data Structures in Geographical Information Systems. Proc. Int. Symp. on Spatial Data Handling, 1984, Zurich, Switz.

Shelton, R.L. & Estes, J.E. 1981 Remote Sensing & Geographical Information Systems : An Unrealised Potential.
Geo - Processing , voll, pp 395-420.

Smith, T.R.& Pazner, M.I. 1984 Knowledge - Based control of Search & Learning in a large scale GIS. Proc. Int. Symp. on Spatial Data Handling, 1984, Zurich, Switz.

Smith,T.R. & Pazner,M.I. 1984 A knowledge based system for answering queries about geographical objects. William T. Pecora Memorial symposium on spatial information technologies for remote sensing today and tomorrow. proc..9th pp 286-289.

Smith, T.R. & Peuquet, D.J. 1985 Control of spatial search for complex queries in a knowledge-based geographic information system. Proc. Int. Remote Sensing Society/CERMA Conf. on Advanced Technology for Monitoring and Processing Global Environmental Data, London, 1985, p439-453.

Stewart,I.P. 1986 Quadtrees: storage and scan conversion. The Computer Journal Vol 29, No 1, 60-75

Tomlinson, D.J. 1984 Geographical Information Systems - A New Frontier. Proc. Int. Symp. Spatial Data Handling, 1984, Zurich.

Townshend J.R.G.(ed) 1981 Terrain analysis and remote sensing. George Allen & Unwin, London, pp231.

Symposium on Remote Sensing for Resources Development and Environmental Management / Enschede / August 1986

A comprehensive LRIS of the Kananaskis Valley using Landsat data

G.D.Lodwick, S.H.Paine, M.P.Mepham & A.W.Colijn
The University of Calgary, Alberta, Canada

ABSTRACT: This paper describes the design and development of a general land-related information system (LRIS) covering the upper Kananaskis Valley of south western Alberta. Landsat data provide information on surface cover, principally vegetation, as well as on land use. In addition, a range of overlays of thematic data obtained from conventional sources, such as geology, topography, snow-cover, hydrology and pedology, is being added to the LRIS. The land information is stored on a VAX 11/750 computer system using VAX/DBMS a CODASYL compliant network data base system. The primary key is the position-based data defined by the geographic coordinates of the various data types. Information retrieval by various secondary keys is possible. As well, topological relationships have been incorporated into the LRIS design allowing data retrieval in both graphical and tabular form.

RÉSUMÉ: Cette contribution décrit la conception et le développement d'un système général d'information foncière (SGIF) couvrant la partie supérieure de la vallée de Kananaskis dans le sud-ouest de l'Alberta. Les données de Landsat fournissent l'information pour la couverture de surface, principalement la végétation, mais aussi l'utilisation de la terre. De plus, une série de recouvrements de données thématiques obtenues de sources conventionnelles telles que la géologie, la topographie, la couverture de neige, l'hydrologie et la pédologie est actuellement ajoutée au SGIF. L'information foncière a été emmagasinée sur un système d'ordinateur VAX 11/750 utilisant VAX/DBMS, un système de banques de données en réseau compatible avec CODASYL. L'adresse primaire est en termes des données de positions définies par les coordonnées géographiques pour les différents types de données. Le retrait de l'information selon des adresses secondaires est aussi possible. De plus, les relations topologiques ont été incorporées dans le SGIF pour permettre le retrait des données sous formes graphique et tabulaire.

1. INTRODUCTION

Satellite borne electro-optical systems are being used extensively for thematic mapping data collection. The primary system is the Landsat Multispectral Scanner (MSS) which is recognized world wide as a useful source of map data. This system was specifically designed for resource mapping, with large area coverage and four spectral bands for discrimination of surface cover detail (Paine, 1984). It has also been used for collection of data for land use mapping (Anderson et al, 1976) and terrain classification (Schreier et al, 1982). In terms of topographic mapping it has been used indirectly for flight planning of photogrammetric missions (Myhre, 1982). Landsat has also been used for direct topographic mapping of uniform surface cover areas, such as ice and snow (Lodwick and Paine, 1985). The Landsat Thematic Mapper (TM) system is also available with roughly the same coverage, but with better ground resolution and seven wavelength bands for better spectral resolution (Holmes, 1984). According to Welch et al (1985) the 20 m positional accuracy of the TM system meets accuracy standards for maps of 1:50,000 or smaller, and is well-suited for image maps of 1:100,000 scale.

Electro-optical systems in the future will have all of the advantageous characteristics of the Landsat series, with even better ground resolution and stereo coverage. In the near future, data will be available from the SPOT satellite, which has the capability for collecting stereo images (Holmes, 1984). A proposed system called Mapsat will have continuous stereo coverage from three overlapping detectors (Colvocoresses, 1982). Welch (1985) suggests that the SPOT data should permit the production of topographic maps of 1:50,000 to 1:250,000 scale with contour intervals of 40 to 50 m or greater. Image maps of 1:25,000 to 1:50,000 scale should prove satisfactory to most users.

The use of database management systems (DBMSs) has increased sharply over the past few years. Important reasons for this are that such systems facilitate the design and implementation of data bases, which have a very complicated structure, the fact that such systems allow data to be treated consistently and with greater confidence of correctness, and also that they make it easier for all or parts of a database to be shared by many applications and/or users. In return for these advantages there is a price, however. Since a DBMS is an additional "layer" of software between the operating system and the applications programs, and because it is general, instead of being tailor-made for a particular application, there is often a significant overhead, resulting in decreased execution speed. There is also usually a penalty in disk storage space, because often a large number of pointers is maintained automatically by a DBMS, even if not all of them are required for a particular application. Nevertheless, the advantages of using a database management system usually outweigh the disadvantages. Recent examples of the application of DBMSs to LRISs are reported by Moore et al (1984) and Dangermond and Burns (1986).

This research aimed to utilize both the features of remote sensing data acquisition techniques and the advantages of database technology to develop an LRIS containing both natural resources and survey data. It was hoped that this prototype LRIS would show the value of such multi-layered information sets in answering a range of questions in surveying and resource mapping, and to indicate the widespread gains possible through the synthesization of remotely sensed and conventionally derived data.

2. REMOTELY SENSED DATA

2.1 Data Acquisition

Techniques have been developed for position-based mapping of surface cover using Landsat digital data and implemented for a study area in the upper Kananaskis Valley of southwestern Alberta, located in part of the Rocky Mountains (see Figure 1) (Lodwick, 1981; Paine, 1984). The research involved approximately one-twentieth of an image of the Calgary scene, taken on September 20, 1975, and comprised natural areas of woodland and rock. Input is in the form of digital computer-compatible tapes of Landsat multispectral scanner data supplied by the Canada Centre for Remote Sensing in Ottawa. The output products are maps at scales and accuracies suitable for a wide range of environmental applications.

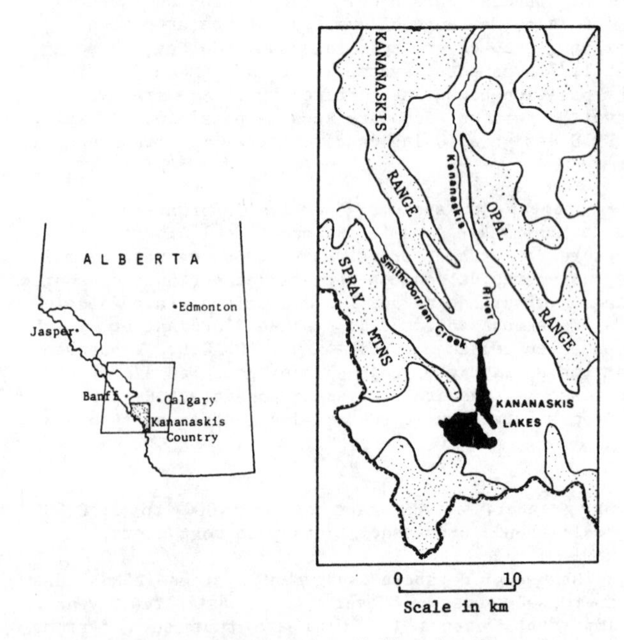

Figure 1. The Kananaskis Valley in Alberta

The analysis procedure comprised four major stages to preprocess, enhance, position and classify the data. Firstly, data from the computer-compatible tapes were reformatted and radiometrically corrected to remove errors in video response values introduced during image acquisition. Secondly, the data were enhanced for investigation and interpretation using principal components analysis. The advantage of this technique is that the original information can be defined in one, or at most two variables, which are readily interpretable in terms of natural surface phenomena. Thirdly, the image was geometrically adjusted, using a second order polynomial model and ground control points to spatially resect it to the earth's surface. Individual pixel values were then resampled to a UTM grid of 50 m spacing using the nearest-neighbour technique. Finally, interpretation of the data was carried out with a supervised classification scheme using a parallellepiped technique.

The results indicated that the first and second principal components contained over 99 percent of the information in the four original data sets. Classification using these data resulted in accurate definition of the surface cover of the test area. The rectification established the positioning accuracy of these data with an RMS error of less than 50 m, which is suitable for mapping at a scale of 1:50,000. The resulting map contained ten distinct surface cover classes, which were used as one input data set for the LRIS. The significant advantage of this approach is that sequential Landsat imagery enables updating on a regular basis.

2.2 Format Conversion (Raster to Vector)

Remotely sensed data are usually collected in a raster format and the decision about storage, manipulation and retrieval of these data often hinges on the question of raster versus vector format. Broadly, the methods of describing the positional extent (or coding) of spatial entities are either vector-based, using coordinates, or raster-based, using scan lines. Arguments have been made in favour of each of the raster and vector formats. Generally, particular formats and format conversions are not of concern to the system user but of the information system itself (Haralick, 1980).

Using vector format, polygons are described by a series of lines or points, lines are also described by a series of lines or points (Peucker and Chrisman, 1975; Edson, 1975) and points are described either by absolute coordinates (relative to the coordinate system origin) or by relative coordinates (relative to a previous point) (Baxter, 1976; Burton, 1979).

Using raster format, polygons, lines and points are all represented by those parts of an overlain grid matrix, or system of scan lines, which cover them. A raster matrix, or grid map may be full, where all cells or pixels are stored, or sparse, where only significant cells of each scan line are stored (Miller, 1980; Barber, 1982). The usual comment made, when discussing the relative merits of vector and raster formats, is that vectors are more efficient in terms of storage space, and rasters are more efficient in terms of computation time (Barber, 1982).

With Landsat data, storage in raster format is an appropriate approach. However, when dealing with an LRIS, where survey data as points and lines are to be stored, positional accuracy is an important consideration. Also efficiency considerations favour storage by polygons (vectors) (Mepham and Paine, 1986). To polygonize the remote sensing data set required that the data be smoothed to a level of generalization which would provide a more efficient storage format. This smoothing/filtering process requires careful selection of a spatial filter which will not degrade the data (Paine and Mepham, 1986). This resulted in a data set represented by polygons formed from line segments based on unique points.

3. CONVENTIONAL MAP DATA

3.1 Selection of Data Types

One of the requirements of this prototype LRIS was for test data that represented a broad range of spatial land information. Thus, three basic data types were selected: 1) primary data (positional), 2) secondary data (thematic), and 3) digital elevation model (DEM) data.

The first two types are fairly standard and have accepted definitions. Primary land-related information comprises three main components: 1) geographic positioning systems established with a range of survey techniques, 2) mapping systems established from the basic survey data, and 3) land registries containing crown land and land titles information (LRIS, 1981).

Secondary land-related information can loosely be defined as thematic data (Kozak, 1980). It is usually represented graphically as a map with the appropriate point, line or areal symbolization. These data include the basic biological and physical land characteristics, man made features and land use. Certain social and economic information is also included, as it is often useful for planning and managerent of the land base. This information usually has a geographic basis at both the collection and utilization stage. The phenomena that secondary data represent are often more dynamic than those of primary data. As a result of these changes periodic assessments for currency are required

The third type (DEM) could be classed as primary data but it has some unique storage, retrieval and utilization characteristics that require special handling in a database. Therefore, it has to be established as a separate data type.

Table 1. Data types for input into the LRIS.

Classes	Subclasses	Source
Survey Contr((Primary)	Cadastral, Boundaries, Control Data	digitized 1:50 000 NTS digitized RT series digitized 1:50 000 NTS
Hydrography (Secondary)	Lakes, Rivers, Swamps, Ice	digitized 1:50 000 NTS
Geology (Secondary)	Surficial, Bedrock	digitized 1:50 000 RE digitized 1:1000 000GSC
Transportation (Secondary)	Roads, Powerlines, Pipelines	digitized 1:50 000 NTS
Surface Cover (Secondary)	Forest, Grass, Rock/Soil, Farm, Urban	derived from digital Landsat data
Digital Elevation Models (DEM)	Airphoto, Digital Stereo	1:40 000 airphotos psuedostereo test data

With these differences in mind there were six classes of the different data types selected for processing in the study (Table 1). These classes have many of the varied subclass types found in land information. The data sources are also varied, with some data collected by digitizing existing maps, some derived from digital remote sensing imagery, and some from published descriptions.

3.2 Primary and Secondary Data

Survey control, hydrography, geology and transportation data were digitized directly from maps. The existing map data were digitized with a six parameter transformation to transfer the positional data to a UTM coordinate system. There is a header for each file, which contains the mounting information, composed of the control points utilized and their input coordinates and adjusted residuals. There is also a record containing the adjusted scale factors, offsets, rotation and non-perpendicularity parameters. The data files follow and contain an ID code, a character string description and the list of point coordinates describing the feature. The digitized features were displayed on a vector graphics screen as they were collected to allow verification of the data. Any blunders or corrections were flagged at the time of data capture and the data files were later edited to remove the errors.

3.3 Digital Elevation Model (DEM) Data

DEM data can be collected in one of three ways: 1) field observations, 2) photogrammetrically (traditional airphotos), or 3) utilizing digital stereo correlation of remotely sensed imagery. For the prototype LRIS, the DEM data were derived by the second and third methods. A digital elevation model was produced from 1:40 000 scale photography of a test area on a Wild AC-1 analytical plotter. An area covering approximately 5.6 by 5.6 km was chosen, and elevations were measured on the same 50 m UTM grid that the Landsat data utilized. The area chosen has a good range of terrain types, with elevations ranging from 1650 m to 2550 m, and various surface cover conditions for testing the algorithms. There is a rugged area of mountainous terrain, some flat and rolling terrain and two lakes.

These traditionally derived data were then used to generate a test digital stereo data set to evaluate the third possible digital elevation model input method. The Landsat image was assumed to be an orthoimage and the grid reference systems for both the image and the digital elevation model were assumed to be the same. A parallactic angle of 30 degrees was used, as this produces a significant elevation change from a horizontal displacement. This angle and the flying height of the satellite produce a typical rate change for all the X, Y, Z coordinates (ie a 50 m change in X produces a 50 m change in Z). Various methods of digital correlation and processing were evaluated and the resulting digital elevation models were compared to the airphoto derived model. The results indicate accuracies of subpixel range and suggest this as a viable method of obtaining DEM data from satellite systems, such as SPOT, which have stereo imagery and appropriate resolution.

There are two methods for the storage and representation of elevation data. The data can be stored in the format in which it is collected, which will typically be either in the form of a regular grid of spot heights or an irregular grid of spot heights (a triangulated irregular network). It can also be stored in the form of contours derived from the collected data. While contours are simpler to interpret visually, they are also less accurate than the original data. In addition, the storage of contour data will generally require much more disk space than the storage of the original digital terrain model. This is particularly true in areas of relatively smooth (not flat) terrain, where only a limited number of spot heights is required to describe the surface. Also, the one advantage that contours have over spot heights (their simple visual interpretation) is rapidly decreasing in importance due to the increasing use of digital processing over manual processing and interpretation.

4. DATA STORAGE

4.1 Conceptual Design

The Kananaskis database is being designed as a prototype of a database suitable for much larger application, such as the entire province of Alberta, Canada. With this in mind, some initial design criteria were established to guide its development.

The first criterion examined was the question of response time (speed). It was decided that "ordinary" requests must be answered quickly in an online system. Such requests might be "Find all occupied dwellings within 5 km of a certain gas well" or "Find all forest that is primarily pine and is on property owned or leased by the Cut-Em Forest Company". In addition, there will also be other "strange" requests, such as "Find all lots larger than 200 sq m that are owned by somebody whose last name starts with a vowel". While such requests are not likely to occur very often, the database design must allow for such an occurrence. However, it is felt that such "strange" requests need not be answered quickly. A time lapse of one or two days would not be unreasonable.

In addition to retrieval time, a second aspect is the time required to make changes (additions, deletions, modifications) to the database. It was considered that immediate updates would not generally be necessary. If the updates could be made on an overnight basis that would be satisfactory.

The third consideration in designing the database was the spatial nature of the data. In all but the very smallest of databases the searches will normally be restricted to limited areas of the region covered by the database. This makes it desirable that some sort of spatial indexing system be designed, which can be used to quickly eliminate from any subsequent search all entities that are not in the area of interest.

A final design consideration is the variation in accuracy of the data that will be entered into the database. It was decided that all data should be entered into the database regardless of accuracy. This implies that all data must have an accuracy qualification associated with it. An effective solution appears to be to use a confidence circle for each point. This can be represented by a single number and can be estimated for any point that it is not actually computed for. A more difficult problem is accuracy estimation for attribute data. An example of this is the classification accuracy of the surface cover for an area, or the degree of certainty regarding who is the owner of a parcel of land. The inclusion of accuracy specifications for each data element in the database will allow users the flexibility of using only that data that is accurate enough and reliable enough for their applications, without restricting the data that can be stored in the database.

4.2 Georeferencing

In positional terms there are only three types of data to be entered into the database. These data types are point, line and polygon. In the design of the database, the more complex data types are redefined in terms of simpler ones, e.g. polygons are encoded as a closed series of connected lines, and lines are defined as a series of connected straight line segments between points. The three files holding these definitions are all accessible from the remainder of the database. For example, the position attribute field of a survey control point will refer directly to the point file, whereas the location field of a surface cover entry will refer to the polygon file. This in turn will refer to the line file, which will in turn refer to the point file. This is illustrated in Figure 2.

There are several advantages to this approach. The first is that all positional information is in a single homogenous file. This will allow global transformations of the coordinates (as will be required as a result of the redefinition of the geodetic reference system) to be performed with a minimum of effort. A second advantage is that common points will only be stored once. This will result in a space saving and will also allow easy boundary adjustment of polygons, and eliminate the introduction of inconsistencies in the data as a result of editing the data in the database.

A further advantage of using this common point approach is that the most important topological relationship between spatial entries in the database is implicitly defined. This relationship is adjacency. The determination of what properties, for example, are adjacent to a particular property is carried out simply by determining what properties have at least one point in common with the property of interest. This direct determination of adjacency is accomplished without having to explicitly encode the information.

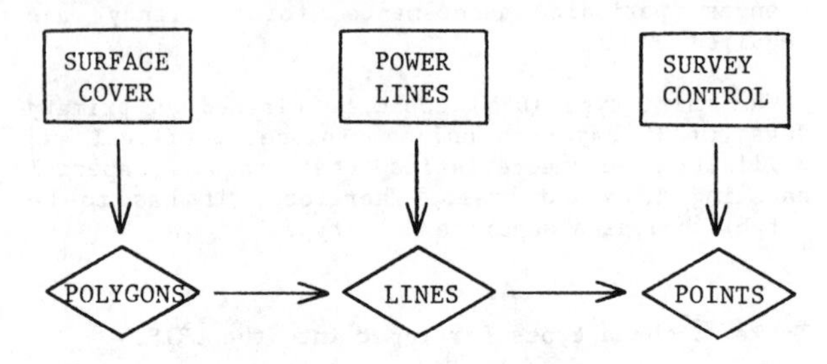

Figure 2. Referencing location files

4.3 Spatial Indexing

The great majority of the inquiries that will be made of the database will be limited to a specific region of the database and are likely to cover only a small area. In order to maintain a quick response time for the requests, it is highly desirable that only those entities that actually exist in the area of interest be examined in satisfying the inquiry. For this reason a spatial index is being established.

The region covered by the database will be divided into a series of rectangles called tiles. These tiles will be mathematically defined in terms of the coordinates of their corners. There is no plan to try to have the tiles coincident with map sheets or any other "standard" division of the earth's surface. The main problem is to optimize the size so that any additional overhead cost is more than compensated for by the saving in retrieval time. Since this is a function of the number of entities that occur in an area covered by a tile, and the distribution of information in the region covered by the database will vary, it has even been proposed that a variable tile size be used.

For each tile a record will be kept listing the entities that touch on the area covered by the tile. This entity list will then be used to restrict the search of the database to only those entities that could possibly be in the area of interest. Typically, a user's search area will be defined by an irregular polygon. The tiles overlapping the polygon will then be used to identify the entities of potential interest. Figure 3 illustrates how the entities in a tile will be listed. Usually, the tile list will be more extensive than required, but should be of manageable proportions.

5. DATA BASE IMPLEMENTATION

5.1 Data Model Alternatives

Relational and network models are typical approaches taken by commercial DBMS's and are discussed in standard texts (e.g. Bradley, 1982). The relational approach has been gaining in popularity, as shown for example by the release of IBM's DB2 (Date, 1986). In the relational approach it is necessary only to define relations, the close analogue of ordinary files in traditional data processing, and it is not necessary to pre-define connections among different relations. In the network approach, on the other hand, one defines record types, the analogue of relations, but it is also necessary to

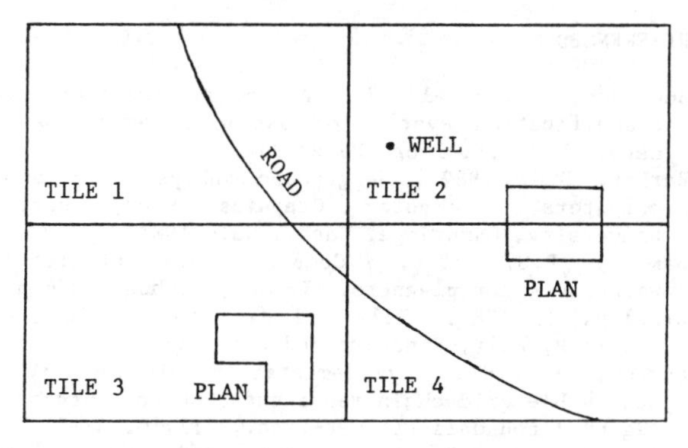

INDEX_1
ROAD_17 PT_921 . . . PT_1004

INDEX_2
WELL_23
PLAN_24 PT_5197 . . . PT_5203

INDEX_3
ROAD_17 PT_1004 . . . PT_1029
PLAN_92

INDEX_4
ROAD_17 PT_1029 . . . PT_1102
PLAN_24 PT_5203 . . . PT_5197

Figure 3. Spatial indexing

define connections between different record types. These connections are defined by means of DBTG (Database Task Group) sets.

The relational approach offers the advantage of flexibility, since there are no pre-defined linkages. Any query that can be formulated in the query language can be accomplished. For this reason, relational DBMSs, such as ARC/INFO (Dangermond and Freedman, 1984), are popular for LRIS design. The network approach, on the other hand, offers the potential advantage of greater efficiency of those queries for which suitable linkages have been pre-defined. Of course, when suitable DBTG sets have not been pre-defined, a query may either be impossible or very inefficient to answer. In a network database one is said to "navigate" through the database, and in such a database system it is necessary to structure the record types and DBTG sets in such a way tnat frequently occurring queries can be satisfied by relatively simple navigation through the database.

As a general rule, the inclusion of more DBTG sets improves retrieval speed, but may make the speed of updating, both insertion of new records and changing values in existing records, slower, because more pointers and indexes must be changed. In the current project much of the data is not volatile; much of it will change rarely, if ever. Even those data that do change will not need to be updated in "real-time". Consequently, the design of the database is oriented toward efficient retrieval at the cost of slower updates.

This prototype LRIS will use the VAX-11 DBMS, which is a "CODASYL compliant" network data base management system, and which runs under the VMS operating system on a DEC VAX-11/750. A good Fortran interface, and the retrieval efficiencies that are possible with a network database system are among the reasons for choosing it. As with all CODASYL network DBMS's, a database is defined by defining a schema, which contains, among other things, definitions of the record types and the DBTG sets. Restricted access (such as read-only access) for certain classes of users, oı different "user views", can be provided by means of different subschemas. And, on the VAX-11 DBMS in particular, retrieval efficiency can be optimized by means of a storage schema, which provides details on such things as indices, and attempts to store related items on the same page of disk storage.

5.2 Infological Model

The infological model is related to human concepts and real world representation. In other words the data are organized in terms of representing reality the way humans perceive it. The most common infological model is a hierarchical tree structure which would break the data into levels as shown in Figure 4. This type of model is feasible to implement in a database but it does have some limitations. If there are a large number of classes near the root of the tree it is difficult to navigate around the data to establish and utilize the various relationships. This type of model also has a problem with data redundancy, which arises from the need for each entity to have its own non-unique attributes. This results in nonstandardized entity/attribute records, where a geology polygon would require fewer record fields than the same polygon representing surface cover.

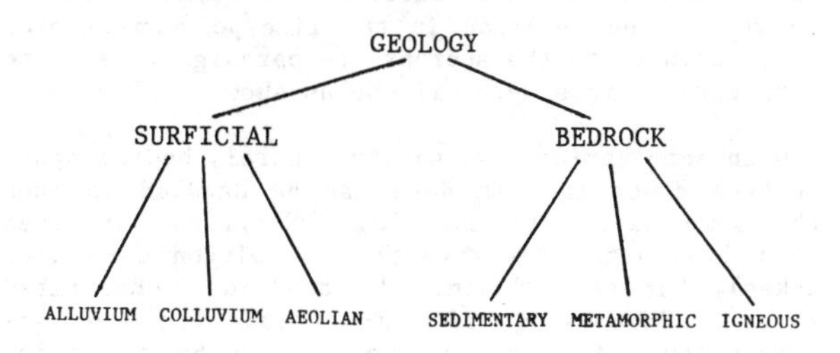

Figure 4. Hierarchical infological model

With these limitations in mind, a different infological model has been designed. This model also takes into account the nature of the network database model, which allows definition of backwards and forwards file pointers. This infological model utilizes paradigm files to reduce redundancy and to allow easier navigation and establishment of relationships. A paradigm file contains a set of standardized entity attribute records as shown in Table 2.

Table 2. Geologic bedrock paradigm file.

Geologic code	Geologic age	Formation name	Rock type	Structure
D	Devonian	undiff	.	.
Dbf	Devonian	Banff	.	.
Dbw	Devonian	Bow	.	.
J	Jurassic	undiff	.	.
Jcl	Jurassic	Coleman	.	.
.	.	.	.	.

These paradigm files are very stable, allowing easy updating and multiple entry to and from various different entities. Thus, it is possible to standardize the basic entity attribute record structure by utilizing the pointers to these files as shown in Figure 5. This reduces data redundancy in many cases and provides a framework for the definition of the relationships between the various data.

ID #	ENTITY TYPE	ENTITY ATTRIBUTES	DATA SOURCE
(unique random)	(point line polygon)	(bedrock surficial)	(Can. Geol. Survey)

Figure 5. Standardized entity file

5.3 Datalogical Model

As explained earlier, the point file (or points record type) is of great importance. Polygonal and line data refer to points, which are stored uniquely. But since several occurrences of polygons (adjacent polygons, for example) will refer to the same points, an auxiliary record type must be used. Lines and polygons can be represented by the same record type, which would carry an indicator as to whether a line or polygon is represented, with the assumption in the latter case that there is a line segment from the last point back to the first.

Taking geological data as an example, the data record for a particular occurrence will contain information for a particular area, including a unique identifier, the source of the data, a pointer to the correct polygon in the line/polygon records and a pointer to the appropriate paradigm file. The structure of this data will be as shown in Figure 5.

With some variations, survey control, hydrography, surface cover and DEM data can be handled in much the same way. For example, hydrography involves both line data (for rivers) and polygon data (for lakes), but can otherwise be handled as described above. Processing of DEM data, due to its complexity, will have to be handled by a special applications program outside of the DBMS. Within the data base it will suffice, therefore, to have some basic information, the DEM type, the source of the data etc., as well as a pointer to the correct data record in a separate file of DEM information.

Transportation data, including such things as roads and power lines, can also be handled as described above, except that an extra "level" of information may be required. Different sections of a highway may have been built by different contractors, for example, and therefore there may have to be a record type which is subordinate to, or "owned by", the record type for roads, which describes a segment of a road that has constant attributes.

6. CONCLUSIONS

Land related information systems are becoming more important as tools to assist in spatial data management. By interfacing such systems with modern data acquisition and storage technology, such as satellite remote sensing and data base management systems, further significant payoffs are possible. This paper has reported on research that is integrating a range of spatial data types into an LRIS. Apart from the normal range of conventionally derived thematic data, also included are remotely sensed data as well as DEM and survey data. This has resulted in design criteria involving a more complex conceptual model than would be typical of less comprehensive geographic information systems. As well, the research has resolved certain practical aspects of data acquisition and storage. What remains to be done is to complete the development of the prototype LRIS in order to undertake an evaluation of the system to assess its potential for more extensive application.

REFERENCES

Anderson, J.R. et al 1976. A land use and land cover classification system for use with remote sensor data. U.S.G.S. Prof. Paper 964.

Barber, D.G. 1982. Geographic analysis for small computers. Computer Graphics Week, Harvard University, Cambridge, Mass., July 1982.

Baxter, R.S. 1976. Computer and statistical techniques for planners. London, Methuen, 336 p.

Bradley, J. 1982. File and data base techniques. New York, Holt, Rinehart and Winston.

Burton, W. 1979. Implementation of the binary searchable grid chain representation for curves and regional boundaries. Geo-Proc., 1:1, p. 37-52.

Colvocoresses, A.P. 1982. An automated mapping satellite system (Mapsat). P.E.R.S., 48:10, p.1585-1591.

Dangermond, J. and T. Burns 1986. A successful case study in geographic information implementation: Anchorage, Alaska. ACSM-ASPRS Annual Convention, Washington, March 1986, 3, p. 258-265.

Dangermond, J. and C. Freedman 1984. Findings regarding a conceptual model of a municipal database and implications for software design. Intern. Symp. on Spatial Data Handling, Zurich, Aug. 1984, II, p. 479-496.

Date, C.J. 1986. An introduction to database systems. Fourth Ed. Reading, Mass., Addison-Wesley.

Edson, D. 1975. Digitial cartographic data base: preliminary description. In John Kavalinas and Frederick Broome (eds.), Auto-Carto II, Intern. Symp. on Comp.-Assist. Cart., Sept.1985, p.523-538.

Haralick, R.M. 1980. A spatial data structure for geographic information systems. In H. Freeman and G.G. Pieroni (eds.), Map Data Processing. New York, Academic Press, 374 p.

Holmes, R.A. 1984. Advanced sensor systems: Thematic Mapper and beyond. R.S.E., 15, p. 213-221.

Kozak, E.L. 1980. Land related information systems current situation analysis (User Survey). Working Paper, LRIS Coord. Proj. Edmonton, Bureau of Statistics Treasury, 40 p.

Lodwick, G.D. 1981. A computer system for monitoring environmental changes in multitemporal Landsat data. C.J.R.S, 7:1, p. 24-33.

Lodwick, G.D. and S.H. Paine 1985. A digital elevation model of the Barnes Ice Cap derived from Landsat MSS data. P.E.R.S., 51:12, p. 1937-1944.

LRIS 1981. Land-related information systems a network concept. Report No. 2, LRIS Coord. Proj. Edmonton, Bureau of Statistics Treasury, 31 p.

Mepham, M.P. and S.H. Paine 1986. An evaluation of storage methods for Landsat-derived raster data. ASPRS-ACSM Fall Convention, Anchorage, Sept. 1986.

Miller, S.W. 1980. A compact raster format for handling spatial data. ACSM Fall Technical Meeting, Niagara, Oct. 1980, 18 p.

Moore, R. et al 1984. Towards a CODASYL database for the U.K. river network. Intern. Symp. on Spatial Data Handling, Zurich, Aug. 1984, II, p. 574-575.

Myhre, R.J. 1982. Satellite photos can aid navigation on aerial photo missions. P.E.R.S., 48:2, p. 275-279.

Paine, S.H. 1984. Using Landsat imagery for position-based surface-cover mapping in the Rocky Mountains. C.J.R.S., 10:2, p. 190-200.

Paine, S.H. and M.P. Mepham 1986. Spatial filtering of digital Landsat data for extraction of mapping information. Tenth Canadian Symposium on Remote Sensing, Edmonton, May 1986.

Peucker, T.K. and N.R. Chrisman 1975. Cartographic data structures. The Amer. Cartog. 2:1, p. 55-69.

Schreier, H. et al 1982. The use of digital multidate Landsat imagery in terrain classification. P.E.R.S., 148:1, p. 111-119.

Welch, R. 1985. Cartographic potential of SPOT image data, P.E.R.S., 51:8, p. 1085-1091.

Welch, R. et al 1985. Comparative evaluations of the geodetic accuracy and cartographic potential of Landsat-4 and Landsat-5 Thematic Mapper image data. P.E.R.S., 51:9, p. 1249-1262.

Symposium on Remote Sensing for Resources Development and Environmental Management / Enschede / August 1986

The CRIES Resource Information System: Computer-aided spatial analysis of resource development potential and development policy alternatives

Gerhardus Schultink
Comprehensive Resource Inventory and Evaluation System (CRIES) Project, Michigan State University, East Lansing, USA

ABSTRACT: This paper addresses analytical procedures and two of the major micro computer-based software modules of the Comprehensive Resource Inventory and Evaluation System (CRIES) Resource Information System (RIS). System objectives, the land evaluation framework and selected outputs of the CRIES Geographic Information System (CRIES-GIS) and the the CRIES Agro-economic Information System (CRIES-AIS) are discussed using selected examples from developing countries.

Specific emphasis is given to the use of the integrated spatial data base, maintained in the GIS system and the AIS-YIELD and AIS-MULBUD model used to assess comparative production advantage. YIELD provides the capability to predict yield for a large number of food and export crops for user-selected locations and agro-ecological zones. MULBUD provides multiple enterprise analysis of short and long season crops to determine economic returns resulting from farming systems options representing Land Utilization Types (LUT's)

In combination with other CRIES-RIS modules and the established data base, the YIELD model provides the user with the analytical framework to evaluate physical and socio-economic attributes by location, and determine the comparative advantage of sites for land use alternatives. Farming systems, and regional or national aggregates can also be evaluated with regards to their optimum performance characteristics and resulting socio-economic benefits derived under alternative land use and development policy scenarios.

1. INTRODUCTION

The Comprehensive Resource Information and Evaluation System (CRIES) Project encompasses a systematic resource analysis approach to evaluate public and private benefits derived from alternative land use options and policy scenarios. Examples in the developing world include the creation of food self-sufficiency while meeting rural employment requirements or meeting balance of payment goals through the expansion and/or intensification of the production of food and cash crops, resulting in increased exports of agricultural commodities and import substitution.

The resource evaluation methodology employed by CRIES represents an effort to develop, adapt, and document general procedures to inventory, classify and analyze current land use, its distribution, extent, and development potential of agricultural and natural resources for development planning. CRIES has three general objectives:

a) to apply a consistent approach to land resource assessment which is adaptable to many countries and suitable for the transfer of appropriate agrotechnology;

b) to provide assistance in integrated surveys, development of a computer compatible resource data base and computer-aided analysis software suitable for the analysis of development options and policy evaluation; and

c) to provide the training and technical assistance necessary to develop indigenous capabilities to inventory and classify renewable resources, to assess crop production potential, and to systematically evaluate development alternatives and derived public and private benefits.

This evaluative framework is supported by computer-aided procedures designed to systematically delineate agro-ecological production zones (the so-called Resource Planning Units or RPU's), representing areas with physical characteristics considered relatively homogeneous at the level of detail supported by the land evaluation. These RPU's can be related to existing land uses, associated farming systems and estimated crop production potential based on existing and alternative land utilization types and associated enterprises identified.

The difference between current land use and resource production potential, expressed in crop yields or economic terms, such as "land rent" is defined as "unrealized production potential," a first approximation of the quantitative magnitude of development potential to meet critical policy objectives.

Selected software modules, developed by the CRIES project or adapted from existing software, provide the user with the option to evaluate land use alternatives on a site-specific basis or for regional and national aggregates representing a mix of existing land use types or alternatives considered.

2. THE SPATIAL AND TEMPORAL DIMENSION IN INFORMATION SYSTEMS AND ASSOCIATED LAND EVALUATION.

Generally, land evaluation studies have been associated with aggregate levels of analysis: e.g. agro-ecological zones, the administrative district, region or nation. The CRIES information system components described in this paper, provide for spatial referencing of resource attribute data using orthogonal coordinate systems. This grid-based reference system permits the analysis of comparative advantage on a locational basis represented by the data cells comprising the integrated resource data base. This feature ensures that at all stages of analysis the spatial identity is incorporated in the final results and, therefore, reflected in the derived information content on which resource development recommendations and resulting land use policies are based. The location specific dimension of the information also permits economic feasibility analyses for alternative projects varying in scope and magnitude.

The temporal dimension is reflected by the average frequency of availability of selected data sets over time and their representation in the data base (e.g., population density data per administrative district over a recent time period at selected time intervals represented by the attribute files in the data base). With the presence of these multi-temporal attribute files, time series analyses can be conducted addressing such issues as the type, location and magnitude of land use change, population growth forecasting or environmental impacts.

3. INTEGRATION AND LINKAGE OF INFORMATION SYSTEM MODULES

The CRIES resource inventory and analysis approach to integrated rural development planning and agricultural sector analysis can be illustrated by the following major Resource Information System (RIS) components: the CRIES-GIS and the CRIES-AIS (Figure 1). As indicated in the figure, the linkage between the GIS and AIS constitutes, in essence, the CRIES approach in resource inventory and evaluation. The specific interactions as represented involve a set of software modules used to systematically evaluate land use options given a complex set of physical and socio-economic data, resource management objectives, carrying capacity considerations, and policy concerns.

The main objective of the GIS is to compile and analyze the integrated data base to provide a systematic assessment of the (unrealized) production potential of the resource base. Subsequently, this information can then be linked with selected AIS modules to provide estimates of the performance characteristics (derived socio-economic benefits) of single or multiple enterprise options for a specific location or an aggregate, thereof, such as a community and its surrounding supply area, i.e., a (sub)region, administrative district, national level or even multi-national regions.

To provide users with an initial orientation toward the functions of the main GIS and AIS modules, the main menu of both systems is provided in this paper, containing short fuction descriptions (Fig. 2 and 3). The automated analytical protocols between the GIS and AIS are under development.

CRIES
Resource Information System (RIS)

consists of

CRIES Geographic Information System (GIS)	CRIES Agro-economic Information System (AIS)
CRIES - GIS: (handles data with an intrinsic spatial component)	CRIES - AIS: (handles data of an aggregate nature)
Information Unit: Grid Cell (e.g. 1, 4, 100 hectares)	Information Unit: Planning Unit (e.g. district, agro-eco. zone)
RESOURCE PRODUCTION BASE and PHYSICAL PRODUCTION POTENTIAL	PERFORMANCE DYNAMICS - BENEFITS/ ALTERNATIVE USE/INPUT SCENARIOS
data examples: soils, precipitation, temperature, bio-climatic zones, elevation, population density, administrative districts	data examples: crop requirements, availability and cost of inputs,food requirements, enterprise budgets agro-physical profiles
MAJOR SOFTWARE MODULES: * GIS-INPUT * GIS-STATISTICS * GIS-ANALYSIS * GIS-TERRAIN * GIS-DISPLAY * GIS-CHOROLINE * GIS-UTILITIES	MAJOR SOFTWARE MODULES: * AIS-WATBUG * AIS-WATER * AIS-YIELD * AIS-MULBUD * AIS-ENTERPRI * AIS-LINPROG * AIS-IN/OUT * AIS-MSTAT

Figure 1. Role and Linkage of Major Software Modules used in the CRIES Land Resource Assessment Approach for Integrated Rural Development Planning.

CRIES - GEOGRAPHIC INFORMATION SYSTEM
(CRIES - GIS) - Version 6.00
MAIN MENU

PLEASE SELECT GIS MODULE :

[F1] GIS-INPUT	these phases are used for primary data entry
[F2] GIS-STATISTICS	these phases provide statistical summaries
[F3] GIS-ANALYSIS	these phases analyze and create raster files
[F4] GIS-TERRAIN	these phases support 3 - D modeling
[F5] GIS-DISPLAY	these phases display the contents of a raster file
[F6] GIS-CHOROLINE	these phases produce dot matrix printer maps
[F7] GIS-UTILITIES	these phases create, edit and reformat files

Figure 2 . Main Screen Menu of the CRIES Geographic Information System.

CRIES - AGRO-ECONOMIC INFORMATION SYSTEM
(CRIES - AIS)
MAIN MENU

PLEASE SELECT AIS MODULE :

[F1]	AIS-WATBUG	provides water balance analysis with minimal data requirements (*)
[F2]	AIS-WATER	provideswaterbalancemodeling (**)
[F3]	AIS-YIELD	provides yield prediction for selected locations or agro-ecological zones
[F4]	AIS-ENTERPRI	conducts single enterprise analysis
[F5]	AIS-MULBUD	conducts multiple enterprise analysis for multiple time periods
[F6]	AIS-LINPROG	optimizes objective function using linear programming
[F7]	AIS-IN/OUT	support economic input/output analysis (**)
[F8]	AIS-MSTAT	provides experimental design, data analysis and data management for agronomic research experiments

[NOTE: AIS modules not yet available for MSDOS are identified by (*) and AIS modules under development by (**).]

Figure 3. Main Screen Menu of the CRIES Agro-economic Information System.

4. THE CRIES - GEOGRAPHIC INFORMATION SYSTEM (GIS)

This section describes the Geographic Information System (GIS) component of the Comprehensive Resource Inventory and Evaluation System (CRIES) - Resource Information System (RIS) as modified for IBM, Model AT micro-computers using the MS-DOS operating system. The CRIES-GIS provides the capability to store, edit and process digital map data and creates the master data base (disk files) for subsequent analysis. The various phases are used to input, retrieve, combine, statistically analyze, overlay, modify and display the results in the form of various statistical summaries or computer maps. Typical products include area measurements and displays of single map attributes or combinations at various map scales.

The IBM-AT (or compatible MS-DOS micros) Version 6.0 requires only 256K of RAM, is menu-driven, supports a data base of 65,535 rows by 65,535 columns (the individual information cells), multiple attributes (information layers) with attribute values (or classes) of 1 to 65,535, depending on the user-selected bit field. The system requires the 80286 processor and uses the 80287 co-processor to accelerate mathematical operations where available. However, a 8088 (with optional 8087) (IBM -XT) processor may be used if lower processing speeds are acceptable. The digitizing is supported using the TECMAR color graphics master board. The software minimizes data storage requirements through the use of, user-selectable, bit-fields based on the complexity of map attributes. (1, 2, 3, 4, 5, 8 or 16 bit fields). The bit size selected (2 to the n.th power - 1, in which n = bit size) determines the number of attribute values per attribute. For instance, for n = 8, 256-1 attributes values may be represented (for n = 16, 65536 -1). A standard, 80 or 132-column printer is required to generate the output for selected phases and computer printer maps. The more attractive maps generated by the Choroline phase require an EPSON, 80 or 132 column, or compatible, dot matrix printer. User-selectable map scaling options in x and y direction increase two-dimensional accuracy of dot matrix printer maps, and increase printer compatibility.

Overview of CRIES- GIS Modules, Phase and Sub-phases.

An short descriptive overview of the CRIES-GIS, which comprises 7 major modules, 30 phases and 8 sub-phases, is provided, below:

The CRIES-GIS-INPUT module, used for primary data manual or electronic data entry, verification and editing, consists of 4 phases and 2 sub-phases :
1. EDITCELL - permits geocoding and editing of a raster file
2. DIGITIZE - converts analog map data into digital form using an X-Y digitizing tablet
   2.1 MONODIG - this sub-phase permits low resolution, monochrome digitizing
   2.2 COLORDIG - this sub-phase permits high resolution, color digitizing
3. POLYFILL - converts converts polygon data created by DIGITIZE into a raster file
4. CUTTER - uses existing raster files to cut outline boundaries into overlapping raster files

The CRIES-GIS-STATISTICS module, used to provide statistical summaries of raster files, consists of 3 phases :
1. HISTGRAM - provides a frequency table and a histogram from a single raster file
2. TALLY - calculates summary statistics for a user-specified rectangular subwindow of a single or multiple raster files
3. CROSSTAB - creates a crosstabulation of two or more raster files of up to 255 attribute values

The CRIES-GIS-ANALYSIS module, used to analyze raster files, consists of 7 phases :
1. EROSION - calculates erosion rates using the Universal Soil Loss Equation using user-specified variable weights
2. GROUP - groups ranges of attribute values into user-selected single values using single or multiple raster files
3. INVERT - inverts the range of attribute values in a raster file
4. MATCH - creates new attribute values for user-specified co-occurrences of existing attribute values of multiple files
5. NORMALIZ - normalizes the range of attribute values based on user-defined limits
6. OVERLAY - permits weighted and unweighted overlay analysis of attribute values of corresponding grid cells with optional overriding values
7. SEARCH - permits proximity analysis from points, lines and areas

The CRIES-GIS-TERRAIN module, used to support three-dimensional terrain analysis, consists of 2 phases:
1. SURFCELL - creates a trend surface in the form of a raster file using X, Y, Z point data as input
2. SURFAREA - (under development) creates a trend surface in the form of a raster file using area data as input. This phase permits via the use of sub phases calculation of slope gradient, slope length and orientation.

The CRIES-GIS-DISPLAY module,used to display the contents of raster files in the form of character maps or digital form, consists of three phases:

```
Prepared by: CRIES-GIS / Michigan State University
             Wednesday April 23, 1986   Time 09:53

File name: CHOLELEV.ras

     0000000000000000000011111111111111111111222222222222222222222223333333333333
     0011223344556677889900112233455667788990011223344556677889900112233345566
     1616272727273838383849494949505050506161616172727272838383839494949494940505

001                                                                              001
010       0000I                                                                  010
018      X0010                                                                   018
027      X00000                                                                  027
036      X000001I..II                                                   0        036
044     XX0II.....III0.0III000                                      I0000        044
053      XXX00.....I000II00000                                   00I000XXXXX     053
061         0II...I0000000XX0                                  0000I0000XXXXXX   061
070         0I.I.I0X00000000X                                 0II00000000XXXXX   070
079          ........I0I0I.I0I0XX                          X0000I000IIII00XXXX   079
087           ..............II0XX                        XXXXX0IIIIII00X0XXXX    087
096            .0III.....III100X00                     XXXX0II10000000XXXXXXX    096
105             000I....I00II000XX0I00I                XX0IIOX00000I00XXXXXXX    105
113              .0I...II0X0000XXXX000II          0X0II00XX00XX0000XXXXXXX       113
122               ....II00000XX0XXXXX00..I0IIII..III000XXXXXXXXXXXXXXXX          122
131               .......II00000000XXX0I......IIII000XX0XXXXXXXXXXXXXXXXX        131
139                I....III000X00I00XX00....I000000XXXXXXXXXXXXXXXXXXXXX         139
148                ..00..000011000.IIXI....00000000XXXXXXXXXXXXXXXXXXXX          148
156                ........0...I0.II00I..0000I0000XXXXXXXXXXXXXXXXXXXXXX          156
165               ......I00I0.0.IIII.I.......000XXXXXXXXXXXXXXXXXXXXXX           165
174             .........II.I.I...........IIII00XXXXXXXXXXXXXXXXXXXXXXXX         174
182           ....II........I.I...........I0X0XXXXXXXXXXXXXXXXXXXXXXX5XXXX       182
191       ...............II..................I00XXXX00XXXXXXXXXXXXXXXXXXXX      191
200      ... ...........I...................I0000XXX00XXXXXXXX5555XXXXXXXX      200
208      . .....I...........................III0XXX0000XXXXXXX   XXXXXXX         208
217   .. .....I.....................I...I0000XX000000000          XXXXX          217
225  . ........................III00X0II00XX0X0000000II                          225
234      ........................0I.I000XXXXX00I000000                           234
243       ...........................II0XX0XX00II00000                           243
251       .............................I0000X000I.IIII                           251
260        ............................I00I0I00IIIIII                            260
269        ............................I000IIII.I.....I.                         269
277         ...........................III..II...III...                          277
286         ...........................II.....I......                            286
295            ...........................I......                                295
303              ..................I.......I.......                              303
312              ...........................                                     312
320               .........................                                      320
329                 ... ......................                                   329
338                   .....................                                      338
346                  ..........          ...                                     346
355                                                                              355

     0000000000000000000011111111111111111111222222222222222222222223333333333333
     0011223344556677889900112233455667788990011223344556677889900112233345566
     1616272727273838383849494949505050506161616172727272838383839494949494940505

CHOLUTECA ELEVATION RANGES

HONDURAS DATA BASE - REGIONAL LEVEL - MAP SCALE SELECTED 1 : 500,000
ARRTIBUTE: ELEVATION.  ATTRIBUTE VALUES: CONTOURS, VARIABLE INTERVAL

        Mapping      Print
        Directive    Symbol      Frequency

           0                        1410
           1           .             765
           2           I             186
           3           0             306
```

Figure 4 -CHARMAP OUTPUT - ELEVATION, CHOLUTECA DEPARTMENT HONDURAS.

Prepared by: CRIES-GIS / Michigan State University
Wednesday April 23, 1986 Time 09:23

Attribute: CHOLELEV.ras — Description: CHOLUTECA DEPARTMENT

Attribute: CHOLRAIN.ras

	Row Totals	Atr Val 1	Atr Val 2	Atr Val 3	Atr Val 4	Atr Val 5
Column Totals	444469	209056	53113	82425	98481	1394
	100.00	100.00	100.00	100.00	100.00	100.00
	100.00	47.04	11.95	18.55	22.16	0.32
	100.00	47.04	11.95	18.55	22.16	0.32
Atr Val 1	89725	1944	10106	25669	51512	494
	20.19	0.93	19.03	31.15	52.31	35.44
	100.00	2.17	11.27	28.61	57.42	0.56
	20.19	0.44	2.28	5.78	11.59	0.12
Atr Val 2	83275	9969	12494	23531	36381	900
	18.74	4.77	23.53	28.55	36.95	64.57
	100.00	11.98	15.01	28.26	43.69	1.09
	18.74	2.25	2.82	5.30	8.19	0.21
Atr Val 3	184900	125856	24069	26556	8419	0
	41.61	60.21	45.32	32.22	8.55	0.00
	100.00	68.07	13.02	14.37	4.56	0.00
	41.61	28.32	5.42	5.98	1.90	0.00
Atr Val 4	86569	71287	6444	6669	2169	0
	19.48	34.10	12.14	8.10	2.21	0.00
	100.00	82.35	7.45	7.71	2.51	0.00
	19.48	16.04	1.45	1.51	0.49	0.00

Crosstabulation Table

Format: Frequency / Col Pct / Row Pct / Tot Pct — Frequency in hectares

Figure 5 - CROSSTAB OUTPUT - PORTION OF TWO-WAY CROSSTABULATION OF RAINFALL AND ELEVATION, CHOLUTECA, HONDURAS.

1. CHARMAP - creates a user-selectable alphanumeric character map from a raster file at user-selectable scales
2. VALUEMAP - displays the actual attribute values of the raster files
3. LOCATE - finds the location of user-specified attribute values of the raster files

The CRIES-GIS-CHOROLINE module, used to produce high quality dot matrix printer maps with user-selectable or user generated patterns, consists of three phases and 6 sub-phases :

1. RASTOMAP - reformats a raster file into a CHOROLINE format
2. PATTERN - creates pattern files
    - 2.1 - CODEPATS - creates new patterns
    - 2.2 - COMPATS - combines patterns
    - 2.3 - PRNTGRID - prints the pattern composition diagram
    - 2.4- SLCTPTRN - selects patterns from existing pattern files
3. DISPLAY - prints patterns and maps
    - 3.1- PRINTMAP- prints the reformatted map files
    - 3.2 - PRNTPTRN - prints the pattern files

The CRIES-GIS-UTILITIES module, used to create, edit, reformat and merge files, consists of 5 phases and two sub-phases.

1. CREATE - creates a new empty raster file
2. HEADER - edits and/or prints a raster's file header information
3. REFORMAT - reformats raster files and converts raster files and standard files
4. MOSAIC - combines contiguous raster files into a larger raster file
5. RESIZE - increase or decrease the grid cell size of a raster data base
    - 5.1 AGGREGAT - aggregates cells in user-specified increments
    - 5.2 DISAGGRE - disaggregates raster files in user-specified increments (under development)

To introduce users to the output of the CRIES - GIS selected examples are provided below:

5. THE CRIES- AGRO-ECONOMIC INFORMATION SYSTEM (AIS)

This section describes two selected modules of the CRIES Agro-economic Information System (AIS) as modified for IBM-XT or AT micro-computers using the MS-DOS operating system and equipped with a floating point processor. They include the YIELD and MULBUD modules. The YIELD model is adapted from "Yield Response to Water" by Doorenbos and Kassam (FAO, 1979). The MULBUD model was developed by Etherington and Matthews (1985) and modified for MSDOS.

5.1 The CRIES - AIS "YIELD" model.

YIELD provides the capability to predict yield response for a large number of food and export crops for user-selected locations and agro-ecological zones. The model, can be linked with a national data base containing agro-climatic data to predict yield response for locations, agro-ecological zones or regional and national aggregates. If the data base is not established, the user is provided with a data management option facilitating data input and editing for selected locations. Data requirements include: location identification, temperature, precipitation, relative humidity, wind velocity, solar radiation, sowing dates, crop selection, length of growing stages, soil texture, relevant water table characteristics on the root zone, and root growth over various growing stages.

If the data base is not established, the user is provided with a data management option facilitating data input and editing for selected locations. Data requirements include: location identification, temperature, precipitation, relative humidity, wind velocity, solar radiation, sowing dates, crop selection, length of growing stages, soil texture, relevant water table characteristics on the root zone, and root growth over various growing stages. In combination with other CRIES-RIS modules and the established data base, the YIELD model provides the user with the analytical framework to evaluate physical and socio-economic attributes by location, and determine the comparative advantage of sites for cropping alternatives.

The CRIES-AIS-YIELD module was developed to provide agronomists, land use planners, resource managers and policy analysts with a low cost, micro computer-based analysis tool in resource assessment studies. A summary description of the model is provided below (Table 1):

Table 1. SUMMARY DESCRIPTION OF THE CRIES-AGRO-ECONOMIC INFORMATION SYSTEM - YIELD

(CRIES-AIS-YIELD)

* MODIFIED AFTER DOORENBOS ET AL. "YIELD RESPONSE TO WATER" FAO
* INCORPORATES THE "WAGENINGEN METHOD" (BASED ON SLABBERS AND THE WIT) FOR ALFALFA, MAIZE, SORGHUM AND WHEAT.
* INCORPORATES THE AGRO-ECOLOGICAL ZONE METHOD FOR LARGE AREA YIELD PREDICTIONS OF 26 CROPS.
* COMPUTER MODEL PREDICTS YIELD RESPONSE TO WATER AVAILABILITY UNDER RAIN-FED OR IRRIGATED CONDITIONS.
* PROVIDES QUANTIFICATION OF VARIOUS CROP YIELD FOR AGRO-ECOLOGICAL ZONES
* MODEL HAS FIVE PHASES (FOUR CURRENTLY IMPLEMENTED)
-PHASE 1: DETERMINATION OF MAXIMUM POTENTIAL YIELD (Ym) OF ADAPTED CROP VARIETY ASSUMING NO LIMITING GROWTH FACTORS (E.G. WATER, FERTILIZER, PESTS, AND DISEASES)
-PHASE 2: CALCULATION OF MAXIMUM EVAPOTRANSPIRATION (ETm-PENMAN) WHEN CROP WATER REQUIREMENTS ARE FULLY MET BY AVAILABLE WATER SUPPLY.
-PHASE 3: COMPUTATION OF ACTUAL EVAPOTRANSPIRATION (ETa) BASED ON FACTORS AFFECTING CROP WATER AVAILABILITY.
-PHASE 4: CALCULATION OF ESTIMATED YIELD (Ye) BY EVALUATING CROP WATER REQUIREMENTS, WATER SUPPLY AND RESPONSE FACTORS AT VARIOUS STAGES OF PLANT DEVELOPMENT.
*INCLUDES THE FOLLOWING CROPS: ALFALFA, BANANA, BEAN, CABBAGE, CITRUS, COTTON, GRAPE, GROUNDNUT, MAIZE, OLIVE, ONION, PEA, PEPPER, PINEAPPLE, POTATO, RICE, SAFFLOWER, SORGHUM, SOYBEAN, SUGARBEET, SUGARCANE, SUNFLOWER, TOBACCO, TOMATO, WATERMELON, AND WHEAT.

5.2 The CRIES - AIS - MULBUD Model.

Farming systems, representing land utilization options and regional or national aggregates, need to be evaluated with regards to their optimum performance characteristics and resulting socio-economic benefits derived under alternative land use and development policy scenarios. Typical production options in developing countries can be characterized as variations of mixed cropping systems, representing multiple enterprises and their

Table 2. SUMMARY DESCRIPTION OF THE CRIES-AGRO-ECONOMIC INFORMATION SYSTEM - MULBUD

(CRIES-AIS-MULBUD)

* ADAPTED FROM ETHERINGTON AND MATTHEWS, 1984
* PERMITS ECONOMIC APPRAISAL OF A SINGLE FARMING SYSTEM CHARACTERIZED BY MULTIPLE AND INTERCROPPING PRACTICES AS TYPICALLY FOUND IN THE (SUB)TROPICS.
* PERMIT THE EVALUATION OF LAND USE/PRODUCTION OPTIONS OVER A PERIOD OF TIME BASED OF ANNUAL AND PERENNIAL CROPS, AGRO-FORESTRY SYSTEMS AND LIVESTOCK PRODUCTION OPERATED AS A MULTIPLE ENTERPRISE ON A SINGLE LOCATION (FARMING SYSTEM UNIT).
* MODEL INPUTS INCLUDE: LABOR OPERATIONS AND COSTS, WAGE RATES AND LABOR AVAILABILITY, DISCOUNT RATES, TERMINAL VALUES, FERTILIZER, PLANT MATERIAL, CHEMICALS, PRODUCT TYPES AND PRICES, LAND USE INDICES.
* MODEL OUTPUTS: GRAPHICAL DISPLAY OF LABOR REQUIREMENTS AND COSTS, MATERIAL REQUIREMENTS AND VARIABLE COSTS, GROSS AND NET REVENUES, NET REVENUE PER LABOR UNIT, SUM OF NET PRESENT VALUES, SENSITIVITY ANALYSIS FOR VARIABLE DISCOUNT RATES, INTERNAL RATE OF RETURNS, PROFILE OF LAND USAGE WITH TIME.

Table 3. CRIES-AIS-YIELD model, Example of Phase I Output for Bananas.

CRIES - AGRO-ECONOMIC INFORMATION SYSTEM
YIELD GENERATOR (CRIES - AIS - YIELD)
MICHIGAN STATE UNIVERSITY, EAST LANSING, MI 48823

COUNTRY NAME: JAMAICA
POL./ADM. DISTRICT: ST. CATHERINE
WEATHER STATION NAME: WORTHY PARK
RESOURCE PROD UNIT OR AGRO-ECOLOGICAL ZONE: 11
CROP TYPE: BANANA TROPICAL
YEAR OF ORIGIN: 1982
GROWTH PERIOD: 2/15/1982 TO 12/30/1982
PROD POTENTIAL UNIT: 1.12
LATITUDE: 18.01(deg)
HEMISPHERE: NORTHERN
AVERAGE ALTITUDE: 59.06(m)
TIME INCREMENTS (dT): 5 (days)

PHASE 1: DETERMINATION OF MAXIMUM POTENTIAL YIELD (Ym).

T (DAYS)	MTH/DAY	GROSS DRY MATTER PRODUCTION STANDARD CROP - (kg/ha) total (absol change)	total (cumm change)	overcast day (absol change)	overcast day (cumm change)	clear day (absol change)	clear day (cumm change)	GROSS DRY MATTER PRODUCTION corrections (prod rate)	(cloud perc)	(leaf area f)	(net prod f)	(harvs index)	POTENTIAL YIELD (kg/ha) (absolute change)	(cumm change)
5	2/15	2341.	2341.	994.	994.	1901.	1901.	50.11	.48	.48	.50	.10	56.	56.
10	2/20	2383.	4724.	1011.	2005.	1929.	3830.	50.78	.49	.48	.50	.10	57.	113.
15	2/25	2426.	7150.	1028.	3032.	1956.	5786.	51.44	.50	.48	.50	.10	58.	172.
20	3/ 2	2469.	9619.	1044.	4077.	1984.	7770.	52.11	.50	.48	.50	.10	59.	231.
25	3/ 7	2512.	12132.	1061.	5138.	2011.	9782.	52.78	.51	.48	.50	.10	60.	291.
30	3/12	2556.	14687.	1078.	6216.	2039.	11820.	53.44	.51	.48	.50	.10	61.	353.
35	3/17	2599.	17286.	1094.	7310.	2064.	13885.	54.11	.52	.48	.50	.10	62.	415.
40	3/22	2641.	19927.	1109.	8419.	2088.	15973.	54.78	.52	.48	.50	.10	63.	478.
45	3/27	2684.	22611.	1124.	9542.	2112.	18085.	55.44	.53	.48	.50	.10	64.	543.
50	4/ 1	2728.	25339.	1138.	10681.	2136.	20220.	56.11	.53	.48	.50	.10	65.	608.
55	4/ 6	2771.	28110.	1153.	11834.	2160.	22380.	56.78	.53	.48	.50	.10	67.	675.
60	4/11	2815.	30925.	1168.	13002.	2183.	24563.	57.44	.54	.48	.50	.10	68.	742.
65	4/16	2824.	33750.	1179.	14181.	2202.	26765.	57.93	.55	.48	.50	.10	68.	810.
70	4/21	2810.	36560.	1186.	15367.	2216.	28981.	58.41	.57	.48	.50	.10	67.	877.
75	4/26	2795.	39354.	1194.	16561.	2231.	31212.	58.89	.59	.48	.50	.10	67.	945.
80	5/ 1	2779.	42134.	1202.	17763.	2245.	33457.	59.37	.61	.48	.50	.10	67.	1011.
85	5/ 6	2763.	44896.	1209.	18972.	2260.	35717.	59.85	.62	.48	.50	.10	66.	1078.
90	5/11	2746.	47643.	1217.	20189.	2274.	37992.	60.33	.64	.48	.50	.10	66.	1143.
95	5/16	2770.	50413.	1222.	21411.	2283.	40275.	60.70	.63	.48	.50	.10	66.	1210.
100	5/21	2821.	53234.	1224.	22635.	2289.	42564.	61.07	.62	.48	.50	.10	68.	1278.
105	5/26	2873.	56108.	1227.	23861.	2294.	44858.	61.44	.61	.48	.50	.10	69.	1347.
110	5/31	2925.	59032.	1229.	25090.	2299.	47157.	61.81	.59	.48	.50	.10	70.	1417.
115	6/ 5	2977.	62009.	1232.	26322.	2305.	49462.	62.19	.58	.48	.50	.10	71.	1488.
120	6/10	3030.	65039.	1234.	27556.	2310.	51772.	62.56	.57	.48	.50	.10	73.	1561.
125	6/15	3080.	68119.	1235.	28791.	2311.	54083.	63.30	.56	.48	.50	.10	74.	1635.
130	6/20	3129.	71248.	1234.	30025.	2309.	56392.	64.04	.55	.48	.50	.10	75.	1710.
135	6/25	3179.	74426.	1234.	31259.	2307.	58699.	64.78	.54	.48	.50	.10	76.	1786.
140	6/30	3221.	77647.	1233.	32492.	2305.	61004.	65.00	.54	.48	.50	.10	77.	1864.

products produced over a period of time. The MULBUD model (Etherington and Matthews, 1985) permits appraisal of land use options over a period of time and their derived socio-economic benefits on a sustained basis.

A summary description of the MULBUD model is provided (Table 2).

Important features of MULBUD include the ability to: specify variable cost of inputs (labor and materials) and total product value by time period/production cycle (season and year) for selected enterprises, produce graphical displays of seasonal labor requirements versus net revenues generated, conduct sensitivity analysis on discount rates and changes in material costs and gross revenue, and provide summaries of variable costs, returns and net present values at user-defined discount rates.

Table 4. CRIES-AIS-YIELD model, Example of Phase IV Output

YIELD GENERATOR (CRIES - AIS - YIELD)
MICHIGAN STATE UNIVERSITY, EAST LANSING, MI 48823

COUNTRY NAME:	JAMAICA	CROP TYPE:	TOBACCO	LATITUDE:	18.01(deg)
POL./ADM. DISTRICT:	ST. CATHERINE	YEAR OF ORIGIN:	1981	HEMISPHERE:	NORTHERN
WEATHER STATION NAME:	WORTHY PARK	GROWTH PERIOD:	2/15/1981 TO 5/15/1981	AVERAGE ALTITUDE:	59.06(m)
RESOURCE PROD UNIT OR AGRO-ECOLOGICAL ZONE:	11	PROD POTENTIAL UNIT:	1.12	TIME INCREMENTS (dT):	5 (days)

PHASE 4: CALCULATION OF ESTIMATED YIELD (Ye)

T (DAYS)	MTH/DAY	MAX EVAPT(ETm) (mm/dT)	ACT EVAPT(ETa) (mm/dT)	YLD RESP FACTOR (ky)	CHANGE P YIELD (kg/ha)	CHANGE E YIELD (kg/ha)	CUMM YIELD (Ye) (kg/ha)
5	2/15	7.04	7.04	.15	161.50	161.50	162.
10	2/20	7.10	7.10	.15	164.02	164.02	326.
15	2/25	7.16	7.16	.15	166.54	166.54	492.
20	3/ 2	15.47	15.46	.00	169.07	169.07	661.
25	3/ 7	15.59	15.59	.00	171.61	171.61	833.
30	3/12	15.72	15.71	.00	174.15	174.15	1007.
35	3/17	15.73	15.73	.00	175.14	175.14	1182.
40	3/22	15.68	15.68	.00	175.12	175.12	1357.
45	3/27	22.90	20.11	.50	175.11	164.44	1522.
50	4/ 1	22.80	19.84	.50	175.11	163.74	1685.
55	4/ 6	22.71	19.57	.50	175.12	163.01	1848.
60	4/11	22.60	19.30	.50	175.14	162.35	2011.
65	4/16	22.70	22.23	.50	175.91	174.09	2185.
70	4/21	22.94	22.95	.50	177.16	177.16	2362.
75	4/26	20.01	20.01	.50	178.42	178.42	2540.
80	5/ 1	20.22	20.22	.50	179.67	179.67	2720.
85	5/ 6	20.43	20.43	.50	180.94	180.94	2901.
90	5/11	20.64	20.64	.50	182.20	182.20	3083.
TOTAL							3083.

40.95
38.95
36.95
34.95
32.95
30.95
28.95
26.95
24.95
-- Rainfed
-- Reported
1965 1967 1969 1971 1973 1975 1977 1979 1981

Jamaica/St. Catherine/Worthy Park
Sugar Cane Performance 1963 - 1982

Figure 6. Graphic Comparison of Predicted (CRIES-AIS-YIELD) and Observed Sugar Cane Production for the Period 1960 - 1982 in Worthy Park Sugar Estate, St. Catherine Parish, Jamaica.

6. SELECTED EXAMPLES OF CRIES-AIS-YIELD

To introduce users to YIELD model which can be employed in quantitative land evaluation studies, selected outputs are provided, using data from Jamaica. First of all, output is provided of a yield simulation for two commodities grown in close proximity to the Worthy Park weather station (Tables 3 and 4). Phase I output indicates the format of maximum potential yield based on genetic crop production potential without input and management constraints.

Phase IV output calculates the estimated yield based on moisture availability for user-specified increments during the cropping cycle.

Estimates can be provided for a "standard year" calculated from time series data and monthly means or for a single year and compared with observed yields. This last capability is especially useful in comparing the predictive ability of the yield model for selected agro-ecological conditions or representative farming systems. As an example, a comparison is provided between estimated and reported yield for sugar cane over the time period 1965 - 1981 for a selected sugar estate in St. Catherine Parish, Jamaica (Figure 6). The reported low in 1977 is the result of harvesting problems experienced during that year.

7. REFERENCES

Doorenbos J. and A.H. Kassam, 1979. Yield Response to Water. FAO Irrigation and Drainage Paper # 33, FAO, Rome.

Etherington, D. and P. Matthews, 1985. MULBUD User's Manual, National Centre for Development Studies, Australian National University.

Power P. et al., 1985. MSTAT, A Micro computer Program for the Design, Management, and Analysis of Agronomic Research Experiments. Michigan State University, East Lansing, Michigan.

Schultink G. et al., 1986. User's Guide to the CRIES Geographic Information System - Version 6.0, CRIES project, Michigan State University, East Lansing, Michigan.

Symposium on Remote Sensing for Resources Development and Environmental Management / Enschede / August 1986

Soils an important component in a digital geographic information system

Carlos R.Valenzuela*, Marion F.Baumgardner & Terry L.Phillips
Purdue University, Laboratory for Applications of Remote Sensing, West Lafayette, Ind., USA
** Present address: ITC, Enschede, Netherlands*

ABSTRACT: There is an increasing use of digital geographic information systems to meet the demand for specific accurate and rapid information of our resources. The degree of usefulness of this information depends on the accessibility and efficiency of the methods utilized for input, storage, analysis, and retrieval of informatio
The demand for accurate and rapid soil information is growing in our society, thus the element soil, because se of its importance, is one of the basic components of a complete geographic information system.
The Indiana soil association map at a scale 1:500,000 was digitized, projected to an Albers equal-area map projection, rasterized, and stored in a geo-referenced database created for the state of Indiana, U.S.A. Using the digital soils data stored in the geo-referenced database, new sets of data were generated by changing the coding of the soil associations or by combining two or more of these new generated products. Among the new digital data sets generated from the soils data are: Prime agricultural lands, Potential erosion, Dominant drainage, Dominant relief, average Corn yields.

1 GENERAL

The complexity and increasing volumes of available information, and the demand for the storage, analysis and display of large quantities of environmental data, has led in recent years, to rapid development in the application of computers to environmental and natural resources data handling and the creation of sophisticated information systems (Tomlinson et al., 1976).

Effective utilization of large spatial data volumes is dependant upon the existance of an efficient geographic handling and processing system that will transform these data into usable information. The major tool for handling spatial data is the geographic information system (Marble and Peuquet, 1983). Increasingly data of all types are being collected and converted to digital format. Extensive digital geographically oriented databases are being developed, and automated spatial information systems are used for storage, retrieval, manipulation, analysis, and display of information (Tom and Miller, 1974; Power, 1975; Knapp, 1978; Jerie et al., 1980; Anderson and Bernal, 1983).

A digital geographic information system (GIS) is a computerized system designed to stored, process and analyse spatial data and their corresponding attribute information. Advances in computer technology and techniques have made it possible to integrate a wide range of information (Gribbs, 1984). Technological advances have increased input techniques, storage, analysis and retrieval capabilities. Furthermore, there has been a reduction in costs and an increase in accessibility, so that a larger user community has been developed (Moellering, 1982). Geographic information systems have provided planners with a readily accessible source of objective earth science related facts, and an inexpensive, rapid and flexible tool for combining these facts with various other products to create decision alternatives (van Driel, 1975; Stow and Estes, 1981; Stoner, 1982).

A digital GIS is an information system which has as its primary source of input a base composed of data referenced by geographic coordinates and in which a major part of the processing is done with a digital computer (Kennedy and Meyers, 1977). It basically performs the following major functions: a) Data input, b) Data storage and retrieval, c) Data manipulation, and d) Data output (Tomlinson et al., 1976; Knapp, 1978; Nagy and Wegle, 1978; Jerie et al., 1980; Marble and Peuquet, 1983; Bartlocci et al., 1983; Valenzuela, 1985).

Basic information on the location, quantity and availability of natural resources is indispensable for planning more retionally their development, use and/or conservation. The demand for specific, accurate, and rapid soil information is growing in our modern society. Soils, because of their importance in agricultural and non-agricultural matters, and their inherent relationships with other environmental resources are a basic and fundamental component of any complete geographic information system.

Johnson (1975) points out that the conventional preparation of soil interpretive maps combining information of the soil resource with other resource information are excessively expensive, especially if various source maps have to be converted to a common scale and if the interpretive requirements are complex.

Automatic data processing systems have create inmense opportunities for storing and disseminating soil data (Bertelli, 1979). As the demand for interpretive maps increases, computers are used to speed up and cut down costs of the process (Bertelli, 1966; Shields, 1976; Bertelli, 1979; Miller and Nichols, 1979; Bie, 1980; Santini et al., 1983; Valenzuela, 1985).

The national Soil Handbook of the USDA Soil conservation Service (1983), indicates that the use of computer generated interpretive soil maps is encouraged where the soil survey has been digitized because they cost less than maps prepared by other means.

The objective of this study was to evaluate the significance of the element soils in the overall context of a digital geographic information system.

2 METHODOLOGY

The digital geographic information system developed at the Laboratory for Applications of Remote Sensing of Purdue University, basically consists of five major subsystems: a) input subsystem, b) database subsystem, c) management subsystem, d) modeling and analysis subsystem, and e) output subsystem. A simplified schematic configuration of the system is illustrated in Figure 1.

The soil association map of Indiana, USA, was prepared by the Indiana Soil Survey Staff of the United States Department of Agriculture Soil Conservation Service and Purdue University Agricultural Experiment

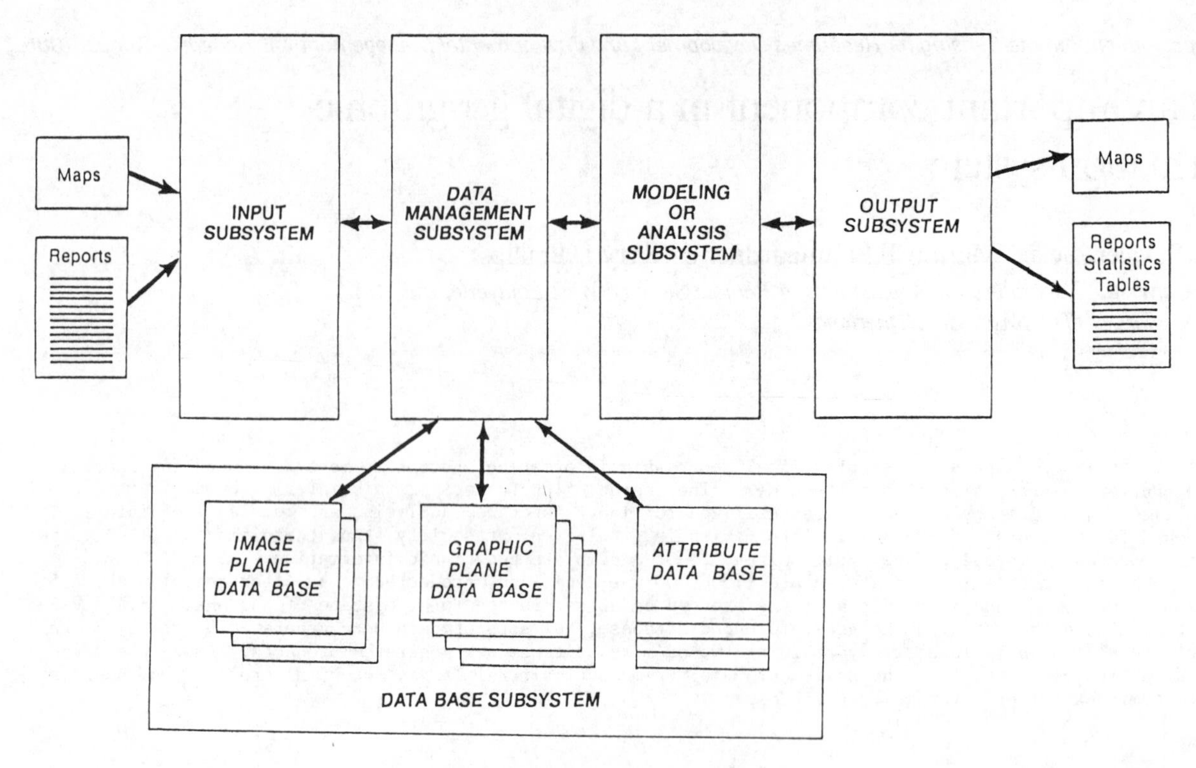

Figure 1. Basic components of a digital geographic information system.

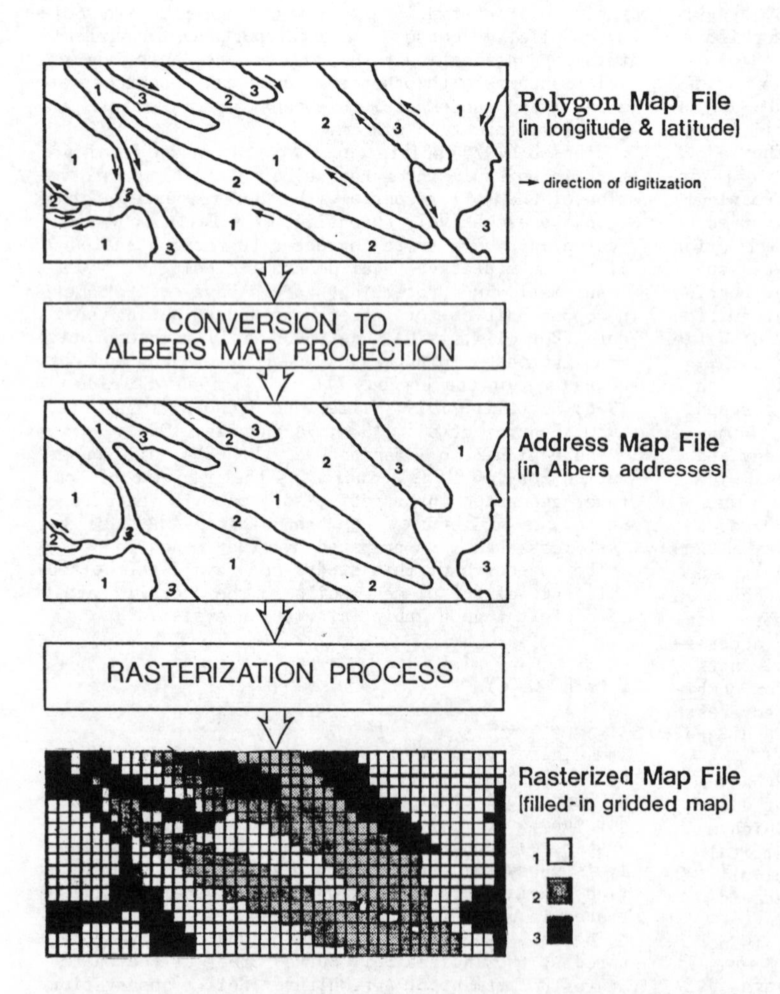

Figure 2. Steps involved in the conversion of a polygonal map file to a rasterized (cellular) map file.

Station, and made available to users as publication AY 209 (1980). The map at a scale 1:500,000 was published by the Cooperative Extension Service of Purdue University in cooperation with the state Soil and Water Conservation Committee of the Indiana Department of Natural Resources and the Soil Conservation Service of the United States Department of Agriculture.

The soil association map was manually digitized using the Purdue University/LARS digitizing system. This system is composed of a Talos table digitizer and an APPLE II Plus microcomputer. A complete documentation of this menu-driven system was prepered by Phillips (1983).

The data capture (map digitization) consisted in the transformation of three map primitives, i.e. control points, boundaries (limits of soil units), and centroids into a format compatible with digital computers. After the process of data capture was completed, the computer compatible data were transferred from the APPLE II Plus microcomputer to the host (main) computer (IBM 370/158) where the data were stored and the activities of editing, coordinate transformation, and rasterization were performed. Editing the digitized data was accomplished by manual and automatic editing routines using a graphics terminal Tektronics 4045.

Twelve control points, as illustrated in Figure 2, were used to derived statistically a biquadratic regression model which was used to transform the digitized values in X and Y into longitude and latitude geographic coordinates. These data were subsequently transformed into an Albers equal-area cartographic project. This was the projection (cartographic) selected for the Indiana geographic information system implemented at the Laboratory for Applications of Remote Sensing (LARS) of Purdue University.

The final step of the map input procedure was the rasterization process. During this process, the boundary and centroid files, stored in addresses corresponding to the Albers equal-area cartographic projection, were converted into an image file. The map units were filled-in with cells according to a predefined grid (500 m x 500 m on the ground) and subsequently each cell was assigned a class code associated with the centroid file (0 - 255). Figure 3 illustrates the different steps performed during the map input procedure. The coding (fill characters) assigned to each of the 55 soil associations existing in Indiana and to the portion of Lake Michigan in the state, is shown in Table 1.

For the construction of the attribute database (hierarchical), extensive use was made of the available information generated for the state soil associations of Indiana (Galloway et al., 1975). Other information not readily available in tables or as maps, at this level of detail, were obtained by interpretation, extrapolation and generalization of the information present in the description of the soil series forming each soil association (Galloway and Stainhardt, 1981; Franzmeier and Sinclair, 1982).

For displaying purposes and generation of color outputs of the computer generated interpretive soil maps, the rasterized image was transferred to the image processing device IBM 7350 "HACIENDA".

3 RESULTS AND DISCUSSION

Once the input of the data is completed and the rasterized data set and the corresponding attribute data set are stored in the database, the spatial information can be easily retrieved, handled, analyzed and displayed. The degree of the analitical capabilities implemented in a system depends on the nature, purpose and general objectives of the user. However, a well thought-out system will be one that is flexible enough to respond to the needs for input, analysis and display of different kinds of data required by the main user of the system.

Regardless of the objective of the principal user, one element seems to be present in almost every digital information system. It is the element soils, depicting soil types as obtained from soil surveys. It occurs because of its relation to the fauna, vegetation and climate, and its strong interaction with other natural resources elements. Soils, landuse and infrastructure constitute the fundamental and basic elements forming part of the database of geographic information systems for natural resources. The nature and types of information available in a soil survey enables the generation of several interpretive soil maps. These maps can be used as new variables for analysis or modeling of resources to predict changes that may occur through time.

The soil associations of Indiana in digital format displayed in the High Level Image Processing System (HLIPS) device IBM 7350 "HACIENDA", is shown in Figure 3. The area estimates and percentage of occurrance of each soil association in Indiana is presented in Table 1. Soil association Crosby-Brookston present on nearly level surfaces of Wisconsinan age glacial till plains in central Indiana, constitutes the largest association in Indiana covering an area of approximately 703,050 ha or 7.4 % of the state, followed by the Morley-Blount-Pewamo association, occurring on end moraines and on rolling areas near streams that dissect till plains. This association covers an area of 636,825 ha or 6.7 % of the state. The smallest associations are Riddles-Tracy-Chelsea on the end moraines in northwesten Indiana and Lyles-Ayrshire-Princenton developed on calcareous outwash sand and eolian fine sand deposited in Wisconsinan time covering an area of 13,500 and 15,600 ha respectively, or, 0.14 and 0.16 percent of the state.

The potential soil erosion was calculated using the Universal Soil Loss Equation. The factors of the USLE for each soil association were estimated by Brentlinger et al.(1979). This information was used to reclassify the digital soil association map into four potential soil erosion groups: low, medium, high and very high. The potential soil erosion map is illustrated in Figure 4, and the area estimates for each erosion group are shown in Table 2. This interpretive information can be used in conjuction with landuse data to predict the erosion hazard or gross erosion in the state. It can also be related to slope, landuse and proximity to streams to determine agricultural pollution due to erosion and to estimate sedimentation hazards and the related dangers of floodings.

Soil maps in Indiana are used in reassessment of agricultural land. The basic aim of any assessment activity is the equal treatment of all individual landowners. Yahner (1979) described the procedures used in agricultural land reassessment using estimates of corn yields. Each state soil association has been assigned an estimated corn yield value. Figure 5 illustrates the corn yield estimate map of Indiana after grouping the values in high, medium and low yield values for each soil association. Because of the resolution (scale) of the data and the generalization involved in the creation of the soil associations, some problems and difficultities may exist in the actual assessment of individual farm evaluation. However, it can be used to obtain rapid information on the approximate value of agricultural land.

The possibility of deriving different interpretive maps from the soil association map can be very useful in creating a set of illustrative material for didactical purposes . One such example is the possibility of showing graphically the influence of the soil forming factors in determining the actual soil characteristics. Figure 6 illustrates the parent material from which the Indiana soils were developed. It depicts the various kinds of materials including old sedimentary rocks in the southern part of the state, defferent thickness of loess deposits over glacial till, alluvial, lacustrine and eolian deposits from which the soils were developed.

Topography or relief has a great influence on the processes of weathering and soil formation. It in-

Figure 3. Soil associations digital map of the state of Indiana.

Figure 4. Digital soil associations illustrating potential erosion in Indiana.

Figure 6. Digital soil associations illustrating parent materials in Indiana.

Figure 5. Digital soil associations illustrating corn yield estimates in Indiana.

Figure 7. Digital soil associations illustrating dominant relief in Indiana.

Table 1. Coding, area estimates and percentage of occurrance of soil associations and Lake Michigan in Indiana.

Soil associations	Map symbol	Code	Area in Ha	Percentage
Genesse-Eel-Shoals	A1	1	202,050	2.13
Fox-Genesse-Eel	A2	2	187,825	1.98
Sloan-Ross-Vincennes-Zipp	A3	3	44,225	0.47
Stendal-Haymond-Wakeland-Nolin	A4	4	400,000	4.23
Wheeling-Huntington-Lindside	A5	5	65,800	0.69
Houghton-Adrian	B1	6	70,600	0.74
Maumee-Gilford-Sebewa	B2	7	333,350	3.51
Rensselaer-Darroch-Whitaker	C1	8	131,450	1.39
Sebewa-Gilford-Homer	C2	9	30,150	0.32
Lyles-Ayrshire-Princeton	C3	10	15,600	0.16
Milford-Bono-Rensselaer	D1	11	23,900	0.25
Patton-Lyles-Henshaw	D2	12	51,350	0.54
Zipp-Markland-McGary	D3	13	38,275	0.40
Tracy-Door-Lydick	E1	14	91,475	0.96
Elston-Shipshe-Warsaw	E2	15	77,325	0.82
Oshtemo-Fox	E3	16	219,100	2.31
Fox-Ockley-Westland	E4	17	208,825	2.20
Parke-Negley	E5	18	22,175	0.23
Oakville-Adrian	F1	19	19,450	0.21
Plainfield-Maumee-Oshtemo	F2	20	198,475	2.09
Princeton-Bloomfield-Ayrshire	G	21	98,100	1.03
Alford	H	22	173,350	1.83
Ragsdale-Raub	I1	23	45,225	0.48
Sable-Ipava	I2	24	17,300	0.18
Fincastle-Ragsdale	I3	25	273,700	2.88
Reesville-Ragsdale	I4	26	99,050	1.04
Iva-Vigo	I5	27	62,200	0.66
Brookston-Odell-Corwin	J1	28	155,300	1.64
Crosier-Brookston	J2	29	135,300	1.43
Crosby-Brookston	J3	30	703,050	7.41
Blount-Pewamo	K1	31	517,925	5.46
Hoytville-Nappanee	K2	32	28,375	0.30
Parr-Brookston	L1	33	102,700	1.08
Riddles-Tracy-Chelsea	L2	34	13,500	0.14
Miami-Crosier-Brookston-Riddles	L3	35	387,700	4.09
Miami-Crosby-Brookston	L4	36	493,400	5.20
Miami-Hennepin-Crosby	L5	37	72,475	0.76
Miami-Russell-Fincastle-Ragsdale	L6	38	490,425	5.17
Russell-Hennepin-Fincastle	L7	39	67,700	0.71
Markham-Elliott-Pewamo	M1	40	42,950	0.45
Morley-Blount-Pewamo	M2	41	636,825	6.71
Bartle-Peoga-Dubois	N1	42	72,000	0.76
Weinbach-Wheeling	N2	43	43,325	0.46
Avonburg-Clermont	N3	44	135,850	1.43
Hosmer	O1	45	152,575	1.61
Zanesville-Wellston-Tilsit	O2	46	125,100	1.32
Cincinnati-Vigo-Ava	O3	47	246,575	2.60
Cincinnati-Rossmoyne	O4	48	383,950	4.05
Wellston-Zanesville-Berks	P	49	461,950	4.87
Crider-Bedford-Lawrence	Q1	50	25,675	0.27
Crider-Hagerstown-Bedford	Q2	51	284,925	3.00
Crider-Baxter-Corydon	Q3	52	63,125	0.67
Berks-Gilpin-Weikert	R1	53	185,675	1.96
Corydon-Weikert-Berks	R2	54	56,725	0.60
Eden-Switzerland	R3	55	138,525	1.46
Lake Michigan		60	62,500	0.66

Table 2. Area estimates and percentage of potential erosion classes in Indiana.

Classes	Hectares	Percent
Very high	865,050	9.1
High	2,085,875	22.0
Medium	2,223,050	23.4
Low	4,250,850	44.8
Lake Michigan	62,500	0.7

Table 3. Area estimates and porcentage of dominant relief classes in Indiana.

Classes	Hectares	Percent
Hilly	906,000	9.5
Rolling	3,736,850	39.4
Undulating	592,275	6.2
Nearly level	4,189,700	44.2
Lake Michigan	62,500	0.7

Table 4. Soil associations and corresponding interpretive information in the State of Indiana.

Code	Symbol	Area in ha	Erosion	Slope	Drainage	Corn	%OM
1	A1	202,050	1	1	1	3	2
2	A2	187,825	1	1	1	3	2
3	A3	44,225	1	1	4	3	2
4	A4	400,900	1	1	1	3	1
5	A5	65,800	2	1	1	3	2
6	B1	70,600	1	1	4	3	4
7	B2	333,350	1	1	4	3	2
8	C1	131,450	1	1	4	4	3
9	C2	30,150	1	1	4	3	2
10	C3	15,600	1	1	4	3	1
11	D1	23,900	1	1	4	3	3
12	D2	51,350	1	1	4	4	3
13	D3	38,275	2	1	3	3	2
14	E1	91,475	1	2	1	3	2
15	E2	77,325	1	1	1	2	2
16	E3	219,100	1	2	1	2	1
17	E4	208,825	2	1	1	3	2
18	E5	22,175	4	3	1	2	1
19	F1	19,450	1	3	1	2	1
20	F2	198,475	1	2	1	2	1
21	G	98,100	2	3	1	2	1
22	H	173,350	3	3	1	3	1
23	I1	45,225	1	1	4	4	3
24	I2	17,300	1	1	4	4	4
25	I3	273,700	1	1	3	4	2
26	I4	99,050	1	1	3	4	2
27	I5	62,200	1	1	3	4	2
28	J1	155,300	1	1	4	4	3
29	J2	135,300	1	1	3	3	2
30	J3	703,050	1	1	3	3	2
31	K1	517,925	1	1	3	3	3
32	K2	28,375	1	1	4	3	3
33	L1	102,700	1	2	1	3	3
34	L2	13,500	1	3	1	3	1
35	L3	387,700	2	3	3	3	2
36	L4	493,400	2	3	1	3	2
37	L5	72,475	3	3	1	2	1
38	L6	490,425	3	3	3	3	1
39	L7	67,700	3	3	1	2	1
40	M1	42,950	2	3	3	3	3
41	M2	636,825	2	3	3	2	2
42	N1	72,000	2	2	3	3	1
43	N2	43,325	2	1	3	3	2
44	N3	135,850	2	1	3	3	1
45	O1	152,575	3	3	1	2	1
46	O2	125,100	3	3	1	1	2
47	O3	246,575	3	3	3	2	1
48	O4	383,950	3	3	1	2	1
49	P	461,950	4	4	1	1	1
50	Q1	25,675	3	3	1	2	1
51	Q2	284,925	3	3	1	2	1
52	Q3	63,125	3	4	1	2	1
53	R1	185,675	4	4	1	1	1
54	R2	56,725	4	4	1	1	1
55	R3	138,525	4	4	1	1	2
60	Lake Michigan						

Where:

Erosion	1(low); 2(medium); 3(high); 4(Very high)
Slope	1(nearly level); 2(undulating); 3(rolling); 4(hilly)
Drainage	1(well); 2(moderately well); 3(somewhat poorly); 4(poorly)
Corn yield	1(low); 2(medium); 3(high); 4(very high)
Org. matter	1(1.5%); 2(1.6-2.5%); 3(2.6-3.5%); 4(3.6%)

fluences internal drainage, surface runoff, and especially erosion. It also affects the use of farm machinery by inhibiting its use in areas of high relief. Information on relief or slope gradient can be very useful in determining the construction of septic systems. The extent of each of the dominant relief classes in Indiana is shown in Table 3. Figure 7 illustrates the spatial location of the dominant relief classes. This information shows a good relationship with potential erosion estimates.

Table 4 presents all the soil associations in the state with their respective tabular information (attributes) used to generate the interpretive maps and statistical information. The production of digital interpretive maps does not involve changes in the original data set. It uses attribute files to regroup the original map units into interpretive classes in real time. These new data (Interpretive maps) can be used in analysis and modeling processes with other resources data sets available in the GIS.

The element soils constitute one of the basic and fundamental component of any natural resources geographic information system. This is shown by the way in which soils are related to other environmental resources. These relationships indicate potential interactions and provide information which could improve planning and decision making activities at different levels of detail.

REFERENCES

Anderson, J.R. & P.F. Bermel 1983. Coordination of digital cartography in the federal government. Technical papers, ACSM-ASP 43rd annual meeting, Washington, D.C.,

Bartolucci, L.A., T.L. Phillips & C.R. Valenzuela 1983. Bolivian digital geographic information system. Proceedings of the 9th international symposium on machine processing of remotely sensed data, Purdue University, West Lafayette, Indiana.

Bertelli, L.J. 1966. General soil maps - A study of landscapes. Journal of soil science and water conservation, 21(1), p.3-6.

Bertelli, L.J. 1979. Interpreting soil data. Planning the use of management of land. ASA-CSSA-SSSA, 677 south segoe road, Madison.

Bie, S.W. 1980. Computer assisted soil mapping. The computer in contemporary cartography, edited by D.R.F. Taylor, J.Wiley & sons ltd.

Brentlinger, F.C., J.B. Hays, R.L. Lauster & C.S. Pierce 1979. Agricultural assessment for the non-designated and designated 208 planning areas of Indiana. State soil and water conservation committee, Indiana Department of Natural Resources.

Franzmeier, D.P. & H.R. Sinclair 1982. Key to soils of Indiana: showing the relationships of soils to each other and to their environment. AY-249, Coop. Ext. Ser., Purdue University, Indiana.

Galloway, H.M., J.E. Yahner, G. Srinivasan & D. P. Franzmeier 1975. Users guide to the general soils map and interpretive data for counties of Indiana. Supplement to AY-50 series, Coop. Ext. Ser., Purdue University, Indiana.

Galloway, H.M. & G.C. Steinhardt 1981. Indiana's soil series and their properties. AY-212, Coop. Ext. Ser. Purdue University, Indiana.

Gribb, W.J. 1984. The combining of geographic and other information systems to create a new type of information system. Technical papers of the 44th annual meeting of the american congress on surveying and mapping, ASP/ACSM convention, Washington, D.C.

Jerie, H.G., J. Kure & H.K. Larsen 1980. A systems approach to improving geo-information systems. ITC journal 1980-4, Enschede, The Netherlands.

Johnson, C.G. 1975. The role of automated cartography in soil survey. Proceedings of the meeting of the ISSS working group in soil information systems, Wageningen, The Netherlands.

Knapp, E. 1978. Landsat and ancillary data inputs to an automated geographic information system: Application for urbanized area delineation. Goddard space flight center and computer science corporation.

Kennedy M. & C.R. Meyers 1977. Spatial information systems: An introduction. Urban studies center, Louisville, U.S.A.

Marble, D.F. & D.J. Peuquet 1983. Geographic information systems and remote sensing. Manual of remote sensing. vol 1, Chapter 22.

Miller F.T. & J.D. Nichols 1979. Soils data. Planning the use and management of land. ASA-CSSA-SSSA, 677 south segoe road, Madison, U.S.A.

Moellering, H. 1982. The challange of developing a set of national digital cartographic data standards for the United States. Proceedings ACSM/ASP, Spring meeting, p. 201-212.

Nagy, G. & S.G. Wagle 1978. Geographic data processing. Computing surveys, vol 11, No. 2.

Phillips T.L. 1983. Operation of the Purdue/Lars digitizing system. LARS/Purdue University, Indiana.

Power, M. 1975. Computerized geographic information systems: An assessment of important factors in their design, operation and success. Washington University, Saint Louis, Missouri.

Santini, J., J.E. Yahner & D.P. Franzmeier 1983. Soil maps and interpretation system. FACTS User guide. FX-71(AY), Coop. Ext. Ser., Purdue University, Indiana. U.S.A.

Shields, R.L. 1976. New generalized soil maps guide land use planning in Maryland. Journal of soil and water conservation, vol 32, No. 6 p. 276-280.

Stoner, E.R. 1982. Agricultural land cover mapping in the context of a geographically referenced digital information system. Report No. 205, NASA, Space technology laboratories, Earth Resources Laboratory, NSTL Station.

Stow, D.A. & J.E. Estes 1981. Landsat and digital terrain data for county level resource management. Photogrammetric Engineering and Remote Sensing, vol 47, No. 2, p. 215-222.

Tom, C. & L.D. Miller 1974. A review of computer-based resource information systems. Land use planning information report 2, Colorado State University, Fort Collins, Colorado, U.S.A.

Tomlinson, R.F., H.W. Calkins & D.F. Marble 1976. Computer handling of geographical data. The UNESCO Press.

USDA 1983. National soils handbook. USDA, soil conservation service, U.S. Government printing office, Washington, D.C.

Valenzuela, C.R. 1985. The element soil as a basic component in a digital geographic information system. Ph.D. thesis, Agronomy department, Purdue University, Indiana, U.S.A.

Van Driel, N. 1975. Geological information in a computer mapping system for land use planning. Proceedings of the international symposium on computer assisted cartography, Bureau of the Census and ACSM Auto Carto II.

Yahner, J.E. 1979. Use of soil maps in Indiana's farmland reassessment. Agronomy guide AY-216, Cooperative Extension Service, Purdue University, Indiana, U.S.A.

Symposium on Remote Sensing for Resources Development and Environmental Management / Enschede / August 1986

Land suitability mapping with a microcomputer using fuzzy string

J.P.Wind & N.J.Mulder
ITC, Enschede, Netherlands

INTRODUCTION: In this report a system is described which allows digitizing, editing and storage of land units in a database. Land suitability for selected crops is mapped on a color raster screen by selecting a color code for suitability mapping. Intervals are defined on the attributes as strings.
The measure for suitability for a certain crop is derived in two possible ways:

1. For each suitability class the required properties)strings) are matched (AND-function of (un)equal property values of crop and land unit) against the land unit property table; an exact match produces the choosen color on the screen.

2. For one crop the ideally required properties are matched against the land unit property table (string) but a measure or (mis)match is given based on the Hamming distance of the property strings. Each distance can by assigned a different color.

1. HOW TO MAKE A LAND SUITABILITY MAP.

To obtain a landsuitability map the following steps have to be taken:

a. digitize/edit land unit boundaries (free digitizing).
b. define/edit list of centroids vs. land units.
c. generate a list of map attributes.

map unit: soil depth: salinity: ph: ....
for each attribute generate a menu of allowed attribute values (text strings):

attrl. menu:
soil depth
0 to 5 cm
5 to 10 cm
10 to 20 cm
20 to 40 cm
40 to 80 cm
> 80 cm

d. Generate a table with lines containing:
map unit, attributel, attr.2,.....

and for each attribute select an attribute value from the current attribute value menu using the cursor, move to the next attribute next menu etc.

Relation table I
Map unit attributes

map unit name	soil depth	ph of soil	attr.3
Aal	5 to 10 cm	7.0	....
Bal	> 80 cm	5.2	....
...	....	...	....

e. Generate a table relating crop suitability to map unit attributes.
For the attributes choose for each attribute a value from the menue for that attribute.

f. For each required crop type and range of suitability classes define a colour code:

Relation table II
Crop requirements:

crop	class	soil depth	ph	attr.3	...
patat	good	> 80cm	7	...	...
"	med.	5 to 10cm	6	...	...
"	bad	0 to 5cm	5	...	...
beets	good	> 80cm	6.8	...	...
"	med.	20 to 40cm	5.0	...	...
"	bad	10 to 29cm	4.8	...	...
.....	good	.. to ..cm	...	...	...

Relation table III
Crop suitability vs colour:

crop	class	red	green	bleu
Patat	good	255	255	000
"	med.	128	128	000
"	bad	32	32	00
Beets	good	000	255	000
"	med.	000	128	000
....				

g. For all crop classes compare the attributes required (strings) with the map unit attributes (strings) available, if all attributes match then link the suitability colour coding table III through to the centroid table IV expanding the proper map units in the proper colours or: map the per pixel map unit number through the colour look up table into the right colour.

Relation table IV

centroid xc,yc	map unit number	map unit code
123,987	0	Aal
345,6241	0	Aal
23,719	1	Ba2
998,34	2	Pq9
....	..	...
321,789	255	ZwO

LAND SUITABILITY MAPS

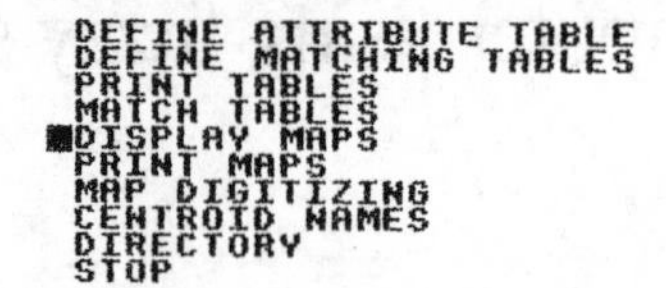

DEFINE ATTRIBUTE TABLE
DEFINE MATCHING TABLES
PRINT TABLES
MATCH TABLES
DISPLAY MAPS
PRINT MAPS
MAP DIGITIZING
CENTROID NAMES
DIRECTORY
STOP

Figure 2.1 Menu choice by cursor movement (black box is cursor).

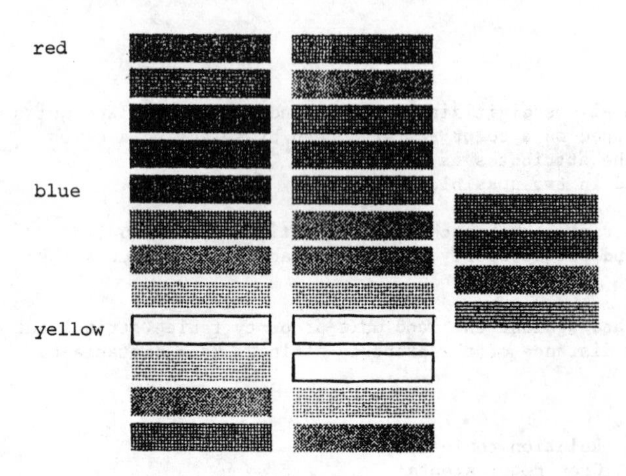

Fig. 2.2 Possible colours and patterns

DEFINE ATTRIBUTE TABLE

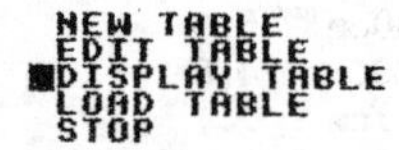

Fig. 2.3 Function choice by cursor

DISPLAY TABLE

SLOPE %

0.0-2.0
2.0-8.0
8.0-16.0
16.0-25.0
25.0-55.0

Fig. 2.4 Attributee from attribute table with its values displayed on a screen.

h. Overlay the land suitability map with any required line or symbol information.

i. Desing the colour legend.

j. Save the colour coded suitability map as an image file for further processing or hardcopy production.

k. Future extension:
   kl: the map unit unit attributes contain numerical values like: soil-depth= 17 (cm)
   k2: the crop requirement attributes contain an expression to be evaluated like:

   IF soil-depth > 12 or soil-depth < 32 THEN condition=TRUE

SLOPE %
VB1 0.0-2.0
VB2 0.0-2.0
VB3 0.0-2.0
TM 0.0-2.0
AA 0.0-2.0
AO 2.0-8.0
AC 25.0-55.0
AE 0.0-2.0
AV 0.0-2.0
EM 2.0-8.0
8.0-16.0
16.0-25.0

Fig. 2.5 Values of attribute 'slope' per landunit, displayed on screen.

Match is TRUE if all attribute requirements evaluate to TRUE (AND of all partial evaluations).

REMARKS: EXPERT SYSTEM.

In putting an expression evaluator in the table of required attributes we have implemented an expert system for land suitability evaluation.

The program evaluating the expression for attributes required is in fact a rule intepretor.

The table with map units and attributes together with the polygon data form the geographical data base.

The table with required attributes for crop suitability classes forms the actual rule base or knowledge base.

2. HOW DOES IT LOOK LIKE.

In this section different screen dumps are shown of the actual presentation.

Every choice that must be made is done by displaying the possible choices on the screen in a menu and indicating the choice by moving the cursor up and down (see fig. 2.1).

2.a Map digitizing and editing.

Map digitizing is done by so called "free digitizing". This means that lines are not stored as vectors, but the entire image is saved pixel by pixel. It also can be stored by run length coding. So there is no line indication. To delete a line or a polygon a square rubber can be put on the screen which can be moved by the cursor on the tablet to what must be deleted. By pressing one of the buttons the area under the rubber is deleted (i.e. becomes black/white). One can choose between three rubber sizes.

To plot a line three colors can be choosen. To fill a polygon with a color, a color fill subroutine is written which fills a polygon from the centroid with a color pattern (for color pattern see figure 2.2). Once a polygon is actually filled it can not be filled with another pattern. The digitized map can be printed on a color inkjet printer in two sizes using the full or the half width of the paper.

2.b Attribute definition.

To make a land unit table (table I, section 1) and a crop requirements table (table II, section 1) first an attribute table has to be made. By cursor movement one of the functions (figure 2.3) can be choosen. A new table can be defined or an existing table edited. For each attribute its values are asked. The values are stored as strings. Each attribute and its values can be shown on the screen (figure 2.4).

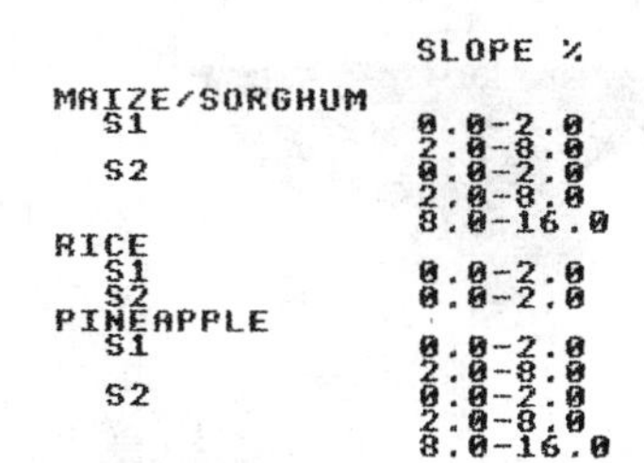

Fig. 2.6 Values of attribute 'slope' per crop class, displayed on screen.Crop requirements for an attribute.

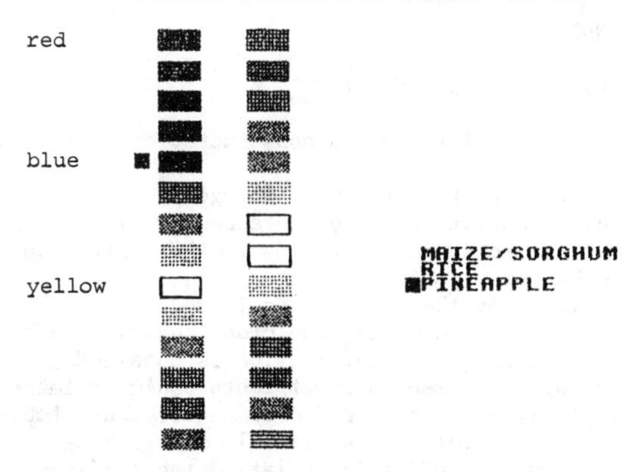

Fig. 2.7 Colour choice for crop subclasses. Cursor can also be moved left or right to change column.

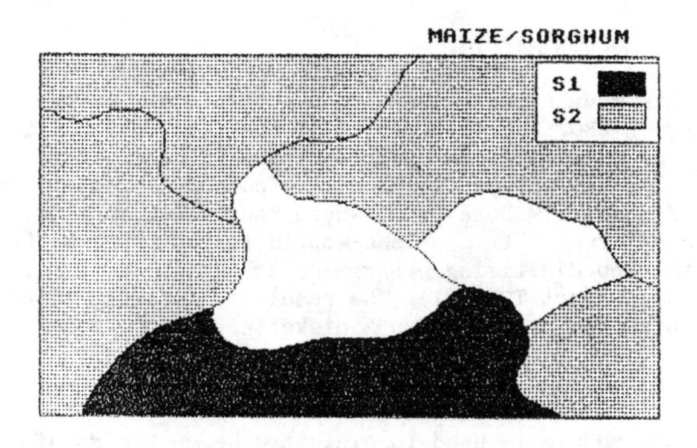

Fig. 2.8 Landsuitabilitymap for maize/sorghum (black/white is not suitable)

2.c Land unit and crop requirements table.

These tables are constructed and edited by using the attribute table. Attributes and values are indicated by cursor movement.
Each land unit or crop can be displayed per attribute with its values (figure 2.6).

2.d Matching.

To obtain a land suitability map for each defined crop the above mentioned tables (section 2.c) have to be matched. Each crop and its subclasses have to be assigned a color. This is done by displaying all possible colors on the screen and choosing by cursor movement (figure 2.7).

2.e Land suitability map.

The matching information is linked with the actual map and a crop suitability map can be displayed (figure 2.8). A hard copy on an inkjet plotter can be made.

2.f Printing tables.

The constructed attribute, land unit and crop requirements tables can be printed on a printer (figure 2.9).

2.g Overlay of maps

A land suitability map can be overlayed with e.g. an infrastructure map (figure 2.10).

ATTRIBUTE TABLE

SLOPE %	SOIL TEXTURE	SOIL DEPTH
1. 0.0-1.9	1. COARSE TEXTURED	1. <50 cm
2. 2.0-7.9	2. MODR. COARSE TEXT.	2. 50-180 cm
3. 8.0-15.9	3. MEDIUM TEXTURED	3. >100 cm
4. 16.0-24.9	4. MODR. FINE TEXT.	
5. 25.0-55.8	5. FINE TEXTURED	

INHER. FERT.	PH RANGE	SOIL DRAINAGE
1. LOW	1. 6.5-7.5	1. VERY POORLY DRAINED
2. MEDIUM	2. 5.5-6.5	2. POORLY DRAINED
3. HIGH	3. 4.5-5.5	3. IMPERFECT. DRAINED
	4. 7.5-8.5	4. MODER.WELL DRAINED
	5. <4.5 or >8.5	5. WELL DRAINED
		6.SOMEWH.EXCES.DRAINED
		7. EXCESSIVELY DRAINED

FLOODING RISK
1. NONE TO SLIGHT
2. MODERATE
3. HIGH

Fig. 2.9 (a) Attribute table

LAND SUITABILITY

	SLOPE %	SOIL TEXT.	SOIL DEPTH	INHER.FERT.	PHRANGE
VB1	1	5	3	2, 3	2
VB2	1	5	3	2, 3	2
VB3	1	5	3	2, 3	2
TM	1	4	3	1	2
AA	1	4	3	1, 2	3
AO	2	3, 4	2	1	3
AC	5	3, 4	1	1	3
AE	1	3	3	1	3
AV	1	3	3	1	3

	SOIL DRAINAGE	FLOODING RISK
B1	2, 3, 4, 5	1
B2	2, 3, 4, 5	2
B3	2, 3, 4, 5	3
M	4	1
AA	5	1
AO	5	1
AC	5, 6, 7	1
AE	2, 3	3
AV	3, 4, 5	2
TM	6, 7	1

Fig. 2.9 (b) Land unit table

CROP REQUIREMENTS

	SLOPE %	SOIL TEXTURE	SOIL DEPTH
MAIZE/SORGHUM			
S1	1, 2	3, 4	3
S2	1, 2, 3	2, 3, 4, 5	2, 3
RICE			
S1	1	4, 5	2, 3
S2	1	3, 4, 5	2, 3
PINEAPPLE			
S1	1, 2	3, 4	3
S2	1, 2, 3	2, 3, 4, 5	2, 3

	INHER. FERT	SOIL DRAINAGE	FLOODING RISK
MAIZE/SORGHUM			
S1	2, 3	4, 5	1
S2	1, 2, 3	3, 4, 5, 6	1, 2
RICE			
S1	2, 3	1, 2	1
S2	2, 3	1, 2, 3, 4	1, 2
PINEAPPLE			
S1	2, 3	4, 5	1
S2	1, 2, 3	3, 4, 5, 6	1

Fig. 2.9 (c) Crop requirements table

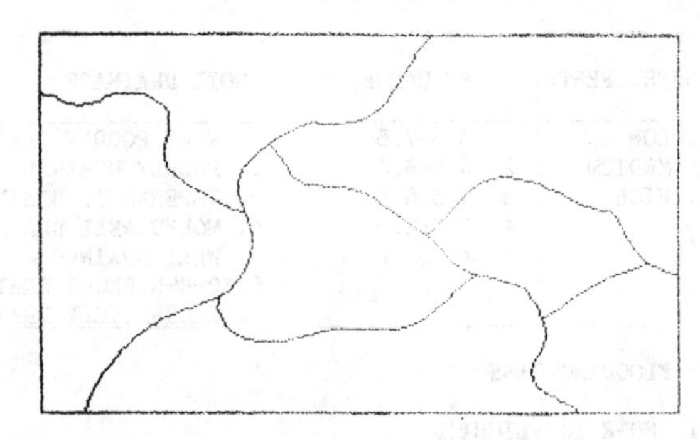

Fig. 2.10 (a) Land unit boundaries (red lines)

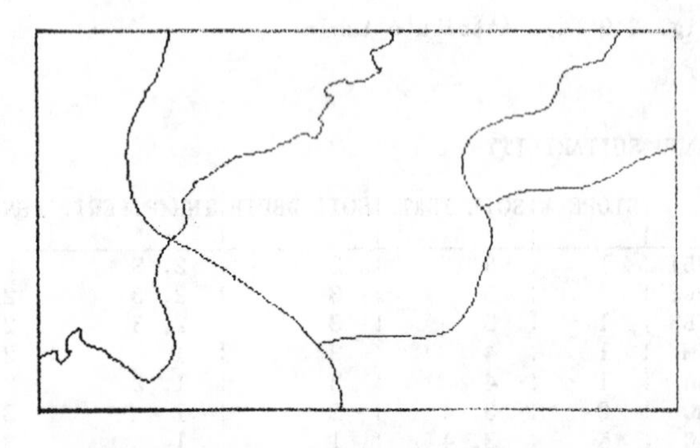

Fig. 2.10 (b) Infrastructure (roads) (blue lines)

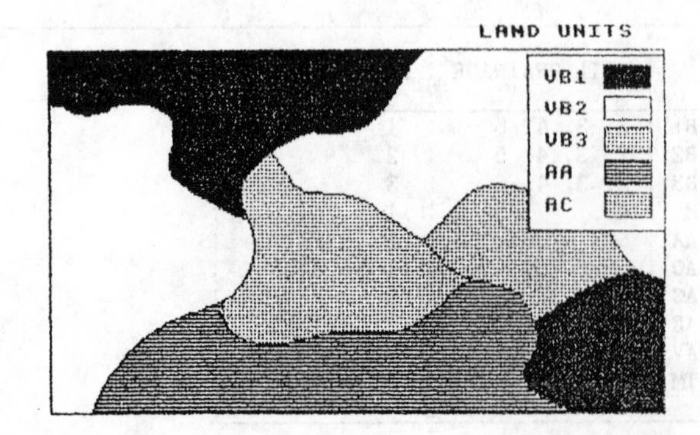

Fig. 2.10 (c) Land unit map

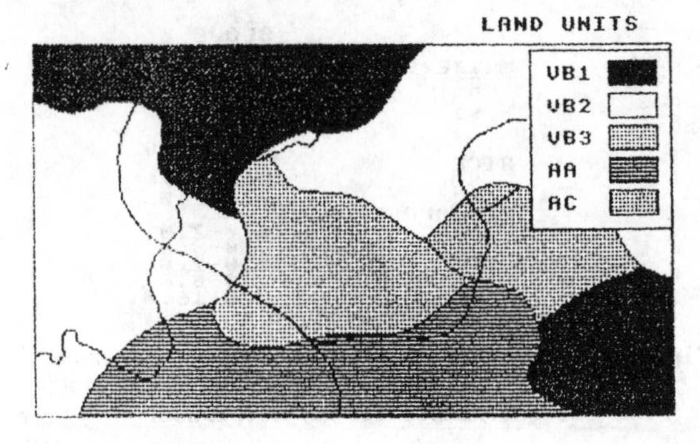

Fig. 2.10 (d) (c) overlayed with (b)

APPENDIX

DESCRIPTION OF THE MODULES.

A land suitability map is constructed by using two kinds of data:

a. a map description: lines, polygons, centroids; this description can be obtained by digitizing a map and assigning centroids to the contructed polygons;
b. colours for the polygons (related to the centroids) indicating the crop (sub)classes for the polygon; this information is obtained by matching the map unit attribute table (relation table I, section 1) and crop requirements table (relation table II, section 1) and linking it with a colour coding table (relation table III, section 1).

To achieve this 7 modules are built.

The listing of the different modules and their in- and outputs are shown in scheme A.1.

A.1 MODULE MAPDIG.

This module can be used to digitize a map on the Acorn BBC. For the purpose of the land suitability maps only lines and polygons have to be digitized (they may not be filled with a colour, because this is done in MODULE DISPL where the result maps are displayed). It is a less sophisticated program than the map digitizing program on the PDP11 and RAMTEK (see report TR85081). The result is an image file which can be stored on a diskette.

A.2 MODULE MAPCEN.

This module is used to classify the centroids of the polygons of a map. Of each polygon a centroid can be defined by cursor movement and after it is indicated its class (map unit name) must be given. The result is a table with the coordinates of the centroids and a pointer to the class name (see section 1, Table IV).

A.3 MODULE ATTTAB.

This module is used to define a table with attribute names and their possible values (see section I.a, attribute menu). This table is used as a menu for constructing a map unit table or a crop requirements table (see section A.4).
This module also has the possibility to edit an already defined attribute table.

A.4 MODULE MUTCRT.

This module is used to define a map unit table (see section 1, relation table I) and a crop requirements table (see section 1, relation table II). Also existing tables can be edited.

Schema A.1 Survey of modules and their interaction

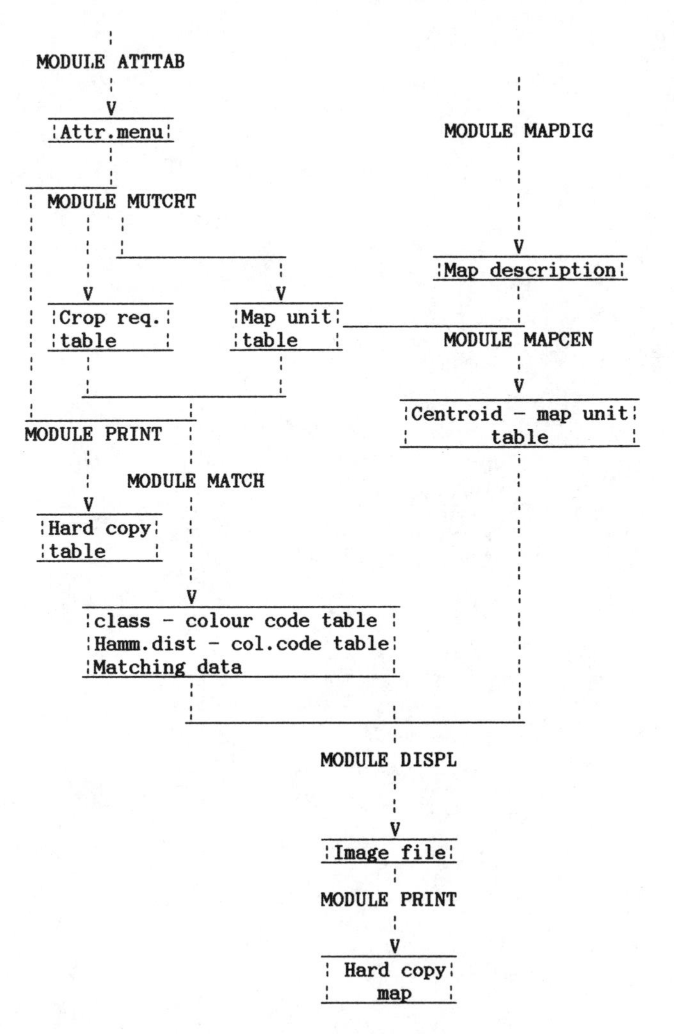

For the map unit table map unit names must be given and for the crop requirements table the class name (crop names) and their subclass names. By cursor movement the attributes and the values for each item in the table(s) can be choosen from the attribute table constructed in module ATTTAB (see section A.3). The map unit names are used in the module MAPCEN (see A.2) to classify the centroids.

A.5 MODULE MATCH.

This module is used to match two tables: a map unit table and a crop requirements table (both constructed by module MUTCRT see section A.4).
For each class (crop) a colour code must be specified for its subclasses by choosing from a menu of colours.
It also registrates the measure of mismatch for each map unit. This measure is the Hamming distance to a full match (if result is TRUE, TRUE, FALSE, FALSE the Hamming distance to a full match - i.e. TRUE, TRUE, TRUE, TRUE - is 2). For each possible distance also a colour must be specified.
The result is a table with map units and colour codes and a table with Hamming distances and colour codes. Further for each map unit and subclass a decimal coded number is given for the (mis)match (for example - 1=TRUE, 0=FALSE -: 1,1,1,0 =12; 0,1,1,1 =7; 1,0,0,1 =9). This information is the input for module DISPL (see section A.6).

A.6 MODULE DISPL.

This module is used to display land suitability maps on the screen. By using a menu the class (crop) to be displayed can be choosen. Further there is a choice between a full match picture or partly match picture. A full match picture shows the possible units to grow different subclasses of a crop, a partly match picture shows the land units and their distance to a full match for the best suitability class of a crop.

Symposium on Remote Sensing for Resources Development and Environmental Management / Enschede / August 1986

Land resource use monitoring in Romania, using aerial and space data

N.Zegheru
Institute of Geodesy, Photogrammetry, Cartography and Land Management, Bucharest, Romania

ABSTRACT: Tendencies in land uses, main causes of land retirement and measures provided for their protection are presented. Considering existing land cadastre based on photogrammetric maps, achievements in automatic cadastral mapping using analitical photogrammetric methods and data storage in a data base are discussed. The importance of these photogrammetric cadastral bases as a support of the land information systems, the only ones able to supply update, accurate and complete data in due time regarding land resource situation is emphasized. Multispectral aerial and space recordings, bisides aerial photography, are used to complete data on land resources. Taking into account the present-day situation of the land information systems and the existing possibilities in this field of activity, expectations related to land resource monitoring and land efficient use in Romania are also given.

RESUME: On présente les tendences de l'utilisation des terrains, les causes principales de leur dégradation et les mesures prises pour leur protection. Dans les conditions de l'existances du cadastre foncier avec base topographique on présente les réalisations concernant l'établissement automatique, par méthodes photogrammétriques numériques, des plans cadastraux et le stockage des données dans des bases des données. On met en évidence la signification de ces bases des données cadastrales comme support des systèmes informationnels du territoire, seuls en mesure de fournir, en temps utile,des données et des informations actuelles exactes et complètes concernant la situation du fonds foncier. Pour compléter les données sur le fonds foncier on utilise non seulement des photographies aériennes mais aussi des enregistrements aérospatiaux multispectraux. On présente, tenant compte de la situation actuelle et des possibilités existantes, les perspectives dans le domaine de la surveillance et de l'utilisation efficiente du fonds foncier en Roumanie.

Considering the land use dynamics, some permanent changes resulting from the built-in area extension and land retirement, concurrently with population and its requirement increases have taken place. To answer such important matters, it is quite a complex problem; on the one hand, an agricultural and forest land protection and extension legislation and, on the other hand,a modern and intensive management should be taken into account. When land-use is a hazard activity, in the course of time, the result is a real calamity, especially considering the future generations. As it is already well-known, some large areas have become waste, desolate regions, in other words, desertification, pollution entailing soil degradation or deforestation, have appeared, which bring about negative consequences related to the land and surrounding microclimate of the respective zone. Irrigated and drainage lands and those situated near non-arranged river banks have also endured a lot of consequences; salinity, swamp formation, floods, etc. could appear.

As regards the national land resources, every country has its own policy on the rational use of this invaluable national good, based on a proper legislation. Romanian legislation makes provisions for an ensemble of measures on the national land resource use and protection, and they are a permanent concern of some governmental organizations.

Data and information collection on the present-day land situation is a prerequisite, in order to accomplish objectives provided by our legislation and other governmental documents. As a consequence, land cadastre containing an ensemble of technical, economic and legal operations to systematically and permanently inventory land resources, to establish surface size, land-use category and landownership of each parcel is obviously engaged. Land cadastre envisages the country land as a whole devided into territorial-administrative units based on the land parcel showing shape, size, area, land-use, ownership or tenant, land quality, and other elements.

The following technical, economic, and legal activities can be accomplished, using land cadastre: cadastral surveyings on parcels, mapping, surface calculations, and cadastral register establishment; qualitative classification of soils; identification and registration of all land owners based on legal documents, regarding their rights on lands and constructions.

Activity sectors holding large surfaces or constructions are establishing specialized cadastre giving all necessary thematic data, based on land cadastre.

Connections among various land cadastre divisions, as well as, its connections with the main specialized cadastres are shown in Figure 1.

1:5,000 scale maps for zones with few topographic details and large size plots, located in plain and mountain regions, 1:2,000 scale maps for zones with many topographic details located in hilly regions and for all villages, and 1:1,000 scale maps for all towns are used in cadastral works. Some other cadastral maps at 1:10,000, 1:25,000, 1:50,000, 1:100,000 scales are also used.

The activity within the national land re-

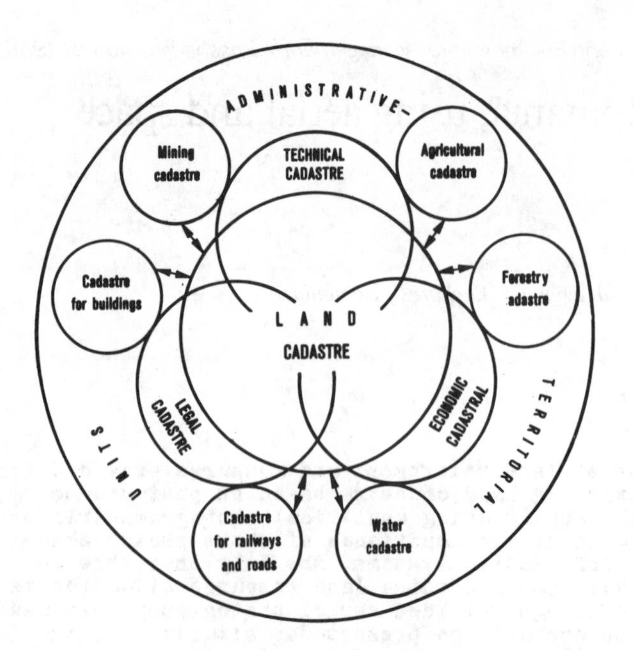

Figure 1. Diagram of the land cadastre structure and their connections with special cadastre.

sources are becoming more and more complex, and thus, the classical cadastre is gradually transformed into a more complex one, that is an ensemble of land information systems, able to answer the many various problems in due time. In this respect, there is a tendency to compile digital maps, an almost new established concept, meaning a digitally stored data collection representing a map contents (Stefanovič 1980). Considering digital map features, we easily can understand the technological process (Figure 2) used to compile it, where data employed are automatically processed (Zegheru 1981).

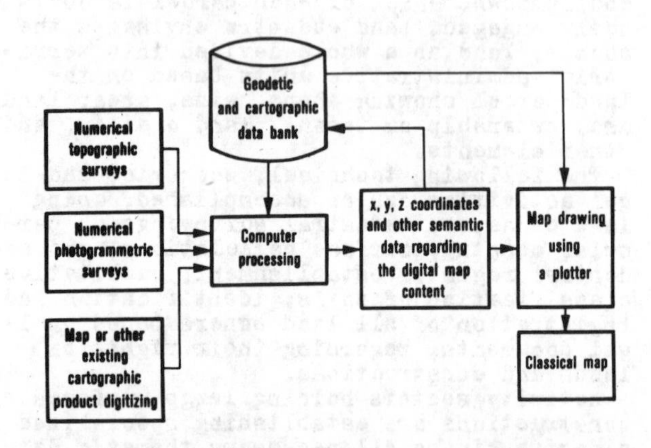

Figure 2. Input data sources, the main technological process stages in digital mapping and final results.

The previous diagram obviously illustrates that input data can be derived from topographic and photogrammetric surveys, some existing maps or document digitizing, photogrammetric and remote sensing data and information. In digital mapping technology, geodetic and cartographic data base is an compulsory element to deliver data during the processing stage, to store these map contents and other information related to the national land resources derived from the aerial and space multispectral imagery.

The cartographic material users are accustomed to have maps as a graphical or photographic representation, which contents could be read, studied and on which they can work. That is why, a plotter is supplied to the necessary hardware to compile a digital map, in order to draw a classical map, as well.

To compile a digital map, therefore, a certain technological development stage is required, namely, the existance of a topographic, photogrammetric, and remote sensing equipment able to record digitally, a computation unit with a proper software to process data, a geodetic and cartographic data bank including a data base for digital maps, and a plotter, as well.

Although the scale has not the same meaning for a digital map as against the classical one, it was established to use a symbol for each conventional sign being in a biunivocal relation (Figure 3), just to make a graphical representation, when required.

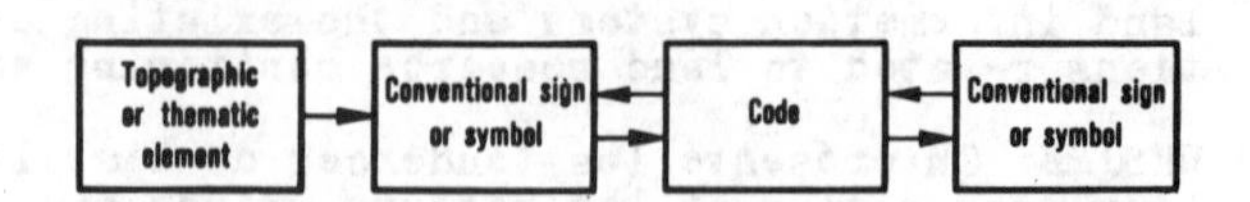

Figure 3. Diagram of connections among topographical elements, conventional signs and codes.

Errors due to drawing and its base are added to the measuring and data processing errors during classical map compilation. In this way, the digital map proofs to be more accurate that the classical one.

The following reasons are in favour of the digital mapping technology: updating map simplification, maintaining the accuracy in new and modified elements introduced in the original map; simplification of the map compilation having various themes derived from the same original map: various thematic maps could be derived from a topographical map, using a display or a plotter; a shorter time between data collection and map editing stages; the possibility for a manifold analysis of information derived from the map contents and their various thematic processings; the possibility to automatically generalize from the digital map to smaller scale classical maps, according to the data base files.

The existing data base was conceived to store 1:5,000, 1:2,000 and 1:1,000 scale digital cadastral maps, using analytical photogrammetric methods (Zegheru et al.1982). A conventional sign atlas suited for the automatic cartographic requirements was developed, to carry out this data base on files corresponding to the above mentioned scales. These conventional signs were coded considering both connections among symbols, representing the same object at various scales, and data retrieval for different themes (Figure 4) (Fusoi et al. 1981).

The digital cadastral map contents are stored on a scale file, according to its content and accuracy. The derived map and plan information is prepared at the drawing scale, and after its cartographic representation, nothing is to be stored in the data base.

Among the important facilities of the digital mapping technology at the same time, being also the main reasons for digital cadastral mapping, we can mention storage, their

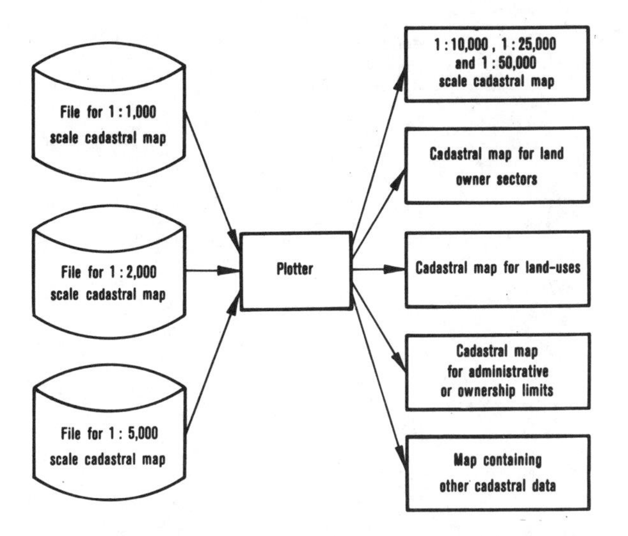

Figure 4. Diagram of data base configuration for digital maps and graphical facilities.

content interpretation and processing, updating, as well as, some land-use data retrieval.

The cadastral maps compiled as we have already mentioned, which are supplemented by aerial and space conventional and multispectral data imagery are used in land-use monitoring. In view of these facts, methodologies to use aerial photographs in land-use monitoring have been established; these photographs are a real and efficient information source supplying a large data volume existing in the field. Obviously, results are depending on many factors, among which we mention: aerial photograph and photosensitive material qualities, the best period selection, emphasize on phenomena to be monitored, the best aerial photograph scale selection, equipment, human operator experience and skill in aerial photograph, data and information interpretations.

Several accomplished themes related to land-use monitoring could be mentioned: study on land evolution within hydromeliorative systems, land-use monitoring within settlements, and monitoring surfaces, on which garbage and industrial wastes are to be found. Some other themes are to be approached, using also other data sources such as multispectral data delivered by aerial and space platforms, aiming at the development of a complex cadastre being able to permanently monitor land-use, in real time.

- On land evolution within hydromeliorative systems

In the last decades, large irrigation and drainage systems have been developed in various country zones. Aerial photograph taking in some special time periods, these aerial photograph interpretations, and conclusions presented to the decision-making units are recommended, to monitor their running effects on the respective fields.

Studies using 1:12,000-1:15,000 scale aerial photographs and photogrammetric products over diked areas in the Danube Delta and irrigation systems in other country zones have been made, in order to develope a proper methodology. The execution project related to canal network, the respective system running efficiencies, and the secondary effects(marsh formation, salinity,a.s.o.), efficient land-use in farming, according to the land management projects, and land cadastre updating were implemented in the above mentioned studies.

The developed methodology recommends : taking aerial photographs during periods of time when the respective systems have a maximum load, carrying out photogrammetric documents emphasizing excess moisture, correlation among excess moisture zones and micro-relief shapes, excess moisture surface mapping and estimation, correlation between the respective zone photographic image and the corresponding map showing the isophreatic lines emphasizing the field excess moisture), densitometric measurements using diagrams and interpretations, as well as, phenomena which cannot be percieved by visual photointerpretation, correlation among types of soil having vegetations covering excess humidity surfaces, according to the photographic image.

- On land-use monitoring within settlements

Towns and villages are permanently and intensively extending, as a result of a new and impressive social, cultural and economic development. The national land resource legislation provides allotment of lands for various fields of activity, without diminishing farming and forest surfaces. Periodical 1:20,000 scale aerial photographs are taken, to supervise these legislation provisions, aiming at: the settlement building limit identifications, field identifications within the settlements to be built, and their use previous to the dwelling development and putting under crop the lands, which temporally were unproper to agriculture because their wrong uses.

- On monitoring surfaces on which garbage and industrial wastes are to be found

Concentrations of wastes, cinder, slag and refuse are permanently carried and deposited in large usual natural holes on non-productive fields. Stockpiling in densely populated towns and large industrial plants have always brought about serious problems, because it must comply with the legal provisions regarding health safety conditions, soil, water and air pollutions. The studies implemented, and which have used aerial photographs recommends: identification of farming lands covered by wastes, stockpiling approval checkings, depositing dynamics monitoring according to the technical documentation, its effect on the neighboured zones, putting lands under crop, according to the schedule, based on the established projects.

Stockpiling zones for garbage and industrial wastes for some large towns with heavy industries have been studied, using 1:4,000 scale periodical aerial photographs used as a base for 1:500 scale topographic mapping; hole stockpiling volume has been calculated according to stockpiling rate; filling diagrams have been developed and, finally, putting these lands under crops.

Some themes on the national land-use monitoring have been briefly presented, using aerial photographs and photogrammetric documents. Methodologies for all approached themes are developed using the above mentioned studies.

Space multispectral imagery taken in the course of the last mid-decade (Zegheru 1976) has been experimentally used to monitor the national land-use, for the first time in our country. Our own system to automatically process remote sensing digital multispectral data (SPADAM) for unsupervised multispectral image classifications, using histograms

(Vass 1980) based on the methodology below has been developed within the Institute of Geodesy, Photogrammetry, Cartography and Land Management (I.G.F.C.O.T.):

- Zone pixel selection in CCT Landsat strip and radiometric measure introduction in a file, which can be processed, using Fortran language.
- The general pixel population histogram development in the first two main components. Histogram is directly made in a file and a printing list is obtained.
- Classes to be separated are established on a histogram. Each class is included in a number of rectangulars parallel to the axes. A card containing the old class number, the new class number and rectangular size is punched for each rectangular.
- A printed histogram for each established class in the first two main components of the respective class is developed. All histograms are compiled and printed at only one programme running. If some histograms give possibility to devide the respective classes, the process should be repeated. If all histograms show class homogeneities, the operation is over.

These classified images are used to obtain a graphical representation at a desired scale, in a plotter.

All package programmes corresponding to the above mentioned process stages have been developed, imagery delivered by Landsat satellites were processed, small scale maps showing land-use were compiled, and other themes were studied, using this system. The findings were good enough for such geographic information systems (Geo-Information Systems).

REFERENCES

Zegheru, N. & D.Gheorghiu 1984. Data bank for land cadastre. Analele I.G.F.C.O.T. 6: 35-46.
Stefanovič, P. 1980. Computer assisted mapping. I.T.C. Journal, Special Issue.4:606-637.
Zegheru, N. 1981. Prospects related to digital mapping in our institute (I.G.F.C.O.T.) Analele I.G.F.C.O.T. 3:37-46.
Zegheru, N., D.Gheorghiu, A.Fusoi, G.Galbură & Z.Doroghy 1982. Data base for digital cadastral maps. Analele I.G.F.C.O.T. 4:43-49.
Fusoi, A., G.Galbură & D.Gheorghiu 1981. Conventional signs for automated topographic mapping. Analele I.G.F.C.O.T. 3:63-69.
Zegheru, N. 1976. Land-use map compilation using the satellite images. Surveys for Development - I.T.C. Enschede, The Netherlands.
Vass, G. 1980. Remote sensing multispectral digital data processing. Analele I.G.F.C.O.T. 2:97-108.